Fundamental Constants

CONSTANT	SYMBOL	VALUE	SI UNITS	cgs UNITS
Speed of light in vacuum	c	2.99792	10^{8} m s^{-1}	10^{10} cm s^{-1}
Avogadro's number	N_A	6.0221	10^{23} mol^{-1}	10^{23} mol^{-1}
Planck's constant	h	6.6262	10^{-34} J s	10^{-27} erg s
Boltzmann's constant	k_B	1.38066	10^{-23} J K^{-1}	10^{-16} erg K^{-1}
Gas constant	R	8.3144	J K^{-1} mol^{-1}	10^{7} erg K^{-1} mol^{-1}
Faraday's constant	F	9.6487	10^{4} C mol^{-1}	
Bohr radius	a_0	5.29177	10^{-11} m	10^{-9} cm
Elementary charge	e	1.60219	10^{-19} C	
		4.80298		10^{-10} cm$^{3/2}$ g$^{1/2}$ s^{-1} mol^{-1}
Electron rest mass	m_e	9.1095	10^{-31} kg	10^{-28} g
Proton rest mass	m_p	1.67265	10^{-27} kg	10^{-24} g
Gravitational constant	G	6.6720	10^{-11} N m^{2} kg^{-2}	10^{-8} dyn cm^{2} g^{-2}
Permittivity of vacuum	ϵ_0	8.85419	10^{-12} kg^{-1} m^{-3} s^{4} A^{2}	
Ice point at 1 atm		2.7315	10^{2} K	
Atomic mass unit		1.6605	10^{-27} kg	10^{-24} g
π		3.14159		
e		2.71828		
$\log_e 10 = \ln 10$		2.30258	($\ln x = 2.30258 \log_{10} x$)	

Conversion Factors for Energy Units

To convert from energy in units of the left column to units of the first row, multiply by the factor at their intersection.

	eV	cm^{-1}	kcal mol^{-1}	kJ mol^{-1}	K	erg
1 ev =	1	8065	23.06	96.48	1.160×10^{4}	1.602×10^{-12}
1 cm^{-1} =	1.240×10^{-4}	1	2.859×10^{-3}	1.196×10^{-2}	1.439	1.986×10^{-16}
1 kcal mol^{-1} =	4.336×10^{-2}	349.8	1	4.184	503.2	6.948×10^{-14}
1 kJ mol^{-1} =	1.036×10^{-2}	83.60	0.239	1	120.3	1.661×10^{-14}
1 K =	8.617×10^{-5}	0.6950	1.987×10^{-3}	8.314×10^{-3}	1	1.381×10^{-16}
1 erg =	6.241×10^{11}	5.034×10^{15}	1.439×10^{13}	6.022×10^{13}	7.243×10^{15}	1

Physical Chemistry

with Applications to the Life Sciences

THE BENJAMIN / CUMMINGS PUBLISHING COMPANY, INC.
Menlo Park, California   Reading, Massachusetts
London   Amsterdam   Don Mills, Ontario   Sydney

Physical Chemistry

with Applications to the Life Sciences

DAVID EISENBERG
University of California, Los Angeles

DONALD CROTHERS
Yale University

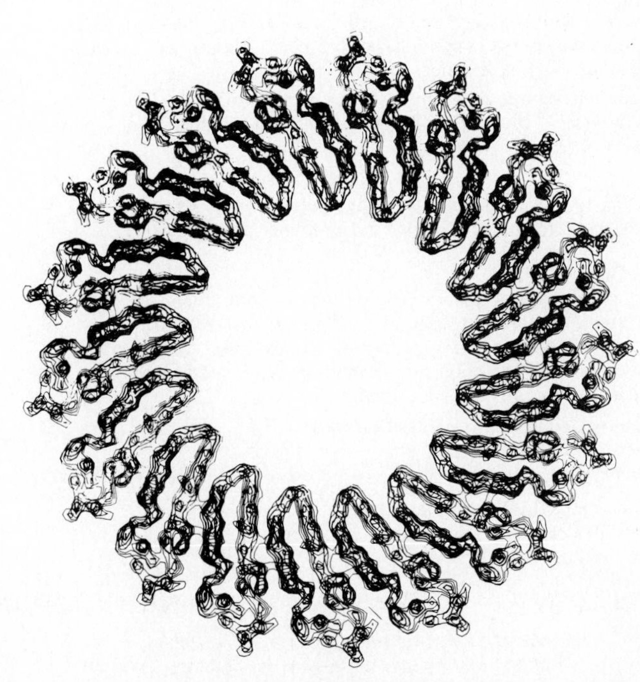

Sponsoring editor: Mary Forkner
Production editors: Betsey Rhame, Margaret Moore
Book designer: Design Office / Peter Martin
Cover designer: Design Office / Peter Martin
Artist: Michael Fornalski

About the cover

The cover shows the protein disk which is the basic unit of the structure of tobacco mosaic virus. When stacked face to face, these disks form a tube closely related to the protein coat of the virus. The disk is composed of 17 polypeptide chains, each of which is shown here as a comma-shaped object by contours of electron density. This picture, determined by the methods of x-ray diffraction described in Chapter 17, is too coarse in resolution to reveal amino acid residues, much less atoms. Further work using the same methods by Drs. A. Klug, A. C. Bloomer, J. N. Champness, G. Bricogne, and R. Staden in Cambridge, England, has established the positions of most of the atoms in the disk, as shown in Figure 16–4. (Figure courtesy of Dr. A. Klug and colleagues.)

About the part opening photographs

p. 1: Photographs from *Snow Crystals* by W. A. Bentley and W. J. Humphreys, McGraw-Hill Book Co., New York, 1931.

p. 269: Electron micrograph (magnification $\cong 10^5$), taken by the freeze fracture replica method, of the inner and outer surfaces of a rabbit ventrical cell membrane. The pits and particles are gap junctions that connect neighboring cells. (Micrograph by Dr. Joy Frank, University of California, Los Angeles.)

p. 399: Electron micrograph of a closed circular DNA molecule, showing formation of the D-loop structure which results from hydrogen bonding of a single stranded DNA fragment to one of the strands in the closed circular molecule. D-loop formation results in unwinding of the superhelical turns, and is catalyzed by the *rec A* protein from *E. coli*, a protein throught to participate in DNA recombination. (Micrograph courtesy of Charles Radding and Chanchal Das Gupta.)

p. 647: Gel electrophoresis pattern (in polyacrylamide in the presence of sodium dodecyl sulfate) of the cell proteins of the bacterium *Bacillus subtilis* infected with bacteriophage SP01, showing the time course of synthesis of proteins. (Photograph by Dr. Michael L. Parker, University of California, Los Angeles.)

p. 747: Electron micrograph of a crystal of the enzyme ribulose biphosphate carboxylase, which initiates the Calvin cycle of photosynthesis. (Micrograph by Dr. T. S. Baker, University of California, Los Angeles.)

Library of Congress Cataloging in Publication Data
Eisenberg, David S
Physical chemistry.

Includes bibliographical references and index.
1. Chemistry, Physical and theoretical. I. Crothers, Donald M., joint author. II. Title.
QD453.2.E37 541′.3 79-12053
ISBN 0-8053-2402-X
KL-MA-8987

THE BENJAMIN/CUMMINGS PUBLISHING COMPANY, INC.
2727 Sand Hill Road
Menlo Park, California 94025

This book is dedicated to our daughters

Jenny and Nell

Nina and Kristina

PREFACE

A New Approach to Physical Chemistry Texts

This book grew out of courses in physical chemistry at UCLA and Yale for students interested in the life sciences, for which we found no current text to be wholly suitable. Our courses, and our book, emphasize those aspects of physical chemistry which find applications to the life sciences. For example, we cover the thermodynamics and spectroscopy of solutions in detail, but devote less space than customary to the physical chemistry of gases. While students may not be exposed extensively to some topics, we believe they are compensated by seeing numerous applications to the areas of biochemistry and biology. Our experience suggests that the majority of students find that it is easier to grasp principles when the principles are illustrated by material they have learned about in other courses.

Our approach to physical chemistry recognizes that many students, particularly those in the life sciences, may not have extensive experience with abstract mathematical description of scientific problems. The logic of the biological sciences is primarily verbal and intuitive, usually relying on qualitative descriptions and models for understanding complex phenomena. In our book we explain physical phenomena in the verbal and intuitive terms that are more familiar to students while at the same time we gradually develop the mathematical formalism which is essential for full mastery and understanding. This text covers the main concepts of physical chemistry and does not evade mathematical rigor in the subjects treated. However, we do try to provide enough conceptual explanations of the models and equations so that students can reinforce quantitative description with qualitative understanding.

To aid students in learning physical chemistry (a subject with a reputation for difficulty) we have employed several devices. These include:

1. Review or development of the required mathematics in the text. (See especially Sections 1–5 through 1–7 for development of the partial derivatives used in thermodynamics and Section 2–10 for review of fundamentals.)
2. The grading of the material of chapters and sections by one of the following symbols:

   ⬡ indicates material appropriate for a basic physical chemistry course.
   ⬢ indicates somewhat more advanced topics in physical chemistry.
   ⬣ indicates biophysical applications, in some cases more appropriate for those students who have special biochemical interests.

3. Marking of important equations by the symbol ◀ in the margins.
4. Gradation of problem difficulty. The problems at the close of each chapter include questions for review, short answer problems, problems that correspond to particular exercises of the chapter, and other problems, some of which are more advanced. Problems based on sections marked ⬢ or ⬡ are so indicated.

To give students more knowledge of the scientists of the past and their influence on our world, most chapters include a biographical sketch of a scientist who contributed to the development of the ideas of that chapter. We have written about these scientists as people, relating not so much the progress of their discoveries, but more the connection of their work to their personal lives and to the world around them. We do not agree with Madame Curie's statement, given to evade questions from journalists, "Science deals with things, not with people."

Each teacher may wish to introduce topics in his or her preferred order. Thus we have attempted to cast the book into relatively independent parts, as explained by the flow diagram on the next page. A connecting line indicates which earlier chapter or section is required for reading a later chapter or section. Notice that Part One contains fundamentals, and any of the other four parts can be taken up next. The book is designed to make selective omission of topics as simple as possible. We have found that in a two-quarter course at UCLA we cover about two-thirds of the material of the book (usually omitting either Part Four or Part Five). In a two-semester course at Yale, nearly all the material is covered. A one-semester course might cover only the basic material in Parts One and Three. A typical two-semester course might cover Parts One, Two, Three, and either Four or Five.

A Note to Students on Units and Dimensions

We use SI (official) units for the most part, but also intersperse them with other units in common use (such as atmosphere, Ångstrom, and kcal) which you will encounter elsewhere. You may not be familiar with the SI quality calculus used in this book; headings of tables and labels on axes are written in fractional form. For example, a table of energies will have the heading Energy/J. This means that a number in the table represents that number of Joules, divided by the unit Joule.

You will find that working problems is easier if you retain units in all computations and equations, and keep in mind that the physical dimensions on both sides of an equation must balance. This procedure of *dimensional analysis* helps to show you if you have omitted a factor, or if your equation is otherwise faulty.

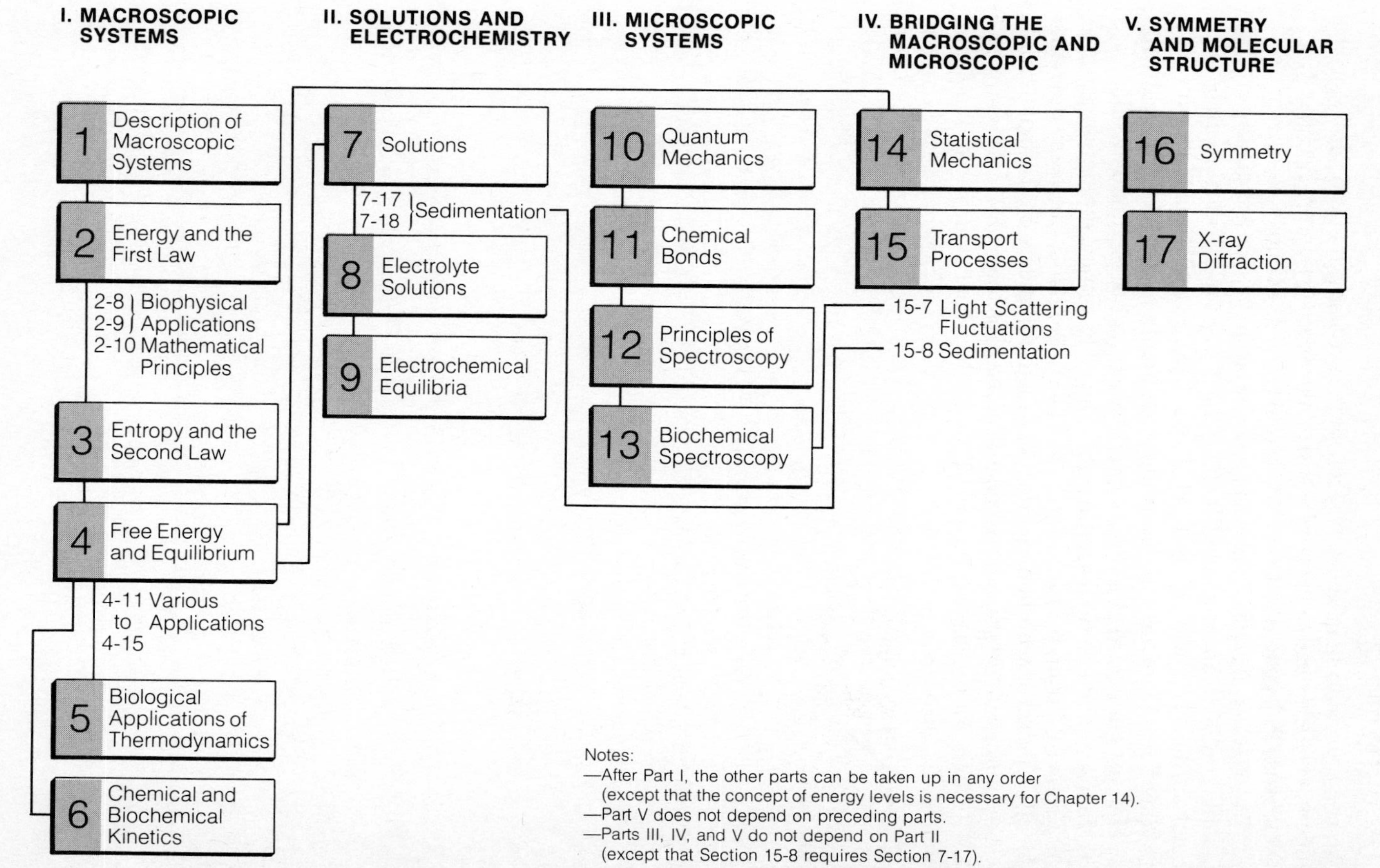
I. MACROSCOPIC SYSTEMS
1 Description of Macroscopic Systems
2 Energy and the First Law
2-8 Biophysical
2-9 Applications
2-10 Mathematical Principles
3 Entropy and the Second Law
4 Free Energy and Equilibrium
4-11 to 4-15 Various Applications
5 Biological Applications of Thermodynamics
6 Chemical and Biochemical Kinetics
II. SOLUTIONS AND ELECTROCHEMISTRY
7 Solutions
7-17
7-18 Sedimentation
8 Electrolyte Solutions
9 Electrochemical Equilibria
III. MICROSCOPIC SYSTEMS
10 Quantum Mechanics
11 Chemical Bonds
12 Principles of Spectroscopy
13 Biochemical Spectroscopy
IV. BRIDGING THE MACROSCOPIC AND MICROSCOPIC
14 Statistical Mechanics
15 Transport Processes
15-7 Light Scattering Fluctuations
15-8 Sedimentation
V. SYMMETRY AND MOLECULAR STRUCTURE
16 Symmetry
17 X-ray Diffraction
Notes:
—After Part I, the other parts can be taken up in any order
(except that the concept of energy levels is necessary for Chapter 14).
—Part V does not depend on preceding parts.
—Parts III, IV, and V do not depend on Part II
(except that Section 15-8 requires Section 7-17).

Acknowledgements

We are grateful to helpful discussions with, or comments from, the following friends and colleagues: D. E. Atkinson, C. Cantor, P. Cole, R. E. Dickerson, P. Lyons, H. Reiss, W. Scanlon, V. N. Schumaker, L. Stryer, R. M. Sweet, I. Tinoco, J. Tuchman, L. Weissman. We are greatly indebted to W. Kauzmann and R. M. Rosenberg for a critical reading of much of the text, to S. W. Suh, G. Kobayashi, and C. Stauffacher for reading portions of the text from the standpoints of advanced students, to J. Hall for editing, M. Moore for production, and to A. Jakubowski, N. Johnson, and D. Nicol for diligent typing. The optical diffraction patterns in Chapter 17 were prepared by T. S. Baker. We thank our many students for their patience with imperfection and their comments on the manuscript.

For comments on the biographical sketches we thank D. L. D. Caspar, J. T. Edsall, L. Eisenberg, R. Eisenberg, K. N. Trueblood, B. W. Tuchman, and W. B. Wood.

David Eisenberg
Donald M. Crothers
Los Angeles and New Haven

INTRODUCTION: A Map of Physical Chemistry

A traveler about to enter unfamiliar territory should be equipped with a map. Like a map, this introduction is designed to show the main subdivisions and boundaries of the territory; and also like a map, it can give no feeling for the inspiring vistas or the possible impediments that the traveler will encounter on his or her way.

Three Theories of Physical Chemistry

Physical chemistry rests on three main conceptual theories: quantum mechanics, statistical mechanics, and thermodynamics.

Quantum mechanics describes the motions and energetics of *micro*scopic particles. These are particles too small to be seen with a light microscope, such as electrons, nuclei, atoms, and molecules. Given an expression for the potential energy of interaction of any group of these particles, quantum mechanics allows us, at least in principle, to compute properties of the particles such as the probabilities that they have certain positions and momenta, and the quantized energy levels accessible to them (see Figure I-1).

The modifying phrase ''in principle'' is crucial to the accuracy of the preceding sentence, because computations on all but the simplest systems bog down within a few steps in fearful mathematical complexity. Some idea of the very limited power of quantum mechanics to make predictions emerges from the list of real systems for which it gives the exact descriptions: the hydrogen atom and the H_2^+ ion. The list ends there. To be fair, excellent approximate descriptions have been given for many small atoms and molecules. Calculations of the structure and properties of a single water molecule can now be made in good agreement with experiment (Figure I-2*a*), but the computation for a group of ten water molecules is a frontier problem, and calculations of the properties of sodium and chloride ions surrounded by water molecules are at the limit of—and perhaps beyond—the capacity of current methods and computers.

In contrast to quantum mechanics, *thermodynamics* deals with the interdependence of *macro*scopic properties, such as the temperature, pressure, and concentrations in an aqueous solution of hemoglobin and oxygen. This general theory has two facets useful in applications to the life sciences. First, thermodynamics places strict limitations on the

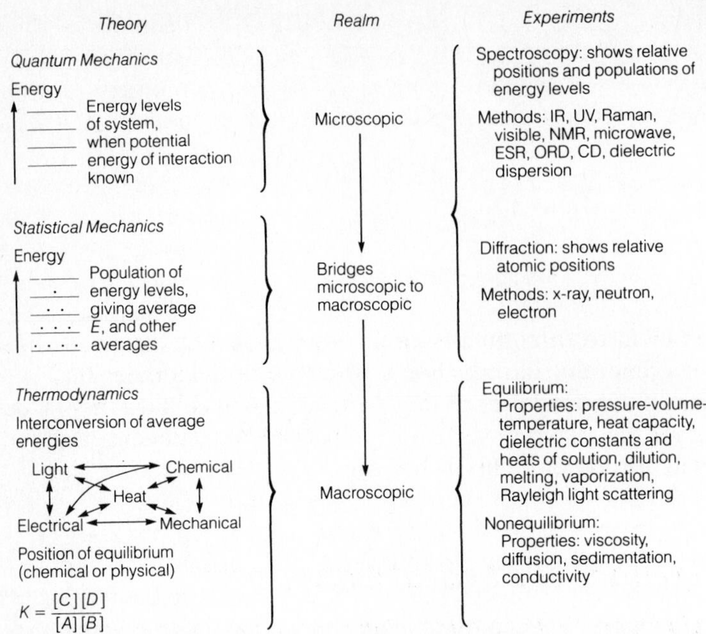

FIGURE I-1
Classification of theories and experiments of physical chemistry

interconversion of different forms of energy. Since the interconversion of heat, chemical energy, electrical energy, mechanical energy, and light is basic to life, it is clear that thermodynamics must be at the center of any physical description of living matter. The second important facet of thermodynamic theory is that it allows us to predict the position of any chemical equilibrium, knowing only the properties of the substances involved in the reaction. This is of great importance in understanding metabolism. In addition to these two facets, thermodynamics is such a familiar theory to physical chemists that its concepts are borrowed for use in other less fundamental theories, such as that of chemical kinetics (Chapter 6). This is why we take up thermodynamics at the start of the book, and treat it thoroughly enough so that before long you may regard it as an old friend.

Statistical mechanics bridges the *micro*scopic realm of quantum mechanics and the *macro*scopic realm of thermodynamics (see Figure I-1). This theory tells us how to use the energies of the individual particles described by quantum mechanics to arrive at the properties of the macroscopic system. An example of an application of statistical mechanics is the computation of the pressure (a macroscopic property) of liquid argon as a function of its temperature and volume. Values of the pressure, in good agreement with experiment, can be calculated if one starts with an accurate potential energy function for the interaction of two argon atoms. Such a potential energy function can be obtained from quantum mechanics. Today, similar calculations on liquid water are a research problem in statistical mechanics. Although it is possible to use statistical mechanics to relate some characteristics of polymer chains to observable properties, to treat a solution of hemoglobin by rigorous methods, and to obtain reasonable answers, is not possible at the present time. Numerous examples of bridging from the microscopic to the macroscopic realm are also given in Chapter 15 on transport processes. These are processes such as diffusion, viscosity, and electrophoresis which find many applications in biochemistry. Because these phenomena depend on the development in time of the system, they are sometimes called *nonequilibrium* properties.

It should be clear from the few examples given for quantum mechanics and statistical mechanics, that biochemists or life scientists must resort to experiments if they are to learn much about the systems that attract their interest. But this does not mean that quantum and statistical mechanics are of little use in their work. On the contrary, they constitute the framework within which nearly every experimental result must be interpreted. It is the interpretive power of these theories, in addition to their predictive power, that makes them so useful to scientists. Our discussions of the principles of quantum mechanics (Chapter 10) and of statistical mechanics (Chapter 14) form the foundation for the discussions of experimental methods in many of the other chapters.

The Divisions of This Book

Part One treats macroscopic systems. The fundamentals of thermodynamics are developed in Chapters 1–4, and are applied to systems related to the life sciences in these chapters, and in the extra illustrations of Chapter 5 to some more complex systems. The experimental data discussed in these chapters are characteristic of a macroscopic system at equilibrium. They include pressure-volume-temperature data, heat capacities and other thermal-energy functions, and equilibrium constants. Part One closes with Chapter 6 on biochemical kinetics, which is an introduction to time-dependent systems.

Part Two (Chapters 7–9) continues the development of thermodynamics for solutions and electrochemical systems. The treatment is more detailed than that of Part One, and is more directly applicable to systems studied by most biochemists.

Part Three (Chapters 10–13) deals with microscopic systems. The basic elements of the theory of quantum mechanics are developed, and spectroscopic methods which yield direct information on molecules are discussed. Spectroscopic experiments can reveal the quantized energy levels of a system, and the relative populations of particles in these various levels. Nuclear magnetic resonance, electron spin resonance, as well as infrared, microwave, visible, ultraviolet, and Raman spectroscopy all are in this group. For relatively simple systems, such as the water molecule, the information on energy levels can be converted rapidly into information on

$r = 0.9572 \pm 0.0003$ Å

$\alpha = 104.52° \pm 0.05°$

(b)

(a)

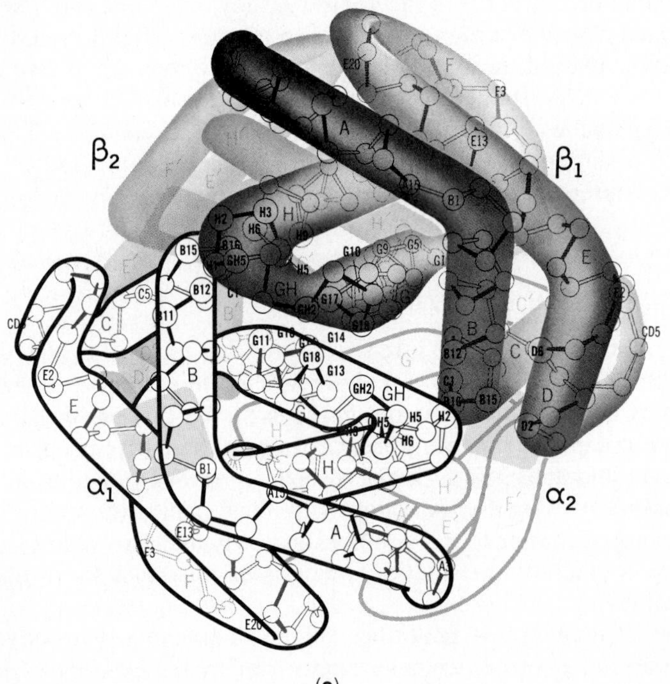

(c)

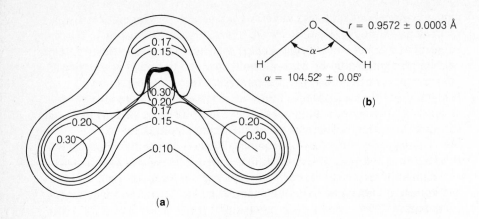

FIGURE I-2
Some structures determined by methods of physical chemistry. **(a)** Density of electrons in the water molecule, as computed from quantum mechanics. The density of electrons is shown by contour lines, just as altitudes are shown by contours on a topographical map. The oxygen nucleus is at the intersection of the two straight lines and the hydrogen nuclei are at the opposite ends of the lines. (Reproduced from Bader and Jones, 1963.) **(b)** Dimensions of the water molecule as determined by infrared spectroscopy. (By Benedict *et al,* 1953.)
(c) Structure of the hemoglobin molecule, as determined by x-ray diffraction. The outlines are shown of the four polypeptide chains of the protein (α_1, α_2, β_1, and β_2). Circles represent some of the amino acid residues. (Reprinted from *The Structure and Action of Proteins* by Richard E. Dickerson and Irving Geis. Menlo Park, Calif.: W.A. Benjamin, Inc. Copyright 1969 by Dickerson and Geis.)

molecular structure. For example, from the infrared spectrum of water vapor the O—H bond length and the H—O—H bond angle have been determined to better than one part per thousand accuracy (see Figure I-2*b*). For larger molecules, the conversion of spectroscopic to chemical information is not as direct, but spectroscopic techniques are extremely powerful for studying molecular motions and changes in structure.

For these larger molecules, structural information is obtained from the diffraction methods, described in Part Five. Diffraction techniques include x-ray, neutron, and electron diffraction, and the related technique, electron microscopy. These techniques give direct information on the structures of molecules as large as enzymes and viruses, and have played an important part in the development of chemistry and molecular biology (see Figure I-2*c*).

Part Four develops the theory of statistical mechanics used to bridge macroscopic properties to the microscopic realm. As noted earlier, this is done in two chapters: Chapter 14 on the principles of statistical mechanics, and Chapter 15 on time-dependent phenomena.

Why Mathematics?

The newcomer to the study of physical chemistry finds one feature common to all its divisions: the language of mathematics. Some newcomers, particularly from the life sciences, wonder whether this is necessary, or whether it is employed as a means to keep down the number of newcomers. We believe that there are three reasons why mathematics forms an essential part of physical chemistry. The first is precise definition. When a quantity is defined by an equation, one may dispute the wisdom of the definition, but rarely is there any dispute about its meaning. This is much less true with verbal definitions, as the thriving activity of our law courts and diplomatic missions attests. The second benefit of mathematics is the possibility of mathematical deduction. Once scientific concepts have been defined in mathematical terms, manipulation of symbols can produce results that are not obvious in the original concepts. The third reason for using mathematics is more subtle, and though vital in science, it sometimes seems unimportant to students. From mathematical expressions, scientists can estimate the probable error in a quantity they are calculating or measuring. This estimate can be crucial, because few scientific experiments lead to a clear "yes" or "no" answer; more often the answers are of the form "maybe yes" or "perhaps no." To be sure of the conclusion, one must be able to judge the size of the error in the measurements and to know just how big is the "perhaps." One method for estimating errors from mathematical relationships is described in Chapter 1.

The authors realize that many newcomers to physical chemistry experience some initial difficulty with the mathematical language, and have endeavored to be careful guides and interpreters. Section 2-10 is designed to help readers whose calculus is rusty or so newly acquired that it needs breaking in.

Model and Reality in Physical Chemistry

Physical chemistry may be defined as "the science that describes the world in terms of atoms, molecules, and energy." As we noted previously, the physical chemist casts his or her descriptions of matter and energy in terms of the three main conceptual theories. Each of these theories is a model of reality. We use these models in thinking about the world because they are consistent with many observations, but no serious scientist would take these models as reality itself; each model is at best a simplified representation of reality.

The relationship of models to reality has concerned thinkers since Plato (428–348 BC). One school of thought, the Idealists, of whom the philosophers Plato and Hegel (1770–1831) are the most notable examples, holds that concepts such as space, order, and causation are attributes of the human mind which organize our perceptions of the world. This school believes that models preexist in our minds. A second school is that of the Empiricists, as exemplified by the philosopher David Hume (1711–1776). They believe that our sense organs record and define the order that exists in the world about us. Space, time, forces, causation, and so forth are attributes of this world that exist independently of us.

Most modern scientists, if forced to state their views on the question, might take a position between these extremes. Experience and observation are essential for understanding the world, but the complexity of external reality is so great that to comprehend it we must describe it in terms of simplified models. Great scientific minds produce these models, and the models are extended and refined by our fitting them to observations. This interplay of model and reality, of our minds with the external world, is part of the excitement of physical chemistry, and of science in general.

References

Bader, R. F. W., and Jones, G. A. 1963. *Canadian Journal of Chemistry, 41,* 586.

Benedict, W. S., Gailar, N., and Plyler, E. K. 1953. *J. Chem. Phys., 21,* 1301.

Dickerson, R. E., and Geis, I. 1969. *The Structure and Action of Proteins.* Menlo Park, Calif.: W. A. Benjamin, Inc.

CONTENTS

PART ONE

Macroscopic Systems 1

PART THREE

Microscopic Systems 399

CHAPTER 10

Principles of Quantum Mechanics 401

CHAPTER 11

Chemical Bonds 468

CHAPTER 12

Principles of Spectroscopy 516

CHAPTER 13

Biochemical Spectroscopy 562

PART FOUR

Bridging the Macroscopic and Microscopic 647

CHAPTER 14

Statistical Mechanics 649

CHAPTER 15

Transport Processes 700

CHAPTER 17

X-ray Diffraction and the Determination of Molecular Structure 798

PART ONE

Macroscopic Systems

CHAPTER 1

Description of Macroscopic Systems

○ 1–1 The Nature of Thermodynamics

How can it be that dissolving one substance in water, such as NaOH, heats the liquid up, while dissolving another, such as urea, cools it down? Why is it that mixing ethanol and butanol yields a final volume greater than the combined volumes of the separated liquids, whereas mixing two other substances, such as ethanol and water, gives a volume smaller than the total of the initial substances? How can it be that compressing ice tends to melt it to water, whereas compressing most liquids tends to make them freeze? Why is it that as you use the books on your desk in a random way they tend to get out of order rather than tending toward some special order, such as alphabetical?

These are a few of the diverse questions that are illuminated by thermodynamics, the branch of physical chemistry that describes matter on the macroscopic level and the physical and chemical changes that it undergoes. This description rests both on a simplified representation, or *model,* of physical reality, and on a small number of *laws*. This chapter discusses the model, the next two chapters introduce the laws, and the following chapters discuss deductions that can be formed from these laws, as well as applications of these deductions to chemical and biochemical problems. For reading this chapter, all you need know about the laws of thermodynamics is that they are statements, in terms of physics, of accumulated empirical experience about the interchange of energy among substances. The core of thermodynamics is the development from the laws and the model of a myriad of useful relationships that describe the interchange of energy and the direction of equilibria in physical processes and chemical reactions.

As useful as thermodynamics is in chemistry, biology, and physics, it has nothing to say about many of the most important and interesting questions of these sciences. The model and laws of thermodynamics make no mention of the atomic properties of matter, and consequently the theory can yield no direct information about the microscopic realm. Only when additional molecular assumptions are introduced can thermodynamics contribute to our understanding of molecular processes. Another basic limitation of thermodynamics is that the theory has nothing to say about rates. Though the laws can be used to determine which direction a process will take of its own accord, they give no clue as to how long it may take for the process to occur. The element of time is in the province of *kinetics,* which is discussed in Chapter 6 of this book.

1–2 The Thermodynamicist's Picture of the World

Thermodynamic theory depicts the world in a highly simplified fashion. All fine details are smeared out, and each object is considered to be characterized by a small number of uniform *properties,* such as temperature, pressure, and energy. The part of the world that is selected for consideration is called the *system.* Depending on one's interest, the system might be a reaction flask, a steam engine, a human being, or the earth. The region around the system is called the *surroundings.* The surface dividing the system from the surroundings is called the *boundary.* It may be real or imaginary. For example, part of the boundary of the reaction flask in Figure 1-1 is the real glass surface, and part is the interface between the liquid and the atmosphere.

All interactions of a system with the surroundings occur across the boundary. Sometimes matter may cross the boundary, as may the two forms of energy, *heat* and *work.* We will not give precise definitions of these quantities until later, but your common experience tells you that they are different. When you heat a pot of water on a stove, energy in the form of heat crosses the boundary of the system (the pot) and warms the water inside. Energy can also cross the boundary of a system in the form of work. Suppose you have a sample of air contained in a cylinder, and you compress it by forcing a piston to move inside the cylinder. The work you do is the force you exert times the distance the piston moves; the energy expended crosses the boundary of the system. A consequence of doing work on a gas, such as air, is that it tends to warm up because of the work energy that crosses the system boundary. Since both heat and work can cause the system's temperature to increase, they must be equivalent in some way. The relationship between the two is a main preoccupation of the study of thermodynamics.

Systems in thermodynamic theory are classified by the permeability of the boundary to matter, heat, and work. An *open system* has a boundary through which matter, heat, and work are free to pass, whereas the boundary of a *closed* system does not permit transport of matter (Fig. 1-1) but may allow passage of heat and work. The boundary of an *adiabatic* system is impermeable to heat and matter. Though no real system is ideally adiabatic, thermal insulation such as styrofoam can be used to approximate the thermodynamic

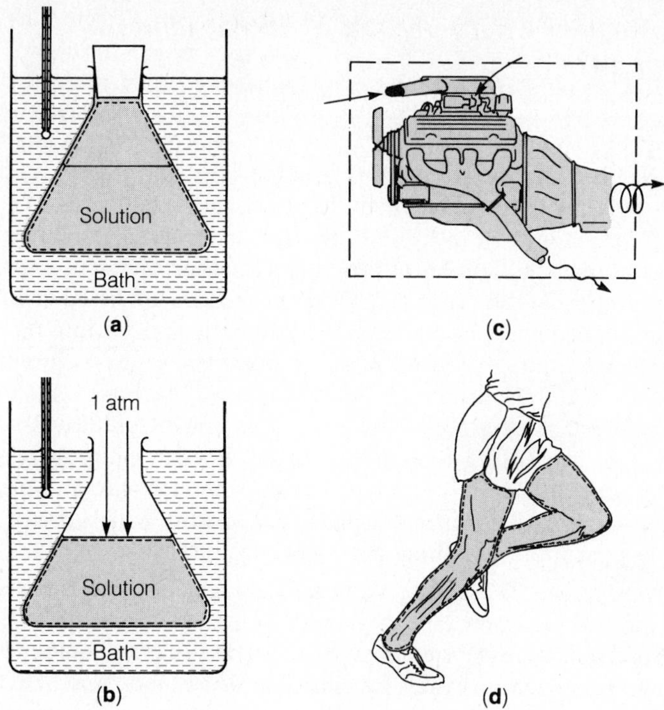

FIGURE 1-1
Some thermodynamic systems. **(a)** A *closed system:* a tightly stoppered flask holding a solution immersed in a heat bath is at constant temperature and volume. **(b)** An *open system:* the same flask in the heat bath, but open to the atmosphere, is at constant temperature and pressure (of 1 atm). **(c)** An automobile is an open system, doing work on the surroundings when in operation; heat also crosses the boundary. **(d)** The muscles in a sprinter's legs may be considered an open system. The system does work on surroundings (bones), as glucose and oxygen enter the system and CO_2 leaves the system.

ideal. An *isolated* system has a boundary that transmits neither heat nor work nor matter.

EXERCISE 1-1

To what extent are the following systems closed, adiabatic, or isolated: (*a*) a bacterium, (*b*) a bird's egg, (*c*) the London underground, (*d*) a space capsule in orbit, (*e*) the earth, (*f*) the solar system, (*g*) the universe.

ANSWER

(*g*) is isolated by definition; (*d*), (*e*), and (*f*) are nearly closed; (*d*) would be isolated were it not for radiation transfer (from the sun and into space); (*b*) is open to diffusion of gases and is certainly not adiabatic (since heat from the mother is essential for hatching); (*a*) and (*c*) are open.

A System is Characterized by its Variables or Properties

In the simplified picture of reality used in thermodynamics, each system is described by a small number of *variables of state* or *properties*. A characteristic of the system is a property only if its value does not depend on how the system was brought to the chosen conditions. A property cannot be a function of the past history of a system; it depends only on the conditions at the time of measurement. Examples of properties are the volume of a system, V, and the temperature of the system, T. Other properties are listed in Table 1-1. Properties are called *intensive* if, when the system is divided into parts, each part has the same value of the property. Thus temperature is intensive because if a system at uniform temperature is divided, each part has the same temperature. A property is called *extensive* if its value is additive, such that when parts of the system are brought together, values of the properties of the parts add together to form the value for the whole. Volume is clearly an extensive property. All *specific* or *molar* properties, which have units of something per gram or something per mole, are intensive.

EXERCISE 1-2

Are the following intensive or extensive properties: (*a*) the number of moles of a solute in a solution; (*b*) the concentration of a solute?

ANSWER

(*a*) Extensive. (*b*) Intensive.

TABLE 1-1 *Some Properties of a System*

TYPE	EXTENSIVE	INTENSIVE
Mass-related	Mass	Density Concentrations of solutes
Pressure-volume-temperature	Volume	Specific volume (volume per gram) Molar volume (volume per mole) Pressure Temperature
Thermal energy	Heat capacity	Specific heat (heat capacity per gram)
	Energy	Molar energy
	Entropy	Molar entropy
	Enthalpy	Molar enthalpy
	Free energy	Chemical potential
Some other properties		Dielectric constant Refractive index Viscosity

A system is said to be in a *defined state* when all its variables or properties have specified values. We would not consider the system of an ice cube just immersed into a cup of hot coffee to be in a defined state, because it is not possible to specify the temperature and concentration of the coffee on a macroscopic level. A uniform solution of hemoglobin at one atmosphere pressure at 37°C at a concentration of 10 mg/ml is in a defined state. A defined state is said to be in *thermodynamic equilibrium* if its properties are independent of time and there is no flow of mass or energy. The hemoglobin solution is a system in thermodynamic equilibrium.

When there is a flow of matter or energy through a system, and yet no change of properties with time, we say it is in a *steady state*. The wall of a heated house could be regarded as a system in a steady state: the temperature of the inside surface is, say, 25°C, and that of the outside surface is 10°C, and the temperature at any given point within the wall has a constant value. But the system is not at equilibrium because of the net flow of energy. A suspension of metabolizing cells is also in a steady state, because of its continual intake of O_2 and its release of CO_2 and H_2O. The suspension may be considered to be at equilibrium if the concentrations of dissolved substances are everywhere the same and not varying appreciably with time, but the interior of the cell is certainly not at equilibrium.

The Properties of a System are Related by Equations of State

A mathematical relationship between some of the properties of a system is called an *equation of state*. These relationships cannot be derived from the laws of thermodynamics. They are usually empirical relationships devised to fit experimental data, or relationships derived from a molecular theory. An example is the ideal gas law

$$PV = nRT \tag{1-1}$$ ◀

in which n is the number of moles of gas at temperature T and pressure P in the volume V, and R is the gas constant (0.082 liter atm K^{-1} mol^{-1}). This equation relates four variables of state for one type of system (a dilute inert gas). Another equation of state is that for the volume of a mole of water as a function of temperature; a possible form for such an equation of state is

$$V = a + bT + cT^2 + fT^3 \tag{1-2a}$$

where a, b, c, and f are constants determined by fitting the equation to the observed volume. This equation gives in mathematical form the dependence of one property on another. Other equations of state, such as Raoult's law, describe solutions (see Tab. 1-2).

Equations of state such as these are descriptions of some of the macroscopic properties of a substance. They are the means by which the thermodynamicist summarizes data on the system of interest. We shall see in the next several chapters that when the laws of thermodynamics are applied to

TABLE 1-2 *Some Equations of State*

EQUATION	PROPERTIES RELATED BY EQUATION	OTHER QUANTITIES	COMMENT
Ideal gas: $PV = nRT$	Pressure, P Volume, V Temperature, T Number of moles, n	Gas constant, R $= 0.0821$ liter atm mol^{-1} K^{-1} $= 1.987$ cal mol^{-1} K^{-1} $= 8.314$ J mol^{-1} K^{-1}	Derived from kinetic theory or from experiments on inert gases.
Van der Waal's equation for a gas: $\left(P + \frac{n^2a}{V^2}\right)(V - nb) = nRT$	Same as ideal gas	a and b are constants for a given gas, determined by fit to experimental data.	This equation represents deviations from ideality semi-quantitatively.
Virial equation for gas: $PV = nRT\left(1 + \frac{nB'}{V} + \frac{n^2C'}{V^2} + \cdots\right)$	Same as ideal gas	B', C', . . . are the *virial coefficients*, determined by fit to experimental data.	This equation can represent deviations from ideality with higher precision.
Tait's equation for a liquid: $\frac{1}{V_0}\frac{dV}{dP} = -\frac{C}{B + P}$	Pressure, P Volume, V Volume at zero pressure, V_0	C and B are constants for a given liquid and temperature, determined by fit to data.	Describes dependence of volume on pressure for many liquids.
Raoult's law for ideal solutions: $P_A = X_A P_A^{\bullet}$	Vapor pressure of component A over solution, P_A Vapor pressure of pure A, $P_A^{\bullet}$ Mole fraction of A in solution, X_A		Describes vapor pressure (and therefore other properties) of solutions in which molecules mix evenly (see Chapter 7).

such equations of state, many predictions about the behavior of that substance can be made. Scientists place a great deal of confidence in such predictions, beçause the laws of thermodynamics have been tested for a century, and conclusions based on these summaries of experience have invariably proved correct. Notice that no single equation of state summarizes all the macroscopic behavior of a substance. The ideal gas equation, for example, tells nothing about the viscosity or refractive index of the gas.

A *state function* is a concept related to an equation of state; it is a mathematical expression for a property in terms of values of other properties used to specify the state of a system. State functions are often written only in functional form, for example

$$V = V(T,P) \tag{1-2b}$$

because the detailed function is unknown. Even such a sparse relationship is useful, because the laws of thermodynamics can tell us a good deal about its behavior. You should note that in contrast to Equation 1-2*b*, Equation 1-2*a* is not a state function, because the independent variables (T) do not include all those necessary to specify the state of the system.

How Many Properties are Needed to Specify the State of a System?

One of the most fundamental discoveries of all thermodynamics, and one that justifies the simple model of reality used in this science, is that the state of a system is fully specified once the values of just a few properties have been specified. For example, if your system is a mole of water, once you have specified the temperature and pressure of the water, the values of *all* other properties are fixed, including volume, viscosity, refractive index, and so forth. Alternatively, if you specify the viscosity and the refractive index, you have fixed the volume, temperature, pressure, and so forth.

The number of properties that have to be specified in order to define the state of a system is given by the *phase rule*. This was first stated by the great American physicist J. Willard Gibbs in his famous paper, *On the Equilibrium of Heterogeneous Substances,* published in 1878. We must wait until Chapter 4 to derive the phase rule, but the result is simple enough to be stated here, where it is helpful to an understanding of the thermodynamic model of reality.

For a single pure and homogeneous substance, such as liquid water, the phase rule tells us that one extensive property (such as mass) and two intensive properties (such as temperature and pressure) are sufficient to specify the state of the system. All other properties are then automatically fixed, or in other words, there are no further *degrees of freedom* that can be varied. For each additional component present in the homogeneous system, one additional intensive property must be stated to specify completely the state of the system. A *component* is defined as a constituent of the system whose concentration can be independently varied.

Hemoglobin dissolved in the water sample would be an additional component and its concentration could be the additional property. For each additional phase present, one intensive property less must be given to specify the state of all intensive variables of the system. However, one more extensive property is needed to specify the mass of each phase. A *phase* is a region of the system that is physically and chemically homogeneous, such as a solid region or a gaseous region.

The significance of the phase rule is that macroscopic systems at equilibrium can be described by a very small number of quantities (which we call *properties* or *variables of state*). This results in a vast simplification over any molecular description of a system, which would require on the order of 10^{23} numbers (approximately the number of molecules in a mole). This enormous reduction in the complexity of description shows the usefulness of the thermodynamicist's picture of the world.

EXERCISE 1-3

How many *intensive* variables of state (properties) are sufficient to specify the state of each of these systems (in addition to one extensive variable per phase): (*a*) one mole of ice; (*b*) liquid water in equilibrium with water vapor; (*c*) an ice cube floating in water in equilibrium with water vapor; (*d*) a solution of hemoglobin in water?

ANSWER

(*a*) two (e.g., pressure and temperature); (*b*) one (because at any temperature the pressure is fixed—for example, at 100°C the pressure must be 1 atm); (*c*) none—the temperature must be 0.01°C and the pressure 4.5 mm of Hg (this is the *triple point* of water, the only condition at which these three phases can exist in equilibrium); (*d*) three (e.g., pressure, temperature, and concentration of hemoglobin).

J. WILLARD GIBBS

1839–1903

J. Willard Gibbs, shown in a bas-relief donated to Yale University by W. Nernst on the occasion of his visit as a Silliman lecturer in 1906. (Photograph courtesy of Arthur M. Ross.)

J. Willard Gibbs received the second Ph.D. degree in science granted in America, but he remains to this day the outstanding American contributor to theoretical physics and chemistry.

Gibbs was born, educated, and spent almost his entire life in New Haven, Connecticut. He came from a family whose members had for generations been educated in America: his father was Professor of Hebrew at Yale, his maternal grandfather had been a physician and professor of chemistry at Princeton, and on the Gibbs side, his grandfather, both great grandfathers, and several great-great, and great-great-great grandfathers had attended Harvard.

Gibbs himself received his Ph.D. from Yale in 1863 for a thesis in engineering, on the meshing of gears. In the same year he was appointed tutor in Yale College. For two years he taught Latin and for a third, mathematics. During this period he invented a hydraulic turbine and patented an improved railway car brake.

From 1866 to 1869 Gibbs studied mathematics and physics at the Universities of Paris, Berlin, and Heidelberg, where he was introduced to the frontiers of physical science of that day. He returned to New Haven and in 1871 was appointed Professor of Mathematical Physics—without salary. During the next several years he developed the fundamentals of chemical thermodynamics and published the theory in his celebrated paper of 1878 in the *Transactions of the Connecticut Academy of Arts and Sciences*.

This achievement brought an offer of a professorship at Johns Hopkins University, carrying a stipend of $3000 a year. Yale shrewdly countered with an offer of $2000, and Gibbs elected to remain in New Haven.

During the period from 1880 until his death in 1903, Gibbs developed the essentials of vector analysis, worked in theoretical optics, and laid the foundations of statistical mechanics. Towards the end of his life he received many of the highest scientific honors of the period. He died in the third year of the Nobel awards without having received one, but at least four later awards were given for work based directly on his concepts.

Gibbs the person was the very model of professorial sobriety, serenity, and restraint. He never married. He shared a home with his sisters. His patience was such that he could carry out a prolonged correspondence with a retired middle-western clergyman who wrote him abusive letters about Gibbs' attempts to explain the velocity of sound in a gas, and with another man who claimed that the true value of π is 3.125. His restraint is illustrated by his brief statement at the end of a prolonged debate of the Yale faculty on whether studying classical language or mathematics is the better discipline for students. Gibbs rose to say, "Mathematics *is* a language."

○ *1–3 A Process is an Event in which a Property of a System Changes*

The essence of the preceding section is that the state of a macroscopic system in equilibrium can be specified by a small number of properties. We must now ask how a system is changed from one state to another. Since by definition the system in the new state will be characterized by a new set of properties, it is obvious that one or more of the variables of state must be changed. In thermodynamic terms, a *process* is a change in some property of the system.

When a process occurs in a system previously at equilibrium, something must cross the boundary of the system. It can be heat, work, or matter, or some combination of these. As one of these entities crosses the boundary, one or more variables of state of the system assumes a new value. For example, if component A is introduced into the system, the concentration of A is increased. Several examples are given in Figure 1-2. Every chemical reaction is an example of a process, because the concentrations of some components are increased and others are diminished.

A type of change of special interest in thermodynamics is a *reversible process*. This is a process that proceeds through a succession of equilibrium states, each differing from the last by an infinitesimal change in a variable of state. Figure 1-3 illustrates the difference between reversible and irreversible processes. In the case of expansion of a gas from a smaller volume V_1 to a larger volume V_2, the expansion can be carried out reversibly as follows: the system is the gas, confined in a cylinder by a plunger that has many tiny weights on it. If one weight is removed, the pressure P is reduced infinitesimally, the plunger rises and the volume is increased infinitesimally to $V_1 + \delta$. Another tiny weight is removed, and so forth, until the volume is V_2. The system is always at equilibrium, and hence the process is reversible. If the plunger instead is suddenly released so that it can spring up to a volume V_2, the system passes through a period in which the pressure in the cylinder cannot be specified (there may, for example, be turbulence in the gas) and

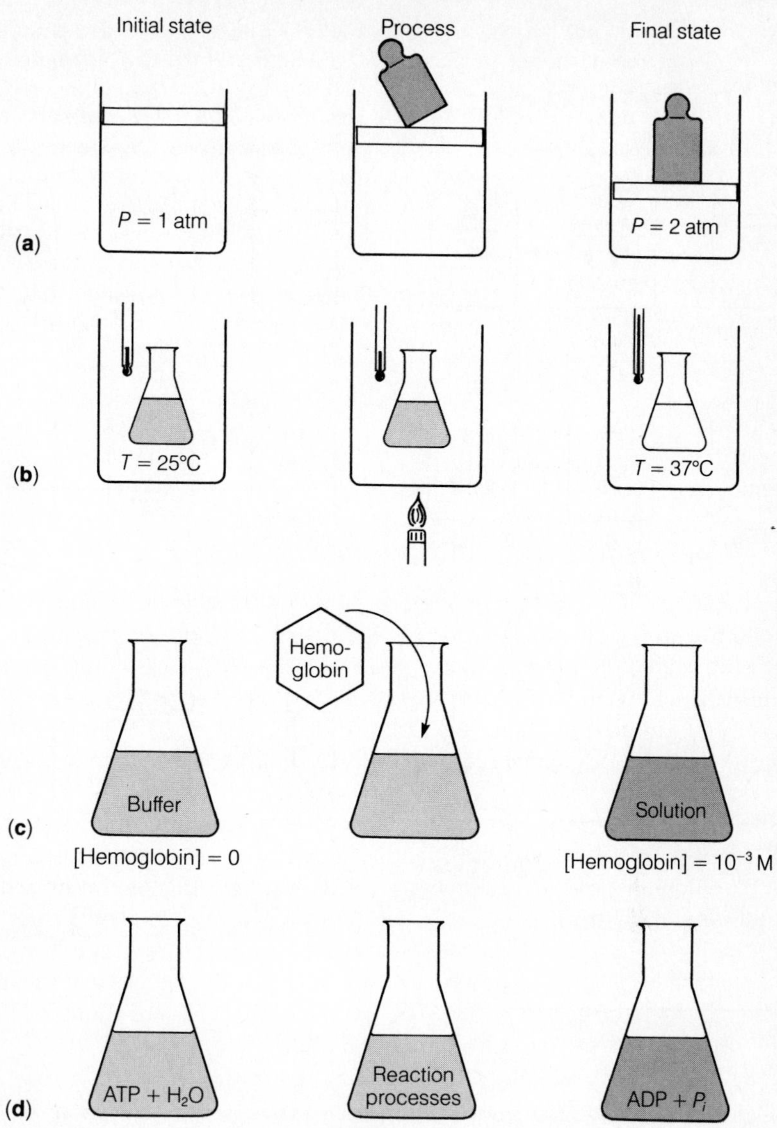

FIGURE 1-2
Examples of processes: **(a)** The compression of a gas in a cylinder by pressing down a piston in the cylinder. **(b)** Heating water in a flask. **(c)** Adding crystalline hemoglobin to a solution. **(d)** A chemical reaction, e.g., hydrolysis of ATP to yield ADP and inorganic phosphate, P_i.

during which the properties are changing with time. Hence the system is not at equilibrium during the process, and we call the process *irreversible*.

Now consider two processes, one irreversible and one reversible, that go from the same initial state to the same final state. An example is the two expansions we have just considered. Two such processes have an important similarity and an important difference. They are similar because the states of

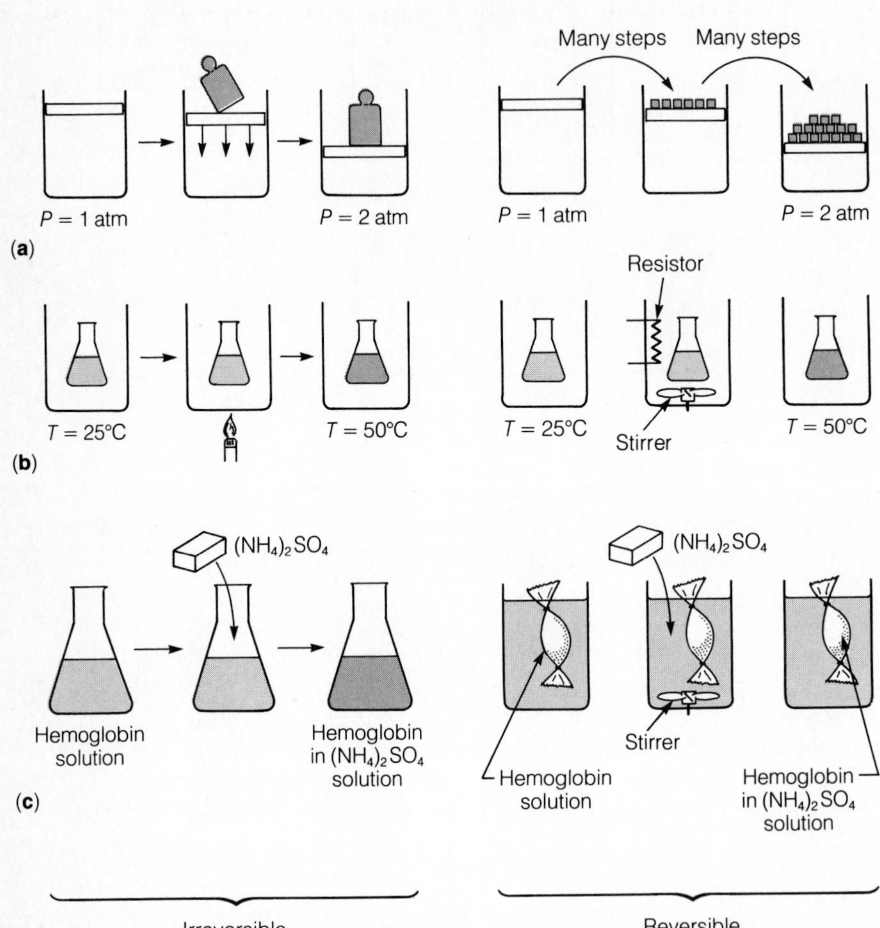

FIGURE 1-3

Some irreversible and reversible processes between the same initial and final states: **(a)** A gas is compressed irreversibly by adding a heavy weight to a piston or reversibly by adding a succession of tiny weights. **(b)** A solution is heated irreversibly by a Bunsen burner or reversibly by passing a small current through a resistor and stirring the solution. **(c)** Ammonium sulfate is added to a solution of hemoglobin in solid form, and then reversibly passed through a membrane that does not allow the hemoglobin molecules to escape, because they are too large for the pores of the membrane.

the system for the two processes are identical before the process, and identical afterwards. This means that all properties of the system are the same in the final state regardless of the process. This is so just by our definition of a property. The important difference is that the amounts of heat and work that cross the boundary during the process are different for the two processes. Heat and work depend on the detailed path of the process, not just on the initial and final states. We shall see numerous examples of this in the following chapters. The fact that heat and work are not determined by the state of a system, but rather by the details of a process means that they are not properties. It is meaningless to speak of the heat or work of a system; rather one must speak of the heat or work that pass between a system and its surroundings in a specific process.

1–4 Definitions of Heat and Work

The essence of the first law of thermodynamics (Sec. 2-1) is that both heat and work are forms of energy. As we shall see, *energy is a property of a system but heat and work are the means by which energy can flow*. Their values depend on how a process is carried out.

Historically, *work* had a precise definition much earlier than did *heat,* since the science of classical mechanics includes a definition of work. In its simplest form, work is the product of a force times the distance moved:

$$\text{Work} = \text{force} \times \text{distance}$$

More precisely, for a small distance displacement (ds), the work (dW) is the product of ds and the force F:

$$dW = F\,ds \qquad (1\text{-}3)$$

Work can be done on the system by the surroundings, or by the system on the surroundings. In order to distinguish these two possibilities, we adopt the convention that *dW is positive when work is done on the system by the surroundings*. (Some textbooks, particularly older ones, use the opposite sign convention.) The consequence of our convention is that a *positive dW* represents a flow of energy *into* the system, supplied by the work done by the surroundings.

An example of work done on a system is the compression of a gas in a cylinder by an external pressure, P_{ext} (Fig. 1-4). Pressure is force per unit area, $P = F/A$, so the force on a piston of area A is

$$F = P_{ext}\,A$$

If ds is the distance the piston moves in the compression, the work done by the surroundings on the system is

$$\begin{aligned} dW &= F\,ds \\ &= P_{ext}\,A\,ds \end{aligned}$$

When the piston moves a distance ds in compressing the gas, the volume of the gas decreases, changing by dV. Therefore, we set $-dV$ equal to the product of area and distance, or

$$A\,ds = -dV$$

and get

$$dW = -P_{ext}\,dV \qquad (1\text{-}4)$$

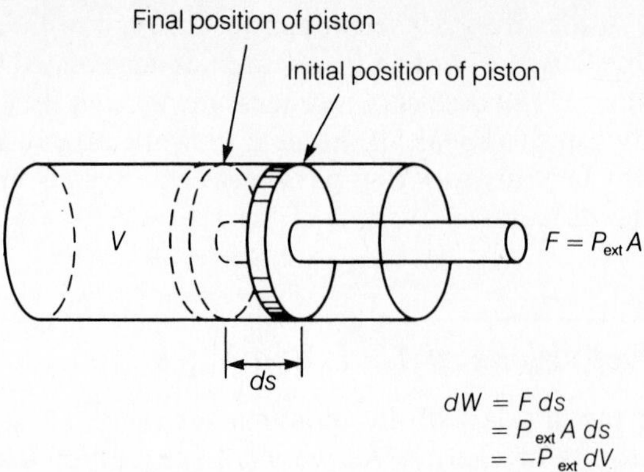

FIGURE 1-4
Calculation of the work done by the surroundings in compressing gas (the system) contained in a cylinder. The piston, which is of area A, moves a distance ds from right to left, under the force $F = P_{ext}A$ supplied by the surroundings. $dW = -P_{ext}\, dV$ is *positive* for compression of the gas (dV negative), because work is done on the system by the surroundings. If the gas were to expand, doing work on the surroundings, dW would be negative.

Remember that dV is negative in a compression, so dW is positive. If the compression of the gas is carried out reversibly, the external pressure (P_{ext}) is always equal to the internal pressure (P), so Equation 1-4 can be replaced by

$$dW = -P\, dV \text{ (reversible process)} \qquad (1\text{-}5) ◀$$

You may wonder in what circumstances P_{ext} would not equal the pressure of the gas. It would not be an equality, for example, if the piston experienced friction against the cylinder. Then we would have $P_{ext} > P_{int}$. In the imaginary reversible process, the reality of friction is ignored. In either case, the work done on the system is defined in terms of the external pressure, because that quantity times dV gives the infinitesimally small energy change with the surroundings.

Work can be mechanical, electrical, or chemical. In each case dW is expressed as the product of an intensive variable times a small change in an extensive variable. For example, electrical work is the product of a voltage difference (intensive variable) times the amount of charge (extensive variable) transferred through the voltage change. Table 1-3 summarizes expressions for the different kinds of thermodynamic work. (The meaning of some of the variables in this table will not be discussed until later chapters.)

The simplest definition of heat is in terms of its ability to change the temperature of an object. Consider Figure 1-5, which shows a system absorbing heat dQ from the surroundings. As a consequence, the temperature of the system rises, a change that can be measured very accurately. If dT is the temperature increase, the heat absorbed by the system is

$$dQ = C\, dT \qquad (1\text{-}6) ◀$$

TABLE 1-3 *Some types of thermodynamic work*

TYPE	INTENSIVE VARIABLE	EXTENSIVE DIFFERENTIAL	EXPRESSION FOR WORK
General	Force, F	Change in distance, ds	$W = \int F\,ds$
Expansion	Pressure, P	Volume change, dV	$W = \int P\,dV$
Elevation in a gravitational field	Gravitational force, mg	Change in height, dh	$W = \int mg\,dh$
Electrical	Voltage difference, $\Delta\Phi$	Change in charge, dq	$W = \int \Delta\Phi\,dq$
Chemical	Chemical potential of component A, μ_A	Change in number of moles of A, dn_A	$W = \int \mu_A\,dn_A$
Surface	Surface tension, γ	Change in surface area, dA	$W = \int \gamma\,dA$
Stretching	Tension, τ	Change in length, dl	$W = \int \tau\,dl$

in which C is a characteristic of the system, called its *heat capacity*. Notice that dQ and dT are positive when heat is absorbed by the system from the surroundings, and negative when the system gives up heat to the surroundings.

The definition of heat in Equation 1-6 introduces another unknown quantity, the *heat capacity,* and for that reason may seem to have little utility. However, this difficulty was avoided in the early years of thermodynamics by defining a system of units based on the heat capacity of water: one *calorie* was defined as the amount of heat required to raise the temperature of one gram of water by one degree from 14.5°C to 15.5°C at atmospheric pressure. This was a very convenient system of units for measuring heat, because of the ease of measuring the temperature change in a water sample that absorbed the heat from some unknown process. It is natural to measure work

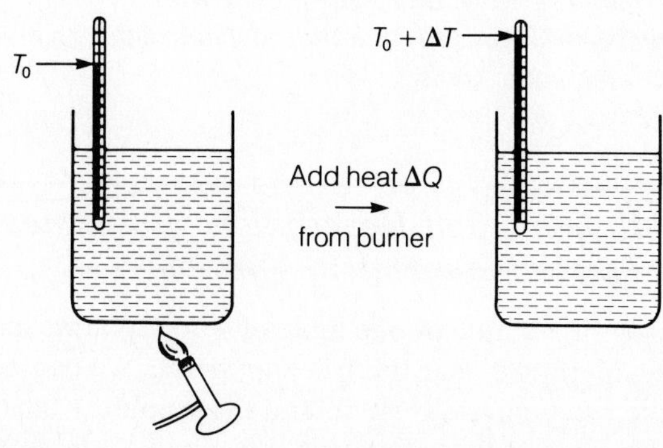

FIGURE 1-5
Relationship between the heat absorbed by a system and its temperature change. The heat is $dQ = C\,dT$, in which C is the *heat capacity* of the system.

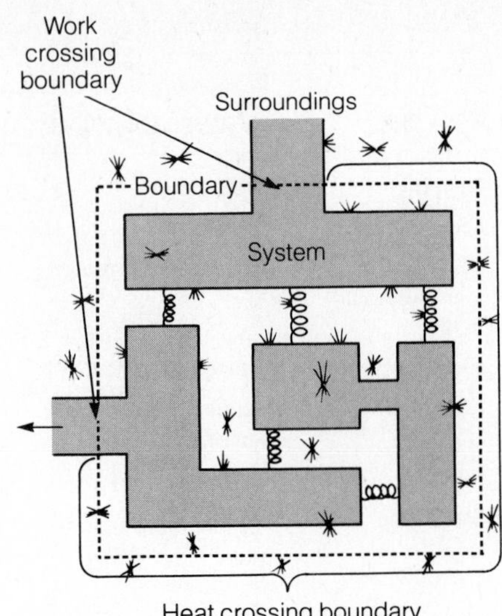

FIGURE 1-6
A highly schematic diagram to illustrate the distinction between heat and work.

in *Joules* (force times distance, or *newton meter*), and heat in calories. The relationship between these two units is discussed in Section 2-3.

You may get some feeling for the difference between work and heat from Figure 1-6, which is a highly schematic representation of a system and its surroundings. There are large handles, connectors, and springs in the system, and the handles cross into the surroundings. Also in the figure are many symbols of the form $*$. These "pows" represent collisions of molecules with each other, and with the handles, connectors, and springs of the system. The number and intensity of "pows" is a measure of the temperature. If the surroundings are warmer than the system, then there are more vigorous "pows" outside the boundary than inside. As collisions transfer energy through the boundary and into the system, the system comes up to the same temperature as the surroundings. The idea of all this is that work is represented in the figure by motions of the handles, and heat by the "pows". *The chief difference between work and heat is that work consists of motions across the boundary that are organized on a macroscopic scale, whereas heat consists in motions on a molecular scale.*

○ *1–5 Mathematical Description of a System with a Single Independent Variable*

Figure 1-7 shows the volume of one mole of water at atmospheric pressure as a function of temperature near the freezing point. We can consider the mole of water as a thermodynamic system, and the volume V as a variable of state that depends on temperature, which is a second variable of state. We can express this mathematically by the equation

$$V = V(T) \tag{1-7}$$

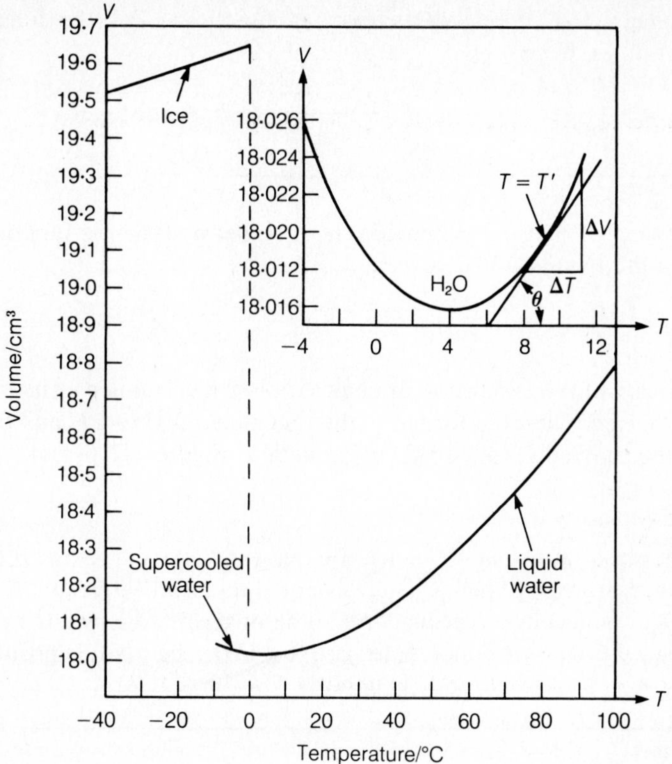

FIGURE 1-7
The volume of one mole of water as a function of temperature. The inset shows in detail the volume around 4°C, the temperature of maximum density. A tangent is drawn to the curve at temperature T', and makes an angle θ with the T axis. The slope of the curve at $T = T'$ is $(dV/dT)_{T'} = \tan\theta$. (From Eisenberg and Kauzmann, 1969.)

in which T is the independent variable, V is the dependent variable, and $V(T)$ represents the functional dependence. In fact, a function of the form of the equation of state 1-2 is:

$$V = a + bT + cT^2 + fT^3 + \cdots \tag{1-8}$$

and can be fitted to the curve of Figure 1-7 by proper selection of the constants $a, b, c, f, \ldots$. The thermodynamic function V of Equation 1-8 satisfies the definition of a mathematical function because it defines for every value of the independent variable T a value of the dependent variable V.

In describing functions, either in graphical form as in Figure 1-7, or in algebraic form as in Equation 1-8, it is useful to consider the derivative of the function. The geometric definition of a derivative is shown in the inset to Figure 1-7. If a tangent is drawn to the curve $V(T)$ at point T', and θ is the angle between this tangent and the T axis, then the derivative of V with respect to T, dV/dT, equals $\tan\theta$. The derivative is thus a measure of the slope of the curve at any given point T'. We are also familiar with the algebraic method of taking derivatives of functions. For the function of Equation 1-8, we have

$$\frac{dV}{dT} = \frac{d(a + bT + cT^2 + fT^3 + \cdots)}{dT} = b + 2cT + 3fT^2 + \cdots$$

EXERCISE 1-4

The coefficient of thermal expansion, α, is a thermodynamic function defined by the expression

$$\alpha = \frac{1}{V}\frac{dV}{dT} \tag{1-9}$$

It is a measure of the expansion or contraction of a substance as its temperature is increased. Sketch α for ice in the region of −5°C to 0°C and for liquid water in the range 0°C to +10°C, using data in Figure 1-7.

ANSWER

The coefficient α is zero at 4°C since the tangent to the curve has zero slope at this temperature, and hence dV/dT is equal to zero. Between 0°C and 4°C the sign of α is negative, because the slope of the $V(T)$ curve is negative. From below 0°C and for liquid water above 4°C the sign of α is positive. The magnitude of α for ice at −5°C is about $1.5 \times 10^{-4}\ °C^{-1}$.

EXERCISE 1-5

Between −4°C and +12°C, the volume of liquid water in $cm^3\ mol^{-1}$ is given roughly by

$$V(t) \cong 18.017 + \left(\frac{0.009}{8^2}\right)(t - 4)^2 - \left(\frac{0.001}{8^3}\right)(t - 4)^3$$

in which t is the temperature in °C. Derive a quantitative expression for α.

ANSWER

$$\alpha \cong \frac{1}{18}\left[\frac{0.018}{8^2}(t - 4) - \frac{0.003}{8^3}(t - 4)^2\right]$$

EXERCISE 1-6

Show that for an ideal gas at constant pressure, the coefficient of thermal expansion α, equals $1/T$.

ANSWER

$$\alpha = \frac{1}{V}\frac{dV}{dT} = \frac{1}{V}\frac{d}{dT}\left(\frac{nRT}{P}\right)$$

$$= \frac{P}{nRT}\frac{d}{dT}\left(\frac{nRT}{P}\right) = \left(\frac{P}{nRT}\right)\left(\frac{nR}{P}\right) = \frac{1}{T}$$

Note from the inset to Figure 1-7 that if dV/dT is known, one can estimate the approximate change in the dependent variable ΔV that corresponds to a small change ΔT in the independent variable. From the triangle in the inset,

$$\frac{dV}{dT} = \tan\theta \cong \frac{\Delta V}{\Delta T} \tag{1-10}$$

and therefore

$$\Delta V \cong \frac{dV}{dT} \cdot \Delta T \tag{1-11}$$

1–6 *Mathematical Description of a System with Two or More Variables*

The volume of a sample of water depends on pressure as well as temperature. Therefore a more complete functional description than Equation 1-7 is of the form

$$V = V(T,P) \tag{1-12}$$

in which there are now two independent variables. If the water contains solutes, then experience tells us that the volume is also a function of the concentrations of the solutes. Hence, we must add these concentrations as additional independent variables in Equation 1-12. In this section we will work only with two independent variables, since extension of all the formulae to other variables is straightforward.

A schematic representation of the volume of liquid water as a function of temperature and pressure is given in Figure 1-8. The volume is represented by a surface. For any chosen values of T and P, the volume is given by the V coordinate of the surface. As with Equation 1-8, some empirical function can be found to represent $V(T,P)$. Let us suppose that this function has the form

$$V = a + bT + cT^2 - eP - fTP + gP^2 \tag{1-13}$$

where $a, b, \ldots, g$ are constants. Note that this function satisfies the definition of a mathematical function, in that for any chosen values of T and P, it gives a value for V.

As for a function of a single variable, it is useful to characterize a function of several variables by its derivatives. Now, however, there are two *partial* derivatives:

$$\left(\frac{\partial V}{\partial T}\right)_P \quad \text{and} \quad \left(\frac{\partial V}{\partial P}\right)_T$$

Each of these gives the rate of change of the function $V(T,P)$ with respect to one of its independent variables, while the other is held constant. The geometric interpretation of these derivatives is shown in Figure 1-8. Suppose the plane $P = P'$, which is parallel to the V and T axes, intersects the

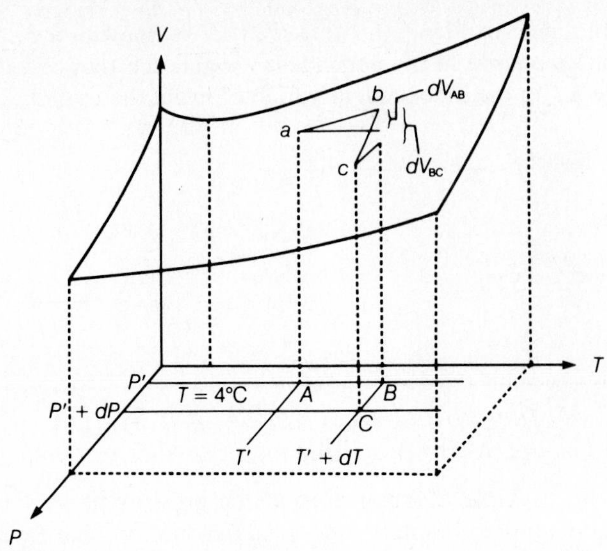

FIGURE 1-8
A curved surface which schematically represents the volume of liquid water as a function of temperature and pressure. A plane at $P = P'$ parallel to the V,T plane intersects the surface along the curve a-b. As the temperature is increased from T' to $T' + dT$ at the constant pressure P', the volume increases from a to b, by an amount dV_{AB}. The tangent to curve a-b at the point $T = T'$ has the slope $(\partial V/\partial T)_{P=P'}$. A second plane parallel to the P,V plane at $T = T' + dT$ intersects the surface along the curve b-c. As the pressure is increased from P' to $P' + dP$, the volume changes by an amount dV_{BC}. The tangent to the curve b-c above the point B has the slope $(\partial V/\partial P)_{T=T'+dT}$.

surface along the curve a-b. Then the slope of the tangent to the curve a-b at the point T' is the derivative:

$$\left(\frac{\partial V}{\partial T}\right)_P$$

The subscript P denotes that the pressure is held constant along the curve a-b. Note that the increase in V as T increases by the small amount from T' to $T' + dT$, which we denote dV_{AB}, is given by

$$dV_{\mathrm{AB}} = \left(\frac{\partial V}{\partial T}\right)_P dT \tag{1-14}$$

Similarly the derivative:

$$\left(\frac{\partial V}{\partial P}\right)_T$$

is the slope of the tangent to the curve b-c, formed by the intersection of the plane $T' + dT$ with the surface $V(T,P)$. Because the volume of water decreases as pressure increases the value of $V(T,P)$ is smaller at point c than at b. Thus this derivative is negative and the change in V along bc, denoted dV_{BC},

$$dV_{BC} = \left(\frac{\partial V}{\partial P}\right)_T dP \tag{1-15}$$

is negative.

EXERCISE 1-7

By differentiation, find the first and second partial derivatives of the function $V(T,P)$ of Equation 1-13.

ANSWER

In taking the partial derivative with respect to T, $(\partial V/\partial T)_P$, we take the ordinary derivative of V with respect to T, treating P as a constant.

First partial derivatives:

$$\left(\frac{\partial V}{\partial T}\right)_P = \frac{\partial}{\partial T}\,[a + bT + cT^2 - eP - fTP + gP^2]_P = b + 2cT - fP$$

$$\left(\frac{\partial V}{\partial P}\right)_T = \frac{\partial}{\partial P}\,[a + bT + cT^2 - eP - fTP + gP^2]_T = -e - fT + 2gP$$

Second partial derivatives:

$$\left(\frac{\partial^2 V}{\partial T^2}\right) = \frac{\partial}{\partial T}\left[\left(\frac{\partial V}{\partial T}\right)_P\right]_P = \frac{\partial}{\partial T}\,[b + 2cT - fP]_P = 2c$$

$$\left(\frac{\partial^2 V}{\partial P^2}\right) = \frac{\partial}{\partial P}\left[\left(\frac{\partial V}{\partial P}\right)_T\right]_T = \frac{\partial}{\partial P}\,[-e - fT + 2gP]_T = 2g$$

Note that in taking the second derivative we simply differentiate again, holding the same variable constant.

EXERCISE 1-8

Show, for the function of Equation 1-13, that

$$\frac{\partial^2 V}{\partial P \partial T} = \frac{\partial^2 V}{\partial T \partial P} \tag{1-16}$$

These quantities are called *mixed second partial* derivatives. They are calculated by differentiating first with respect to the variable on the right of the denominator, holding the one on the left constant, and then with respect to the variable on the left, holding the one on the right constant. Equation 1-16 shows that the *order* of differentiation does not affect the result. This is a property of any of the well-behaved functions that we are likely to encounter in thermodynamics.

ANSWER

$$\frac{\partial^2 V}{\partial P \partial T} = \frac{\partial}{\partial P}\left[\left(\frac{\partial V}{\partial T}\right)_P\right]_T = \frac{\partial}{\partial P}\,[b + 2cT - fP]_T = -f$$

$$\frac{\partial^2 V}{\partial T \partial P} = \frac{\partial}{\partial T}\left[\left(\frac{\partial V}{\partial P}\right)_T\right]_P = \frac{\partial}{\partial T}\,[-e - fT + 2gP]_P = -f$$

Therefore:

$$\frac{\partial^2 V}{\partial T \partial P} = -f = \frac{\partial^2 V}{\partial P \partial T}$$

The Total Differential Gives the Change in a Function for Combined Changes in the Independent Variables

Consider the change in the volume V as the temperature is increased from T' to $T' + dT$, and the pressure is increased from P' to $P' + dP$. This change is represented on the surface of Figure 1-7 as the change dV_{AB} as the temperature is increased, and the change dV_{BC} as the pressure is increased. The total small change in V, denoted dV, is given by

$$dV = dV_{AB} + dV_{BC} = \left(\frac{\partial V}{\partial T}\right)_P dT + \left(\frac{\partial V}{\partial P}\right)_T dP \qquad (1\text{-}17)$$

where the values for dV_{AB} and dV_{BC} have been substituted from Equations 1-14 and 1-15. This change in V is given on the surface as the difference in the V coordinates of points a and c. The combined change dV is called the *total differential* of the function V.

For a general function $Z = Z\ (x,y,u,\ .\ .\ .\ )$, the total differential dZ is defined as

$$dZ = \left(\frac{\partial Z}{\partial x}\right)_{y,u\ldots} dx + \left(\frac{\partial Z}{\partial y}\right)_{x,u\ldots} dy + \left(\frac{\partial Z}{\partial u}\right)_{x,y\ldots} du + \cdot\cdot\cdot \qquad (1\text{-}18)$$ ◀

It can be thought of as the combined change in Z created by small changes dx, dy, du, . . . in the independent variables.

EXERCISE 1-9

Find the total differential of the function

$$Z = \tfrac{1}{2}x^2y + c$$

where c is a constant.

ANSWER

$$dZ = xy\ dx + \tfrac{1}{2}x^2\ dy \qquad (1\text{-}19)$$

○ 1–7 Exact Differentials Represent Small Changes in Properties in a Process; Inexact Differentials Represent Small Changes of Heat and Work

We have seen above that the small change in a function which results from combined small changes in its independent variables is given by the total differential of the function. Since properties are represented by functions, a small change in a property can be represented by the total differential of the function. The total differential of a function is also called an *exact differential*. Thus a small change in a property is given by an exact differential.

An example of an exact differential is $dZ = xy\,dx + (x^2/2)dy$. You showed in Exercise 1-9 that this is the total differential of $Z = \frac{1}{2}x^2y + c$. Similarly the differential

$$dV = (b + 2cT - fP)dT + (-e - fT + 2gP)dP$$

is an exact differential because it is the total differential of the mathematical function of Equation 1-13.

There is another class of differentials, known as *inexact differentials*, which are not the total differentials of any mathematical functions. An example is $dZ = xy\,dx + xy\,dy$. There is no function $Z(x,y)$ whose total differential is given by this dZ. An interesting characteristic of inexact differentials is that the change in Z, ΔZ, caused by changes in the independent variables x and y, depends not only on the initial and final values of x and y, but also on the detailed path along which x and y change. This is demonstrated for one inexact differential in Figure 1-9.

A moment's thought shows that inexact differentials are not suitable for representing small changes in properties. Properties are represented as functions, and changes in them depend only on the initial and final states; the changes must be independent of path.

We shall see presently that changes in heat and work depend on the details of the process that a system undergoes in passing from a given initial to a given final state. That is, heat and work depend on the path. Therefore small changes in heat dQ and in work dW are inexact differentials, and heat and work are not properties. There is no such concept as the "heat of a system" or the "work of a system." In contrast, changes in volume and energy are independent of the details of path, and can be represented by exact differentials. Thus the "volume of a system" or the "energy of a system" has meaning, and therefore there are functions that correspond to these properties. In general, though we know such functions exist, we do not know their precise form (the equation of state) until experiments have been performed.

Euler's Criterion for Exact Differentials

There is a method for determining whether any given differential is exact or inexact, known as *Euler's criterion of exactness*. Any differential of two independent variables, whether exact or inexact, can be written in the form

$$dZ = M(x,y)\,dx + N(x,y)\,dy \tag{1-20}$$

where M and N are some functions of x and y. Now if dZ is an exact differential, it must be the total differential of some functions $Z(x,y)$. We can write the total differential of $Z(x,y)$ in the form

$$dZ = \left(\frac{\partial Z}{\partial x}\right)_y dx + \left(\frac{\partial Z}{\partial y}\right)_x dy \tag{1-21}$$

	EXACT	INEXACT
1. Function:	$Z = Z(x,y) = x^2y + c$	None
2. Differential:	$dZ = 2xy\,dx + x^2dy$	$dZ = xy\,dx + xy\,dy$
3. Change in Z from state 1 to state 2:	$\Delta Z = 2\int_{x_1}^{x_2} xy\,dx + \int_{y_1}^{y_2} x^2\,dy$	$\Delta Z = \int_{x_1}^{x_2} xy\,dx + \int_{y_1}^{y_2} xy\,dy$

4. *Limits:* Initial state: $x_1 = 0, y_1 = 0$
Final state: $x_2 = 1, y_2 = 1$
Paths: I: Along $x = y$
II: Along $y = x^2$
IIIA: Along $y = 0$ to $x = 1, y = 0$
IIIB: Then along $x = 1$ to $x = 1, y = 1$

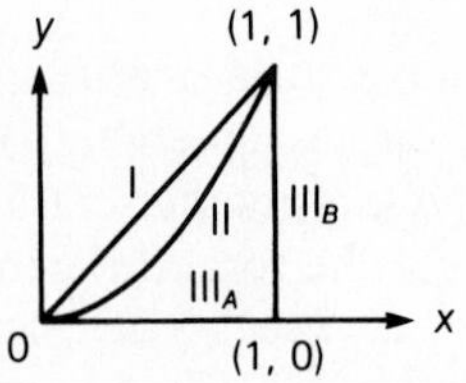

y (1, 1) I II III$_B$ III$_A$ 0 (1, 0) x

5. ΔZ for path I: $dZ = 2x \cdot x\,dx + y^2\,dy$

$$\Delta Z = 2\int_{x=0}^{x=1} x \cdot x\,dx + \int_{y=0}^{y=1} y^2dy = \left[\frac{2x^3}{3}\right|_0^1 + \left[\frac{y^3}{3}\right|_0^1 = \frac{2}{3} + \frac{1}{3} = 1$$

$$\Delta Z = \int_{x=0}^{x=1} x \cdot x\,dx + \int_{y=0}^{y=1} y \cdot y\,dy = \left[\frac{x^3}{3}\right|_0^1 + \left[\frac{y^3}{3}\right|_0^1 = \frac{1}{3} + \frac{1}{3} = \frac{2}{3}$$

6. ΔZ for path II: $dZ = 2x \cdot x^2\,dx + y\,dy$

$$\Delta Z = 2\int_{x=0}^{x=1} x \cdot x^2dx + \int_{y=0}^{y=1} y\,dy = \left[\frac{2x^4}{4}\right|_0^1 + \left[\frac{y^2}{2}\right|_0^1 = \frac{1}{2} + \frac{1}{2} = 1$$

$$\Delta Z = \int_{x=0}^{x=1} x \cdot x^2\,dx + \int_{y=0}^{y=1} y^{3/2}\,dy = \left[\frac{x^3}{3}\right|_0^1 + \left[\frac{2y^{5/2}}{5}\right|_0^1 = \frac{1}{3} + \frac{2}{5} = \frac{11}{15}$$

7. ΔZ for path III:

$$\Delta Z = 2\int_{x=0}^{x=1} x \cdot 0\,dx + \int_{y=0}^{y=0} x^2dy\Big\}\ \text{III}_A + 2\int_{x=1}^{x=1} xy\,dx + \int_{y=0}^{y=1} 1^2dy\Big\}\ \text{III}_B = 0 + 0 + 0 + 1 = 1$$

$$\Delta Z = \int_{x=0}^{x=1} x \cdot 0\,dx + \int_{y=0}^{y=0} x \cdot 0\,dy\Big\}\ \text{III}_A + \int_{x=1}^{x=1} 1 \cdot y\,dx + \int_{y=0}^{y=1} 1 \cdot y\,dy\Big\}\ \text{III}_B = 0 + 0 + 0 + \tfrac{1}{2} = \tfrac{1}{2}$$

FIGURE 1-9
Demonstration that changes associated with inexact differentials depend on the detailed path of the process, whereas changes associated with exact differentials depend only on the initial and final stages. The differential $dZ(x,y)$ on the left side of line 2 is exact; the differential on the right is inexact. Line 3 gives general expressions for changes in Z corresponding to changes in the independent variables, from their initial values x_1 and y_1 to their final value x_2 and y_2. Line 4 shows three paths between $x_1 = 0, y_1 = 0$ and $x_2 = 1$, $y_2 = 1$. Lines 5, 6, and 7 evaluate the changes for the three paths for both differentials. Note that ΔZ for the exact differential is the same for all paths, but ΔZ for the inexact differential depends on the path.

By comparing Equations 1-20 and 1-21 we see that

$$M(x,y) = \left(\frac{\partial Z}{\partial x}\right)_y \quad \text{and} \quad N(x,y) = \left(\frac{\partial Z}{\partial y}\right)_x \tag{1-22}$$

You will recall that the mixed second partial derivatives of a function must be equal, $\partial^2 Z/\partial y\,\partial x = \partial^2 Z/\partial x\,\partial y$. *Therefore, if a differential of the form 1-20 is exact it must obey the relationship*

$$\left[\frac{\partial}{\partial y}\,M(x,y)\right]_x = \frac{\partial^2 Z}{\partial y\,\partial x} = \frac{\partial^2 Z}{\partial x\,\partial y} = \left[\frac{\partial}{\partial x}\,N(x,y)\right]_y \qquad \text{(1-23)} ◀$$

This is Euler's criterion. If a differential dZ obeys the equality of Equation 1-23, it is exact; if not, it is inexact and there is no mathematical function corresponding to Z.

EXERCISE 1-10

Show for an ideal gas that the small volume change dV caused by small changes in temperature and pressure is given by

$$dV = \frac{nR}{P}\,dT - \frac{nRT}{P^2}\,dP \qquad \text{(1-24)}$$

ANSWER

For an ideal gas,

$$V = V(T,P) = \frac{nRT}{P}$$

Taking the total differential of V,

$$\begin{aligned} dV &= \left(\frac{\partial V}{\partial T}\right)_P dT + \left(\frac{\partial V}{\partial P}\right)_T dP \qquad \text{(1-25)}\\ &= \frac{\partial}{\partial T}\left(\frac{nRT}{P}\right)_P dT + \frac{\partial}{\partial P}\left(\frac{nRT}{P}\right)_T dP \\ &= \frac{nR}{P}\,dT - \frac{nRT}{P^2}\,dP \end{aligned}$$

EXERCISE 1-11

Prove that dV is an exact differential for an ideal gas (or, in other words, that volume is a property of an ideal gas).

ANSWER

Using Euler's criterion (Equation 1-23),

$$M(T,P) = \frac{nR}{P}; \qquad N(T,P) = -\frac{nRT}{P^2}$$

and

$$\left[\frac{\partial}{\partial P}\left(\frac{nR}{P}\right)\right]_T = -\frac{nR}{P^2} = \left[\frac{\partial}{\partial T}\left(-\frac{nRT}{P^2}\right)\right]_P = -\frac{nR}{P^2}$$

Since Euler's criterion is satisfied, dV is exact.

EXERCISE 1-12

We have seen above that the work done by a gas during a reversible expansion can be expressed $dW = -PdV$. Combining this relationship with Equation 1-24, we obtain a relationship for the work done by an ideal gas during a reversible expansion:

$$dW = -nRdT + \frac{nRT}{P}\,dP \tag{1-26}$$

Prove that dW is an inexact differential (or, in other words, that work is not a property of an ideal gas).

ANSWER

Again using Euler's criterion,

$$M(T,P) = -nR \qquad N(T,P) = \frac{nRT}{P}$$

and

$$\frac{\partial}{\partial P}[-nR]_T = 0 \qquad \neq \qquad \frac{\partial}{\partial T}\left[\frac{nRT}{P}\right]_P = \frac{nR}{P}$$

1–8 The Thermodynamicist's Description of Nature: A Summary

The thermodynamicist's model for reality centers on a portion of the world called the *system*, characterized by its properties. When the system is in an equilibrium state, properties have uniform and unchanging values. All molecular structures and motions of which modern physical chemists are aware are averaged when describing a thermodynamic system, and these averages are represented by a few macroscopic properties such as temperature, pressure, and energy. The number of properties that we must specify to define the state of the system depends on the number of components and phases of the system, as dictated by the phase rule.

When heat, work, or matter cross the boundary that separates the system from its surroundings, the system changes from one equilibrium state to another. If the process is reversible, the system is always at equilibrium, and all properties retain uniform values during the process. If the process is irreversible, then some properties, such as temperature and pressure, are not uniform throughout the system and do not have defined values until equilibrium is restored. A few properties, such as volume and energy, have meaning even during irreversible processes.

Notice that if a system undergoes a series of processes that return it to its original state, then all properties must assume their original values. An overall process in which a system is returned to its original state is called a *cyclic process*. Cyclic processes are of interest in nearly every practical application of thermodynamics, because all machines (including enzymes and living cells) operate by them. A machine that did not operate by a cyclic process would involve some never-ending displacement, or never-ending change of concentration, and it could not continue to function.

From the definition of a property, the sum of the changes in any property during a cyclic process must be zero. This is so because the system is restored to its original state, and so it must again be characterized by its original properties. We can represent this for pressure by the summation over all changes of the property,

$$\sum_{\text{all steps}} \Delta P = 0$$

Similar expressions can be written for any property. We can also sum the work or heat or specific forms of matter that cross the boundary during a cyclic process, but there is no reason that these entities should sum to zero. We can express this as follows:

for work: $$\sum_{\text{all steps}} W \neq 0$$

and for heat: $$\sum_{\text{all steps}} Q \neq 0$$

These mathematical statements are equivalent to saying that whereas pressure is a property of a system, heat and work are not. They are also equivalent to saying that dP is an exact differential, whereas dQ and dW are inexact differentials.

QUESTIONS FOR REVIEW

1. Classify four types of thermodynamic systems according to their boundaries.
2. Name three extensive thermodynamic properties and three intensive properties. Name three quantities used in thermodynamics that are not properties.
3. What are equations of state, and how are they determined? What is a state function?
4. Explain the phase rule and give an example.
5. Give several examples of processes. What variables of state are changed in each process?
6. Is it correct to state that the heat of a system changes during a process?
7. Define: (a) Total differential. (b) Mixed second partial derivative. (c) Exact differential.
8. What is Euler's criterion of exactness?

PROBLEMS

1. Are the following statements true or false?

   (a) A system undergoing a reversible process may be regarded as being at equilibrium.

   (b) The energy of an isolated system is constant.

   (c) A closed system cannot be in a steady state.

   (d) An open system must be in a steady state.

   (e) The energy of an adiabatic system is constant.

   (f) Equations of state are derived from laws of thermodynamics.

   (g) The right-hand side of Equation 1-13 is a state function.

Answers: (a) T, by definition. (b) T. (c) F, there can be a flow of energy through a closed system. (d) F. (e) F, work done on, or by, the system can change the energy. (f) F, they are empirical or are based on microscopic theories. (g) T.

2. Insert the appropriate words:

   (a) The sign of heat is defined as being positive when it crosses the boundary ________ (into, out of) a system.

   (b) The sign of work is defined as being ________ when work is done on the system.

   (c) A ________ process is one that proceeds by a succession of equilibrium states.

   (d) The total differential of a function is an ________ differential.

   (e) Changes associated with ________ differentials depend on the path between a given initial state and a given final state.

   (f) The change in the value of a property in a cyclic process is ________.

Answers: (a) Into. (b) Positive. (c) Reversible. (d) Exact. (e) Inexact. (f) Zero.

PROBLEMS RELATED TO EXERCISES

3. (Exercise 1-1) Fill in the following table with a + if heat, work, or matter is able to cross the boundary of the corresponding system, and a − if passage is not permitted.

System	*Heat*	*Work*	*Matter*
Open			
Closed			
Adiabatic			
Isolated			

4. (Exercise 1-2) State whether the following are intensive or extensive properties: (a) The total weight of the students in your class. (b) The average weight of the students in your class.

5. (Exercise 1-3) In addition to one extensive variable per phase, how many intensive variables are sufficient to specify the state of the following systems: (a) A solution of NaCl. (b) A solution of NaCl in equilibrium with ice. (c) A solution of NaCl in equilibrium with ice and water vapor. (d) A solution of NaCl and hemoglobin in equilibrium with ice and water vapor.

6. (Exercise 1-7) Find the first and second partial derivatives of the function:

$$V = aP + \frac{bT^2}{P} + \frac{cP^2}{T}$$

in which a, b, and c are constants.

7. (Exercise 1-8) Find the mixed second partial derivatives of the function in Problem 1-6.

8. (Exercise 1-9) Find the total differential of the function $Z = xy + xy^2 + x^2y$.

9. (Exercise 1-10) Show for an ideal gas that the total differential dT due to changes in the pressure, volume, and number of moles is

$$dT = \frac{V}{nR}\,dP + \frac{P}{nR}\,dV - \frac{PV}{n^2R}\,dn$$

10. (Exercise 1-11) When a differential contains three terms,

$$dZ = Ldw + Mdx + Ndy$$

Euler's criterion for exactness is that

$$\frac{\partial^2 L}{\partial x\,\partial y} = \frac{\partial^2 M}{\partial w\,\partial y} = \frac{\partial^2 N}{\partial w\,\partial x}$$

Use this criterion to show that the differential dT in Problem 1-9 is exact.

OTHER PROBLEMS

11. Determine whether each of the following properties is intensive or extensive:

(a) The coefficient of thermal expansion, α, defined

$$\alpha = \frac{1}{V}\left(\frac{\partial V}{\partial T}\right)_P$$

(b) The coefficient of isothermal compressibility, β, defined

$$\beta = -\frac{1}{V}\left(\frac{\partial V}{\partial P}\right)_T$$

(Note that β is a measure of how easily a substance is compressed at a given temperature.)

(c) Gibbs free energy, G, defined

$$G = E + PV - TS$$

where E and S are the energy and entropy, respectively.

(d) Mole fraction of the ith component in a solution, X_i, defined

$$X_i = \frac{n_i}{n_1 + n_2 + n_3 + \cdots}$$

in which n_i is the number of moles of the ith component in solution.

Answers: (a) and (b) are intensive because they both involve a ratio of extensive properties (V). (c) is extensive because every term includes an extensive quantity (see Tab. 1-1). (d) is intensive, because it involves a ratio of extensive quantities.

12. (a) Describe the geometric interpretation of the isothermal compressibility β (see Prob. 11b) in terms of the *P-V-T* surface (Fig. 1-8). (b) Is β a function of temperature? (c) Is β a function of pressure?

Answers: (b) Yes, (c) Yes

13. Considering volume as a function of temperature and pressure, $V = V(T,P)$, prove that

$$\left(\frac{\partial \alpha}{\partial P}\right)_T = -\left(\frac{\partial \beta}{\partial T}\right)_P$$

Hint: Mixed second partial derivatives of a property are equal.

14. In 1889, P.G. Tait analyzed data that had been collected on the oceanographic research vessel *H.M.S. Challenger.* He found that sea water obeys the relationship

$$\frac{1}{V_0}\left(\frac{\partial V}{\partial P}\right)_T = -\frac{C}{B + P} \tag{1-27}$$

in which C and B are constants and V_0 is the volume at $P = 0$. Since then many liquids have been found to obey this equation, known as *Tait's equation.*

(a) Is Tait's equation an equation of state, even though it does not include temperature? Is the right-hand side a state function?

(b) State the physical meaning of Tait's equation.

(c) Show that the volume difference predicted by Tait's equation at pressures P_1 and P_2 can be expressed by

$$\frac{V(P_1) - V(P_2)}{V_0} = C \ln \left[\frac{B + P_2}{B + P_1}\right]$$

Answers: (a) Yes, because it relates properties (pressure and compressibility); no. (b) At a given temperature T, the rate of decrease of volume with applied pressure declines as pressure is increased. In other words, as a liquid is compressed, each further increment of pressure produces a smaller volume decrement. (c) At constant temperature, we may treat the partial derivative as an ordinary derivative.

$$\frac{1}{V_0}\frac{dV}{dP} = -\frac{C}{B + P}$$

Then we may integrate between State 1 and State 2:

$$\frac{1}{V_0}\int_{V_1}^{V_2} dV = -\int_{P_1}^{P_2} \frac{CdP}{B+P}$$

$$\frac{V(P_2) - V(P_1)}{V_0} = -C[\ln(B+P)]\Big|_{P_1}^{P_2}$$

$$= -C\ln(B+P_2) + C\ln(B+P_1) = -C\ln\left[\frac{B+P_1}{B+P_2}\right]$$

15. Consider the number of moles of an ideal gas to be a function of the independent variables pressure, temperature, and volume. Find the expression for the total differential dn.

Answer:

$$dn = \frac{V}{RT}\,dP + \frac{P}{RT}\,dV - \frac{PV}{RT^2}\,dT$$

16. Total differentials are useful for estimating the error in a quantity that has been derived from several other quantities. For example, suppose you wish to estimate the error in determining the volume of a rectangular box for which you have measured the lengths of its edges a, b, and c. You calculate V from the relationship $V = a \cdot b \cdot c$ (This box might be the unit cell of a crystal; see Chap. 17). Suppose also that you have estimated that the error in measuring edge a is da, that is measuring edge b is db and that in edge c is dc. Then the total differential of V, dV, is a measure of the maximum error in V produced by the errors in a, b, and c:

$$dV = \left(\frac{\partial V}{\partial a}\right)_{b,c} da + \left(\frac{\partial V}{\partial b}\right)_{a,c} db + \left(\frac{\partial V}{\partial c}\right)_{a,b} dc \qquad (1\text{-}28)$$

Since the errors in a, b, and c are not likely to be all in the direction to increase V, or all in the direction to decrease V, the probable error is likely to be smaller than that predicted by Equation (1-28). The probable error is frequently assumed to be of the form

$$dV = \sqrt{\left(\frac{\partial V}{\partial a}\right)_{a,c}^2 (da)^2 + \left(\frac{\partial V}{\partial b}\right)_{a,c}^2 (db)^2 + \left(\frac{\partial V}{\partial c}\right)_{a,b}^2 (dc)^2} \qquad (1\text{-}29)$$

(a) Estimate the probable error dV if the edges are of lengths 10, 15, and 20 cm, and if the estimated error in measuring each edge is ± 0.1 cm.

(b) Referring to Problem 1-15, estimate the probable error in determining the number of moles of 50 liters of an inert gas confined at 2.0 atm pressure and 298 K. The estimated errors in volume, pressure, and temperature are ± 0.3 1, ± 0.1 atm, and ± 1°C.

Answers: (a) $V = 3000 \pm 39$ cm^3
(b) $n = 4.1 \pm 0.2$ moles.

17. Which of the following differentials are exact and which inexact?

(a) $dZ = xy^2\,dx + yx^2\,dy$

(b) $dZ = xy^2\,dx + xy^2\,dy$

(c) $dZ = \cos x \cos y\, dx - \sin x \sin y\, dy$

(d) $dZ = \ln y\, dx + \frac{x}{y}\, dy$

Answer: (a), (c), and (d) are exact; (b) is inexact.

18. The volume of a right cone with base of radius r and altitude h is given by $V = \pi r^2 h/3$

    (a) Consider V as a function of the variables r and h, and determine the total differential of V.

    (b) Prove that dV is exact.

19. Prove that the differential on the left-hand side of Figure 1-9 is exact, and that the differential on the right-hand side is inexact.

20. For a reversible expansion of an ideal gas, the heat absorbed can be expressed by

$$dQ_{\text{rev}} = C_V\, dT + P dV$$

in which C_V may be regarded as a constant (called the *heat capacity at constant volume*).

    (a) Show that dQ_{rev} is an inexact differential (i.e., that Q_{rev} is not a property).

    (b) Show that dQ_{rev}/T is an exact differential. (Note that this implies $\int dQ_{\text{rev}}/T$ is a property. We shall see in Chap. 3 that this property is called *entropy*.)

21. According to the chain rule of differentiation, when f is a function of y, and y is a function of x, then

$$\frac{df}{dx} = \frac{df}{dy}\frac{dy}{dx}$$

For example, the variation of pressure with temperature at constant number of moles can be expressed

$$\left(\frac{\partial P}{\partial T}\right)_n = \left(\frac{\partial P}{\partial V}\right)_n \left(\frac{\partial V}{\partial T}\right)_n$$

(Notice that the variable held constant must be the same in all the partial derivatives in this expression.) Assume Equation 1-13, and suppose that the number of moles is proportional to the volume, at constant temperature:

$$n = kV$$

Calculate $(\partial n/\partial P)_T$, using the chain rule.

22. The phase rule can be used to interpret the solubility of proteins. The data shown below were measured for chymotrypsinogen by J. A. V. Butler [*J. Gen. Physiol., 24,* 189 (1940)]. The ordinate gives the number of mg nitrogen in 1 ml of filtrate (a measure of protein in solution) as a function of the number of mg of nitrogen in 1 ml of suspension (a measure of total protein in the system, both in solution and solid form). For low protein concentrations, all data lie on a straight line of slope unity, showing that all protein is in solution at these concentrations.

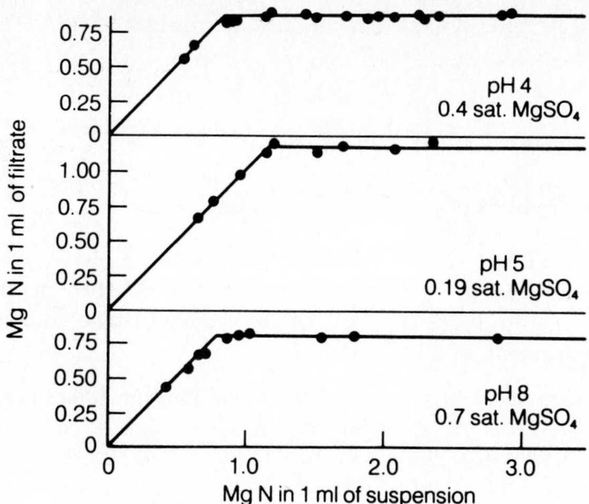

Interpret the shapes of these curves, given that these solutions contain four components (chymotrypsinogen, water, hydrogen ion, and $MgSO_4$) and that the experiments are carried out at constant temperature and pressure.

Answer: Since four components are present in solution, six properties are needed to specify the state of the system (four concentrations plus T and P), minus the number of phases present. The conditions of each experiment are constant temperature, pressure, pH, and concentration of $MgSO_4$. Thus one property must be specified if a single phase is present and none if two phases are present. The solution at low protein concentrations constitutes a single phase, and the one property to be specified is the concentration of protein. This property varies as protein is added. But when the protein concentration reaches the bending point in the curves, a second phase appears (solid protein), and no properties must now be specified (or can be varied). Thus, at concentrations above this point, the protein concentration in the solution remains constant, and the curves are horizontal. Thus the phase rule applied to the data clearly shows that the protein chymotrypsinogen is a single component, not a mixture of substances.

23. The property $\gamma = (\partial P/\partial T)_V$ is sometimes called the *coefficient of thermal pressure*. Prove that

$$\alpha = \beta\gamma$$

where α is the coefficient of thermal expansion and β is the coefficient of isothermal compressibility.

Answer: If $V = V(T,P)$, then

$$dV = \left(\frac{\partial V}{\partial T}\right)_P dT + \left(\frac{\partial V}{\partial P}\right)_T dP$$

Dividing this expression by a small temperature change at constant volume $(dT)_V$ we get

$$\left(\frac{\partial V}{\partial T}\right)_V = 0 = \left(\frac{\partial V}{\partial T}\right)_P \left(\frac{\partial T}{\partial T}\right)_V + \left(\frac{\partial V}{\partial P}\right)_T \left(\frac{\partial P}{\partial T}\right)_V$$

The left-hand side is zero, since V cannot vary at constant V, and $(\partial T/\partial T)_V$ is unity. Therefore, we have

$$\left(\frac{\partial V}{\partial T}\right)_P + \left(\frac{\partial V}{\partial P}\right)_T \left(\frac{\partial P}{\partial T}\right)_V = 0$$

Substituting the definitions of α, β, and γ, we find $\alpha = \beta\gamma$.

24. Analyze the ideal gas equation of state in terms of the phase rule. How many components and how many phases are in the system? How many properties need be specified to define the state of the system? Is the equation $PV = nRT$ consistent with this number?

25. Taking the example of Figure 1-9 of exact and inexact differentials, select a fourth path, and evaluate Z for both the exact and inexact differentials.

QUESTIONS FOR DISCUSSION

26. (a) In a recent publication on the thermodynamics of the automobile industry, the authors define their system as the total of all the metal in operating automobiles in the United States. Matter enters the system during the assembly of new automobiles and leaves when old automobiles are junked. Is this a reasonable system to select for thermodynamic analysis?

    (b) You wish to apply thermodynamics to the metabolism of a living human subject. What would you select as your system? What quantities would you measure? What variables would you have to control?

    (c) A psychology major accidentally enrolls in a course on thermodynamics. In attempting to make the best of the situation, the psychologist develops a theory: states of mind (anger, greed, suspicion, etc.) are thermodynamic states of a region of the brain that can be considered a system. Discuss the theory.

    (d) A houseperson notes that soaking dishes first makes dishwashing easier, and wants to understand the physical-chemical basis of this observation. Would you recommend a course in thermodynamics or one in kinetics?

    (e) In the absence of a spark or catalyst that would create H_2O, a mixture of O_2 and H_2 can exist for years. Is the mixture in equilibrium?

FURTHER READING

Kauzmann, W. 1967. *Thermodynamics and Statistics*. Menlo Park, Calif.: W. A. Benjamin, Inc. An exceptionally clear introductory text that stresses the thermodynamic model and its mathematical description.

Denbigh, K. 1963. *The Principles of Chemical Equilibrium*. Cambridge: Cambridge University Press. An advanced but readable development of thermodynamics.

Margenau, H., and Murphy, G. M. 1961. *The Mathematics of Physics and Chemistry*. Princeton, N.J.: Van Nostrand. A full development of the mathematics of thermodynamics, useful for many practical applications.

CHAPTER 2

Energy and the First Law of Thermodynamics

> A theory is the more impressive the greater the simplicity of its premises, the more different are the kinds of things it relates, and the more extended is its range of applicability. Therefore, the deep impression which classical thermodynamics made upon me. It is the only physical theory of universal content which I am convinced, that within the framework of applicability of its basic concepts, will never be overthrown.
>
> Albert Einstein

2–1 The First Law of Thermodynamics Relates Heat, Work, and Energy

Classical thermodynamics is not an easy science to learn. It developed gradually over a period of more than a century, beginning with the early experiments of Lavoisier and Laplace on heat in the late eighteenth century followed by the refinement of the concept of entropy in the mid nineteenth century, and culminating with the idea of free energy introduced by Gibbs and Helmholtz in the late nineteenth century. The early work was stimulated by the invention of the steam engine by James Watt in 1769, which made an understanding of the relationship between heat and work of clear practical importance. From its beginnings in a specific engineering problem, thermodynamics gradually developed into a general theory, unifying many observations that spanned both engineering and the physical sciences.

Thermodynamics is an abstract science whose laws do not rely on any specific composition of matter, or even on the molecular nature of matter. However, we will see that it is much easier to visualize thermodynamic quantities such as heat when one realizes that matter consists of atomic or molecular particles. In the early years of the development of thermodynamics the particulate nature of matter was not yet generally accepted, so the conceptual problems were especially acute. We now know that heat is related to molecular motion, but it is not surprising that the early investigators thought that heat, or *caloric,* was a substance that could be added to or taken away from matter under certain conditions. The relationship between heat, work, entropy, and free energy occupied some of the best scientific minds during a century of thought, and there were many sharp debates among the scientists involved. Therefore, you should not despair if the results they obtained are not obvious to you at first reading. Thermodynamics is traditionally a subject that requires repeated exposure for mastery; facility in adapting it to any problem at hand requires practice, for which there is no substitute.

The First Law of Thermodynamics is a Statement of the Conservation of Energy

The first law states that the total energy of a system is a property of the system, and that changes in the energy E (dE) result from the sum of energy added in the form of heat (dQ) and work (dW):

$$dE = dQ + dW \tag{2-1}$$ ◀

In this equation dQ and dW are inexact differentials, but dE is exact. This means that dQ and dW depend on how the process is carried out, but their sum dE does not. For quantities of heat that are not infinitesimally small, the first law can be stated as

$$\Delta E = Q + W \tag{2-2}$$ ◀

Conservation of energy, described by the first law of thermodynamics, is a universal principle, as illustrated by a variety of experimental systems in Figure 2-1.

In Figure 2-1(a) work is done on a gas enclosed in a cylinder by moving the

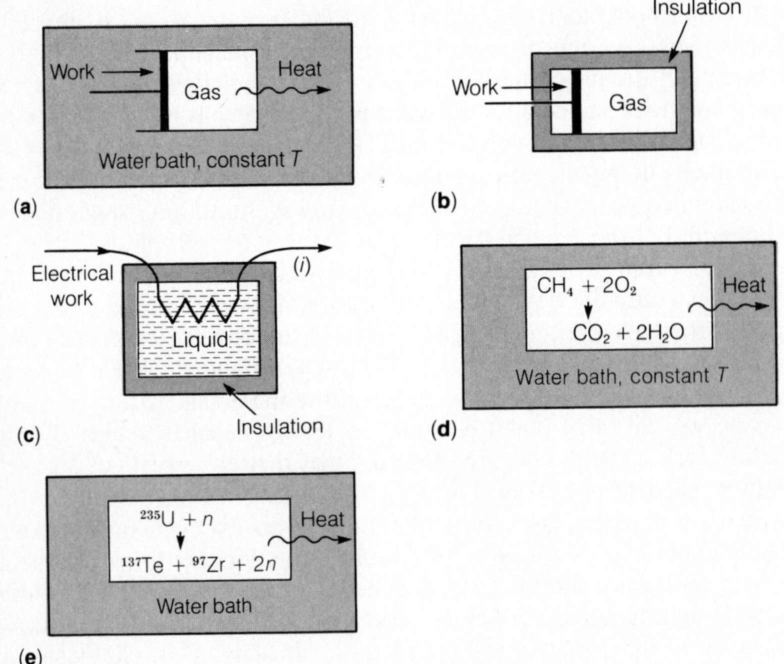

FIGURE 2-1
Examples of the application of the first law of thermodynamics to a variety of systems. In **(a)** a gas is compressed by doing work on it, and heat is given off to the surroundings, the water bath. In **(b)** heat flow is blocked so that $dE = dW$. In **(c)** the energy change of the system dE is equal to the electrical work done on it. In **(d)** the energy change of the system (ΔE) is equal to the negative of the heat released by the burning methane. In **(e)** the energy change (ΔE) is due to conversion of matter to heat which is given up to the surroundings.

piston. Compressing a gas produces heat, which is released through the cylinder wall to the surrounding constant-temperature water bath. For a small increment in the compression, the energy change of the gas, dE, is the sum of the work done on the gas, dW, plus the heat absorbed by the gas, dQ. Notice that in the process as written, heat is released, and dQ is a negative quantity.

Figure 2-1(*b*) shows an adiabatic system, in which no heat is allowed to flow, so dQ is zero. In this case, dE is simply equal to dW. We will see later that for the idealized state of matter called the *ideal gas,* dW and dQ can be easily calculated for both the isothermal and adiabatic volume changes shown in (*a*) and (*b*).

Figure 2-1(*c*) illustrates the introduction of energy into an adiabatic system by means of electrical work: a current (*i*) is passed through a resistor immersed in a liquid contained in a constant-volume, insulated container. Because the system is adiabatic, $dQ = 0$, and because there is no volume change, the work of expansion ($dW = -P_{\text{ext}}\,dV$) is zero. Hence, only electrical work (dW_{el}) contributes to the energy change, from which we conclude that $dE = dW_{\text{el}}$. The increased energy of the liquid in the container is reflected by an increase in its temperature. A system such as that shown in (*c*) is convenient for determining the relationship between electrical work, usually expressed in joules (the product of voltage and charge, Table 1-3), and the traditional unit of heat, the calorie (Section 1-4). The proportionality constant between the two units is established by first measuring the amount of heat in calories required to raise the temperature of the liquid sample (when no work is done on it) by a certain amount. Since the calorie is defined as the heat needed to raise the temperature of 1 g of water from 14.5° to 15.5°C, it would be convenient to choose the liquid sample to be 1 g of water at 14.5°C. The amount of electrical work in joules (4.184 J) required to raise the temperature to 15.5°C is then equivalent to one calorie.

Figure 2-1(*d*) shows a system which is convenient in the measurement of the heat of a chemical reaction. A constant-volume vessel (called a bomb calorimeter) contains methane and oxygen, which are mixed and ignited. A large, constant-temperature water bath surrounding the calorimeter absorbs the reaction heat as it passes out of the system (the calorimeter is the system.) Because the volume is constant, no work is done, and $\Delta E = Q$. The energy change ΔE is the difference between the energy of the substances present after the reaction ($CO_2 + 2H_2O$) and the energy of the reactants ($CH_4 + 2O_2$). Because the first law tells us that $\Delta E = Q$ we need only measure the amount of heat given off to the water bath ($= -Q$) in order to determine the energy change produced by the chemical reaction.

In Figure 2-1(*e*) the chemical reaction of (*d*) is replaced by a nuclear reaction, in which the uranium isotope ^{235}U captures a neutron and undergoes fission to ^{127}Te and ^{97}Zr, plus two neutrons. The heat produced by this reaction—and others like it—is the source of the electrical power generated by nuclear power plants. Viewed from the perspective of the first law, the system, held at constant volume, undergoes an energy change $\Delta E = Q$, in which $-Q$ is the heat released to the water bath. The energy of the system changes because some of the matter it contains is converted to heat by the nuclear fission process. Obviously, for understanding nuclear reactions, we must consider the matter in a system to be part of its potential energy content.

However, it is very important to notice that *the first law does not rely on, nor specify in any way, the manner in which energy is stored in a system.* The change in the energy of the systems in Figure 2-1 arises from several different mechanisms, such as change in the internal kinetic energy due to motion of the gas or liquid molecules, or change in the chemical nature of the system, or conversion of matter to energy. In spite of these important differences, each system is described by the first law.

Calculations Using the First Law

The first law provides a simple relationship between heat, work, and energy change which can be used to solve a number of quantitative problems. Since there are three variables and only one equation, the values of two of the variables must be known to calculate the third. The following exercise illustrates such a calculation.

EXERCISE 2-1

The energy of a gas increases by 0.4 J (joules) when it is compressed by a force of one newton acting through 0.5 meter. Calculate the heat change.

ANSWER

$$W = 0.5 \text{ meter} \times 1 \text{ newton} = 0.5 \text{ J}$$
$$Q = \Delta E - W = 0.4 - 0.5 = -0.1 \text{ J}$$

Because the sign of Q is negative, heat is released when the gas is compressed.

A Caution Concerning the Conversion of Heat to Work

The first law considers heat and work to be forms of energy, and makes no distinction between the two. However, as we will see in Chapter 3, heat and work are not completely equivalent. For example, imagine a system in which a paddlewheel is immersed in water. Doing work on the system by turning the paddlewheel will cause the temperature of the water to rise, but the reverse does not happen: such a system never spontaneously drops in temperature, giving rise to motion of the paddlewheel. Hence, nature imposes a restriction on the conversion of heat to work. This is the subject of the second law of thermodynamics, from which we will see that, in symbolic form,

$$W \underset{\leftarrow - - - - - - - -}{\longrightarrow} Q \qquad (2\text{-}3)$$

which indicates that work can freely be converted to heat, but the reverse conversion of heat to work is subject to constraints.

2–2 Molecular Interpretation of Energy Changes

Even though the first law does not depend on any particular interpretation of the energy of a system, our intuitive understanding of energy changes will benefit greatly from a closer look at the meaning of the energy of a system, and the molecular mechanisms by which energy can be stored.

You might find it helpful to think of the energy of a system as analogous to the balance in your bank account. Money (energy) may be deposited (plus sign) or withdrawn (minus sign) in the form of cash (work) or checks (heat). The daily change in your balance (ΔE) is the sum of all the transactions that day ($W + Q$). The bank statement records your balance (E), which indicates the amount of money (energy) stored in your account. With modern banking practices, balances can be positive or negative, and so can the energy of a system.

Added heat and work increase the energy of a system, but how is the incremental energy stored? (If you ask the same question about the money you deposit in your bank account, you will find that the answer is also complicated, but not obviously analogous to the molecular mechanism of energy storage.) You can visualize the meaning of the energy E that characterizes a state by considering the molecular nature of matter, a topic not strictly part of thermodynamics. The energy supplied to a system by work or heat is stored in the kinetic energy K_E of the particles, and in the potential energy U of their interaction,

$$E = K_E + U \tag{2-4}$$

in which we assume for simplicity that the substance itself is at rest, and that there is no energy due to gravitational or applied electromagnetic fields. The kinetic energy due to movement of a single particle through space, called *translational motion,* is given by classical mechanics as $mv^2/2$, where m is the particle mass and v the velocity. The total translational kinetic energy is the number of particles N multiplied by the mean translational kinetic energy, or $Nm\langle v^2\rangle/2$, in which the brackets indicate the average of the square of the velocity. If the gas particles contain more than one atom, they also can have energy due to rotation and vibration.

The Ideal Monatomic Gas has only Kinetic Energy

An *ideal monatomic gas* is a substance that consists of point atoms that do not interact with each other, and which obeys the equation of state $PV = nRT$. Because there is no interaction between the particles, $U = 0$ and

Equation 2-4 becomes $E = K_E$. It can be shown (see Chap. 14) that the energy E and the temperature T of an ideal monatomic gas are related by

$$E = \tfrac{3}{2}\, nRT \qquad (2\text{-}5)$$ ◀

in which R, the gas constant, is a proportionality constant whose value depends on the units of the temperature and energy scales. Table 2-1 lists some values. The energy of an ideal gas which is not monatomic is larger than $\frac{3}{2}\, nRT$ because of the energy stored in rotation and vibration. This topic is considered in detail in Chapter 14.

TABLE 2-1 *Values of the gas constant R*

UNITS	VALUE
J K^{-1} mol^{-1} (SI)	8.31431
cal K^{-1} mol^{-1}	1.98717
cm^3 atm K^{-1} mol^{-1}	82.0575
l atm K^{-1} mol^{-1}	0.0820575

Equation 2-5 is one of several possible equivalent definitions of *temperature*. It states that T is a quantity proportional to the energy of an ideal gas, and implies that there is a minimum temperature at which $E = 0$ and all motion ceases. This temperature is called *absolute zero* and is −273.15°C, which is defined to be 0 K on the Kelvin scale. Notice that the energy of one mole of an ideal gas does not change unless the temperature changes.

All energy supplied to a monatomic ideal gas, either as heat or as work, increases the kinetic energy, which is the only form of energy the noninteracting gas particles can have. When heat is added, the molecular mechanism of energy transfer begins with increased motion of the particles in the gas container, because the temperature at its boundary is increased. The motion of the boundary particles is transferred to the gas particles by collision, and the gas molecules move faster, with increased energy and temperature as a consequence.

Work also can increase the temperature of an ideal gas. If the gas is compressed by a piston, the motion of the piston imparts added velocity to the gas molecules, thereby increasing their temperature, unless the energy increase is compensated for by heat release. The following exercises will give you some practice calculating heat, work, and energy changes for ideal monatomic gases.

EXERCISE 2-2

One mole of an ideal monatomic gas is compressed adiabatically by a force of 1.000 newton acting through 0.500 meter. Calculate the temperature rise.

ANSWER

In an adiabatic process, $Q = 0$. Therefore $\Delta E = W = 0.500$ J $mole^{-1}$. From Equation 2-5, $\Delta T = (2/3R)\Delta E$, or

$$\Delta T = \frac{2 \times 0.500 \text{ J mole}^{-1}}{3 \times 8.314 \text{ J K}^{-1} \text{mole}^{-1}} = 0.0401 \text{ K}$$

EXERCISE 2-3

An ideal gas is compressed isothermally by a force of 1.0 newton acting through 0.5 meter. Calculate the heat change.

ANSWER

In an isothermal process for an ideal gas, $\Delta E = 0$, since $\Delta T = 0$ (see Equation 2-5). Hence $Q = -W = -0.5$ J. The negative sign means that heat is released when the gas is compressed isothermally.

EXERCISE 2-4

One mole of an ideal monatomic gas is at 300 K. (a) Calculate the energy E. (b) Describe two ways of increasing the energy by 1 J.

ANSWER

(a) $E = (3/2) \times 1 \text{ mole} \times 8.314 \text{ J K}^{-1} \text{ mole}^{-1} \times 300 \text{ K} = 3.74 \times 10^3$ J.
(b) Either add 1 J of heat to the gas held at constant volume or do 1 J of adiabatic work on the gas, or do any combination of these two that adds up to 1 J.

EXERCISE 2-5

The energy of one mole of an ideal monatomic gas is 5000 J. What else do we need to know to describe the state?

ANSWER

Either the pressure or the volume. The energy tells us only the product of pressure and volume ($PV = nRT = 2E/3$). The answer is in agreement with the phase rule, which demands two variables in order to specify the state of a homogeneous phase of a given amount of pure substance.

Real Substances Possess Potential Energy of Interaction

No real substance is an ideal gas. Real atoms and molecules interact, and this interaction produces potential energy. The origin of such interaction energies will be the subject of Chapters 10 and 11; for the moment we will content ourselves with the definition of the potential energy.

If two particles interact, they exert a force on each other. The force $F(R)$, which is a function of the distance R between the particle centers as shown in Figure 2-2(a), is defined to be negative when the particles attract and positive when they repel each other. Moving them farther apart by a distance dR requires work dW:

$$dW = -FdR$$

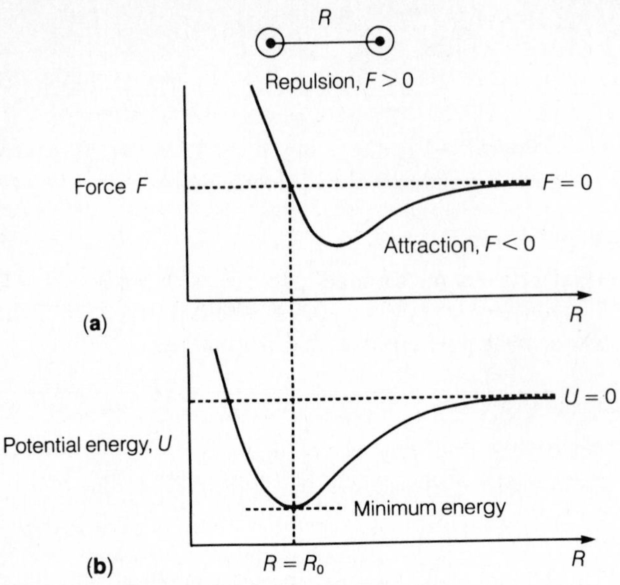

FIGURE 2-2
Schematic diagram of the force F between two interacting particles at different distances of separation R. The equilibrium separation is the point of minimum energy, $R = R_0$, because work is required for either an increase or a decrease in R. At that point, $F = -dU/dR = 0$. Note that $F = -dU/dR$ is the negative of the slope of the plot of U versus R.

The negative sign is required so that dW will be positive when two attracting particles (F negative) are moved farther apart (dR positive). *The total work required to move the particles from infinite separation ($R \to \infty$) to a separation of distance R is called the potential energy U.* This quantity is a sum of all the work terms dW from $R' = \infty$ to R, which we can write as an integral (see Section 2-10 if your calculus is rusty):

$$U = \int_{\infty}^{R} dW$$

Substituting $-FdR'$ for dW, we get

$$U = -\int_{R'=\infty}^{R'=R} FdR' \qquad (2\text{-}6)$$

(A prime is added here simply to distinguish the integration variable R' from the final separation R.) This is an important equation that relates the potential energy U to the force F between two particles. Figure 2-2(*b*) illustrates the variation of U with R.

Equation 2-6 can be differentiated to yield

$$F = -\frac{dU}{dR} \qquad (2\text{-}7)$$

which is another way of expressing the relation between force and potential energy. Notice that Equation 2-7 states that the force F is the negative of the slope of the curve that gives the dependence of U on R. Furthermore, when $F = 0$, dU/dR must also be zero, so U must be a minimum or maximum when $F = 0$. Finally, when we define U by Equation 2-6, we find that $U = 0$ when R becomes very large. In effect, we have defined a reference point ($R \to \infty$) from which the potential energy is measured. *Unlike kinetic energy, which can be expressed in terms of molecular velocity, potential energy always has an arbitrary zero to which all energy values refer.* Notice that the potential energy U is negative at the minimum energy point, $R = R_0$, in Figure 2-2(*b*).

Energy is Stored in Potential and Kinetic Form

You can readily imagine that the molecules in a substance such as ice or liquid water exert a force on each other and therefore have potential energy due to their interaction. Consequently, when energy in the form of heat or work is added to water, it can change both the kinetic energy and the potential energy. Figure 2-3 summarizes the conversion of heat to work and the storage of energy in molecular motion and potential energy. A sample at 0 K (absolute zero) in State (*a*) is heated by adding heat Q at constant pressure to produce State (*b*). The first law tells us that the change in the energy ΔE is the heat absorbed Q plus the work done on the system, $W = -P\Delta V$. Since the sample expands, $\Delta V > 0$ and $W < 0$. Therefore, $\Delta E = Q + W$ is less than Q. Hence, we can say that of the heat added, $Q = \Delta E - W$, some is used to do work of expansion $-W$ on the atmosphere, and the remainder is used to increase the internal energy E of the sample. The energy change ΔE ($\Delta E < Q$) in turn is used to increase both the kinetic energy of molecular motion and the potential energy of intermolecular interactions, as illustrated in the figure.

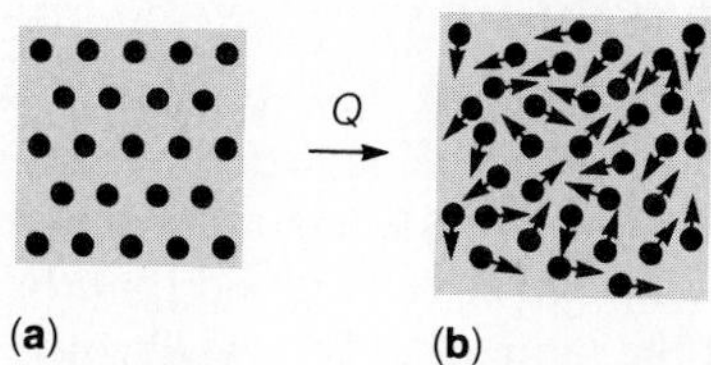

FIGURE 2-3
Storage of energy in molecular motion and potential energy. In **(a)**, the temperature is absolute zero, so there is no kinetic energy and the molecules are all at the equilibrium separation R_0, in which the energy is a minimum. In **(b)** energy has been added by heating at constant pressure so there is molecular motion and $T > 0$. Some of the energy is used to increase the kinetic energy K_E of the molecules. In addition, the system has expanded, so that some energy is used to increase the potential energy U, because the molecules are no longer at the minimum energy separation R_0. The total increase in energy ΔE is less than Q because some of the energy added is used to do work of expansion against the constant pressure that the surroundings (for example, the atmosphere) exert on the system.

The quantitative importance of intermolecular interactions can be recognized by considering the melting of ice. Suppose enough heat is added to convert a sample of ice at 0°C to water at 0°C. The kinetic energy of molecules depends primarily on the temperature, as is the case for an ideal gas, so the kinetic energy of water molecules in two phases in equilibrium is essentially equal (except for quantum mechanical constraints on the excitation of vibration or rotation—see Chapters 10 and 14). Virtually all the energy added to ice to convert it to water goes into increasing the potential energy of the molecular interactions, thus weakening the bonds between water molecules to convert the solid to liquid.

EXERCISE 2-6

The heat required to melt one mole of ice (0°C) is 5980 J mol^{-1}. Calculate the increase in temperature this would produce if converted to kinetic energy in one mole of an ideal monatomic gas.

ANSWER

$$\Delta T = \frac{2\Delta E}{3R} = \frac{2 \times 5980 \text{ J mol}^{-1}}{3 \times 8.3143 \text{ J K}^{-1} \text{ mol}^{-1}} = 479 \text{ K}$$

This example shows that in condensed materials (liquid and solids) the magnitude of the potential energy term is large compared to the kinetic energy at room temperature (≈295 K).

2–3 *Heat, Energy Change, Enthalpy Change, and Heat Capacity are Readily Measured Quantities*

With some feeling for the molecular basis of energy, we are now prepared to return to thermodynamics proper. The measurement of the heat absorbed in a process is one of the basic experiments in thermodynamics. This may seem contradictory at first, since we have stressed that dQ is an inexact differential, which implies that the amount of heat Q absorbed in a process depends on the path by which the process is carried out. However, it is easy to show that there are two simple paths for which Q depends only on the initial and final values of a state function. When the process occurs at *constant volume*, the heat absorbed Q_V is equal to ΔE. At *constant pressure* the heat Q_P is equal to the change ΔH in a new property, H, which is called the *enthalpy*. In each case the path is constrained in such a way that the heat change equals a change in a property.

We first take the condition of constant volume. The work term dW in the first law, $dE = dQ + dW$, can be replaced by $dW = -PdV$ (see Tab. 1-3), if we assume that the only work is that of expansion-compression. Applying the condition of constant volume ($dV = 0$) to the equation

$$dE = dQ - PdV$$

leads us to the expression,

$$dE = dQ_V \tag{2-8a}$$

in which dQ_V denotes that the heat change is at constant volume. When the process is not infinitesimally small we can write

$$\Delta E = Q_V \tag{2-8b}$$ ◀

This important equation tells us how to measure changes in the energy of a system. We simply determine the heat Q_V under conditions of constant volume (see Fig. 2-4).

When the heat is measured at constant pressure, and dW again is replaced by $-PdV$, the first law becomes

$$dE = dQ_P - PdV$$

in which dQ_P denotes the heat change at constant pressure. For a finite change, assuming $P = P_{\text{ext}}$ is constant throughout the process, we write:

$$\Delta E = Q_P - P\Delta V$$

This equation can be rearranged to

$$Q_P = \Delta E + P\Delta V \tag{2-9}$$

Since E, P, and V are properties, the right side of Equation 2-9 must be equal to the change in a property.

The property whose change is equal to Q_P is called the *enthalpy, H:*

$$H = E + PV \tag{2-10a}$$ ◀

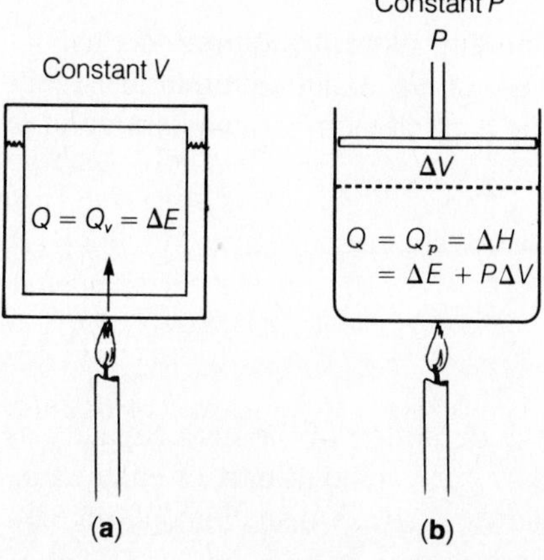

FIGURE 2-4
Heating systems at **(a)** constant volume as in a pressure bomb, and **(b)** constant pressure as in a flask open to the atmosphere.

At constant pressure the enthalpy change, ΔH, is

$$\Delta H = \Delta E + P\Delta V \tag{2-10b}$$

The right side of this expression is identical to Equation 2-9, so we can conclude that

$$Q_P = \Delta H \tag{2-11}$$

Equation 2-11 states that the *heat absorbed at constant pressure is equal to the enthalpy change of a process.* Restricting the path to constant pressure, like constant volume, makes heat absorbed equal to a change in a property of the system. Equation 2-9 tells us that part of the heat absorbed at constant pressure increases the energy of the system, and part goes to expand the system against an external pressure. The expansion term $P\Delta V$ is a small fraction of the total except in processes involving gases and in processes at pressures of many thousands of atmospheres. With these exceptions in mind, we can write

$$\Delta H \cong \Delta E \qquad \text{(except gases or at high pressure)}$$

It often is easier to measure heat changes at constant pressure (usually 1 atm) than at constant volume, so ΔH is a frequently measured quantity. The enthalpy change is an important characteristic of chemical reactions and physical processes, as we shall see in subsequent sections.

EXERCISE 2-7

Express the enthalpy of an ideal monatomic gas in terms of temperature.

ANSWER

$H = E + PV = \frac{3}{2}nRT + nRT = \frac{5}{2}nRT$. Notice that H, like E, for the ideal gas depends only on T, not on P or V.

Heat Capacity Measures the Ability to Store Energy

The *heat capacity* of a substance is the amount of heat required to raise its temperature by one degree. Since dQ is not an exact differential except under prescribed conditions, we define the heat capacity at constant volume C_V and at constant pressure C_P by

$$C_V = \frac{dQ_V}{dT} = \left(\frac{\partial E}{\partial T}\right)_V \tag{2-12}$$

$$C_P = \frac{dQ_P}{dT} = \left(\frac{\partial H}{\partial T}\right)_P \tag{2-13}$$

The first part of these two equations is the definition of the heat capacity as heat dQ divided by temperature change dT. The second part of each equation is obtained by setting $dQ_V = dE$ and $dQ_P = dH$. We shall make frequent use of Equations 2-12 and 2-13 in their differential forms

$$(dE)_V = C_V(dT)_V$$
$$(dH)_P = C_P(dT)_P$$

in which the subscripts V and P make it clear that V or P must be held constant.

The heat capacities C_V and C_P express the amount of heat required to increase the temperature of one mole of a substance by one degree under conditions of either constant volume or constant pressure. They are different because in the case of constant pressure some heat is required for work of expansion, as illustrated in the following exercise.

EXERCISE 2-8

Show that for an ideal monatomic gas $C_P - C_V = nR$.

ANSWER

$E = \frac{3}{2}nRT$ so $C_V = \frac{3}{2}nR$; $H = E + PV = \frac{3}{2}nRT + nRT$, so $C_P = \frac{3}{2}nR + nR$. Hence, $C_P - C_V = nR$. It can be shown that this result is true for all ideal gases, not just monatomic gases (see Problem 2-8).

The *specific heat* of a substance is the amount of heat required to raise the temperature of 1 g of a substance by 1°C. Thus the heat capacity of a mole of a substance is the product of its specific heat and its molecular weight. Unless otherwise specified, the specific heat as usually tabulated refers to conditions of constant pressure.

The Molecular Basis of Heat Capacities

The physical meaning of Equation 2-12 is that a large value of C_V indicates how much heat energy must be put into a substance to warm it by a given amount. The energy can be kinetic (increasing vibrations of molecules in a crystal, rotations and vibrations in a liquid, or translations and other motions in a gas), or the energy can be potential, as in the weakening of interactions among molecules.

We noted earlier that, except for reactions of gases and for substances under great pressure, $H \approx E$. Therefore, with the same exceptions, C_P is roughly equal to C_V (within 10%–20%) and its magnitude is governed by the same factors.

Units of Energy

A readily measurable effect of heat is to increase the temperature of a substance. This led to the definition of a unit of heat called the *calorie,* which is historically defined as the heat required to increase the temperature of 1 g of water from 14.5°C to 15.5°C at atmospheric pressure. Since by Equation 2-13

$$C_P \Delta T = Q_P$$

and $\Delta T = 1°C = 1$ K, the value of $Q_P = 1$ cal defines the specific heat of water at 15°C as 1 cal $K^{-1}g^{-1}$. The calorie and kilocalorie (1 Kcal = 1000 calories) are sometimes employed in biochemical literature, but are no longer standard units. The "calorie" cited for foods and beverages is the kilocalorie used by scientists.

The standard unit of work, energy, and heat is the *joule,* which is one volt-coulomb (electrical work) or one newton-meter (mechanical work). The establishment of the equivalence between heat and work, and the proportionality between calories and joules, is one of the fascinating stories in the development of thermodynamics. In 1798 Count Rumford of the Holy Roman Empire noticed while boring cannon barrels in Bavaria that heat was produced by the work done against frictional forces and in proportion to the work (see following biographical sketch). It was years after his death before it was generally accepted that work could be converted to heat, since the prevailing view was that "caloric" was a substance. In retrospect, Rumford's experiments proved his point, and his value of the proportionality constant between work and heat (about 5.5 joules per calorie in modern terms) is surprisingly close to the present standard of 4.1840 joules cal^{-1}. The precise experiments of James Prescott Joule gave the value 4.154 J cal^{-1} in 1849, which has been modified by less than 1% since that time.

Sensitive modern calorimeters use the equivalence of heat and electrical work. In general, they measure heat capacity by measuring the amount of electrical heating required to produce a given temperature increase, or they measure enthalpy changes by determining the electrical energy required to produce the same temperature increase as is caused by the process under study. Temperature changes on the order of microdegrees can be measured in solutions as small as one milliliter; with care the electrical energy can be measured with a sensitivity better than a microjoule—or roughly a microcalorie. This makes possible the direct measurement of heat changes on biochemical materials available only in small quantities. One example is the study by Hunt and colleagues (1972) of the heat absorbed by the enzyme glutamine synthetase as ions bind to it. They found that 3 kcal were absorbed per mole of Mn^{2+} ion bound. By observing the temperature change as the system came to equilibrium after Mn^{2+} was added, they found that the process has two steps. These correspond to initial binding of a Mn^{2+} ion in which a proton of the enzyme is released, and then to slower release of a second proton.

○ 2-4 An Example: *W* and *Q* for an Ideal Gas Expansion

No thermodynamics course is complete without a calculation of the heat and work for expansion of an ideal gas. The reason—in our case at least—is not so much to obtain the specific results, but rather to illustrate the application of the first law and the related concept of heat capacity. The mathematical

BENJAMIN THOMPSON, COUNT RUMFORD

1753-1814

Benjamin Thompson's progress from poor Massachusetts farmboy to Imperial Count of the Holy Roman Empire and husband of the wealthy and socially prominent widow of the great chemist Lavoisier seems more like the plot of a comic opera than history. Thompson, like Gibbs, was an American-born contributor to the theory of heat. But the similarity ends there. Whereas Gibbs was content to contemplate the nature of matter and energy and to participate quietly in academic life, Thompson was driven by ruthless ambition towards the centers of power, prestige, and wealth. His scientific and technological discoveries, although undeniably important, were often intertwined with his political intrigues and frequently were stimulated by a practical military or administrative problem that he faced. Thompson's qualities of energy, ambition, and arrogance propelled him to discovery and power, but they were also limitations. In the end they turned men against him and diminished his scientific and political impact.

Thompson was born to a poor farming family in Woburn, Massachusetts, and starting at the age of 13 he was successively apprenticed to two dry goods merchants and a doctor. He failed to please his employers because he was less interested in learning a trade than in studying science and carrying out experiments. Consequently, in 1772, at the age of 19, he took a job as schoolmaster in Concord, New Hampshire (originally called Rumford, New Hampshire). Within four months he made the move that started his career on its incredible path: he married the wealthiest and most prominent widow of the colony, a woman eleven years his senior. The schoolmaster instantly became gentleman farmer, and soon received from the British Royal Governor the commission of major in the New Hampshire militia.

This was, of course, the time during which the American Revolution was brewing. Local patriots had no illusions about which camp Thompson's sympathies would lie in, and when he heard a rumor of approaching tar and feathers, he fled back to Woburn, abandoning his wife and infant daughter. Thompson first worked for the British as a spy, and then joined them openly in Boston. When the city fell in 1776, he sailed for London where he acted as an expert on the colonial war. By exaggerating his own importance and by the persuasive power he seemed to exert over men, he soon became private secretary to Lord Germain, who was Secretary of State for Colonies and the man in charge of pursuing the war against the revolutionaries. By 1780 he had secured for himself a position of independent power as Undersecretary of State for the Northern Department, which entailed the responsibility of recruiting, equipping, and transporting British forces to America.

In 1780 Thompson also carried out a series of experiments on the forces produced by exploding gunpowders of various compositions. He designed an improved ballistic pendulum for this purpose. The experiments were judged important enough scientifically that he was elected Fellow of the Royal Society in the following year, and the results were thought to be of sufficient practical importance that he was invited on maneuvers with the British fleet. It was typical of Thompson to turn to fundamental science to solve a practical problem which he encountered in his work. It was also typical that his results led to a strengthening of his political position or prestige.

Thompson was forced from his position with the British government when it was rumored that he was a spy for the French. At the close of the American Revolution he was a soldier of fortune, seeking a new arena for his talents and ambitions. His talents in science and administration gained for him a post as adviser to the Elector of Bavaria, in charge of reorganizing the country's inefficient army. He secretly maintained his contacts with the British Foreign Office, however, and sent back reports on the Bavarian army.

Thompson noted that clothing and feeding the soldiers were the principal items of the army budget. To find ways to trim these costs, he turned again to basic science. He devised a simple instrument to measure the thermal conductivity of materials, and set out to discover which fabrics were the best insulators for uniforms. He discovered that an insulating cloth is one in which air is prevented from moving. When local manufacturers were reluctant to produce cloth by the weave of his design, he opened "workhouses," in which beggars and other poor people did weaving in return for food and clothing. To educate children in the workhouses (and also the adults who desired it) he initiated what was one of the earliest free schools in Europe. Scientific advisers and innovators in government are not a development of the twentieth century.

In order to provide maximum nutrition at minimum cost for the army and for his workhouses, Thompson began to experiment with foods. He found that a "Rumford soup" of peas, barley, and potatoes was both nutritious and inexpensive. He set up gardens, including Munich's famous English garden, as experimental agricultural centers, and also to produce foods for his establishments. Thompson's interest in efficient preparation of food led him to devise the kitchen range to replace the open hearth, and then to invent the baking oven, the double boiler, the pressure cooker, a portable field kitchen for the army, and the drip coffee pot. To study the efficiency of heating fuels he designed the combustion calorimeter, and to improve lighting in his workhouses he invented a photometer and used it to study the luminosity of various candles and lamps. Other Thompson inventions include central heating and the modern efficient fireplace with smoke-shelf and damper.

It was during this period that, as superintendent of the Munich arsenal, Thompson performed his famous cannon boring experiments. He discovered that the intense heat produced during the boring of a cannon was proportional to the mechanical work of boring. This made it unlikely that heat was a substance (then called "caloric") squeezed from the metal as boring took place, because the caloric in any substance would be exhausted after extended boring. Thompson's idea that work and heat are energetically equivalent foreshadowed the first law of thermodynamics.

Thompson's genius for administration and innovation was soon apparent to the Elector, who promoted him from adviser and army colonel to Minister of War, Minister of Police, Major General, Chamberlain of the Court, and State

Councillor, and then to Count of the Holy Roman Empire. Thompson never again referred to himself by his family name, but only by his title, Count Rumford.

Holding all of these positions simultaneously was a drain on his health, and created numerous enemies. He decided that it was time to return to England, and the Elector cooperated by naming him Bavaria's Minister Plenipotentiary to the Court of St. James. This proposal angered King George's government, first because Thompson had supposedly been acting as a spy for Britain but had obviously held back information, and second because Rumford was a subject of the King, and hardly could act as ambassador from another country.

This setback was a disappointment to Rumford, who then began to cast about for a new position. It took a person of his incomparable arrogance to devise the next plan: he proposed to the new United States government that he return to America to set up a military academy. Fortunately his role during the Revolution was discovered before the appointment was arranged.

Rumford then busied himself with the foundation of the Royal Institution of Great Britain. It was started as a showcase for his inventions and discoveries, but when he brought in a young and talented chemist, Humphrey Davy, as professor, the future of the institution as a great research laboratory was established. Rumford also founded two prizes in his name, to be awarded periodically by the Royal Society of London and the American Academy of Arts and Sciences for research on heat and light. These were the first international awards for science, predating the Nobel awards by a century. Gibbs was a later recipient of the Rumford Medal.

Rumford returned to the Continent. where he courted and eventually married Madame Lavoisier. The marriage was not a happy one, and before long Rumford moved into his own villa to pursue his experiments. He died in 1814, leaving his estate to Harvard College. In a final act of intrigue, to insure that his benefaction would be accepted in a quarter that might have reason to reject it, Rumford arranged to have the signing of his will witnessed by Lafayette, the French hero of the American Revolution.

expression of the thermodynamic laws is very simple, but their application requires careful and systematic consideration of the conditions of each experiment. Let us compare the expansion of an ideal gas under isothermal and adiabatic conditions, considering reversible and irreversible paths. Our objective will be to derive equations for the heat and work in each case. We will also find that the reversible path yields greater work of expansion than do irreversible paths.

Isothermal (Constant Temperature) Expansion

A schematic diagram of an experiment to measure isothermal expansion is shown in Figure 2-5.

The work done on the system is

$$dW = -P_{ext}\,dV \tag{2-14}$$

in which the sign is negative because when the system does work, $dV > 0$. Work dW is defined as positive for the work done *on* the system. (Work always is expressed in terms of the external pressure P_{ext} because that quantity times ΔV defines the energy exchange with the surroundings.) The difference between reversible and irreversible paths is that for the reversible path the external and internal pressures are equal,

$$P_{ext} = P_{int} \qquad \text{(reversible)} \tag{2-15}$$

because the system is always very close to equilibrium, whereas for an irreversible expansion the external pressure is reduced so that

$$P_{ext} < P_{int} \qquad \text{(irreversible)} \tag{2-16}$$

This would be the case if, for example, you yanked up the piston in Figure 2-5, so that expansion of the gas lagged behind movement of the piston surface.

For a reversible expansion, $P_{ext} = nRT/V$, so Equation 2-14 becomes

$$dW = -nRT\,\frac{dV}{V} \tag{2-17}$$

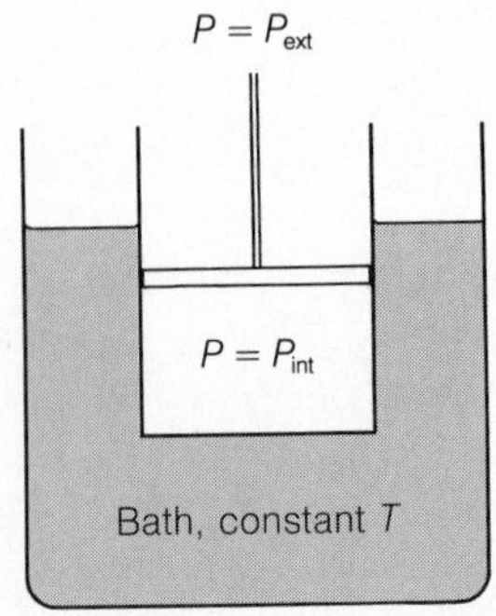

FIGURE 2-5
Experimental system for measuring the work of isothermal expansion of an ideal gas. The work $dW = -P_{ext}\,dV$ is always defined in terms of P_{ext} because $P_{ext}\,dV$ determines the energy exchange with the surroundings. For a reversible process, $P_{ext} = P_{int}$, while for an irreversible expansion, $P_{ext} < P_{int}$.

Since the temperature is constant we can obtain W by integrating dV/V (see Section 2-10), giving

$$W = -nRT \int_{V_1}^{V_2} \frac{dV}{V}$$

$$= -nRT \ln \frac{V_2}{V_1} \qquad \text{(reversible)} \qquad (2\text{-}18a)$$

in which ln represents the natural logarithm function (see Section 2-10).

The heat change Q can be readily calculated from W, because the temperature of the ideal gas is constant in an isothermal process. Hence $\Delta E = 0$, and the first law allows us to set $Q = -W$. Therefore

$$Q = nRT \ln \frac{V_2}{V_1} \qquad \text{(reversible)} \qquad (2\text{-}18b)$$

The heat and work terms for an irreversible expansion cannot be expressed by a general equation, because they depend on the path that is followed. However, we can show that the gas always does less work when it expands irreversibly than when it expands reversibly. Equation 2-18*a* for W is equal to the negative of the area under the solid curve in Figure 2-6. Since P_{ext} is decreased below P_{int} for all irreversible expansions, the area under the corresponding dotted curve will be less. Hence

$$|W_{irrev}| < |W_{rev}| \qquad (2\text{-}19)$$

where we use absolute values to avoid confusion arising from the negative sign. Equation 2-19 states that the work energy exchanged with the surroundings is smaller for an irreversible than for a reversible expansion path.

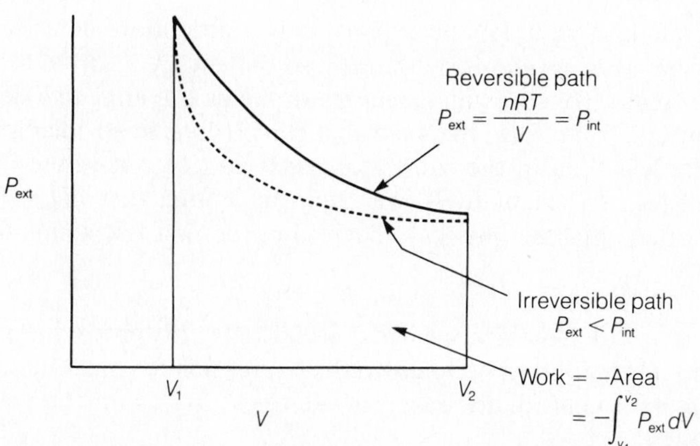

FIGURE 2-6
Reversible and irreversible work of isothermal expansion of an ideal gas. This diagram shows that the work term for irreversible expansion is smaller in absolute value than for a reversible process. Work is the negative of the area under the curve of P_{ext} versus V.

The maximum work that can be obtained by expansion is by a reversible path; that expansion, however, will occur more slowly than by an irreversible path because of the vanishingly small net pressure difference, $P_{int} - P_{ext}$, on the piston. There are many cases in biochemical systems where the thermodynamic efficiency of a reversible reaction is sacrificed for the speed gained when the process is irreversible.

EXERCISE 2-9

Calculate the work for reversible isothermal expansion of one mole of an ideal gas from 10 atm pressure to 1 atm at 300 K.

ANSWER

$V_2/V_1 = P_1/P_2 = 10$, so $W = -nRT \ln (V_2/V_1) = -8.3143 \text{ J K}^{-1} \text{ mole}^{-1} \times 1 \text{ mole} \times 300 \text{ K} \times \ln 10 = -8.3143 \times 300 \times \log 10 \times 2.303 \text{ J} = -5743 \text{ J}$.

EXERCISE 2-10

According to Exercise 2-4, the energy of one mole of an ideal gas at 300 K is 3741 J. Where does the energy come from to do 5743 J of work on the surroundings in Exercise 2-9?

ANSWER

From the bath. Since T is constant, $\Delta E = 0$ for the gas and $Q = -W$, so $Q = 5743$ J. The gas absorbs heat from the bath and uses it to do work of expansion. (Notice that in this case there is no problem in the conversion of heat to work; see Chap. 3.)

Adiabatic Expansion

A consideration of an adiabatic expansion is a little more complicated. No heat is absorbed or given off by the gas, so the energy to drive the work of expansion comes from the kinetic energy of the gas, leading to a decrease in its temperature. Therefore, the work that can be done in an adiabatic expansion will be less than in the same expansion carried out isothermally (see Exercise 2-10). To calculate W we begin by noting that $dQ = 0$, by the definition of an adiabatic process. Therefore, the first law states that

$$dE = dW$$

Thus if we can calculate dE, we can calculate dW. In approaching problems of this kind it is helpful to write down the expression for the total differential of the quantity sought. In this case, considering E to be a function of T and V, we obtain

$$dE = \left(\frac{\partial E}{\partial T}\right)_V dT + \left(\frac{\partial E}{\partial V}\right)_T dV \tag{2-20}$$

(See Equation 1-18.) We see immediately that this was a good choice of independent variables, because $(\partial E/\partial V)_T$ is zero for an ideal gas (Why?), whereas $(\partial E/\partial T)_V$ is defined by Equation 2-12 to be C_V. Hence, for expansion from state T_1V_1 to T_2V_2,

$$dE = C_V\,dT$$

$$\Delta E = W = \int_{T_1}^{T_2} C_V\,dT = C_V(T_2 - T_1) \tag{2-21}$$

in which we have used the fact that C_V is independent of T for an ideal gas in order to perform the integration (see Section 2-10). Notice that Equation 2-21 is valid for both reversible and irreversible paths, because W is equal to ΔE, a property that depends only on $T_2 - T_1$. However, we will see in the next paragraphs that calculation of $T_2 - T_1$ requires that we assume a reversible path.

We have not quite finished the derivation, because those who set such problems are perverse enough to give us not the final temperature we need in Equation 2-21 but rather the final volume or pressure. We must then find a relation between the volume change and the temperature change. To get an expression involving the volume, we return to the relationship, valid when $dQ = 0$, that $dE = dW = -P_{\text{ext}}\,dV$, and set $-P_{\text{ext}}\,dV$ equal to $dE = C_V\,dT$.

$$C_V\,dT = -P_{\text{ext}}\,dV \tag{2-22}$$

We now assume a reversible process, so that $P_{\text{ext}} = nRT/V$ at all times. Hence, dividing by $T = P_{\text{ext}}V/nR$,

$$C_V\frac{dT}{T} = -nR\frac{dV}{V} \qquad \text{(reversible)} \tag{2-23}$$

Integrating, we get

$$C_V \ln\left(\frac{T_2}{T_1}\right) = nR \ln\left(\frac{V_1}{V_2}\right) \tag{2-24}$$

Rearranging terms gives

$$\ln\left(\frac{T_2}{T_1}\right) = \frac{nR}{C_V}\ln\left(\frac{V_1}{V_2}\right)$$

Taking the exponential (see Section 2-10 for a review of exponentials) yields

$$\frac{T_2}{T_1} = \left(\frac{V_1}{V_2}\right)^{nR/C_V} \qquad \text{(reversible)} \tag{2-25}$$

This expression tells us by what fraction the temperature will be reduced when the volume expands from V_1 to V_2. Notice from Exercise 2-8 that nR/C_V is 2/3 for an ideal monatomic gas. Table 2-2 summarizes the discussion in this section by giving the thermodynamic changes for reversible isothermal or adiabatic volume changes of an ideal gas. Be sure that you understand every entry in the table.

TABLE 2-2 *First Law Quantities and Temperature Changes in Reversible Volume Changes of an Ideal Gas*

PROCESS	W	Q	ΔE	T_2/T_1
Reversible isothermal expansion or contraction	$-nRT_1 \ln \frac{V_2}{V_1}$	$nRT_1 \ln \frac{V_2}{V_1}$	0	1
Reversible adiabatic expansion or contraction	$C_V(T_2 - T_1)$	0	$C_V(T_2 - T_1)$	$(V_1/V_2)^{nR/C_V}$

EXERCISE 2-11

One mole of an ideal monatomic gas is expanded reversibly and adiabatically from 1 liter to 10 liters, starting at 300 K. Calculate W.

ANSWER

$T_2/T_1 = (0.1)^{2/3} = 0.215$; $T_2 = 0.215 \times 300$ K $= 64.6$ K (Equation 2-25). Hence, $W = 3/2 \times 1$ mole $\times$ 8.3143 J K^{-1} mole^{-1} (64.6 − 300) K = −2936 J (Equation 2-21). The work for this process carried out irreversibly will be less in absolute value than for the reversible path. This can be shown with a diagram such as that in Figure 2-6. Consequently, the final temperature will not drop as much in the irreversible process as in the reversible one.

EXERCISE 2-12

Write the total differential, dE, using (V,P) and (T,P) as independent variables. Explain why Equation 2-20 is a better choice.

ANSWER

$$dE = \left(\frac{\partial E}{\partial V}\right)_P dV + \left(\frac{\partial E}{\partial P}\right)_V dP$$

In this case, neither of the partial derivatives is zero, so a simple expression such as (2-21) is not obtained. Alternatively,

$$dE = \left(\frac{\partial E}{\partial T}\right)_P dT + \left(\frac{\partial E}{\partial P}\right)_T dP$$

In this case $(\partial E/\partial P)_T$ is zero for an ideal gas, but $(\partial E/\partial T)_P$ is a quantity not yet defined. However, you should be able to show that $(\partial E/\partial T)_P = C_P - nR = C_V$ for an ideal gas, so that the equation $dE = C_V\, dT$ can be obtained by this approach, as well as from Equation 2-20.

○ 2–5 *Enthalpy Changes: A General Principle*

The measurement of enthalpy changes constitutes the heart of the application of the first law of thermodynamics to chemical and biochemical systems. In the next section we take up thermochemistry, the most important of those applications. There is one easily stated general principle that many applications have in common. Before stating it, however, we should be sure that we understand the definition of an enthalpy change. ΔH for the process

$$\underset{H = H_1}{\text{State 1}} \xrightarrow{\Delta H} \underset{H = H_2}{\text{State 2}}$$

in which the system changes from State 1 to State 2 is equal to the enthalpy of the final state minus the enthalpy of the initial state, or

$$\Delta H = H_2 - H_1$$

The important general principle of this section can be stated in several equivalent ways. For example, ΔH does not depend on the path by which State 1 is converted to State 2. This is simply another way of saying that H is a property or function of state, so the principle is not a new one. However, it is equivalent to what is sometimes called *Hess's law of heat summation*.

Hess's law states that the total enthalpy change of a process is equal to the sum of the enthalpy changes of the individual steps of the process. For example, if we consider the process

$$A \underset{\Delta H_{AB}}{\longrightarrow} B \underset{\Delta H_{BC}}{\longrightarrow} C$$

the enthalpy change, ΔH_{AC}, for converting A to C must be equal to

$$\begin{aligned} \Delta H_{AC} = H_C - H_A &= (H_C - H_B) + (H_B - H_A) \\ &= \Delta H_{BC} + \Delta H_{AB} \end{aligned}$$

In other words, ΔH_{AC} is the sum of the heats for the two steps, $\Delta H_{AB} + \Delta H_{BC}$.

The utility of this general principle is that it allows us to calculate the enthalpy change for processes we may not be able to measure directly. For example, suppose we can measure ΔH_{AC} and ΔH_{BC}, but that the step $A \rightarrow B$ cannot be carried out experimentally without converting B to C. Hess's law allows us to set

$$\Delta H_{AB} = \Delta H_{AC} - \Delta H_{BC}$$

and to calculate the unknown from measurable quantities.

○ 2–6 *Thermochemistry Relates Heat to the Enthalpy Change of Chemical Reactions*

Thermochemistry is the study of the enthalpy or energy changes of chemical reactions. It provides a fundamental knowledge of the nature of chemical and biochemical compounds and reactions, and is probably the most important biochemical application of the first law of thermodynamics. Specifically, measurement of the heat change of chemical reactions provides:

1. The relative energies of compounds, which is of great usefulness in chemistry, biochemistry, nutritional medicine, and other sciences.
2. The energy of chemical bonds, which is essential to understanding the nature and strength of chemical bonding.

Many years of thermochemical experiments have produced standard enthalpy tables from which the heat of a wide variety of chemical reactions can be calculated. In this section we will consider how those tables were generated and how they are used. In Section 2-7 we will examine what the results tell us about the energetics of chemical bonds.

Heats of Reaction

Reaction heats are measured either in a bomb calorimeter, which is a closed system in which $Q_V = \Delta E$, or in an open calorimeter at constant pressure in which $Q_P = \Delta H$. The reaction mixture, which constitutes the thermodynamic system, is immersed in a heat bath (Figure 2-7).

As the reaction proceeds, heat is absorbed or released by the system. This is detected by a temperature change of the bath, ΔT. The change is expressed by using the definition of heat capacity, $C_P = Q_P/\Delta T$. Hence

$$\Delta T = \frac{Q_P(\text{bath})}{C_P(\text{bath})} = -\frac{Q_P(\text{system})}{C_P(\text{bath})} = -\frac{\Delta H(\text{reaction})}{C_P(\text{bath})}$$

When $Q(\text{system}) < 0$ we say a reaction is *exothermic* (heat is produced), and when $Q(\text{system}) > 0$ it is *endothermic* (heat is absorbed).

The enthalpy change ΔH of any process, including a chemical reaction, is the total enthalpy of the products minus the enthalpy of the reactants. Suppose our reaction is the combination of hydrogen and oxygen to form water:

$$H_2 + \tfrac{1}{2} O_2 \rightarrow H_2O \tag{2-26}$$

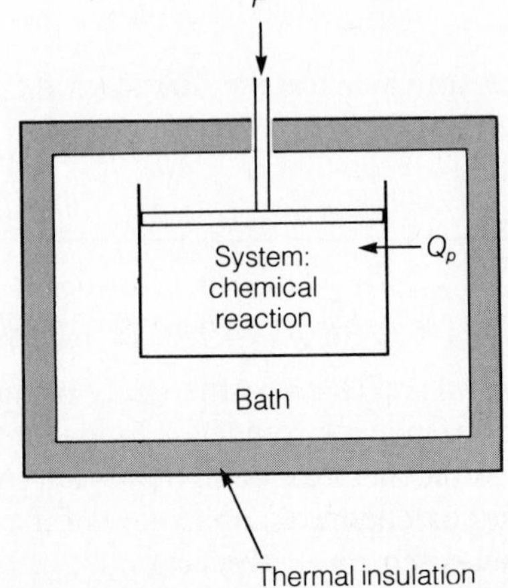

FIGURE 2-7
Schematic diagram of an open calorimeter to measure the heat Q_P of a chemical reaction. A very small temperature change of the bath, ΔT, is produced by the reaction. The heat capacity of the bath can be determined by measuring the amount of electrical energy introduced as heat that is required to produce a given value of ΔT.

Then the enthalpy change is the difference of the enthalpy of products and reactants,

$$\Delta H = H(H_2O) - H(H_2) - \tfrac{1}{2} H(O_2) \qquad (2\text{-}27)$$

In general we can write

$$\Delta H = \Sigma\, H(\text{products}) - \Sigma\, H(\text{reactants}) \qquad (2\text{-}28)$$

The energy change ΔE of a reaction is described by an analogous equation

$$\Delta E = \Sigma\, E(\text{products}) - \Sigma\, E(\text{reactants}) \qquad (2\text{-}29)$$

Alternatively, if the enthalpy change has been measured, ΔE can be calculated from the volume change at constant pressure, ΔV_P, from

$$\Delta E = \Delta H - P\Delta V_P \qquad (2\text{-}30)$$

and the definition

$$\Delta V_P = \Sigma\, V(\text{products}) - \Sigma\, V(\text{reactants})$$

Thermochemical Equations and Standard Conditions

A few conventions are used in compiling and using tables of enthalpies of compounds.

The first is that thermochemical equations always give the physical state of reactants and products of the reaction. For example, to contain complete information, Equation 2-26 must be written in the form

$$H_2(g) + \tfrac{1}{2}O_2(g) \rightarrow H_2O(g) \qquad (2\text{-}31)$$

in which g signifies that all substances are gases. Similar equations might indicate that some of the reactants are solid (s), liquid (l), in aqueous solution (aq), or in some particular crystal form (c). The following exercise shows how ΔH differs if the water produced were in the liquid state rather than in the vapor state.

EXERCISE 2-13

The enthalpy change of Reaction 2-31 is -57.7979 kcal/mole of H_2O = -241.8 kJ mol^{-1} at 25°C (298.15 K) and 1 atm pressure. The enthalpy of vaporization of H_2O at 25°C and 1 atm pressure is 44 kJ mol^{-1}. Calculate ΔH for production of $H_2O(l)$ from $H_2(g) + \tfrac{1}{2}O_2(g)$ at 25°C and 1 atm pressure.

ANSWER

For $H_2O(g) \rightarrow H_2O(l)$, $\Delta H = 44$ kJ mol^{-1}. Therefore, $\Delta H = -241.8$ kJ mol^{-1} -44 kJ mol^{-1} $= -285.8$ kJ mol^{-1}.

The second convention is that tabulated values of enthalpies are for a set of *standard conditions,* so that the temperature and pressure are the same for all compounds. The usual convention is 25°C and one atmosphere pressure. The superscript ° emphasizes that H° refers to specific conditions, which must be stated along with the results.

A third convention pertains to the point of zero enthalpy. We define the enthalpy of the chemical elements as being equal to zero when they are in their most stable physical state (solid, liquid, or gas) under the standard conditions of temperature and pressure. Note that the elements are not necessarily in atomic form. For example, oxygen is in the form O_2 at the standard conditions.

Finally, the quantity tabulated is the *standard enthalpy of formation*, denoted ΔH_f°, and defined as the enthalpy of formation of the compound from its elements, whose enthalpy is set equal to zero when they are in their most stable form at the standard temperature and pressure. You can see now that the enthalpy change of -241.8 kJ mol^{-1} for Reaction 2-31 is actually a standard enthalpy of formation of $H_2O(g)$. This is true because the reactants are the elements in their most stable form under the standard conditions. Therefore, $\Delta H_f^\circ = \Delta H^\circ = -241.8$ kJ mol^{-1}.

Examples of the Uses of Thermochemical Data

With these conventions and a table of enthalpies of formation such as Table 2-3 below, we can calculate the enthalpy change for many reactions, including some that are difficult or impossible to measure directly, and even for reactions that have never been run. Some examples follow.

First, consider how the values in the table were obtained. The enthalpy of formation of methane (CH_4) is given by ΔH_f° for

$$\text{C(graphite)} + 2\text{H}_2\text{(g)} \rightarrow \text{CH}_4\text{(g)} \tag{2-32}$$

in which the standard form of solid carbon is specified as graphite. Reaction 2-32 will not occur in stoichiometric amounts under any conditions. However, one can measure the heat of combustion (burning, or reaction with oxygen) of C, H_2, and CH_4 and combine the data to obtain ΔH_f° for Reaction 2-32. The individual reactions are:

$$\begin{array}{ll}
-\ [\text{CH}_4\text{(g)} + 2\text{0}_2\text{(g)} \rightarrow \text{CO}_2\text{(g)} + 2\text{H}_2\text{O(g)} & \Delta H^\circ = -804.1\text{ kJ}] \\
+\ [\text{C(graphite)} + \text{O}_2\text{(g)} \rightarrow \text{CO}_2\text{(g)} & \Delta H^\circ = -395.3\text{ kJ}] \\
+\ 2 \times [\text{H}_2\text{(g)} + \tfrac{1}{2}\text{O}_2\text{(g)} \rightarrow \text{H}_2\text{O(g)} & \Delta H^\circ = -241.8\text{ kJ}] \\
\hline
\text{C(graphite)} + 2\text{H}_2\text{(g)} \rightarrow \text{CH}_4\text{(g)} & \Delta H_f^\circ = -74.8\text{ kJ}
\end{array}$$

When combined with the sign and coefficients indicated at the left, the separate reactions yield the desired total reaction; the heats are combined in the same way. This procedure uses the general principle discussed above in Section 2-5; since enthalpy is a property, ΔH adds exactly as the reactions do, with the same sign and multiplicity. Thus, the standard enthalpy of formation is $\Delta H^\circ = -(-804.1) - 395.3 - 2 \times 241.8 = -74.8$ kJ.

TABLE 2-3 *Standard Enthalpy of Formation of Selected Compounds (1 atm pressure, 25°C)*

INORGANIC AND SIMPLE ORGANIC COMPOUNDS[a]	ΔH_f°/kcal mol^{-1}	ΔH_f°/kJ mol^{-1}
H^+(aq)	0.0 (by convention)	0
LiCl(c)	−97.7	−408.8
NaCl(c)	−98.2	−410.9
NaF(c)	−136.0	−569.0
NaBr(c)	−86.0	−359.8
NaI(c)	−68.8	−287.9
KCl(c)	−104.2	−436.0
$MgCl_2$(c)	−153.4	−641.8
CO(g)	−26.42	−110.5
CO_2(g)	−94.05	−393.5
CO_2(aq)	−98.69	−412.9
CH_4(g)	−17.89	−74.85
HCOOH(g)	−86.67	−362.6
H_2CO_3(aq)	−167.0	−698.7
HCO_3(aq)	−165.18	−691.1
CO_3^{2-}(aq)	−161.63	−676.2
NO(g)	21.6	90.4
NO_2(g)	8.09	33.8
NH_3(g)	−11.04	−46.2
NH_4^+(aq)	−31.74	−132.8
PO_4^{3-}(aq)	−306.9	−1284.1
HPO_4^{2-}(aq)	−310.4	−1298.7
$H_2PO_4^-$(aq)	−311.3	−1302.5
H_3PO_4(aq)	−308.2	−1289.5
O_3(g)	34.0	142.3
OH^-(aq)	−54.96	−230.0
H_2O(g)	−57.798	−241.8
H_2O(l)	−68.317	−285.84
SO_2(g)	−70.6	−295.4
SO_3(g)	−94.45	−395.2
H_2S(g)	−4.82	−20.2
Cl^-(aq)	−40.02	−167.4
HCl(aq)	−40.02	−167.4

SUBSTANCES OF BIOCHEMICAL IMPORTANCE	ΔH_f°/kcal mol^{-1}	ΔH_f°/kJ mol^{-1}
Carbohydrates[b]		
Glucose(s)	−304.3 (−303)[c]	−1273.2
Galactose(s)	−307.4 (−306)[c]	−1286.2
Sucrose(s)	−532.0 (−534)[c]	−2225.9
Alcohols and carboxylic acids[b]		
Acetic acid(l)	−115.8	−484.5
Citric acid(s)	−369.0	−1543.9
Ethanol(l)	−66.4	−277.8
Glycerol(l)	−159.8	−668.6
Lactic acid(s)	−165.9	−694.1
Pyruvic acid(l)[d]	−138.8	−580.7
Nitrogen-containing compounds[e]		
Alanine(s)	−134.5	−563

Continued

TABLE 2-3 *(Continued)*

SUBSTANCES OF BIOCHEMICAL IMPORTANCE	ΔH_f°/kcal mol^{-1}	ΔH_f°/kJ mol^{-1}
Nitrogen-containing compounds[e] (contd).		
Cysteine(s)	−126.7	−530
Glutamic acid(s)	−241.3	−1010
Glycine(s)	−128.4	−537
Leucine(s)	−152.4	−638
Serine(s)	−173.6	−726
Tryptophan(s)	−99.2	−415
Tyrosine(s)	−160.5	−672
Urea(s)	−79.6	−333
DL-Alanylglycine(s)	−186.0	−778
Glycylglycine(s)	−178.1	−745
Glycyl-DL-Tryptophan(s)	−148.8	−623
DL-Serylserine(s)	−281.5	−1178
Fatty acids[b]		
Palmitic acid(s)	−213.1	−891.6
Stearic acid(s)	−226.5	−947.7

[a] Dickerson, R. E. 1969. *Molecular Thermodynamics*. Menlo Park, Calif.: W. A. Benjamin, Inc.
[b] Cox, J. D., and Pilcher, G. 1970. *Thermochemistry of Organic and Organometallic Compounds*. New York: Academic Press.
[c] Values in parentheses were taken from *International Critical Tables*, Vol. V (1929. New York: McGraw-Hill) to illustrate how little thermochemical quantities have had to be revised in the past 50 years.
[d] Wagman, E. D., ed. 1965. *Selected Values of Chemical Thermodynamic Properties* (National Bureau of Standards Circular 500). Washington, D.C.: U.S. Government Printing Office.
[e] Sober, H. A., ed. 1968. *Handbook of Biochemistry*. Cleveland, Ohio: The Chemical Rubber Co.

Table 2-3 above lists the standard enthalpies of formation of selected compounds; the book by Dickerson (1969) cited in the references to the table contains a more extensive listing, and the National Bureau of Standards Circular 500 (Wagman, 1965) contains a very comprehensive compilation. These values can be combined to yield ΔH for any reaction involving tabulated compounds. The principle again is Hess's law. A reaction may formally be considered to follow a path in which reactants are dissociated to elements, then recombined to products:

$$\text{Reactants} \xrightarrow[\text{(Reactants)}]{-\Sigma\Delta H_f^\circ} \text{Elements} \xrightarrow[\text{(Products)}]{\Sigma\Delta H_f^\circ} \text{Products}$$

Since ΔH must be the same for all paths, it must equal the difference between the heats of formation, which equal the heats of the individual steps above,

$$\Delta H^\circ = \Sigma\Delta H_f^\circ(\text{products}) - \Sigma\Delta H_f^\circ(\text{reactants}) \tag{2-33}$$

This important equation tells us how to calculate ΔH° for any reaction if we know the standard heats of formation of reactants and products.

Standard heats of formation are useful for estimating the enthalpy available from biochemical oxidations. For example, the heat of oxidizing glucose to carbon dioxide and water, which is described by the reaction

$$C_6H_{12}O_6(s) + 6O_2(g) \rightarrow 6CO_2(g) + 6H_2O(g)$$

is the sum of the heats of formation of the products, minus those of the reactants. From Table 2-3, ΔH_f° for $6CO_2(g) + 6H_2O(g)$ is $-2361 - 1325.5$ kJ mol^{-1}, and ΔH_f° for glucose is -1273.2 kJ mol^{-1}. The heat of formation of $O_2(g)$ is zero, since the gas is the stable form of the element at 1 atm and 298 K. Hence ΔH° for oxidation of glucose under standard conditions is $-3686.5 - (-1273.2) = -2413.3$ kJ mol^{-1}. This means that 2413.3 kJ of heat is released in the reaction at 298 K and 1 atm pressure. When organisms metabolize glucose to CO_2 and H_2O, they are able to utilize roughly 50% of that energy to do chemical and mechanical work.

EXERCISE 2-14

Using the data in Table 2-3, calculate ΔH° for $NO(g) + CO(g) \rightarrow CO_2(g) + \frac{1}{2} N_2(g)$ at 298 K.

ANSWER

$\Sigma\Delta H_f^\circ$(Products) $= -393.5$ kJ mol^{-1} + 0; $\Sigma\Delta H_f^\circ$(Reactants) $= 90.4 - 110.5 = -20.1$ kJ mol^{-1}. $\Delta H^\circ = -393.5 - (-20.1) = -373.4$ kJ mol^{-1}.

Reaction Heats at Other Temperatures can be Calculated if the Heat Capacities are Known

It often may be the case that a reaction cannot be measured at 298 K, or that we need to know the heat of a reaction at some temperature other than 298 K. The effect of temperature on ΔH can be calculated if the heat capacity of products and reactants is known. Since, from Equation 2-13,

$$(dH)_P = C_P \, dT$$

we can write the same equation for ΔH, because $(dH_1)_P - (dH_2)_P = (d\Delta H)_P$:

$$d(\Delta H)_P = \Delta C_P \, dT \qquad (2\text{-}34)$$

where ΔC_P is the difference in heat capacity of products and reactants.

$$\Delta C_P = \Sigma C_P \text{ (products)} - \Sigma C_P \text{ (reactants)} \qquad (2\text{-}35)$$

Therefore,

$$\Delta H_T^\circ = \Delta H_{298}^\circ + \int_{298}^{T} \Delta C_P \, dT \qquad (2\text{-}36)$$

which states that the enthalpy change of a reaction at any temperature T can be determined from the 298 K value if ΔC_P is known over the temperature interval from 298 to T.

2–7 Molecular Interpretation of Enthalpy Changes

Chemical reactions often have large enthalpy changes because of the contribution of the energies of chemical bonds. One of the main objectives of thermochemistry is to determine these bond energies. To see how ΔH is related to the difference in bond energies we must correct for all other contributions to ΔH. These are summarized in Figure 2-8 and discussed below.

A minor difference arises from $P\Delta V$ work. Since $\Delta H = \Delta E + P\Delta V$, the difference in energy of products and reactants at constant pressure is the enthalpy change minus the work term, $P\Delta V$:

$$\Delta E = \Delta H - P\Delta V$$

In most reactions $P\Delta V$ is small compared to ΔH, even if gases are involved. For example, in the reaction of H_2 and O_2 to produce H_2O,

$$H_2(g) + \tfrac{1}{2}\, O_2(g) \rightarrow H_2O(g)$$

the measured ΔH at 298 K is -241.8 kJ mol^{-1}. The reaction consumes $\frac{1}{2}$ mole of gas. By the ideal gas law $P\Delta V = -(\frac{1}{2})RT = -(8.3143/2)$ J K^{-1} mol^{-1} $\times$ 298 K $= -1.24$ kJ mol^{-1} at 298 K. This is small compared to ΔH, leaving us with $\Delta E = -241.8 - (-1.2) = -240.6$ kJ mol^{-1}.

The major contribution to the energy of reactions in the gas phase is usually the difference between the bond energies of products and reactants.

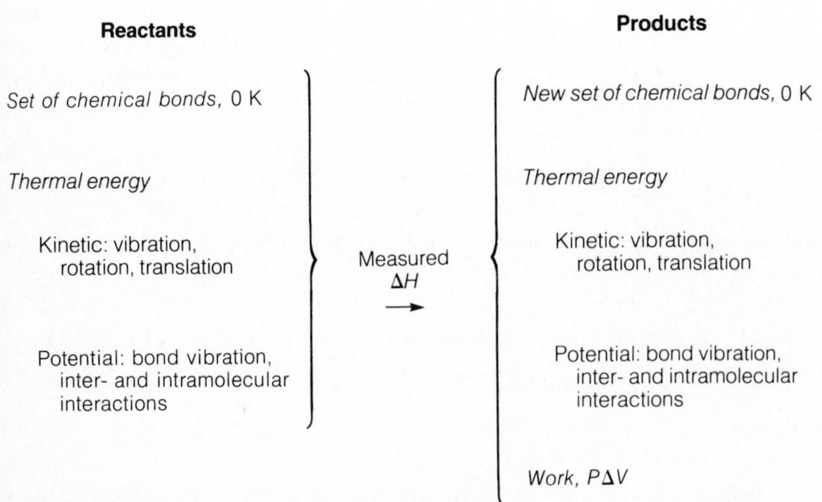

FIGURE 2-8
Sources of the enthalpy change of a gas phase chemical reaction. The kinetic energies of reactants and products may differ, as may the potential energies. Work ($P\Delta V$) may be done during the reaction. Thus the observed ΔH will differ from the change in bond energies.

This can be shown quantitatively by imagining the reaction of H_2 and O_2 to occur by the idealized path

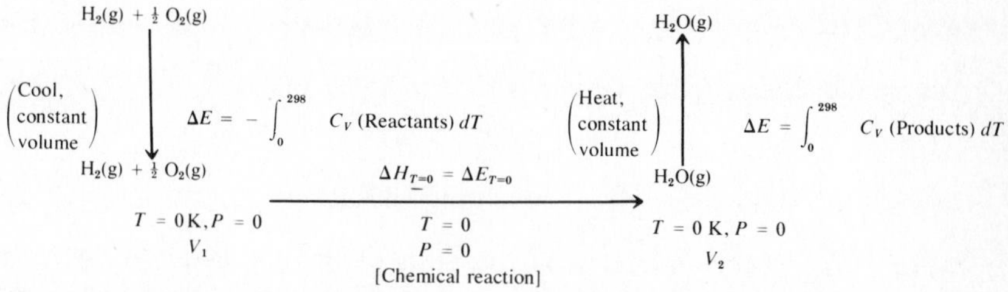

$$
\begin{array}{ccccc}
H_2(g) + \tfrac{1}{2}O_2(g) & & & & H_2O(g) \\
\left(\begin{array}{l}\text{Cool,}\\ \text{constant}\\ \text{volume}\end{array}\right)\downarrow & \Delta E = -\int_0^{298} C_V\,(\text{Reactants})\,dT & & \left(\begin{array}{l}\text{Heat,}\\ \text{constant}\\ \text{volume}\end{array}\right)\uparrow & \Delta E = \int_0^{298} C_V\,(\text{Products})\,dT \\
H_2(g) + \tfrac{1}{2}O_2(g) & \xrightarrow{\Delta H_{T=0} = \Delta E_{T=0}} & & H_2O(g) & \\
T = 0\,\text{K}, P = 0 & T = 0 & & T = 0\,\text{K}, P = 0 & \\
V_1 & P = 0 & & V_2 & \\
 & [\text{Chemical reaction}] & & &
\end{array}
$$

The reacting gases are imagined to be cooled to 0 K at constant volume, giving zero pressure. The energy change is $\Delta E = \int_{298}^{0} C_V dT$. With $P = 0$, the reaction enthalpy ΔH is equal to the energy change $\Delta E_{T=0}$, which is the energy due to rearranging the chemical bonds. The products are then reheated at constant volume to 298 K. The total energy change at the higher temperature is

$$\Delta E = \Delta E_{T=0} + \int_0^{298} \Delta C_V dT \tag{2-37}$$

in which the difference in heat capacities ΔC_V is

$$\Delta C_V = C_V\,(\text{products}) - C_V\,(\text{reactants}) \tag{2-38}$$

We call $\int_0^{298} \Delta C_V\, dT$ the *thermal energy difference*. It is the difference between the energy required to heat the product and reactant gases.

The integral $\int_0^{298} \Delta C_V dT$ for the product and reactant gases cannot be *measured* directly because the gases condense to liquids and solids at lower temperatures. However, the *idealized* thermal energy difference can be *calculated* from the heat capacity difference ΔC_V of the gases at higher temperature, using the methods of statistical mechanics (see Chap. 14), along with spectroscopic properties which describe the rotation and vibration of the reactant and product molecules (see Chap. 13).

Since $\Delta E_{T=0}$ is the difference in bond energies, Equation (2-37) tells us that ΔE equals the difference in bond energies plus the difference in thermal energies. The thermal energy difference for the reaction of H_2 and O_2 is found to be -1.65 kJ mol^{-1} by calculating the integral over $\Delta C_V\, dT$. Hence, the difference in bond energies is $-240.6 - (-1.6) = -239$ kJ mol^{-1}. This says that when the chemical bonds in one mole of H_2 and one-half mole of O_2 are broken in an idealized gas phase at 0 K and rejoined to form one mole of H_2O, 239 kJ of energy is released. In other words, the bond energy in one mole of H_2O is larger than that in the constituent $H_2(g)$ and $\frac{1}{2}$ $O_2(g)$ by 239 kJ mol^{-1}.

This number can be combined with other data to calculate the energy required to break the two O—H bonds in water without reforming any other bonds. By spectroscopic methods discussed in Chapter 10, it is possible to measure the minimum energy needed to dissociate simple diatomic molecules such as H_2 and O_2 into their constituent atoms:

$$H_2(g) \rightarrow 2H\cdot \qquad \Delta E = 432 \text{ kJ mol}^{-1}$$
$$\tfrac{1}{2}\,O_2(g) \rightarrow O\cdot \qquad \Delta E = 247 \text{ kJ (mol of O}\cdot)^{-1}$$

We can consider the reaction between H_2 and O_2 to occur by the imaginary path

$$H_2(g) + \tfrac{1}{2}\,O_2(g) \xrightarrow{\Delta E_1} 2H\cdot + O\cdot \xrightarrow{\Delta E_2} H_2O(g)$$

The energy ΔE_1 of the first step is the sum of the dissociation energies, $\Delta E_1 = 432 + 247 = 679$ kJ mol^{-1}. Since the total energy change is the sum of ΔE for the two steps,

$$\Delta E_T = \Delta E_1 + \Delta E_2 \tag{2-39}$$

we can use the result that $\Delta E_T = -239$ kJ mol^{-1} to calculate $\Delta E_2 = \Delta E_T - \Delta E_1 = -239 - 679 = -918$ kJ mol^{-1}. The energy released when two H atoms combine with an O atom to yield H_2O is 918 kJ mol^{-1}. Therefore, we can say that the average H—O bond energy in H_2O is half this value, or 459 kJ mol^{-1}. This compares with 432 kJ mol^{-1} for the bond energy of H_2, and 494 (= 2 × 247) kJ mol^{-1} for the O═O bond in O_2, given by the dissociation energies above. Bond dissociation energies range from about 900–1100 kJ mol^{-1} for strong triple bonds such as N≡N, C≡O, and C≡C, to 100–500 kJ mol^{-1} for single bonds. The book *Molecular Thermodynamics* by Dickerson (1969) cited in Table 2-3 contains a more detailed discussion of this subject.

In condensed media such as solutions, the reaction heat can be dominated by intermolecular interactions rather than by bond energies. For example, the enthalpy of hydrolysis of ATP to ADP in aqueous medium, which is about −7.3 kcal per mole (−30.5 kJ mol^{-1}) for the reaction

$$\text{ATP (aq)} \rightarrow \text{ADP (aq)} + \text{inorganic phosphate (aq)}$$

includes the difference in enthalpy of all interactions of water with the products and reactants. In such cases the actual bond energy change may be very different from the measured enthalpy change.

2-8 *Transition Enthalpies and Heat Capacities*

Physical changes of state generally provide smaller enthalpy changes than do chemical reactions. Examples of physical changes are the melting of ice to liquid water and the conversion of the liquid water to water vapor. The phase rule tells us that when there are two phases of a single homogeneous component, the phase change will occur at a single temperature if we hold the pressure at 1 atm. The heat absorbed to melt the solid to liquid is called ΔH_{fus}, the *latent heat of fusion*. Similarly, the heat required to convert liquid to vapor is the *heat of vaporization* (ΔH_{vap}), and the sum of these two, at a given temperature, is the *heat of sublimation* (ΔH_{sub}), or the heat required to convert solid to vapor,

$$\Delta H_{sub} = \Delta H_{fus} + \Delta H_{vap} \tag{2-40}$$

(Notice that Equation 2-40 is another example of the law of heat summation.)

The heat absorbed in the melting or vaporization transitions is required to overcome the intermolecular interactions that stabilize the crystal or the liquid phase. Macromolecules such as proteins are also stabilized in their *native* (or active form) structure by noncovalent interactions. These include hydrogen bonds, interactions between oppositely charged groups, and other interactions that we will consider in later chapters.

Cooperative Transitions Display Large Heat Capacities

The heat capacity of water at the melting temperature is infinite, since heat is taken up with no change in temperature. This is an example of what is called a *cooperative transition,* because a large number of water molecules must cooperate, or act together, in order to change from one state to another with a negligible change in temperature.

There are a number of physical transitions in which the heat capacity is anomalously large, but not infinite, and which involve a smaller degree of cooperativity. For example, biological macromolecules can be converted by heat from their ordered native structure to a more disorganized form. This process is analogous to the melting of ice to water, and the heat absorbed is used to break the hydrogen bonds and other interactions that stabilize the native macromolecular structure. Like melting, the process happens near a

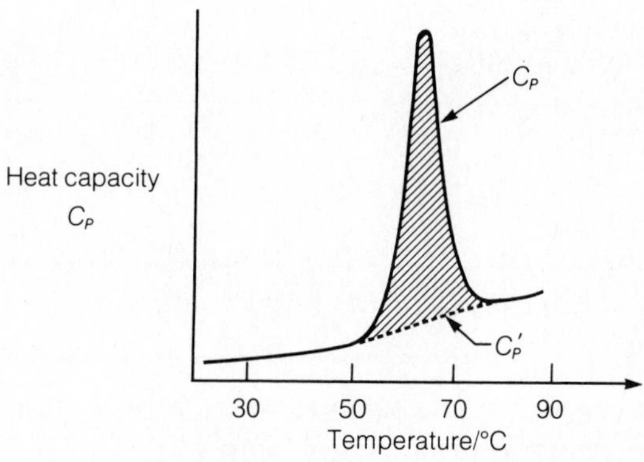

FIGURE 2-9
Heat capacity of a solution of lysozyme at pH 2.5 corrected for the heat capacity of the buffer. The shaded area is ΔH for the denaturation transition. (From P. L. Privalov and N. N. Khechinashvili. 1974. *J. Mol. Biol. 86,* 665–684, fig. 2. Copyright by Academic Press Inc. (London) Ltd.

single temperature. Figure 2-9 shows an experimental measurement of the amount by which the heat capacity of a solution containing the enzyme lysozyme exceeds the heat capacity of the buffer. The narrow temperature region in which C_P is large corresponds to the unfolding of the macromolecule.

The magnitude of ΔH for unfolding can be calculated from the experimental values of C_P. Since $(\partial H/\partial T)_P = C_P$, we can write

$$dH = C_P\, dT$$

for measurements at constant pressure. Let us denote the heat capacity expected if there were no cooperative transition by C'_P. We can estimate C'_P by making a smooth interpolation between C_P at high and low temperatures (dashed line in Fig. 2-9). Then the increment of heat absorbed due to the unfolding transition is $dH = (C_P - C'_P)\, dT$. Hence, the unfolding enthalpy is

$$\Delta H = \int_{T_1}^{T_2} (C_P - C'_P)\, dT$$

which is the shaded area in Figure 2-9.

The heat absorbed in the unfolding transition is used to break the noncovalent interactions that stabilize the native structure, except for a small contribution to the work of expansion. Protein molecules typically have denaturation (unfolding) enthalpies in the range 200–600 kJ mol^{-1}, as illustrated in Table 2-4.

TABLE 2-4 *Enthalpy of Protein Unfolding at 50°C**

PROTEIN	ΔH/kJ mol^{-1}
α-chymotrypsin	560
Ribonuclease *A*	400
Lysozyme	370
Metmyoglobin	285
Cytochrome *c*	210

* From Privalov and Khechinashvili (1974).

2–9 *The Unusual Thermal Properties of Water*

Thales of Miletus in about 580 BC believed water to be the origin of all things. Later Greek philosophers regarded water merely as one of the four elements. Then in the 1790s the English experimentalist Henry Cavendish reduced the status of water further to that of a mere chemical compound when he found that it is composed of hydrogen and oxygen.

Despite this decline in the fortunes of water, most scientists regard water as the matrix of life. Living tissues other than bone contain at least 50%

water by weight, and some, such as brain and lung tissue, contain 80%. Modern biochemists, notably L. J. Henderson (1913), have argued that the connection between water and life is unique, and that there is no other chemical that could assume the central role of water in living matter, even for life on another planet.

These arguments are based on the thoroughly studied thermodynamic properties of water. These properties must be determined from experimental measurements since there is no theory that allows us to calculate all the thermodynamic properties of real substances. Only in the case of dilute gases near the ideal gas limit is theory accurate enough to be relied on in the place of experiment.

One property that can be measured directly is the *molar volume,* denoted $\overline{V}^{\bullet}$.[1] This is the volume V divided by the number of moles in the sample. Figure 1-7 shows the molar volume of solid and liquid water between $-40°C$ and 100°C. The major feature is an 8% decrease in the volume, $\Delta\overline{V}_{\text{fus}}$, at the freezing (or fusion) point. In addition, as the inset in the figure shows, the molar volume is a minimum (the density is a maximum) at 4°C. At the boiling (or vaporization) point (100°C, 1 atm) there is a large increase ΔV_{vap} in the molar volume, and $\overline{V}^{\bullet}$ rises to a value (3.014×10^4 cm^3 mol^{-1}) near the ideal gas limit of $\overline{V}^{\bullet} = RT/P = 3.062 \times 10^4$ cm^3 mol^{-1}.

Another property that can be measured directly is the heat capacity at constant pressure. We determine the amount of heat energy, Q_P, required to increase the temperature by ΔT, and calculate $C_P = Q_P/\Delta T$ from the ratio. Figure 2-10 shows this result. The measured heat capacity approaches zero as T approaches zero, as is found for all substances. Liquid water has a considerably larger heat capacity than either the solid or the vapor, because there are many interactions between water molecules in the liquid state which are broken gradually as the temperature is increased. Energy is absorbed to increase the potential energy, accounting for the large amount of heat that must be absorbed to produce a given temperature increase ΔT.

The molar enthalpy $\overline{H}^{\bullet}$ of water can be calculated from the measured heat capacity. Since $dH = C_P\,dT$, the enthalpy change from absolute zero, $\overline{H}^{\bullet} - \overline{H}^{\bullet}_0$, is

$$\overline{H}^{\bullet} - \overline{H}^{\bullet}_0 = \int_{T'=0}^{T} \overline{C}_P\,dT' + \Delta H_{\text{latent heat}}$$

This quantity is the area under the dotted curve in Figure 2-10; it is also plotted directly as $\overline{H}^{\bullet} - \overline{H}^{\bullet}_0$ by the solid curve. The latent heat term accounts for the heat absorbed in melting or vaporization, provided these transitions occur below the temperature T to which $\overline{H}^{\bullet} - \overline{H}^{\bullet}_0$ refers. Notice that the enthalpy of the vapor is much larger than that of solid or liquid. The reason for this is that energy is required to overcome the potential energy of interaction of water molecules. Notice also that $\overline{C}_P$ is largest in the liquid region,

[1] We will reserve the symbol $^{\bullet}$ to specify thermodynamic properties of *pure* substances. The symbol $^{\circ}$ refers to a substance in a *defined* or *standard* state, which need not be the pure substance. The line above V indicates a molar or partial molar (see Chap. 6) property.

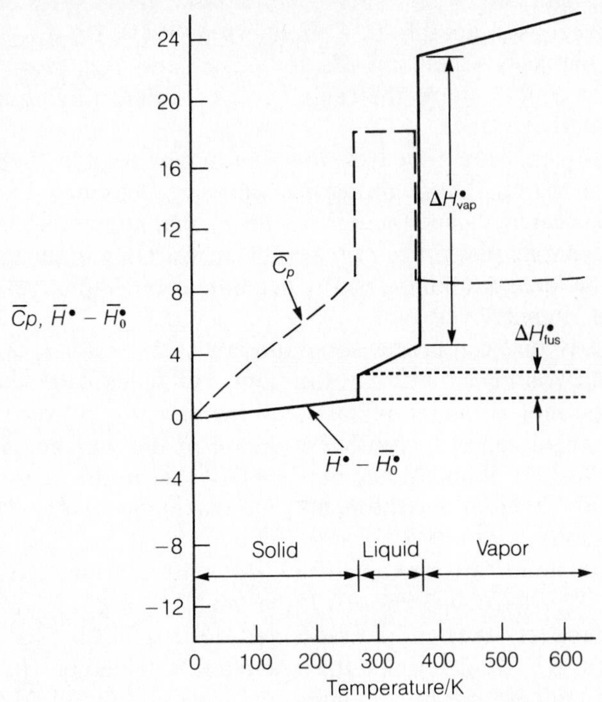

FIGURE 2-10

Heat capacity C_P and molar enthalpy, $\bar{H}^{\bullet} - \bar{H}_0^{\bullet}$, of water. $\bar{C}_P$ is given in cal mol^{-1} deg^{-1} and $\bar{H}^{\bullet} - \bar{H}_0^{\bullet}$ in kcal mol^{-1}. $\bar{H}_0^{\bullet}$ is the molar enthalpy when $T = 0$ K. (From Eisenberg and Kauzmann, 1969).

where the enthalpy is increasing most rapidly with temperature [recall that $C_P = (\partial H/\partial T)_P$]. The energy E is calculated readily from H and V, since $E = H - PV$. In the solid phase, H and E are not very different, because PV is small. The following exercises will help you see the relation between ΔH and ΔE.

EXERCISE 2-15

The heat of fusion of ice at 0°C (1 atm pressure) is 5.98 kJ mol^{-1}, and $\Delta\bar{V}_{fus} = -1.621$ cm^3 mol^{-1}. Calculate ΔE_{fus} at 1 atm pressure.

ANSWER

$\Delta\bar{E}_{fus} = \Delta\bar{H}_{fus} - P\Delta\bar{V}_{fus} = 5.98 \times 10^3$ J mol^{-1} + 1 atm $\times$ 1.621 cm^3 mol^{-1} $\times$ 0.1013 J cm^{-3} atm^{-1} = 5980 J mol^{-1} + 0.164 J mol^{-1}, or 5.98×10^3 J mol^{-1}. This example shows that $\Delta\bar{H}_{fus}$ and $\Delta\bar{E}_{fus}$ are essentially identical for water, because $\Delta\bar{V}$ is very small. However, when the material is converted to a gas, the work of expansion is sufficient to make $\Delta\bar{E}_{vap}$ and $\Delta\bar{H}_{vap}$ appreciably different.

EXERCISE 2-16

The heat of vaporization of H_2O at 100°C (1 atm) is 40.6 kJ mol^{-1}, and the volume increase is $\Delta\bar{V}_{vap} = 3.01 \times 10^4$ cm^3 mole^{-1}. Calculate $\Delta\bar{E}_{vap}$.

ANSWER

$\Delta\bar{E}_{vap} = \Delta\bar{H}_{vap} - P\Delta\bar{V}_{vap} = 40.6 \times 10^3$ J mol^{-1} − 1 atm $\times$ 3.01 $\times 10^4$ cm^3 mol^{-1} $\times$ 0.1013 J cm^{-3} atm^{-1} = 40.6 $\times 10^3$ J mol^{-1} − 3.05 $\times 10^3$ J mol^{-1} = 37.5 $\times 10^3$ J mol^{-1}.

Note from Exercises 2-15 and 2-16 that the energy of vaporization is considerably larger than the energy of fusion. The reason is that in conversion to the gas, virtually all the interactions between water molecules must be broken.

The Biological Significance of the Thermodynamic Properties of Water

The thermal properties of water have considerable importance for the regulation of temperature both within organisms and in the environment. All organisms generate heat from the biochemical reactions in their metabolism. For example, an average 60 kg person generates about 2500 kcal = 2.5 $\times 10^6$ cal (about 10^7 J) of heat per day from the food he or she eats. If the body were adiabatic, the temperature increase (assuming the specific heat is 1 cal g^{-1} °C^{-1}, as for water) would be $\Delta T = \Delta H/C_p = 2.5 \times 10^6$ cal/(1 cal g^{-1} °C^{-1} $\times$ 6 $\times 10^4$ g) = 42°C, which would be fatal.

Several features of water make it particularly effective for temperature regulation. First, as shown in Table 2-5, it has a large specific heat compared

TABLE 2-5 *Specific Heats of Various Liquids*

SUBSTANCE (LIQUID)	SPECIFIC HEAT/cal g^{-1} °C^{-1}
H_2	3.41
NH_3	1.23
H_2O	1.0
CH_3CH_2OH	0.6
$CHCl_3$	0.24
NaCl	0.21
Fe	0.1
Hg	0.03

to most other liquids. This means that temperature changes, given by $\Delta T = \Delta H$/specific heat, will be smaller in water for a given amount of heat, ΔH, absorbed per gram of liquid than in other liquids. The large specific heat of water makes it an effective buffer against temperature changes both in organisms and in the general environment. The moderating effect on climate exerted by large bodies of water is due partly to this substantial specific heat of H_2O.

Second, the large heat of vaporization of water provides an effective mechanism for dissipating the heat generated in an organism by metabolism. Using the relationship $Q_P = \Delta H_{fus}$, we can see that a large heat of fusion implies that evaporation of a small amount of material requires a large amount of heat. Table 2-6 shows the heat of vaporization of several liquids in cal/gram.

TABLE 2-6 *Heat of Vaporization of Various Liquids*

SUBSTANCE	T/°C	ΔH_{vap}/cal g^{-1}
Water	40	574
Water	100	540
Ethanol	78.3	204
Acetic acid	118	97
Hexane	68	79

EXERCISE 2-17

Calculate the amount of water that would have to be vaporized at 40°C to expend the 2.5×10^6 cal of heat generated by a person in one day.

ANSWER

Q_P = heat of vaporization × wt H_2O, so wt $H_2O(g) = 2.5 \times 10^6$ cal/(574 cal g^{-1}) = 4.4×10^3 g, or about 4.4 liters. This is the amount of water that would be required to cool the body if the temperature of the surroundings was kept at body temperature so that evaporation was the only mechanism for cooling. Most of us do not sweat that much, although volumes of nearly that magnitude (5 qts) can be lost (at the cost of severe dehydration) by vigorous exercise in hot weather, or a protracted stay in a sauna bath. Note that there is also evaporative cooling from the vaporization of water in the lungs.

The large heat of vaporization of water has an important moderating effect on the Earth's climate. Much of the radiant energy from the sun is used to vaporize water rather than heat the atmosphere or the Earth's surface. When this radiant energy disappears at night or in cloudy weather, condensation of water vapor to liquid water releases the stored energy and prevents the temperature from dropping rapidly. Areas without water, such as deserts or the moon, are consequently subject to much more extreme temperature variations than are humid regions.

Finally, the heat of fusion of water is also relatively large on a weight basis (see Table 2-7). The freezing of ice in winter and its thawing in spring moder-

TABLE 2-7 *Heats of Fusion of Various Substances*

SUBSTANCE	T_{fus}/°C	ΔH_{fus}/cal g^{-1}
Water	0	80
Ammonia	−75	108
Nitric acid	−47	9.6
Benzene	5.4	30

ate climatic changes in a manner similar to condensation and vaporization of water.

The contraction of ice when it melts is unusual; this property is shared only by diamond, silicon, germanium, and a few alloys of antimony, bismuth, and tin. The significance of this contraction was expressed by Henderson (1913) as follows:

> There is an old experiment of [Count] Rumford's which well illustrates what conditions must have been had the contraction of water been normal and ice denser than water. He found that in a vessel filled with water, which contains ice confined at the bottom, it is possible to heat and even boil the superficial portion of the water without melting the ice. And so it would be with lakes, streams and oceans were it not for the anomaly and the buoyance of ice. The coldest water would continually sink to the bottom and there freeze. The ice, once formed, could not be melted, because the warmer water would stay at the surface. Year after year the ice would increase in winter and persist through the summer, until eventually all or much of the body of water, according to the locality, would be turned to ice. As it is, the temperature of the bottom of a body of fresh water cannot be below the point of maximum density; on cooling further the water rises; and ice forms only on the surface. In this way the liquid water below is effectually protected from further cooling, and the body of water persists. In the spring the first warm weather melts the ice, and at the earliest possible moment all ice vanishes.

It was Henderson's overall thesis that water's properties of specific heat, heat of fusion, heat of vaporization, and contraction of fusion—in addition to its remarkable properties as the best of all solvents—combine to make it the chemical uniquely fit to sustain life.

○ 2–10 *Mathematical Principles: Integrals, Logarithms, and Exponentials*

In thermodynamics we have to make frequent use of the *integral* of a function. Integration is a process in which we add together a very large number of very small quantities. More specifically, let y be a quantity that depends on x (Fig. 2-11). The expression $\int_{x_1}^{x_2} y\,dx$ means that we divide the x axis

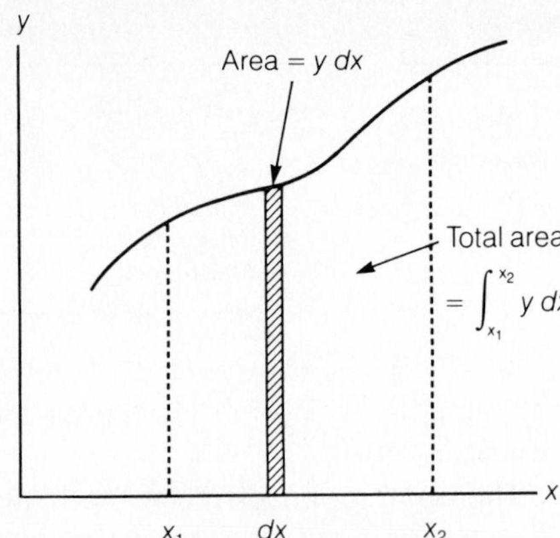

FIGURE 2-11
The integral of the function y between x_1 and x_2, $\int_{x_1}^{x_2} y\,dx$, is the area under the curve bounded by x_1, x_2, y, and $y = 0$.

between x_1 and x_2 into small increments, dx, multiply each of these by the value of y in that increment, and add all of the products, $y\ dx$, together. Notice that $y\ dx$ is the area of the rectangular section whose sides are dx and y, and that the sum of all such products between x_1 and x_2, $\int_{x_1}^{x_2} y\,dx$, is the area under the curve between x_1 and x_2.

As a more direct physical example, consider the integral, Equation 2-6, which defines the potential energy U. U is the work required to move two particles from infinite separation to a distance R apart. The total work is the sum of work elements dW between $R' = \infty$ and $R' = R$, which we can express as an integral

$$U = \int_{R'=0}^{R'=R} dW$$

in which the limits on the integral sign indicate the range of values of R' over which the values of dW are to be added. The differential dW is a product of the force $-F$ and another differential dR', so

$$U = -\int_{R'=\infty}^{R'=R} F\,dR'$$

This integral also can be interpreted as the area under a curve.

A simple integral was evaluated in Equation 2-21. We start with the equation

$$dE = C_V\,dT \tag{2-41}$$

The change in E, $\Delta E = E_2 - E_1$, is the integral or sum over all increments dE between State 1 and State 2:

$$\Delta E = \int_{(1)}^{(2)} dE$$

The sum of all energy increments dE is the total energy change ΔE between the two states. If we integrate over the left side of Equation 2-41, we must integrate over the right side between the same limits, so

$$\Delta E = \int_{T_1}^{T_2} C_V \, dT$$

If C_V is a constant, as it is for an ideal gas, it can be removed from the integral sign because it does not vary when T is changed. Hence we are left with

$$\Delta E = C_V \int_{T_1}^{T_2} dT$$

The sum of dT values between T_1 and T_2 is $T_2 - T_1$, so

$$\Delta E = C_V(T_2 - T_1)$$

as we obtained in Equation 2-21.

A general way to evaluate integrals is based on recognizing that integration and differentiation are inverse operations. Let the derivative of the function f be g:

$$g = \frac{df}{dx} \tag{2-42a}$$

We rewrite this as an equation in differentials and integrate to get

$$\int_{(1)}^{(2)} g \, dx = \int_{(1)}^{(2)} df$$

The right side is just $f_2 - f_1 = \Delta f$, so

$$\int_{(1)}^{(2)} g \, dx = f_2 - f_1 \tag{2-42b}$$

Equations 2-42a and b tell us that if we can find the function f of which g is the derivative, we can easily evaluate integrals over g. The following exercises give some simple examples.

EXERCISE 2-18

Calculate $\int_0^5 x \, dx$.

ANSWER

$g = x$, $f = x^2/2$ (you should confirm for yourself that $g = df/dx$). Therefore $\int_0^5 x \, dx = 5^2/2 - 0 = 12.5$.

EXERCISE 2-19

Calculate $\int_{-1}^{1} x^2 \, dx$.

ANSWER

$g = x^2, f = x^3/3$. Hence $\int_{-1}^{1} x^2 \, dx = (1)^3/3 - (-1)^3/3 = 2/3$.

Logarithms

A very important function in thermodynamics, and one which does not have a simple algebraic integral, is $1/x$. An example is Equation 2-17,

$$dW = -nRT\frac{dV}{V}$$

The integral of dV/V cannot be expressed by a simple expression involving V^n. We therefore define a new function called the (natural) logarithm, $\ln(x)$, which has the property that its derivative is $1/x$

$$\frac{d\ln x}{dx} = \frac{1}{x}$$

Therefore the integral of $1/x$ must be $\ln x$. We are now in a position to integrate Equation 2-17:

$$\begin{aligned} W &= -nRT\int_{V_1}^{V_2}\frac{dV}{V} \\ &= -nRT\,[\ln V_2 - \ln V_1] \end{aligned} \tag{2-43}$$

We note three important properties of the logarithm:

$$\ln x_1 + \ln x_2 = \ln (x_1x_2) \tag{2-44a}$$
$$\ln (1/x) = -\ln x \tag{2-44b}$$
$$\ln (x^a) = a\ln x \tag{2-44c}$$

Figure 2-12 shows $\ln x$ as a function of x. Common logarithms (to the base 10) can be obtained by dividing $\ln x$ by 2.303. The natural logarithm is larger because its base is smaller, and so the exponent (logarithm) must compensate by being larger.

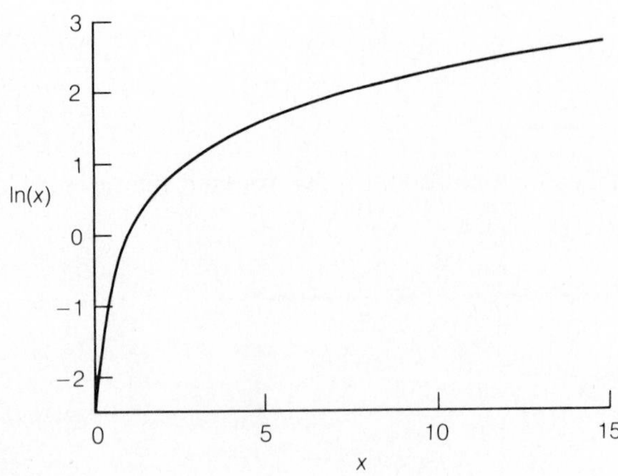

FIGURE 2-12
A plot of $\ln x$ against x. Notice that $\ln(1) = 0$.

EXERCISE 2-20

Show that Equation 2-43 is equivalent to $W = -nRT \ln (V_2/V_1)$.

EXERCISE 2-21

Use Equation 2-44*a* to show that ln(1) must be zero.

Exponentials

The inverse operation of taking the logarithm of a number is exponentiation, which consists of raising the number e to the power x, where e is the base of the natural logarithms, and equal to 2.71828 . . . :

$$\text{exponential}\,(x) = e^x$$

Some important properties of the exponential function are:

$$e^{\ln x} = x \qquad (2\text{-}45a)$$

$$e^{a \ln x} = x^a \qquad (2\text{-}45b)$$

$$e^{x_1 + x_2} = e^{x_1} e^{x_2} \qquad (2\text{-}45c)$$

$$\ln (e^x) = x \qquad (2\text{-}45d)$$

$$e^{-x} = 1/e^x \qquad (2\text{-}45e)$$

EXERCISE 2-22

Confirm using Equation 2-45*b* that Equation 2-25 is obtained from $\ln (T_2/T_1) = (nR/C_V) \ln (V_1/V_2)$.

QUESTIONS FOR REVIEW

1. State the first law of thermodynamics. Describe its physical meaning and that of each of the quantities in it.
2. Define heat capacity, specific heat, and enthalpy. How are these quantities related?
3. What is the molecular basis for increase in the energy E of a system as it is heated?
4. Why can one substance have a large heat capacity and another a small one? Why does liquid water have a greater heat capacity than either ice or water vapor?
5. What is the standard enthalpy of formation of a compound? Why is this quantity tabulated? How can its values be used to calculate the heat of a reaction? What additional information is required to calculate the heat of reaction at a nonstandard temperature?
6. How is a measured heat of reaction related to the bond energies of the reacting substances?

7. How does the existence of a phase transition affect the thermal properties of a substance?

8. How can the increases in energy and enthalpy of a substance be measured as the substance is heated?

PROBLEMS

1. Are the following statements true or false?

   (a) Kinetic energy cannot be negative.

   (b) Potential energy cannot be negative.

   (c) $-PdV$ is an exact differential.

   (d) In an exothermic reaction, ΔH is negative.

   (e) Applications of thermodynamics rest on the molecular nature of matter.

   (f) More work is done when a system expands reversibly than when it expands irreversibly to the same final state.

   (g) A wet object is cooled if the water on its surface can evaporate.

   (h) C_V is independent of volume for a fixed mass of substance.

Answers: (a) T. (b) F. (c) F. (d) T. (e) F. (f) T. (g) T. (h) F (this is true only for an ideal gas at constant temperature).

2. Add the appropriate words:

   (a) The force between two particles is proportional to the rate of change of their ________ with distance.

   (b) The standard enthalpy of formation of an ________ is defined as zero.

   (c) The heat capacity is a property of state when either ________ or ________ is held constant.

   (d) The energy units commonly associated with mechanical work and thermochemistry are the ________ and the ________ respectively.

   (e) In an isothermal expansion the ________ is constant.

   (f) No heat crosses the boundary in an ________ process.

Answers: (a) Potential energy. (b) Element. (c) Pressure or volume. (d) Joule, calorie. (e) Temperature. (f) Adiabatic.

PROBLEMS RELATED TO EXERCISES

3. (Ex. 2-1) Calculate the energy change of a gas when it is compressed by a force of 1 newton acting through 2 meters, and heat energy of 1.6 J flows to the surroundings.

4. (Ex. 2-2) One mole of ideal monatomic gas expands adiabatically through a distance of 2.0 meters against an external force of 1.0 newton. Calculate the temperature change.

5. (Ex. 2-3) Calculate the distance moved when an ideal gas expands isothermally against a force of 1.0 newton, absorbing 0.5 calorie in the process.

6. (Ex. 2-4) One mole of an ideal monatomic gas has an energy of 5000 J. Calculate the temperature.

7. (Ex. 2-6) The heat required to vaporize 1 mole of H_2O at 100°C is 9720 cal. Calculate the increase in temperature of one mole of an ideal monatomic gas which this energy would produce.

8. (Ex. 2-8) Use the relationship $H = E + PV$ to show that for all ideal gases, $C_P = C_V + nR$. (This problem will require some independent thought.)

9. (Ex. 2-9) Calculate the work for reversible isothermal compression of one mole of an ideal gas from 2 atm to 5 atm at 400 K.

10. (Ex. 2-10) Calculate Q for the process in Problem 2-9.

11. (Ex. 2-11) Calculate W for the reversible adiabatic compression of one mole of an ideal gas from 2 atm to 5 atm, starting at 400 K.

12. (Ex. 2-12) Write expressions for the total differential dH in terms of the variables (V,T), (V,P), and (T,P). Which of these expressions simplifies to $dH = C_P\,dT$ for an ideal gas?

13. (Ex. 2-14) Calculate $\Delta H°$ for the reaction $CO(g) + \frac{1}{2}\,O_2(g) \rightarrow CO_2(g)$.

OTHER PROBLEMS

14. Use the following heats of combustion (reaction with O_2 to produce CO_2 and H_2O): pyruvic acid ($CH_3COCOOH$) (l), -277 kcal mol^{-1}; and acetic acid (CH_3COOH) (l), -207 kcal mol^{-1} (both under standard conditions) to calculate the enthalpy change for oxidative metabolism of pyruvic acid (l) to acetic acid (l) and CO_2 under standard conditions.

15. A particle is held in place by a spring which exerts a force $F = -kx$ on it; k is the *force constant* and x the displacement from the equilibrium position $x = 0$. If $k = 10^3$ dynes/cm, calculate the work required to displace the particle by $x = 20$ cm. Express your answer in (a) ergs, (b) joules, (c) calories, (d) kcal.

16. Calculate the average kinetic energy of a particle in an ideal monatomic gas at 300 K. If the particles have atomic weight 4, what is their mean square velocity? The *root-mean-square* velocity is the square root of this quantity.

17. Calculate $\left(\dfrac{\partial E}{\partial V}\right)_T$

    for an ideal monatomic gas (*Hint*: use $E = \frac{3}{2}\,RT$).

18. Calculate $\left(\dfrac{\partial H}{\partial V}\right)_T$

    for an ideal monatomic gas (*Hint*: use the definition of H and the results of Problem 2-17).

19. Calculate

$$\left(\frac{\partial H}{\partial T}\right)_V$$

for an ideal monatomic gas (Answer: $\frac{5}{2}nR$).

20. Begin with the expression

$$dH = \left(\frac{\partial H}{\partial P}\right)_T dP + \left(\frac{\partial H}{\partial T}\right)_P dT$$

and show that

$$C_P = -\left(\frac{\partial H}{\partial P}\right)_T \left(\frac{\partial P}{\partial T}\right)_H$$

21. Solid sulfur exists in two crystal forms, rhombic and monoclinic. The rhombic crystal is the more stable, and its $\Delta \bar{H}_f^\bullet$ is set equal to zero. $\Delta \bar{H}_f^\bullet$ for the monoclinic crystal is 0.071 kcal/mole. The densities at 25°C are 1.96 g cm^{-3} for the monoclinic and 2.07 g cm^{-3} for the rhombic form. Calculate $\Delta \bar{E}^\bullet$ for converting rhombic sulfur to monoclinic at 25°C and 1 atm pressure.

22. One mole of an ideal gas is compressed reversibly and adiabatically from 1 atm to 10 atm starting at 25°C. Calculate ΔE and the temperature change ΔT. Will the temperature change be greater, smaller, or the same if the compression is carried out irreversibly?

23. (a) The enthalpy of unfolding cytochrome *c* is 50 kcal/mole at 50°C and 88 kcal/mole at 70°C. Estimate ΔC_P for the unfolding of this protein, assuming ΔC_P is constant over this temperature range. (b) The heat capacity of denatured (unfolded) proteins is generally found to be higher than for the same proteins in their native structure. Can you think of any reasons for this observation?

24. A biochemist decides to measure the heat of oxidizing sucrose (s) ($C_{12}H_{22}O_{11}$) to CO_2 (aq) and water (l) by allowing the solid sugar to be metabolized by bacteria. He finds a heat evolution of 400 kcal mol^{-1} of sucrose. Compare this value with the prediction from Table 2-3, and identify the erroneous assumption in the design of the experiment.

25. Calculate the enthalpy change for conversion of glucose (s) ($C_6H_{12}O_6$) to ethanol (l), and CO_2 (aq) under standard conditions (1 atm pressure, 25°C).

● 26. (a) The enthalpy change for atomization of graphite at 298 K, C (graphite) → C (g), is $\Delta H° = 171.7$ kcal mol^{-1}, and the enthalpy change for dissociation of H_2, H_2 (g) → 2H (g), is $\Delta H° = 104.2$ kcal mol^{-1}. Combine this data with the heat of formation of methane ($\Delta H_f°$ of $CH_4 = -17.9$ kcal mol^{-1}) to calculate the total enthalpy change for dissociation of the 4 C—H bonds in methane.

(b) Calculate the average C—H bond enthalpy.

● 27. Combine the following data for ethane

$$\Delta H_f° (CH_3CH_3) = -20.2 \text{ kcal mol}^{-1}$$

with the data for atomization of graphite and dissociation of H_2, and the calculated C—H bond enthalpy in Problem 2-26 to calculate the C—C bond enthalpy in ethane.

28. A gas at pressure P_1 is forced through a porous partition, with pressure P_2 maintained on the other side of the partition. The gas on both sides of the partition is enclosed adiabatically, except for the porous partition. Prove that there is no change in the enthalpy of the gas under these conditions.

29. Calculate the work done in the reversible adiabatic expansion of one mole of an ideal monatomic gas from 10 atm to 1 atm, starting at 300 K.

QUESTIONS FOR DISCUSSION

30. (a) A tiny mass of steam emerging from the spout of a tea kettle can cause a burn much worse than an equal mass of liquid water at the same temperature. How can this be?

    (b) A thermodynamicist at your dining table asks why the cooked tomatoes always seem hotter than other vegetables. Can you explain?

    (c) Why does Los Angeles have a temperate climate while Needles, on about the same latitude and 200 miles to the east, is very hot in summer and often cold in winter?

FURTHER READING

In learning a subject as difficult as thermodynamics, it often helps to read a second treatment of the material. In addition to the text by Kauzmann referred to in the Further Readings of Chapter 1, the following are among the many good treatments of the subject.

Daniels, F., and Alberty, R. A. 1975. *Physical Chemistry,* 4th ed. New York: John Wiley & Sons. Perhaps the simplest treatment of thermodynamics of those suggested here.

Dickerson, R. E. 1969. *Molecular Thermodynamics.* Menlo Park, Calif.: W. A. Benjamin, Inc. A clear thermodynamics text, with a fascinating chapter on Thermodynamics and Living Systems.

Edsall, J. T., and Wyman, J. 1958. *Biophysical Chemistry.* New York: Academic Press. Chapter 2 on Water is a classic. The treatment of thermodynamics is more advanced.

Klotz, I. M., and Rosenberg, R. M. 1972. *Chemical Thermodynamics,* 3rd ed. Menlo Park, Calif.: W. A. Benjamin, Inc. A full treatment of chemical potentials and solutions.

Moore, W. J. 1972. *Physical Chemistry,* 4th ed. Englewood Cliffs, N.J.: Prentice-Hall.

REFERENCES

Eisenberg, D., and Kauzmann, W. 1969. *The Structure and Properties of Water.* New York: Oxford University Press.

Henderson, L. J. 1913. *Fitness of the Environment.* New York: Macmillan. An examination of the biological fitness of the properties of matter. Still stimulating reading.

Haley, A. R., and Snaith, J. W. 1968. *Biopolymers 6,* 1355.

Hunt, J. B., Ross, P. D., and Ginsburg, A. 1972. *Biochemistry 11,* 3716.

Kuntz, J. D., Jr., and Kauzmann, W. 1974. *Adv. Prot. Chem. 28,* 239.

Mrevlishvili, G. M., and Privalov, P. L. 1969. In *Water in Biological Systems,* ed. L. P. Kayuskin. p. 63. New York: Plenum.

Privalov, P. L., and Khechinashvili, N. N. 1974. *J. Mol. Biol. 86,* 665–684.

Wagman, E. D., ed. 1965. *Selected Values of Chemical Thermodynamics Properties* (National Bureau of Standards Circular 500). Washington, D.C.: U.S. Government Printing Office. A useful compendium of thermodynamic data.

CHAPTER 3

Entropy and the Second Law of Thermodynamics

3–1 Spontaneous Changes Imply the Need for a Second Law

The first law of thermodynamics does not provide a complete description of natural processes. It simply says that energy is conserved, that heat absorbed and work done on a system are stored as the energy E. But it does not tell us if a given process is *spontaneous,* that is, if it proceeds of its own accord in a given direction. The *second law of thermodynamics* gives us a way of determining whether a process is spontaneous, and thus leads to a description of equilibrium such as we find in a chemical reaction.

The need for a criterion for spontaneous processes can be illustrated by a simple example. Suppose we have two identical blocks of silver, in contact with each other and isolated from the environment [see Fig. 3-1(a)]. At the outset of the experiment, we fix the temperature of one block at 75°C and the other at 25°C. Common intuition correctly tells us, and experiment verifies, that when sufficient time is allowed for equilibration, the temperature will become the same in both blocks. The heat energy Q_P absorbed by the colder block is given up by the warmer block, so that the energy change for the system of two blocks is $\Delta E = 0$. We also expect from experience that the reverse process does not occur. When we begin with two blocks in thermal contact and isolated from the environment, one of them will *not* spontaneously become warmer than the other. Thus, even though the reverse of the spontaneous process in Figure 3-1(a) also has $\Delta E = 0$, it does not occur.

Mechanical systems reach equilibrium when their potential energy is minimized. For example, the interacting particles in Figure 2-2(b) adopt the equilibrium distance R_0, which minimizes the potential energy, when they are allowed to come to *mechanical* equilibrium (no motion). The example in Figure 3-1(a) shows that minimization of energy is not an adequate principle for defining *thermal* equilibrium, since the initial and final states have the same energy.

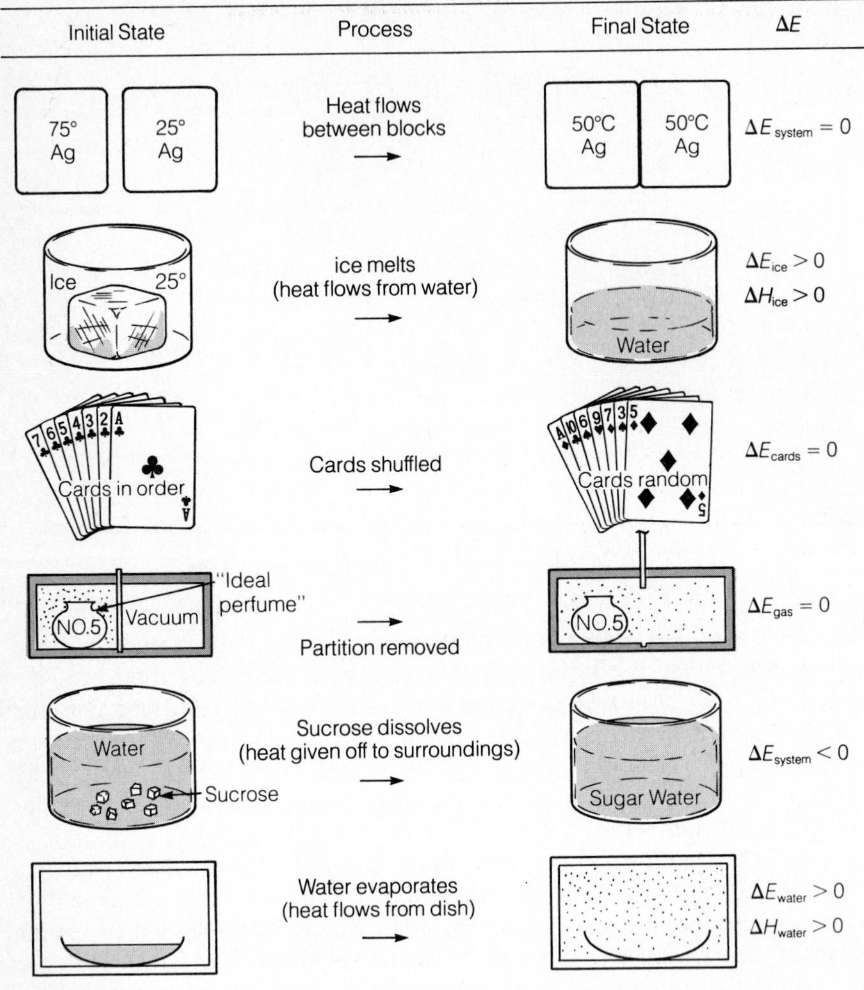

FIGURE 3-1
Some spontaneous processes.

Several other spontaneous processes are illustrated in Figure 3-1. These include the melting of ice at 25°C, the randomizing of a deck of ordered cards as they are shuffled, the diffusion of a volatile perfume into a larger volume as a partition is removed, the dissolving of sugar in water, and the evaporation of water from a dish. There are common features exhibited by all these processes. They include:

1. A direction of spontaneous change. The reverse processes never proceed spontaneously. Water never freezes at 25°C, releasing heat to the surroundings. Shuffling a deck of cards never puts them in perfect order (at least in the hands of an honest dealer). A gas never concentrates itself in half its container. Sugar never concentrates and precipitates from a solution. The fact that most processes are so strongly directional suggests that there is some deep principle of nature at work.

2. None of the reverse, unspontaneous processes are forbidden by the first law of thermodynamics. This was noted above for the two silver blocks. The same is true in all other examples of Figure 3-1, since energy is conserved in all. Thus the first law is inadequate to deal with the *direction* of change, and it is obvious that we need a new principle. From the last column of Figure 3-1, it is obvious that minimization of neither energy nor enthalpy can dictate the direction of spontaneous change. Thus the new principle must consider another property of a system.
3. Each of the spontaneous processes illustrated here is accompanied by an increase in disorder. Shuffled cards are less well ordered than those in a new pack. Liquid water is less well ordered than crystalline ice. A gas whose molecules have a greater volume to move in is more disordered. Indeed, it has been found that *all spontaneous processes in isolated systems are accompanied by increasing disorder.* This suggests that the new property which indicates a spontaneous change is a measure of disorder. Since we have already noted that heat is a less well organized form of energy than work, it seems reasonable that the new property might have some connection with heat.

In 1850 Rudolf Clausius, Professor of Physics at the Artillery and Engineering School of Berlin, found a property of state that is related to spontaneous change. He called it *entropy, S,* from the Greek words *en* (in) and *trope* (turning), meaning "to give direction." In time it was realized that S is a measure of disorder.

○ 3–2 *The Second Law of Thermodynamics*

The essence of the second law is to define a new property, the entropy, S, which is a function of state that reaches a maximum at equilibrium in an isolated system.

The second law may be stated as follows. Suppose a spontaneous process occurs in a system (see Fig. 3-2). Suppose also that the system plus its surroundings are isolated. We call this isolated region the *isolated enclosure.* Then

1. For a small amount of heat dQ_{rev} delivered to the system in a reversible process at temperature T, the differential

$$dS = \frac{dQ_{\text{rev}}}{T} \qquad (3\text{-}1a) \blacktriangleleft$$

is an exact differential.

2. For a spontaneous process, the total entropy change of the isolated enclosure is positive:

$$\Delta S_{\text{total}} = \Delta S_{\text{system}} + \Delta S_{\text{surr}} > 0 \qquad (3\text{-}1b) \blacktriangleleft$$

in which ΔS_{surr} is the entropy change of the surroundings.

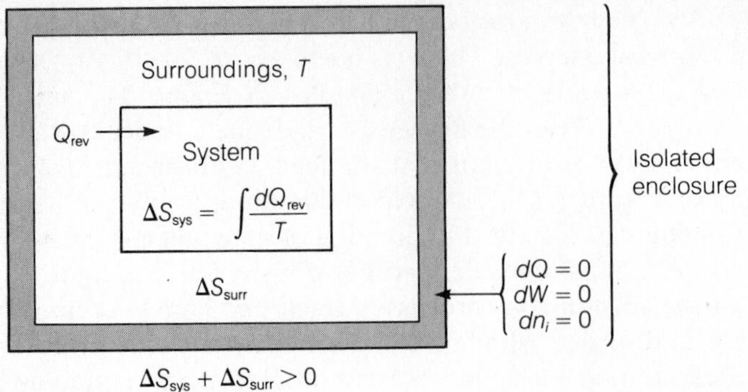

FIGURE 3-2
The basic elements of the second law of thermodynamics. The isolated enclosure consists of system plus surroundings and is isolated (no heat, work, or matter can enter or leave). For a spontaneous process in the system, resulting in entropy changes ($\Delta S_{sys} = \int dQ_{rev}/T$ and ΔS_{surr}), the sum of ΔS_{sys} and ΔS_{surr} is positive.

Several comments should be made before we apply the second law. The first is that statement (1) of the law is equivalent to stating that entropy is a property of a system. The second is that entropy has the dimensions of energy/temperature, or J K^{-1}, so that when multiplied by temperature, the product has dimensions of energy. The third is that the increase in entropy for spontaneous processes holds only for the *isolated* enclosure. Recall that ΔE, Q, and W are zero for any isolated system. Finally, we should note that although the law is mathematically very simple, virtually everybody finds it difficult to grasp conceptually at first. The common complaints are that the definition of dS seems arbitrary, and that it is not obvious why S should increase in a spontaneous process. Working through the examples of the following sections is a good way to convince yourself that the entropy, S, does in fact increase in spontaneous processes and that it behaves as a property. The molecular interpretation of entropy as disorder given in Section 3-4 and more fully in Chapter 14 will show that S is a measure of disorder. The remainder of this section will show that dQ_{rev}/T is a plausible choice for a function of state of a system.

How Clausius May Have Seen that Entropy is a Property

To appreciate how the function $dS = dQ_{rev}/T$ was discovered, put yourself in the place of Clausius. You know that heat and work are forms of energy and that work can be converted freely to heat, but there are limits on the conversion of heat to work. You recognize that work represents concerted motion under a macroscopic force, whereas you suspect that heat is expressed in chaotic molecular motion. Once energy is "degraded" to heat, only a fraction of the energy can be converted back to work in a cyclic process. You might then ask whether you can construct from the heat Q a property whose value reflects the degree of chaos or disorder in matter.

Consider the problem in the context of an ideal gas in a cyclic process. We choose a cyclic process because the change in any property in such a process must equal zero. Figure 3-3 shows a schematic diagram of an ideal gas subjected to cyclic expansion and compression. The first step is an isothermal expansion, with heat absorbed and work done by the system. You will recall that for an ideal gas at constant temperature, $\Delta E = 0$, $Q = -W$. According to Table 2-2, for a reversible process the heat and work are

$$Q_{\text{rev}}\,(1) = -W_{\text{rev}}\,(1) = nRT \ln \frac{V_2}{V_1} \tag{3-2}$$

If the process is irreversible, W_{irrev} and Q_{irrev} are smaller in absolute value. This tells us immediately that if we are to develop a property involving the heat, it will have to refer to the heat change in processes carried out *reversibly*, or Q_{rev}, since otherwise the indeterminacy of Q will render inexact all differentials involving dQ.

The other three steps in Figure 3-3 are also easily described. Steps 2 and 4 are adiabatic so $Q_{\text{rev}}\,(2) = Q_{\text{rev}}\,(4) = 0$. For Step 3, by analogy with Step 1,

$$Q_{\text{rev}}\,(3) = nRT' \ln \frac{V_4}{V_3} \tag{3-3}$$

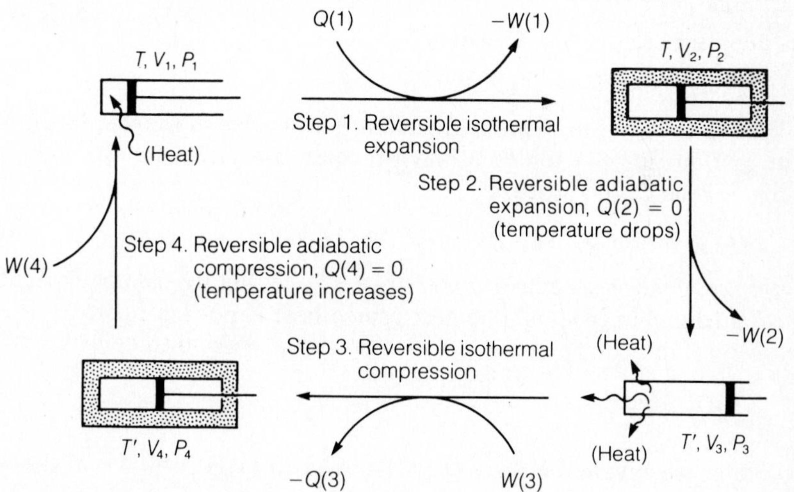

FIGURE 3-3
Heat and work changes in the cyclic expansion and compression of an ideal gas. In each step, Q is the heat *absorbed* by the system and W the work done *on* the system. If the direction of the arrow shows heat *produced* or work *done by* the system, then a negative sign is attached. In the text it is shown that $Q_1 = nRT \ln (V_2/V_1)$ and $Q_3 = -nRT' \ln (V_2/V_1)$, with $Q_2 = Q_4 = 0$. If entropy is defined by $\Delta S = Q_{\text{rev}}/T$, the entropy change for the cycle is $\Delta S_T = \Delta S(1) + \Delta S(2) + \Delta S(3) + \Delta S(4) = Q(1)/T + Q(3)/T' = nR \ln (V_2/V_1) - nR \ln (V_2/V_1) = 0$. A property must have the characteristic that it is unchanged by any cyclic process. Division of Q_{rev} by T to yield ΔS gives us a function with that characteristic.

in which $T' < T$ because of the adiabatic expansion, Step 2. Now we can combine Equations 3-2 and 3-3 to obtain for the total heat change in the cyclic process, Q_{total},

$$\begin{aligned} Q_{total} &= Q_{rev}(1) + Q_{rev}(2) + Q_{rev}(3) + Q_{rev}(4) \\ &= nRT \ln (V_2/V_1) + nRT' \ln (V_4/V_3) \end{aligned} \qquad (3\text{-}4)$$

Equation 3-4 has too many variables in it (four different volumes and two different temperatures) to be directly useful to us. However, the volumes and temperatures are related by equations that describe the reversible adiabatic steps, as derived in Section 2-4. Applying the results in Table 2-2 to Figure 3-3, we see that

$$\underset{\text{(Step 2)}}{T'/T = (V_2/V_3)^{nR/C_V}} \qquad \underset{\text{(Step 4)}}{T/T' = (V_4/V_1)^{nR/C_V}}$$

Inverting the second of these equations gives $T'/T = (V_1/V_4)^{nR/C_V}$, whose right side can be equated to the right side of the equation for Step 2, to give

$$(V_2/V_3)^{nR/C_V} = (V_1/V_4)^{nR/C_V}$$

Raising both sides of this equation to the power C_V/nR (which multiplies the exponent by C_V/nR) yields

$$V_2/V_3 = V_1/V_4$$

or

$$V_4/V_3 = V_1/V_2$$

This allows us to convert Equation 3-3 to

$$Q(3) = -nRT' \ln (V_2/V_1) \qquad (3\text{-}5)$$

Now we are able to express Equation 3-4 in a more useable form. Substituting Equation 3-5 for $Q(3)$ into Equation 3-4 yields for the total heat change for the cyclic process

$$Q_{total} = nRT \ln (V_2/V_1) - nRT' \ln (V_2/V_1) \qquad (3\text{-}6)$$

This expression is *not* equal to zero, since $T' \neq T$, but we did not expect that the total heat change should be zero, since heat is not a property.

However, the entropy change for each step is defined to be

$$\Delta S = \frac{Q_{rev}}{T} \qquad (3\text{-}7)$$

Therefore, we divide the term $Q_{rev}(1) = nRT \ln (V_2/V_1)$ by T and the term $Q_{rev}(3) = -nRT' \ln (V_2/V_1)$ by T'. Recalling that $Q_{rev}(2)$ and $Q_{rev}(4)$ are zero, instead of Equation 3-6 we get

$$\begin{aligned} \Delta S_T &= \Delta S(1) + \Delta S(2) + \Delta S(3) + \Delta S(4) \\ &= nR \ln (V_2/V_1) - nR \ln (V_2/V_1) = 0 \end{aligned} \qquad (3\text{-}8)$$

The cyclic process of Figure 3-3 has the entropy changes shown in Figure 3-4. The crucial point to understand is that *with ΔS defined by Equation 3-7, we find no change in S ($\Delta S_T = 0$) for the cycle of Figures 3-3 and 3-4. This is a necessary characteristic of a property.*

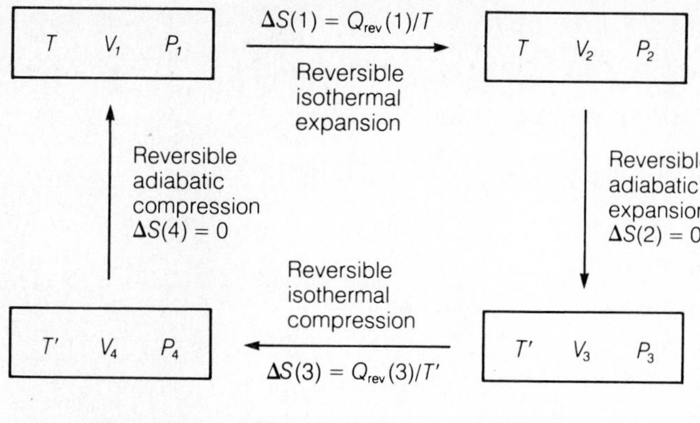

FIGURE 3-4
Entropy changes in the cyclic process shown in Figure 3-3.

The reversible expansion and compression cycle of an ideal gas is a limited example, and demonstrating that ΔS defined by Equation 3-7 is a property for that cycle does not constitute a general proof. However, *it does eliminate all other functions involving* Q_{rev}, because only the choice Q_{rev}/T leaves S unchanged in the cycle we considered. A true property cannot fail in *any* example. Thus we are compelled to try a function of the form dQ_{rev}/T. A century of experiments have confirmed that S so defined is indeed a property for all systems. The path to this understanding of nature was not simple or rapid, as illustrated by the sketch of William Thomson's life on pages 92–94.

EXERCISE 3-1

Explain why $\Delta S = 0$ for Steps 2 and 4 in Figures 3-3 and 3-4.

ANSWER

Because $Q_{rev} = 0$ for both steps, and $\Delta S = \int dS = \int dQ_{rev}/T = 0$ for both steps. An entropy change occurs only if there is a heat change when the process is carried out *reversibly*.

EXERCISE 3-2

Calculate the entropy change for the reversible isothermal expansion of one mole of an ideal gas from 10 atm to 1 atm at 300 K.

ANSWER

Combining Equations 3-7 and 3-2, we get

$$\Delta S = nR \ln\frac{V_2}{V_1} \quad \text{(isothermal ideal gas)} \tag{3-9}$$

So $\Delta S = 8.314 \ln 10 = 8.314 \times 2.303 \times 1 = 19.14\ \text{J K}^{-1}$.

WILLIAM THOMSON, BARON KELVIN

1824–1907

Lord Kelvin with his compass for use on board iron ships. (The Bettmann Archive.)

William Thomson entered university at age 10. He was one of the deepest thinkers of all the brilliant founders of thermodynamics and was author of 661 scientific papers, yet he stands as a refreshing contrast to the stereotype of the great scientist as a tortured neurotic introvert. Thomson was gregarious, fond of sports and travel, and dedicated almost as much to the advance of practical instrumentation as to the development of fundamental theory. His 70 patents, many on telegraphic and navigational instruments, brought him a fortune, part of which he spent on a 126-ton sailing yacht, the *Lalla Rookh*. The importance of his leadership in the growing industrial society of England led Queen Victoria to honor him, first with a knighthood and then with a peerage of the realm. As a title Thomson selected Kelvin, after the river near his university.

Thomson was the son of a Scottish writer of mathematical textbooks. His mother died when he was six, and his father devoted himself to the education and advancement of his children. William matriculated three months past his tenth birthday in the University of Glasgow, where his father held the chair of mathematics. Speaking later in life of his early education, Thomson said, "A boy should have learned by the age of twelve to write his own language with accuracy and some elegance; he should have a reading knowledge of French, should be able to translate Latin and easy Greek authors, and should have acquaintance with German."

Thomson was introduced to Fourier's analytical theories (see Chap. 17) at Glasgow, well before they were known in England. Not only were his first two publications (at age 17) on Fourier's ideas, but Fourier's methods were important in much of his later work. He moved from Glasgow to Cambridge University when he was 17, earned a B.A. degree with honors, rowed competitively, and played second horn in the orchestra.

When he was 22, the chair of natural philosophy (professorship of physics) at Glasgow became open. Thomson's father carefully managed the candidacy of his son, and William was unanimously elected to the post, which he held for his entire career. As a lecturer he was not methodical, and was hard for poorer students to follow. One complained, "Well, I listened to the lectures on the pendulum for a month, and all I know about the pendulum yet is that it wags."

Thomson contributed to many areas of physics, but his most profound work was in the development of thermodynamics. After graduating from Cambridge he spent several months in Paris, where he read the works of Carnot and Clapeyron. Carnot had analyzed the working of heat engines in terms of the flow of heat, then regarded as an indestructible substance. Thomson's great work was to reconcile Carnot's analysis with Joule's demonstration that work can be transformed into heat by revolving paddles in a fluid.

Joule was from a wealthy brewing family in Manchester. His papers and talks on the equivalence of heat and work had been greeted with cold indifference until Thomson pointed out their significance. Joule recalled the moment that this took place: at the conclusion of his talk at the 1847 meeting of the British Association for the Advancement of Science in Oxford. Joule related, ". . . discussion not being invited, the comunication would have passed without comment if a young man had not risen in the section, and by his intelligent observations created a lively interest in the new theory. The young man was William Thomson, who had two years previously passed the University of Cambridge with the highest honor. . . ." Thomson's recollection of the same moment is the following: "I heard his paper read at the section, and felt strongly impelled to rise and say that it must be wrong . . . But as I listened on and on I saw that (though Carnot had vitally important truth not to be abandoned) Joule had certainly a great truth and a great discovery, and a most important measurement to bring forward."

Two weeks afterwards Thomson journeyed to the Alps. He was astounded to run into Joule there, who was combining his honeymoon with an attempt to measure the mechanical conversion of heat of falling water in waterfalls. Thomson recalled, ". . . about a fortnight later I was walking down from Chamounix to commence the tour of Mount Blanc, and whom should I meet walking up but Joule, with a long thermometer in his hand, and a carriage with a lady in it not far off. He told me that he was going to try for elevation of temperature in waterfalls."

Thomson's thinking about combining the ideas of Carnot and Joule culminated in 1851 in a clear statement of the first and second laws of thermodynamics. Clausius, who developed the notion of entropy at about the same time, grumbled over the question of priority, but Thomson seemed little concerned. He wrote, "Questions of personal priority, however interesting they may be to the persons concerned, sink into insignificance in the prospect of any gain in deeper insight into the secrets of nature."

By the mid-1850s Thomson's eminence as a scientist and his interest in instruments brought him to the attention of a consortium of British industrialists who wanted to lay a submarine telegraph cable across the Atlantic. Telegraphy was already a profitable business, and a line between the hemispheres would open new possibilities for rapid communication. Thomson was made a director of the company, but the technical details were entrusted to Mr. O. E. W. Whitehouse, a retired physician. The cable was manufactured in 1200 pieces of two miles each, then joined into lengths of 300 miles each and coiled in tanks. Because Whitehouse was ill at the critical time, Thomson supervised the paying-out from on board the H.M.S. *Agamemnon.* Earlier Thomson had applied Fourier's methods to analyze the nature of currents transmitted through miles of cable, and had found that rapid communication would be possible only at the small voltages produced by batteries. Accordingly he invented a mirror galvanometer that could detect miniscule currents. Whitehouse, however, insisted that his own apparatus, relying on the higher voltages of induction coils, was superior. When Whitehouse found that his instruments could transmit only very slowly, exactly as Thomson had predicted, he secretly substituted Thomson's galvanometer. When his deception was discovered, Whitehouse was fired and Thomson became a hero to the British financial community and to the Victorian public.

In 1862 Thomson made an ingenious calculation of the geological age of the earth by applying Fourier's theory of heat conduction to the cooling of the sun and the earth. He assumed that the sun's heat was not inexhaustible, and that it must have cooled over geological time. From reasonable estimates of the solar specific heat and the rate of solar radiation, he concluded that the sun had not illuminated the earth for more than 500 million years. From Fourier's laws of heat transfer and underground temperature, he also concluded that the earth solidified more than 20 million years ago but not more than 400 million years ago. These estimates brought him into conflict not only with geologists but also with Darwin and his followers, whose theories required an older earth and more gradual changes. Of course the flaw in Thomson's analysis was that the sun is not a cooling object, but one in which heat is continually produced through nuclear reactions. Radioactivity was discovered only later in Thomson's life, and he found it impossible to accept that vast energy can emerge from the atom. In a 1903 letter to Lord Rayleigh (see Chap. 12) Thomson wrote, "It seems to me utterly out of the question to suppose, as Rutherford and others have done, that heat emitted by radium is generated by a self-contained store of energy." Even the greatest minds are often unconvinced by the revolutionary thoughts of the new generation.

EXERCISE 3-3

Calculate the entropy change of the ideal gas if the expansion of Exercise 3-2 is carried out from the same initial state to the same final state, but by an irreversible path. One such possible irreversible path is that of a *free expansion,* in which a partition is removed, allowing the gas to expand to ten times its initial volume.

ANSWER

The answer is the same as for Exercise 3-2, 19.14 J K^{-1}. An entropy change cannot be *measured* along an irreversible path, but since S is a property, *ΔS for an irreversible path must be the same as ΔS for a reversible path,* for which Q_{rev} can be measured. This is an important point that will help avoid confusion about entropy changes.

3-3 $\frac{1}{T}$ *Is an Integrating Factor for* dQ_{rev}

An *integrating factor* in mathematics is a factor which when multiplied by an inexact differential converts it to an exact differential. We have just seen that $1/T$ is an integrating factor for dQ_{rev} since multiplication of the two yields the exact differential dS. Again the ideal gas supplies a simple illustrative example. Combining Equation 2-5 for the energy of an ideal monatomic gas, $E = \frac{3}{2}nRT$, with the ideal gas law, $PV = nRT$, we obtain

$$E = \tfrac{3}{2}PV$$

Hence the differential dE for an ideal gas is

$$dE = \tfrac{3}{2}PdV + \tfrac{3}{2}VdP$$

Euler's criterion (Equation 1-23) verifies that dE is exact because

$$\frac{\partial}{\partial P}\left(\tfrac{3}{2}P\right)_V = \frac{\partial}{\partial V}\left(\tfrac{3}{2}V\right)_P = \tfrac{3}{2}$$

In a reversible process $dW_{\text{rev}} = -PdV$; hence $dQ_{\text{rev}} = dE - dW_{\text{rev}}$ is

$$\begin{aligned} dQ_{\text{rev}} &= \tfrac{3}{2}PdV + \tfrac{3}{2}VdP + PdV \\ &= \tfrac{5}{2}PdV + \tfrac{3}{2}VdP \end{aligned}$$

In this case Euler's criterion says that dQ_{rev} is inexact, because

$$\frac{\partial}{\partial P}\left(\tfrac{5}{2}P\right)_V = \tfrac{5}{2} \neq \frac{\partial}{\partial V}\left(\tfrac{3}{2}V\right)_P = \tfrac{3}{2}$$

However, if we now divide dQ_{rev} by $T = PV/nR$, dQ_{rev}/T is

$$\frac{dQ_{\text{rev}}}{T} = \frac{5}{2}\frac{nR}{V}dV + \frac{3}{2}\frac{nR}{P}dP$$

Once again using Euler's criterion, we now can show that dQ_{rev}/T for an ideal gas is an exact differential by calculating

$$\frac{\partial}{\partial P}\left(\frac{5}{2}\frac{nR}{V}\right)_V = \frac{\partial}{\partial V}\left(\frac{3}{2}\frac{nR}{P}\right)_P = 0$$

This example shows that dS is an exact differential in the specific case of an ideal monatomic gas. Of course it does not prove the general rule, but it does supply some insight into the mathematical reason why S is a property (dS exact) whereas Q_{rev} is not (dQ_{rev} inexact).

3-4 *The Spontaneity Criterion*

Using the ideal gas as an example, we have seen that $dS = dQ_{\text{rev}}/T$ is a reasonable definition for a property S, which is called the entropy. We now must confront the question that is at the heart of the second law: Why does a positive entropy change, $\Delta S > 0$, always accompany a spontaneous process in an isolated enclosure? It is only through this criterion that entropy can "give a direction" to chemical and physical changes.

States that can be Achieved in more Ways are more Probable and have Greater Entropy

As is usual in thermodynamics, a conceptual grasp of the nature of entropy and the origin of the spontaneity criterion comes most easily from consideration of the molecular nature of matter. We will see throughout this book that entropy is a measure of disorder on a molecular scale. Your understanding of this point will grow with exposure, including some examples in this chapter, and the discussion of statistical mechanics in Chapter 14. As an introduction, the spontaneity criterion can be visualized by the following example.

Figure 3-5 shows a simple, idealized system in which an ideal gas is confined by a partition to the left half of the volume of a cylindrical container. Insulation surrounds the system, so no heat flows in, and the system volume is held constant, so no work is exchanged with the surroundings. (Notice that changing the volume of the gas does not change the volume of the system, because the system boundary is the insulated box.) Now, imagine removing

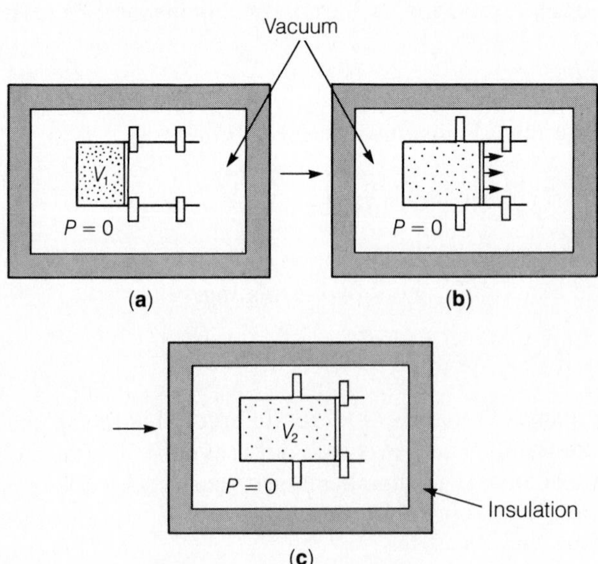

FIGURE 3-5
Expansion of an ideal gas to double its volume in an isolated system. In (a) the gas is constrained by a partition to half the final cylinder volume by restraining stops. In (b) these have been moved in an (idealized) step in which no work is done, in order to allow the partition to move by. The (massless) partition moves unopposed by the vacuum which surrounds the cylinder, until it reaches the final state (c). The entire system is insulated, and no work is done on the surroundings; $V_2 = 2V_1$.

the (massless) stops restraining the partition, while still doing no work on the system. This part of the process is especially idealized to avoid complicating the problem. The gas will expand to fill the whole cylinder volume once the partition is allowed to move to the next set of stops. Why?

Heat, work, and energy do not help us answer the question. Q and W are zero, so the energy change (ΔE) also must be zero, by the first law. Furthermore, since the gas is ideal and its energy is constant, the temperature does not change. (Recall that the energy of an ideal gas depends only on T, not on P or V.) Obviously, the quantities involved in the first law do not lead to an understanding of the direction of the spontaneous change.

The general reason that the gas expands is that its entropy increases. By Equation 3-9, $\Delta S = nR \ln (V_{\text{final}}/V_{\text{initial}}) = nR \ln 2$, which is a positive quantity. The entropy increase in turn is due to the increased "disorder" of the gas after expansion. But what does "disorder" mean in this context?

For simplicity, suppose that our sample of ideal gas contains just two molecules, A and B. Before the expansion, both molecules must be in the left half of the cylinder, but after the expansion each molecule is free to move about in the final volume, and has equal probability of occupying the left or the right half of the container. Let us define a *substate* of the system as a particular assignment of the two molecules to the two half-volumes. For example, both molecules A and B could be in the left half-volume (L), a substate which we designate by A(L)B(L). The three other possible substates are A(L)B(R), A(R)B(L), and A(R)B(R), depending on whether the molecules are in the left (L) or right (R) half-volumes.

We are now prepared to understand why the gas expands. Before the expansion, it can only be in the state A(L)B(L), but once the partition is allowed to move, this state becomes only one substate out of a larger number of possibilities. All of these possible substates have equal energy, and all are equally probable. Doubling the volume has increased the number of substates by a factor of four (when the sample contains two molecules), so the state with increased volume is more probable than the original state. In summary, *the gas expands because there are more substates that correspond to occupation of the total volume than to occupation of the original volume*. The gas is more "disordered" after expansion because it has four times as many substates than before, and we do not know which of these it might be in at any time. In general, the less precisely we can specify the state of a system, the more disordered it is.

With only two molecules in our sample, we cannot appreciate the immense number of substates in a macroscopic system; their very large numbers lead to an overwhelming preponderance for states in which the whole volume is occupied by the gas, compared to ocupancy of only the original half-volume. Notice that when the sample contained only two molecules, the probability was 1/4 that both molecules would occupy the original volume, because the substate A(L)B(L) accounted for 1/4 of the total number of substates. On the other hand, if the sample is one mole of ideal gas, it contains Avogadro's number, $N_A = 6.022 \times 10^{23}$, of molecules. Each molecule has two choices, to occupy the left or the right half-volume, so there are $2 \times 2 \times 2 \times 2 \times \cdots \times 2 = 2^{N_A}$ possible substates. (Check this expression by writing down and counting the substates when there are 2, 3, or 4 molecules in the sample.) When the one-mole sample of gas expands to double its volume, the substate that corresponds to occupation of only the original volume accounts for just one of the 2^{N_A} total. Therefore, the probability of finding the gas exclusively in the left half-volume after the partition is allowed to move is $1/2^{N_A}$, an extremely small number, as we shall see.

This example can give some idea of the enormous predictive power of the second law of thermodynamics. Since the law actually is based on the laws of probability, you might think that the spontaneity criterion should more rigorously be stated in probabilistic terms. For example, you might prefer to say that it is extremely likely that a process in an isolated system will proceed in a direction such that the entropy increases. The second law is generally not stated this way because numbers the size of $1/2^{N_A}$ are vanishingly small. With $N_A = 6 \times 10^{23}$, we have $1/2^{N_A} = 2^{-N_A} = 10^{-1.8\times10^{23}}$, which is a decimal point followed by 1.8×10^{23} zeroes before the first significant figure.[1] Even if the one mole sample of gas could change its substates (by motion of its molecules) once every 10^{-10} sec, the probability remains vanishingly small that it would by a sudden fluctuation have occupied only half of its available volume at any time during the known lifetime of the universe.

It is a general rule that fluctuations in the *extensive* properties of a system are significant and measurable when the number of particles is small, say a hundred or a thousand, but not significant for samples the size of a mole. Fluctuations in the *local* value of *intensive* properties can be significant for any size system. For example, we will see in Chapter 13 that local fluctuations in the refractive index give rise to scattering of light by a liquid.

○ 3–5 *The Boltzmann Equation Relates the Degeneracy of a State to its Entropy*

The quantitative relationship between entropy and disorder was first derived by Boltzmann. In Chapter 14 we will examine his equation in much more detail, but because the mathematical expression is simple and the consequences of the concept are far-reaching, we will present the equation briefly here.

Consider a macroscopic sample of matter which has constant energy and volume (an isolated system). The molecules in the sample can store the same total energy in many different arrangements. For example, the total kinetic energy of a gas can be distributed in many different ways among the particles of the gas. The number of ways of achieving a given energy state is called its

[1] In the type style used for this book, a line 2.4×10^{17} miles long would be needed to write this number. A beam of light would require about 40,000 years to traverse that distance. Hence you would never have time to read the number.

degeneracy, Ω. Unfortunately, before we are ready to calculate Ω for even so simple a material as an ideal gas, we must learn the elements of both quantum mechanics (Chap. 10) and statistical mechanics (Chap. 14). However, if for the moment we accept the idea that the number Ω exists and can be calculated, the Boltzmann equation is simple. It states that the entropy S is given by

$$S = k_B \ln \Omega \tag{3-10a}$$

in which k_B is called the *Boltzmann constant,* which is related to the gas constant and Avogadro's number by

$$k_B = R/N_A = 1.380 \times 10^{-23} \text{ J K}^{-1} \tag{3-10b}$$

Equation 3-10*a* concisely expresses the relationship between entropy and disorder. According to the second law, the entropy of an isolated system reaches a maximum at equilibrium. Equation 3-10*a* reveals that maximizing S is equivalent to maximizing Ω, implying that *the most probable macroscopic state of an isolated system is the state of maximum degeneracy.*

Even though we are not yet prepared to calculate Ω rigorously, an example may help you see the utility of the Boltzmann equation. It turns out (as seems plausible, but which you will have to accept for the moment without proof) that Ω for an ideal gas is proportional to the number of substates calculated for the system in Figure 3-5. Therefore, we can state that the ratio of the final and initial values of Ω, for a one-mole sample of gas, is

$$\frac{\Omega_{\text{final}}}{\Omega_{\text{initial}}} = \frac{2^{N_N}}{1} = 2^{N_A}$$

Consequently, the entropy change when a mole of an ideal gas doubles its volume should be

$$\begin{aligned}\Delta S &= k_B \ln \Omega_{\text{final}} - k_B \ln \Omega_{\text{initial}} \\ &= k_B \ln \left(\frac{\Omega_{\text{final}}}{\Omega_{\text{initial}}}\right) \\ &= k_B \ln 2^{N_A} = k_B N_A \ln 2 = R \ln 2\end{aligned}$$

Notice that this is precisely the result you get by substituting $n = 1$ and $V_2/V_1 = 2$ into Equation 3-9, which was derived earlier from the second law of thermodynamics.

3–6 General Principles and Examples of Spontaneous Processes

With the general concept that spontaneous changes happen because the final state has higher probability than the initial state, we now look at specific examples to see the mathematical reason why the function S increases for a spontaneous process in an isolated system. To do this we will calculate entropy changes for several processes. The principles that will emerge from the examples are the following:

1. For a reversible process, the entropy changes of system and surroundings are equal in magnitude and opposite in sign. Thus the total entropy change for the isolated enclosure is zero.
2. For a spontaneous (irreversible) process, the entropy changes of system and surroundings are not equal. The total entropy of the isolated enclosure increases.
3. Though entropy changes can be measured only for reversible processes, the entropy change of a system can be calculated for irreversible processes. The reason for this is that the entropy of a system is a property of state and depends only on the initial and final states. Thus as long as a process begins and ends in equilibrium states, we can always construct a reversible process with the same initial and final states as the spontaneous process, and can calculate or measure the entropy change for this process.

The Total Entropy Change in a Reversible Process is Zero

The total entropy change of a system plus surroundings always is zero in a reversible process. This relationship is demonstrated in Figure 3-6. The heat, dQ_{rev}, absorbed by the system at temperature T gives $dS_{\text{system}} = dQ_{\text{rev}}/T$; the heat absorbed by the surroundings is $-dQ_{\text{rev}}$, giving $dS_{\text{surroundings}} = -dQ_{\text{rev}}/T$. Clearly $dS_T = dS_{\text{system}} + dS_{\text{surroundings}} = dQ_{\text{rev}}(1/T - 1/T) = 0$. The reason the entropy changes are the same is that T is the same in the system and surroundings. If the two temperatures were not equal, then dS_T would not be zero. Of course, if the temperatures were not equal, the process would not be reversible, because of the finite temperature difference.

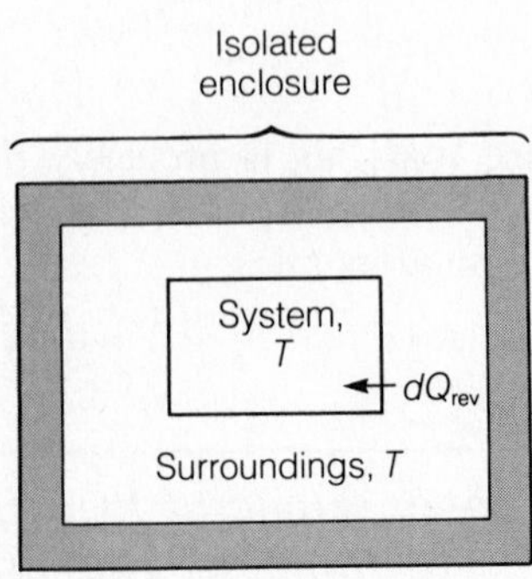

FIGURE 3-6
Reversible heat exchange between system and surroundings at the same temperature T. The entropy change always is zero.

$$dS_{\text{sys}} = \frac{dQ_{\text{rev}}}{T};\ dS_{\text{surr}} = \frac{-dQ_{\text{rev}}}{T}$$

$$dS_{\text{total}} = dS_{\text{sys}} + dS_{\text{surr}} = 0$$

EXERCISE 3-4

Can a reversible process in an isolated system be spontaneous?

ANSWER

No. $\Delta S_{\text{total}} = 0$ for a reversible process, whereas $\Delta S_{\text{total}} > 0$ for a spontaneous process. A system undergoing a reversible process is always at equilibrium, whereas a system undergoing a spontaneous process is moving towards equiiibrium. Of course, *nearly* reversible processes *do* happen, but they occur because of outside intervention that gradually changes the properties of a system. (A *truly* reversible process happens infinitely slowly.) "Outside intervention" means that the system is not isolated; see the diagram of reversible processes in Figure 1-3.

The Flow of Heat from a Hotter to a Colder Body: Establishing Temperature Equilibrium

Consider in detail what happens when the temperatures of the system T and of the surroundings T' are not equal (Fig. 3-7). Experience tells us that heat should flow from the hot body to the cold one so that $T = T'$ when equilibrium is established. The total entropy change dS_{total} accompanying heat flow dQ_{rev} is

$$dS_{\text{total}} = \frac{dQ_{\text{rev}}}{T} + \frac{-dQ_{\text{rev}}}{T'}$$
$$= dQ_{\text{rev}}\left(\frac{1}{T} - \frac{1}{T'}\right) \tag{3-11}$$

Notice that dS_{system} and $dS_{\text{surroundings}}$ are calculated for heat flow at the temperature of the system T and surroundings T' respectively. At no point in the *calculation* does heat flow between different temperatures T and T'. Of course this hypothetical reversible path for transferring the heat does not correspond to the real, irreversible path, but entropy changes cannot be expressed in terms of dQ for irreversible paths.

Equation 3-11 predicts that the direction of heat flow depends on whether T is greater or less than T'. We consider the two cases separately:

Case 1. When $T < T'$, the surroundings are hotter than the system. Since $1/T > 1/T'$, Equation 3-11 indicates that dS_{total} will be positive (process spontaneous) when $dQ_{\text{rev}} > 0$, which implies heat flow *into* the system. Hence heat flows as expected from high to low temperature.

Case 2. When $T > T'$, dS_{total} is positive when $dQ_{\text{rev}} < 0$, implying heat flow *out of* the system.

These two cases show that the *heat flow changes sign when $1/T - 1/T'$ changes sign*. The crucial mathematical feature that causes this sign reversal is inclusion of the factor $1/T$ in the definition of the entropy. *Heat flow from*

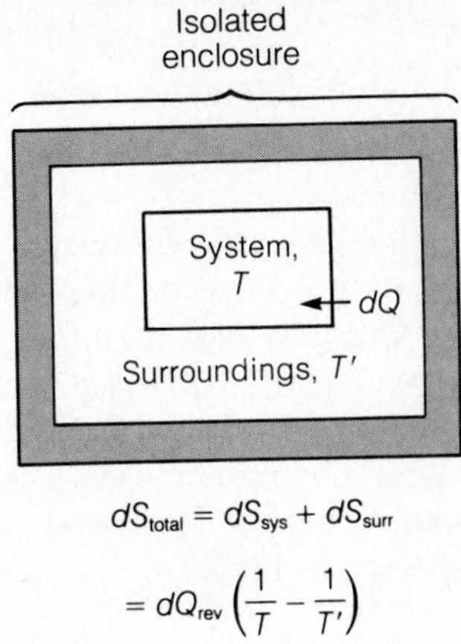

FIGURE 3-7
Spontaneous heat flow between the surroundings and the system when the two are at different temperatures. We expect heat to flow spontaneously from the high to the low temperature. This is demonstrated by calculating the entropy change when an amount of heat $dQ = dQ_{rev}$ is removed from the surroundings at temperature T' by a reversible path, and $dQ = dQ_{rev}$ is added to the system at temperature T, again by a reversible path. By the definition of entropy, $dS_{total} = dQ_{rev}/T - dQ_{rev}/T' = dQ_{rev}\,(1/T - 1/T')$. In Case 1, $T < T'$ and dS_{total} is greater than zero when $dQ = dQ_{rev} > 0$. Heat flows from high temperature (surroundings) to low temperature (system). In Case 2, $T > T'$ and $dS_{total} > 0$ when $dQ = dQ_{rev} < 0$. Heat flows from the system (high temperature) to the surroundings (low temperature). The important point is that in order for dS_{total} to be positive, the sign of dQ reverses when $(1/T - 1/T')$ changes sign.

high to low temperature is predicted by the second law of thermodynamics, but not by the first law.

Figure 3-8 shows how the entropy of the system plus surroundings in Figure 3-7 varies as the temperature T of the system moves away from the equilibrium value $T = T'$. Cases 1 and 2 show that dS is positive as T approaches T' from either side. Therefore the maximum entropy occurs when $T = T'$. This entropy maximization condition is analogous to the minimization of potential energy in mechanical systems at equilibrium [Figure 2-2(b)]. We stress again that the principle of entropy maximization applies only to a

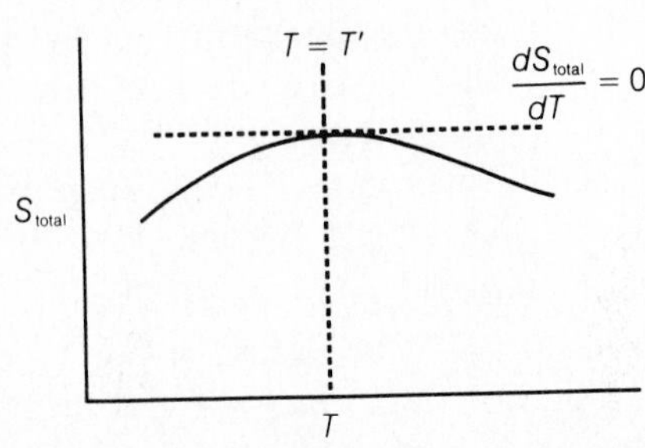

FIGURE 3-8
Maximization of the entropy of the system plus surroundings (isolated enclosure) in Figure 3-7, when $T = T'$. Spontaneous heat flow from high to low temperature ($dS > 0$) brings the temperature of the system to that of the surroundings at equilibrium. Maximization of the total entropy at equilibrium (with constant total energy in an isolated system) is analogous to minimization of the energy at equilibrium in mechanical systems (see Figure 2-2).

process in an isolated system (which we also call the isolated enclosure), in which the total energy is constant and no work is done on the system.

The Melting of Ice: Establishing Phase Equilibrium

Let us consider a nonequilibrium physical state which is not at temperature equilibrium. We will see that again the system initially contains a nonequilibrium which disappears spontaneously according to the entropy maximization criterion. Suppose we add 1 g of ice at 0°C to 4 g of water at 10°C (Fig. 3-9). We can use the first law to calculate how much ice at 0°C must be converted to water at 0°C, where it should be in equilibrium with ice. Using a specific heat of 1 cal g^{-1} $°C^{-1}$, the cooling process removes $Q = 1$ cal g^{-1} $°C^{-1} \times$ 4 g $\times$ 10°C = 40 cal. Because the heat of fusion of ice is 80 cal/g, 0.5 g, or half the ice, must be melted to take up the heat released on cooling the water.

We next ask whether the entropy change for the process in Figure 3-9 is positive. To calculate ΔS we must follow an imaginary reversible path in which heat is added to the ice to give water at 0°C, but the same amount of heat is removed from the water at temperatures ranging from 10°C to 0°C as the rest of the sample is cooled. First, for melting 0.5 g of ice at 0°C,

$$\Delta S_{\text{ice}} = Q_{\text{rev}}/T = \frac{0.5\,\text{g} \times 80\,\text{cal g}^{-1}}{273\,\text{K}}$$
$$= 0.1465\ \text{cal K}^{-1}$$

Next, to cool 4 g of water from 283 K to 273 K, the entropy change is the integral over $dS = dQ_{\text{rev}}/T = \dfrac{C\,dT}{T}$ from 283 K to 273 K,

$$\Delta S_{H_2O} = \int_{283}^{273} \frac{dQ_{\text{rev}}}{T} = -\int_{273}^{283} \frac{C\,dT}{T} = -C \ln\left(\frac{283}{273}\right)$$
$$= -4 \ln \frac{283}{273} = -0.1439\ \text{cal K}^{-1}$$

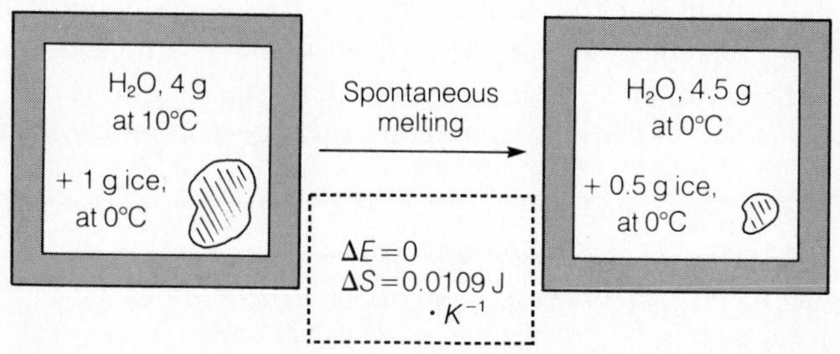

FIGURE 3-9
Ice melts spontaneously to bring the temperature of the water phase down to 0 °C, as shown by a positive value of ΔS. Exercise 3-5 shows that ice will not melt to cool the water phase *below* 0 °C.

(The heat capacity C is 4 cal K^{-1} for the 4 g sample.) Hence the total entropy change is

$$\begin{aligned}\Delta S_{\text{total}} &= \Delta S_{\text{ice}} + \Delta S_{H_2O} = 0.1465 \text{ cal K}^{-1} - 0.1439 \text{ cal K}^{-1} \\ &= 0.0026 \text{ cal K}^{-1} = 0.0109 \text{ J K}^{-1}\end{aligned}$$

which is positive, so the process is spontaneous.

If you examine this derivation closely you will see that the reason ΔS_{total} is positive is that ΔS_{H_2O} is slightly smaller in absolute value than ΔS_{ice}. The reason for this is that the heat removal from liquid H_2O occurs at temperatures at and above 273 K, whereas the same heat is added to the ice at 273 K. Thus the spontaneity of the process is driven by the requirement for temperature equilibrium of the two phases. The following exercise shows that ice will not melt to cool the water phase *below* 0°C.

EXERCISE 3-5

Calculate the entropy change for a hypothetical process in an isolated system in which 0.5 g of ice at 0° melts to cool 4 g of water from 0°C to −10°C.

ANSWER

$$\Delta S_{\text{ice}} = 0.5 \text{ g} \times 80 \text{ cal g}^{-1}/273 = 0.1465 \text{ cal K}^{-1}. \ \Delta S_{H_2O} = -\int_{263}^{273} C\,\frac{dT}{T}$$

$$= -1 \text{ cal g}^{-1} \times \left(\ln \frac{273}{263}\right) \times 4 \text{ g} = -0.1493 \text{ cal K}^{-1}$$

$\Delta S_{\text{total}} = 0.1465 - 0.1493 = -0.0028$ cal K^{-1}. Since ΔS_{total} is negative, the *reverse* of the stated process is spontaneous. If we begin with 0.5 g ice at 0°C and 4 g H_2O at −10°C, the final state will contain 1 g ice + 3.5 g H_2O at equilibrium.

The cases illustrated by Figure 3-9 and Exercise 3-5 show that two phases in equilibrium always have the same temperature. This is consistent with the criterion of maximum disorder or maximum entropy, because if different phases had different temperatures, the higher-energy (higher-temperature) molecules would have to be restricted to one phase and the lower-energy molecules to another. Such arbitrary restrictions violate the principle of maximum disorder and never are seen in systems at equilibrium. In summary, all of the examples in this section share the characteristic that an entropy increase accompanies a spontaneous process in an isolated system.

3–7 *The Efficiency of a Heat Engine is Less than One*

Generations of chemists have learned their thermodynamics in terms of the ideal heat engine. This tradition has strong historical roots, because the development of thermodynamics, and especially the formulation of the second law, were closely tied to analysis of the theoretical efficiency of the steam engine by Carnot in 1824. Of course the steam engine now is just about as far from reality for most modern students as is the second law of thermodynamics.

However, the question of the efficiency of heat engines in general remains of great importance for modern society, since most of our methods for generating electric power rely on heat engines. A heat engine is a device that converts heat into mechanical energy. As with all machines, it must be based on a cyclic process, so that there is no infinite extension of a part, as would be necessary in a continuous, noncyclic process. The efficiency of such an engine (see Fig. 3-10) is limited by the ability to convert the heat Q_1 that is taken from a high temperature (T_1) source into work $-W$ in a cyclic process. Some of the heat must be discharged into a low temperature (T_2) sink, as we will now see.

The *efficiency* ϵ is defined as the ratio of work done to heat absorbed,

$$\epsilon = \frac{-W}{Q_1} \tag{3-12}$$

in which Q_1 is the heat absorbed by the engine at high temperature and $-W$ is the work done by the engine. The waste heat produced is $-Q_2$. The signs are

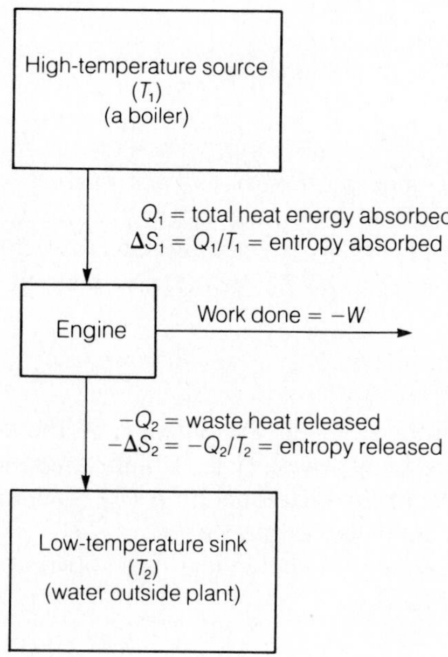

FIGURE 3-10
Schematic diagram of a reversible heat engine. Heat energy Q_1 is absorbed from the high-temperature source, such as a flame. An amount of work $-W$ is done *by* the engine (W is the work done *on* the engine), and heat $-Q_2$ is released from the engine to the low-temperature sink, or coolant. The engine operates in a *cyclic* process, so $\Delta E = Q_1 + Q_2 + W$ must be zero. Hence $Q_1 = -Q_2 - W$, which tells us that the energy absorbed by the engine equals the energy released from the engine. The fraction of energy absorbed that is converted to work is the efficiency $\epsilon = -W/Q_1$. It is demonstrated in the text that $\epsilon = 1 - (T_2/T_1)$.

defined so that W is work done *on* the engine and Q is heat absorbed *by* the engine; $-Q_2$ and $-W$ are positive quantities.

To derive the constraints that thermodynamics places on engines which convert heat to work, we must (1) recognize that any engine operates by a cyclic process, requiring that the total change in the engine's properties be zero for a complete cycle. If this were not so, the engine would be constantly changing with time. Specifically, we can set ΔS and ΔE equal to zero for a complete cycle of the heat engine shown in Figure 3-10.

In addition (2), we must apply the first and second laws of thermodynamics to the engine. The energy change of the engine for a cycle in which it absorbs heat Q_1, does work $-W$, and releases heat $-Q_2$ is

$$\Delta E = Q_1 + Q_2 + W = 0 \quad \text{(first law)} \tag{3-13}$$

in which ΔE is equated to zero because the process is cyclic. Equation 3-13 expresses the conservation of energy, but places no restrictions on how the energy released from the engine $(-Q_2 - W)$ is divided between heat and work.

It is through application of the second law to the heat engine that we perceive how thermodynamics limits the conversion of heat to work by a cyclic engine. Assuming a reversible engine so that the heat flow at each temperature is reversible, the total entropy change of the engine is

$$\begin{aligned} \Delta S &= \Delta S_1 + \Delta S_2 \\ &= \frac{Q_1}{T_1} + \frac{Q_2}{T_2} = 0 \quad \text{(second law)} \end{aligned} \tag{3-14}$$

According to Equation 3-14, *the entropy increase of the reversible engine* (Q_1/T_1) *must equal the entropy decrease of the engine* $(-Q_2/T_2)$. You may regard this as a balance between the entropy flows into and out of the engine, because the total entropy is constant in a reversible process. Assuming definite values for Q_1, T_1, and T_2, then Q_2 also must have a definite value:

$$-Q_2 = Q_1\left(\frac{T_2}{T_1}\right)$$

In other words, in order to keep the entropy of the engine constant, a fraction T_2/T_1 of the heat (Q_1) absorbed at high temperature (T_1) *must* be given off as heat $(-Q_2)$ at lower temperature (T_2); this fraction of the heat absorbed is therefore unavailable to do work.

To calculate the efficiency of the engine, we substitute $Q_2 = -Q_1(T_2/T_1)$ into Equation 3-13, obtaining

$$\begin{aligned} -W &= Q_1 + Q_2 \\ &= Q_1\left(1 - \frac{T_2}{T_1}\right) \end{aligned}$$

Therefore the efficiency ϵ is

$$\epsilon = -\frac{W}{Q_1} = 1 - \frac{T_2}{T_1} \tag{3-15} ◀$$

which is always less than 1, unless $T_2 = 0$ K, or $T_1 = \infty$. This result applies to all reversible heat engines because we have made no assumptions about the nature of the engine. It can be shown that irreversible engines can have efficiencies less than, but not greater than, the efficiency of a reversible engine, as given by Equation 3-15.

EXERCISE 3-6

A 1000 megawatt (electric) power plant operates with the high-temperature stage at 500°C; the cooling stage is a river at 10°C. If regulations permit no more than a 5°C temperature rise in the water, calculate the minimum cooling water flow necessary to operate the plant. Recall that 1 watt = 1 J s^{-1}.

ANSWER

$-W$ = 1000 megawatts = 10^9 J sec^{-1}. $\epsilon = 1 - \frac{283}{773} = 0.633$. Total heat required = $Q_1 = 10^9$ J $sec^{-1}/\epsilon = 1.58 \times 10^9$ J sec^{-1}. Waste heat = $(1 - \epsilon) \times Q_1 = 0.58 \times 10^9$ J sec^{-1}. Specific heat of water = 4.18 J °C^{-1} g^{-1}. Amount of water required = 0.58×10^9 J sec^{-1}/(4.18 J °C^{-1} g^{-1} × 5°C) = 2.8×10^7 g sec^{-1} or 2.8×10^4 liters/sec (approximately 6500 gallons per second, roughly the amount of water in a circular swimming pool 18 feet across and 4 feet deep).

A consequence of Equation 3-15 is the inefficiency of heat engines that rely on the flow of heat between two nearly equal temperatures, T_1 and T_2 in Figure 3-10. A practical example makes the importance of this thermodynamic principle clear. One process which has been proposed to help solve the energy shortage facing modern society is to make use of the heat at the surface of tropical oceans. For example, it might be possible to design a heat engine in which the heat at the surface flows to the cooler water at the bottom, producing work. Unfortunately, the temperature difference between the surface and the bottom is not great enough to make this process very practical. Suppose that the bottom temperature is 7°C (280 K), with the surface at 27°C (300 K); the efficiency ϵ is $1 - 280/300 = 0.0667$, so *at most* 6.67% of the heat that flows from the surface can be converted to work. When the inevitable losses due to friction are included, such an engine appears relatively unattractive.

Another consequence of the thermodynamic analysis of heat engines is that *heat pumps* (engines which utilize work to pump heat from low temperature to high temperature) become highly effective when the two temperatures are nearly equal, as shown by the following exercise.

EXERCISE 3-7

Consider a reversible heat pump (refrigerator) in which all arrows in Figure 3-10 are reversed in direction: The engine absorbs work W to pump heat Q_2 from the low temperature T_2, releasing heat $-Q_1$ to the high temperature T_1.

Show that the ratio of the heat Q_2 pumped out of the refrigerator to the work absorbed by the engine is

$$\frac{Q_2}{W} = \frac{T_2}{T_1 - T_2}$$

ANSWER

The engine operates in a cyclic process, so by the first law,

$$\Delta E = Q_1 + Q_2 + W = 0$$

Similarly, by the second law

$$\Delta S = \frac{Q_1}{T_1} + \frac{Q_2}{T_2} = 0$$

Substituting $Q_1 = -Q_2(T_1/T_2)$ from this equation into the first law expression yields

$$W = Q_2 \left(\frac{T_1}{T_2} - 1\right)$$

which, upon rearrangement, leads to the desired expression. Notice that as T_2 approaches T_1, Q_2/W becomes very large. Hence, a large amount of heat can be pumped reversibly between two objects of similar temperature for a small expenditure of work. This suggests a plausible use of solar energy for heating purposes. During the winter, solar heat could be used to heat a water reservoir to ~16°C (60°F). A heat pump could then be used very efficiently to pump heat from T_2 = 16°C to T_1 = 20°C (68°F) inside a building. In this way the solar heating system would not have to heat water directly to a temperature high enough to allow rapid, spontaneous heat flow to the interior space.

Our analysis of the heat engine is an example of one of the main subjects of the next chapter: What is the maximum amount of work that can be extracted from a spontaneous process? One always finds that the work is a maximum when the process is carried out reversibly. For example, if we simply brought the high and low temperature baths in Figure 3-10 into contact, heat would flow spontaneously and irreversibly, but no work would be done. If we insert a real engine and extract work, the heat flow is slowed down. In the limit of a reversible heat engine, the heat flow and the production of work occur infinitely slowly, but the amount of work produced per unit of heat flow is maximized.

Why Energy is "Degraded" when Converted to Heat

Now you can understand why we sometimes speak of energy being "degraded" to heat, in contrast to storage as mechanical or electrical energy. For example, compare energy generation by a hydroelectric plant to the use of a heat engine. The water behind a dam has potential energy because of its height above the base of the dam. In principle, *all* of the potential energy of the water can be converted to work, or to electrical energy, if the generating plant operates reversibly. (Of course, in reality there are inevitable frictional losses.) In contrast, if the potential energy stored in the chemical bonds in oil is converted to heat by burning, only a fraction of the heat collected by the boiler can be used to do work; no heat engine can be more efficient than specified by Equation 3-15. The lower the temperature at which the heat energy is stored, the more "degraded" is the energy involved, because low temperature sources have low maximum efficiencies of conversion to work.

This principle makes it evident that it is vital for plants, and ultimately for all living organisms, that the sun's energy be captured to do chemical work *before* it is degraded to heat at the earth's surface. If the role of the sun were simply to heat the leaves of a tree, it is unlikely that the observed efficiencies of the order of 1% of the incident energy could be achieved in utilizing solar energy to do biosynthesis. Plants have solved this challenge by evolving molecules such as chlorophyll which are able to trap photons (Chapter 10), the packets of energy in the light which the sun emits, before the concentrated photon energy is spread around over a large number of molecules in the form of molecular motion or heat. The photons in visible light are able to accomplish chemical and physical changes in the leaf which heat energy would be too dispersed to do. Because of the laws of probability which lie behind the second law of thermodynamics, it is exceedingly unlikely that the molecules in a leaf would concentrate their heat energy sufficiently to cause the chemical rearrangements (such as the splitting of water molecules) required in the primary photosynthesis steps.

○ 3-8 The First and Second Laws Combined

The first and second laws can be combined mathematically to yield a number of important thermodynamic relations. Using the now-familiar equations $dE = dQ + dW$, $dS = dQ_{rev}/T$, and $dW_{rev} = -PdV$ (only PV work), we get by simple substitution

$$dE = TdS - PdV \qquad \text{(reversible path)} \tag{3-16}$$ ◀

in which dE is the energy increment accompanying reversible changes dS in entropy and dV in volume. We can get a similar equation for dH by noting that

$$\begin{aligned} dH &= dE + d(PV) \\ &= dE + PdV + VdP \end{aligned}$$

Combining this with Equation 3-16, we find

$$dH = TdS + VdP \qquad \text{(reversible path)} \tag{3-17}$$ ◀

The simple expressions for the exact differentials dE and dH imply that the *natural independent variables* for E are S and V, whereas S and P are the corresponding natural variables for H. A function of state, for example, could be written for the energy, $E(S,V)$, or enthalpy, $H(S,P)$.

Equations 3-16 and 3-17 provide an abundant harvest of useful equations. For example, if we hold the volume constant in Equation 3-16, we find that

$$(dE)_V = T(dS)_V$$

where the subscript V indicates constant volume. This can be rewritten

$$\left(\frac{\partial E}{\partial S}\right)_V = T \tag{3-18}$$

This relationship says that the slope of a plot of energy against entropy, with volume held constant, is the temperature T. The following equations are similarly derived from 3-16 and 3-17:

$$\left(\frac{\partial E}{\partial V}\right)_S = -P \tag{3-19a}$$

$$\left(\frac{\partial H}{\partial S}\right)_P = T \tag{3-19b}$$

$$\left(\frac{\partial H}{\partial P}\right)_S = V \tag{3-19c}$$

EXERCISE 3-8

The entropy of a substance at a constant pressure of 1 atm and 300 K is increased by 0.001 J K^{-1}. Calculate the enthalpy change, assuming that the temperature change is negligible.

ANSWER

From Equation 3-19*b*, $(\Delta H)_P = T(\Delta S)_P$, so $(\Delta H)_P = 300 \text{ K} \times 0.001 \text{ J K}^{-1} = 0.3 \text{ J}$.

EXERCISE 3-9

A substance occupies 1 liter at 1 atm pressure. Calculate the enthalpy change due to a reversible adiabatic increase of pressure to 1.001 atm.

ANSWER

The entropy is constant in a reversible adiabatic process ($dQ = 0$), so $(\Delta H)_S = V(\Delta P)_S = 1 \text{ liter} \times 0.001 \text{ atm} = 0.001 \text{ liter} \cdot \text{atm} = 0.1013 \text{ J}$.

Maxwell Relations

Equations 3-16 and 3-17 are extremely versatile, providing not only Equations 3-18 and 3-19, but also the first two *Maxwell relations* (which will be mentioned further in Chapter 4). Consider the differential for dE in Equation 3-16. Euler's criterion (Equation 1-23) says that because dE is exact,

$$\left(\frac{\partial T}{\partial V}\right)_S = -\left(\frac{\partial P}{\partial S}\right)_V \tag{3-20}$$

and a similar treatment of Equation 3-17 yields

$$\left(\frac{\partial T}{\partial P}\right)_S = \left(\frac{\partial V}{\partial S}\right)_P \tag{3-21}$$

EXERCISE 3-10

Show that

$$dS = -\left(\frac{\partial V}{\partial T}\right)_S dP + \left(\frac{\partial P}{\partial T}\right)_S dV \tag{3-22}$$

ANSWER

Write the total differential (Equation 1-21) for $S = S(P,V)$

$$dS = \left(\frac{\partial S}{\partial P}\right)_V dP + \left(\frac{\partial S}{\partial V}\right)_P dV$$

and substitute the Maxwell relations 3-20 and 3-21 (after inverting them) to get Equation 3-22.

EXERCISE 3-11

Combine Equations 3-19*b* and 3-21 to show that

$$\left(\frac{\partial P}{\partial \ln T}\right)_S = \left(\frac{\partial H}{\partial V}\right)_P$$

ANSWER

First, rewrite $T = (\partial H/\partial S)_P$ as

$$T = \left(\frac{\partial H}{\partial S}\right)_P = \left(\frac{\partial H}{\partial V}\right)_P \left(\frac{\partial V}{\partial S}\right)_P$$

(Notice that this is equivalent to the chain rule of differentiation. It is important to note that the same variable, the pressure in this case, is held constant for all of the partial derivatives.) Next, use Equation 3-21 and substitute $(\partial T/\partial P)_S$ for $(\partial V/\partial S)_P$:

$$T = \left(\frac{\partial H}{\partial V}\right)_P \left(\frac{\partial T}{\partial P}\right)_S$$

Rearrange this to

$$T\left(\frac{\partial P}{\partial T}\right)_S = \left(\frac{\partial H}{\partial V}\right)_P$$

or

$$\left(\frac{\partial P}{\partial \ln T}\right)_S = \left(\frac{\partial H}{\partial V}\right)_P$$

EXERCISE 3-12

How many differential terms dx_i must be present on the right side of an equation for a total differential such as Equation 3-22?

ANSWER

The number of differential terms is equal to the number of independent variables required to specify the state of the system over the applicable range of experimental conditions. Two variables are enough to describe the state of a fixed amount of a single substance (see the phase rule, Chap. 1). However, if more substances are present, more variables and hence more differential terms are required.

○ 3–9 *Calculation of Entropy Changes*

Calculation of the entropy change for a process involves integration of the differential dS. The simplest total differential for the entropy is either in terms of E and V, from Equation 3-16:

$$dS = \frac{dE + PdV}{T} \tag{3-23}$$

or H and P, from Equation 3-17:

$$dS = \frac{dH - VdP}{T} \tag{3-24}$$

However, these equations are awkward for problems such as the ideal gas, where the natural experimental variables are any two of the set T, P, V. To calculate entropy changes we therefore need a new expression for dS which eliminates dH or dE. This involves replacing dE in Equation 3-23. The total differential dE, expressed in terms of V and T (see Equation 1-18) is

$$\begin{aligned} dE &= \left(\frac{\partial E}{\partial T}\right)_V dT + \left(\frac{\partial E}{\partial V}\right)_T dV \\ &= C_V\, dT + \left(\frac{\partial E}{\partial V}\right)_T dV \end{aligned} \tag{3-25}$$

Substituting for dE in 3-23 yields

$$dS = C_V \frac{dT}{T} + \frac{1}{T}\left[\left(\frac{\partial E}{\partial V}\right)_T + P\right] dV \tag{3-26}$$

(Complicated expressions like 3-26 are obtained when one expresses the differential of a thermodynamic quantity in terms of differentials of anything except the natural variables, in this case E and V or H and P.)

For an ideal gas, Equation 3-26 can be simplified. Since $(\partial E/\partial V)_T = 0$ and $P = nRT/V$, we obtain

$$dS = \frac{C_V\,dT}{T} + \frac{nR\,dV}{V} \qquad \text{(ideal gas)} \tag{3-27}$$

Integration of Equation 3-27 yields for the entropy change $\Delta S = S_2 - S_1$, when the state variables of an ideal gas are shifted from V_1, T_1 to V_2, T_2,

$$\Delta S = C_V \ln \frac{T_2}{T_1} + nR \ln \frac{V_2}{V_1} \qquad \text{(ideal gas)} \tag{3-28}$$ ◀

This is a more general expression than Equation 3-9, which allows volume change only at constant temperature. Both T and V can change in Equation 3-28.

Entropy of Mixing

Equation 3-28 is the basis for derivation of an important equation for the *entropy of mixing* ideal gases. Suppose we have n_i moles of gas i, with a total of n moles of gas,

$$n = \Sigma n_i \tag{3-29}$$

At the beginning, each gas is in a different container (Fig. 3-11), with all containers at the same pressure P and temperature T. The volume available to gas i before the stopcocks are opened is

$$V_i = \frac{n_i RT}{P}$$

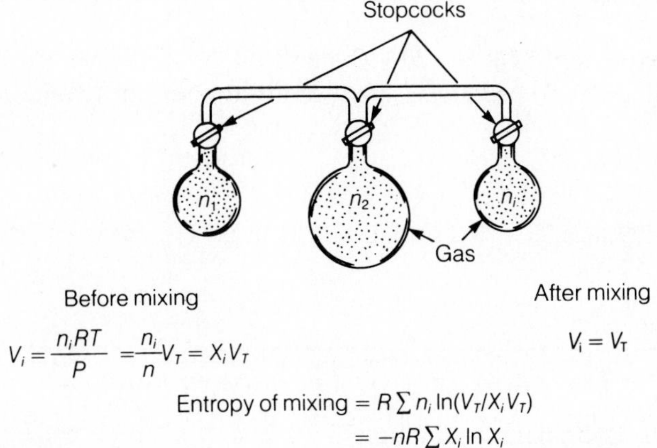

FIGURE 3-11
The entropy of mixing ideal gases. Before mixing, each gas, i, is confined to a volume V_i at pressure P. When the stopcocks are opened, each gas can occupy the total volume $V_T = \Sigma V_i$. The entropy increase of each gas is given by Equation 3-32 as $\Delta S_i = n_i R \ln[V_T/(V_T X_i)]$, so the total entropy increase is $-nR\, \Sigma X_i \ln X_i$.

After mixing, the volume is V_T:

$$V_T = \Sigma V_i = n\frac{RT}{P}$$

The ratio of V_i to V_T is

$$\frac{V_i}{V_T} = \frac{n_i}{n} = X_i \tag{3-30}$$

in which X_i is called the *mole fraction* of gas *i*, the fraction of the total moles of gas accounted for by component *i*. Equation 3-28 gives for the entropy increase at constant temperature due to the increased volume available to gas *i*

$$\Delta S_i = n_i R \ln\frac{V_T}{V_i} = -n_i R \ln\frac{V_i}{V_T}$$
$$= -n_i R \ln X_i \tag{3-31}$$

Because entropy is an extensive property, we may write $\Delta S_{\text{mix}} = \sum_i \Delta S_i$, and so the total entropy of mixing is

$$\Delta S_{\text{mix}} = -\Sigma n_i R \ln X_i$$
$$= -nR\, \Sigma X_i \ln X_i \qquad \text{(ideal gas)} \tag{3-32}$$ ◀

An expression identical to 3-32 also will appear for the entropy of mixing ideal solutions in Chapter 7. The significance of both expressions is that when substances are mixed at constant pressure, they are allowed to share a common volume at random (in this case the volume V_T instead of V_i for component *i*). The consequent increase in volume available to each constituent is a measure of the disorder produced by mixing ("mixed-up-ness").

EXERCISE 3-13

Air is approximately 79% N_2, 20% O_2, and 1% Ar. Calculate the entropy of mixing these gases to produce 1 mole of air at constant temperature and pressure.

ANSWER

$\Delta S = -R[0.79 \ln 0.79 + 0.2 \ln 0.2 + 0.01 \ln 0.01] = 4.61 \text{ J K}^{-1}\text{ mol}^{-1}$.

○ 3–10 The Third Law of Thermodynamics

According to Boltzmann's equation (3-10*a*), $S = k_B \ln \Omega$, entropy is a measure of the degeneracy of a macroscopic state. If this is so, it should be possible to measure entropy on an absolute scale. This is the principle of what sometimes is called the *third law of thermodynamics,* according to which the entropy of a perfectly ordered material (such as a perfect crystal or a perfect superfluid such as liquid helium) approaches zero as the temperature approaches zero,

$$\lim_{T \to 0\ \mathrm{K}} S = 0 \qquad \text{(perfectly ordered material)} \tag{3-33}$$

The entropy approaches zero because there is only one arrangement or substate of a perfectly ordered material, that is, $\Omega = 1$, so $S = k_B \ln \Omega = 0$. A consequence of Equation 3-33 is that in all reactions between perfectly ordered materials, the entropy change $\Delta S = S_{\text{product}} - S_{\text{reactant}}$ must also approach zero as $T \to 0$.

To check the assumption that $S = 0$ at absolute zero, we compare the values of the entropy obtained by two different approaches. On one hand we use statistical mechanical methods to evaluate Ω (see Chap. 14), and calculate S from the Boltzmann equation. Alternatively, the entropy can be calculated as follows from the thermal properties of the material between 0 K and T. At constant pressure, Equation 3-24 gives

$$dS = \frac{(dH)_P}{T} \tag{3-34}$$

With $(dH)_P = C_P dT$ we have

$$dS = \frac{C_P dT}{T} \tag{3-35}$$

If the system is heated from T_1 to T_2 at constant pressure, ΔS can be determined by integrating Equation 3-35:

$$S_2 - S_1 = \int_{T_1}^{T_2} \frac{C_P dT}{T} \qquad \text{(all substances)} \tag{3-36}$$

This important equation tells us how to use heat capacity data to calculate the entropy change between temperatures T_2 and T_1. If $T_1 \to 0$, then $S_1 \to 0$ by the third law. Therefore

$$S = \int_0^{T_2} \frac{C_P dT}{T} \tag{3-37}$$

If there is a phase change at temperature T' which lies between 0 and T_2, and if $\Delta H_{\text{latent heat}}$ is the heat absorbed during this change, then this equation must be modified to

$$S = \int_0^{T_2} \frac{C_P dT}{T} + \frac{\Delta H_{\text{latent heat}}}{T'} \qquad \text{(perfectly ordered at 0 K)} \tag{3-38}$$

There is one term $\Delta H_{\text{latent heat}}/T'$ for each phase transition temperature T', and the infinity in C_P at $T = T'$ is ignored in the integral from 0 to T_2.

Residual Entropy: An Apparent Exception to the Third Law

The meaning of the third law can be appreciated best by considering cases where it does *not* apply. For example, the entropy of crystals of NNO is found not to be zero. This conclusion is reached by comparing the entropy as measured by Equation 3-38 with values calculated by statistical mechanics, based on spectroscopic measurements on NNO gas (see Fig. 3-12). The calculated value is larger than the calorimetrically measured value by 5.73 J K^{-1} mol^{-1}. We conclude that the crystals have a *residual entropy* of $S(0) =$ 5.73 J K^{-1} mol^{-1} at 0 K. This entropy appears because NNO is a linear molecule which fits into the crystal lattice virtually as well with its axis pointed in either direction (NNO or ONN). The crystal is disordered in the sense that each molecule has two orientations, independent of the orientation of its neighbors. Consequently, for one mole, $\Omega = 2 \cdot 2 \cdot 2 \ldots = 2^{N_A}$, where N_A is Avogadro's number. We can then predict the residual entropy to be

$$
\begin{aligned}
S(0) &= k_B \ln 2^{N_A} \\
&= N_A k_B \ln 2 \\
&= R \ln 2 = 5.76 \text{ J K}^{-1}
\end{aligned}
$$

which is in very close agreement with the experimental result (see Fig. 3-12).

For H_2O, the entropy calculated from statistical mechanics and spectroscopic measurements on the vapor at 298 K is 188.7 J K^{-1} mol^{-1}, whereas that determined from calorimetry is 185.3 + $S(0)$ J K^{-1} mol^{-1}. The difference gives $S(0) = 3.4$ J K^{-1} mol^{-1} as the residual entropy at 0 K. A simple and surprisingly accurate explanation of this result was provided by Linus Pauling, who estimated the number Ω of equivalent substates of ice at 0 K. Each water molecule has two protons engaged in hydrogen bonds with other water molecules (see Fig. 3-13). In a crystal with N_A water molecules there are two N_A hydrogen bonds. If each proton could be located on either

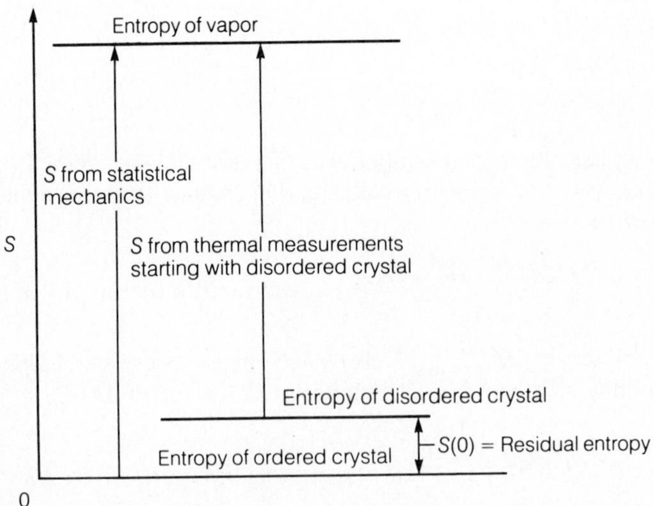

FIGURE 3-12
Apparent exceptions to the third law of thermodynamics occur for disordered crystals. Then the entropy of the gas of the substance, as determined by thermal measurements, is smaller than the entropy calculated by statistical mechanics. The difference is $S(0)$, the residual entropy.

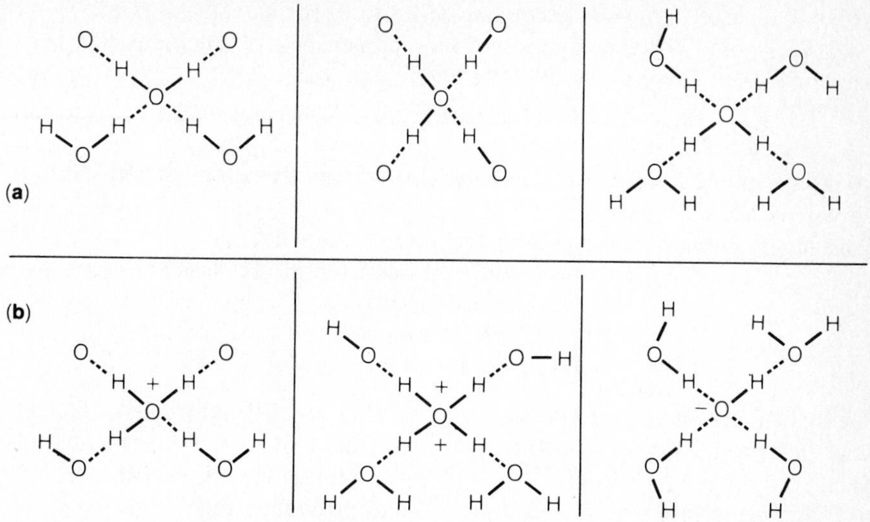

FIGURE 3-13
Hydrogen bonds of a water molecule in ice. Each oxygen atom is tetrahedrally coordinated by hydrogen bonds to four neighboring molecules. Each hydrogen bond contains one proton that can sit nearer either to the central molecule or to its neighbor. There are thus $2^4 = 16$ arrangements. Pauling noted that of these 16 arrangements, only six (of the type in row *a*) satisfy electrical neutrality. The remaining ten lead to charged molecules (of the type shown in row *b*). Draw the other arrangements as an exercise.

side of the hydrogen bond, there would be 2^{2N_A} substates. However, many of these arrangements would leave positive and negative charges in the lattice because there would be more or fewer than two protons on each oxygen. These substates with charge separation have much higher energy and do not occur at 0 K. Hence only arrangements with two protons bonded to each oxygen can be counted. There are only six ways of choosing exactly two protons to be assigned to the central oxygen in Figure 3-13—four for those pairs on each side of the square (with oxygen in the middle of the square) and two for pairs located diagonally across the square. (The square is a planar representation of the tetrahedral arrangement of protons around the oxygen). Without the restriction of electrical neutrality, there are $2^4 = 16$ different ways of positioning the four protons. Hence the 2^{2N_A} ways of arranging the protons must be reduced by a factor $(6/16)^{N_A}$ for the fraction of arrangements that satisfy the constraint of electrical neutrality. Therefore Ω is

$$\Omega = \left(\frac{6}{16}\right)^{N_A} (2)^{2N_A} = \left(\frac{3}{2}\right)^{N_A}$$

and

$$\begin{aligned} S(0) &= k_B \ln \Omega \\ &= R \ln\left(\frac{3}{2}\right) = 3.37 \text{ J K}^{-1} \text{ mol}^{-1} \end{aligned}$$

which is a result in close agreement with the experimental value of 3.4 J K^{-1} mol^{-1}. Other examples of the enumeration of substates will be encountered in Chapter 14.

EXERCISE 3-14

Write a five-line limerick on the subject of thermodynamics or physical chemistry.

Printable Examples:

A chemistry student named Crass
was bored and bemused in his class
With air from the room
he filled a balloon
and labeled it Ideal Gas

A physical chemist named Doar
was eager to transmit his lore
When his notes uninspired
as a text were required
the students all hollered for Moore

The Laws of Rumford and Clausius
to a student were not very obvius
But on further inspection
and prolonged reflection
they made her decidedly nausius

3–11 Entropy and Life

A living organism is not an equilibrium system. It contains many finite gradients, including nonequilibria of pressure, temperature, concentration, and electrical potential. When the organism dies these gradients disappear, some rapidly and some slowly. Our discussion of the second law showed that the spontaneous disappearance of finite nonequilibria is always accompanied by an entropy increase, so we know that the entropy of a living organism in an isolated system is lower than the state into which it decays. However, it is very difficult to measure the entropy of an organism, because of the requirement for following a reversible path to make use of the second law, $dS = dQ_{rev}/T$. In principle, the entropy of the organism could be determined by measuring the heat required to convert it reversibly into its components, each in a state in which its entropy is known. However, such assaults on the structural integrity of living organisms have so far resulted in their irreversible death, so we must at present be content with the knowledge that the entropy of living matter is lower than the entropy of the products to which it decays.

The general problem of maintenance of low entropy in the living state was discussed in an influential book entitled *What is Life* (1945) by E. Schrödinger (of whom more will be said in Chap. 10). To keep itself ordered (low entropy), an organism feeds on low entropy foods such as glucose, and converts these to a higher entropy state (CO_2, H_2O, etc.) using the energy thereby released to do chemical, mechanical, and electrical work. This is an example of the general principle that nonequilibrium steady states, such as life, can be kept low in entropy only in the presence of an energy source.

The ultimate energy source for sustaining life on Earth is the sun. The whole process of biological evolution, working to produce more and more complex life forms, relies on energy flux from the sun. Since the earth is clearly not an isolated system, there is no requirement that the entropy of our planet increase with time. Even less is there a requirement that the entropy of all animate life should increase with time. In principle, the increase of entropy applies to the universe as a whole, but the mysteries of its creation, past, and future are sufficiently great that we probably should not include the statement, "The entropy of the universe increases," in the list of immutable principles of classical thermodynamics.

QUESTIONS FOR REVIEW

1. What is entropy? How can one measure the increase in entropy of a substance as it is heated?
2. State the second and third laws of thermodynamics and indicate the physical meaning of each.
3. What is Boltzmann's equation? What is its physical meaning?
4. Explain in words why an increase in entropy is associated with the approach to equilibrium in an isolated system.
5. What is the relationship of entropy to mixing?
6. Compare reversible and irreversible processes in terms of (1) the amount of heat absorbed, (2) the amount of work performed, and (3) the entropy change.
7. Define the term "ideal monatomic gas." What are the thermodynamic characteristics of an ideal gas, including energy, heat capacities at constant pressure and constant volume, and entropy?
8. Do all substances have the same entropy at 0 K? Which do not?

PROBLEMS

1. Are the following statements true or false?

   (a) The entropy of a closed system cannot decrease spontaneously.

   (b) The entropy of an open system cannot decrease spontaneously.

   (c) The entropy of an isolated system cannot decrease spontaneously.

   (d) When water freezes spontaneously its entropy must increase.

   (e) dQ/T is always an exact differential.

   (f) dS is zero in a reversible process in an isolated system.

Answers: (a) F. (b) F. (c) T. (d) F. (e) F (dQ_{rev}/T is exact). (f) F. (g) T.

2. Add the appropriate words:

   (a) The second law of thermodynamics states that entropy always increases for a spontaneous process in an ______ system.

   (b) The third law of thermodynamics states that all ______ substances have the same entropy at 0 K.

   (c) The entropy change in an isolated system for a ______ process is zero.

   (d) The second law of thermodynamics states that dQ_{rev}/T is an ______ differential.

   (e) The change of a property in any ______ process must be zero.

Answers: (a) Isolated. (b) Perfectly ordered. (c) Reversible. (d) Exact. (e) Cyclic.

PROBLEMS RELATED TO EXERCISES

3. (Exercise 3-2) Calculate the entropy change for the *reversible* isothermal compression of one mole of an ideal gas from 1 atm to 5 atm at 400 K.

4. (Exercise 3-3) Calculate the entropy change for the *irreversible* isothermal compression of one mole of an ideal gas from 1 atm to 5 atm at 400 K.

5. (Exercise 3-5) Calculate the entropy change for the hypothetical process in which 0.5 g of ice at 0°C melts to water at 0°C and 0.5 g water at −10°C freezes to ice at −10°C. Assume $\Delta H_{fus} = 80$ cal g^{-1}, independent of temperature.

● 6. (Exercise 3-6). A 1000 megawatt power plant has available a cooling stream of 10^4 liters/sec at 10°C. Given that the high temperature stage operates at 400°C, calculate the minimum temperature increase in the cooling stream.

● 7. (Exercise 3-7) Calculate the minimum electrical energy required to pump 1 J of heat from 15°C to 20°C.

OTHER PROBLEMS

8. Write expressions in terms of the temperature T_1, T_2, T_3 and T_4 for the entropy change in each step of the following process for an ideal monatomic gas:

   (a) Reversible adiabatic expansion from T_1P_1V to T_2P_2V'. (b) Irreversible heating at constant volume from T_2P_2V' to T_3P_3V'. (c) Reversible adiabatic compression from T_3P_3V' to T_4P_4V. (d) Irreversible cooling at constant volume from T_4P_4V to T_1P_1V. Prove that the total entropy change for the cycle is zero.

9. The cycle of Problem (3-8) can operate as a refrigerator, that is, heat is absorbed in the low temperature compartment and used to heat the gas from T_2 to T_3. After the adiabatic compression, heat is removed from the gas in cooling it from T_4 to T_1. Prove that the ratio of the net work done on the gas to the heat $Q(2)$ absorbed in step (2) is given by $W_T/Q(2) = (V'/V)^{2/3} - 1$. [$W_T$ is $W(1) + W(2) + W(3) + W(4)$.]

10. Suppose you have two symbols A and two symbols B. Calculate the number of distinguishable ways Ω that the four symbols can be arranged in a line, for example, ABAB, BAAB, etc. Calculate the entropy of the state with the symbols A and B mixed.

11. Show that if there are $N/2$ symbols A and $N/2$ symbols B, the number of distinguishable ways Ω that the symbols can be arranged in a line is $N!/[(N/2)!]^2$. ($N!$ is $N(N-1)(N-2) \ldots 3 \times 2 \times 1$.) Show that the entropy of this state is identical to the result given by Equation 3-32 for the entropy of mixing half a mole of A with half a mole of B. (*Note:* you will have to use *Stirling's approximation*, $\ln(N!) = N \ln N - N$.)

12. Explain how the entropy of protein denaturation can be determined from the anomaly in the heat capacity C_P as shown in Figure 2-9. How would you calculate ΔS_{denat} approximately from ΔH_{denat}? Calculate ΔS_{denat} for α-chymotrypsin from $\Delta H = 560$ kJ mol^{-1} at 50°C.

13. Two one-mole samples of ideal monatomic gas at different temperatures are brought into thermal contact and isolated from the surroundings (but not mixed). Their individual volumes are held constant. One sample is initially at 0°C and the other at 50°C. Calculate the final temperature and the entropy change in the spontaneous process of thermal equilibration.

14. Calculate the entropy change when 1 mole of N_2 is mixed with 1 mole of N_2, at constant pressure and temperature.

15. Calculate the entropy change when 1 mole of N_2 at 2 atm and 300 K is mixed with 1 mole of air at 1 atm and 300 K (79% N_2, 20% O_2, 1% Ar). The final pressure is 1.5 atm at 300 K. Assume all gases are ideal.

16. A one-mole sample of a monatomic ideal gas is initially at 1 atm and 300 K. A constant external pressure of 5 atm is suddenly applied, compressing the gas irreversibly to a final state in which P is 5 atm, $T = 400$ K. Calculate: (a) ΔE, (b) W, (c) Q, (d) ΔS, (e) ΔH.

17. A reversible heat engine absorbs heat (Q_1) from a high-temperature source at temperature T_1, and releases heat ($-Q_2$) to the low-temperature sink at temperature T_2. Apply the first and second laws to the engine, and combine these with the fact that the engine operates in a cyclic process to show that the work the engine can do is given by

$$-W = \Delta S_1(T_1 - T_2)$$

in which ΔS_1 is the entropy increase of the engine in the high-temperature heat-absorption step.

18. The cyclic process in Figure 3-3 can operate as a reversible heat engine, absorbing heat $Q(1)$ from the high-temperature source at T, doing work $-W = \sum_{i=1}^{4} -W(i)$, and releasing heat $-Q(3)$ to the low-temperature sink at T'.
    (a) Show that $W(1) = -nRT \ln (V_2/V_1)$ and $W(3) = nRT' \ln (V_2/V_1)$.
    (b) Show that $W(2) = -W(4) = C_V (T' - T)$.
    (c) Combine these results to show that $W = -nR(T - T') \ln (V_2/V_1)$.
    (d) Show that $Q(1) = nRT \ln (V_2/V_1)$.
    (e) Combine these results to show that the efficiency of the engine is $\epsilon = 1 - T'/T$.

19. Prove that

$$\left(\frac{\partial E}{\partial S}\right)_V = \left(\frac{\partial H}{\partial S}\right)_P$$

(*Hint: Begin with Equations 3-16 and 3-17.*)

20. Prove that

$$(\partial E/\partial T)_S = (\partial S/\partial \ln P)_V.$$

(*Hint:* use the Maxwell relation, Equation 3-20, and an equation from 3-19.)

21. Predict the residual entropy per mole at 0 K of a crystal containing the ion $S_2O_3^=$. Assume a tetrahedral arrangement of three oxygen and one sulfur atoms around the central sulfur atom, and assume that the energy of the different possible rotational orientations is the same, with the sulfur able to substitute for any of the three oxygen atoms.

22. (a) Show that Equation 3-26 implies that

$$\left(\frac{\partial S}{\partial T}\right)_V = \frac{C_V}{T} \qquad \text{and} \qquad \left(\frac{\partial S}{\partial V}\right)_T = \frac{1}{T}\left[\left(\frac{\partial E}{\partial V}\right)_T + P\right]$$

(b) Equate the mixed partial derivatives $\left(\frac{\partial^2 S}{\partial T \partial V}\right)$ obtained from the expression in (a) above to show that

$$\left(\frac{\partial E}{\partial V}\right)_T = -P + T\left(\frac{\partial P}{\partial T}\right)_V$$

(c) Insert the ideal gas law into the equation derived in (b) above to show that $\left(\frac{\partial E}{\partial V}\right)_T = 0$ for an ideal gas.

QUESTIONS FOR DISCUSSION

23. (a) Some authors have maintained that the second law of thermodynamics is inconsistent with the notion of biological evolution, since the second law predicts increasing disorder, yet evolution presents an increase in order and complexity. Discuss.

(b) The astronomer Sir Arthur Eddington called entropy "time's arrow," because it is the property that determines the forward direction of time. Discuss.

(c) Pollution of air and water can be thought of as a mixing process that increases entropy. Any process which is carried out to remove pollutants must increase entropy still further. Discuss.

(d) Describe recycling of used materials in terms of thermodynamics.

(e) Would you agree with Clausius: "The energy of the universe is a constant. The entropy of the universe tends towards a maximum."?

FURTHER READING

Morowitz, H. 1970. *Entropy for Biologists*. New York: Academic Press.

Schrödinger, E. 1945. *What is Life?* Cambridge: Cambridge University Press. Chapter 6 is an essay with the same title as the book. This small volume had an important philosophical influence on post-1945 molecular biology.

CHAPTER 4

Free Energy and Equilibrium

This and several later chapters are devoted to the development of thermodynamics and its applications to chemical and biochemical problems. During this exposition we will frequently ask two general questions.

The first question is, what relationships exist between the properties of a system at equilibrium? For example, how does the concentration of hydrogen ions in aqueous solution depend on the concentration of weak acids and bases? How much pressure should develop inside a membrane which is permeable to water but not to the protein dissolved in the water? What should be the voltage drop across a membrane permeable to potassium ions but not to chloride ions? Such wide-ranging problems can be solved by the methods of equilibrium thermodynamics.

The second general question is, when is a process spontaneous? Living systems are not at thermodynamic equilibrium, nor are many of the experimental systems studied by biochemists. The total of all processes occurring in such systems must be spontaneous, but many of the individual steps taken by themselves are not. For example, one of the results of biochemical metabolism is conversion of ADP and inorganic phosphate to ATP, a ubiquitous "energy storage" compound in living organisms. However, if we had only a solution of ADP and inorganic phosphate, quantitative conversion to ATP would *not* occur spontaneously. The role of thermodynamics in such problems is to determine the extent to which reactions such as the formation of ATP are thermodynamically unfavorable and to find how such unfavorable steps may be combined with spontaneous reactions to make the entire process spontaneous. An important biochemical example is the electron transfer chain in mitochondria, in which passage of electrons from one intermediate to another, and ultimately to molecular oxygen, results in oxidation of the metabolites that a cell uses for its energy needs. As we will see in Chapter 9, thermodynamics can help us know which of the electron transfer steps is sufficiently favorable thermodynamically to be able to force the formation of ATP.

The thermodynamic answer to questions of equilibrium and spontaneity proceeds first by establishing a "currency" for energy transactions. This is called the *free energy*. We will see that any process whose total free-energy change is negative can occur spontaneously, assuming the rates are fast enough and that temperature and pressure are the same at the beginning and the end. A favorable individual step with a large negative free-energy change can force occurrence of an unfavorable process that has a smaller positive free energy. We cannot overemphasize the importance of the free energy—and a closely related quantity called the *chemical potential*—for the successful application of thermodynamics to real problems.

4–1 The Gibbs and Helmholtz Free Energies

In previous chapters we stated the laws of thermodynamics and described a criterion for determining whether or not a process can occur spontaneously. According to this criterion, the entropy of an isolated system should be at its maximum when the system reaches equilibrium. In principle that solves the problem of spontaneity, but in practice the criterion is awkward to apply because we rarely study isolated systems. For example, suppose that we carry out a reaction in a flask that is in a water bath to hold the temperature constant and open to the atmosphere to keep the pressure constant. (These are the conditions for the vast majority of biochemical experiments.) Our system consists of the contents of the flask, but it is not isolated because it can (at the least) exchange heat and work with the surroundings. You can immediately see that it would be incorrect to apply the criterion of maximum entropy to a system of this kind.

We can better recognize the grounds for the difficulty with the entropy maximization criterion by noting that the differential in Equation 3-23, $dS = (dE + P\,dV)/T$, implies that S is a natural function of the variables E and V. Fixing E and V ensures that no heat or work can cross the boundary of the system, so the system must be isolated. Therefore, we can paraphrase the spontaneity criterion by saying that entropy is maximized at fixed values of its natural variables. The trouble is that fixing those variables is very awkward for many experiments, especially in biochemistry.

With this prelude it seems plausible that we should seek a new property whose natural variables are convenient to fix and which reaches an extremum (i.e., maximum or minimum) at equilibrium with fixed values of its natural variables. It turns out that there are two such properties, the *Gibbs free energy* (G) and the *Helmholtz free energy* (A). We will find that the natural variables of the Gibbs free energy are temperature and pressure, and that equilibrium is reached at fixed T and P when G attains its minimum value. The Helmholtz free energy has the natural variables T and V, and it reaches a minimum at equilibrium when these natural variables are fixed.

Gibbs Free Energy

The Gibbs free energy is defined by

$$G = H - TS \tag{4-1}$$

Notice that G must be a property, because H, T, and S are properties. We can show that the natural variables for G are T and P by deriving an expression for the differential dG. First, we take the differential of Equation 4-1:

$$dG = dH - T\,dS - S\,dT$$

Remembering that $H = E + PV$, we can replace dH by $dH = dE + P\,dV + V\,dP$ to get

$$dG = dE + P\,dV + V\,dP - T\,dS - S\,dT \qquad (4\text{-}2)$$

We next use the first law for a reversible change,

$$dE = dQ_{\text{rev}} + dW_{\text{rev}}$$

and substitute $dQ_{\text{rev}} = T\,dS$ from the second law, along with $dW_{\text{rev}} = -P\,dV$ (assuming only work of expansion) to get

$$dE = T\,dS - P\,dV$$

When this value is substituted for dE in Equation 4-2, the result is

$$dG = T\,dS - P\,dV + P\,dV + V\,dP - T\,dS - S\,dT$$

Several like terms cancel, leaving us with

$$dG = V\,dP - S\,dT \quad \text{(reversible process, only expansion work)} \qquad (4\text{-}3)$$ ◀

which gives us an expression for the change dG accompanying small pressure (dP) and temperature (dT) changes, *assuming the system remains at equilibrium* (because we used equations for a reversible process), and *only work of expansion occurs*. This important equation shows that P and T are the natural variables for G, because dG can be expressed simply in terms of dP and dT.

Helmholtz Free Energy

The Helmholtz free energy is defined by

$$A = E - TS \qquad (4\text{-}4)$$ ◀

To obtain an expression for dA in terms of two other differentials, we take the differential of Equation 4-4,

$$dA = dE - T\,dS - S\,dT \qquad (4\text{-}5)$$

and, as before, substitute $dE = T\,dS - P\,dV$ to get

$$dA = -P\,dV - S\,dT \quad \text{(reversible process)} \qquad (4\text{-}6)$$ ◀

which confirms that A is a natural function of T and V.

The Gibbs and Helmholtz free energies complete the list of primary thermodynamic properties, or *state functions* (Sec. 1-2). Table 4-1 summarizes these properties and their differentials.

TABLE 4-1 *Thermodynamic Properties*

PROPERTY	NATURAL VARIABLES	DIFFERENTIAL
Energy, E	$E(S,V)$	$dE = T\,dS - P\,dV$
Enthalpy, H	$H(S,P)$	$dH = T\,dS + V\,dP$
Entropy, S	$S(E,V)$	$dS = \frac{1}{T}\,dE + \frac{P}{T}\,dV$
Gibbs free energy, G	$G(T,P)$	$dG = V\,dP - S\,dT$
Helmholtz free energy, A	$A(T,V)$	$dA = -P\,dV - S\,dT$

In the same way that the equations for dE (3-16) and dH (3-17) provided many useful relationships, we can develop similar equations based on the differentials dG and dA. First, holding each variable in turn constant in Equation 4-3 yields

$$\left(\frac{\partial G}{\partial P}\right)_T = V \qquad \left(\frac{\partial G}{\partial T}\right)_P = -S \tag{4-7}$$

and from Equation 4-6

$$\left(\frac{\partial A}{\partial T}\right)_V = -S \qquad \left(\frac{\partial A}{\partial V}\right)_T = -P \tag{4-8}$$

● Two Maxwell relations, obtained by applying Euler's criterion (Equation 1-23) to Equations 4-3 and 4-6 (compare Equations 3-20 and 3-21) are

$$\left(\frac{\partial V}{\partial T}\right)_P = -\left(\frac{\partial S}{\partial P}\right)_T \qquad \left(\frac{\partial S}{\partial V}\right)_T = \left(\frac{\partial P}{\partial T}\right)_V \tag{4-9}$$

○ 4–2 The Free Energy Spontaneity Criterion

Our next objective is to show that the second law leads to the conclusion that the Gibbs free energy is minimized at equilibrium, when T and P are unchanged. The entropy maximization criterion applies only to an isolated system (which we called the isolated enclosure in Chap. 3) whereas the Gibbs free energy minimization requires a system open for heat and work exchange to keep T and P constant. Therefore, we need a combined system of the kind shown in Figure 4-1. The system of interest is enclosed in a heat bath large enough so that any energy flow which occurs will not appreciably affect the temperature or pressure of the bath. The entire bath is isolated from its surroundings. We will refer to the bath as the surroundings for the system of interest, but *we will apply the entropy maximization criterion to the total entropy change of the system plus the bath.* This is allowed because the system plus the bath form a new, isolated enclosure.

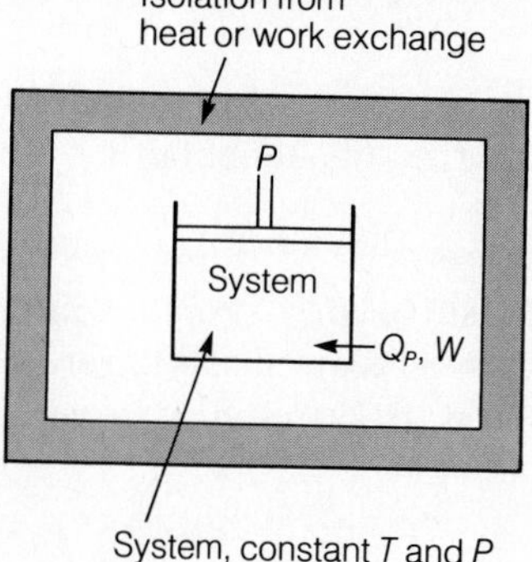

FIGURE 4-1
A system at constant T and P absorbs $Q_p = \Delta H_{sys}$ from the surroundings. The entropy change of the surroundings is $-Q_p/T = -\Delta H_{sys}/T$. Hence the total entropy change is $\Delta S_T = \Delta S_{surr} + \Delta S_{sys} = -\Delta H_{sys}/T + \Delta S_{sys}$. Because $T\,\Delta S_T$ must be positive for a spontaneous process, we conclude that $\Delta H_{sys} - T\,\Delta S_{sys} = \Delta G_{sys}$ must be negative for a spontaneous process.

The Gibbs Free Energy of a System at Constant T and P Decreases during a Spontaneous Process

Imagine a reversible flow of heat Q_P at constant pressure from the bath into the system. According to the second law ($dS = dQ_{rev}/T$) the entropy change of the bath (the surroundings) is

$$\Delta S_{surr} = \frac{-Q_P}{T} \tag{4-10}$$

Because the pressure is constant, $Q_P = \Delta H_{sys}$ (see Eq. 2-11), and Equation 4-10 becomes

$$\Delta S_{surr} = -\frac{\Delta H_{sys}}{T} \tag{4-11}$$

The total entropy change of the system plus the bath is

$$\Delta S_{total} = \Delta S_{surr} + \Delta S_{sys}$$

Substituting for ΔS_{surr} from Equation 4-11, we get

$$\Delta S_{total} = -\frac{\Delta H_{sys}}{T} + \Delta S_{sys} \tag{4-12}$$

At this point we have an equation whose right side depends *only on the properties of the system,* with the sign of the left side determined by the second law.

The right side of Equation 4-12 is $-\Delta G_{sys}/T$. You can see this by writing $G = H - TS$, and taking the difference when T is held constant, or

$$\Delta G = \Delta H - T\,\Delta S \qquad \text{(constant temperature)} \tag{4-13}$$ ◄

Division of Equation 4-13 by $-T$ yields

$$-\frac{\Delta G}{T} = -\frac{\Delta H}{T} + \Delta S$$

We can now replace the right side of Equation 4-12 by $-\Delta G_{sys}/T$ to get

$$\Delta S_{total} = -(\Delta G_{sys}/T)_{T,P}$$

in which we emphasize that temperature and pressure must be unchanged for this relationship to be valid. Another way to write this equation is

$$-T\,\Delta S_{total} = (\Delta G_{sys})_{T,P} \tag{4-14}$$

The second law tells us that ΔS_{total} must be positive for a spontaneous process, and T must always be positive (temperatures below absolute zero have no meaning). Therefore Equation 4-14 states that $(\Delta G_{sys})_{T,P}$ *must be negative for a spontaneous process,*

$$(\Delta G_{sys})_{T,P} < 0 \qquad \text{(spontaneous)} \tag{4-15}$$ ◀

when the temperature and pressure are held constant. This is the free energy spontaneity criterion that we sought.

EXERCISE 4-1

How do the requirements that T and P are constant enter the derivation of $\Delta G < 0$ for a spontaneous change?

ANSWER

Pressure must be constant to set $Q_P = \Delta H$, and temperature must be constant to derive $\Delta G = \Delta H - T\,\Delta S$. Notice that our derivation assumed that pressure is constant *throughout* the process in order to replace Q_P by ΔH. However, because G is a property, ΔG is independent of the path of a process. Therefore, Equation 4-15 applies to any process in which T and P are unchanged at the end from their values at the beginning of the process.

The Helmholtz Free Energy of a System at Constant T and V Decreases in a Spontaneous Process

The derivation of the spontaneity criterion for the Helmholtz free energy follows exactly the lines indicated for the Gibbs free energy, except that volume instead of pressure is constant, Q_V replaces Q_P, and E replaces H. The analogs to Equation 4-13 and 4-15 are

$$\Delta A = \Delta E - T\,\Delta S \qquad \text{(constant temperature)} \tag{4-16}$$ ◀

at constant temperature, and

$$(\Delta A_{sys})_{T,V} < 0 \qquad \text{(spontaneous)} \tag{4-17}$$ ◀

for a spontaneous change at constant temperature and volume.

With Equations 4-15 and 4-17 we have powerful criteria for addressing the problem of the direction of spontaneous changes and the nature of chemical equilibria in many real experimental systems. The great convenience of the free energy criteria is that they rely only on the properties of the system, not the surroundings. The surroundings serve only to fix the temperature and pressure (or volume), and need not be considered further in the thermodynamic analysis. This means, for example, that we should be able to predict whether a given chemical reaction will proceed, knowing only the properties of the chemicals involved.

4–3 *What is Free Energy?*

In spite of its simple definition, free energy is not a simple quantity to grasp conceptually. The Gibbs and Helmholtz free energies are defined mathematically by Equations 4-1 and 4-4 respectively. The Helmholtz free energy is the total energy E of the system minus a quantity TS that depends on the entropy or disorder and on the temperature. Similarly, the Gibbs free energy is the enthalpy H minus the same quantity TS. Both the term "free energy" and the definition as energy or enthalpy minus an entropy-dependent term imply that the free energy must be related somehow to the energy available to carry out a process. This surmise is verified by the following derivation, which shows that *the negative of the free energy change of a process is the maximum work that can be done by that process under specified conditions*. For example $-\Delta A = -\Delta E + T\,\Delta S$ turns out to be the maximum work that can be done by an isothermal process. In a mechanical system, the work done would be $-\Delta E$, but in systems containing many molecules this is replaced by $-\Delta E + T\,\Delta S$. To calculate the "free energy" available to do work, we have to add to the energy change $-\Delta E$ the term $T\,\Delta S$, which can be either positive or negative. We will see that this term reflects the ability of heat to be converted to work, a conversion which cannot be accomplished by a purely mechanical, isothermal system.

$-\Delta A$ is the Capacity to do Isothermal Work

Of the Gibbs and Helmholtz functions, the Helmholtz free energy has the simpler relationship to the maximum work. We consider a reversible, spontaneous process, and seek a relationship between dA and the maximum work that the process can do. When the temperature is constant, $dT = 0$ and the differential $dA = dE - T\,dS - S\,dT$ (Equation 4-5) becomes

$$dA = dE - T\,dS$$

For a reversible change, $T\,dS$ can be replaced by dQ_{rev}, yielding

$$dA = dE - dQ_{\text{rev}} \tag{4-18}$$

According to the first law, the right side of this equation is dW_{rev}, from which we conclude that

$$dA = dW_{rev} \qquad \text{(constant } T\text{)} \tag{4-19a}$$

or, for a finite process,

$$\Delta A = W_{rev} \qquad \text{(constant } T\text{)} \tag{4-19b}$$ ◀

The reversible work done *on* the system is ΔA, and the reversible work done *by* the system is $-\Delta A$. When the process is irreversible, for example in the expansion of a gas, the work it can do is less than $-dW_{rev} = P\,dV$, because $P_{ext} < P_{int}$ (see Fig. 2-4). Consequently, $-\Delta A$ *is the maximum work that can be done by an isothermal process.*

EXERCISE 4-2

A collagen fiber, bathed in a salt solution, can be stretched reversibly. The restoring force which the fiber exerts is $F = -kx$, in which x is the distance, and the force constant $k = 10^4$ dyne cm^{-1}. Calculate the change in the Helmholtz free energy of the fiber when it contracts isothermally from $x = 10$ to $x = 0$.

ANSWER

The work done by the fiber $(-W)$ is $-W = \int_{10}^{0} F\,dx = -k\int_{10}^{0} x\,dx = \frac{k(10)^2}{2} =$ 5×10^5 erg. Since $\Delta A = W$, $\Delta A = -5 \times 10^5$ erg. Notice that when the fiber does work, its Helmholtz free energy decreases.

EXERCISE 4-3

A rubber band exerts a restoring force $F = -kx$, in which $k = 10^4$ dyne cm^{-1}. When it contracts reversibly and isothermally from $x = 10$ to $x = 0$, it absorbs heat $Q_{rev} = 5 \times 10^5$ erg. Calculate ΔA, $T\,\Delta S$ and ΔE.

ANSWER

As in Exercise 4-2, $\Delta A = -5 \times 10^5$ erg. Setting $Q_{rev} = T\,\Delta S = 5 \times 10^5$ ergs, we get $\Delta E = \Delta A + T\,\Delta S = 0$. This exercise illustrates the properties of real rubber: to a first approximation, there is no change in the energy of rubber when it contracts. Rubber resists stretching because the polymer molecules which it contains must be stretched into unlikely shapes (see Chap. 14). The contractile force which rubber exerts arises because the polymer molecules have higher entropy in their more compact states. (Notice that $T\,\Delta S$ is positive for contraction.)

The maximum work that can be done by a system in an isothermal process is $-\Delta A = -\Delta E + T\,\Delta S$. Because of the entropy term, which can be positive or negative, ΔA can be smaller or larger than ΔE. As shown in Exercise 4-3, work can even be done in processes that have $\Delta E = 0$. Another example is the reversible isothermal expansion of an ideal gas, for which $\Delta E = 0$, but $\Delta S = nR \ln (V_2/V_1)$ (Equation 3-28) and $-\Delta A = nRT \ln (V_2/V_1)$. However, you should be very careful to understand that *energy is still required to do work.* The energy needed for the isothermal expansion of an ideal gas is supplied as heat by the bath that keeps the gas isothermal.

Why Free Energy Equals Reversible Work

If our system were a mechanical spring, it would not expand and do work unless it could reduce its energy in the process. The important difference characterizing a collection of molecules is that *entropy increase can make expansion or contraction spontaneous even when the energy of the system does not change.*

Of course, expansion requires a force. The force exerted by a mechanical system such as a spring or two molecules interacting is the derivative of the energy with respect to the distance *x:*

$$F = -\frac{dE}{dx}$$

When the energy is constant, there is no force. For a collection of many molecules at constant temperature the force depends on the derivative of the *free energy* with respect to distance. We can derive this relationship by beginning with the differential

$$dA = -P\,dV - S\,dT$$

At constant temperature

$$\left(\frac{\partial A}{\partial V}\right)_T = -P$$

Remembering that pressure is force per unit area, and setting the volume equal to the area a of a cylinder times its length x, so that $P = F/a$ and $dV = a\,dx$, we get

$$\frac{1}{a}\left(\frac{\partial A}{\partial x}\right)_T = -\frac{F}{a}$$

or

$$F = -\left(\frac{\partial A}{\partial x}\right)_T$$

Comparing this with the expression for the force exerted by a spring, we see that the *free energy replaces the mechanical energy*. Therefore the variation of the term $-TS$, as well as E, with distance gives rise to a mechanical force.

We can now understand why $-\Delta A$ is the maximum work that can be done by an isothermal process. $-\Delta A$ contains two terms: $-\Delta E$ is the contribution to the maximum work arising from reducing the energy stored in the system;

$T\,\Delta S$ is the maximum heat (Q_{rev}) that enters the system during the isothermal process. Because of conservation of energy, the sum of these must be the maximum work that can be done by the system. The term $T\,\Delta S$ does not apply to a simple spring, which is unable to convert heat ($T\,\Delta S$) to work ($-\,\Delta A$) at constant temperature. Evidently, the function A contains an elegant combination of the first and second laws.

ΔG is the Capacity to do Nonexpansion Work at Constant T and P

The Gibbs free energy change also has an important relation to work. Consider Equation 4-3, $dG = V\,dP - S\,dT$, which is valid for a reversible process in which only work of expansion occurs. Clearly, $\Delta G = 0$ for this process at constant temperature and pressure, because dP and dT are both zero. However, suppose that there is work in addition to $dW = -P\,dV$. For example, the chemical reaction in a battery can be used to do electrical work $-W$(other); the term "other" includes all work except work of expansion or compression. Then, instead of substituting $dE = T\,dS - P\,dV$ in Equation 4-2, we substitute $dE = T\,dS - P\,dV + dW(\text{other})$. Therefore Equation 4-3 becomes

$$dG = V\,dP - S\,dT + dW(\text{other}) \qquad (4\text{-}20)$$

in which dW(other) is all work besides work of compression done on the system. As always, the work $-W$(other) done by the system has the largest magnitude if the process is reversible. If T and P are constant, Equation 4-20 states:

$$\Delta G = W(\text{other})_{rev} = W(\text{other})_{max} \qquad (\text{constant } T \text{ and } P) \qquad (4\text{-}21)$$

If a process is irreversible, the magnitude of the work it can do is reduced:

$$|W(\text{other})|_{irrev} < |W(\text{other})|_{rev} \qquad (4\text{-}22)$$

Hence $-\Delta G$ *is the maximum work done by a process at constant T and P, exclusive of work of expansion.*

The Electrical Work of a Battery

A simple example of the use of Equation 4-21 is a galvanic cell or battery, diagramed schematically in Figure 4-2. The chemical reactions A → B and C → D cause a circulation of electrons, as we will discuss in Chapter 9. The maximum electrical work $-W$(other) that can be obtained is $-\Delta G$ for the reaction A + C → B + D. From the laws of electrostatics, electrical work is $nF\mathcal{E}$, in which nF in coulombs per mole is the charge circulated per mole of the reaction A + C → B + D and $\mathcal{E}$ is the potential difference between the two battery electrodes in volts. Consequently, W(other)—the work done *on* the system—is $-nF\mathcal{E}$, and Equation 4-21 gives

$$\Delta G = -nF\mathcal{E} \qquad (4\text{-}23)$$

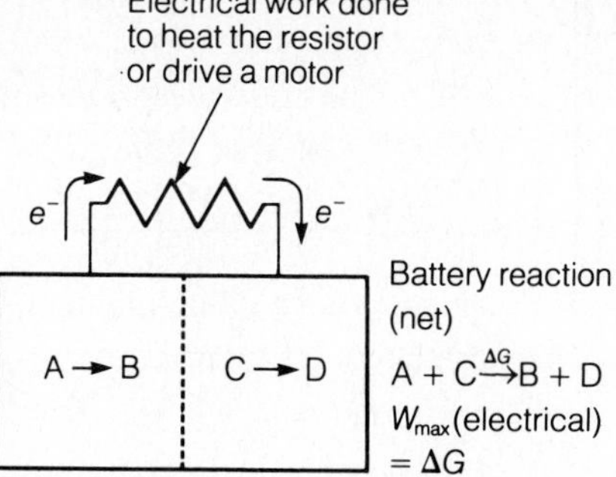

FIGURE 4-2
The free-energy change of the chemical reaction in a battery is equal to the maximum electrical work that can be done when the reaction is carried out reversibly, which means a very small current flow. Electrical work results from the flow of electrons e^- through a resistor or the windings of a motor.

for the relationship between $\mathscr{E}$ and ΔG when the battery is operated reversibly (very small current flow). This equation is central to the analysis of electrochemical reactions, Chapter 9.

In summary we see that both the Gibbs and Helmholtz free-energy changes have simple meanings in terms of the work that can be extracted from a process under defined conditions.

4-4 Free Energy and the Principle of Equilibrium

Chemical or physical equilibrium occurs when there is no time dependence in the amount of each substance present in a mixture. Suppose that a conversion between A and B

$$A \rightleftarrows B \tag{4-24}$$

is at equilibrium. This means that the amount of A and the amount of B do not change with time. It does *not* mean that no A is converted to B nor B to A, but rather that the rates of these processes are equal. All true equilibria are dynamic in that sense.

The principle of minimization of the Gibbs free energy at equilibrium means that if the process $A \rightleftarrows B$ is kept at constant temperature and pressure, and a very small number ($dn_{A \to B}$) of moles of A is converted to B, then the corresponding free-energy change, dG, is zero. The basis for this can be seen by referring to Figure 4-3, which shows the free energy schematically as a function of n_A. Because the free energy is minimized at equilibrium,

$$\frac{dG}{dn_{A \to B}} = 0$$

so dG must be zero for this equation to be obeyed. Be sure you understand why ΔG must be zero for a very small change in a system at equilibrium. The rest of the chapter depends on this principle.

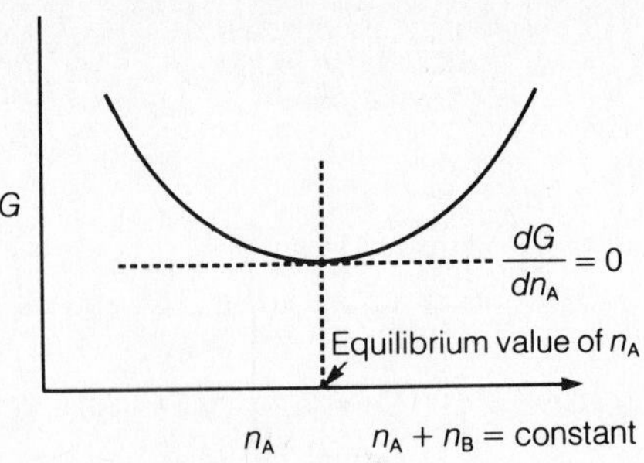

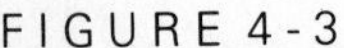
FIGURE 4-3
Substances A and B are in equilibrium through the reaction A $\rightleftarrows$ B. When the total number of moles $n_A + n_B$ is held constant, equilibrium occurs when the free energy is a minimum. This means that $dG/dn_A = 0$ at equilibrium.

The Stable Phase is the One of Lowest Free Energy

Before we can apply the free-energy minimization criterion to chemical reactions in gases or solutions, we must derive an expression for the dependence of the free energy on the pressure or concentration of A and B molecules. However, we can illustrate the principle with a particularly simple case, the melting of ice, where the free energy does not depend on concentration, because ice and water are both pure phases. The free energy for melting one mole of ice at constant T, by Equation 4-13, is

$$\Delta G = \Delta H_{\text{fus}} - T\,\Delta S_{\text{fus}} \tag{4-26}$$

in which ΔH_{fus} and ΔS_{fus} are the molar enthalpy and entropy changes for the reaction $H_2O(s) \rightarrow H_2O(l)$. Combining Equation 4-26 with $\Delta G = 0$ at equilibrium, we see that

$$\Delta H_{\text{fus}} - T_m\,\Delta S_{\text{fus}} = 0$$

T_m is the temperature at which ice and water are in equilibrium. Consequently

$$T_m = \frac{\Delta H_{\text{fus}}}{\Delta S_{\text{fus}}} \tag{4-27}$$

This equation shows that there is only one temperature (at a given pressure) at which ice and water will be in equilibrium, with a value given by the ratio of the heat of fusion to the entropy of fusion.

Figure 4-4 shows the temperature dependence of the free-energy change $\Delta G_{A\rightarrow B} = \Delta H - T\,\Delta S$. You can see graphically why there is only one temperature at which A (ice) and B (water) coexist. At all other temperatures, either A converts spontaneously to B ($\Delta G_{A\rightarrow B}$ is negative) or B converts spontaneously to A ($\Delta G_{A\rightarrow B}$ is positive). The general point of the figure is that changing a variable of state (temperature in this case) alters the relative stability of the phases present. Then a new phase increases at the expense of the old one.

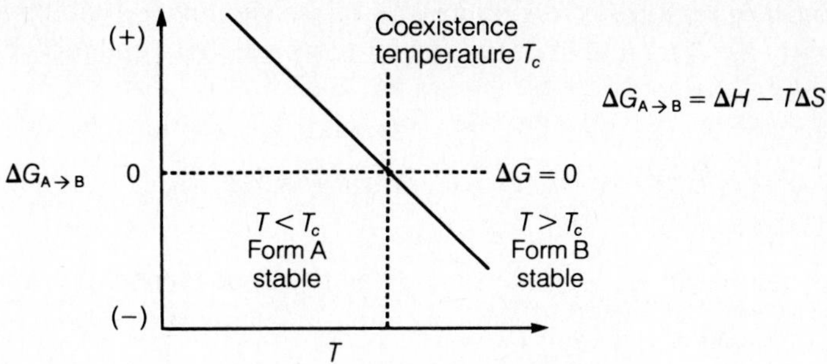

FIGURE 4-4
Temperature dependence of the free energy of the reaction A → B. At temperatures below the coexistence point T, the free energy $\Delta G_{A\to B}$ is positive and only A is stable. Above T_c only B is stable because $\Delta G_{A\to B}$ is negative. At $T = T_c$: $\Delta G_{A\to B} = 0$.

EXERCISE 4-4

The molar heat of fusion of ice is 5980 J mole^{-1} at T_m. Calculate the entropy of fusion at T_m. By what factor would the volume of one mole of an ideal gas have to be increased to achieve the same entropy increase?

ANSWER

$\Delta S_{fus} = \Delta H_{fus}/T_m = 5980$ J/273 K $= 21.9$ J K^{-1}. For an ideal gas, $\Delta S = nR \ln V_2/V_1$, so $V_2/V_1 = \exp(21.9/8.314) = 13.9$.

EXERCISE 4-5

Is the entropy of converting A to B positive or negative in Figure 4-4?

ANSWER

Positive. ΔS is minus the slope of ΔG versus T at constant P, as shown by Equation 4-7.

4–5 Free Energies of Ideal Gases and Ideal Solutions

Suppose that in the reaction A ⇄ B, both A and B are gases, whose partial pressures are P_A and P_B. (The *partial pressure* of one gas in a mixture is the portion of the total pressure contributed by that gas.) As we will shortly see, the free energy of an ideal gas depends on its pressure, because its entropy (not energy or enthalpy) depends on pressure. Therefore the free-energy change of the reaction A → B depends on the partial pressures of the two gases. This fact allows us to express the chemical equilibrium condition in terms of partial pressures, but requires that first we develop an expression for free energy as a function of pressure before we can treat chemical equilibrium.

The Free Energy of a Gas Depends on its Pressure

At constant temperature, Equation 4-3 gives the dependence of G on P as

$$dG = V\,dP$$

For an ideal gas, with $V = nRT/P$,

$$dG = nRT\,\frac{dP}{P}$$

so the free energy-change accompanying a pressure change is

$$\Delta G = nRT \ln \frac{P_2}{P_1} \tag{4-28}$$

It is often convenient to express not the *change* in free energy but rather the free energy itself. In such a case the free energy must be given relative to the free energy of some *standard state*—a state of defined T, P and concentration.

If $\overline{G}_i^\circ$ is the free energy per mole at the chosen standard pressure condition (P°) of one atmosphere, then the free energy of the ideal gas i at partial pressure P_i, with P measured in atmospheres and ΔG given by Equation 4-28 is

$$G_i = n_i\overline{G}_i^\circ + \Delta G_i$$

$$G_i = n_i\overline{G}_i^\circ + n_iRT \ln (P_i/P^\circ) = n_i\overline{G}_i^\circ + n_iRT \ln P_i \qquad \text{(ideal gas)} \tag{4-29}$$

◀

EXERCISE 4-6

Using Equations 3-9 and 4-28 show that the isothermal pressure dependence of ΔG is accounted for by the effect of P on ΔS.

ANSWER

According to Equation 3-9, at constant T, $\Delta S = nR \ln V_2/V_1 = -nR \ln P_2/P_1$. Hence $\Delta G = \Delta H - T\,\Delta S$ is equal to $nRT \ln P_2/P_1$, as predicted by Equation 4-28. This is as expected since H is independent of pressure at constant T, so $\Delta H = 0$ for isothermal pressure changes.

This exercise shows that *the free energy of an ideal gas varies with pressure because the entropy varies with pressure*. The result can also be expressed in terms of the mole fraction of the gas in a mixture. Suppose P_T is the total pressure of all the gases i,

$$P_T = \Sigma P_i$$

where the P_i's are the partial pressures. The mole fraction of gas i, $n_i/\Sigma n_i = X_i$, is

$$X_i = P_i/P_T$$

because each gas exerts a partial pressure P_i proportional to the number of moles it contributes (remember that $P_i = n_iRT/V$; $P_T = \Sigma n_i\ RT/V$). Equation 4-28 shows that the free energy change for diluting the gas from pressure P_T to P_i is

$$\Delta G_i = n_iRT \ln P_i/P_T$$

and, using $X_i = P_i/P_T$, we get for the free energy of dilution

$$\Delta G_i = n_iRT \ln X_i \tag{4-30}$$

The Free Energy of a Solute Depends on its Concentration

Ideal solutions are analogous to ideal gases in terms of physical model and the resulting thermodynamic equations. The dissolved molecules (the *solute*) are assumed to interact with each other exactly as they do with the *solvent* in which they are dissolved. In contrast to the absence of forces in the ideal gas, there is a uniformity of forces in the ideal solution. It turns out (as we will derive formally in Chap. 7) that Equation 4-30 gives the dependence of free energy on mole fraction for an ideal solution. (Since the term $n_iRT \ln X_i$ arises from the free energy of diluting one gas by another, or solute by solvent, it is not surprising that the same equation applies both to gases and solutions.) Letting ΔG_i in Equation 4-30 be the free-energy change to dilute the solute from the state with mole fraction 1 to mole fraction X_i, we can write the free energy of component i, G_i, as

$$G_i = n_i\overline{G}_i^\bullet + n_iRT \ln X_i \qquad \text{(ideal solution)} \tag{4-31}$$ ◀

in which $\overline{G}_i^\bullet$ is the molar free energy of pure substance i (in the state with unit mole fraction).

EXERCISE 4-7

Show that Equation 4-31 gives Equation 4-30.

ANSWER

When $X = 1$, $G = n\overline{G}^\bullet$. $\Delta G = G(X) - G(X = 1) = n\overline{G}^\bullet + nRT \ln X - n\overline{G}^\bullet = nRT \ln X$. In other words, one subtracts Equation 4-31 with $X = 1$ from the same Equation with X retained as a variable.

EXERCISE 4-8

With Equation 4-31, show that the total free energy of mixing pure components to make a solution with mole fraction X_i of each component i is given by $-T\,\Delta S_{\text{mix}}$, in which ΔS_{mix} is given by Equation 3-32 for the entropy of mixing ideal gases at constant T.

ANSWER

By definition, the free energy of mixing is the free energy of the solution minus the free energy of the pure compounds, or

$$\Delta G_{\text{mix}} = \sum_i G_i - \sum_i n_i \bar{G}_i^\bullet$$

Therefore

$$\begin{aligned}\Delta G_{\text{mix}} &= \Sigma(n_i\bar{G}_i^\bullet + n_iRT\ln X_i - n_i\bar{G}_i^\bullet)\\ &= \Sigma\, n_iRT\ln X_i\end{aligned}$$

or, with $n_i = X_i n$ ($n = \Sigma n_i$),

$$\Delta G_{\text{mix}} = nRT\,\Sigma X_i \ln X_i \qquad (4\text{-}32)$$ ◀

This expression is $-T$ times the value of ΔS_{mix} in Equation 3-32. Notice that ΔG_{mix} must be negative because each X_i is a fraction, and hence all the logarithms are negative. Thus all ideal solutions mix spontaneously at constant T and P.

These examples should be sufficient to convince you that the *free energy of an ideal gas or ideal solution varies with partial pressure or mole fraction because the entropy depends on partial pressure or mole fraction.*

The Standard State has Concentration Equal to One

Equations 4-29 and 4-31 show that free energy varies with n_iRT times the natural logarithm of a concentration, either partial pressure or mole fraction. The equations also contain an important convention, which we will repeatedly meet in subsequent sections. When the concentration unit X_i or P_i is one, the natural logarithm term is zero. At that concentration the free energy is $G = n_i\bar{G}_i^\circ$, and $\bar{G}_i^\circ$ is the free energy per mole in the special state with unit concentration. This special concentration is called the *standard state;* $\bar{G}_i^\circ$ is the molar free energy of compound i in the standard state. We will later encounter other thermodynamic properties of the standard state, such as S° and H°. Also, the free energy change of chemical reactions is often expressed as the difference (ΔG°) in the free energy of products and reactants in their standard states. When the standard state happens to be a pure solid or liquid, we represent the thermodynamic properties by the special symbols $G^\bullet$, $H^\bullet$, $S^\bullet$, etc.

It is important that you realize that the standard state for a compound can be chosen arbitrarily, but that the choice is equivalent to the choice of concentration units. For example, for liquid water you might choose mole fraction as a concentration unit, so the standard state would be pure water, with

$X_{H_2O} = 1$. On the other hand, for water vapor you might choose partial pressure as a concentration unit, and the standard state would be water vapor at 1 atm partial pressure. The concentration unit fixes the standard state, because the latter is defined to be the state of unit concentration. The following paragraphs describe other possible choices of concentration units.

Units of Concentration

Most studies of dilute aqueous solutions do not express concentrations in mole fractions, using instead the *molarity,* defined as the moles of solute per liter of solution. If the solution is dilute, consisting of solvent A and solute B, then the mole fraction of B, $X_B = n_B/(n_A + n_B)$ is approximately

$$X_B \approx \frac{n_B}{n_A} \tag{4-33}$$

The molar concentration of B, C_B, is the number of moles of B per liter of solution,

$$C_B = \frac{n_B}{\text{liter of solution}} \tag{4-34}$$

Replacing n_B in Equation 4-34 by $n_B = X_B n_A$ obtained from Equation 4-33, we find

$$C_B = \frac{n_A}{\text{liter of solution}} X_B \tag{4-35}$$

The number of moles of A per liter of *solution* in the right side of Equation 4-35 can be approximated for dilute solution by the number of moles per liter of *solvent.* Letting ρ_A be the solvent density in grams per cm³ and M_A its molecular weight, we have

$$\begin{aligned} 1 \text{ liter solvent} &= 1000 \text{ cm}^3 \times \rho_A \frac{g}{\text{cm}^3} \times \frac{1}{M_A \text{g mol}^{-1}} \\ &= \frac{1000\ \rho_A \text{ mol}}{M_A} \end{aligned} \tag{4-36}$$

Inserting this result in Equation 4-35 provides us with the relationship

$$C_B = \frac{1000\ \rho_A}{M_A} X_B \tag{4-37}$$

Thus in dilute aqueous solutions, the *molar concentration is proportional to the mole fraction.*

EXERCISE 4-9

Calculate the mole fraction in a 0.01 M aqueous solution.

ANSWER

$$\begin{aligned} X_B &= \frac{M_A}{1000\ \rho_A} C_B = \left(\frac{18 \text{ g/mol}}{1000 \times 1 \text{ g/liter}}\right) \times 0.01 \text{ mol liter}^{-1} \\ &= 1.8 \times 10^{-4} \end{aligned}$$

Clearly, the number of moles of solute (B) is much smaller than the number of moles of solvent (A), so the approximations made to derive Equation 4-37 are valid in this concentration range.

With the fact that the molar concentration is proportional to the mole fraction at high dilution, we can convert Equation 4-31, which involves the mole fraction, to its equivalent in terms of molar concentration *C*:

$$G_i = n_i\overline{G}_i^\circ + n_iRT \ln (C_i/C^\circ) = n_i\overline{G}_i^\circ + n_iRT \ln C_i \quad (4\text{-}38)$$ ◀

In this equation $\overline{G}^\circ$ is the free energy per mole in a standard state of $C^\circ = 1$ mole per liter concentration.

Equation 4-38 is central to our further development of the thermodynamic theory of chemical equilibria in solution. It reveals that the free energy of a dissolved substance depends not only on the number of moles *n*, but also on the concentration *C* of the dissolved material.

○ 4–6 The Chemical Potential

To summarize our development thus far, we have found that the free energy of a system is minimized at equilibrium, so that ΔG for a small change at constant *T* and *P* must be zero in a system at equilibrium. In the preceding section we derived a simple formula for the dependence of free energy on the concentration of a substance. We are nearly ready to apply these results to the study of systems at equilibrium.

However, before proceeding we must introduce one more important property, the *chemical potential*. Free energy is an *extensive* quantity, because it depends on the number of moles of substance present. The chemical potential is an *intensive* quantity based on the free energy; it tells us how much the free energy changes per mole of substance added to the system. In this section we will see that equilibrium can be described by equations involving the chemical potential. It is convenient to use an intensive quantity, the chemical potential, rather than the free energy for the same reason that it is convenient to use the intensive quantity pressure to describe equilibration with the atmospheric force rather than the force itself, which is an extensive quantity. The chemical potential of a substance depends on its nature and concentration, not on properties irrelevant for equilibrium such as the size of the reaction flask.

The Chemical Potential is Equal in Two Phases at Equilibrium

Suppose we have two immiscible liquids, such as water and chloroform ($CHCl_3$). They form two separate phases when placed in the same container. Now, let us add solute A that is soluble in both liquids (Figure 4-5). At equilibrium, some of A will be in the water phase, and some in the organic

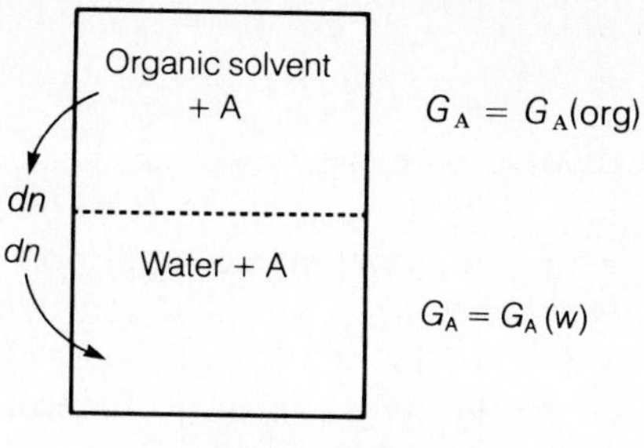

Condition for equilibrium:
$\Delta G = 0$ for transfer of *dn*

FIGURE 4-5
Solute A is soluble in both water and an organic solvent. $\Delta G(\text{org})$ is the free energy change when $\Delta n(\text{org})$ moles of solute A are added to the organic phase, and $\Delta G(w)$ is the same quantity for the water phase. When *dn* moles of A are transferred from the organic phase to the water phase, the total free energy change is $-[\Delta G(\text{org})/\Delta n(\text{org})]dn + [\Delta G(w)/\Delta n(w)]dn$ when T and P are held constant. At equilibrium this quantity must be zero. Hence $\Delta G(\text{org})/\Delta n(\text{org}) = \Delta G(w)/\Delta n(w)$ at equilibrium. Calling $\Delta G/\Delta n$ the *chemical potential,* μ, of the solute, we conclude that the chemical potential must be the same in both phases. When it is not, transfer of *dn* moles can increase or decrease the free energy G, depending on whether the transfer moves the system away from or toward equilibrium (see Fig. 4-3).

phase; our problem is to describe the equilibrium quantitatively. The condition of minimum free energy requires (at constant T and P) that when we transfer a small number of moles, dn, of solute A from the organic phase to the water phase (or the other way around), then the free energy change must be zero. Adding dn moles of solute to the water (w) phase will change its free energy by $[\Delta G(\text{w})/\Delta n(\text{w})]dn$ where $\Delta G(\text{w})/\Delta n(\text{w})$ is the change in the free energy of the solute in the water phase per mole of solute. Similarly, removing dn moles of solute from the organic (org) phase will change its free energy by $-[\Delta G(\text{org})/\Delta n(\text{org})]dn$. $\Delta G_T = 0$ is the sum of these terms, or

$$\Delta G_T = 0 = \left[\frac{\Delta G(\text{w})}{\Delta n(\text{w})}\right] dn - \left[\frac{\Delta G(\text{org})}{\Delta n(\text{org})}\right] dn$$

Consequently, at equilibrium, with constant T and P,

$$\left[\frac{\Delta G(\text{w})}{\Delta n(\text{w})}\right]_{T,P} = \left[\frac{\Delta G(\text{org})}{\Delta n(\text{org})}\right]_{T,P} \tag{4-39a}$$

Converting to the notation of derivatives, which we will employ from now on, we may write

$$\left[\frac{\partial G(\text{w})}{\partial n(\text{w})}\right]_{T,P} = \left[\frac{\partial G(\text{org})}{\partial n(\text{org})}\right]_{T,P} \tag{4-39b}$$

These equations state that at equilibrium the free energy change per mole of solute added is the same in both phases; this quantity is like a "potential" that is balanced between phases at equilibrium.

The derivative $(\partial G/\partial n)_{T,P}$ is of central importance in describing chemical equilibria. It is called the *chemical potential,* μ. In more general notation,

$$\mu_i = \left(\frac{\partial G}{\partial n_i}\right)_{T,P,n_j} \tag{4-40}$$ ◀

which is a partial derivative with the following meaning: μ_i, the chemical potential of component i, is the change in the total free energy per mole of component i added, when the temperature, total pressure, and number of moles n_j of all components besides i are held constant. In the experiment diagramed in Figure 4-5, n_i is the number of moles of solute and n_j is the number of moles of solvent in the water or in the organic phase.

Equation 4-39 shows that at equilibrium the chemical potential of solute A must be the same in both phases,

$$\mu_A(\mathrm{w}) = \mu_A(\mathrm{org}) \tag{4-41}$$

It is a very important general principle of phase equilibria that the chemical potential of a given substance must be the same in all phases.

The Chemical Potential Varies with Partial Pressure or Concentration

Equation 4-29, $G_i = n_i\overline{G}_i^\circ + n_iRT \ln P_i$, expresses the free energy of an ideal gas as a function of pressure. Assuming for the moment that there is only one component in the gas, this equation becomes

$$G = n_i\overline{G}_i^\circ + n_iRT \ln P$$

because the only contributor to G is G_i, and $P = P_i$. Now, holding T and P constant as prescribed by Equation 4-40, and differentiating with respect to n_i, we obtain

$$\left(\frac{\partial G}{\partial n_i}\right)_{T,P} = \mu_i = \overline{G}_i^\circ + RT \ln P_i \tag{4-42}$$

Giving $\overline{G}_i^\circ$ (the free energy per mole in the standard state of $P_i = 1$) the symbol μ_i°, we obtain an important equation for the chemical potential (or free energy per mole) of an ideal gas

$$\mu_i = \mu_i^\circ + RT \ln P_i \qquad \text{(ideal gas)} \tag{4-43a}$$ ◀

This equation is also valid for a mixture of ideal gas components, as is demonstrated formally in Problem 4-31. The effect of pressure on *real* substances must be investigated experimentally, as illustrated by the life's work of Percy Bridgman (page 144).

Figure 4-6 shows Equation 4-43a for μ_i plotted graphically with the constraints of Equation 4-40. As more moles n_i of gas i are added to a gas mixture, holding constant the total pressure, the temperature, and the number of moles of all other components n_j, the partial pressure P_i rises. Therefore the slope of the plot of G versus n_i, which is $\mu_i^\circ + RT \ln P_i$, must increase. The chemical potential μ_i is the value of the slope at a particular partial pressure P_i. Notice that the chemical potential can be positive or negative, depending on μ_i° and P_i, just as an electrical potential can be positive or negative.

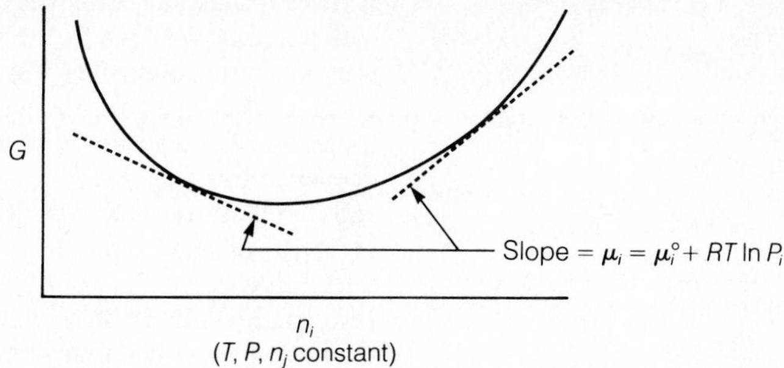

FIGURE 4-6
Schematic diagram of the free energy of an ideal gas. Equation 4-43*a* gives $G_i = n_i\mu_i^\circ + n_iRT \ln P_i$. As the number of moles of *i* is increased in the mixture, with *T, P,* and the number of moles n_j of all other components held constant, the free-energy contribution G_i changes because of the increase in both n_i and P_i. The slope $\left(\frac{\partial G}{\partial n_i}\right)_{T,P,n_j}$ is the chemical potential μ_i. Differentiating *G* (Problem 4-31) gives $\mu_i = \mu_i^\circ + RT \ln P_i$. The chemical potential μ_i can be positive or negative depending on the values of μ_i° and P_i. However, it must increase as P_i increases, so the slope must increase as P_i increases.

For an ideal solution, Equations 4-31 and 4-40 can be combined to yield

$$\mu_i = \mu_i^\bullet + RT \ln X_i \qquad \text{(ideal solution)} \qquad (4\text{-}43b)$$ ◀

in terms of the mole fraction, or

$$\mu_i = \mu_i^\circ + RT \ln C_i \qquad \text{(dilute ideal solution)} \qquad (4\text{-}43c)$$ ◀

for dilute solutions (C_i is the molar concentration). In each of these cases, μ_i° is the chemical potential in the standard state, which is always the state in which the concentration variable P_i, X_i, or C_i is unity. (When $C_i = 1$, ln $C_i = 0$, so $\mu_i = \mu_i^\circ$). Since P_i, X_i, and C_i are unity under different conditions the value of μ_i° depends on the concentration variable chosen.

Equation 4-43*c*, which is plotted graphically in Figure 4-7, indicates that for a (dilute) ideal solution the chemical potential of solute *i* varies linearly with ln C_i. The slope of the line is *RT*, and μ_i crosses the value $\mu_i = \mu_i^\circ$ for the standard state with $C_i = 1$.

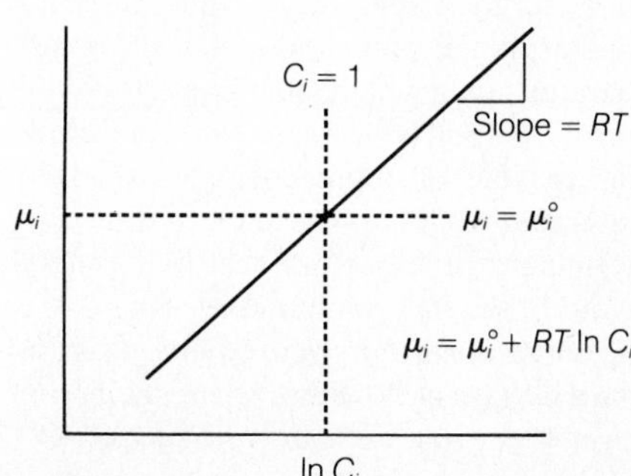

FIGURE 4-7
Graphical illustration of the linear variation of μ with ln C_i. μ_i takes on the value μ_i° in the *standard state* in which $C_i = 1$. The equation $\mu_i = \mu_i^\circ + RT \ln C_i$ applies to solutions only in the ideal limit of a very dilute solution.

PERCY W. BRIDGMAN

1882–1961

With much of the singleminded determination and the practical resourcefulness of a Columbus or a Magellan, Percy Bridgman charted a new world—the world of matter under high pressure. As a young faculty member at Harvard, Bridgman began his explorations of properties of elements and compounds at pressures far above the one atmosphere of our everyday experience. Though his first papers in 1908–1909 described experiments at pressures not much greater than had been attained by other investigators (6500 atmospheres or 6700 kg cm^{-2}), they mentioned a practical discovery that led shortly to the attainment of much higher pressures—a self-sealing gasket of rubber or soft metal. The gasket was restrained on the low-pressure (outer) side of the pressure vessel by a fixed surface whose area was somewhat smaller than that on the inner side. The effect was that the high pressure itself tightened the gasket. Only the strength of the steel casing of the vessel limited the ultimate pressure. This advance, like the support given by the Spanish crown to her admirals, opened a voyage of discovery.

Among Bridgman's discoveries were new polymorphs such as the high-pressure forms of ice (see Fig. 4-13), and many unexpected thermodynamic, electrical, and viscoelastic properties, which deepened our understanding of the solid and liquid states. His work led to the synthesis of new materials, including artificial diamonds. For over 50 years Bridgman published an average of more than five papers per year, many crammed with data and detail on new experimental methods. Typical titles were, "The Resistance of 72 Elements, Alloys, and Compounds to 100,000 kg cm^{-2}," "The Effect of Pressure on the Viscosity and Resistance of 19 Metals to 30,000 kg cm^{-2}," and "Polymorphic Transitions of 35 Substances to 50,000 kg cm^{-2}." Through continued development of his techniques and the availability of stronger steels, Bridgman extended attainable pressures first to 30,000 atmospheres, and then to 100,000 atmospheres, and finally in some systems to an estimated 400,000 atmospheres. Later work by others, using shock waves, has produced transient pressures of several million atmospheres.

Bridgman was a determined individualist. He would arrive in almost any weather on his bicycle when the laboratory opened at 8 A.M. He worked on his current experiment either alone or only with his machinist and research assistant. Over the years a total of 14 doctoral students worked with him, but only two of his more than 260 papers and 13 books were written with a co-author. Bridgman derived great enjoyment from personally operating the pumps and measuring apparatus, and from machining parts, blowing glass, purifying chemicals, and growing crystals of unprecedented size. His attention to detail was such that he determined the trajectory that flying metal would take in his laboratory if his pressure apparatus were to give way; he marked this path with a string, and leapt over the string each time he passed so that he would never be in the zone of danger.

In his singleness of purpose, Bridgman avoided nearly all service on faculty committees, and a long-time colleague reported that during Bridgman's tenure in two chairs of physics at Harvard he was never once seen at a meeting of the faculty. He took little interest in classroom teaching. In odd moments and during stays at his summer home in New Hampshire, he enjoyed music, chess, handball, photography, and gardening; all were pursued with concentration, perfectionism, and a spirit of fierce competition. Bridgman was a wild driver, presumably in part because he wanted nothing to go faster or ahead of him.

In addition to his studies of high pressure, Bridgman contributed ideas to the philosophy of science. Today these form the conceptual framework used by many scientists in their work. Bridgman was concerned that ambiguities are often inherent in defining scientific ideas. These ambiguities of definition can lead to great confusion. The concept of length, for example, has different meanings when measurements are made on earth and when they are made from astronomical observations. Thus concepts such as length, defined either verbally or by an equation, can be ambiguous or obscure. Bridgman's solution to this problem was to insist on what he called "operational definition" of concepts:

> "In general, we mean by any concept nothing more than a set of operations; the concept is synonymous with the corresponding set of operations . . . If a specific question has meaning, it must be possible to find operations by which answers may be given to it. It will be found in many cases that the operations cannot exist, and the question therefore has no meaning."

Bridgman's idea becomes concrete if we consider an example such as heat. In the eighteenth and nineteenth centuries, scientists argued over whether heat is a fluid (caloric) that can be squeezed from matter by work, or whether it consists of rapid atomic motions. But neither "caloric" or "atomic motions" are operational definitions. A possible operational definition of heat is the following: Heat crosses a boundary of a system if the temperature of the surroundings changes. This definition is operational because it consists in an operation (measuring the temperature change of the surroundings).

As useful as operational definitions are for avoiding ambiguities, they fail to give us a feeling for what the essence of a concept is. The operational definition of heat does not satisfy our yearning to understand macroscopic phenomena in terms of atoms. Similarly, the operational definition of an electron as a current from cathode to anode in a given potential difference does not correspond to our feelings of what an electron really is. It is obvious that working scientists do not think of heat, electrons, and many other concepts only in terms of such operational definitions. As fuller knowledge is acquired in a given area (such as thermodynamics) scientists tend to think more in terms of a picture or model of the concept and less in terms of operational definitions. Bridgman's warning to us, however, is that if we use scientific concepts that are not operationally defined, we stand the chance of drifting into statements and interpretations without meaning.

Bridgman also pioneered a social idea now gaining wider acceptance: the right of a person to a dignified death when faced with irreversible or painful degeneration. After he was diagnosed to have Paget's disease in his 80th year he wrote,

> ". . . the disease has run its normal course, and has now turned

into a well-developed cancer for which apparently nothing can be done . . . In the meantime there is considerable pain, and the doctors here do not offer much prospect that it can be made better . . . I would like to take advantage of the situation in which I find myself to establish a general principle, namely, that when the ultimate end is as inevitable as it now appears to be, the individual has a right to ask his doctor to end it for him."

Bridgman was unable to make any such arrangement. Finding that he was losing control of his limbs, he left a note:

> "It isn't decent for Society to make a man do this thing himself. Probably this is the last day I will be able to do it myself. P.W.B."

Then, walking into the garden of his summer home, he killed himself with his own gun.

EXERCISE 4-10

Combine Equations 4-43*c* and 4-41 to obtain an expression for the ratio of concentrations of solute A in the organic and aqueous phases in Figure 4-5.

ANSWER

$$\mu_A(\mathrm{w}) = \mu_A(\mathrm{org}), \text{ so } \mu_A^\circ(\mathrm{w}) + RT \ln C_\mathrm{w} = \mu_A^\circ(\mathrm{org}) + RT \ln C_\mathrm{org} \tag{4-44a}$$

Therefore

$$RT(\ln C_\mathrm{w} - \ln C_\mathrm{org}) = -[\mu_A^\circ(\mathrm{w}) - \mu_A^\circ(\mathrm{org})] \tag{4-44b}$$

Dividing by RT and taking the exponential

$$\frac{C_\mathrm{w}}{C_\mathrm{org}} = \exp\left\{-\frac{\mu_A^\circ(\mathrm{w}) - \mu_A^\circ(\mathrm{org})}{RT}\right\} \tag{4-44c}$$

The quantity on the right side depends on chemical potentials in the standard state and is a constant at a given temperature and pressure. It is called the *partition coefficient*, K_P

$$\frac{C_\mathrm{w}}{C_\mathrm{org}} = K_P \tag{4-45}$$

thus

$$K_P = \exp\left\{-\frac{\mu_A^\circ(\mathrm{w}) - \mu_A^\circ(\mathrm{org})}{RT}\right\} \tag{4-46}$$

The Physical Meaning of the Chemical Potential

We can compare the chemical potential of a solute to the chemical potential of a mass in a gravitational field. The difference $\mu_A^\circ(\mathrm{w}) - \mu_A^\circ(\mathrm{org})$ in Equation 4-46 is a measure of the difference in chemical potential of the solute in the two phases. It is analogous to the difference in potential energy of the mass at two heights above the ground. The larger the difference, the more strongly the solute is attracted to the organic phase, just as the mass releases more energy when dropped from greater heights. This can be seen in Equation 4-44, in which a large difference in the potentials means that the ratio of c_w to c_{org} is small.

Students encountering equations like 4-46 for the first time often wonder why the constant K_P should depend on the chemical potentials in the standard states (μ°'s) rather than on the μ's themselves. One answer to this question is the following: At equilibrium, the chemical potentials $\mu_A(w)$ and $\mu_A(org)$ that describe the distribution of A are equal (see Eq. 4-41). This means that the free-energy change per mole of solute added is the same in both phases. But if the solute A were at unit concentration in both solvents, then the free-energy change per mole added would not be the same in both phases. The difference would be equal to the right-hand side of Equation 4-44*b*. This difference in standard chemical potentials is a measure of the tendency of the solute to distribute unevenly in the solvents, and is thus related to K_P by Equation 4-46.

A second feature of Equation 4-46 that often puzzles students is that the partition coefficient K_P is expressed in terms of the chemical potential of a standard state, which has been set *arbitrarily* at unit concentration. Does this mean that the value of K_P is also arbitrary? Wouldn't one scientist measuring K_P always find the same experimental results no matter what he assumed about standard states? The answer is that K_P and μ° are both arbitrary for the same reason: the choice of concentration units is arbitrary. For example, suppose that you decide to express concentration in the aqueous phase in mg per ml, and in mol per liter in the organic phase. Then, K_P, the ratio of the two concentrations, will be different from the value you would get if you used mg per ml in both cases. Similarly, the standard state, dictated by the concentration units, would be 1 mg per ml in the aqueous phase and 1 mol per liter in the organic phase. Two scientists could measure different values of K_P for the same system, and therefore calculate different values of $\mu_A^\circ(w) - \mu_A^\circ(org)$, since they used different concentration units. However, *if they convert to a common language by using an agreed set of concentration units, their results should be the same.*

The Total Differential of G for Systems with more than one Component

The total differential for the free energy involves the chemical potentials of all components present. We can see this by writing the functional form of G

$$G = G(P, T, n_1, n_2, \ldots)$$

and taking the total differential

$$dG = \left(\frac{\partial G}{\partial P}\right)_{T,n_1,n_2\ldots} dP + \left(\frac{\partial G}{\partial T}\right)_{P,n_1,n_2\ldots} dT + \left(\frac{\partial G}{\partial n_1}\right)_{T,P,n_2\ldots} dn_1 + \cdots$$

If we recall that $(\partial G/\partial P)_T = V$ and $(\partial G/\partial T)_P = -S$ for systems with fixed composition n_i, we can substitute V and $-S$ for the partial derivatives in the first two terms. We can then substitute for the other partial derivatives using Equation 4-40, which gives

$$dG = V\,dP - S\,dT + \sum_i \mu_i\,dn_i \tag{4-47}$$

We will make extensive use of this equation in this and subsequent chapters.

● EXERCISE 4-11

Derive a Maxwell relation that corresponds to Equation 4-47.

ANSWER

Using the expression $G = G(P,T,n_i)$, the Maxwell relations are obtained by equating the mixed third partial derivatives of G obtained by different orders of differentiation. Thus, with

$$dG = \left(\frac{\partial G}{\partial P}\right) dP + \left(\frac{\partial G}{\partial T}\right) dT + \left(\frac{\partial G}{\partial n_i}\right) dn_i,$$

we can, to take an example, equate

$$\frac{\partial^3 G}{\partial n_i \partial T \partial P} = \frac{\partial^3 G}{\partial n_i \partial P dT} = \frac{\partial^3 G}{\partial T \partial P \partial n_i}$$

Since

$$\left(\frac{\partial G}{\partial P}\right) = V, \ \left(\frac{\partial G}{\partial T}\right) = -S \text{ and } \left(\frac{\partial G}{\partial n_i}\right) = \mu_i$$

(see Equation 4-47) we have

$$\frac{\partial}{\partial n_i}\left(\frac{\partial V}{\partial T}\right) = \frac{-\partial}{\partial n_i}\left(\frac{\partial S}{\partial P}\right) = \frac{\partial}{\partial T}\left(\frac{\partial \mu_i}{\partial P}\right)$$

Other relations are possible. All partial derivatives are taken with the other variables constant.

○ 4–7 *Chemical Equilibrium*

The concept of the chemical potential is extremely useful for solving equilibrium problems, including chemical equilibrium. A chemical reaction at constant temperature and pressure is at equilibrium when the Gibbs free energy reaches a minimum. This will happen when the free-energy change due to consumption of a small amount of the reactants exactly balances the free-energy change due to appearance of a small amount of the products.

Let a general chemical reaction be represented by

$$\Sigma \nu_i r_i \rightleftarrows \Sigma \nu_i p_i \tag{4-48}$$

in which r_i are reactants, p_i are products, and ν_i are the *stoichiometric coefficients* of the reaction. For the specific example,

$$NO + \tfrac{1}{2}O_2 \rightleftarrows NO_2$$

$\nu_{NO} = 1$, $\nu_{O_2} = \frac{1}{2}$, and $\nu_{NO_2} = 1$. The free-energy change (dG), when a small amount of reactants is consumed at constant T and P, is, by Equation 4-47,

$$dG(\text{reactants}) = -\Sigma\,\mu_i(\text{reactants})(\nu_i dn_i)$$

in which dn_i is the number of moles consumed of a reactant with $\nu_i = 1$. A similar equation with opposite sign applies to production of products,

$$dG(\text{products}) = \Sigma\,\mu_i(\text{products})(\nu_i dn_i)$$

The free-energy change of the reaction per mole is

$$\Delta G = \frac{dG(\text{products}) + dG(\text{reactants})}{dn_i}$$

Therefore we conclude that

$$\Delta G = \Sigma\nu_i\mu_i(\text{products}) - \Sigma\nu_i\mu_i(\text{reactants}) \qquad (4\text{-}49)$$ ◄

At equilibrium $\Delta G = 0$. Equation 4-49 applies to all reactions at equilibrium in a single phase. It states the condition that a very small transfer of substance from one form to another shall produce zero free-energy change at equilibrium.

The Equilibrium Constant is Related to the Standard Free-Energy Change: $\Delta G^\circ = -RT \ln K$

Building on the groundwork of the last two-and-one-half chapters, we can now derive one of the most remarkable results of all thermodynamics: *the equilibrium constant for a reaction is related to the standard free-energy change of the reacting species*. For the present we will derive this result only for a reaction among ideal gases. We will generalize it to molecules in solution and to nonideal substances in the rest of this chapter and in Chapter 7. The significance of this result is that *the equilibrium constant for a reaction can be computed from tabulated values of standard free energies*. Thus the position of a chemical equilibrium can be computed in terms of properties of the isolated chemicals. It is possible even to compute the equilibrium constant for reactions that have never been examined experimentally, and for reactions difficult or impossible to carry out.

When the reactants and products are ideal gases, Equation 4-49 combines with Equation 4-43 to become

$$\Sigma\nu_i\mu_i^\circ\,(\text{products}) - \Sigma\nu_i\mu_i^\circ\,(\text{reactants}) + RT\,\Sigma\nu_i \ln P_i\,(\text{products}) - RT\,\Sigma\nu_i \ln P_i\,(\text{reactants}) = 0 \qquad (4\text{-}50a)$$

We give the special symbol ΔG° to the difference in free energy of products and reactants under standard conditions,

$$\Delta G^\circ = \Sigma \nu_i \mu_i^\circ(\text{products}) - \Sigma \nu_i \mu_i^\circ(\text{reactants})$$

and call ΔG° the *standard free-energy change* of the reaction. Because μ_i° is the free energy per mole in the standard state, $\Sigma \nu_i \mu_i^\circ$ is the total free energy of the products or reactants in their standard state of unit pressure or concentration. For example the standard free-energy change of the reaction of NO with $\frac{1}{2}O_2$ to form NO_2 is the free-energy difference between products and reactants under standard conditions:

$$\Delta G^\circ = \mu^\circ_{NO_2} - \mu^\circ_{NO} - \tfrac{1}{2}\mu^\circ_{O_2}$$

If we insert the expression for the standard free-energy change into Equation 4-50*a*, and recall Equation 2-44*c* on logarithms, we obtain

$$\Delta G^\circ + RT \ \Sigma \ln P_i(\text{products})^{\nu_i} - RT\,\Sigma \ln P_i(\text{reactants})^{\nu_i} = \Delta G = 0 \qquad (4\text{-}50b)$$

Moreover, the logarithm of a product is the sum of the logarithms, so that

$$\ln P_1^{\nu_1} + \ln P_2^{\nu_2} + \cdots = \Sigma \ln P_i^{\nu_i} = \ln \prod_i P_i^{\nu_i} = \ln[P_i^{\nu_1} \times P_2^{\nu_2} \times \cdots]$$

where the symbol Π means "product over" $P_1^{\nu_1} \times P_2^{\nu_2} \times \cdots$. Thus Equation 4-50*b* can be written

$$\Delta G^\circ + RT \ln \prod_i P_i^{\nu_i}\,(\text{products}) - RT \ln \prod_i P_i^{\nu_i}\,(\text{reactants})$$

$$= \Delta G^\circ + RT \ln \left[\frac{\prod_i P_i^{\nu_i}\,(\text{products})}{\prod_i P_i^{\nu_i}(\text{reactants})} \right] = \Delta G = 0 \qquad (4\text{-}51)$$

where the last step uses the property of logarithms that $\ln x - \ln y = \ln (x/y)$ (Equation 2-44).

If we examine Equation 4-51, we see that the expression in brackets must be a constant at any given T and P. The reason is that ΔG° is a constant, and for the sum ΔG to be equal to zero under all conditions, the bracketed expression must be constant, or

$$K = \frac{\prod_i P_i^{\nu_i}\,(\text{products})}{\prod_i P_i^{\nu_i}\,(\text{reactants})} \qquad (4\text{-}52) \blacktriangleleft$$

Substituting K from Equation 4-52 into Equation 4-51, we have

$$\Delta G^\circ = -RT \ln K \qquad (4\text{-}53) \blacktriangleleft$$

K is a very important quantity, called the *equilibrium constant* for the reaction. For example, in the case of the reaction $NO + \frac{1}{2}O_2 \rightleftarrows NO_2$,

$$K = \frac{P_{NO_2}}{P_{NO} \times P_{O_2}^{1/2}} \tag{4-54}$$

All chemical reactions of gases and solutions can be described by an equilibrium constant. Equating a ratio of concentrations or pressures such as that in Equation 4-54 to a constant means that the number of independent concentration variables is reduced. Given fixed pressures of NO_2 and NO, only one value of the O_2 pressure can be in equilibrium with the mixture.

If all the partial pressures are measured in the equilibrium reaction of NO and O_2 to form NO_2, K can be calculated from Equation 4-54. Equation 4-53 can then be used to calculate the free-energy change of the reaction under standard conditions ($\Delta G°$). You should be sure you understand the close relationship between the equilibrium constant of a reaction and its standard free-energy change.

EXERCISE 4-12

How can one take the logarithm of K in Equation 4-53 when Equation 4-52 indicates that K has dimensions, for example (pressure)$^{-1/2}$ in Equation 4-54?

ANSWER

K in fact is dimensionless. To see this you have to trace the development of the expression $\mu_i = \mu_i^\circ + RT \ln P_i$ from Equation 4-28. The molar free-energy change to convert a gas from its standard state to pressure P_i is $\Delta G = RT \ln (P_i/P_{i,\text{standard}})$. Therefore, the chemical potential is actually $\mu_i = \mu_i^\circ + RT \ln (P_i/P_{i,\text{standard}})$. As a consequence, K in principle should be written in terms of the ratios $P_i/P_{i,\text{standard}}$, which are dimensionless. Because the pressure is unity in the standard state, $P_{i,\text{standard}}$ is dropped from the expression for K. The numerical value of an equilibrium constant depends on the concentration units chosen, so K is frequently given units, such as atm$^{-1/2}$, for convenience. However, it should be realized that K actually is dimensionless in equations such as 4-53.

EXERCISE 4-13

Derive an expression for the equilibrium constant of the reaction

$$CaCO_3(s) \rightarrow CaO(s) + CO_2(g)$$

ANSWER

The chemical potential of a solid is a constant; there is no concentration-dependent term. Therefore at equilibrium Equation 4-49 gives

$$\mu_{CaO}^\circ + \mu_{CO_2}^\circ - \mu_{CaCO_3}^\circ + RT \ln P_{CO_2} = 0$$

or

$$\Delta G^\circ + RT \ln P_{CO_2} = 0$$

hence $K = P_{CO_2}$.

Reactions in Solution

Application of Equation 4-49 to reactants and products in solution, whose chemical potential is given by Equation 4-43*c*, yields (in exact analogy to Equation 4-51):

$$\Delta G^\circ + RT \ln \frac{\prod_i C_i^{\nu_i} \text{ (products)}}{\prod_i C_i^{\nu_i} \text{ (reactants)}} = \Delta G = 0 \tag{4-55}$$

so that the solution equilibrium constant in terms of molar concentration is

$$K = \frac{\prod_i C_i^{\nu_i} \text{ (products)}}{\prod_i C_i^{\nu_i} \text{ (reactants)}} \tag{4-56}$$ ◀

As with gases, $\Delta G^\circ = -RT \ln K$.

Equation 4-53 provides a simple method for determining the standard free-energy change of a reaction. Measurement of the equilibrium concentrations or partial pressures allows calculation of K from Equation 4-56 or 4-52 respectively. ΔG° is then simply $-RT \ln K$.

EXERCISE 4-14

Acetic acid dissolved in water dissociates according to

$$\text{HAc} \rightleftarrows \text{H}^+ + \text{Ac}^-$$

At equilibrium, 25°C, and 1 atm pressure, it is found that $C_{\text{H}^+} = 4.2 \times 10^{-4}$M, $C_{\text{Ac}^-} = 4.2 \times 10^{-4}$M, and $C_{\text{HAc}} = 0.01$M. Calculate K and ΔG°.

ANSWER

$K = C_{\text{H}^+}C_{\text{Ac}^-}/C_{\text{HAc}} = 1.76 \times 10^{-5}$M. $\Delta G^\circ = -RT \ln K = 27.12$ kJ mol^{-1}.

EXERCISE 4-15

The solubility of benzene in water is 0.70 g per liter at 20°C. Calculate the free-energy change for the hypothetical process of dissolving benzene in water at a concentration of 2.0 g per liter, assuming that Equation 4-43*c* is accurate.

ANSWER

A solubility of 0.7 g per liter implies that liquid benzene is in equilibrium with an aqueous solution containing 0.7 g per liter. Therefore, ΔG is zero for the reaction benzene liquid ⇄ benzene solution (0.7 g per liter). According to Equation 4-55, $\Delta G = \Delta G^\circ + RT \ln (0.7) = 0$. (Notice that there is no concentration term for the reactant, liquid benzene.) Hence, $\Delta G^\circ = -RT \ln 0.7 = -1.98 \times 293 \times \ln 0.7 = 207$ cal mol^{-1}. Again using Equation 4-55, we find for the free-energy change when $C = 2$ g liter^{-1}, $\Delta G = 207$ cal mol^{-1} + $RT \ln 2 = 609$ cal mol^{-1}.

EXERCISE 4-16

The solubility of compounds that dissociate into ions is described by the *solubility product,* which is the product of the concentration of the separate ions, each raised to an exponent given by the stoichiometric coefficient of the ion in the salt, for example $K_{SP} = [Cu^+]^2[S^=]$ for Cu_2S. Show that this expression is consistent with Equation 4-56.

ANSWER

For the reaction, Cu_2S (solid) $\rightarrow 2Cu^+ + S^=$, there is no concentration term for the pure solid. Therefore, by Equation 4-56, $K = [Cu^+]^2[S^=]$, which is defined to be the solubility product.

EXERCISE 4-17

The solubility product of BaF_2 is 1.7×10^{-6}, in molar units, at 20°C. Calculate the concentration of F^- in equilibrium with solid BaF_2.

ANSWER

Let $x = [F^-]$. Because of conservation of charge, $[Ba^{2+}] = x/2$. Therefore $x^3/2 = 1.7 \times 10^{-6}$ and $x = (3.4 \times 10^{-6})^{1/3} = 0.015$M.

A Large K can Arise from a Large Negative ΔH° or a Large Positive ΔS°

Equation 4-53 reveals how a chemical equilibrium depends on enthalpy and entropy changes. For the general reaction

$$\nu_A A + \nu_B B + \cdot\cdot\cdot \overset{K}{\rightleftarrows} \nu_X X + \nu_Y Y + \cdot\cdot\cdot$$

the equilibrium constant $K = (C_X^{\nu_X} C_Y^{\nu_Y} \ldots)/(C_A^{\nu_A} C_B^{\nu_B} \ldots)$ will be large if the equilibrium favors the products and small if the reactants predominate. (By convention, the products are written on the right and the reactants on the left of a reaction equilibrium.) Because $\Delta G° = -RT \ln K$, a value of K larger than 1 means $\Delta G°$ is negative, and if K is less than 1, then $\Delta G°$ is positive. Taking account also of $\Delta G° = \Delta H° - T\,\Delta S°$, we can see that a large value of K can result either from a large negative $\Delta H°$ (reaction exothermic) or a large positive $\Delta S°$ (favorable entropy change), or both. Similarly, unfavorable reactions (K small) have a large positive $\Delta H°$ (endothermic), a large negative $\Delta S°$ (unfavorable entropy change), or both.

EXERCISE 4-18

As urea is dissolved in water in a 1M solution, the solution grows perceptibly colder. What statement can you make about $\Delta S°$?

ANSWER

$\Delta S°$ is positive. We know that $\Delta G° = \Delta H° - T\,\Delta S° < 0$, since solution of urea is spontaneous. The reaction is endothermic, hence $\Delta H° > 0$. Thus $-T\,\Delta S° < 0$, and $\Delta S° > 0$.

This is an example of a reaction which is driven by an entropy increase in the system. The entropy increase more than compensates for the enthalpy increase.

A Small Decrease in $\Delta G°$ Causes a Large Increase in K

Because of the logarithmic relationship between $\Delta G°$ and K, a small (relative to RT) change in $\Delta G°$ for a reaction appreciably influences the position of equilibrium. At body temperature, a decrease in $\Delta G°$ by 5.7 kJ mol^{-1} causes a tenfold increase in K. Since 5.7 kJ mol^{-1} is an energy smaller than even one hydrogen bond, we see that small effects can shift the position of biochemical equilibria.

pH and pK also are Logarithmic Functions

You probably are familiar with the use of pH as a measure of the acidity of a solution, defined as

$$\text{pH} = -\log_{10}[\text{H}^+]$$

Like the chemical potential, pH depends on the logarithm of concentration, but in this case on the logarithm using the base 10. However, since $\ln x = 2.303 \log_{10} x$ it is evident that both pH and the chemical potential of H^+ vary linearly with $\ln[H^+]$.

Another logarithmic function is pK, defined as

$$\text{pK} = -\log_{10} K$$

In this case, a parallel can be drawn to $\Delta G°$, which, according to Equation 4-53, also varies linearly with the logarithm of K. You easily can show that

$$\Delta G° = 2.303\, RT\, \text{pK}$$

○ 4–8 *Standard Free-Energy Changes*

The standard free-energy change of a reaction is ΔG for the reaction when all reactants and products are in their standard states. For example, for the reaction

$$\text{NO(g)} + \text{CO(g)} \rightarrow \text{CO}_2\text{(g)} + \tfrac{1}{2}\text{N}_2\text{(g)} \qquad (4\text{-}57)$$

ΔG° refers to the following process: Take separate samples of NO and CO in their standard states, form CO_2 and $\frac{1}{2}N_2$, and convert each of these to its standard state. ΔG° is the free energy per mole of NO or CO for this process; it is analogous to the standard enthalpy change for a reaction.

One way that standard free-energy changes can be calculated is from the *standard free energy of formation*, ΔG_f°, of the reactants and products. The standard free energy of formation is analogous to the standard enthalpy of formation (see Sec. 2-7). It is ΔG for formation of the compound from its elements

$$\text{Elements (298 K, 1 atm)} \xrightarrow{\Delta G_f^\circ} \text{Compound (298 K, 1 atm)} \qquad (4\text{-}58a)$$

where the elements are in their most stable form, and the physical state of the compound must be specified. Table 4-2 lists standard free energies of formation for selected compounds.

Values of ΔG_f° can be used to calculate ΔG°. Any chemical reaction may be imagined to occur by a path in which the reactants are dissociated to their separate elements and reassociated into product:

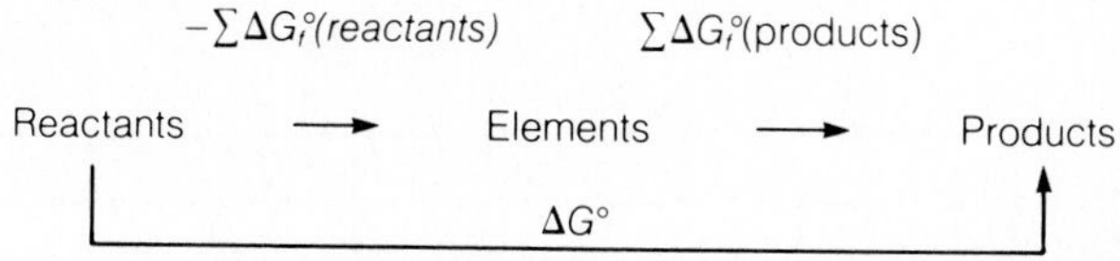

Therefore, because G is a property and ΔG must be the same for all paths,

$$\Delta G^\circ = \Sigma \Delta G_f^\circ(\text{products}) - \Sigma \Delta G_f^\circ(\text{reactants}) \qquad (4\text{-}58b)$$

EXERCISE 4-19

Calculate ΔG° and K at 298 K for reaction 4-57, using the data in Table 4-2.

ANSWER

In kJ mol^{-1}, $\Sigma\Delta G_f^\circ(\text{products}) = -394.4 + 0$; $\Sigma\Delta G_f^\circ(\text{reactants}) = 86.7 - 137.3 = -50.6$; $\Delta G^\circ = -394.4 - (-50.6) = -343.8$ kJ mol^{-1}. (Compare with the calculated enthalpy change, $\Delta H^\circ = -373.4$ kJ mol^{-1}.) $K = \exp-(\Delta G^\circ/RT) = \exp[343.8 \times 10^3/(8.314 \times 298)] = \exp(138.8)$, or 10^{60}. The reaction is thus overwhelmingly on the side of $\frac{1}{2}N_2 + CO_2$ at 298 K.

An Example: Base Pairing in Oligonucleotides

Standard free-energy changes also are useful for interpreting differences in the equilibrium constants of related reactions. Consider the formation of a double helix by the self-complementary oligonucleotide A_nU_n, whose sequence has nA (adenosine) residues followed by nU (uridine) residues. In the double helix, $2n$ A · U base pairs [see Fig. 4-8(a)] are formed, with the

TABLE 4-2 *Standard Free Energy of Formation of Various Pure Substances (1 atm, 298 K)*

INORGANIC AND SIMPLE ORGANIC COMPOUNDS[1]		ΔG_f°/kJ mol^{-1}
LiCl	(c)	−383.7
NaCl	(c)	−384.2
NaF	(c)	−541.0
NaBr	(c)	−347.7
NaI	(c)	−237.2
KCl	(c)	−408.3
$MgCl_2$	(c)	−592.3
CO	(g)	−137.3
CO_2	(g)	−394.4
CH_4	(g)	−50.8
HCOOH	(g)	−335.7
NO	(g)	86.7
NO_2	(g)	51.8
NH_3	(g)	−16.6
O_3	(g)	163.4
H_2O	(g)	−228.59
H_2O	(l)	−237.19
SO_2	(g)	−300.4
SO_3	(g)	−370.4
H_2S	(g)	−33.0

SUBSTANCES OF BIOCHEMICAL IMPORTANCE		
Carbohydrates[2]		
Glucose	(s)	−910
Sucrose	(s)	−1554
Alcohols and carboxylic acids		
Acetic acid	(l)	−392
Ethanol	(l)	−175
Succinic acid	(s)	−748
Amino acids[3]		
Alanine	(s)	−370
Cysteine	(s)	−340
Glutamic acid	(s)	−731
Glycine	(s)	−378
Leucine	(s)	−348
Serine	(s)	−509
Tryptophan	(s)	−119
Tyrosine	(s)	−386
Nitrogen-containing compounds[3]		
Urea	(s)	−197
DL-Alanylglycine	(s)	−490
Glycylglycine	(s)	−488
Glycyl-DL-tryptophan	(s)	−229
DL-Serylserine	(s)	−805

1. Dickerson, R. E. *Molecular Thermodynamics*. Menlo Park, Cal.: W. A. Benjamin, Inc., 1969.
2. Edsall, J. T. and Wyman, J. *Biophysical Chemistry*. New York: Academic Press, 1958.
3. Sober, H. A., ed. *Handbook of Biochemistry*. Cleveland, Ohio: The Chemical Rubber Co., 1968.

two strands running in opposite directions [Fig. 4-8(b)]. The equilibrium constant for double helix formation is given by

$$K = \frac{C_D}{C_M^2}$$

in which C_D is the concentration of the (dimeric) double helix, and C_M the concentration of the single strand. K has been estimated for these reactions at 25°C. For example, when $n = 5$, $K = 5 \times 10^3 M^{-1}$ and when $n = 6$, $K = 2 \times 10^5 M^{-1}$. K is larger for $n = 6$ because there are two more base pairs to stabilize the double helix [Fig. 4-8(c)].

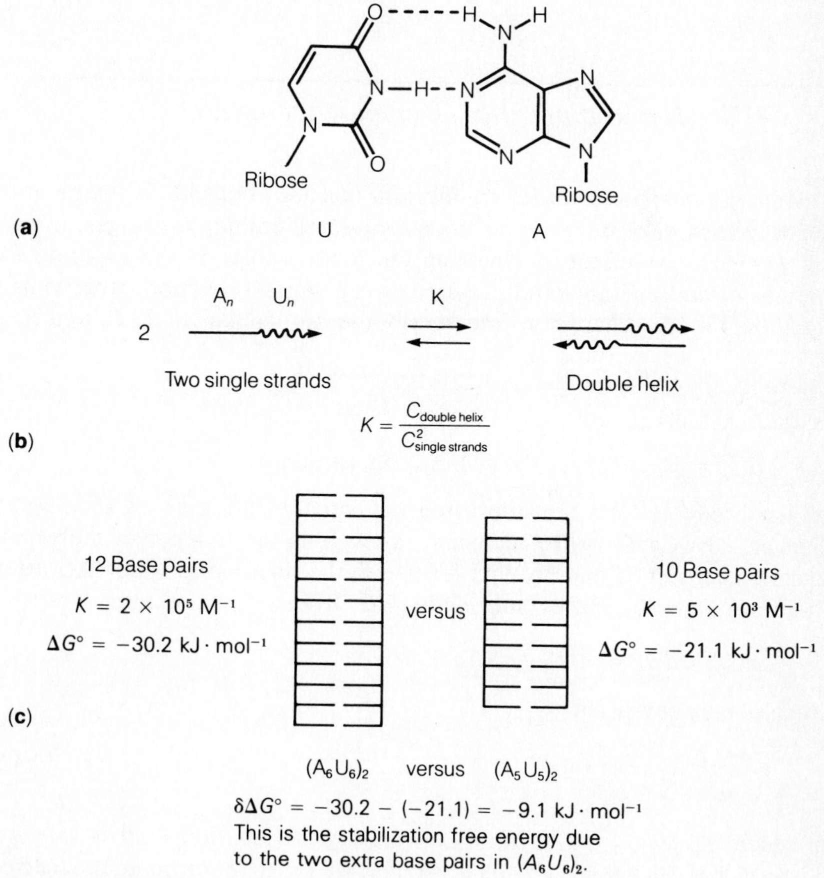

FIGURE 4-8
(a) A base pair formed between uridine (U) and adenosine (A). (b) Double-helix formation between single strands containing nA residues followed by nU residues. Such sequences are said to be *self-complementary* because they can form dimeric double helices with antiparallel strands, as required by the double-helix geometry. (c) Comparison of the equilibrium constant and standard free-energy change of forming a double helix from single strands of different lengths. Under standard concentration conditions, helix formation by the longer oligomer is more favorable because more base pairs are formed. The difference, -9.1 kJ mol^{-1}, is the free-energy change for the two extra A · U base pairs, so -4.6 kJ mol^{-1} is the free-energy change per base pair added to the double helix.

The quantitative stabilizing effect of additional base pairs can be determined by comparing the standard free-energy changes. Because $\Delta G^\circ = -RT \ln K$, we obtain $\Delta G^\circ = -21.1$ kJ mol^{-1} when $n = 5$ and $\Delta G^\circ = -30.2$ kJ mol^{-1} when $n = 6$. The difference in these two reactions *under standard conditions* is due only to the two extra base pairs. Therefore we can ascribe the difference in ΔG°, $\delta\,\Delta G^\circ = -30.2 - (-21.1) = -9.1$ kJ mol^{-1}, to the two base pairs, or -4.6 kJ mol^{-1} of base pairs. In this way the contribution to stabilization of nucleic acid double helices by different base pairs and other structural features has been determined (see Bloomfield *et al*, p. 349).

4–9 The Temperature Dependence of ΔG and K

Free-energy changes and the equilibrium constant depend on temperature. The way they vary depends on the entropy and enthalpy changes. We will now consider the effect of changing the temperature on the separate free energies of the reactants and products in a chemical reaction. According to Equation 4-47 the change in free energy due to changes in T, P, and n_i is

$$dG_r = V_r\,dP - S_r\,dT + \Sigma\mu_i\,dn_i \text{ (reactants)}$$

for the reactants r, and

$$dG_p = V_p\,dP - S_p\,dT + \Sigma\,\Delta\mu_i\,dn_i \text{ (products)}$$

for the products. We are interested only in the influence of temperature changes, so we will hold P constant, as well as the number of moles n_i of reactants consumed and product formed in the reaction to which ΔG refers. Therefore $dn_i = 0$, and by subtracting dG_r from dG_p we obtain

$$dG_p - dG_r = -(S_p - S_r)\,dT$$

which can be rewritten

$$\left(\frac{\partial\,\Delta G}{\partial T}\right)_P = -\Delta S \qquad (4\text{-}59)$$

in which $\Delta G = G_p - G_r$, and $\Delta S = S_p - S_r$. Equation 4-59 is one form of the *Gibbs-Helmholtz equation*. It allows us to determine the entropy change of a reaction from the temperature variation of ΔG, or alternatively, allows us to predict how ΔG will vary with temperature if ΔS is known.

EXERCISE 4-20

Show that an alternative form of the Gibbs-Helmholtz equation is

$$\left(\frac{\partial(\Delta G/T)}{\partial T}\right)_P = \frac{-\Delta H}{T^2}$$

ANSWER

By the chain rule of differentiation,

$$\left(\frac{\partial(\Delta G/T)}{\partial T}\right)_P = \frac{1}{T}\left(\frac{\partial \Delta G}{\partial T}\right)_P + \Delta G \frac{\partial}{\partial T}\left(\frac{1}{T}\right) = \frac{1}{T}\left(\frac{\partial \Delta G}{\partial T}\right)_P - \frac{\Delta G}{T^2}$$

Because

$$\left(\frac{\partial \Delta G}{\partial T}\right)_P = -\Delta S$$

we get

$$\left(\frac{\partial(\Delta G/T)}{\partial T}\right)_P = \frac{-T\,\Delta S - \Delta G}{T^2}$$

so, with $\Delta G + T\,\Delta S = \Delta H$ at any chosen T,

$$\left(\frac{\partial(\Delta G/T)}{\partial T}\right)_P = \frac{-\Delta H}{T^2} \tag{4-60}$$

We can now combine Equation 4-60 with $\Delta G° = -RT \ln K$ to determine the effect of temperature on the equilibrium constant. Dividing Equation 4-56 by T, we find that

$$\frac{\Delta G°}{T} = -R \ln K$$

Substituting $-R \ln K$ for $\Delta G°/T$ in Equation 4-60, we get

$$-R\left(\frac{\partial \ln K}{\partial T}\right)_P = \frac{-\Delta H°}{T^2}$$

or

$$\left(\frac{\partial \ln K}{\partial T}\right)_P = \frac{\Delta H°}{RT^2} \tag{4-61}$$ ◀

This is called the *Van't Hoff equation*, and is used widely to determine reaction heats ($\Delta H°$) from the temperature variation of the equilibrium constant. It allows one to determine enthalpy changes without resorting to calorimetric measurements. Notice that this equation shows that the temperature dependence of the equilibrium constant depends on $\Delta H°$ but not on $\Delta S°$.

EXERCISE 4-21

Show that an alternative form of the Van't Hoff equation is $[\partial \ln K/\partial(1/T)]_P = -\Delta H°/R$.

ANSWER

Because $d(1/T) = -dT/T^2$, $[\partial \ln K/\partial(1/T)]_P = -T^2\,(\partial \ln K/\partial T)_P$. Therefore, with Equation 4-61

$$\left[\frac{\partial \ln K}{\partial(1/T)}\right]_P = \frac{-\Delta H^\circ}{R} \qquad (4\text{-}62)$$

EXERCISE 4-22

Use the heats and free energies of formation in Tables 2-6 and 4-1 to estimate the partial pressure of NO in equilibrium with 0.8 atm N_2 and 0.2 atm O_2 at 25°C and at 1000°C. Assume that ΔH° is independent of temperature.

ANSWER

The reaction is

$$N_2 + O_2 \rightleftarrows 2NO$$

At 298 K, $\Delta H^\circ = 2\Delta H_f^\circ$ (NO) = 180.7 kJ per mole N_2, and $\Delta G^\circ = 2\Delta G_f^\circ$ (NO) = 173.4 kJ per mole N_2. Therefore, at 298 K, $K = \exp(-\Delta G^\circ/RT) = \exp[-173.4 \times 10^3/(8.314 \times 298)] = \exp(-70) = 4 \times 10^{-31}$. The equilibrium constant K is

$$K = \frac{P_{NO}^2}{P_{N_2}P_{O_2}}$$

so, at 298 K, $P_{NO} = (P_{N_2}P_{O_2}K)^{1/2} = 2.5 \times 10^{-16}$ atm.

At 1000°C = 1273 K, we use $[\partial \ln K/\partial (1/T)] = -\Delta H^\circ/R$ to estimate K. Therefore

$$\Delta \ln K = \frac{-\Delta H^\circ}{R}\Delta\left(\frac{1}{T}\right) = \frac{-180.7 \times 10^3}{8.314} \times (-2.57 \times 10^{-3}) = 55.9$$

$$\ln K_{1273} = \ln K_{298} + \Delta \ln K = \ln(4 \times 10^{-31}) + 55.9$$

$$\ln K_{1273} = -14.1;\ K_{1273} = 7.6 \times 10^{-7}$$

Hence, at 1273 K, $P_{NO} = (P_{N_2}P_{O_2}K)^{1/2} = 3.5 \times 10^{-4}$ atm.

Exercise 4-22 shows that the pressure of NO in equilibrium with atmospheric pressures of N_2 and O_2 increases from a negligible amount (2.5×10^{-16} atm) at 25°C to an appreciable pressure (3.5×10^{-4} atm) at 1000°C. This reaction is the origin of the NO produced by internal combustion engines. Some NO is formed at high temperatures, and the exhaust gases cool more rapidly than the thermodynamically favorable reconversion of NO to N_2 and O_2 can occur. The NO emitted is an important participant in the photochemical reactions (activated by sunlight) that produce smog.

Unfolding of Macromolecules

Equation 4-61 implies that a reaction with a large standard enthalpy change can shift preferred direction over a narrow temperature range. There are many examples of this in the *denaturation* of biological macromolecules. This is the process considered in Chapter 2 in which the native structure unfolds in some unspecified manner to a form that is no longer biologically active:

$$\text{Native} \underset{\text{folding}}{\overset{\text{unfolding}}{\rightleftharpoons}} \text{Denatured} \tag{4-63}$$

The equilibrium constant for this reaction is

$$K = \frac{\text{concentration denatured}}{\text{concentration native}}$$

As shown in Figure 2-3 and Table 2-2, the enthalpy change for such reactions can be larger than 10^5 J mol^{-1}, which makes K exceedingly temperature sensitive. For example, for the unfolding of chymotrypsin at pH = 2 the reaction is strongly endothermic, with $\Delta H° = 418$ kJ mol^{-1}. This means that a very large amount of enthalpy is required to denature one mole of chymotrypsin, but since the reaction does occur, it must be very favorable entropically. The measured value of $\Delta S°$ is 1.32 kJ K^{-1} mol^{-1}. If we assume that these values are independent of temperature over a narrow range, we can write

$$K = \exp \frac{-\Delta G°}{RT} = \exp \frac{-\Delta H° + T\,\Delta S°}{RT} \tag{4-64}$$

or, substituting numerical values for $\Delta H°$ and $\Delta S°$,

$$K = \exp \left[\frac{-418 \times 10^3}{8.314\,T} + \frac{1.32 \times 10^3}{8.314} \right] \tag{4-65}$$

In terms of concentrations of native C_N and denatured C_D forms, K is

$$K = \frac{C_D}{C_N} \tag{4-66}$$

We let θ be the fraction of molecules in the native form

$$\theta = \frac{C_N}{C_N + C_D} \tag{4-67}$$

Equation 4-66 gives $C_D = KC_N$, which when substituted into 4-67 yields

$$\theta = \frac{1}{1 + K} \tag{4-68}$$

With Equations 4-65 and 4-68 we can calculate θ as a function of T. The results are shown in Figure 4-9, which is a denaturation or "melting" curve for the protein molecule.

Denaturation reactions that occur within a very small temperature range like that in Figure 4-9 are said to be *cooperative*. You can understand that terminology in the following sense: Each of the many noncovalent interactions (see Chap. 11) that stabilize the folded structure of a macromolecule contributes only a small enthalpy, entropy, or free-energy change to the folding reaction. However, there are many such interactions in a protein or nucleic acid, and if they all "cooperate" by breaking together, the many small contributions add up to a large reaction enthalpy and entropy. The consequence is a sharp, cooperative melting transition, as in Figure 4-9.

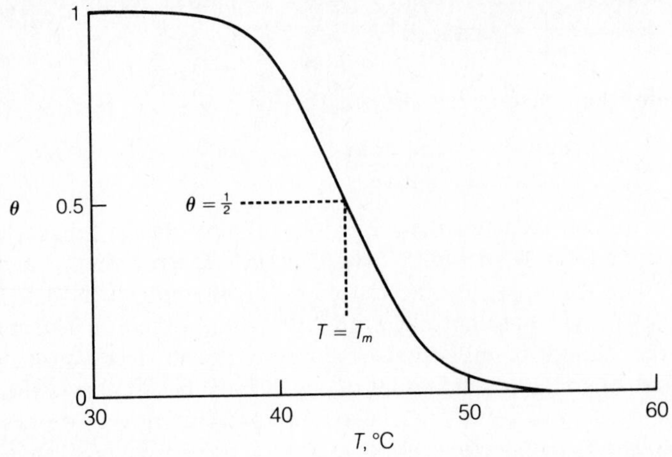

FIGURE 4-9
Calculated curve for the denaturation, or unfolding, of chymotrypsin at pH = 2. The curve is that described by Equations 4-65 and 4-68. θ is the fraction of molecules that have the native structure. The transition midpoint, $T = T_m$, occurs when $K = 1$, $\theta = \frac{1}{2}$. This is the temperature at which there is equal probability that a molecule will have the native or the denatured structure.

EXERCISE 4-23

Using Equation 4-65, calculate K and ΔG° for denaturation of chymotrypsin (pH = 2) at 30°C.

ANSWER

$$K = \exp\left[\frac{-418 \times 10^3}{8.314 \times 303} + \frac{1.32 \times 10^3}{8.314}\right] = 7.76 \times 10^{-4}$$

$$\Delta G^\circ = -RT \ln K = 18 \text{ kJ mol}^{-1}$$

This example shows that even though chymotrypsin (pH = 2) at 30°C is only 13.7°C away from its denaturation temperature (43.7°C), it is highly stable, with less than 0.1% denatured molecules, and 18 kJ mol^{-1} unfavorable free energy of denaturation. This behavior is a consequence of the very large denaturation enthalpy and entropy.

○ 4–10 *A Summary of the Methods for Determining ΔG°*

In Section 4-9 we saw that for reactions occurring at constant temperature and pressure (as do virtually all biochemical reactions) the central quantity for predicting and describing equilibrium is ΔG°. We have already seen a number of methods for determining ΔG°, but since they have been scattered

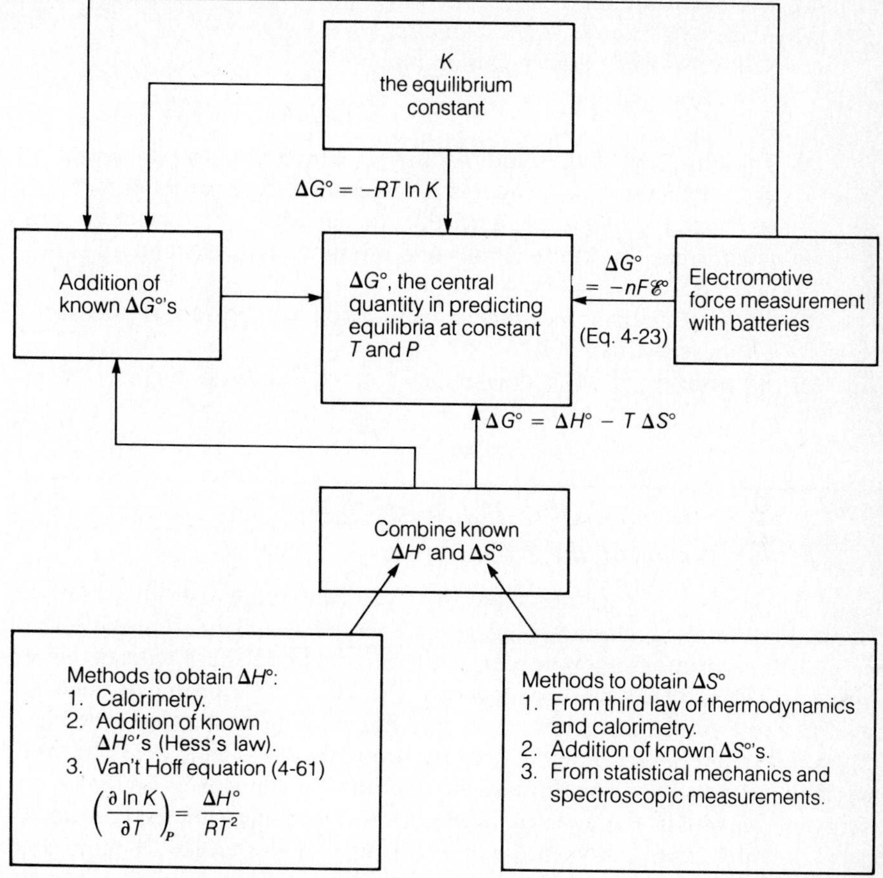

FIGURE 4-10
Summary of the methods to determine ΔG° for reactions.

through several chapters it may be helpful to review them here. All the methods are summarized in Figure 4-10, which is largely self-explanatory.

The main points of the figure are that the ΔG° for a given reaction at a given temperature and pressure can be determined by:

1. Measuring the equilibrium constant for the reaction and using $\Delta G^\circ = -RT \ln K$.
2. Measuring the electromotive force (voltage) for a battery that uses the reaction of interest, and using $\Delta G = -nF\mathscr{E}$. For the ΔG to be ΔG°, the chemicals in the reaction must be at standard conditions.
3. Adding known ΔG°'s to yield the ΔG° of interest. The ΔG's that are added have previously been determined from measurements of other equilibrium constants or from batteries, etc.
4. Combining ΔH° and ΔS° to give ΔG° by $\Delta G^\circ = \Delta H^\circ - T\,\Delta S^\circ$. The ΔH's and ΔS's can in turn be determined by the methods noted in the figure. ΔH° can be measured by:

(a) Direct calorimetry.
(b) Addition of known $\Delta H°$'s, that is, by Hess's law.
(c) From the Van't Hoff equation (4-61).

$\Delta S°$ can be determined by:

(a) The third law of thermodynamics, in which the heat capacities of both reactants and products are measured and the entropies determined by integration according to Equation 3-37. The difference in entropies for the reactants and products in their standard states gives $\Delta S°$.
(b) From statistical mechanics, employing spectroscopic measurements (see Chap. 14).
(c) By addition of $\Delta S°$'s determined from methods (a) and (b) above.

4–11 Hydrophobic Bonding Results from Minimization of the Free Energy

If you recall your elementary chemistry you will remember that simple chemical bonds occur because of minimization of the potential energy; the energy results from interaction between atoms that can share valence electrons (see Chap. 11). As a consequence, the length of a chemical bond is equal to the distance at which the energy has its minimum value (see Fig. 2-2). Many of the interactions that cause proteins and nucleic acids to fold into their native, active conformations can also be thought of in terms of energy minimization. For example, *hydrogen bonds* result from the favorable energy of interaction between a relatively positively charged H atom attached to N or O and a relatively negative atom such as N or O. Thus a significant portion of the energy required to vaporize liquid water is needed to break the O—H---O hydrogen bonds. Similar hydrogen bonds contribute to the stability of proteins and nucleic acids. Other important stabilizing factors include charge-charge (electrostatic) interaction (Chap. 8), interactions between local dipole moments (Chap. 11), and *dispersion forces* (Chap. 11), which result from correlations in the motion of electrons in adjacent parts of the macromolecule.

In addition to these forces, whose magnitude we usually calculate from the derivative of the energy ($F = -dU/dR$, Eq. 2-7), there is another factor that influences macromolecular conformation, which we must calculate from the derivative of the *free* energy. The interaction that gives rise to this additional factor was called the *hydrophobic bond* by Kauzmann in 1959. It differs from the other bonding interactions because it does not arise from pairwise interaction of atoms or parts of a molecule, but rather involves a considerable number of solvent (water) molecules. Because the entropy of the solvent is significantly affected by the hydrophobic interactions, it is necessary to consider the free energy in calculating the effective force in a hydrophobic bond.

The basic idea of a hydrophobic bond, illustrated in Figure 4-11, is that the majority of the nonpolar amino acid side chains are sequestered in the interior of a native, globular protein. In this conformation they are kept from contact with the aqueous solvent. The interaction is called "hydrophobic" because the nonpolar groups tend to avoid water, clustering together instead. When a protein is denatured or unfolded, the polypeptide backbone is unwound and the nonpolar side chains are exposed to the aqueous solvent.

Kauzmann noted that nonpolar groups tend to organize the water molecules about them in hydrogen-bond cages. Water-cage compounds called *clathrates* had been known since the early nineteenth century, and physical chemists had noted in the 1940s and 1950s that the low solubility of nonpolar compounds in water must be due at least in part to the formation of incipient clathrates, sometimes called "icebergs," around the nonpolar side chains. "Icebergs" may not be a very good term since it implies a more organized and extensive network of hydrogen-bonded water molecules than probably exists around the nonpolar groups, but the idea of lower solubility stemming from ordering of water is almost certainly correct.

The Stability of the Denatured State is Reduced by Ordering of Water

The ordering of water around the exposed nonpolar side chains lowers the stability of the denatured structure because the entropy of water is reduced as it becomes more ordered. If we think of the free energy change during denaturation as having contributions from both entropy and enthalpy

$$\Delta G_{\text{denat}} = \Delta H_{\text{denat}} - T\,\Delta S_{\text{denat}}$$

then the ordering of water will make a negative contribution to ΔS_{denat}. Thus the ordering will tend to make $-T\,\Delta S_{\text{denat}}$ positive, and will tend to make ΔG_{denat} positive. If this quantity is positive, then at constant pressure and temperature, the native structure of the protein will be favored. This is, of course, what is observed. Thus it is the incipient clathrate or "iceberg" formation that would take place if the protein unfolded that tends to hold the protein in its organized, native structure.

Data from Model Compounds Support the Idea of the Hydrophobic Bond

Thus far we have discussed Kauzmann's conclusion about the hydrophobic bond, but have not examined the data that led him to the conclusion. Table 4-3 shows the changes in thermodynamic quantities for the transfer of small nonpolar molecules such as methane and benzene to water from nonpolar solvents, including benzene and CCl_4. These transfers may be thought of as crude models for the transfer to the aqueous medium of nonpolar amino acid side chains from the interior of a protein, where they are in contact with other nonpolar side chains (see Fig. 4-11). Thus the transfer processes of Table 4-3 mimic one aspect of the denaturation of a protein: the transfer of nonpolar side chains from a nonpolar environment to water.

TABLE 4-3 *Thermodynamic Changes for the Transfer of Hydrophobic Hydrocarbons from a Nonpolar Solvent to Water*[1]

PROCESS	TEMP/K	$\Delta S°$/CAL MOL^{-1}	$\Delta H°$/CAL MOL^{-1}	$\Delta G°$/CAL MOL^{-1}
CH_4 in benzene → CH_4 in H_2O	298	−18	−2800	+2600
CH_4 in ether → CH_4 in H_2O	298	−19	−2400	+3300
CH_4 in CCl_4 → CH_4 in H_2O	298	−18	−2500	+2900
C_2H_6 in benzene → C_2H_6 in H_2O	298	−20	−2200	+3800
Liquid benzene → C_6H_6 in H_2O	291	−14[2]	0	+4070[2]
Liquid toluene → C_7H_8 in H_2O	291	−16[2]	0	+4650[2]

[1] Compiled by Kauzmann (1959).
[2] Entry corrected to represent transfer from benzene at a mole fraction that corresponds to 1M concentration, into H_2O where the benzene concentration is 1M. This is the same transfer process as is represented by the other table entries.

For all transfers listed, ΔS of transfer to water is negative. This is what we would expect if significant ordering of the water takes place. The sign of ΔH is negative for transfer of aliphatic non-aromatic compounds into water from hydrocarbon solvents. This means that heat is given off, and that the enthalpy contributes to a negative free energy of transfer. Yet the overall free energy of transfer is positive. This is because the entropy term more than compensates for the enthalpy term. Therefore the preference of nonpolar molecules for the nonpolar solvent comes not from a greater *energy* of interaction with the nonpolar solvent (they interact more strongly with water),

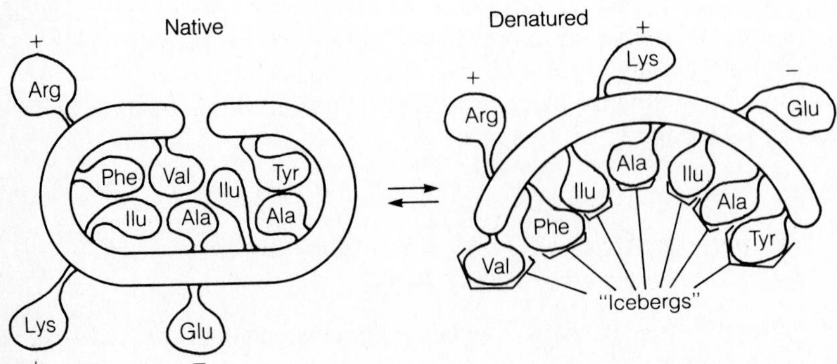

FIGURE 4-11
Schematic representation of the denaturation of a protein. Nonpolar side chains in the interior are exposed to the aqueous solvent. Water molecules tend to be more ordered ("icebergs") about the nonpolar side chains, contributing a negative component to ΔS of denaturation. The nonpolar side chains shown here are for the amino acids phenylalanine (Phe), valine (Val), isoleucine (Ilu), alanine (Ala), and tyrosine (Tyr). The charged side chains on arginine (Arg), lysine (Lys), and glutamic acid (Glu) are pictured in contact with the solvent in both the native and denatured states.

but from the unfavorable (lower) *entropy* in water. This is consistent with the idea of "iceberg" formation about the nonpolar groups in water.

To check that the same thermodynamic effects operate with amino acids, Tanford and his co-workers carried out similar transfer studies for amino acids. Their results are shown in Table 4-4. Amino acids are too insoluble in solvents such as benzene for the same transfers to be studied, so the Tanford group chose to study the transfer from alcohol to water. The free energy of transfer turns out to be negative (favorable) for the movement of all amino acids into water from the alcohol. However, much of this decrease in free energy must come from the favorable interaction of the ionic carboxylate and ammonium ions of each amino acid residue of structure $+NH_3$-CH-COO^-. To assess this contribution, the investigators subtracted from each $\Delta G°$ of transfer the $\Delta G°$ of transfer of glycine (the amino acid in which the side chain is simply a hydrogen atom). Thus the differences should show the effects of the more complex hydrophobic side chains. In all cases $\Delta G°$ of transfer is positive (see the right-hand column of Table 4-4). Hence the data on amino acids support the idea that the low water-solubility of nonpolar side chains is a factor in the stability of native protein structures.

Finally, what about data on proteins themselves? Do these support the idea of the hydrophobic bond? Here the situation is more complicated. The structures of proteins as determined by x-ray diffraction (Chap. 17) show that the majority of the nonpolar side chains are clustered in the interior, away from solvent. The thermodynamic data, however, are harder to interpret (see Table 4-5). First, notice that for three of the four entries the sign of ΔS of denaturation is positive. This is the opposite of the sign for the transfers of model compounds, and seems at first glance to contradict the hypothesis that hydrophobic bonding is an important force in holding proteins in the native state. But there is another contribution to ΔS of denaturation: Each protein has only a single native conformation, yet many possible conformations when denatured. Thus there is a large increase in the number of conformations possible for the protein when denatured. This provides a positive component to the ΔS of denaturation. In the same process, the ordering of

TABLE 4-4 *Free-Energy Changes for Transfer of Amino Acids to Water at 25°C*[1]

AMINO ACID	$\Delta G°$/CAL MOL^{-1} FOR TRANSFER	$\Delta G°$/CAL MOL^{-1} FOR TRANSFER OF SIDE CHAIN
Glycine	−4630	(0)
Alanine	−3900	730
Valine	−2940	1690
Leucine	−2210	2420
Isoleucine	−1690	2970
Phenylalanine	−1980	2650
Proline	−2060	2600

[1] Compiled by Klotz (1967).

TABLE 4-5 *Changes in Thermodynamic Quantities for Proteins During Denaturation*[1]

PROTEIN	CONDITIONS FOR DENATURATION	ΔG/KCAL MOL^{-1}	ΔH/KCAL MOL^{-1}	ΔS/CAL MOL^{-1} °C^{-1}	ΔC_P/CAL MOL^{-1} °C^{-1}
Ribonuclease	pH 2.5, 30°C	0.9	57	185	2000
Chymotrypsinogen	pH 3, 0.01M Cl^-, 25°C	7.3	39	105	2600
Myoglobin	pH 9, 25°C	13.6	42	95	1900
β-lactoglobulin	5M urea, pH 3, 25°C	0.6	−21	−72	2150

[1] From Tanford (1968).

water gives a negative component, and the sign of the total depends on which effect is larger. One bit of evidence that significant ordering of water does occur is the very large values observed for ΔC_p of denaturation (last column of Table 4-5). A large value for ΔC_p means that the heat capacity of the denatured protein solution is much greater than the heat capacity of the native protein solution. One likely reason for this observation is that as the denatured solution is heated, the "icebergs" around the nonpolar chains are melted. This additional contribution to the heat capacity in the denatured (but not native) solution increases ΔC_p. Another possible contribution to the heat capacity increase on denaturation will be discussed in Section 14-6.

In summary, the data clearly indicate that ordering of the water molecules around nonpolar side chains is an important factor in hydrophobic bonding. The large entropy change of this process requires that one consider the free energy rather than just the energy in describing the bond. However, it would be a mistake to conclude that ordering of water molecules is the *only* factor involved in hydrophobic bonding. For example, Sinanoglu has developed a theory that emphasizes the large free energy required to form the surface of a cavity in water into which a solute is placed. In this view, clustering of solute groups reduces the total surface area required, and hence minimizes the free energy. Water is unusual because its surface free energy, measured by the surface tension, is unusually large. More theoretical and experimental work will be required before the hydrophobic forces are fully understood.

○ 4–12 *Phase Equilibria*

According to the classical definition, a single phase consists of matter that is uniform throughout, both in chemical and physical composition. For example, liquid water is a single phase, as are ice, water vapor, a solution of sucrose in water, or the mixture of gases in a sample of air. Two or more phases occur when there is a macroscopic surface of demarcation between the homogeneous phases. Ice, water, and air form a three-phase system—the surfaces that separate the phases are clearly visible to the eye. Liquids that are immiscible form two phases, as do mixtures of solids or different crystal forms of the same solid.

In more modern terms, however, the number of identifiable phases depends on how closely we scrutinize the sample. For example, a solution of DNA appears homogeneous to the unaided eye, but if viewed (after drying) with an electron microscope, one can see inhomogeneities identifiable as DNA molecules. Do these constitute a separate phase? Or suppose we have, in order of increasing particle size, a suspension of ribosomes, mitochondria, bacteria, or white blood cells. When is the particle large enough to be considered a separate phase? There is a sound thermodynamic reason for asking these questions (as well as a sound thermodynamic answer), because the classical analysis of heterogeneous or multiple-phase equilibria provides us with powerful thermodynamic techniques for characterizing the state of biochemical materials. In this and the following sections we will consider Gibbs's treatment of heterogeneous equilibria, which is now a century old, and its consequences for macroscopic phase equilibria. Section 4-15 will consider the application of these principles to macromolecules and submacroscopic molecular aggregates of the kind found so often in biological systems.

The state of a sample of matter is specified by *intensive variables* such as temperature, pressure, and concentration, which do not depend on how much material is present, and by *extensive variables* such as mass and volume, which are proportional to the amount of substance. Generally, we require knowledge of one extensive variable per phase in order to specify the mass of each phase; the remainder of the properties of the system can be determined by the intensive variables.

The *Gibbs phase rule* tells us how many intensive variables are sufficient to specify the state of the system in addition to one extensive variable per phase. *We define the degrees of freedom, f, as the minimum number of intensive variables sufficient to describe the system; their values can be changed independently without altering the number of phases.* As in Chapter 1, p is the number of separate phases and c the number of chemical components whose concentration must be specified to determine the chemical composition of the system. Thus, if component A is in equilibrium with B, the concentration of A determines B and $c = 1$. According to the phase rule (which we will derive shortly),

$$f = c - p + 2 \tag{4-69}$$ ◀

which relates the number of degrees of freedom in the intensive variables to the number of components and number of phases. (In its rearranged form, $p + f = c + 2$, the phase rule is easy to remember as an acronym of: *P*olice *F*orce = *C*ops + 2.) Notice that the phase rule does not prevent us from replacing an intensive variable, such as concentration, by an extensive variable, such as the number of moles. One extensive variable per phase is *necessary* to describe the system, and the phase rule gives us the number of additional variables required. It is *sufficient* if all of these are chosen as intensive, but they may also be chosen to be extensive.

EXERCISE 4-24

One gram of H_2O is a mixture of liquid water and ice at 1 atm pressure. What else do we need to know to specify the state of the system?

ANSWER

$c = 1$ and $p = 2$, so $f = 1$. With the pressure specified, no degrees of freedom remain in the intensive variables. The temperature *must* be 0°C. However, we do not know the relative amounts of ice and water; full specification of the state of the system requires knowing one more extensive variable, such as the mass of ice. Remember that we need one extensive variable per phase, and $p = 2$.

EXERCISE 4-25

Gases in a mixture react according to the following equilibria:

$$NO + CO \rightleftarrows CO_2 + \tfrac{1}{2}N_2 \qquad (1)$$
$$CO + \tfrac{1}{2}O_2 \rightleftarrows CO_2 \qquad (2)$$

Calculate f.

ANSWER

The main problem here is the calculation of c. The simplest way to solve problems of this kind is to count the number of chemical species (five in this case) and subtract from that the number of distinct chemical reactions relating them. (A distinct reaction cannot be a sum of other reactions in the system.) You can see the reason for this counting procedure if you write the equilibrium constant expressions for the two reactions in this exercise. The five partial pressure variables are related by two equations: $K_1 = P_{CO_2} \cdot P_{N_2}{}^{1/2}/(P_{CO} \times P_{NO})$ and $K_2 = P_{CO_2}/(P_{CO} \times P_{O_2}{}^{1/2})$. Two equations with five variables leaves three variables undetermined. Hence $c = 5 - 2 = 3$, and $f = c - p + 2 = 3 - 1 + 2 = 4$. Knowledge of four intensive variables, such as the temperature, total pressure, and two of the partial pressures determines the state of all the intensive variables.

EXERCISE 4-26

The differential of dG is $V\,dP - S\,dT + \Sigma\mu_i\,dn_i$ (see Equation 4-47). How many terms dn_i must be retained for the system in Exercise 4-25?

ANSWER

The total number of differential terms must equal the degrees of freedom, f. In principle, any choice of four will do, but in practice it is usually simplest to retain T and P as independent variables, in which case two terms dn_i would be needed. These correspond to the two partial pressures appearing as independent variables in the answer to Exercise 4-25.

Derivation of the Phase Rule

To derive the phase rule, we begin by counting the total number of variables in the system, then subtracting one for each independent equation that can be written relating the variables. (An independent equation is one that cannot be arrived at by algebraic operations on the other independent equations.) In a system containing p phases ($a, b, c \ldots n \ldots p$) and c components ($1, 2, 3 \ldots i \ldots c$), the state of all intensive variables is specified if we know the temperature, the pressure, and the mole fraction $X_i^{(n)}$ of each component i in every phase n. The total number of variables $X_i^{(n)}$ is $c \times p$, since there are c components for each phase. Adding T and P, the total number of intensive variables is $c \times p + 2$. However, for each phase the sum of the mole fractions is one:

$$\sum_i X_i^{(n)} = 1, n = a, b, \ldots p \tag{4-70}$$

which gives us p restrictive equations. Furthermore, the chemical potential of each component i must be equal in every phase (see Section 4-6). Thus we have the equations

$$\begin{aligned}
\mu_1^a &= \mu_1^b = \mu_1^c = \cdots = \mu_1^p \\
\mu_2^a &= \mu_2^b = \mu_2^c = \cdots = \mu_2^p \\
&\cdot \\
&\cdot \\
&\cdot \\
\mu_c^a &= \mu_c^b = \cdots = \mu_c^p
\end{aligned} \tag{4-71}$$

There are $c(p - 1)$ equations in this set (count the number of equals signs). Therefore, taking account of Equations 4-70 and 4-71, the total number of independent variables is

$$\begin{aligned}
f &= c \times p + 2 - p - c(p - 1) \\
&= c - p + 2
\end{aligned}$$

which is in agreement with the previous statement of the phase rule.

4–13 Phase Diagrams

The state of a system may be represented by a *phase diagram,* which must have as many coordinate axes as there are degrees of freedom, f. For example, a pure substance may exist in one or more phases; the maximum number of degrees of freedom required occurs when $p = 1$, for which $f = 1 - 1 + 2 = 2$. The usual choice of independent variables is temperature and pressure.

Figure 4-12 shows a phase diagram for water, as a function of T and P. The lines on a phase diagram correspond to conditions of temperature and pressure where two phases are observed. Where three lines intersect, three

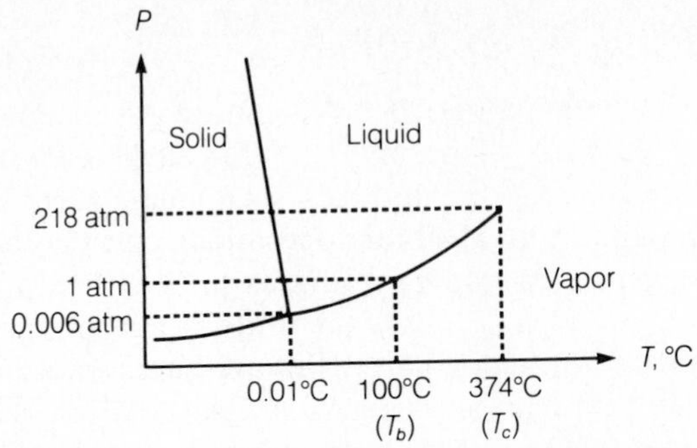

FIGURE 4-12
Phase diagram for H_2O. The solid lines represent conditions of T and P for which two phases are present. The intersection of two lines is a triple point, with three phases present. At the normal boiling point $T_b = 100°C$, and the vapor pressure is 1 atm. The critical temperature T_c is the highest temperature at which two phases exist; it occurs at 374°C and 218 atm.

phases are in equilibrium; this is called a *triple point*. For pure water, the triple point occurs at 0.01°C and 0.006 atm pressure. The three lines leading from this point correspond to the three possible two-phase equilibria between solid, liquid, and vapor. The liquid-vapor equilibrium line crosses 1 atm pressure at 100°C; this is the normal boiling point of water. If the temperature exceeds 100°C, all the liquid is converted to vapor unless the pressure is increased above 1 atm.

It is important to realize that *for any condition of temperature and pressure not on a line in the phase diagram, only one phase is observed*. In those regions of the phase diagram, there is only one phase, and P and T can be varied independently ($p = 1$, $c = 1$, and $f = 2$). On any one of the lines there are two phases, so $p = 2$ and $f = 1$. Hence determination of T (or P) fixes P (or T). The intersection of lines at the triple point occurs at a unique temperature and pressure and $p = 3$ while $f = 0$.

The *critical point* occurs at the highest temperature T_c and pressure P_c at which two phases can be observed for a vapor-liquid equilibrium. Above T_c, change of the pressure produces a continuous change in the density, without a sudden transition to a more condensed phase.

EXERCISE 4-27

Water in a dish in a room at 20°C and 1 atm evaporates completely, yet the phase diagram indicates that the stable phase is liquid. What is wrong?

ANSWER

The system is not at equilibrium and contains more than one component (water plus air). If the room is closed and enough water is provided to reach equilibrium, there will be two phases (liquid water and air plus water vapor) and two components (air plus water, assuming air to contain a fixed ratio of gases). Thus there are $f = 2 - 2 + 2 = 2$ degrees of freedom. Fixing the temperature and total pressure determines the system.

Solids at high pressure may have complex phase diagrams. Figure 4-13 shows the phase diagram of water at pressures up to 25,000 atm. There are at least ten solid phases of water, eight of which are shown on this diagram. Two others are metastable phases at low temperatures. Ordinary ice is called ice I. In all phases, water molecules are hydrogen-bonded to four neighboring molecules (as in Fig. 3-13), but in most high-pressure phases, hydrogen bonds are bent, permitting closer approach of nonbonded neighbors. Using ultra–high-pressure presses, investigators have followed the curve between ice VII and liquid water to over 200,000 atm. At this pressure, ice VII melts at about 400°C, well above the critical point of water vapor.

EXERCISE 4-28

There are numerous triple points in the H_2O phase diagram, but no quadruple point, at which four phases meet. Prove that quadruple points are impossible for this system.

ANSWER

Computing the degrees of freedom at a quadruple point from the phase rule, we have $f = c - p + 2$, with $c = 1$ (H_2O) and $p = 4$ (quadruple point), giving $f = -1$, which is impossible.

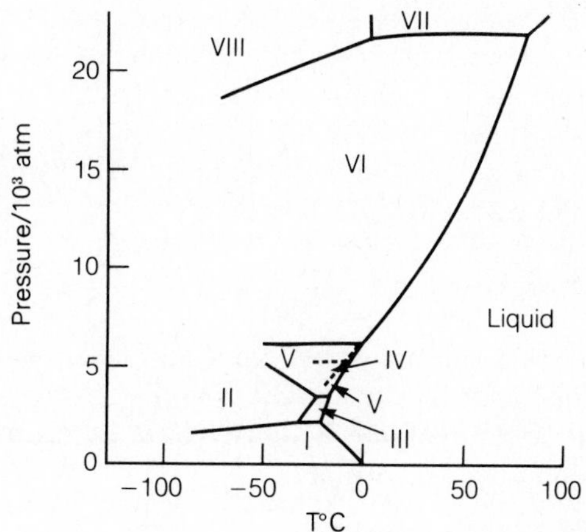

FIGURE 4-13

Phase diagram of solid ice at high pressures. The different forms of ice crystal are numbered I-VIII. Ice IV is a metastable form, not observed at equilibrium. From Eisenberg and Kauzmann (1969).

○ 4–14 The Clausius-Clapeyron Equation Gives the Slope of Lines on a P-T Phase Diagram

The slope of the lines on a phase diagram, dP/dT, is related to thermodynamic properties through what is called the *Clausius-Clapeyron equation*. Suppose we have one mole of Phase A in equilibrium with one mole of Phase B (for example, a liquid-solid equilibrium) and we change the temperature. How much must the pressure change in order to maintain equilibrium? According to Equation 4-47 the change, $d\mu_A$, in the free energy per mole is

$$d\mu_A = V_A\, dP - S_A\, dT \tag{4-72a}$$

in which V_A and S_A are the molar volume and entropy in Phase A. A similar equation applies to Phase B:

$$d\mu_B = V_B\, dP - S_B\, dT \tag{4-72b}$$

By definition, the system remains at equilibrium, $\mu_A = \mu_B$, so $d\mu_A = d\mu_B$. Subtracting 4-72*a* from 4-72*b* yields

$$0 = \Delta V_t\, dP - \Delta S_t\, dT \tag{4-73}$$

in which $\Delta V_t = V_B - V_A$, and $\Delta S_t = S_B - S_A$ are the molar volume and entropy changes in the phase transition. Rearranging, we obtain

$$\frac{dP}{dT} = \frac{\Delta S_t}{\Delta V_t} \tag{4-74}$$ ◄

which is one form of the Clausius-Clapeyron equation.

Notice that the phase rule states that one component and two phases, as in this problem, allow only one degree of freedom, $f = 1$. Consequently only one of the two differentials in Equation 4-72 is necessary to specify dG. If we include two, dP and dT, we get a definite relationship between them. Generally, $f + 1$ differentials are required when a relationship between two of them is sought.

Equation 4-74 can be cast in terms of the latent heat by noting that $\Delta G_t = \Delta H_t - T\,\Delta S_t = 0$, so $\Delta S_t = \Delta H_t/T$. Therefore

$$\frac{dP}{dT} = \frac{\Delta H_t}{T\Delta V_t} \tag{4-75}$$

When the phase transition involves vapor and either liquid or solid, the molar volume of the vapor is much larger than that of the condensed phase. Approximating the molar volume V_{vapor} by RT/P, with $\Delta V_t = V_{\text{vapor}}$, and substituting into Equation 4-75 we get

$$\frac{dP}{dT} = \frac{P\Delta H_t}{RT^2}$$

Substituting $dP/P = d \ln P$ yields

$$\frac{d \ln P}{dT} = \frac{\Delta H_t}{RT^2} \tag{4-76}$$

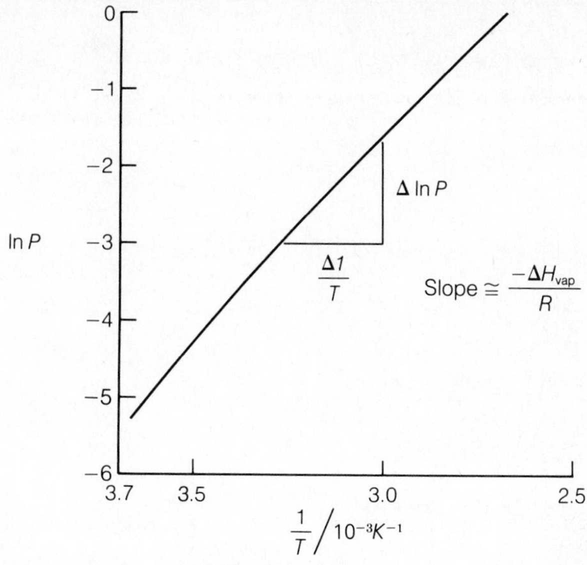

FIGURE 4-14
Variation of ln P (P in atm) with $1/T$ for water. The slope at a given temperature is a good approximation to $-\Delta H_{\text{vap}}/R$ (see Equation 4-77).

which is a useful equation for determining the heat of vaporization or sublimation by measuring the vapor pressure as a function of temperature. If we remember that $d(1/T) = -dT/T^2$, we can convert Equation 4-76 to

$$\frac{d \ln P}{d(1/T)} = \frac{-\Delta H_t}{R} \tag{4-77}$$

which has the virtue that a plot of ln P against $1/T$ is nearly linear because ΔH_t varies only slowly with temperature. Figure 4-14 shows such a plot for the vapor pressure of H_2O. The slope is a good approximation to $-\Delta H_{\text{vap}}/R$.

Equations 4-74 and 4-75 describe the slope of the lines on a phase diagram such as those in Figures 4-12 and 4-13. They can also be used to estimate the influence of pressure on the phase transition temperature.

EXERCISE 4-29

Why does the melting temperature of ice decrease as the pressure increases (see Fig. 4-12)?

ANSWER

Because the volume decreases on melting, $\Delta V_t < 0$. In a melting transition, the entropy always increases, $\Delta S_t > 0$, because heat is absorbed in the process. Therefore ΔS_t and ΔV_t have different signs, and Equation 4-74 indicates that $dP/dT < 0$. Therefore a pressure increase produces a decrease in T_m.

EXERCISE 4-30

The densities of ice and water at 0°C and 1 atm are 0.9917 g cm^{-3} and 0.9998 g cm^{-3} respectively. The molar heat of fusion of ice is 5980 J mol^{-1}. Calculate the change in the melting temperature if the pressure is changed to 2 atm.

ANSWER

From Equation 4-75,

$$\frac{dT}{T} = \frac{\Delta V_t}{\Delta H_t}\, dP$$

Assuming ΔV, ΔH constant,

$$\ln \frac{T_2}{T_1} = \frac{\Delta V_t}{\Delta H_t}(P_2 - P_1)$$

The volume change per mole is

$$\left(\frac{1}{0.9998} - \frac{1}{0.9917}\right) \text{cm}^3\ \text{g}^{-1} \times 18\ \text{g mol}^{-1} = -0.147\ \text{cm}^3\ \text{mol}^{-1}$$

Therefore

$$\ln\left(\frac{T_2}{T_1}\right) = \frac{-0.147\ \text{cm}^3\ \text{mol}^{-1}}{5980\ \text{J mol}^{-1}} \times 1\ \text{atm} \times 0.1013\ \text{J cm}^{-3}\ \text{atm}^{-1}$$
$$= 2.5 \times 10^{-6}$$

Hence $T_2 = T_1 \exp(-2.5 \times 10^{-6})$, and $\Delta T = T_2 - T_1 = T_1(\exp[-2.5 \times 10^{-6}] - 1) = -7 \times 10^{-4}$ °C.

4–15 *Transitions in Biological Macromolecules and Molecular Aggregates*

Biological macromolecules such as proteins and nucleic acids, and molecular aggregates such as ribosomes and lipid vesicles, have highly ordered structures that frequently can be altered by a change in external conditions. A simple example is the unfolding, or denaturation, of chymotrypsin, shown in Figure 4-9. There are many useful parallels between these structural changes and the phase changes covered by the classical theory of heterogeneous equilibria. A macromolecule or molecular aggregate is analogous to an ordered solid phase, and the change in shape is like a solid-solid phase transition. If the analogy can be firmly based thermodynamically, then the theory of phase equilibria can be applied to macromolecules or organelles as separate phases.

Let us consider a solution or suspension of macromolecules or molecular aggregates, *M*. Our problem is to decide whether we should consider the system to be one phase with two homogeneously mixed components or whether the large molecule can constitute a separate phase. The answer depends on how the system behaves, and what is convenient for our analysis. If there are two components in one phase, $f = c - p + 2 = 3$, and we require three intensive variables to specify the system, the natural choices would be temperature, pressure, and macromolecular concentration, C_M. However, if we consider that there are two components (water and macromolecule), and two phases (water and macromolecule), then $f = c - p + 2 = 2$. In this case only two variables, T and P, are needed to describe the system. The concentration of macromolecules has been lost as a variable. This leads us to a general principle: *We can consider the aggregate or macromolecule to be a separate phase only when none of the intensive properties of interest depends on its concentration.*

For example, if we are studying the unfolding of chymotrypsin and find that the transition curve (see Fig. 4-9) is independent of chymotrypsin concentration, we do not need protein concentration as a separate variable and can consider the protein to be a separate phase.

The same principle applies to macroscopic phase equilibria as well. For example, if ice in equilibrium with water is broken into small pieces, it can be considered a separate phase only if the interactions between pieces of ice have no consequences for the thermodynamic properties of the system. *Therefore the smallest possible collection of matter that can be called a phase is that which contains within it all the interactions relevant to the properties under study.*

However, the relatively small size of macromolecules or aggregates leads to an important contrast with macroscopic phase equilibria. Macroscopic phases contain vast numbers of molecules and intermolecular interactions. This allows completely cooperative transitions from one phase to another. However, the transitions of macromolecules or molecular-scale aggregates always show a finite transition breadth. Consider the example of the unfolding of chymotrypsin in Figure 4-9. The transition enthalpy is large enough to give a narrow thermal transition zone, but to yield a phase transition the enthalpy would have to be virtually infinite. That would imply a molecule of macroscopic size.

Transitions in Lipid Bilayers

The influence of particle size on transition width can be seen in the order-disorder transition exhibited by lipid bilayers. The lipid molecules (see Fig. 4-15) in bilayers are oriented with their hydrocarbon "tails" clustered together, and their charged "heads" in contact with the water phase (Fig. 4-16). These bilayers can form spherical particles of 25 nm diameter called *vesicles,* with water trapped inside. They also form large folded structures of less well defined form, called *multilayers*.

As the temperature of a suspension of vesicles or multilayers is increased, the physical properties change within a narrow temperature range, and there is an anomaly in the specific heat, like that seen in Figure 2-3. This phenome-

```
          H
          |                              O
       H—C—O—R                           ‖
          |                   R = —C—(CH2)12—CH3
  R—O—C—H   O                       (Side chains)
          |    ‖
       H—C—O—P—O—CH2—CH2—N(CH3)3
          |    |
          H   O—
          Dimyristryl-L-α-lecithin
```

FIGURE 4-15
A typical lipid capable of forming bilayers in either vesicle (spherical) or multilayer form. The polar "head" (charged end) of the molecule is in contact with water, and the nonpolar hydrocarbon, $—(CH_2)_{12}—CH_3$, side chains form a nonpolar phase. A bilayer structure is shown schematically in Figure 4-16.

non is due to a transition in the bilayer phase in which increased entropy is acquired by the hydrocarbon tails. As diagrammed in Figure 4-15, below the transition all the hydrocarbon chains are fully extended, and above the transition some are bent in a random manner.

Figure 4-17 shows a comparison of the transition in suspensions of vesicles and multilayers. The larger multilayer aggregates clearly have a sharper transition, corresponding to the larger total energy available per particle. This dependence of transition width on the size of the phase is not seen in macroscopic phase equilibria. The finite transition width forces us to define more carefully the transition temperature, which is usually taken as the midpoint of the transition, called T_m, at which half the material has been converted to the high-temperature form.

With this definition of the transition temperature, the results of the preceding sections can be applied directly to transitions involving macromolecules.

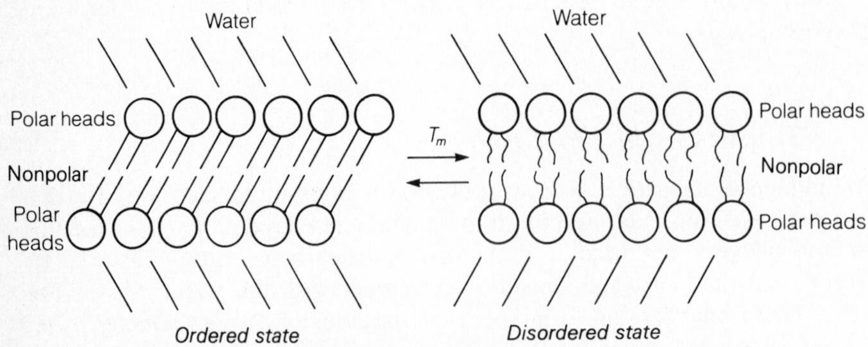

FIGURE 4-16
Schematic diagram of the order-disorder transition in a lipid bilayer. The hydrocarbon side chains (nonpolar region) are extended and packed together in an ordered array at low temperature. Above T_m, the side chains become kinked, and the packing becomes disordered. The transition is analogous to melting a solid to a fluid, or a gel to a liquid.

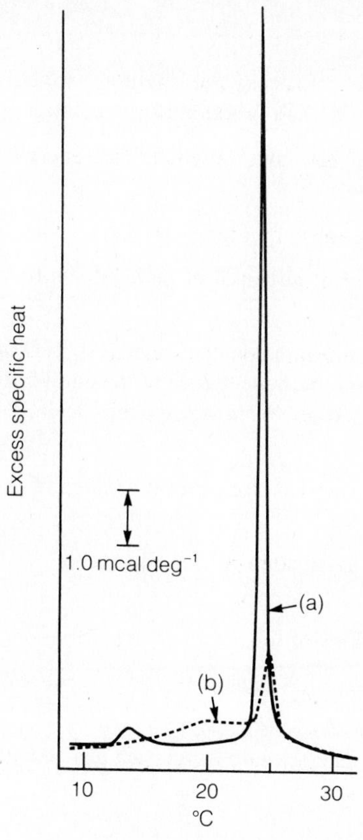

FIGURE 4-17
Comparison of heat uptake due to the "melting" transition (a) in multilayers and (b) in vesicles, of dimyristryl-L-α-lecithin. The total heat uptake is larger for multilayers (6.26 kcal mol^{-1} of lipid) than for sonicated vesicles (4.30 kcal mol^{-1} of lipid). Of special interest is the greater width of the transition for vesicles because of their smaller size. (Figure courtesy of S. Mabrey; see Mabrey and Sturtevant, 1976).

For example, Figure 4-18 shows the effect of pressure on the thermal transition temperature T_m for unfolding metmyoglobin (Zipp and Kauzmann, 1973). The slope $dP/dT_m = \Delta S_t/\Delta V_t$ (Equation 4-74) is negative when $P = 1$ atm at 80°C. It is known that ΔS_t is positive under these conditions, so ΔV_t must be negative. Zipp and Kauzmann calculated values of ΔV_t of about -100 cm³ mol⁻¹ for denaturation of metmyoglobin at pH 5, T = 20° to 40°C, and P = 3000 to 4000 kg cm⁻².

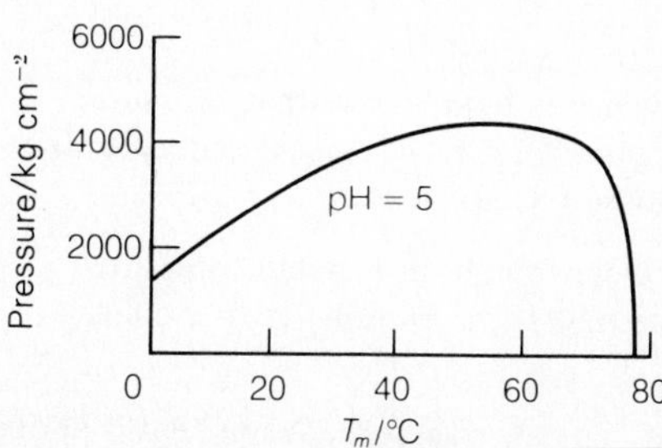

FIGURE 4-18
Variation of T_m for metmyoglobin with pressure at pH = 5. The native state is more stable than the denatured state inside the contour. (From Zipp and Kauzmann, 1973. Reprinted with permission from *Biochemistry 12,* 4217. Copyright by the American Chemical Society.)

QUESTIONS FOR REVIEW

1. Define the Gibbs and Helmholtz free energies. What are the "natural variables" for these functions? Under what conditions is each function at a minimum?
2. What are the relationships to work of the Gibbs and Helmholtz free energies? What is the physical meaning of free energy?
3. Define *mole fraction* and *molarity*. How are these quantities related?
4. Define *chemical potential*. How is the chemical potential of an ideal solution related to the mole fraction and the molarity?
5. Define *equilibrium constant*. How is the equilibrium constant related to the standard free-energy change? How is the temperature dependence of the equilibrium constant related to the standard enthalpy change? What is the Gibbs-Helmholtz equation?
6. Describe four methods of determining the standard free-energy change of a reaction.
7. Define the following terms: triple point, critical point, phase, degree of freedom.
8. State the phase rule and give two applications of it.
9. What is a phase diagram? In what way does the Clausius-Clapeyron equation describe phase diagrams?
10. ● Under what circumstances can one regard an aggregate in solution as a separate phase?

PROBLEMS

1. Are the following statements true or false?
   (a) For a spontaneous process at constant T and P, ΔG is always negative.
   (b) For a spontaneous process at constant T and P, ΔG° is always negative.
   (c) For a spontaneous process at constant T and P, ΔS is always positive.
   (d) The equilibrium constant for a reaction can be calculated from ΔG.
   (e) The temperature dependence of an equilibrium constant can be determined from ΔH° for the reaction.
   (f) The free-energy change of the surroundings needs to be considered in determining spontaneity of a reaction.
   (g) An equilibrium mixture of ice and water is held at constant pressure. When heat is added so that some ice melts, the Gibbs free energy of the system decreases. (*Hint:* Use Equation 4-3, $dG = V\,dP - S\,dT$).
   (h) An equilibrium mixture of ice and water is held at constant pressure. When heat is added so that some ice melts, the Helmholtz free energy decreases.
   (i) A mixture of ice and water is held at constant pressure. When heat is added so that some ice melts, the entropy increases.

Answers: (a) T. (b) F, ΔG determines spontaneity, and even with a positive $\Delta G°$, ΔG can be negative if the concentrations of the reactants are large (Equation 4-51). (c) F, only for an isolated system. (d) F, from $\Delta G°$. (e) T. (f) F. (g) F, the temperature remains constant and $dG = 0$. (h) F, $dV < 0$ so that $dA > 0$. (i) T.

2. Add the appropriate word:

   (a) The Gibbs free energy of a system at constant temperature and ______ decreases during a spontaneous process.

   (b) The negative change in Helmholtz free energy during an ______ process is the maximum work available from the process.

   (c) The amount of work available from a reversible process is ______ than from an irreversible process.

   (d) In a dilute ideal solution, the chemical potential of a solute is proportional to the ______ of its concentration.

   (e) The change with temperature of an equilibrium constant depends on the change in standard ______.

   (f) The product of the melting point and the entropy of fusion is the ______ of fusion.

Answers: (a) Pressure. (b) Isothermal. (c) Greater. (d) Logarithm. (e) Enthalpy. (f) Enthalpy.

PROBLEMS RELATED TO EXERCISES

3. (Exercise 4-1) How do the requirements that T and V are constant enter the derivation of $\Delta A < 0$ for a spontaneous process?

4. (Exercise 4-2) Calculate the change in the Helmholtz free energy of the collagen fiber described in text Exercise 4-2 when it is stretched isothermally from $x = -5$ to $x = 5$.

5. (Exercise 4-3) Explain why a stretched rubber band which is suddenly allowed to contract becomes colder.

6. (Exercise 4-4) The molar heat of vaporization of water is 40.64 kJ mol^{-1} at 100°C. Calculate the molar entropy of vaporization.

7. (Exercise 4-8) Calculate the free energy of mixing 1 mole each of two pure liquids to form an ideal solution at 25°C.

8. (Exercise 4-9) Calculate the molarity of an aqueous solution of sucrose whose sucrose mole fraction is 10^{-4}.

9. (Exercise 4-10) The concentration of a substance A is found to be 100 times higher in a benzene phase than in an aqueous phase equilibrated with the benzene phase at 25°C. Calculate the difference between the standard chemical potentials of A in the two solvents.

10. (Exercise 4-13) Write an expression for the equilibrium constant for the vaporization of water, $H_2O(l) \rightleftarrows H_2O(g)$.

11. (Exercise 4-17) Calculate the concentration of Ba^{2+} in equilibrium with a solution whose F^- concentration is 0.1M. The solubility product of BaF_2 is 1.7×10^{-6} M^3.

12. (Exercise 4-19) Calculate $\Delta G°$ and K at 298 K for the reaction $SO_2(g) + \frac{1}{2}O_2(g) \rightarrow SO_3(g)$, using the data in Table 4-2.

13. (Exercise 4-22) Estimate the partial pressure of NO in equilibrium with 5 atm N_2 and 1 atm O_2 at 100°C.

14. (Exercise 4-23) Calculate K and $\Delta G°$ for denaturation of chymotrypsin (pH = 2) at 70°C, using Equation 4-65.

15. (Exercise 4-25) Acetic acid dissociates in water according to the equation

$$HAc + H_2O \rightleftarrows H_3O^+ + Ac^-$$

Calculate the degrees of freedom f for an aqueous solution of HAc.

OTHER PROBLEMS

16. A 10 g piece of ice is placed in an enclosed box at −1°C and 1 atm pressure. The box also contains a dish of 100 ml of a 1 N solution of NaCl. The ice gradually disappears and the volume of the solution grows. Which of the following is the reason for this?

    (a) The total free energy of the solution is lower than the total free energy of the ice.

    (b) The mean free energy of the solute NaCl and the solvent water is lower than the free energy of the ice.

    (c) The free energy per mole of the H_2O in the solution is lower than the free energy per mole of the H_2O of the ice.

    (d) The free energy per gram of solution is lower than the free energy per gram of ice.

Answer: (c).

17. Prove that $(\partial P/\partial T)_G = S/V$.

18. Use $\Delta A = \Delta E - T\,\Delta S$ to calculate the maximum work that can be done by one mole of an ideal gas in expanding from 10 atm to 1 atm at 300 K.

19. *Trouton's rule* says that the boiling temperature T_b of many liquids is proportional to the enthalpy of vaporization: $T_b \approx \Delta H_{vap}/21$ when T_b is in K and ΔH_{vap} is in calories. What does this imply about the entropy of vaporization?

20. Calculate the change in the Gibbs free energy of one mole of an ideal gas during isothermal expansion from 10 atm to 1 atm at 300 K.

21. An ideal gas is carried through the following reversible cycle: (1) isothermal expansion, (2) heating at constant P, (3) isothermal compression, and (4) cooling at constant P to the starting condition. Draw diagrams showing the qualitative variation of (a) G as a function of T, and (b) G as a function of P.

22. Derive an expression for the isothermal variation of the Helmholtz free energy of an ideal gas with (a) volume, and (b) pressure.

23. Derive an expression for the Helmholtz free-energy change when ideal gases i at pressure P are mixed in mole fractions X_i and constant temperature. Let the volume after mixing be the sum of the individual volumes before mixing.

24. Calculate the difference in the chemical potential of sucrose at 0.01 M and at 0.0001 M concentrations. Assume ideal solutions and $T = 300$ K.

25. The partition coefficient of the drug actinomycin between water and a mixed organic solvent is 0.5 [ratio $C(\text{w})/C(\text{org}) = 0.5$]. If the maximum solubility of actinomycin in water is 1 mg/ml, calculate the maximum solubility in the organic solvent. Justify your calculation with a thermodynamic argument based on chemical potentials. Assume ideal solutions.

26. Calculate the standard free-energy change for the hydrolysis of solid DL-alanylglycine by liquid H_2O to produce solid alanine and solid glycine.

27. Calculate the standard free-energy change for conversion of glucose (s) to 2 ethanol (l) and $2CO_2$(g). What else would you need to know to calculate the free-energy change in solution?

28. Show that Equation 2-18 for the work and 3-28 for ΔS of reversible isothermal expansion of an ideal gas are consistent with $W = \Delta A$.

● 29. Apply Equation 4-19 to the cycle in Figure 3-3 and show that the efficiency of the reversible heat engine is $\epsilon = 1 - T'/T$.

30. Intercalation of the planar drug ethidium between the base pairs in DNA causes unwinding of the double helix by 26° per drug molecule bound. When both strands in the double helix are covalently closed circles, the molecule cannot unwind, and intercalation of the drug causes a change in the *tertiary twist* or *superhelix* content in the molecule. (Superhelical turns are produced when a circular molecule is twisted into a figure-8.) Unwinding of one turn of double helix in a closed circular molecular produces one tertiary turn of positive sign. The number of tertiary turns in the molecule is given the symbol τ.

   (a) Naturally occurring closed circular DNA molecules contain intact double helix and tertiary turns. The closed circular molecules bind the drug ethidium more tightly than do linear (noncircular) molecules when only a few ethidium molecules are bound per DNA. Explain this observation, and predict whether the tertiary turns have a positive or negative sign.

   (b) Assume that the free energy of twisting closed circular molecules depends on τ according to $G = k\tau^2/2$. Show that the standard free energy of binding ethidium to closed circular DNA is (per mole of ethidium)

$$\Delta G^{\circ\prime} = \Delta G^{\circ} + k\tau\left(\frac{26}{360}\right) \tag{1}$$

   in which ΔG° is the standard free energy of binding to the *linear* DNA. *Hint:* The extra free-energy change per ethidium bound to closed circular molecules is $(\partial G/\partial n) = (\partial G/\partial \tau)(\partial \tau/\partial n)$ in which n is the number of ethidium molecules.

   (c) Show that the force constant k in Equation 1 above for twisting closed circular molecules is

$$k = -\frac{RT}{\tau}\left(\frac{360}{26}\right)\ln\left(\frac{K'}{K}\right)$$

in which K is the binding constant of ethidium to linear DNA, and K' is the binding constant to closed circular DNA containing τ superhelical turns. *Hint:* $\Delta G^\circ = -RT \ln K$; use the results of (b).

(d) When large numbers of ethidium molecules are bound to closed circular DNA, τ changes sign. Which should bind ethidium more tightly under these conditions, a closed circular molecule or a linear molecule? Explain your reasoning.

(e) Which should show a greater tendency for local, spontaneous unwinding of double helix, a natural closed circular DNA or a linear DNA? Explain your reasoning.

31. The free energy of a mixture of ideal gases can be written

$$G = \sum_j n_j \mu_j$$
$$= \sum_j n_j \mu_j^\circ + \sum_j n_j RT \ln P_j$$

(This is an example of the *sum rule,* which we will discuss in Chapter 7.) Show by differentiating this expression with respect to the number of moles of one component (n_i) that the equation above is consistent with the more familiar equation

$$\mu_i = \left(\frac{\partial G}{\partial n_i}\right)_{T,P,n_{j\neq i}} = \mu_i^\circ + RT \ln P_i$$

Hint: You will have to use the chain rule for differentiating the second summation. You will get one term, $RT \ln P_i$, plus a term $\sum_j n_j RT \frac{\partial \ln P_j}{\partial n_i}$. Rearrange the term $n_j RT\, d\ln P_j$ so that you can take advantage of the fact that $\Sigma\, dP_j = 0$ (because the total pressure is constant).

32. Show that the chemical potential μ_i can be defined by

$$\mu_i = \left(\frac{\partial A}{\partial n_i}\right)_{T,V,n_j}$$

Hint: Begin with $dG = V\,dP - S\,dT + \Sigma \mu_i\, dn_i$ and $A = G - PV$. Take the differential of the latter, substitute for dG, and use the fact that the result is the complete differential of A.

33. Use an approach similar to that in Problem 32 to show that

$$\mu_i = \left(\frac{\partial H}{\partial n_i}\right)_{S,P,n_j}$$

$$\mu_i = \left(\frac{\partial E}{\partial n_i}\right)_{S,V,n_j}$$

34. When a lipid bilayer undergoes transition from the ordered or "solid" phase to a "liquid" phase, the area of the bilayer increases. This area change is analogous to the volume change in other phase transitions. Pursue this analogy through the following steps.

(a) Write an expression analogous to the three-dimensional expression $-P\,dV$ for the work done on the two-dimensional system when the area changes. Define appropriate substitutes for P and dV, and give units in which they might be expressed.

(b) By analogy with the Clausius-Clapeyron equation, derive an expression for the variation of the transition temperature T_m with the two-dimensional analogue of pressure.

(c) Define the analogue of the three-dimensional isothermal compressibility (Chap. 1) for the two-dimensional system.

(d) Make a *schematic* plot of the quantity defined in (c) against temperature, through the temperature range of the order-disorder transition in the bilayer. Can you see any mechanical advantage to a cell that might arise from operating within the temperature range of the transition?

QUESTIONS FOR DISCUSSION

35. (a) Heat can be converted to work with 100% efficiency in the reversible isothermal expansion of an ideal gas, yet the efficiency of a heat engine is always less than 1. Explain the differences between the two processes that account for the facts.

(b) We have considered the enthalpy change of phase transitions (latent heat) and the entropy change of phase transitions, but not the free-energy change of phase transitions. Why is this?

(c) Reactions that are enormously exothermic are usually—but not always—spontaneous. Why?

(d) In Kurt Vonnegut's novel *Cat's Cradle*, life on earth comes to an end when a crystal of "ice IX" nucleates into a blue solid all water (including that in living organisms) which comes into contact with it. Suppose you are one of the most thermodynamically knowledgeable survivors (having survived by knowing that you must not touch the stuff), but that you are fascinated by the intellectual challenge of the new substance. Analyze its properties.

FURTHER READINGS

See the texts referenced at the ends of Chapter 1 and 2, especially those by Dickerson, Kauzmann, Klotz and Rosenberg, and Moore.

Denbigh, K. 1963. *The Principles of Chemical Equilibrium*. Cambridge, England: Cambridge University Press. A more advanced yet very clear exposition of thermodynamics, including much material on chemical potentials and phase equilibria.

Klotz, I. M. 1967. *Energy Changes in Biochemical Reactions*. New York: Academic Press. An informative and clear exposition of biological thermodynamics, with many examples.

OTHER REFERENCES

Bloomfield, V., Crothers, D. M., and Tinoco, I., Jr. 1974. *Physical Chemistry of Nucleic Acids.* New York: Harper & Row.

Eisenberg, D., and Kauzmann, W. 1969. *The Structure and Properties of Water.* New York: Oxford University Press.

Zipp, A., and Kauzmann, W. 1973. *Biochemistry 12,* 4217.

Kauzmann, W. 1959. *Adv. Prot. Chem. 14,* 1.

Mabrey, S., and Sturtevant, J. 1976. *Proc. Nat. Acad. Sci. U.S.A. 73,* 3862.

Tanford, C. 1968. *Adv. Prot. Chem. 23,* 121.

CHAPTER 5

Biological Applications of Thermodynamics

The main elements of chemical thermodynamics were developed in Chapters 1 through 4. With these in mind it is possible to interpret many of the phenomena in the world around us that involve the transformation of energy. The present chapter is intended to give you some practice in this, particularly with biological examples. Following this excursion and Chapter 6 on kinetics, in Chapters 7–9 we will resume the development of thermodynamics and its application to solution chemistry and to electrochemistry.

The historical sketch for this chapter describes two pioneers in the application of physical concepts in biology, J. D. Bernal and Max Delbrück. Their importance was not in thermodynamic applications but more generally in holding the conviction that organisms can be understood and described by the laws of physics. They were also leaders in interesting other physical scientists in biology, and, in fact, were major influences in starting the field now known as molecular biology.

○ 5–1 Solar Energy

Solar radiation falling on the upper atmosphere amounts to about 1.7×10^{14} kw, and varies only by 3% as the earth moves in its elliptical orbit about the sun. This energy is equivalent to about 2 cal min^{-1} cm^{-2}. Of course the energy that strikes a given point on the earth depends on the latitude, time of day, and prevailing weather. Much of the solar radiation is scattered by water droplets in clouds and by dust. Some is reflected back into space, and some is absorbed by gases of the atmosphere. These processes reduce the solar radiation reaching earth to between 1.5 cal cm^{-2} min^{-1} for a desert area near the equator down to 0 cal cm^{-2} min^{-1}. On a clear day about 10% of radiation reaching the earth comes from scattered light rather than directly from the sun; on a cloudy day most of the light is scattered light.

J. D. BERNAL

1901–1971

AND

MAX DELBRÜCK

BORN 1906

The origins of molecular biology can be traced to two groups of workers, one in England concerned with the structures of biological molecules, and one in the United States concerned with the transfer of genetic information. Each of these groups had an intellectual leader trained in physics and possessing a vision that the details of biological processes could be described by physical laws. The pioneering experiments were performed by these leaders in the 1930s. Each attracted a group of workers around him, but progress was slowed in the 1940s by World War II. Work resumed late in the decade, and then in 1952 a collaboration between members from the two groups produced a great discovery, the DNA double helix. Watson had been trained in the genetics group in the United States and Crick in the structural group in Cambridge, England. Their finding of the double helix—and the flood of molecular biology that followed—can be regarded as a confluence of the streams of thought of these two groups.

More than any other person, J. D. Bernal established the field of structural molecular biology. His lightning-fast scientific understanding and encyclopedic knowledge earned him the nickname "Sage" among his friends. He is said to have read a book a day and to have had so tenacious a memory that facts rarely escaped. His range of interests both in science and outside was immense, and his abilities as a raconteur were fabled. One colleague, A. Mackay, recalled that Bernal ". . . could captivate his audience, whether a scientific meeting or a pretty woman by his discourses on almost any topic. He gave his ideas away freely, never doubting that he had plenty more. He was generous, considerate, and Bohemian in his private life . . ." This Bohemian streak included an unabashed taste for many women, decidedly left-wing political views, and friendships with writers and artists. When Picasso stayed with Bernal in London, he left a drawing on Bernal's wall as a gesture of thanks. The novelist C. P. Snow modeled on Bernal the lively scientist named Constantine in his novel *The Search*.

Bernal was trained in classical crystallography at Cambridge. In 1923 he joined Sir William Bragg's research group in x-ray crystallography of organic compounds at the Royal Institution in London. As an undergraduate at Cambridge he had already written a paper on the derivation of the 230 space groups. When he arrived at the Royal Institution he was confronted with the job of measuring the intensities of x-ray diffraction by the accurate but tedious spectrometer method devised by Bragg. Lacking the patience for this, he invented a simple, rapid x-ray camera from a clock mechanism and spare parts, and then showed how the resulting rotation photographs could be systematically interpreted. Bernal recalled later, "These years at the Royal

Institution were the most exciting and most formative of my scientific life." Another crystallographer training with him there recalled the day when the impatient experimentalist, Bernal, was on his knees looking for his one and only crystal of γ-brass.

In 1927 Bernal returned to Cambridge as a lecturer in crystallography, and for the next eleven years made his most outstanding contributions. One was a monumental paper with R. H. Fowler in 1933 on the structure of water and aqueous solutions. This discussion of the arrangement of water molecules in ice, liquid water, and solutions was the first to conclude that molecules in the liquid are tetrahedrally coordinated by hydrogen bonds, as they are in the solid. The following year Bernal, along with Dorothy Crowfoot (later Dorothy Hodgkin), demonstrated that it might be possible to determine the arrangement of atoms within a protein using x-ray diffraction. This discovery depended on having large single crystals of a protein. Such crystals of the digestive enzyme pepsin had been grown accidentally by a researcher at Uppsala, Sweden, who had left his preparation in the refrigerator while he went on a skiing trip. On his return he was astonished to find crystals some 2 mm in length. A visitor from Cambridge told him, "I know a man in Cambridge who would give his eyes for those crystals," and carried some back to Bernal in the vial in which they were growing in their mother liquor. Bernal soon made the critical observation that a crystal was birefringent under the polarizing microscope—demonstrating the periodicity required to diffract x-rays—but it lost birefringence when removed from its liquid of crystallization. He reasoned that to observe x-ray diffraction from a protein crystal, he would have to mount it in a thin-walled glass capillary which would maintain it in the wet state by equilibrium with some mother liquor. The experiment worked, revealing that the crystal diffracted x-rays very well. That night Bernal walked the streets of Cambridge full of excitement about the future when it might be possible to deduce the atomic structures of protein from their x-ray diffraction patterns. It was some 26 years before Bernal's postdoctoral student, Max Perutz, and Perutz's colleague, John Kendrew, were able to achieve this goal.

Bernal also initiated the studies of viruses by x-ray diffraction with his study of tobacco mosaic virus in 1936. From this beginning some 40 years elapsed before the structure of this virus was determined to near atomic detail. Quicker to yield results, but perhaps no less important, was Bernal's preliminary work in small-angle diffraction, which was carried out when he returned to London as Professor at Birkbeck College in 1938. This technique has become a tool of great usefulness in studying the organization of biological material at the level of cellular and subcellular aggregates, such as the myelin sheath of nerves and the structure of ribosomes.

Bernal's social views were forged during his youth in Ireland at the time of the rebellion against England. He became a devoted Marxist, and later authored several books on the Marxist view of the history and function of science. His left-wing views, however, did not keep him from service on behalf of England during World War II. When the Minister of Home Security, Sir John Anderson, needed a scientific mind to devise ways to protect civilian buildings against bombing, Bernal was recommended to him by several people, but with the qualification that he was a Communist. After two such reports, and still desperately in need of a brilliant mind, Sir John is reported to have said, "He may be as red as the flames of hell, but we must use him." Bernal's analyses of bombing and other military problems became increasingly important during the course of the war. Much later, some military experts gave the opinion that Bernal was the British scientist who had helped most in the English war effort. This is high praise when it is considered that Bernal was presumably being compared to the developers of radar, and also that those making the comparison were utterly opposed to his political views.

Max Delbrück was the leader of the group that was interested in genetics rather than structure. He was brought up in Berlin down the street from Max Planck, from whose garden he picked and ate cherries. Delbrück's father, like Planck, was a professor at Berlin University and his great-grandfather was the renowned chemist von Liebig. Six of Delbrück's relatives by blood or marriage perished during World War II as members of the German resistance to Nazism. Delbrück himself studied astronomy, and later physics. Because he had little background in experimental physics, he failed his Ph.D. oral exam on the first attempt, a "devastating embarrassment" to him. After advanced training with Niels Bohr in Copenhagen and Wolfgang Pauli in Zürich, he became assistant to Meitner and Hahn at the Kaiser Wilhelm Institute for Chemistry. On the side, he formed a very small discussion group of a few theoretical physicists, biochemists and geneticists, who began to consider the nature of genes. Because some members were Jews who had been dismissed from scientific work by the Nazi government, they met at Delbrück's home, and spread out their papers on his mother's sewing board. A few years earlier H. J. Muller had found that ionizing radiations produce mutations. Further work in Berlin had shown that these mutations were caused by either single pairs of ions or small clusters of them. The discussion group reasoned that genes thus have a stability comparable to that of molecules, and might well be molecules. That this could be an open question in the mid-1930s illustrates that it required vision then even to hold the conviction that physical laws describe genes.

In 1937, a year and a quarter before Hahn and Meitner discovered uranium fission, Delbrück left to become a research fellow at California Institute of Technology, and to search for a way to apply his quantitative background to the understanding of genetics. Arriving at Caltech, Delbrück met another research fellow, Emory Ellis, who showed him the "one-step growth" curves which chart as a function of time the number of bacterial viruses (or *phage*) that are present as the phage infect bacteria. These curves indicated that within a half hour of infection the bacteria are broken, and that 100 daughter phage are released for each original phage. Delbrück's first comment on seeing the curves was, "I don't believe it." He and Ellis repeated the experiments and Delbrück put to work his analytical power to describe the results. The main significance of this meeting was that Delbrück had found an organism, phage, that he regarded as being sufficiently simple in its hereditary mechanism to be capable of description and understanding in terms of physical principles. He devoted the next 20 years to this goal, and through his charismatic personality and his insistence on clear thinking, he stimulated the revolution in genetics.

In 1940 Delbrück took a job at Vanderbilt University in Tennessee. The group of investigators that were attracted to his orbit, several of them ex-physicists, became known as the Phage Group. His influence on them went beyond his insight of fundamental importance that phage would prove to be the central tool in deciphering genetics at the molecular level. His papers described elegant experiments in a sparse and logical style. He insisted on clear-cut logic from others, as well as from himself. Numerous stories are told about Delbrück preventing a seminar from proceeding until the speaker clarified a point to Delbrück's satisfaction. Speakers regarded by Delbrück as illogical or obscure were often treated afterwards to a private comment from him to the effect that the seminar was the worst he had ever heard. Most everyone understood that devotion to rigor and passionate rejection of vagueness, rather than cruelty, were his motives for such candid behavior. Members of the Phage Group knew that Delbrück would insist that their results fit logically into a hypothesis, and this goaded them to think clearly about their results before they checked with him for his opinion. But they had to check. For many years nearly every worker in the field of molecular genetics sent each finished manuscript to Delbrück before it was sent on to a journal. Max's approval was the best sign that things made sense. Watson was no exception. He reproduces in his book *The Double Helix* the letter to Max that reported that he and Crick had arrived at the structure of DNA. Delbrück's high standards, as much as any factor, catalyzed the rapid progress of the study of genes at the molecular level.

Yet despite his demanding attitude about seminars and papers, Delbrück extended a warm welcome to generations of graduate students and postdoctoral fellows at Caltech, the institution to which he returned as professor in 1947. He eagerly sought out young arrivals to ask about their work at their previous institutions. What could have been more exciting to a postdoctoral fellow than to have this famous scholar questioning him about the fine points of his Ph.D. thesis? Many of these immigrants to southern California were invited to join in the Delbrücks' frequent weekend camping trips to the southern desert. Saturday in the desert invariably involved a long hike through winding canyons. In the evenings astronomy was a favorite pastime. Delbrück recorded his observations of planetary positions in the night sky for several years, "to see how long it would take to catch up with Copernicus, or at least Ptolemy." On more than one occasion he entertained campers by demonstrating motions of the solar system using the campfire as the sun, and campers, circulating about it according to his shouted directions, as planets and moons.

EXERCISE 5-1

An estimate for the total energy consumed on earth by burning of coal, oil, gas, and wood in the year 1957 was 3.2×10^{16} kcal. Show that if the solar energy falling on a relatively small desert area of 150 miles^2 could be collected and fully utilized, it would suffice for world needs. Make the conservative assumptions that the sun shines on the desert for 500 min day^{-1} ($8\frac{1}{3}$ hours) and that the average solar energy during this period is 1.0 cal cm^{-2} min^{-1}.

ANSWER

The solar energy falling on this desert amounts to $(150)^2$ mile^2 $\times$ [5.280 $\times$ 10^3 feet mile^{-1} $\times$ 12 inches foot^{-1} $\times$ 2.54 cm inch^{-1}]^2 $\times$ [1.0 cal cm^{-2} min^{-1} $\times$ 500 min day^{-1} $\times$ 365 day year^{-1}] = 1.1×10^{20} cal year^{-1} = 1.1×10^{17} kcal year^{-1}.

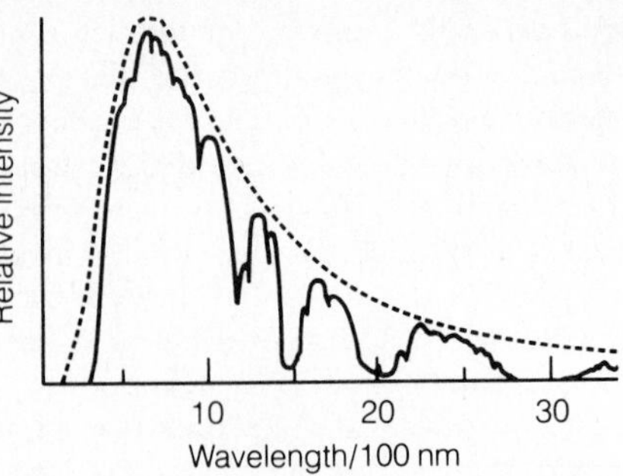

FIGURE 5-1
The spectrum of solar energy. Dashed line: spectrum falling on the upper atmosphere. Solid line: spectrum reaching the surface of the earth. The spectrum on the upper atmosphere is essentially that of a perfect radiator or blackbody heated to 6000°C. The spectrum reaching earth differs because of the absorption of the atmosphere. (From Daniels, 1964.)

Exercise 5-1 illustrates that a major challenge facing modern scientists is to advance methods for the collection, conversion, and distribution of solar energy.

The spectrum of solar energies reaching earth is shown in Figure 5-1. The dashed line represents the spectrum reaching the outer atmosphere, and the solid curve shows the spectrum reaching the surface of the earth. We will see in Chapter 12 that the light in the very left-hand (short wavelength) part of the dashed curve consists of x-rays and ultraviolet radiation. Radiation at wavelengths shorter than 200 nm is absorbed by O_2 and N_2, and most of the radiation from 200 to 300 nm is absorbed by the ozone of the outer atmosphere. At some longer wavelengths, substantial fractions of the solar radiation are removed by absorption of water, ozone, and carbon dioxide.

Collectors of solar energy work on simple principles. In flat-plate collectors, a blackened surface—often sheet aluminum—absorbs much of the radiation that falls on it. The light is converted to heat (by processes discussed in Chap. 12 and 13), and is transferred to moving air or water below. To prevent reradiation and loss of heat, the collector is covered by a sheet of transparent plastic or glass. The plastic or glass transmits much of the solar energy but reflects back inside the infrared radiation (heat) that is emitted by the warmed black surface.

Such flat-plate collectors work well for heating houses or swimming pools in sunny climates, but are not usually effective for boiling water or for other needs which require temperatures above 90°C. However, with parabolic focusing collectors, temperatures up to 3500°C can be obtained. The disadvantages of focusing collectors include (a) their greater cost, (b) the need for moving them to track the sun, and (c) the fact that they can use only the radiation that arrives unscattered by clouds, whereas the flat-plate collectors use scattered as well as direct radiation.

EXERCISE 5-2

What is the minimum time that water of a depth of 2 inches must be held in a flat-plate collector to raise its temperature from 20°C to 50°C on a day when the solar radiation is 1 cal cm^{-2} min^{-1}?

ANSWER

Since 2 inches is about 5 cm, each cm^2 area of the collector represents about 5 cm^3 of water, or about 5 g. From $\Delta H = C_P \Delta T$, we know that about 5 g $\times$ 1 cal g^{-1} °C^{-1} $\times$ 30°C of energy is required = 150 cal. At a rate of 1 cal min^{-1} for each cm^2 of area, 150 min = 2.5 hours are required. In fact, because of heat losses, 3 to 4 hours are required.

5–2 *The CO_2-O_2 Cycle of the Biosphere*

Energy from the sun supports the main matter and energy cycle of life on our planet, the CO_2-O_2 cycle of the biosphere. This cycle is depicted schematically in Figure 5-2. In the upward stage of the cycle, sunlight is absorbed by chlorophyll and other pigments of green plants, and by the photosynthetic microorganisms of the soil and seas. This radiant energy is converted in the process of *photosynthesis* to chemical energy by the combination of carbon dioxide and water into glucose. The overall reaction in green plants can be summarized:

$$6\,CO_2 + 6\,H_2O \rightarrow \underset{\text{(Glucose)}}{C_6H_{12}O_6} + 6\,O_2$$

$$\Delta G^* = 686 \text{ kcal/mole of glucose} = 2870 \text{ kJ/mole of glucose} \qquad (5\text{-}1)$$

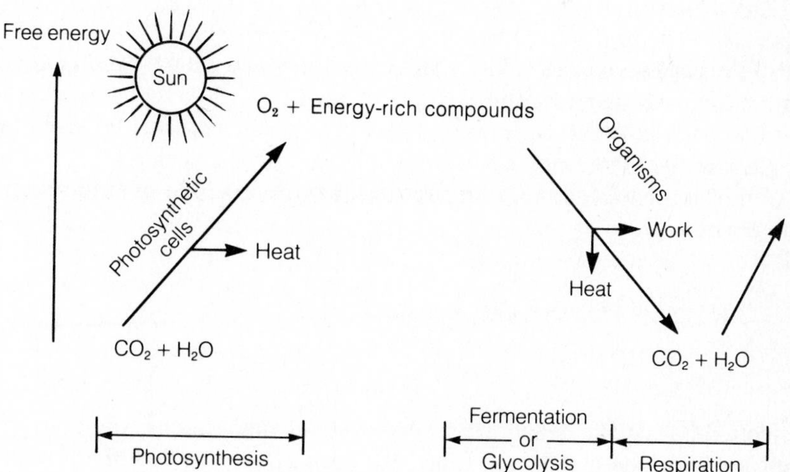

FIGURE 5-2

A schematic representation of the CO_2-O_2 cycle, showing the synthesis of high–free-energy compounds in photosynthetic cells, and their utilization for work in respiring cells. On the far right of the figure the cycle begins its second swing.

This equation summarizes a *very* complicated process by which the sun's energy is converted into high–free-energy molecules in a sequence of steps which are still understood only in part. As in any real process, much of the absorbed energy cannot be utilized for work, and escapes as heat. Lehninger (1971) estimates that the equivalent of about 2000 kcal of absorbed light energy is required to form one glucose molecule, so that the process may be regarded as having an efficiency of about one-third. A vastly greater amount of sunlight escapes absorption by biochemical pigments altogether.

Animals feed on glucose and other energy-rich compounds derived from photosynthesis. These compounds are broken down or *catabolized* to CO_2 and H_2O in the downward leg of the cycle. This yields free energy for the needs of biological organisms. This leg represents the catabolic processes of fermentation and respiration, also termed *biological oxidation*. Fermentation is a process that takes place in cells in the absence of oxygen (anaerobic cells). These cells break glucose into compounds of lower free energy. For some microorganisms—and also our muscle cells under conditions of vigorous exercise (see Sec. 5-4)—the end product is lactate (lactic acid), in others it is acetone, and in still others—the yeast cells used in baking and brewing—it is ethanol. For cells that convert glucose to lactate the overall process of many steps can be summarized:

$$\underset{\text{Glucose}}{C_6H_{12}O_6} \longrightarrow \underset{\text{2 Lactic acid}}{2CH_3CHOHCOOH}$$

$$\Delta G^* = -47 \text{ kcal/mole of glucose} = -197 \text{ kJ/mole of glucose} \qquad (5\text{-}2)$$

Some of the energy released in this conversion can be used for cellular needs, and as in any real process, some is lost as heat.

Aerobic cells, such as those of our body, can derive much more energy from glucose. In the combined processes of glycolysis (which is all but the final step of fermentation) and respiration, they reverse the overall reaction of photosynthesis.

$$\begin{aligned} &C_6H_{12}O_6 + 6\,O_2 \rightarrow 6\,CO_2 + 6\,H_2O \\ &\Delta G^* = -686 \text{ kcal/mole of glucose} \end{aligned} \qquad (5\text{-}3)$$

in which ΔG^* is the standard free-energy change under standard biochemical conditions (*pH* 7, see Chap. 9). If you have studied biochemistry, you know that this equation is actually a summary of about 25 reactions. Many of these steps couple to other reactions in which "high-energy" molecules are made. One of these high–free-energy molecules is adenosine triphosphate (ATP) (Fig. 5-4), the main molecular storage for biological free energy. We can include the results of the coupled reactions in Equation 5-3 by writing

$$\text{Glucose} + 6\ O_2 + 36\ \text{ADP} + 36\ \text{Phosphate} \rightarrow$$
$$6\ CO_2 + 42\ H_2O + 36\ \text{ATP}$$
$$\Delta G^* = -423\ \text{kcal/mole of glucose} \quad (5\text{-}4)$$

This reaction written this way indicates that the breakdown of glucose is coupled to the formation of 38 "high–free-energy" phosphodiester bonds. The difference between 686 kcal in Equation 5-3 for the breakdown of glucose and the 423 kcal in Equation 5-4 represents 263 kcal for the formation of 36 phosphodiester bonds in the ATP molecules, each of which yields considerable free energy when it undergoes hydrolysis.

To explore this topic further we must ask what is the meaning in molecular terms of "coupling" one reaction to another, and what is the meaning of "high–free-energy bond." These two topics are the subject of the following section.

However, before leaving the CO_2-O_2 cycle, a little more needs to be said about the arrow in Figure 5-2 that is labeled "Work."

The kinds of work performed by a cell include (a) mechanical work, such as muscular contraction and other movements; (b) transport work, such as moving Na^+ ion out of a cell against a concentration gradient; and (c) biosynthesis, in which molecules needed by the cell are built up. These molecules include, among others, amino acids, proteins, DNA, RNA, and lipids. All three types of work utilize the stored chemical free energy of ATP. Thus, in summary, the breakdown of glucose yields ATP in coupled reactions, and then this ATP can be used in other coupled reactions to do biosynthesis and other types of work. This summary is shown schematically in Figure 5-3.

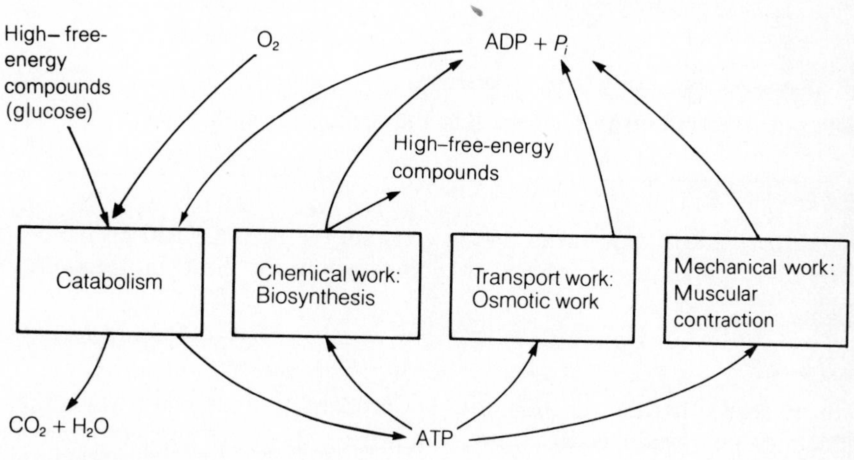

FIGURE 5-3
A block diagram illustrating the overall energetics of respiring cells. The catabolism block represents the oxidation of high–free-energy compounds to CO_2 and H_2O, and the coupling of this oxidation to the synthesis of ATP. As indicated in the other blocks, ATP is then utilized for biosynthesis and for mechanical and transport work. This work is achieved by coupling the energy-yielding hydrolysis of ATP to the work-performing reactions. The hydrolyzed products, such as ADP, are recycled for resynthesis into ATP.

5–3 Coupled Reactions and Group Transfer Potentials: Endergonic Reactions are Coupled to Exergonic Reactions by an Enzyme

We noted in the preceding section that if the free energy released from one process (say, the oxidation of glucose) is to be utilized for another process (say, biosynthesis of an amino acid), there must be some means by which the two processes are coupled. In the absence of coupling, the energy released in the first process would not be harnessed for biosynthesis and would eventually appear as heat.

Let us consider a specific example of coupling: the synthesis of the amino acid glutamine and the simultaneous hydrolysis of ATP to ADP. This is one of many examples that could be chosen. It shares three characteristics with others: (a) an endergonic (free-energy requiring) reaction is coupled with an exergonic (free-energy yielding) reaction so that the combined coupled reaction is exergonic overall and therefore spontaneous; (b) the exergonic reaction is the hydrolysis of ATP to ADP and phosphate; and (c) an enzyme is the coupling factor which unites the two reactions into one so that the exergonic part drives the endergonic part.

The amino acid glutamine is a source of nitrogen in the biosynthesis of many compounds, including amino acids, nucleic acids, and amino sugars. Glutamine is formed by the condensation of NH_3 with glutamate:

$$\underset{\text{Glutamate}}{\begin{array}{l} {}^{+}NH_3{-}CH{-}COO^- \\ \qquad\;\; | \\ \qquad CH_2 \\ \qquad\;\; | \\ \qquad CH_2 \\ \qquad\;\; | \\ O{=}C{-}OH \end{array}} + \underset{\text{Ammonia}}{NH_3} \rightleftarrows \underset{\text{Glutamine}}{\begin{array}{l} {}^{+}NH_3{-}CH{-}COO^- \\ \qquad\;\; | \\ \qquad CH_2 \\ \qquad\;\; | \\ \qquad CH_2 \\ \qquad\;\; | \\ O{=}C{-}NH_2 \end{array}} + \underset{\text{Water}}{H_2O} \qquad \begin{aligned} \Delta G^* &= +3.4\,\text{kcal mol}^{-1} \\ &= +14\ \text{kJ mol}^{-1} \end{aligned} \tag{5-5}$$

The positive sign of ΔG^* shows that the reaction is endergonic.

EXERCISE 5-3

Does the positive ΔG^* mean that at constant temperature and pressure Reaction 5-5 can never proceed spontaneously in the direction from left to right?

ANSWER

No. It shows that the reaction does not proceed spontaneously when the reactants and products are in their standard states (1 M concentrations or unit activities for all species except H^+, which is at 10^{-7} M; we will use concentrations in the following examples).

EXERCISE 5-4

Suppose the concentration of NH_3 is at the physiologically reasonable level of 10 mM (10^{-2} M). What is the ratio of glutamate to glutamine necessary for Reaction 5-5 to proceed spontaneously at 25°C?

ANSWER

For the reaction to proceed spontaneously, ΔG must be negative. ΔG is given by

$$\Delta G = \Delta G^* + RT \ln \frac{[\text{Glutamine}][H_2O]}{[\text{Glutamate}][NH_3]} \tag{5-6}$$

When $\Delta G = 0$, we have

$$RT \ln \frac{[\text{Glutamine}][1]}{[\text{Glutamate}][10^{-2}]} = -\Delta G^* \tag{5-7}$$

where we have recalled the convention that $[H_2O] = 1$.
Solving for the ratio of concentrations of glutamate to glutamine we find

$$\frac{[\text{Glutamate}]}{[\text{Glutamine}]} = 3 \times 10^4 \tag{5-8}$$

This shows that at a reasonable concentration of NH_3, the concentration of glutamate must be enormously greater than that of glutamine for the reaction to move towards the right. This example demonstrates that Reaction 5-5 cannot be utilized *on its own* for the biosynthesis of glutamine.

Nature's method for overcoming the natural direction of an endergonic reaction like 5-5 is to couple it to an exergonic reaction. "Couple" means that as the exergonic reaction is allowed to proceed by an enzyme, the resulting free energy is used to drive the appropriate endergonic reaction. This is achieved often by an enzyme-bound transitory molecule that is an intermediate in both reactions.

The exergonic reaction coupled to the biosynthesis of glutamine (and to many other biosynthetic reactions) is the hydrolysis of ATP (see Fig. 5-4).

$$\text{ATP} + H_2O \rightarrow \text{ADP} + \text{Phosphate} \qquad \Delta G^* = -7.3 \text{ kcal mol}^{-1} = -30 \text{ kJ mol}^{-1} \tag{5-9}$$

The value of ΔG^* for this reaction is so negative that at any physiologically reasonable phosphate concentration, the equilibrium lies far to the right. If we assume a phosphate concentration of 10 mM and proceed as in Exercise 5-4, we find that at equilibrium the ratio of concentrations of ADP to ATP is 2×10^7. Thus if this reaction were to come to equilibrium in a cell, there would be virtually no ATP and the reaction would be useless for cellular needs. Thus thermodynamic analysis by itself tells us that there can be no normally accessible cellular catalyst (that is, no enzyme) that permits Reaction 5-9 to come to equilibrium.

The type of catalyst required is one that permits the hydrolysis of ATP to proceed as it drives an endergonic reaction. An enzyme that couples Reactions 5-5 and 5-9 would catalyze the reaction that is the sum of these two:

$$\text{Glutamate} + NH_3 + \text{ATP} \rightleftarrows \text{Glutamine} + \text{ADP} + \text{Phosphate}$$
$$\Delta G^* = 3.4 - 7.3 = -3.9 \text{ kcal mol}^{-1} \tag{5-10}$$

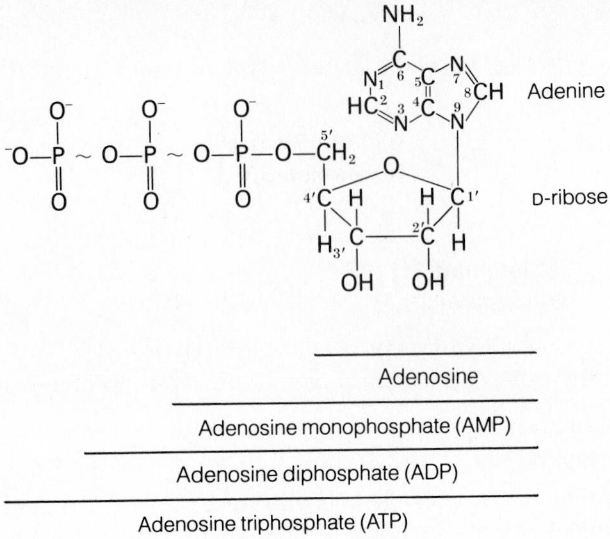

Adenine

D-ribose

Adenosine

Adenosine monophosphate (AMP)

Adenosine diphosphate (ADP)

Adenosine triphosphate (ATP)

FIGURE 5-4
The covalent structure of ATP and its contributing substructures. The symbol ~ between the two pairs of phosphate groups indicates that a large free energy is released when these bonds are hydrolyzed.

This coupled reaction is spontaneous ($\Delta G < 0$) both for the standard conditions (since $\Delta G^* < 0$) and for any reasonable concentrations of ammonia and phosphate. This ensures that when glutamate is available, it will be converted to glutamine to supply the needs of the cell for nitrogen-containing metabolites.

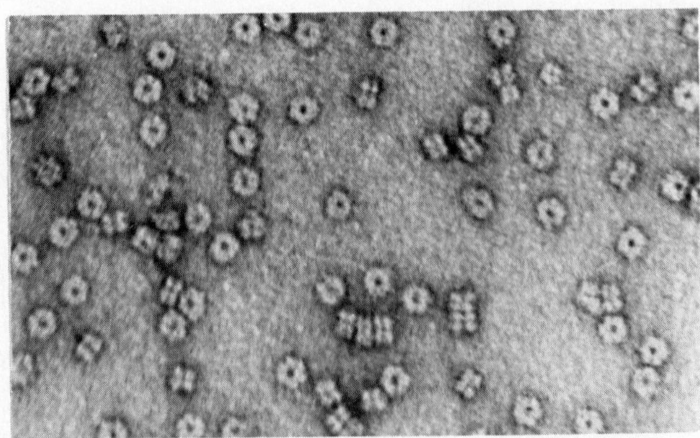

FIGURE 5-5
An electron micrograph of molecules of glutamine synthetase from a bacterium. This enzyme catalyzes the reaction of Equation 5-10. The 12 identical subunits are arranged in two hexagonal rings. The distance across the hexagonal face is about 14 nm and the height of the double ring is about 9 nm.

The enzyme that couples the two half-reactions into Reaction 5-10 is called glutamine synthetase. In the bacterium it consists of 12 identical, nearly spherical polypeptide chains, arranged like the carbon atoms of two benzene rings, eclipsed face to face (see Fig. 5-5). Each of these polypeptide chains has a molecular weight of 50,000 and contains one active site at which Reaction 5-10 is catalyzed. No reaction can take place until ammonia, glutamate, and ATP are in the active site. Thus the enzymatic functions of catalysis and proper coupling of reactions are carried out together.

There is also another function of the enzyme, called *regulation*. If nitrogen-containing metabolites synthesized from glutamine are available in abundance, it is then desirable that Reaction 5-10 supply less glutamine, that is, that it be switched off. The reaction will slow to some extent simply because there is an increasing concentration of glutamine present, and the free energy for the reaction becomes increasingly less negative. But a more sensitive control is desirable. This is achieved by direct interaction of the nitrogen-containing metabolites with glutamine synthetase, thereby making glutamine synthetase a less effective catalyst. This is called *allosteric regulation*.

Summary of the Significance of ΔG and ΔG^* for Biochemical Reactions and Pathways

For an individual reaction such as the formation of glutamine from glutamate (Equation 5-5) to be spontaneous, ΔG for the reaction must be negative. The value of ΔG can be negative even though ΔG^* may be positive, provided the concentrations of the reactants are much higher than those of products. But we have seen that this situation is unlikely even if ΔG^* is only moderately positive. In reality, for a reaction with positive ΔG^* to become spontaneous, it must be coupled to an exergonic reaction, such as the hydrolysis of ATP, and the sum of the ΔG^*s must be negative or close to zero.

The situation is slightly more complicated when we consider a sequence of reactions that constitutes a biochemical pathway. Let us consider a sequence that might represent, say, the synthesis of histidine from glutamate via glutamine

$$\mathrm{A} \overset{\Delta G_1^*}{\rightarrow} \mathrm{B} \overset{\Delta G_2^*}{\rightarrow} \mathrm{C} \overset{\Delta G_3^*}{\rightarrow} \mathrm{D} \tag{5-10a}$$

in which the ΔG^*s give the standard free-energy change for each step. We can write the free-energy change for each step:

$$\begin{aligned} \mathrm{A} \rightarrow \mathrm{B}: \quad & \Delta G_{\mathrm{AB}} = \Delta G_{\mathrm{AB}}^* + RT \ln \frac{[\mathrm{B}]}{[\mathrm{A}]} \\ \mathrm{B} \rightarrow \mathrm{C}: \quad & \Delta G_{\mathrm{BC}} = \Delta G_{\mathrm{BC}}^* + RT \ln \frac{[\mathrm{C}]}{[\mathrm{B}]} \\ \mathrm{C} \rightarrow \mathrm{D}: \quad & \Delta G_{\mathrm{CD}} = \Delta G_{\mathrm{CD}}^* + RT \ln \frac{[\mathrm{D}]}{[\mathrm{C}]} \end{aligned} \tag{5-10b}$$

Then by adding the free energies, we can obtain the free energy for the synthesis of D from A:

$$\Delta G_{\mathrm{AD}} = \Delta G^*_{\mathrm{AB}} + \Delta G^*_{\mathrm{BC}} + \Delta G^*_{\mathrm{CD}} + RT \ln \frac{[\mathrm{B}][\mathrm{C}][\mathrm{D}]}{[\mathrm{A}][\mathrm{B}][\mathrm{C}]}$$

$$= \Sigma\Delta G^* + RT \ln \frac{[\mathrm{D}]}{[\mathrm{A}]} \qquad (5\text{-}11)$$

The point to note is that the total free energy depends on the sum of the standard free energy changes and on the concentrations of A and D, but *not* on the concentrations of any of the intermediate metabolites. In adding these equations we assume that the concentrations of B and C are uniform in the system. Thus in a biochemical pathway in which the intermediary metabolites are at uniform concentrations, the direction of the flux of matter is governed by the sum of the standard free-energy changes, $\Sigma\Delta G^*$, and by the concentrations of the initial and final compounds.

The sign of the overall ΔG for a biochemical pathway must be negative if material is to move through the pathway. It is the continual supply of the first metabolite, A, and the continual removal of the last, D, that keeps the material in flux. Any living system must be in such a steady state.

The actual value of ΔG depends on two factors: (a) an environmental factor, or the rate at which A is supplied and D is utilized; and (b) an evolutionary factor—the sum of the ΔG^*s of the constituent reactions. Evolution has selected reactions and stoichiometries of chemicals that yield an overall ΔG that is negative under normal environmental conditions of concentration and temperature.

EXERCISE 5-5

Calculate the equilibrium constants at 37°C for Reaction 5-3 for the oxidation of glucose, and Reaction 5-4 for oxidation of glucose coupled to the synthesis of ATP with the observed stoichiometry. Comment on the magnitudes.

ANSWER

$K = \exp[-\Delta G^*/RT]$, and $\ln K = -\Delta G^*/RT$. For Reaction 5-4 $\ln K = 423$ kcal mol^{-1}/(1.987 $\times$ 10^{-3} kcal mol^{-1} K^{-1} $\times$ 310 K) = 687. Thus $\log K$ = 687/2.303 = 298. Therefore $K = 10^{298}$. For Reaction 5-3, $K = 10^{483}$.

These equilibrium constants are enormous; that for Reaction 5-4 is much smaller because of the ATP's produced, but is still vast. Thus at any reasonable concentrations of the participating metabolites, the overall reactions are highly irreversible. Possible explanations for this are that irreversibility gives speed or that metabolism is forced in the proper direction even in the most unfavorable values of concentrations.

A "High-Energy Bond" is One Having a Large Negative Free Energy of Hydrolysis

The term "high energy," as applied to various phosphate bonds in biochemistry, tends to obscure the important point about these bonds: they have an unusually large free energy of hydrolysis which can be coupled to endergonic reactions. Klotz (1967) emphasized this point by referring to this property as a group-transfer potential. *The group transfer potential of a chemical species is defined as its standard free energy of hydrolysis,* ΔG_h^*. The concept of group transfer potential is analogous to the proton transfer potential of an acid, or the electron potential (redox potential) of an electron-donating group. For example, we may write the equation

$$AH \rightarrow A^- + H^+ \tag{5-12}$$

for the dissociation of an acid. But we really mean

$$AH + H_2O \rightarrow A^- + H_3O^+$$

since the proton is always hydrated. The free energy per mole for the transfer reaction of 5-12 may be called the *transfer potential.*

The concept of the group transfer potential is similar. The cleavage of a "high-energy bond" is often written

$$A - PO_4^= \rightarrow A + PO_4^{\equiv}$$

but what is really meant is the hydrolysis

$$A - PO_4^= + H_2O \rightarrow A - OH + HPO_4^=$$

The standard free energy of this hydrolysis can be considered the group transfer potential, in analogy to the free energy of proton or electron transfer (see Tab. 5-1).

TABLE 5-1 *A Comparison of Three Types of Transfer Potentials*[a]

	PROTON-TRANSFER POTENTIAL (ACIDITY)	ELECTRON-TRANSFER POTENTIAL (REDOX POTENTIAL)[b]	GROUP-TRANSFER POTENTIAL (HIGH-ENERGY BOND)
Concise equation	$AH \rightarrow A^- + H^+$	$A \rightarrow A^+ + e$	$A \sim PO_4 \rightarrow A + PO_4$
Equation with acceptor	$AH + H_2O \rightarrow A^- + H_3O^+$	$A + H^+ \rightarrow A^+ + \frac{1}{2}H_2$	$A \sim PO_4 + H_2O \rightarrow A{-}OH + HPO_4$
Measure of transfer potential	$pK_a = \dfrac{\Delta G^\circ}{2.303\,RT}$	$\mathscr{E}^\circ = -\dfrac{\Delta G^\circ}{nF}$	ΔG_h^*
Nature of transfer potential	$\propto \Delta G^\circ$ per mole H^+ transferred	$\propto \Delta G^\circ$ per mole e transferred	$\propto \Delta G^\circ$ per mole PO_4 transferred

[a] From Klotz, 1967.
[b] This topic forms a large part of Chapter 9.

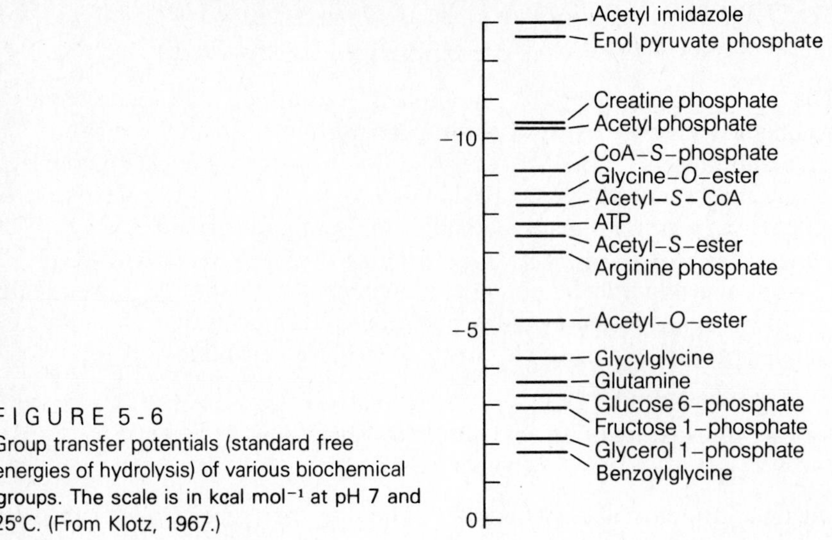

FIGURE 5-6
Group transfer potentials (standard free energies of hydrolysis) of various biochemical groups. The scale is in kcal mol^{-1} at pH 7 and 25°C. (From Klotz, 1967.)

The group transfer potentials for several biochemical molecules are shown in Figure 5-6. There is an almost continual spread from near 0 to 13 kcal/mole-of-transferred group.

5-4 *Metabolic and Thermodynamic Strategies of Bicycle Racing*

Bicycle racing illustrates much about exercise physiology and human energy conversion. Let us start with a brief analysis of the physical factors of bicycle racing and then turn to muscle physiology.

Air Friction Dictates Racing Strategies

At speeds above 10 miles $hour^{-1}$, most of a cyclist's energy is dissipated by friction of the air against the rider and machine. This far exceeds the internal rubbing of the bicycle, which is almost negligible for a well-maintained racing bicycle. A third source of friction, the tires rolling on the road, is only about 0.02 horsepower (15 watts) for a firmly inflated tire at 10 miles $hour^{-1}$ and only double this at 20 miles $hour^{-1}$. Air friction, however, while also only 15 watts at 10 miles $hour^{-1}$, increases with about the third power of the speed, and at 30 miles $hour^{-1}$ is 0.6 horsepower (450 watts). This is illustrated in Figure 5-7. Now, it is a fact that the maximum power that can be applied by a cyclist is about 700 watts for the duration of one minute, and about 300–400 watts for the duration of a 1–4 hour bicycle race. Since this is comparable to the power dissipated by air friction, it is clear that air friction determines the strategy of bicycle racing.

Bicycle racers combat air friction by riding in packs, with each racer riding in the slip stream of the bicycle in front. In this shielded pocket the air friction is reduced by as much as 20%, cutting the energy a rider must exert to keep a given speed. The air friction effect is illustrated in Figure 5-8. Each rider takes his turn "pulling pace" at the front of the pack, shielding the air for those behind, and then rotates back into the pack where he can "sit in" and ride with only 80%–90% his previous exertion. Of course, as the race nears the finish line, this atmosphere of cooperation erodes, and each racer sprints for the winning place.

From this you might conclude that the best sprinter will always win. While good sprinting ability is certainly an advantage, there are other ways to win a bicycle race. An alternative strategy is *attacking*. A racer attacks by increasing the speed of the pack when he rotates into the front position. When this happens, other riders work to keep up because they fear being separated from the pack. Even a space of 100 cm of intervening air exposes a rider to full air friction and may increase his energy dissipation so much that he cannot regain contact with the pack. Riders also fear a *breakaway group*. This is a smaller group of riders that moves ahead of the pack. Once there is even a little intervening space, air friction makes it difficult to cover the open space between the pack and the breakaway group.

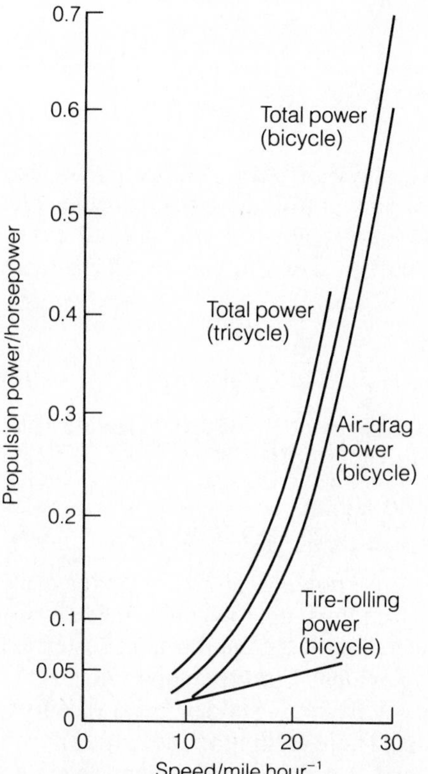

FIGURE 5-7
The power required to propel a racing bicycle at various speeds, and the fractions of this power that go into air friction and rolling friction. The additional power required to propel a tricycle offsets the friction of the third wheel. (Reprinted from *Bicycling Science: Ergonomics and Mechanics* by F. R. Whitt and D. G. Wilson by permission of The MIT Press, Cambridge, Massachusetts. Copyright © 1974 by The MIT Press.)

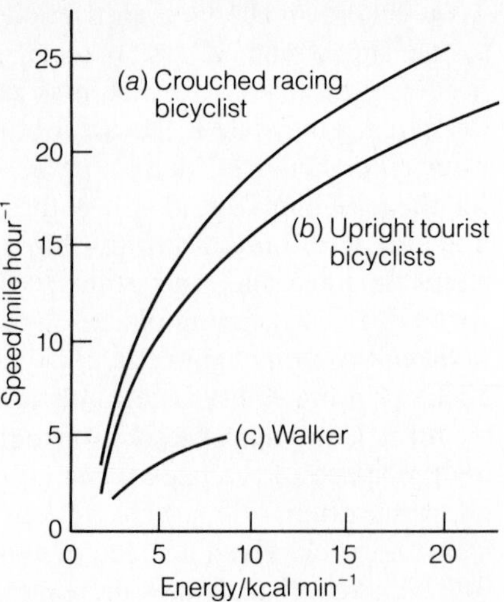

FIGURE 5-8
The energy required for bicycling and walking. The racing bicyclist goes faster for the same input of energy that does the touring bicyclist, because the former presents a smaller cross-section to the air and consequently dissipates less energy into air friction. (Reprinted from *Bicycling Science: Ergonomics and Mechanics* by F. R. Whitt and D. G. Wilson by permission of The MIT Press, Cambridge, Massachusetts. Copyright © 1974 by The MIT Press.)

For these reasons racers push hard to keep up with the pace of the attacker, and a strong attacker can exhaust his opponents. Strong sprinters, as we shall see below, may be especially susceptible to tiring if the race proceeds at a constant fast pace. Thus, towards the finish a successful attacker may have neutralized the sprinting superiority of his competition.

Spriting is Essentially Anaerobic; Attacking is Essentially Aerobic

Muscular contraction is driven by the hydrolysis of ATP. ATP is replenished largely by the breakdown of glycogen, a polymer of glucose in muscles and liver. This process has an anaerobic (airless) segment that is carried out in coupled reactions in which glycogen is broken down to lactate. The overall summary of several steps is given by:

$$\text{(Glucose)}_n + 3\ \text{HPO}_4^{=} + 3\ \text{ADP} \rightarrow 2\ \text{Lactate} + 3\ \text{ATP} + \text{(Glucose)}_{n-1} \qquad (5\text{-}13)$$

Then there is an aerobic (oxygen-using) segment in which pyruvate (one reaction step from lactate) is oxidized to CO_2 and H_2O:

$$2\ \text{Pyruvate} + 5\ \text{O}_2 + 30\ \text{ADP} + 30\ \text{HPO}_4^{=} \rightarrow 6\ \text{CO}_2 + 26\ \text{H}_2\text{O} + 30\ \text{ATP} \qquad (5\text{-}14)$$

The reaction sequence summarized in 5-13 is known as *glycolysis,* takes place in the cytoplasm of muscle cells, and requires no oxygen. The reaction sequence summarized in 5-14 takes place in self-contained organelles called *mitochondria,* and requires oxygen, as is evident from the equation.

During the first moments of exercise, or during a sudden burst of effort, the source of ATP is anaerobic glycolysis. There is no time for extra oxygen to be carried by the blood to the tissues and to diffuse to mitochondria where it is required for the steps of oxidative phosphorylation that are summarized in Reaction 5-14. As anaerobic exercise takes place the concentration of lactic acid increases. Lactate is a high–free-energy compound compared to CO_2, and for the main synthesis of ATP to be carried out, lactate must diffuse toward the mitochondria, be converted to pyruvate, and finally be oxidized in the citric acid cycle. The diffusive steps involved with this oxidative metabolism are relatively slow. This is the reason that lactate piles up under conditions of vigorous exercise. Physiologists say that under these conditions the athlete incurs an "oxygen debt," to be repaid as pyruvate is slowly oxidized.

The application of this to bicycle racing is as follows. Sprinting is a short-term, extremely vigorous exertion, the ATP for which must be supplied by the rapid process of anaerobic glycolysis. After this transitory effort, the excess lactate can be oxidized at the slower pace of the citric acid cycle. The excess lactate has also another effect. It reduces the pH of the muscle, leading to muscle fatigue—a feeling that athletes know well.

Not surprisingly, good sprinters are often large-muscled people, whose tissues hold much glycogen and glucose for anaerobic glycolysis, and who can tolerate large amounts of lactate in the muscles. They can generate on the order of 700–900 watts in a sprint situation, such as in the final moments of a race. The best attackers are often of a different body type. Their exertion is made over a relatively long period of time, and the greater part of this ATP must come from aerobic metabolism. Therefore, they must be particularly efficient at supplying oxygen to their tissues. A slender body with relatively small muscle mass per volume of blood vessel would seem to be optimal for an attacker. In fact, good hill climbers and attackers are often thin. They can probably generate on the order of 400–600 watts over periods of 1 to 10 minutes.

The strategy of an attacker is clear. He can supply oxygen to his tissues with great efficiency and thus can carry out aerobic metabolism at a relatively high rate. His goal is to force the pace of the race to a point that just permits him to metabolize aerobically, but that forces his sprinter opponents into anaerobic glycolysis. Their oxygen debt will increase during these periods, perhaps to the level where they can no longer keep up with the pack. Once the strong sprinters have been dropped, the attacker can relax the pace, knowing that he stands as good a chance as the others to win.

EXERCISE 5-6

Show that 1 kcal min^{-1} = 69.73 watt.

ANSWER

Recall that 1 watt = 1 J s^{-1}, and 1 cal = 4.184 J. Then

$$\begin{aligned} 1 \text{ kcal} &= 4.184 \times 10^3 \text{ J} \\ 1 \text{ kcal min}^{-1} &= 4.184 \times 10^3 \text{ J min}^{-1} \times 60 \text{ min s}^{-1} \\ &= 69.73 \text{ J s}^{-1} = 69.73 \text{ watt} \end{aligned}$$

EXERCISE 5-7

Let us define the efficiency of cycling as the rate of work actually applied to the bicycle divided by the rate of work expended by the cyclist. Calculate the efficiency of cycling at 20 miles hour^{-1}, using Figure 5-7 for the rate of work applied and Figure 5-8 for the rate of work energy expended.

ANSWER

From Figure 5-7 the power expended is 0.23 horsepower × 746 watt-horsepower^{-1} = 172 watt. From Figure 5-8, curve *a*, at 20 miles hour^{-1} the rate of work expended is 12 kcal min^{-1} = 837 watt. Therefore, the efficiency is (172/837) 100 = 21%. This figure is close to that of 25% arrived at by detailed studies of the mechanical efficiency in aerobic muscular exercise (Margaria, 1976).

EXERCISE 5-8

Why is muscular exercise only 20%–25% efficient?

ANSWER

The nonexpansion work available from any process (such as glycolysis and respiration) is as great as the Gibbs free-energy change of the process only when the process is perfectly reversible (Sec. 4-3). But the production of ATP by the oxidation of glucose is highly irreversible, as is the hydrolysis of ATP to drive muscular contraction. If these processes were reversible, then the concentrations of all the hundreds of chemicals involved would have their equilibrium values, and human movement would be infinitely slow.

EXERCISE 5-9

The world's record for distance traveled by an unaided bicycle rider on a level track in one hour is 49.408 km, set by E. Merckx in Mexico City in 1972. Given the data in Figure 5-7 and the above estimate of 21% for efficiency of bicycle transportation, determine the number of pounds of sugar that Merckx would have had to ingest at the conclusion of his achievement to replenish his energy. Note that carbohydrate yields about 4 kcal g^{-1} of free energy.

ANSWER

49.4 km hour^{-1} = 30.7 mile hour^{-1}, which from Figure 5-7 suggests that his bicycle required about 0.7 horse power (5.2×10^2 watt) to propel it. Assuming an efficiency of 21% (this may be too low for the greatest cyclist of modern times), Merckx's mean requirement for power would have been $5.2 \times 10^2/0.21 = 2.5 \times 10^3$ watt. In one hour he would have expended 3600 s × 2.5×10^3 J s^{-1} = 9×10^6 J = 2.2×10^3 kcal. This is equivalent to 2.2×10^3 kcal × $\frac{1}{4}$ g-of-carbohydrate kcal^{-1} = 5.4×10^2 g = 1.2 lb. of carbohydrate.

A physiological study of running, which illustrates some of the same points, is summarized in Figure 5-9. The solid lines plot the consumption of oxygen and the production of lactate in normal subjects constrained to run on a treadmill such that they require energy in the amount shown on the horizontal axis. At energy requirements above 220 cal kg^{-1} min^{-1}, lactate builds up in the blood. At energies below this value, respiration can keep pace with glycolysis and there is no buildup of this oxygen debt. The point at which an oxygen curve turns horizontal represents the maximum rate of respiration of the subject. The dashed lines in the figure are for long-distance runners, who have enhanced aerobic capacity. This enhanced capacity means that their oxygen curves turn horizontal only at a greater energy requirement, and thus the curve for their buildup of lactate is displaced toward greater energy requirement. The slope of the lactate line (230 cal g^{-1}) is the same for both normal and athletic subjects because it represents the energy liberated per gram of glucose that undergoes glycolysis.

A measure of aerobic capacity used by physiologists is V_{O_2}, the rate of oxygen consumption of the subject in ml min^{-1} per kg of body mass. An athlete can consume oxygen at a rate of 40 ml min^{-1} kg^{-1}.

One other point about the science of bicycle racing that deserves comment is the preference of racers for pedaling at high rates. Rates of 60–120 revolutions per minute of the pedals are common in races. Two reasons for this can be given, one a matter of physics and the other a matter of physiology. The first is that the maximum force that can be applied to the pedals is not much greater than the product of the mass of the rider and the acceleration of gravity. Since a rider cannot increase his force, the only way for him to increase the power he puts into the machine is to increase the *rate* of pedaling. The second reason for high rates is that rapid pedaling is accompanied by frequent contractions of the leg muscles. These contractions help to push blood up the veins to the heart and to perfuse blood into the tissues. This more efficient circulation makes pedaling at high rates more comfortable for experienced riders.

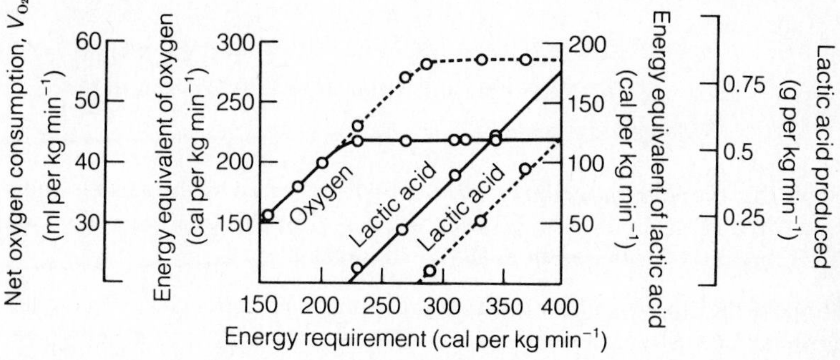

FIGURE 5-9
Rate of oxygen consumed (left ordinate) and net lactate produced (right ordinate) in human subjects on treadmills. Solid line: normal subjects. Broken line: athletes. (From Margaria, 1976. Used with permission of the American Physiological Society.)

PROBLEMS

1. Are the following statements true or false?

   (a) The flux of solar radiation on the outer atmosphere of the earth is a function of the season of the year, time of day, and weather.

   (b) The spectrum of solar radiation reaching the earth differs at short wavelengths from that impinging on the upper atmosphere.

   (c) ΔG^* for a reaction in a biochemical pathway must be negative.

   (d) ΔG for a reaction in a biochemical pathway must be negative.

   (e) ΔG^* for an overall biochemical pathway is likely to be negative.

   (f) Biological cells are nearly in a state of thermodynamic equilibrium so that they can derive maximum work from high–free-energy compounds.

   (g) Humans can convert more than 50% of the free energy of the oxidation of glucose into useful work.

   (h) The biological oxidation of one gram of ethanol yields more free energy than oxidation of one gram of carbohydrate, but less than the oxidation of one gram of fat (see Problem 7).

   (i) The biological oxidation of glucose is nearly reversible.

Answers: (a) F. (b) T. (c) F. (d) T. (e) T. (f) F. (g) F. (h) T. (i) F.

2. Add the appropriate word.

   (a) ________ collectors of solar energy can achieve high temperatures but are not useful for collecting scattered light.

   (b) ________ processes are those that proceed in the absence of oxygen.

   (c) A "high-energy bond" is one with a large free energy of ________ .

   (d) The build-up of lactate during vigorous exercises is called the creation of an ________ ________.

Answers: (a) Focusing. (b) Anaerobic. (c) Hydrolysis. (d) Oxygen debt.

3. Show that the solar radiation of 1.7×10^{14} kw received by the earth is equivalent to 2.4×10^{15} kcal min^{-1} or 2.0 cal cm^{-2} min^{-1} for the area receiving radiation. Take the radius of the earth as 3.8×10^3 miles.

4. Suppose the biochemical pathway represented in Equation 5-10a is coupled at its first step to the hydrolysis of ATP to ADP. How would this change the expression for ΔG of the pathway?

5. Answer the following questions on the basis of the free energies of formation at 25°C shown in the table at the top of the next page:

   (a) Does the peptide bond between alanine and glycine form spontaneously under standard conditions?

SUBSTANCE	ΔG_f^0/kcal mol^{-1}
DL-alanine (aq)	−89.11
DL-alanylglycine (aq)	−114.57
Glycine (aq)	−89.14
H_2O (l)	−56.69

(b) Suppose alanine and glycine have the same concentration in aqueous solution. At equilibrium, what would be the ratio of concentrations of peptide to one of the amino acids?

6. Estimate the amount of additional exercise that you would have to perform to lose $\frac{1}{2}$ lb. of fat a week. Use the following table for the energy expenditures of various activities. (Notice that thinking is not listed as an activity that requires energy.) To use this table you must know your surface area in m². You can obtain a good estimate from the relationship

$$\text{Area/m}^2 = 0.202 \times W^{0.425} \times H^{0.725}$$

in which W is your weight in kg and H is your height in m. One g of fat corresponds to 9 kcal of free energy.

ACTIVITY*	ENERGY/kcal m^{-2} hour^{-1}
REST	
Sleeping	35
Lying awake	40
Sitting upright	50
LIGHT ACTIVITY	
Writing, clerical work	60
Standing	85
MODERATE ACTIVITY	
Washing, dressing	100
Walking (3 mph)	140
Housework	140
HEAVY ACTIVITY	
Bicycling	250
Shivering	up to 250
Swimming	350
Lumbering	350
Skiing	500
Running	600

* (From Brown, 1973.)

7. In 1960, R. F. Randall bicycled 1000 miles in 2 days, 10 hours, and 40 minutes. Assume that 10 hours of this time was devoted to rest and refreshment off the bicycle.

   (a) Estimate the energy Randall expended during his feat.

   (b) Calculate the number of grams of carbohydrate Randall would have had to consume to replenish expended energy (1 g carbohydrate corresponds to 4 kcal of free energy).

(c) Make the same calculation for fat and protein (which yield respectively 9 and 4 kcal g^{-1} of free energy).

(d) Suppose instead that one elects to replenish the energy with a 50% solution of ethanol. Ethanol yields 7 kcal g^{-1} when oxidized to CO_2 and H_2O. How many liters of solution would have to be imbibed?

8. Some exercise physiologists believe that muscle glycogen must be nearly exhausted before the body derives energy from oxidation of fats from adipose tissue. Suppose your muscles hold enough glycogen for you to cycle 30 miles at 20 miles per hour. How many more miles would you have to cycle at the same rate to lose one pound of fat?

9. During one stage (a single day's racing) of the 1976 Tour de France, competitors climbed a total of over 20,000 feet on their bicycles. Assuming a rider weighs 165 lb. and his bicycle 20 lb., calculate the minimum work (in J and g of carbohydrates) he performed that day.

QUESTIONS FOR DISCUSSION

10. Why is the efficiency of running or walking so much less than that of bicycling (try to run 25 miles if you have any doubts about this proposition)? How is it that dogs can run 20–25 mile $hour^{-1}$? (*Hint:* consider the mass of the foot in your analysis.)

11. Why is it that sprinters require only 1 or 2 breaths as they run the 100 m dash?

12. Why did E. Merckx select Mexico City for his successful attempt to break the one-hour distance record?

FURTHER READINGS

Alberty, R. A. 1969. *J. Biol. Chem., 244,* 3290–3302; and *J. Am. Chem. Soc., 91,* 3899–3903. These two papers present a thorough analysis of the thermodynamics of the hydrolysis of ATP.

Atkinson, D. E. 1977. *Cellular Energy Metabolism and Its Regulation.* New York: Academic Press. The authoritative monograph on the subject.

Dickerson, R. E. 1969. *Molecular Thermodynamics.* Menlo Park, Cal.: W. A. Benjamin, Inc. Chapter 7 on Thermodynamics and Living Systems makes stimulating scientific reading.

Daniels, F. 1964. *Direct Use of the Sun's Energy.* New York: Ballantine Books. A handbook describing progress in this area up to the early 1960s.

Klotz, I. M. 1967. *Energy Changes in Biochemical Reactions.* New York: Academic Press. An informative and clear exposition of biological thermodynamics. Many examples.

Lehninger, A. L. 1971. *Bioenergetics,* 2nd ed. Menlo Park, Cal.: W. A. Benjamin, Inc. A summary at an elementary level of the thermodynamics and biochemistry of biological energy transformation.

Margaria, R. 1976. *Biomechanics and Energetics of Muscular Exercise*. Oxford: Clarendon Press.

Whitt, F. R., and Wilson, D. G. 1974. *Bicycling Science*. Cambridge, Mass.: M.I.T. Press. An intriguing collection of information from physics, engineering, and sports, but weak in its integration of biological science. What is the flaw in the demonstration on page 26 that the second law of thermodynamics does not apply to humans?

OTHER REFERENCES

Brown, A. C. 1973. *Physiology and Biophysics*, 12th ed., ed. T. C. Ruch and H. D. Patton. Philadelphia: W. B. Saunders Co.

CHAPTER 6

Chemical and Biochemical Kinetics

6–1 Rates, Not Energy Differences, Control the Course of Biochemical Reactions

In the preceding chapters we considered the properties of systems at equilibrium, for which we did not need to include time as a variable. Of course, the real world is not at equilibrium: the flow of energy from the sun and from processes within the earth's core provides the driving force for natural phenomena ranging from the biochemical to the geological. A physical chemist unfamiliar with the study of rates and mechanisms of natural processes would be ill-equipped to study biochemistry—or any other part of the real world.

The study of thermodynamics taught us that reactions proceed spontaneously if there is a decrease in free energy when temperature and pressure are held constant. We were cautioned that the term "spontaneous" does not imply that a reaction necessarily proceeds at a measurable rate, but we were given little opportunity to ponder the serious limitation this sets on the thermodynamic description of biochemistry. For example, consider a bowl of sugar (sucrose) in contact with air. Oxidation of this sucrose,

$$C_{12}H_{22}O_{11} + 12\ O_2 \rightarrow 12\ CO_2 + 11\ H_2O$$

proceeds with a free-energy change of −5693 kJ per mole when all gases are present at 1 atm pressure. Thermodynamics allows that the reaction is spontaneous, but our experience tells us that a flash fire in a sugar bowl is an event rarely, if ever, seen. You may sense an even greater level of personal concern when informed that the carbohydrates, proteins, and nucleic acids in your body are thermodynamically unstable with respect to hydrolysis to sugars, amino acids, and nucleotides. Yet life continues, so these essential substances evidently are not unstable in an immediate sense.

The study of kinetics teaches us that processes differ enormously in their rates. Many biochemical reactions that are thermodynamically spontaneous proceed at negligible velocities. The selected few that do occur with measurable rates are those catalyzed by enzymes. Thus enzymes, by accelerating reaction rates, organize the myriad of thermodynamically permitted reactions between the biochemicals in an organism into a meaningful metabolism. In this sense *rates*, not energy differences, direct the course of biochemical metabolism.

The importance of relative rates in dictating reaction products is not confined to biochemistry. An organic chemist usually cannot predict a product by calculating which of the possible compounds has the lowest free energy. Instead, he or she must run the reaction and analyze the products to discover those that are kinetically favored. On a grander scale, if thermodynamic principles dominated our world the earth would be incinerated by nuclear fusion of the deuterium atoms in the oceans to form helium by the process that actually occurs in a hydrogen bomb. It is clear that the principles of kinetics must be considered alongside those of thermodynamics in determining the properties of macroscopic systems.

○ 6–2 *What Kineticists Do—A Summary*

We can define a kineticist as a scientist who studies the rate of a process in order to increase our understanding of a chemical, biochemical, or physical transformation. Usually, a simple measurement of the rate is only one of the experiments the kineticist would do. For example, an organic chemist studying a molecular rearrangement might analyze the product distribution. She or he might also use isotopic substitution of a particular atom in the structure to allow identification of the source of atoms in newly formed bonds. Another relevant question would be whether optical configuration at a chiral center is retained, inverted, or lost. Similarly, a biochemist studying oxidation reactions might suspect the presence of a free-radical intermediate. An electron paramagnetic resonance measurement (Chapter 13) could directly verify this.

The kineticist's general purpose in these experiments is to infer for the process under study a plausible *reaction mechanism,* which describes in detail the transient intermediates that intervene between reactants and products. Because the intermediates are unstable, they often cannot be isolated and identified. Hence kineticists must place an unusual reliance on indirect evidence for their conclusions. This accounts for the variety of experimental approaches taken, in addition to simple measurement of the rate of a process.

When the kineticist measures the reaction rate, the result is expressed as the rate of change of the amount of a substance present in the reaction

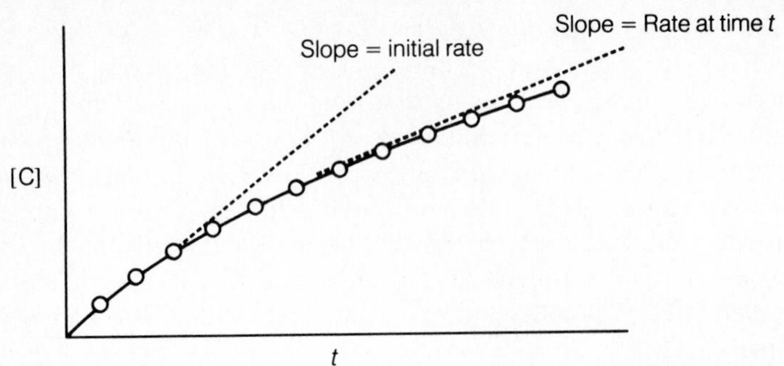

FIGURE 6-1
Determination of the rate of a chemical reaction from the slope of a plot of concentration of the product C against time. The *initial rate* of a reaction is the rate at $t = 0$ when the product is present in negligible amounts. For most chemical reactions the rate decreases as t increases.

mixture. Usually the expression is given in terms of an intensive variable, the concentration, or

$$\text{Rate} = \frac{d[\mathrm{C}]}{dt} \tag{6-1}$$

in which $d[\mathrm{C}]/dt$ is the rate of change with time t of the product concentration [C]. As shown in Figure 6-1, the rate is thus the slope of a plot of concentration against time.

Of course, the chemical reaction rate can be expressed as the time rate of change of any of the concentrations. For example, in the general reaction $\mathrm{A} + \mathrm{B} \rightarrow \mathrm{C}$, one molecule of B disappears for each molecule of A that reacts, and one molecule of C appears as a result. Therefore the rate of appearance of the product equals the rate of disappearance of the reactants, or

$$\frac{d[\mathrm{C}]}{dt} = \frac{-d[\mathrm{A}]}{dt} = \frac{-d[\mathrm{B}]}{dt}$$

The only time this expression is not valid is when the transient build-up of intermediates delays appearance of the products relative to disappearance of the reactants.

When the stoichiometric coefficients are not all unity, more care must be taken to express the rate in a consistent fashion. For example, in the reaction $2\mathrm{A} \rightarrow \mathrm{C}$, two molecules of A disappear for each molecule of product C that appears. Therefore, the rate of change of [C] will be half as great as the rate of change of [A], $-\frac{1}{2}(d[\mathrm{A}]/dt) = d[\mathrm{C}]/dt$. It is customary to express this fact by dividing each derivative by its stoichiometric coefficient in the reaction.

$$\nu_\mathrm{A}\mathrm{A} + \nu_\mathrm{B}\mathrm{B} \rightarrow \nu_\mathrm{C}\mathrm{C} \tag{6-2a}$$

$$\text{Rate} = \frac{1}{\nu_\mathrm{C}}\frac{d[\mathrm{C}]}{dt} = -\frac{1}{\nu_\mathrm{A}}\frac{d[\mathrm{A}]}{dt} = -\frac{1}{\nu_\mathrm{B}}\frac{d[\mathrm{B}]}{dt} \tag{6-2b}$$ ◀

EXERCISE 6-1

Express the rate of the reaction $2A \rightarrow 3C$ in terms of the time rate of change of the concentrations.

ANSWER

$$\text{Rate} = \frac{1}{3}\frac{d[\text{C}]}{dt} = -\frac{1}{2}\frac{d[\text{A}]}{dt}$$

The kineticist has a number of methods available for determining $d[\text{C}]/dt$. For example, if the reaction is slow one might remove small samples of the reaction mixture at various times and analyze chemically for the concentration of a reactant or product. Faster reactions can often be followed by spectroscopic methods. For example, the oxidation or reduction of heme-containing proteins such as cytochrome *c* leads to a change in the transmission of light by the solution. Therefore, the time rate of change of transmission at an appropriate wavelength can be used to infer $d[\text{C}]/dt$. Other methods used by kineticists include changes in the conductivity of a solution, and more complicated techniques that we will discuss later, such as broadening of an NMR resonance (Chapter 13) or absorption of energy by a solution due to resonance matching between the frequency of an applied perturbation (a sound wave) and the response frequency of the chemical reaction.

Once the reaction rate has been measured, the kineticist's next step is to establish the *rate law*. This is done by showing how the observed rate depends on the concentration of reactants (and products, if the reverse of the reaction occurs at an appreciable rate). For example, it might be found that the rate of a reaction $A + B \rightarrow C$ is linear in the concentrations of A and B:

$$\text{Rate} = \frac{d[\text{C}]}{dt} = k[\text{A}][\text{B}] \tag{6-3}$$

in which k is called the *reaction rate constant*. This equation gives the rate law for the process; in general, the rate law for a reaction must be determined by experiment and *cannot* be inferred from the reaction stoichiometry.

A more general form of Equation 6-3 is

$$\text{Rate} = \frac{d[\text{C}]}{dt} = k[\text{A}]^{\alpha}[\text{B}]^{\beta} \tag{6-4}$$

The superscripts α and β are used to define the *order* of a reaction: if α is 1, the reaction is said to be "first-order in A," if it is 2 the reaction is "second-order in A," etc. The sum of all the exponents in the rate law, $\alpha + \beta$ in this case, is called the *total reaction order*. In simple cases α and β are integers, but this is not always the case.

EXERCISE 6-2

Atkinson *et al* (1965) found that at very low concentrations the rate law for the enzyme-catalyzed oxidation of isocitrate by the NAD-dependent isocitrate dehydrogenase of yeast is given by

$$\text{Rate} = k[E][S]^4[A]^2[N]^2[Mg^{2+}]^2$$

in which E, S, A, and N represent respectively enzyme, substrate, AMP, and NAD. What is the total reaction order?

ANSWER

11.

Having established the rate law, the kineticist can calculate a value for the reaction rate constant. Notice that the constant k must have units and that its value may depend on the concentration units chosen. For example, according to the rate law 6-3, the rate has units concentration $\times$ time^{-1}; in order that the product $k[A][B]$ have the same units or dimensions, k must have units concentration^{-1} $\times$ time^{-1}.

EXERCISE 6-3

What are the units of the rate constant for a reaction with overall order of n?

ANSWER

Concentration^{1-n} time^{-1}.

The value of a rate constant has implications for the reaction mechanism. Some molecules react every time they collide; an example is the reaction of H^+ and OH^- in solution. Such reactions are said to be *diffusion limited* because of their dependence on the rate of molecular motion (diffusion) which brings the partners together. Most reactions occur more slowly, which implies that not every collision is reactive.

Failure of colliding partners to react on every collision can be due to two general features:

1. An intermediate species formed during the reaction may have higher energy than readily available from the kinetic or potential energy of the reactants. The "energy barrier" to reaction that results is called the *activation energy* of the process.
2. The reacting partners may have to orient in a certain way before reaction can occur. This requirement can be expressed quantitatively in terms of the *activation entropy* for the process.

The kineticist can use the temperature dependence of the reaction rate constant to help dissect the reaction barrier into its energetic and entropic components. As temperature increases the molecular kinetic energy increases, so the reacting partners are more likely to have sufficient energy to form the necessary intermediate of high energy. In Section 6-11 we will investigate the relationship between the rate constant and the energy and entropy barriers to the reaction.

6–3 A Reaction Mechanism Consists of a Set of Elementary Steps

A primary objective of kinetic studies is to infer from the data a plausible reaction mechanism, which consists of a series of what are called *elementary steps*. These have the following four important features, to which an additional three will be added in subsequent sections.

1. Elementary reaction steps result from processes in a single molecule, or from reactive collisions between two or sometimes three molecules.
2. No chemically stable intermediate can be detected between reactants and products in an elementary step.
3. The rate law for an elementary step can be deduced from its stoichiometry.
4. The sum of the elementary steps is the net reaction.

According to the first two of these conditions, the elementary steps account for all the intermediates for which the experimenter has evidence. For example, in the binding of the flat drug proflavine (P) by DNA (D) to form a complex $(PD)_{in}$ with the drug intercalated between the base pairs, the net reaction is

$$P + D \rightleftarrows (PD)_{in}$$

However, there is kinetic evidence for an intermediate, a complex $(PD)_{out}$ in which the drug is attached more loosely to the outside of the double helix. In this case the two elementary reaction steps are

$$P + D \underset{k_{-1}}{\overset{k_1}{\rightleftarrows}} (PD)_{out} \qquad (1)$$

$$(PD)_{out} \underset{k_{-2}}{\overset{k_2}{\rightleftarrows}} (PD)_{in} \qquad (2)$$

The first of these steps results from a *bimolecular* collision between the drug and DNA, and the second results from a *unimolecular* process that involves an intramolecular rearrangement of the drug-DNA complex. At present, no evidence exists for further intermediates in the process, and each of these reactions is postulated to be an elementary step. However, it is usually not possible to prove the nonexistence of other intermediates, so reaction mechanisms can generally be revised by doing further experiments that disclose additional intermediates and hence additional elementary steps.

Condition (3) that an elementary step must obey follows from conditions (1) and (2). Stated in more detail, an elementary step that involves ν_A molecules A will be ν_Ath order in A. For example, let the step be a unimolecular decay of A to B, with rate constant k:

$$A \xrightarrow{k} B \qquad (6\text{-}5a)$$

Since the A molecules are independent, the fraction $d[A]/[A]$ of molecules that decay in the time interval dt will be simply proportional to dt:

$$\frac{d[\mathrm{A}]}{[\mathrm{A}]} = -k\,dt \tag{6-5b}$$

or

$$\frac{d[\mathrm{A}]}{dt} = -k\,[\mathrm{A}] \tag{6-5c}$$

(The minus sign is needed because [A] decreases.) Hence *a unimolecular elementary reaction step has a first-order rate law.*

The rate of bimolecular elementary reaction steps is proportional to the frequency of collisions per unit volume between the reacting partners, such as A and B in that most famous of chemical reactions

$$\mathrm{A} + \mathrm{B} \xrightarrow{k} \mathrm{C} \tag{6-5d}$$

The frequency of collisions per unit volume that a given A molecule experiences with B molecules is proportional to the concentration of B; the total collision frequency per unit volume is obtained by multiplying this expression by the concentration of A. Hence

$$\frac{d[\mathrm{A}]}{dt} = -k[\mathrm{A}][\mathrm{B}] \tag{6-5e}$$

Thus, *a bimolecular elementary step has a second-order rate law.* The generalization of this result is that stated above: an elementary step that involves ν_{A} molecules A is ν_{A}th order in A.

EXERCISE 6-4

Write the rate laws for the elementary reaction steps (1) and (2) for proflavine reacting with DNA (see page 217).

ANSWER

$$\frac{d[(\mathrm{PD})_{\mathrm{out}}]}{dt} = k_1[\mathrm{P}][\mathrm{D}] - k_{-1}[(\mathrm{PD}_{\mathrm{out}})] \qquad (1)$$

$$\frac{d[(\mathrm{PD})_{\mathrm{in}}]}{dt} = k_2[(\mathrm{PD})_{\mathrm{out}}] - k_{-2}[(\mathrm{PD})_{\mathrm{in}}] \qquad (2)$$

Notice that both reactions are *reversible,* meaning that a rate term must be included for both forward and reverse reactions.

The fourth condition that constrains the set of elementary reaction steps is that their sum must yield the net reaction. All transient intermediates produced during the reaction must also be consumed—hence they appear as products in one reaction and reactants in another. In taking the sum of elementary steps the transient intermediates must therefore cancel. Failure of a reaction mechanism to satisfy this condition means that it does not accurately describe the reaction.

6-4 An Equilibrium Constant Equals a Ratio of Rate Constants

A very important concept in physical chemistry is the relationship between the reaction rate constants and the equilibrium constant. Consider the elementary reaction steps

$$\begin{aligned} A + B &\underset{k_{-1}}{\overset{k_1}{\rightleftarrows}} C && (1) \\ C &\underset{k_{-2}}{\overset{k_2}{\rightleftarrows}} D && (2) \\ \hline A + B &\rightleftarrows D \end{aligned} \qquad (6\text{-}6a)$$

The rate laws for the individual elementary steps are

$$\frac{d[C]}{dt} = k_1[A][B] - k_{-1}[C] \qquad (1)$$

$$\frac{d[D]}{dt} = k_2[C] - k_{-2}[D] \qquad (2)$$

When a given step is at equilibrium, the concentration is time-independent. Therefore

$$\begin{aligned} k_1[A][B] - k_{-1}[C] &= 0 && (1) \\ k_2[C] - k_{-2}[D] &= 0 && (2) \end{aligned} \qquad (6\text{-}6b)$$

These equations can be rearranged to

$$\begin{aligned} \frac{[C]}{[A][B]} &= K_1 = \frac{k_1}{k_{-1}} && (1) \\ \frac{[D]}{[C]} &= K_2 = \frac{k_2}{k_{-2}} && (2) \end{aligned} \qquad (6\text{-}6c)$$

in which K is the equilibrium constant for the elementary step. Because the rate law for an elementary step can be deduced from its stoichiometry (as can the equilibrium constant) this result is readily generalized to the fifth constraint on the elementary steps:

5. The equilibrium constant for an elementary step is the ratio of its forward and reverse rate constants.

A sixth condition that applies to the set of elementary reaction steps is also easily derived. For the same reason that transient intermediates cancel in taking the sum of elementary reaction steps, the transient intermediate concentrations vanish in taking the product of equilibrium constants for all elementary steps. (Every time an intermediate appears in the numerator as a product it will appear again in the denominator as a reactant). For the example of Equations 6-6:

$$K_1K_2 = \frac{k_1k_2}{k_{-1}k_{-2}} = \frac{[C][D]}{[A][B][C]} = \frac{[D]}{[A][B]} = K_T$$

in which K_T is the equilibrium constant for the total reaction. Hence, in general:

6. The equilibrium constant for the net reaction (K_T) is the ratio of products of forward (k_i) and reverse (k_{-i}) rate constants for the elementary steps, or

$$K_T = \frac{\Pi k_i}{\Pi k_{-i}} \tag{6-7}$$ ◀

6–5 *The Principle of Microscopic Reversibility Requires that a Reaction Have the Same Intermediates in Forward and Reverse Directions*

Beginning practitioners of kinetics are sometimes tempted to postulate that a chemical reaction proceeds through different intermediates in forward and reverse directions, for example, in the hypothetical cyclic mechanism

(6-8)

in which it is proposed that A is converted to C through the intermediate B, and C is converted back to A through the intermediate D. This can be expressed quantitatively by neglecting both the back reactions from C to A through B, and the forward reactions from A to C through D. However, the implication of assuming a value of zero for these rate constants is that the equilibrium constant

$$K_{AC} = \frac{k_{AB}k_{BC}}{k_{BA}k_{CB}}$$

is infinity when calculated through intermediate B, whereas the equilibrium constant for the reverse process

$$K_{AC}^{-1} = \frac{k_{CD}k_{DA}}{k_{DC}k_{AD}}$$

is also infinity when calculated through intermediate D. Obviously it is impossible that both K_{AC} and $1/K_{AC}$ be infinity. Hence one cannot neglect the forward reaction in one path and the reverse reaction in the other path.

A general principle that applies to this problem is called the *principle of microscopic reversibility,* which can be rigorously derived by the methods of statistical mechanics for a system at equilibrium. The principle states that *the forward and reverse transformation rates must be equal for any step in a process at equilibrium.* Applied to chemical reactions such as Equation 6-8, it is sometimes called the *principle of detailed balancing*. For example, when Reaction 6-8 is at equilibrium, it is necessary that

$$\text{Rate}(A \rightarrow B) = \text{Rate}(B \rightarrow A) \tag{6-9}$$
(detailed balance at equilibrium)

(Notice that we implicitly used this principle in writing Equations 6-6*b* for Reaction 6-6*a* at equilibrium.)

You can get a general idea of the origin of the principle of microscopic reversibility by considering the consequences if it were not valid. If the rate $A \rightarrow B$ were greater than $B \rightarrow A$ at equilibrium, each of the rates $B \rightarrow C$, $C \rightarrow D$, and $D \rightarrow A$ would also have to be greater than their reverse rates in order to prevent build-up of the concentration of any species, which is not permitted at equilibrium. In this case there would be a preferred direction of operation of the reaction cycle. Such a spontaneous cycle in a system at equilibrium is not consistent with the principles of thermodynamics.

The consequences of the principle of detailed balance for a reaction mechanism can be summarized in a seventh (and final) constraint on the elementary steps:

7. The equilibrium constant (calculated as the ratio of the product of forward rate constants to the product of reverse rate constants) must be unity for any *cyclic* path in the reaction mechanism.

As you can recognize from the equation $\Delta G° = -RT \ln K$, this condition restates the requirement that ΔG be zero for any cyclic process, utilizing also the relationship of rate and equilibrium constants.

○ 6–6 *How Reaction Mechanisms are Inferred*

A mechanism proposed for a given chemical reaction is usually the result of several lines of evidence. In studying this subject, you should realize that it is not possible to prove that a particular reaction mechanism is *correct,* but it is possible to demonstrate that another mechanism is *incorrect*. For example, one can show that the formation of gaseous phosgene from carbon monoxide and chlorine gases

$$CO + Cl_2 \rightarrow COCl_2 \tag{6-10}$$

does not proceed by a simple elementary bimolecular reaction step as might be erroneously inferred from the overall reaction. Three general lines of experimentation can be followed to support this conclusion. These apply to the study of reaction mechanisms in general:

1. Independent evidence is sought for the existence of specific intermediates or molecular rearrangements that are predicted only by certain reaction mechanisms.

2. The reactions proposed must be consistent with general chemical knowledge, including experimental evidence on the reactivity of analogous systems.
3. The rate law must be consistent with the proposed mechanism.

The first of these approaches might be applied to the phosgene system by using, for example, isotopically labeled chlorine gas *Cl—Cl* in a mixture that also contains Cl_2 in another isotopic form. Mass spectroscopy of the products would reveal whether the Cl atoms that were bonded together in Cl_2 remain bonded to the same molecule in $COCl_2$. The negative answer that is found implies that a single Cl can be transferred independently at some stage in the reaction and argues strongly against bimolecular reaction of Cl_2 with CO.

The second approach to constructing a reaction mechanism emphasizes chemical knowledge and intuition. For example, it is found in many reactions of Cl_2 that radical chain reactions occur which involve Cl· atoms. One might therefore propose that a step in the reaction mechanism is

$$Cl_2 \rightleftarrows 2Cl\cdot$$

This hypothesis would encourage the investigator to try an experiment in which Cl· radicals would be trapped, perhaps by reaction with an added radical. A positive result would encourage further exploration of mechanisms involving Cl·.

The third line of evidence used to support a proposed reaction mechanism is comparison of the observed rate law with that predicted by the mechanism. One of the major preoccupations of chemical kinetics is the calculation of rate laws from proposed reaction mechanisms. This problem can be examined in terms of the specific example of phosgene synthesis, Equation 6-10. The rate of phosgene formation is found experimentally to follow the initial rate law (when the Cl_2CO concentration is so small that the back reaction can be neglected):

$$\frac{d[Cl_2CO]}{dt} = k[Cl_2]^{3/2}[CO] \tag{6-11}$$

Our problem is to devise a mechanism and show that it is consistent with the experimental observations. With the chemical data pointing toward a chain reaction mechanism involving Cl·, an appropriate set of elementary steps to explore is

$$Cl_2 \underset{k_{-1}}{\overset{k_1}{\rightleftarrows}} 2Cl\cdot \tag{6-12a}$$

$$Cl\cdot + CO \underset{k_{-2}}{\overset{k_2}{\rightleftarrows}} Cl\dot{C}O \tag{6-12b}$$

$$Cl\dot{C}O + Cl_2 \underset{k_{-3}}{\overset{k_3}{\rightleftarrows}} Cl_2CO + Cl\cdot \tag{6-12c}$$

Notice that Step (*a*) initiates the reaction by producing Cl· radicals, but it contributes only insignificantly to the net reaction stoichiometry. Step (*b*) consumes Cl· , but Step (*c*) leads to production of another Cl· radical, which can cycle through Step (*b*) again. Hence Step (*a*) is needed to start the reaction, and helps determine the rate by influencing the amount of Cl· present.

Our next problem is to derive a rate law for the mechanism in Equations 6-12. There are three general approaches or assumptions of interest to us in this connection:

1. Direct computation can be used to solve for the time dependence.
2. A rate-limiting step can be assumed, with the transient intermediate concentrations determined by an assumed *pre-equilibrium* condition.
3. A rate-limiting step can be assumed, with the transient intermediate concentrations determined by an assumed *steady-state* condition. This assumption is less restrictive than (2).

1. *Direct Computation to Plot the Time Development of the System*
Differential equations can be written for the concentration of each species present, which can be solved by computer methods. The number of independent variables is limited by conservation of matter, as the following exercise shows.

EXERCISE 6-5

How many independent variables are there in the reaction mechanism 6-12?

ANSWER

(See the discussion of the number of components in the treatment of the phase rule, Section 4-12). The chemical species present are Cl_2, Cl· , CO, $Cl\dot{C}O$, and Cl_2CO. Three conservation equations can be written, one each for Cl, C, and O. For example, assuming the total amount of Cl is constant,

$$2[Cl_2] + [Cl\cdot] + [ClCO] + 2[Cl_2CO] = \text{constant}$$

Five species minus three conservation equations leaves two independent variables for the time dependence.

EXERCISE 6-6

Write rate equations that describe the time development of Mechanism 6-12.

ANSWER

Only two independent rate equations can be written, because the system has only two independent variables. For example, one might choose

$$\frac{d[Cl_2CO]}{dt} = -k_{-3}[Cl_2CO][Cl\cdot] + k_3[Cl\dot{C}O][Cl_2]$$

$$\frac{d[Cl\cdot]}{dt} = 2k_1[Cl_2] - 2k_{-1}[Cl\cdot]^2 - k_2[Cl\cdot][CO] + k_{-2}[Cl\dot{C}O] - k_{-3}[Cl_2CO][Cl\cdot] + k_3[Cl\dot{C}O][Cl_2]$$

(The factor of 2 is needed for k_1 and k_{-1} because the stoichiometric coefficient is 2 in Reaction 6-12*a*.)

With modern computers one can rapidly solve this coupled set of first-order nonlinear differential equations, varying the assumed values of the rate constants and repeating the calculation in an attempt to get agreement with experiment. This might be called the "brute force computer approach," which unfortunately does not usually provide much intuitive enlightenment.

The other two methods generally used rely on the simplification of identifying a *rate-limiting step,* the one reaction in the set whose rate is the slowest, and which therefore determines the rate. Using that approach, the overall rate can be set equal to the expression for the rate law of the rate-limiting step. For example, in the mechanism 6-12, suppose that Step *c* is rate-limiting; then the rate of product formation is (assuming that the back reaction from the product can be neglected)

$$\frac{d[\mathrm{Cl_2CO}]}{dt} = k_3[\mathrm{Cl\dot{C}O}][\mathrm{Cl_2}] \tag{6-13}$$

However, this expression is not useful until one can decide on a method for calculating the concentration of the transient species $\mathrm{Cl\dot{C}O}$. This is usually done by either the pre-equilibrium (2) or steady-state (3) approximation.

2. Pre-equilibrium

In this case, the reaction Steps 6-12*a* and 6-12*b* which precede the rate-limiting step are assumed to be at equilibrium:

$$K_1 = \frac{[\mathrm{Cl}\cdot]^2}{[\mathrm{Cl_2}]} \qquad K_2 = \frac{[\mathrm{Cl\dot{C}O}]}{[\mathrm{Cl}\cdot][\mathrm{CO}]}$$

Eliminating $[\mathrm{Cl}\cdot]$ from these two equations yields

$$[\mathrm{Cl\dot{C}O}] = K_2[\mathrm{Cl}\cdot][\mathrm{CO}] = K_2K_1^{1/2}[\mathrm{Cl_2}]^{1/2}[\mathrm{CO}] \tag{6-14}$$

Substituting this result into Equation 6-13 gives

$$\frac{d[\mathrm{Cl_2CO}]}{dt} = k_3K_2K_1^{1/2}[\mathrm{Cl_2}]^{3/2}[\mathrm{CO}] \tag{6-15}$$

Notice that this expression is identical to the experimental rate law (Equation 6-11) if one sets

$$k = k_3K_2K_1^{1/2}$$

thus allowing us to interpret the experimental rate constant k in terms of the properties of the elementary steps. As shown in Problem 6-12, assuming that the second step in Reaction 6-12 is rate-limiting yields the rate law, rate = $k_2K_1^{1/2}[\mathrm{Cl_2}]^{1/2}[\mathrm{CO}]$, which is not in agreement with experiment. It is often found that the choice of the rate-limiting step can be made on the basis of the experimental rate law.

3. The Steady-State Assumption

This is a more general approach to the problem of solving for the rate law. In the simplest case one still assumes a single rate-limiting step, so that Equation 6-13 remains valid for the initial rate. However, we now calculate the concentration $[\text{Cl}\dot{\text{C}}\text{O}]$ by assuming that $\text{Cl}\dot{\text{C}}\text{O}$ is present as a transient intermediate whose concentration is small, with a time derivative that is negligible. At the very beginning of the reaction there is a brief period during which the concentration $[\text{Cl}\dot{\text{C}}\text{O}]$ builds up. Once the *steady-state* is reached, $[\text{Cl}\dot{\text{C}}\text{O}]$ varies only slowly with time.

The steady-state assumption is used by writing the rate law for the concentration of the steady-state intermediate and setting that derivative equal to zero. This approach is valid as long as the rates of transformations into and out of the steady-state intermediate are large compared to the change in its concentration. Hence, adding the rates for forming and decomposing $\text{Cl}\dot{\text{C}}\text{O}$ gives

$$\frac{d[\text{Cl}\dot{\text{C}}\text{O}]}{dt} = k_2[\text{Cl}\cdot][\text{CO}] - k_{-2}[\text{Cl}\dot{\text{C}}\text{O}] - k_3[\text{Cl}\dot{\text{C}}\text{O}][\text{Cl}_2] = 0 \qquad (6\text{-}16)$$

The terms on the right must be large compared to $d[\text{Cl}\dot{\text{C}}\text{O}]/dt$, but they effectively cancel. Notice again that we are neglecting the back reaction from the final product Cl_2CO. Equation 6-16 can be solved for $[\text{Cl}\dot{\text{C}}\text{O}]$, yielding

$$[\text{Cl}\dot{\text{C}}\text{O}] = \frac{k_2[\text{Cl}\cdot][\text{CO}]}{k_{-2} + k_3[\text{Cl}_2]} \qquad (6\text{-}17)$$

For simplicity we eliminate $[\text{Cl}\cdot]$ by assuming that the first reaction step is at equilibrium (a steady-state condition could also be used on $[\text{Cl}\cdot]$). Substituting $[\text{Cl}\cdot] = K_1^{1/2}[\text{Cl}_2]^{1/2}$ into Equation 6-17 gives

$$[\text{Cl}\dot{\text{C}}\text{O}] = \frac{k_2 K_1^{1/2}[\text{Cl}_2]^{1/2}[\text{CO}]}{k_{-2} + k_3[\text{Cl}_2]} \qquad (6\text{-}18)$$

Inserting this equation into Equation 6-13 provides the result

$$\frac{d[\text{Cl}_2\text{CO}]}{dt} = \frac{k_3 k_2 K_1^{1/2}[\text{Cl}_2]^{3/2}[\text{CO}]}{k_{-2} + k_3[\text{Cl}_2]} \qquad (6\text{-}19)$$

When $k_{-2} \gg k_3[\text{Cl}_2]$, this result is identical with Equation 6-15 (remember that $k_2/k_{-2} = K_2$). Thus the steady-state approach tells us that $k_3[\text{Cl}_2]$ must be much smaller than k_{-2} in order for the experimental rate law to be valid. It also suggests that if we examine the kinetics at very large $[\text{Cl}_2]$ we might find deviations from the experimental rate law (Equation 6-11). We will make further use of the steady-state approximation in considering the kinetics of enzyme-catalyzed reactions.

6–7 First-Order Reactions Follow an Exponential Time Course

First-order reactions have a simple mathematical form. Let us suppose that the irreversible reaction

$$A \rightarrow B$$

can be described by the rate equation

$$-\frac{d[A]}{dt} = k[A] \tag{6-20}$$

so that it is a first-order reaction by our definition. We can solve this equation for the concentration of A at any time t in terms of the concentration of A at time zero, A_0. To do so we first separate variables

$$\frac{d[A]}{[A]} = -k\,dt$$

and then integrate both sides to get

$$\ln [A] = -kt + c$$

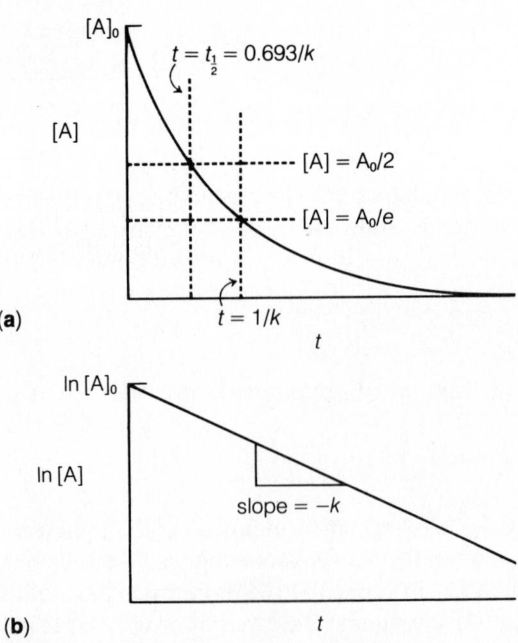

FIGURE 6-2
(a) Exponential decay of [A] in a first-order reaction according to the equation $[A] = A_0 \exp(-kt)$. At $t = t_{1/2} = 0.693/k$, [A] has decayed to half its initial value A_0; $t = 1/k$ when $[A] = A_0/e$. **(b)** Linearization of the decay curve by plotting ln [A] against t. The rate constant can be evaluated readily because the slope of the line is $-k$.

in which c is an integration constant. We can evaluate c by noting that when $t = 0$, $\ln[A] = \ln[A]_0 = c$. Therefore,

$$\ln[A] = -kt + \ln[A]_0 \qquad (6\text{-}21a)$$

or, on taking the exponential of both sides,

$$[A] = A_0 \exp(-kt) \qquad (6\text{-}21b)$$

Equation 6-21*b* reveals that the concentration of A decays exponentially to zero, starting from the value A_0 when the reaction is initiated, as shown in Figure 6-2(*a*). The time required for A to reach $1/e$ of its initial value is $1/k$.

Equation 6-21*a* provides a convenient expression for evaluating k because it is linear in t. A plot of ln[A] against t gives a straight line with slope $-k$ [Fig. 6-2(*b*)].

Another measure of the time constant for a first-order reaction is the *half-life* ($t_{1/2}$), defined as the time necessary for [A] to decay to half its initial value [Fig. 6-2(*a*)]. At $t_{1/2}$, $[A] = A_0/2$. Substituting in Equation 6-21*a* we find that

$$\ln(1/2) = -kt_{1/2}$$

Therefore

$$t_{1/2} = 0.693/k \qquad (6\text{-}22)$$

6–8 Radioactive Decay is a First-Order Reaction

Radioactive labeling, so important in biochemistry and medicine, and radiocarbon dating used in archeology, are based on a first-order reaction—radioactive decay. Many of the unstable isotopes used in biochemistry decay by emitting an electron (a β-particle, or e^-) and a neutrino, thus converting a neutron in the nucleus to a proton. For example, tritium (^3H) decays to the stable nucleus ^3He by β-decay

$$^3H^+ \rightarrow {}^3He^{2+} + e^- + \text{neutrino}$$

The energy released by the nuclear transition is carried away by the kinetic energy of the electron and the neutrino. Since the energy is not partitioned between the two in a fixed ratio, there is a distribution of electron kinetic energies. In addition, different nuclei differ widely in the average kinetic energy of emitted electrons, as shown in Table 6-1.

It is the kinetic energy of the emitted electron that allows these isotopes to be detected. High-energy electrons, such as those emitted by ^{32}P, are able to pass through the window of a Geiger counter and cause ionization of the gas inside, resulting in flow of an electric current. Weak β-emitters such as ^3H cannot be detected this way because the electrons are absorbed by the window. However, all β-emitters can be detected by a *liquid scintillation*

TABLE 6-1 *Average Energy of Emitted Electrons, and Half-Lives of Common Radioisotopes*

ISOTOPE	E_{avg}/MeV	$t_{1/2}$
^{3}H	0.0055	12.3 years
^{14}C	0.050	5580 years
^{32}P	0.700	14.3 days
^{35}S	0.0492	88 days
^{45}Ca	0.077	165 days
^{131}I	0.188	8.05 days

E_{avg} is the average kinetic energy of emitted electrons in millions of electron volts.
$t_{1/2}$ is the isotope half-life (see Eq. 6-23*b*).

counter in which a special fluid to which the isotope is added emits light in response to the electrons produced in the nuclear decay process.

It has been found experimentally that the rate of decrease in the number of unstable nuclei, $-dN/dt$, follows a first-order rate law:

$$-\frac{dN}{dt} = k^*N$$

in which k^* is called the *decay constant*. By analogy with Equations 6-21*b* and 6-22 we can write

$$N = N_0 \exp(-k^*t) \tag{6-23a}$$

and the half-life is

$$t_{1/2} = 0.693/k^* \tag{6-23b}$$

Table 6-1 gives the half-lives of some radioactive nuclei used in biochemical experiments. The relatively long half-life of ^{14}C makes this nucleus useful for dating archeological artifacts. ^{14}C is created by the action of cosmic rays on atmospheric nitrogen. The radioactive isotope enters the food chain in the form of $^{14}CO_2$, and its abundance is such that the radioactivity due to ^{14}C in plants and animals while alive is about 12.5 disintegration/minute/g carbon. After the animals die, no new ^{14}C is added, so the level of ^{14}C decays according to Equation 6-23*a*; the approximate time of death can be calculated from the remaining activity. The method works well for dating within the past 30,000 years, but assumes that the level of atmospheric ^{14}C has been constant over that period.

The amount of a radioactive substance is often expressed in terms of the curie (Ci) (see the historical sketch in this chapter), which is defined as the amount of material that produces 3.7×10^{10} disintegrations per second (dps). Notice that the number of disintegrations per second is $-dN/dt$, or from Equation 6-23*a*,

$$\begin{aligned} -\frac{dN}{dt} &= k^*N_0 \exp(-k^*t) \\ &= k^*N \end{aligned} \tag{6-24}$$

Therefore $-dN/dt$ is proportional not only to N, the number of nuclei present, but also to the decay constant k^*.

EXERCISE 6-7

Calculate the number of ^3H atoms in 1 Ci of ^3H.

ANSWER

From Table 6-1, $t_{1/2} = 12.3$ years.

$$12.3 \text{ years} \times 365 \frac{\text{days}}{\text{year}} \times 24 \frac{\text{hrs}}{\text{day}} \times 60 \frac{\text{min}}{\text{hr}} \times 60 \frac{\text{sec}}{\text{min}} = 3.88 \times 10^8 \text{ sec.}$$

$k^* = 0.693/t_{1/2} = 1.79 \times 10^{-9}$ sec^{-1}. With $-dN/dt = 3.7 \times 10^{10}$ dps $= k^*N$, we obtain $N = 3.7 \times 10^{10}$ dps/1.79×10^{-9} sec^{-1} $= 2.07 \times 10^{19}$ atoms (or 3.44×10^{-5} moles).

A common measure of the dosage of radiation absorbed by a substance is the *rad,* defined as the absorption of energy per unit mass. One rad corresponds to absorption of 100 ergs/g, or 10^{-2} J/kg. Notice that the dose defined this way depends on the number of disintegrations per unit time (or electrons absorbed by the substance irradiated) times the average energy times the time of exposure, divided by the sample mass:

$$\text{rads} = \frac{dN}{dt} \text{(absorbed)} \times E_{\text{avg}} \times \frac{\Delta t}{\text{mass}} \tag{6-25}$$

EXERCISE 6-8

All of the energy of a 10 mCi sample of ^{32}P is absorbed in 10 g of tissue for 1 hour. Calculate the average dose in rads.

ANSWER

$$E_{\text{avg}} = \frac{0.7 \times 10^6 \text{ ev} \times 96{,}487 \text{ J mol}^{-1}\text{ev}^{-1}}{6.023 \times 10^{23} \text{ mol}^{-1}} = 1.12 \times 10^{-13} \text{ J}$$

$$\frac{\text{energy absorbed}}{\text{mass}}$$

$$= \frac{10 \times 10^{-3} \times 3.7 \times 10^{10} \text{ sec}^{-1} \times 1.12 \times 10^{-13} \text{ J} \times 60 \times 60 \text{ sec}}{0.01 \text{ kg}}$$

$$= 14.9 \text{ J kg}^{-1}/10^{-2} \text{ J kg}^{-1}\text{rad}^{-1}$$

$$= 1493 \text{ rads}$$

Allowed levels of human exposure to radiation are expressed in *rems,* equal to the dose in rads times a *quality factor,* Q. For short-wavelength radiation (x-rays and γ-rays) and electrons, $Q = 1$; Q ranges from 1 to 20 for heavier particles (protons, neutrons, α particles). Presently allowed human occupational exposure to radiation ranges from 0.5 rem (pregnant women) to 75

MARYA SKLODOVSKA—MARIE CURIE

1867–1934

Pierre and Marie Curie shortly after their marriage. (Photograph courtesy of Laboratoire Curie.)

When the Inspector of private schools in Warsaw entered the classroom, the little girls knew he had come to make sure that Russian language and history were being taught, as required. Much of Poland had been under Russian domination for a century, and Polish language and culture were suppressed in the schools.

"Please call on one of these young people," the Inspector asked of the schoolmistress. As in nearly all inspections, Marya Sklodovska was the student selected. Her Russian was flawless, her memory remarkable, and her emotions under sufficient control to withstand antagonistic questioning.

The Inspector demanded, "Name the tsars who have reigned over our Holy Russia since Catherine II." The child replied, "Catherine II, Paul I, Alexander I, Nicholas I, Alexander II."

Then came a series of further questions that emphasized Polish subjugation. Finally the Inspector asked, "Who rules over us?" Marya hesitated and he demanded, "Who rules over us!"

"His Majesty Alexander II, Tsar of All the Russians."

The Inspector nodded and moved on to the next classroom. The Polish schoolmistress who only minutes before had been giving an illegal lesson on Polish history in the Polish language, said to Marya, "Come here, my little soul." Marya went up to the schoolmistress who kissed her on the forehead, and the child burst into tears.

This girl, from a poor family in an oppressed nation, grew up not only to be the first person to receive two Nobel Prizes, but also to raise a daughter who herself became a Nobel Prize winner. Her father had been a high-school teacher of mathematics and physics, but a series of misfortunes had reduced him to supporting his family of six by boarding boys from the school. Because there were as many as ten boarders at a time, Marya had to give up her room and sleep in the living room. One of her older sisters contracted a fatal case of typhus from a boarder, and two years later, when Marya was eleven, her mother died of tuberculosis. The father had hoped to send his daughters to foreign universities, but despite their gold medals for brilliant high-school careers, he could not afford it.

Marya supported herself for five years as a governess with well-to-do families in the provinces and in Warsaw. She sent half of her meager monthly salary to her older sister, who had gone to Paris to study medicine. They agreed that once the sister was established as a doctor, she would send for Marya, who would then start university. In Poland higher education was closed to women.

Marya was 24 by the time her sister sent for her. She traveled by third-class train, except in Germany where there was a fourth class. Marya Gallicized her name to Marie and registered as a student in the Faculty of Science in the University of Paris. Because her brother-in-law's ceaseless chatter interrupted her study, she moved to an unheated garret. For weeks on end she was so absorbed in work that she ate only buttered bread and tea. At the end of the year she passed the exams in physics ranking first, and then the next year in mathematics, ranking second. Starting research, she met Pierre Curie, a physicist, eight years her senior, who had established a scientific reputation for studies of the magnetic properties of crystals, and who shared her serious approach to science. His first gift to her was a copy of his paper on the symmetry of electromagnetic fields.

They were married in 1895. An in-law of Marie's offered her a wedding dress as a present. Marie responded, "I have no dress except the one I wear every day. If you are going to be kind enough to give me one, please let it be practical and dark, so that I can put it on afterwards to go to the laboratory." With their only money, a gift from a cousin, they bought two bicycles.

The year 1897 was an important one. It brought both their first daughter (the future Nobel Prize winning physicist) and Marie's first scientific paper. Marie had placed first in her exams for certification as a physics teacher and was now looking for a thesis topic for her Ph.D. degree. Among physicists there was much excitement about Roentgen's discovery of x-rays, and in 1896 Henri Becquerel in Paris had found that uranium salts emit rays related in their effects to x-rays.

She chose this topic for research, and was given as a laboratory a rough glassed-in studio, previously used as a storeroom on the ground floor of the School of Physics in which Pierre was on the faculty. It was not really suitable for laboratory work: her research notebook shows that on February 6, 1898 the room temperature was 6.25°C. Using a sensitive electroscope built by Pierre Curie and his brother, Marie detected the radiation emitted by uranium. She called this property *radioactivity*.

Marie found that that radioactivity was proportional to the mass of uranium in a sample, and that it was independent of temperature and exposure to light. This led to an insight of great importance: radioactivity is an *atomic property*. With this hypothesis she tested every other element to find if other atoms displayed this property. One did—thorium. Continuing her systematic tests, she found that several minerals in the collection of the School of Physics, particularly pitchblende ore, registered greater radioactivity than could be accounted for by their uranium or thorium contents. Since she had already tested every known element, logic forced her to conclude that this ore contained a new, highly radioactive element.

Pierre dropped his research to share the excitement of Marie's work. She remained an unpaid researcher and supplemented his small salary by lecturing in physics at a girls' school. In setting out to isolate the new element, radium, from ore, the Curies had to pay for pitchblende from Pierre's salary. Fortunately, through the intervention of the

Austrian Academy of Sciences, the Austrian government donated a ton of pitchblende from the state-owned mine in Bohemia, and the Curies had to pay only for its transportation.

Marie underestimated the work of isolation. Originally she guessed that pitchblende might contain 1% radium, whereas in fact it contained less than one part per million. She later recalled, "And yet it was in that miserable old shed that the best and happiest years of our life were spent. I sometimes passed the whole day stirring a mass in ebullition, with an iron rod nearly as big as myself. In the evening I was broken with fatigue." Along the way, the radioactivity separated into two fractions, and it was evident that there was another new element in the ore. "You will have to name it," said Pierre to his young wife. "Could we call it polonium?" she asked, thinking her work could honor her oppressed homeland.

In 1902, four years after the start of the work, Marie succeeded in isolating radium and in determining its atomic weight. Recognition followed almost immediately. The following year the Curies shared the Nobel Prize with Becquerel. Pierre and Marie skipped the official ceremony, saying that it would take them away from their teaching for too long. About this time it became evident that radium would be useful in medicine, and that to patent the process for its isolation would make the Curies rich. Pierre raised this possibility to Marie, who needed to consider it only a few seconds. "It is impossible," she said, "It would be contrary to the scientific spirit."

Just as Marie's life appeared to have reached a point of ease, tragedy stepped in. First Pierre died in a horrible accident. His head was crushed beneath the wheels of a horsedrawn wagon in a street of Paris. Marie firmly rejected collections on her behalf but did accept the professorial chair that Pierre had occupied for eighteen months at the Faculty of Science. She was the first female professor at the Sorbonne and in the whole of France.

Having triumphed in her first move alone in the male-dominated scientific world, she suffered in the next a highly publicized defeat. She announced in the fall of 1910 that she would stand for election to the vacancy in the general physics section of the French Academy of Sciences. Since its reestablishment in 1795, this prestigious body had admitted no woman. Sensing drama in this situation, the popular press began to build up the election. The respectable paper *Le Temps* opened its columns to the permanent secretary of the Academy who documented the case for Madame Curie's election. But the right-wing press hotly opposed her election, with arguments to the effect that she had ridden Pierre's coattails to the Nobel Prize and a professorship. Just then Edouard Branly announced he would oppose Marie. Branly had done creditable research on aerials for electromagnetic waves, but his work was in no way the equal of Marie's fundamental discoveries. His main qualifications were that he was male, gentlemanly, devoutly Catholic, and might not live until another vacancy occurred. The right-wing press lined up behind him, and the election was called "the War of the Sexes." In front of a crowd of journalists and photographers, the members cast 30 votes for Branly and 28 for Curie.

As distressing as was this public defeat, there was worse to come. The uproar over the election had made every move of Madame Curie's a subject for public interest, and late in the same year came another round of intense publicity. It was started by the headline of *Le Journal,* "A Story of Love. Mme Curie and Professor Langevin." Paul Langevin had been Pierre Curie's pupil and protégé and was his successor at the School of Physics.

Langevin had married a girl of his own working-class background, and as he advanced in education, she raised their four children. She could not understand his reluctance to leave his university post for better pay in industry. He sought the sympathy of university friends, and his distress particularly touched Marie, with her ethic of pure science. Langevin took a small apartment near the School of Physics, where Marie visited him frequently. What made matters worse for Marie was that she also put her feelings in letters. Aided by her brother-in-law, Madame Langevin broke the lock of Langevin's study and took the letters, which she planned to use to get a separation.

These letters were the basis of attacks of the most unpleasant sort on Marie from the popular press, which gradually built the situation into a great scandal. One article for example said that Marie Curie, "the vestal virgin of radium," had used vile suggestions to detach Langevin from his wife and children. Others hinted that she had driven Pierre to suicide. Crowds gathered outside the Curie house shouting, "Get the foreign women out! Husband stealer!" and a rock was thrown at the house. The Curies were taken to a friend's apartment where the fourteen-year-old daughter, having seen a copy of the main press attack, lay numbed. Langevin challenged the author of the attack to a pistol duel, but it was concluded without wounds.

In the midst of all this came a telegram revealing that Marie had received a second Nobel Prize, this time for the chemistry of radium. She took her older daughter to Stockholm for the award ceremony, as interest in Paris in her private life gradually died away. When she returned she collapsed and was carried to a nursing home. It took a year for her to return fully to work.

Before her death from leukemia, Marie Curie suffered from cataracts. Her extensive exposure to atomic radiation probably contributed to both afflictions.

rems (exposure to hands) per year. For comparison, the whole-body radiation absorbed from background natural sources is roughly 0.1 rem per year, and the average exposure due to medical diagnosis is about 0.07 rem per year.

○ 6–9 *Second-Order Rate Laws can be Integrated*

Rate equations for reactions of second and higher orders can be integrated—as we have done for first-order rate laws—to give explicit expressions for the concentrations of reactants and products as a function of time. Numerous books, including those by Frost and Pearson (1952) and Moore (1972), derive and discuss these expressions. Let us consider one example, a reaction of the form

$$A + A \rightarrow B$$

which is described by a rate equation

$$-\frac{1}{2}\frac{d[A]}{dt} = k[A]^2 \tag{6-26}$$

Separating variables

$$-\frac{d[A]}{[A]^2} = 2kt\ dt$$

and integrating, we get

$$\frac{1}{[A]} = 2kt + c$$

in which c is an integration constant. As with Equation 6-21*a*, we can evaluate c by noting that at $t = 0$, $c = 1/A_0$, in which A_0 is the initial concentration of A. Therefore the time-dependence of [A] is given by

$$\frac{1}{[A]} = \frac{1}{A_0} + 2kt \qquad (6\text{-}27)$$

Equations 6-21 and 6-27 provide the basis for deciding whether a reaction in which A is converted to B follows a first-order or a second-order rate law. The data are plotted as prescribed by both equations. If ln [A] is linear in t, the reaction is first-order and the experimental slope is $-k$. On the other hand if ln [A] is not linear in t, but 1/[A] is, the reaction is second-order and the slope is twice the second-order reaction rate constant k. If neither plot is linear the rate law is neither first-order nor second-order. If both plots are linear, the reaction has not been observed long enough.

6–10 DNA Renaturation is a Second-Order Reaction

An important application of second-order reaction kinetics is to the characterization of DNA by measurement of the rate of renaturation. The two strands, A and B, of a double helix can be separated by brief exposure to high temperature, and then cooled to allow the complementary base pairs to reform, or *renature*, the double helix, H:

$$A + B \rightarrow H \qquad (6\text{-}28)$$

This reaction is second-order, so the rate depends on the concentration of complementary strands A and B. A DNA sample consists of fragments of the large piece or pieces of DNA originally present in the organism. Some genes, or base sequences, may be present in only one copy per cell, whereas others may be repeated many times in "redundant" copies. One would expect the repeated sequences to renature more rapidly because their concentration is higher. Quantitative treatment of the renaturation rate allows one to estimate how many copies of a particular gene are present per cell.

The rate law for the second-order reaction in Equation 6-28 is

$$\frac{d[A]}{dt} = -k[A][B] \qquad (6\text{-}29)$$

Since complementary strands A and B are present in equal numbers, [A] = [B] and Equation 6-29 simplifies to

$$\frac{d[\mathrm{A}]}{dt} = -k[\mathrm{A}]^2$$

which is identical in form to Equation 6-26, and which may consequently be integrated to give Equation 6-27, with $2k$ replaced by k. We let the fraction of strands dissociated be f; thus

$$f = \frac{[\mathrm{A}]}{\mathrm{A}_0} \tag{6-30}$$

in which A_0 is the total concentration of A strands. Equation 6-30 can be combined with $1/[\mathrm{A}] = 1/[\mathrm{A}_0] + kt$ to yield

$$f = \frac{1}{1 + \mathrm{A}_0 kt}$$

The strand concentration A_0 is not necessarily known, but is related to the base-pair concentration C_0 by

$$\mathrm{A}_0 = \mathrm{C}_0/X$$

in which X is the number of base pairs that must be taken in order to obtain one copy of the base sequence on strand A (Fig. 6-3). Therefore the fraction dissociated is

$$f = \frac{1}{1 + \mathrm{C}_0 tk/X} \tag{6-31}$$

Half the base pairs will be renatured ($f = \frac{1}{2}$) when $\mathrm{C}_0 tk/X = 1$, or

$$\mathrm{C}_0 t = X/k \qquad (f = \tfrac{1}{2}) \tag{6-32}$$

Figure 6-4 shows a plot of f versus $\mathrm{C}_0 t$, on a logarithmic scale, for several nucleic acid samples with different values of X. The value of $\mathrm{C}_0 t$ when $f = \frac{1}{2}$ varies from 2×10^{-6} for poly A · poly U to 3×10^3 for calf DNA. For poly A · poly U, $X = 1$ because the repeated sequence or "gene" is just one A · U base pair. Therefore $k = 1/(2 \times 10^{-6}) = 5 \times 10^5 \mathrm{M}^{-1}\mathrm{s}^{-1}$. This allows calculation of X for all other samples from their values of $\mathrm{C}_0 t$ at $f = \frac{1}{2}$. The scale at the top of Figure 6-4 shows the X-values obtained in this way.

The mechanism of DNA renaturation can be described to a first approximation by the pre-equilibrium assumption, in which the fast equilibrium step is formation of a short helix containing about two base pairs (Fig. 6-5). This two–base-pair helix is unstable, and dissociates more rapidly than it grows ($k_{-1} > k_2$), allowing us to assume that the first reaction step occurs many times before the second step has time to happen. The rate-limiting step is addition of a third base pair to form a stable helix *nucleus*, which grows more rapidly than it dissociates.

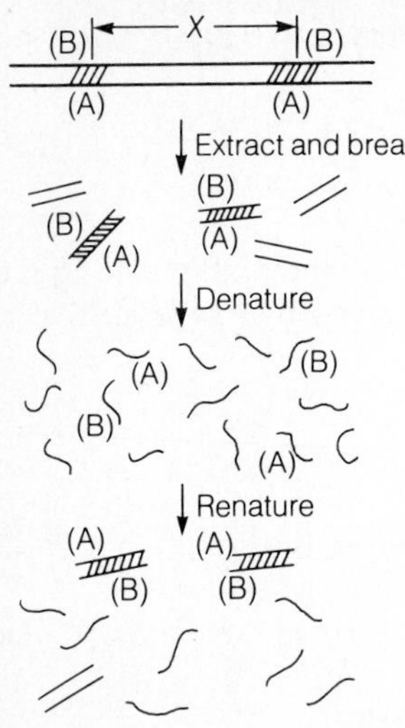

(**a**) Original long DNA molecule in genome.

(**b**) Shear-degraded small duplex molecules.

(**c**) Denatured single strands.

(**d**) Partially renatured sample. Complementary strands A and B renature relatively rapidly because gene A is repeated, and therefore present in higher concentration.

FIGURE 6-3

DNA renaturation experiment. The original long DNA molecule in **(a)** has an average distance X (in base pairs) between repeats of the gene A. X can range in length from one for a simple sequence such as poly A · poly U, up to the full length of the DNA content of one cell if the gene is present in only one copy in the cell. In **(b)** the DNA molecule has been broken into smaller fragments. Since there is one copy of A per X base pairs in **(a)**, the concentration of A in **(b)** is $1/X$ times the total base pair concentration. **(c)** The sample is denatured by heat treatment. **(d)** Restoration of lower temperature results in the renaturation reaction, shown here as only partially complete. Notice that gene A renatures relatively rapidly because it was present in multiple copies in the original genome **(a)**.

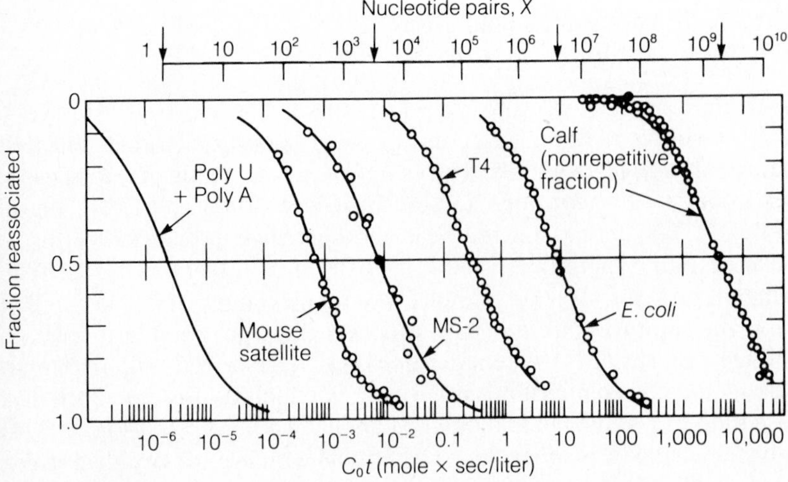

FIGURE 6-4

"C_0t curves" for reassociation of double stranded nucleic acids from various sources. The fraction dissociated (f) is plotted as a function of initial concentration times time (C_0t, lower scale) and number of base pairs (X) required to get one copy of the renaturing sequence. (From Britten, R. J., and Kohne, D. E. 1968. *Science 161,* 529. Copyright 1968 by the American Association for the Advancement of Science.)

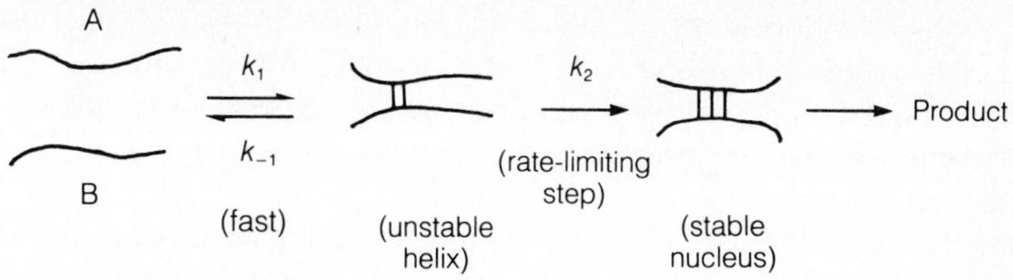

A

B

$\underset{k_{-1}}{\overset{k_1}{\rightleftharpoons}}$ (fast) (unstable helix) $\xrightarrow{k_2}$ (rate-limiting step) (stable nucleus) $\longrightarrow$ Product

FIGURE 6-5
Simplified mechanism for renaturation of two DNA strands A and B. The unstable helix dissociates much more rapidly than it undergoes growth to the stable three–base-pair helix, so one can assume pre-equilibrium for the first step. Dissociation of the stable nucleus is neglected in this simplified mechanism.

We now show that the mechanism in Figure 6-5 leads to a bimolecular rate law. The rate of production of double-helical product P is

$$\frac{d[\mathrm{P}]}{dt} = k_2\,[\text{unstable helix}]$$

By the pre-equilibrium assumption

$$K_1 = \frac{[\text{unstable helix}]}{[\mathrm{A}][\mathrm{B}]}$$

and we obtain

$$\frac{d[\mathrm{P}]}{dt} = k_2 K_1 [\mathrm{A}][\mathrm{B}]$$

which allows us to conclude that the measured rate constant for DNA renaturation is $k = k_2K_1$. From the experimentally estimated values of $k \simeq 10^6\mathrm{M}^{-1}\mathrm{sec}^{-1}$ and $\mathrm{K}_1 \simeq 10^{-1}\mathrm{M}^{-1}$, it has been concluded that k_2, the rate constant for base-pair formation, is about $10^7\mathrm{sec}^{-1}$. This value is important because base-pair formation is the elementary step of genetic information transfer in biological systems. Another important point you should notice from this example is that a second-order rate law can result, even though the rate-limiting step is first-order.

6–11 Elementary Rate Constants Depend on the Energy and Entropy of Activation

So far we have expressed the rate of an elementary chemical reaction step in terms of the rate constant—an experimentally determined quantity that can be used to describe quantitatively the rate of a chemical or biochemical reaction. But what determines the magnitude of a rate constant? In this section we consider theoretical answers to this question.

The size of a rate constant for an elementary step depends on the detailed molecular interactions that occur during the *reactive encounters* which take individual molecules through the transition from products to reactants. Even for very slow reactions the reactive encounters are brief; slow reactions result because only a few encounters are reactive. As we will see here, the probability that an encounter is reactive can be expressed in terms of the potential energy and entropy changes that occur during the course of a reactive encounter.

Molecular encounters that lead to chemical reaction require several spatial variables for description, even for a reaction as simple as the exchange of one hydrogen atom for another in H_2:

$$H + H_2 \rightarrow H_2 + H$$

In this case an H atom (a) can approach the molecule $H_b - H_c$ at various angles θ:

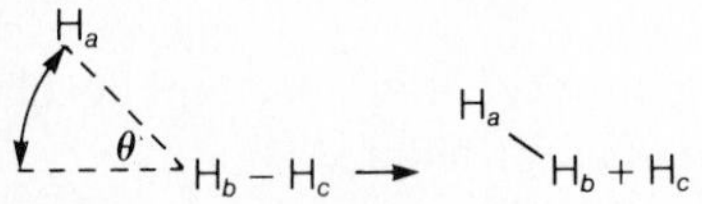

$$H_a \overset{\theta}{\cdots} H_b - H_c \rightarrow H_a \diagdown H_b + H_c \qquad (6\text{-}33)$$

The relative positions can be described sufficiently by three variables, the distances $H_a - H_b$ and $H_b - H_c$, along with the angle θ. However, keeping track of the energy and entropy changes of this reacting system as functions for all three variables leads to complicated diagrams.

A general simplification of this problem is to describe the reaction in terms of the *reaction coordinate,* which measures the progress of the encounter along the most probable reaction path. A simple analogy is the measurement of your progress across a mountain range. In principle there may be many ways to cross the mountains, but suppose that the only one of any practical consequence is a winding path that traverses a single pass. One variable, the distance you have moved along the path, is sufficient to describe your progress. That distance variable has the same meaning as the reaction coordinate.

The Reaction Profile Describes Energy Changes along the Reaction Coordinate

The potential energy of a reacting system can be thought of as a surface in multidimensional space. If we restrict ourselves to two spatial variables, this surface can be visualized as the equivalent of a topographical map. For example, detailed calculations show that the most probable direction for reactive attack in Reaction 6-33 is when H_a is on line with H_b and H_c ($\theta = 0$). Therefore, we are left with only two variables to describe the reaction—the distances $H_a - H_b$ and $H_b - H_c$. Figure 6-6 shows schematically the potential energy of the system by plotting equipotential contours when these two distances are varied independently. Notice that there are two energy minima, one each for the equilibrium bond lengths of H_2. The reaction coordinate moves along a trough between these two points over the lowest possible energy-barrier height. The top of the energy barrier is described mathematically as a *saddle point,* at which the energy is a maximum for motion along the reaction coordinate, but a minimum for motion at right angles to

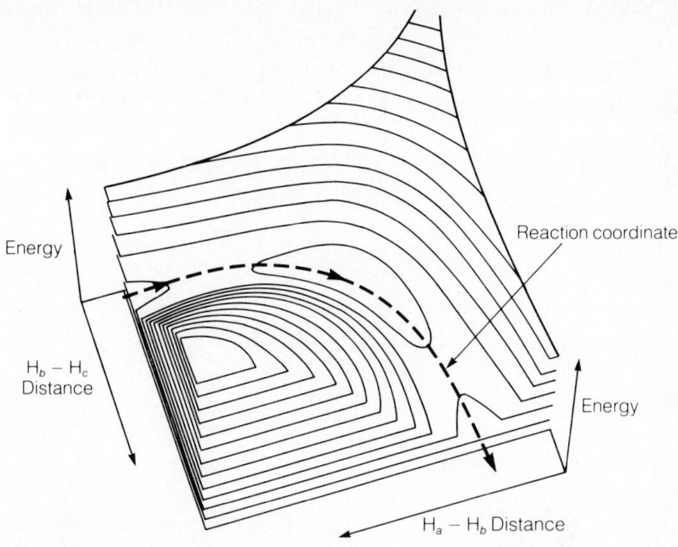

FIGURE 6-6
The potential-energy surface for the reaction $H_a + H_b - H_c \rightarrow H_a - H_b + H_c$. The surface represents energy in the vertical direction, as a function of the $H_a - H_b$ and $H_b - H_c$ separations, which increase in the direction of the arrows. Notice that $H_a - H_b$ is large at the beginning of the reaction, and decreases toward the equilibrium bond distance as the system moves along the reaction coordinate.

the reaction coordinate. The reacting atoms at the saddle point are said to be an *activated complex,* also called the *high-energy transition state.*

Plotting the energy of the system as a function of the reaction coordinate gives a *reaction profile* of the kind shown in Figure 6-7. The energy difference between the reactants and the transition state is called the *activation energy* E_A, which can also be defined for the reverse reaction. For the reaction in Equation 6-33 and Figure 6-6, the energy of reactants and products is zero, so $\Delta H = 0$.

Arrhenius Expressed the Rate Constants in Terms of the Temperature, Activation Energy, and the Frequency Factor of a Reaction

In 1889, some 45 years before quantum-mechanical computations of potential-energy surfaces had been carried out, Arrhenius realized that the activation energy was an important parameter in determining the rate of a reaction. Arrhenius reasoned that because an equilibrium constant may be viewed as the ratio of two rate constants (Equation 6-6*c*), the temperature dependence of a rate constant should be similar to that of an equilibrium constant. It was known then that the change with temperature of an equilibrium constant can be expressed by the van't Hoff equation

$$\left(\frac{d \ln K}{dT}\right)_P = \frac{\Delta H^\circ}{RT^2}$$

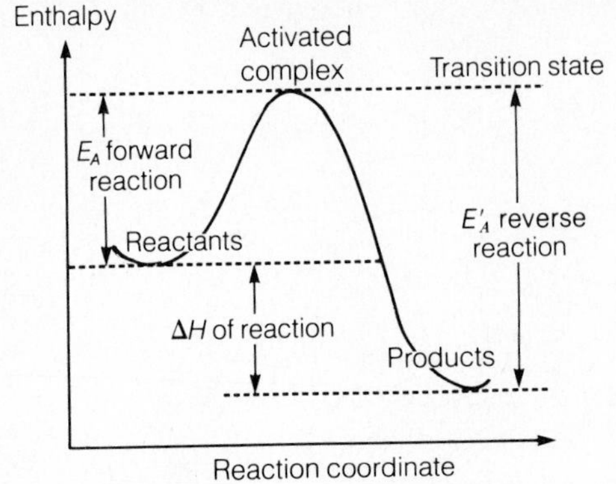

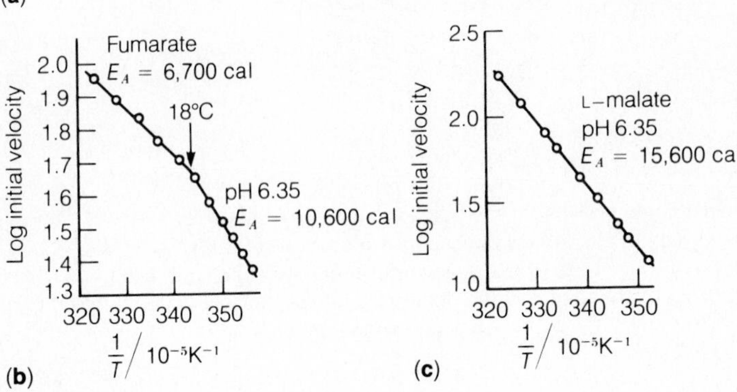

FIGURE 6-7
Arrhenius theory of rate constants. **(a)** A reaction profile showing the relationship of the energy of activation and the thermodynamic enthalpy change. **(b)** Arrhenius plots of the log of the initial reaction velocity against $1/T$, for the fumarase reaction. (From Dixon and Webb, 1964, based on work by Massey, 1953.) **(c)** Same as **(b)**, for the reverse reaction.

in which ΔH° is the standard enthalpy change for the reaction and R is the gas constant. In analogy to this relationship, Arrhenius expressed the temperature dependence of the rate constant k by

$$\left(\frac{d \ln k}{dT}\right)_P = \frac{E_A}{RT^2} \tag{6-34}$$

in which E_A is the activation energy. Then, by separation of variables,

$$d \ln k = \frac{E_A}{R} \frac{dT}{T^2}$$

This expression can be integrated to give

$$\ln k = \frac{-E_A}{RT} + \ln A \tag{6-35a}$$ ◀

The quantity A is the integration constant, and for reasons we will discuss presently, is called the *frequency factor*. Taking the exponential of both sides of Equation 6-31, we get Arrhenius's equation,

$$k = A \exp(-E_A/RT) \tag{6-35b}$$

The physical meaning of this equation can be seen by referring to Figure 6-7(*a*). For the forward reaction, from reactants to products, E_A is the difference in energy between the transition state and the reactants. For the reverse reaction, from products to reactants, E'_A is the activation energy. The difference $E_A - E'_A$ is the thermodynamic enthalpy change for the reaction. The factor $\exp(-E_A/RT)$ in Equation 6-35*b* represents the fraction of reagent molecules having the critical energy E_A for the reaction to occur. The frequency factor, A, may be thought of as the frequency of collisions with the proper orientation to produce a chemical reaction. This factor can be as large as about 10^{13}sec^{-1}, which is about the frequency of collision of molecules in liquids. Thus, in Arrhenius' theory, the rate constant is determined by the ratio of the activation energy to the temperature and by the frequency of collisions that produce a reaction. Notice that at higher temperatures the ratio E_A/RT is smaller, so that $\exp(-E_A/RT)$ is greater, and the rate of reaction is greater.

It is clear from Equation 6-35*a* that for any reaction that obeys Arrhenius' theory, a plot of $\ln k$ against $1/T$ will be linear, with slope $-E_A/R$. Linear plots are found for many reactions, including many enzymatic reactions with a limited temperature range. Figure 6-7*b* shows these so-called *Arrhenius plots* for the enzymatic hydration of fumaric acid. The plot for the forward reaction shows a break at 18°C, indicating a change in reaction mechanism above that temperature.

Before leaving Arrhenius' description of rate constants, it is worth noting that even this simple idea gives some insight into the special catalytic power of enzymes. It is evident from Equation 6-35*b* that an enzyme can increase the reaction rate constant either by lowering E_A through binding to the substrate, or by increasing A through properly orienting the reactants, or by both effects. One example (Jencks, 1969) is that of urease, which catalyzes the hydrolysis of urea to ammonia and carbon dioxide. The nonenzymatic hydrolysis at 21°C requires passage over an energy barrier of about 30 kcal mol^{-1}, whereas the activation energy in the presence of the enzyme is 11 kcal mol^{-1}. Therefore the enzyme accelerates the reaction by a factor of $[\exp(-11 \times 10^3/RT)]/[\exp(-30 \times 10^3/RT)] = 1.5 \times 10^{14}$.

EXERCISE 6-9

Calculate the temperature that would be required for the nonenzymatic hydrolysis of urea to proceed as rapidly as the enzymatic hydrolysis at 21°C.

ANSWER

$$\exp(-11 \times 10^3 \text{ cal mol}^{-1}/1.98 \text{ cal K}^{-1}\text{mol}^{-1} \times 294 \text{ K}) = \exp(-30 \times 10^3 \text{ cal mol}^{-1}/1.98 \text{ cal K}^{-1}\text{mol}^{-1}\text{T})$$

or

$$T = \frac{30}{11} \times 294 \text{ K} = 802 \text{ K or } 529°\text{C}$$

Of course, this temperature cannot be reached in aqueous solution. Thus a catalyst may bring about an increase in reaction rate that would be impossible to achieve by heating.

Eyring's Transition-State Theory Relates the Rate Constant to Quasi-Thermodynamic Parameters

In 1935, Eyring published the *transition-state* theory, which is still the most widely used conceptual scheme for discussing reaction rates. The central idea of this theory is that at any given temperature, the rate of a reaction depends only on the concentration of the high-energy *activated complex* which is in equilibrium with the reactants. All properties of the activated complex are denoted by a double cross, ‡.

Suppose we are interested in the reaction of A with B to give C. According to the Eyring theory, the reactants exist in equilibrium with the activated complex,

$$A + B \rightleftarrows A - B^{\ddagger} \rightarrow C$$

and the equilibrium can be described by a constant:

$$K^{\ddagger} \rightleftarrows \frac{[A - B^{\ddagger}]}{[A][B]} \tag{6-36}$$

The reaction rate, $-d[A]/dt$, is given by the combined product of the concentration of the activated complex, the rate at which the complex moves across the barrier, and the fraction κ, called the *transmission coefficient*, of complexes that pass over without returning. Thus we can write

$$-\frac{[dA]}{dt} = [A - B^{\ddagger}] \times (\text{rate of crossover}) \times \kappa \tag{6-37}$$

Calculating the "rate of crossover" requires some results from a subsequent chapter (Chapter 10) on quantum mechanics. You can visualize the process that occurs in the activated complex as a vibrational motion of one or more atoms in the complex along the direction specified by the reaction coordinate. However, instead of reversing direction at the extreme of the vibration as expected for oscillatory motion, the atom continues along its path, resulting in the breakage of one bond and formation of another. In the Eyring theory, the rate of crossover is set equal to the vibrational rate, as expressed by its frequency (ν) in cycles per second. Chapter 10 shows that the minimum energy E required to excite vibrational motion is $E = h\nu$, in which h is Planck's constant. Therefore, the rate of crossover can be expressed by

$$\text{Rate of crossover} = \nu = \frac{E}{h}$$

The average energy E of the atom moving along the reaction coordinate is then replaced by the average energy of a particle vibrating in one dimension. Statistical mechanical methods are used in Chapter 14 to show that $E = k_B T$, in which k_B is Boltzmann's constant and T is the temperature. Therefore,

$$\text{Rate of crossover} = \frac{k_B T}{h} \tag{6-38}$$

Substituting this result into Equation 6-37 we find

$$-\frac{d[\mathrm{A}]}{dt} = \frac{k_B T}{h}[\mathrm{A} - \mathrm{B}]^{\ddagger}\kappa$$

We can eliminate $[\mathrm{A} - \mathrm{B}^{\ddagger}]$ from this equation using Equation 6-36:

$$-\frac{d[\mathrm{A}]}{dt} = \frac{k_B T}{h} K^{\ddagger}[\mathrm{A}][\mathrm{B}]\kappa$$

Since we are considering a second-order reaction of the form of Equation 6-3, the rate constant in Eyring's theory is given by

$$k = \frac{k_B T}{h} K^{\ddagger}\kappa \tag{6-39}$$ ◀

Building on the idea of a quasi-thermodynamic equilibrium, as expressed in Equation 6-36, Eyring was able to relate $K^{\ddagger}$ to other thermodynamic properties of the activated complex. From the relationships $\Delta G^\circ = -RT \ln K$ and $\Delta G^\circ = \Delta H^\circ - T\,\Delta S^\circ$, it is possible to write

$$K^{\ddagger} = \exp(-\Delta G^{\ddagger\circ}/RT) = \exp(\Delta S^{\ddagger\circ}/R - \Delta H^{\ddagger\circ}/RT)$$

Substituting this result into 6-39 we can express the rate constant as

$$k = \frac{k_B T\kappa}{h}\exp(-\Delta G^{\ddagger\circ}/RT) = \frac{k_B T\kappa}{h}\exp(\Delta S^{\ddagger\circ}/R - \Delta H^{\ddagger\circ}/RT) \tag{6-40}$$ ◀

Relationship of the Arrhenius and Eyring Theories and Evaluation of Quasi-Thermodynamic Quantities

We can relate the parameters E_A and A of the Arrhenius theory to the quasi-thermodynamic quantities of the Eyring theory by examining the temperature dependence of the logarithm of the rate constant in both formulations. For the Eyring theory, Equation 6-40 gives

$$\frac{d\ln k}{dT} = \frac{1}{T} + \frac{\Delta H^{\ddagger\circ}}{RT^2} = \frac{\Delta H^{\ddagger\circ} + RT}{RT^2} \tag{6-41}$$

Comparing this equation to Equation 6-34, we see that

$$E_A = \Delta H^{\ddagger\circ} + RT \tag{6-42}$$ ◀

If we substitute this relationship for E_A into Arrhenius' equation (6-35*b*), and compare the result to Eyring's equation (6-40), we find

$$A = \left(\frac{k_B T \kappa e}{h}\right) \exp(\Delta S^{\ddagger\circ}/R) \quad (6\text{-}43)$$

This equation shows explicitly that the Arrhenius parameter A contains factors for the frequency of vibration in the reaction coordinate, $k_B T/h$, and for orientation, $\exp(\Delta S^{\ddagger\circ}/R)$.

There are several routes for evaluating the quasi-thermodynamic parameters of the Eyring theory. If we have measured the dependence of the rate constant on temperature, then we can use a plot such as Figure 6-7(*b*) to find E_A. Equation 6-42 then yields $\Delta H^{\ddagger\circ}$ from E_A. Then if it is assumed that the transmission coefficient is unity (as is almost always done), the rate constant k at any given temperature yields $\Delta G^{\ddagger\circ}$, using Equation 6-40. Once both $\Delta G^{\ddagger\circ}$ and $\Delta H^{\ddagger\circ}$ are known, $\Delta S^{\ddagger\circ}$ can be evaluated from $\Delta G = \Delta H - T\Delta S$.

The Meaning of Eyring's Equation: A Summary

Eyring's expression for the rate constant k contains three factors: (a) a pre-exponential, $(k_B T/h)$, (b) the exponential containing the quasi-thermodynamic constants, and (c) the transmission coefficient. Usually, little or nothing is known about the transmission coefficient, and its value is assumed to be unity. The pre-exponential factor gives the estimated rate of vibration of the activated complex, which is assumed to be determined by the kinetic energy of translation and is therefore essentially a measure of the collision rate between adjacent particles in solution. Its value is about 10^{13}sec^{-1} at room temperature. The pre-exponential factor increases linearly with temperature, but has relatively little effect on the temperature dependence of the rate constant because temperature also enters into the exponential term, where it exerts a more profound influence.

The quantity that depends on the nature of the reaction is the free energy of activation, which in turn depends on the enthalpy and entropy of activation. A large $\Delta G^{\ddagger\circ}$ means that the free energy of the transition state is far above that of the reactants, and as you can see from Equation 6-40, this means that the rate constant is small. $\Delta G^{\ddagger\circ}$ can be large either because of a large positive enthalpy of activation, $\Delta H^{\ddagger\circ}$, or a large negative entropy of activation, $\Delta S^{\ddagger\circ}$, or both. The enthalpy and entropy of activation are measures of the heat and entropy changes when the reactants (at standard-state concentrations) are converted to the activated complex which is also at its standard-state concentration. The enthalpy of activation always is positive for an elementary reaction step because energy is required to create the partially broken bonds of the activated complex. The entropy of activation can be positive or negative. Simple molecules are generally more constrained in the activated complex than when they are able to move separately in solution, so $\Delta S^{\ddagger\circ}$ is negative. In more complex molecules, however, new degrees of freedom may appear in the activated complex, such as freedom for a segment of the molecule to move independently. In such cases the entropy of the activated complex can be more favorable than the reactants, so $\Delta S^{\ddagger\circ}$ is positive.

From this short summary it is clear that for a catalyst to accelerate a reaction at a given temperature, it must decrease $\Delta G^{\ddagger\circ}$, either by decreasing the enthalpy of activation or by increasing the entropy of activation. Recent evidence, summarized in Section 6-17, indicates that enzymes do both.

6-12 In a Diffusion-Limited Process, Reaction Occurs at Every Collision

The Eyring theory does not give an adequate conceptual description of reactions in which the molecules react every time they collide. In such a case we cannot consider that the reactants are in equilibrium with the activated complex, since it is converted to product every time it is formed. The theory describing these *diffusion-limited* reactions was developed by Smoluchowski in 1917, and later refined by Debye. The theoretical model assumes that two particles react every time they come within a distance r_{12}, the *reaction radius,* of each other, and that they move by random Brownian motion. The rate of random motion of molecules is proportional to the diffusion constant D (see Chap. 15), which consequently enters the expression for k. For uncharged particles Smoluchowski found that

$$k = \frac{4\pi N_A}{1000}(D_1 + D_2)r_{12} \tag{6-44}$$

in which D_1 and D_2 are the diffusion constants (Chap. 15) for particles 1 and 2, respectively. When the particles are charged, Debye showed that at low ionic concentration this expression is modified to

$$k = 4\pi N_A r_{12}(D_1 + D_2)\frac{\phi_{12}}{\exp \phi_{12} - 1} \tag{6-45}$$

in which ϕ_{12} is the ratio of electrostatic energy between the particles at a distance r_{12}, to the thermal energy $k_B T$ (Chap. 8):

$$\phi_{12} = \frac{z_1 z_2 e^2}{\epsilon r_{12} k_B T}$$

The charge on ion i is $z_i e$, and ϵ is the dielectric constant (Chap. 8). If the particles have opposite charges, their diffusion-limited reaction is accelerated by the electrostatic attraction; if they have like charges, it is slowed down.

For typical values of small-molecule diffusion constants and reaction radii, diffusion-limited rate constants are found to be between about $10^9 M^{-1} sec^{-1}$ and $10^{11} M^{-1} sec^{-1}$.

The Lac Repressor Reacts with Operator at a Diffusion-Limited Rate

A repressor is a protein that can react with a particular DNA sequence, called the *operator,* and block expression of adjacent genes. In the presence of lactose, the *lac* repressor does not bind to the operator, and the proteins necessary for utilization of lactose are produced. When lactose is removed, binding of repressor to the operator occurs (Fig. 6-8*a*). Since only a few copies of the repressor normally are produced per cell, the reaction with the operator must be efficient. In 1970, Riggs, Bourgeois, and Cohn showed that the rate constant for the reaction can exceed 10^{10} M^{-1} sec^{-1}, and decreases as the salt concentration is increased. The diffusion coefficient for the repressor should be about 5×10^{-7} cm^2 sec^{-1}. Assuming with Riggs *et al* that r_{12} is about 5 Å, Equation 6-44 indicates that the diffusion-limited rate constant should be about 10^8 M^{-1} sec^{-1}. Since this calculated value is smaller than the observed rate constant we conclude that the reaction is speeded up by one or more factors. One possibility, suggested by Riggs *et al* (1970) is that the bound region on the protein has a charge opposite to that on the DNA, causing the reaction to be accelerated by electrostatic interactions. Richter and Eigen (1974) have argued that the rate is too large to be accounted for on this basis, and have suggested the alternative model shown in Figure 6-8(*b*). Repressor is known to bind with moderately high affinity to nonoperator DNA; Eigen and Richter proposed that DNA within a few hundred base pairs of the operator captures the repressor, which then migrates by Brown-

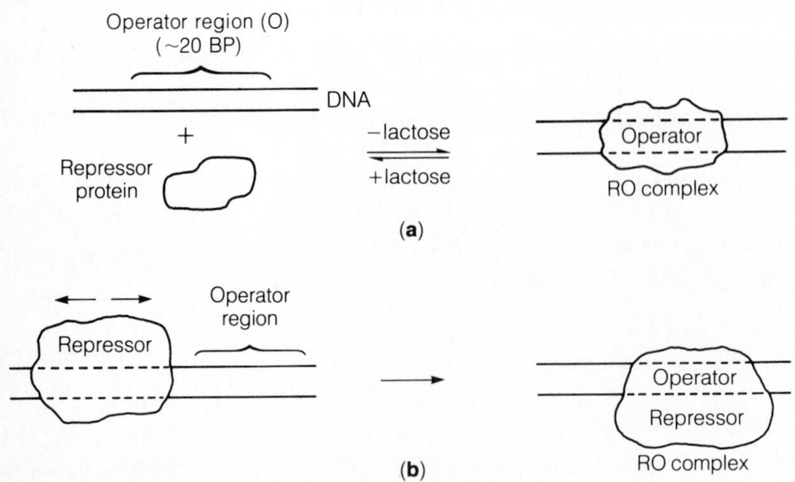

FIGURE 6-8
(a) Removal of lactose causes binding of repressor protein to the operator region of a DNA molecule containing the lactose operon, and addition of lactose causes dissociation. The binding reaction occurs at a rate even greater than predicted for a diffusion-limited reaction. **(b)** A possible explanation for the rapid reaction rate is linear motion of a captured repressor along the DNA to the operator.

ian motion or diffusion (Chap. 15) along the DNA to the operator site. Capture by a larger DNA region speeds up the reaction by increasing the reaction radius, r_{12} in Equation 6-44.

6–13 *Experimental Methods for Measuring Reaction Rates Cover a Wide Range of Times*

Many different techniques are available for measuring the rates of processes of biochemical importance. Each technique has its advantages and limitations, and each covers only part of the vast time domain from 10^{-18} sec (interaction of a photon with a chlorophyll molecule) to millions of years (evolutionary phenomena) that is important for biochemistry. Figure 6-9 shows a summary of the methods and their effective time ranges. Some of the techniques listed in the figure are designed to measure molecular motion. Magnetic resonance techniques are very important in kinetics, and will be considered in Chapter 13.

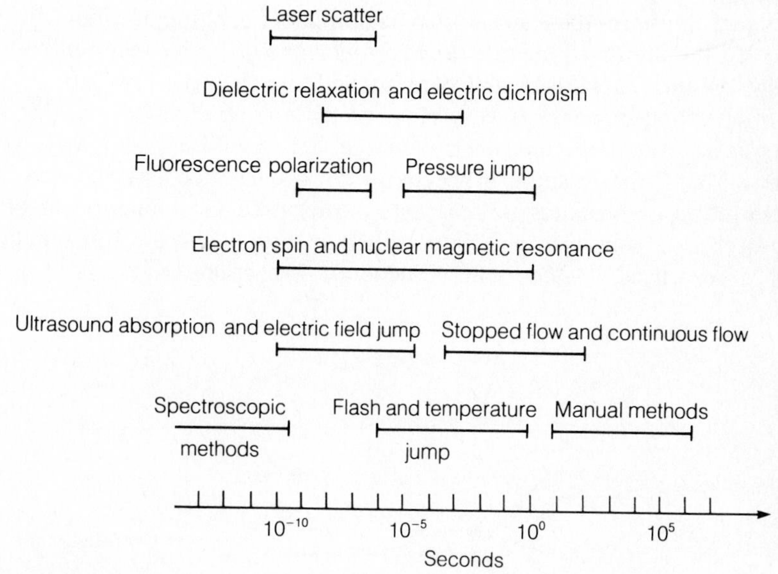

FIGURE 6-9
Summary of the time scale of processes that can be measured by particular techniques. Laser scattering, electric dichroism, dielectric relaxation, and fluorescence polarization are used primarily to study molecular motions such as translational diffusion and rotational motion. Spectroscopic methods refer to techniques such as UV or IR spectroscopy. Manual methods are the simple mixing techniques (limited on the slow side by the lifetime of the observer). Relaxation techniques include temperature, electric field, and pressure jump, and ultrasound absorption; some of these are considered in subsequent sections. Magnetic resonance methods are considered in Chapter 13. (From Schechter, A. N. 1970. *Science 170,* 273. Copyright 1970 by the American Association for the Advancement of Science.)

The simplest way to measure the reaction rate is to mix the reactants together, then measure the concentration of product or remaining reactants at various times after mixing. For example, substrate and enzyme can be mixed in a spectrophotometer cell, and the reaction then followed by measuring the spectral change produced by the reaction. The result is a plot of product concentration as a function of time, with a slope that gives the reaction rate, $d[C]/dt$, as shown in Figure 6-1. The slope, or rate, varies with time, and usually can be determined at the initial condition when $t = 0$, $[C] = 0$ and the reactants are at their starting concentrations.

Until 1923, the fastest time achieved for measurement of solution reaction rates was about 10 sec, the time required to mix the reactants and to sample for extent of reaction. In 1923, Hartridge and Roughton introduced the flow reactor, shown schematically in Figure 6-10. Reactants A and B flow constantly through the mixing chamber, and are observed a distance d down the flow tube. The time reactants have been together is proportional to d, and a time-profile of the reaction can be obtained by measuring concentrations at different distances. In this way Hartridge and Roughton were able to study reactions after a mixing time of about 1 msec (10^{-3} sec).

The continuous-flow reactor has the serious disadvantage for biochemical reactions of requiring large amounts of solution to maintain the flow. With the advent of phototubes and electronic amplifiers capable of following fast reactions, the preferred method now is *stopped flow*. In this technique small amounts of the reactants A and B are mixed and then the flow is stopped. The observation chamber is very close to the mixing chamber, so that the reactants arriving there have been in contact only a millisecond or less. After the flow stops, the reaction proceeds in the observation chamber and the concentration of products or reactants is observed as a function of time. This method is simple in principle, although often elaborate in execution, and has been used to study many biochemical reactions.

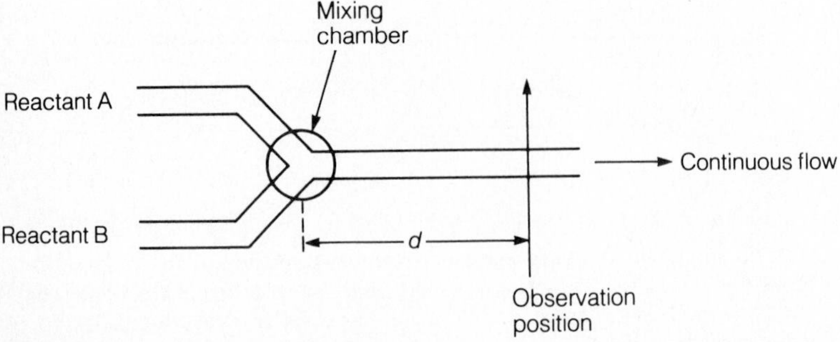

FIGURE 6-10
Schematic diagram of a continuous-flow reactor. Reactants A and B flow continuously through the mixing chamber at a constant rate. The extent of reaction is measured at a variable distance *d* downstream from the mixing chamber, usually by determination of the optical absorption.

6–14 Relaxation Kinetics Relies on Small Perturbations of Reactions at Equilibrium

Mixing techniques are limited in time resolution to about 10^{-4} sec because of the physical problem of rapidly combining two solutions. Relaxation methods circumvent this difficulty. The reactants are present at equilibrium in a solution, to which a perturbation is applied. This perturbation can be a change in pressure, temperature, or electric field, which causes a shift in the amount of the various reactants and products in the mixture. If the perturbation is applied quickly, the equilibrium is not re-established as rapidly as the temperature or pressure is changed. The process of attaining the new equilibrium is called *relaxation*. The time required for relaxation depends on the values of the kinetic constants for the reaction.

An important advantage of relaxation methods is that they result in only small displacement from equilibrium, so that complicated kinetic equations can be linearized. Two physical laws are used to derive the *relaxation time* for a reaction: (1) the rate law for the reaction and (2) conservation of mass in the reaction. Consider a second-order reaction in which A and B combine to yield C:

$$\begin{array}{lcccc} & \mathrm{A} & + \quad \mathrm{B} & \underset{k_{-1}}{\overset{k_1}{\rightleftarrows}} & \mathrm{C} \\ \text{Concentration} = & (\overline{\mathrm{A}} + x) & (\overline{\mathrm{B}} + x) & & (\overline{\mathrm{C}} - x) \end{array}$$

The concentration of each species is expressed in terms of the equilibrium concentrations $\overline{\mathrm{A}}$, $\overline{\mathrm{B}}$, and $\overline{\mathrm{C}}$, plus or minus a *displacement x from equilibrium*. Because of conservation of mass, A and B are displaced from equilibrium by the same amount, and with opposite sign by the same amount as C.

Relaxation involves decay of the displacement x to zero, and the objective of the formal analysis is to find an expression for dx/dt that describes the decay. The rate law for the second-order reaction is

$$\frac{d[\mathrm{C}]}{dt} = k_1[\mathrm{A}][\mathrm{B}] - k_{-1}[\mathrm{C}] \tag{6-46}$$

Substituting concentrations in terms of x, we get

$$\frac{d\overline{\mathrm{C}}}{dt} - \frac{dx}{dt} = k_1(\overline{\mathrm{A}} + x)(\overline{\mathrm{B}} + x) - k_{-1}(\overline{\mathrm{C}} - x) \tag{6-47}$$

At this point we use the assumption that x is small so that terms in x^2 can be dropped. Rearranging Equation 6-47 and neglecting terms in x^2 yields

$$\frac{d\overline{\mathrm{C}}}{dt} - \frac{dx}{dt} = k_1\overline{\mathrm{A}}\,\overline{\mathrm{B}} - k_{-1}\overline{\mathrm{C}} + k_1(\overline{\mathrm{A}} + \overline{\mathrm{B}})x + k_{-1}x$$

The term $d\overline{\mathrm{C}}/dt$ is zero because the temperature or pressure is held constant following its increase or decrease, so the equilibrium concentration does not depend on time. On the right side of the equation, $(k_1\overline{\mathrm{A}}\,\overline{\mathrm{B}} - k_{-1}\overline{\mathrm{C}})$

also equals zero because forward and backward reaction rates are equal at equilibrium. Thus

$$\frac{dx}{dt} = -[k_1 (\bar{A} + \bar{B}) + k_{-1}]x$$

This is an equation of the form (6-20), with the corresponding solution

$$x = x_0 e^{-t/\tau} \tag{6-48}$$

in which

$$1/\tau = k_1(\bar{A} + \bar{B}) + k_{-1} \tag{6-49}$$

The quantity τ is called the *relaxation time* for the reaction. Equation 6-48 shows that x relaxes to zero exponentially; the relaxation time τ is the time required for x to decay to $1/e$ of its original value x_0. It is important to notice that *both forward and reverse rates contribute to the relaxation rate, $1/\tau$.* As shown by Equation 6-49, a plot of $1/\tau$ versus equilibrium concentration, $\bar{A} + \bar{B}$, yields k_1 as slope and k_{-1} as intercept.

EXERCISE 6-10

Show that the relaxation time τ for a reversible first-order reaction $A \underset{k_{-1}}{\overset{k_1}{\rightleftarrows}} B$ is given by $1/\tau = k_1 + k_{-1}$.

When there is more than one step in the reaction mechanism, more than one relaxation time appears in the relaxation or decay curve. It can be shown rigorously that a small displacement from equilibrium of any observed variable x (such as the light absorption) decays to zero with a sum of exponential terms

$$x = \sum_i x_i \exp(-t/\tau_i) \tag{6-50}$$

The number of terms x_i is in general equal to the number of independent variables in the kinetic equations (see Exercise 6-5). Equation 6-50 can be derived from a *normal mode* analysis of the set of differential equations that describes the reaction mechanism. The set of time constants τ_i is called the *relaxation spectrum,* and is analogous to the set of vibrational frequencies that results from normal mode analysis of molecular vibration (Chapter 13).

Periodic Perturbations Yield a Phase Shift at the Relaxation Frequency

In addition to the methods in which a perturbation is applied suddenly and maintained, such as temperature, pressure, and electric-field jump, relaxation kinetics also can be studied with a periodic perturbation. An example is measurement of *sound absorption.* When a sound wave passes through a solution, the pressure at a point oscillates. If a chemical reaction has a volume change ΔV, the equilibrium constant depends on pressure according to the expression (derived from $d\,\Delta G = \Delta V\,dP - \Delta S\,dT$):

$$\left(\frac{\partial \ln K}{\partial P}\right)_T = \frac{-\Delta V^\circ}{RT}$$

Therefore the equilibrium concentration of reactants oscillates with the variation of pressure in the sound wave. When the sound frequency matches the relaxation rate of the chemical reaction, energy is absorbed from the sound wave because of the phase shift between the pressure perturbation and the chemical relaxation response. The phenomenon is analogous to absorption of electromagnetic radiation when its frequency is in resonance with some molecular or electronic motion (Chap. 12). As a consequence, the relaxation time can be determined from the frequency at which maximum sound absorption occurs. For practical reasons this technique is limited to relaxation times between about 10^{-5} sec and 10^{-9} sec. The sound frequencies involved are higher than the detection limit of the human ear (~20 kHz) and the method is often called *ultrasonic* absorption.

6–15 Proton Transfer Reactions are Important in Biochemistry

Virtually all of the biochemical reactions that involve carbon compounds are catalyzed by enzymes, but that should not make us think there are no nonenzymatic reactions that occur in organisms. In particular, there are many acid-base or proton-transfer reactions, and many reactions involving complex formation by metal ions that occur very rapidly in solution and are essential for biochemical functions.

Proton-transfer reactions are ubiquitous in living systems and play a crucial role in the catalytic function of many enzymes. The proton in solution is always bonded to an oxygen of a water molecule, so that it often is written as H_3O^+. The hydrogen-bonded structure of water provides a special path for migration of the protonic charge, as shown in Figure 6-11. The water molecule hydrogen bonded to a basic group X^- may be hydrogen bonded to H_3O^+. Movement of the protons along the hydrogen-bond axis results in transfer of a proton to X^-, without any contact between the original H_3O^+ molecule and

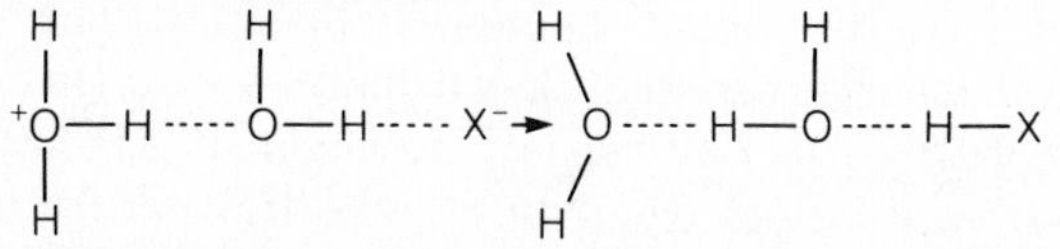

$$\begin{array}{c} \mathrm{H} \\ | \\ {}^{+}\mathrm{O{-}H\cdots\cdots} \\ | \\ \mathrm{H} \end{array}\begin{array}{c} \mathrm{H} \\ | \\ \mathrm{O{-}H\cdots\cdots X^-} \\ \\ \end{array} \longrightarrow \begin{array}{c} \mathrm{H} \\ \backslash \\ \mathrm{O\cdots\cdots H{-}} \\ / \\ \mathrm{H} \end{array}\begin{array}{c} \mathrm{H} \\ | \\ \mathrm{O\cdots\cdots H{-}X} \\ \\ \end{array}$$

FIGURE 6-11
Migration of the protonic charge to X^- by transfer of protons from one water molecule to another.

X^-. In this way protonic charge can move rapidly through the solution, and is not associated with any single proton. (See the relative ion mobilities in Table 8-3.)

EXERCISE 6-11
Devise a path for migration of OH^- in water solution that is analogous to the migration of protons. *Hint.* Treat OH^- as a "missing" proton.

As a result of their special mobilities, H_3O^+ and OH^- recombine at a diffusion-limited rate that is the largest ever measured in aqueous solution. In 1955, Eigen and DeMaeyer used relaxation kinetics to show that the second-order rate constant is 1.4×10^{11} M^{-1} sec^{-1}. This means that if 1 M acid and base solutions could be mixed instantaneously, the neutralization reaction would be nearly complete in 10^{-11} sec.

Proton Transfer is Diffusion-Limited when the Acceptor has a Greater Proton Affinity than the Donor

The kinetics of most proton-transfer reactions can be approximated by the simple general rules developed by Eigen. Recall thàt the pK of a substance is the negative log of the acid dissociation constant:

$$AH \underset{k_{-1}}{\overset{k_1}{\rightleftarrows}} A^- + H^+$$

$$K = \frac{[A^-][H^+]}{[AH]}$$

$$pK = -\log_{10} K$$

The larger the pK, the greater the affinity of the conjugate base A^- for protons. If a proton-transfer reaction occurs between two substances,

$$AH + B^- \underset{k_R}{\overset{k_F}{\rightleftarrows}} A^- + BH$$

the reaction rate constant k_F depends on the difference in pK values between AH and BH. If BH has the larger pK, B^- has greater affinity for protons than A^- has, and proton transfer occurs at every collision between AH and B^-. In this case the transfer rate constant, k_F, usually is at or near the diffusion-controlled limit of 10^9 to 10^{11} M^{-1} sec^{-1}. However, if AH has the larger pK, A^- has greater affinity for protons than B^-, and transfer from AH to B^- will occur on only a small fraction of collisions. In this case the reaction rate is smaller than the diffusion-limited rate and can be approximated by

$$\log_{10} k_F = \log_{10}(k_{\text{diffusion-limited}}) - (pK_{AH} - pK_{BH}) \tag{6-51}$$

EXERCISE 6-12

Calculate the standard free-energy change, ΔG°, of proton transfer between AH and B^-, and show that setting ΔG° equal to the free energy of activation for the transfer reaction leads to Equation 6-51. *Hint:* $\Delta G^\circ = -RT \ln K$.

OH^- is a very strong base ($pK_{H_2O} = 15.75$) and reacts at the diffusion-controlled limit with most acids. Similarly, H_3O^+ is a very strong acid and reacts with most bases at the diffusion-controlled limit. Correspondingly, proton transfer reactions to and from H_2O are relatively slow. Table 6-2 lists values of the rate constants for reactions of H_3O^+ and OH^- with bases B^- and acids BH respectively.

6–16 Enzymes Profoundly Affect Reaction Rates and Reaction Rate Laws

An enzyme is a catalyst that increases the rate of a biochemical reaction, and is unchanged at the conclusion of the reaction. Even though it is unchanged, the enzyme does not act by some mysterious force, but rather by participating in the reaction. The only restriction on its participation is that the enzyme be regenerated with the products of the reaction. With the enzyme playing an active role, it is not surprising that the rate laws for catalyzed and uncatalyzed reactions frequently are very different.

Suppose that an enzyme (E) catalyzes the conversion of a reactant called a *substrate* (S) to a product P:

$$S \xrightarrow{E} P$$

TABLE 6-2 *Rate Constants for Reactions of H_3O^+ and OH^- with B^- and BH, Respectively*

REACTION	pK_A OF BH	$k/M^{-1}\,sec^{-1}$
(a) $H_3O^+ + B^-$		
$H_3O^+ + OH^-$	15.75	1.4×10^{11}
$D_3O^+ + OD^-$	16.5	8.4×10^{10}
$H_3O^+ + F^-$	3.15	1.0×10^{11}
$H_3O^+ + HS^-$	7.24	7.5×10^{10}
$H_3O^+ + NH_3$	9.25	4.3×10^{10}
(b) $OH^- + BH$		
$OH^- + NH_4^+$	9.25	3.4×10^{10}
$OH^- + ROH$	10	1.5×10^{10}
$OH^- + RSH$	9.5	$\sim 10^9$

From Eigen (1963).

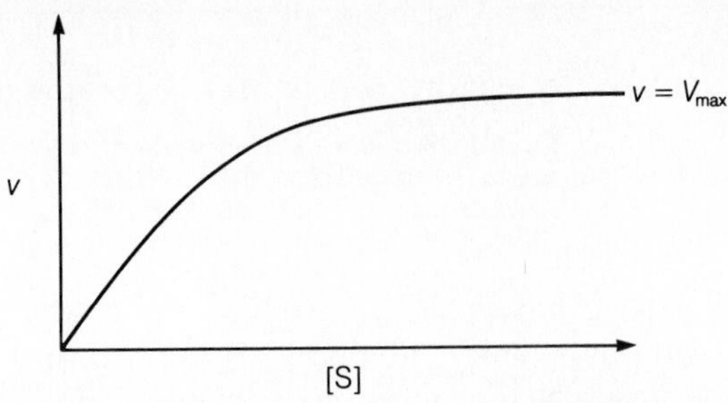

FIGURE 6-12
Schematic variation of the reaction velocity v of an enzymatic reaction as a function of substrate concentration [S]. The reaction is initially linear or first-order in [S], but the rate is limited to the value V_{max}. When that value is reached the reaction rate is constant, or zero-order in [S].

If we plot the rate of this reaction as a function of the concentration of S, we do not find a monotonically increasing velocity v, but rather a velocity that approaches a limit, V_{max}, when [S] is large (Fig. 6-12). Thus the rate of an enzyme-catalyzed reaction increases as substrate is added, until a limiting velocity is reached, whereupon further addition of substrate fails to produce any greater velocity. This behavior is consistent with a saturation of the catalytic site of the enzyme by added substrate, after which no further rate-enhancement can result from adding more substrate.

The Michaelis-Menten Model Describes Hyperbolic Enzyme Kinetics

In 1913, Michaelis and Menten described a model for enzymatic catalysis that accounts for the velocity curve shown in Figure 6-12. Their model was modified by Briggs and Haldane in 1925. The basic idea is that an enzyme and substrate form an enzyme-substrate complex (ES) which is converted to an enzyme-product (EP), which in turn decays rapidly to product and regenerated enzyme. The regeneration of the enzyme is necessary for it to be considered a catalyst. Their model can be represented in general by

$$\mathrm{E} + S \underset{k_{-1}}{\overset{k_1}{\rightleftarrows}} \mathrm{ES} \underset{k_{-2}}{\overset{k_2}{\rightleftarrows}} \mathrm{EP} \underset{k_{-3}}{\overset{k_3}{\rightleftarrows}} \mathrm{P} + \mathrm{E} \tag{6-52}$$

in which k_2 is the rate constant for the catalytic conversion, and the constants k_1, k_{-1}, k_3, and k_{-3} are rate constants for the association and dissociation of the enzyme-substrate complex and the enzyme-product complex. The central assumption of the Michaelis-Menten-Briggs-Haldane model is that the ES complex is present in small amounts, and that its rates of formation and breakdown nearly balance so that its concentration is effectively constant. Therefore, the steady-state approximation applies. To simplify the problem further we neglect the back-conversion of EP to ES by stipulating that Reaction 6-52 is effectively irreversible in the initial reaction stages before the concentration of product has had time to accumulate. (This is why measurements in enzyme kinetics usually are of initial velocities.)

We can express the rate of product formation by multiplying k_2 by the concentration of ES, assuming that Step 2 is rate-limiting and the rate of dissociation of EP to product and E is fast by comparison:

$$v = \frac{d[\mathrm{P}]}{dt} = -\frac{d[\mathrm{S}]}{dt} = k_2[\mathrm{ES}] \tag{6-53}$$

The mathematical consequence of the steady-state assumption is

$$\frac{d[\mathrm{ES}]}{dt} = 0 = k_1[\mathrm{E}][\mathrm{S}] - k_{-1}[\mathrm{ES}] - k_2[\mathrm{ES}] \tag{6-54}$$

The first term on the right side of this equation gives the rate of formation of ES from E and S, and the next two terms give the rate of breakdown to E and S, and to EP. Equation 6-54 embodies the idea that the concentration of ES is constant because of the steady flux of material through ES. The concentration of free enzyme [E] that appears in this equation is a quantity that we know very little about. We can substitute for it, however, by noting that E must be the difference of the total enzyme in the reaction $\mathrm{E_0}$ (which normally is a quantity determined by the experimenter) and the occupied enzyme:

$$[\mathrm{E}] = [\mathrm{E_0}] - [\mathrm{ES}] - [\mathrm{EP}] \tag{6-55}$$

If we again restrict our attention to the case in which EP can be neglected because it dissociates rapidly, we may write

$$[\mathrm{E}] = [\mathrm{E_0}] - [\mathrm{ES}] \tag{6-56}$$

Substituting 6-56 into 6-54 and solving for ES, we get

$$[\mathrm{ES}] = \frac{k_1[\mathrm{E_0}][\mathrm{S}]}{k_{-1} + k_2 + k_1[\mathrm{S}]} \tag{6-57}$$

Then, substituting this value for [ES] into Equation 6-53 for the velocity of the reaction, we find

$$v = \frac{k_2[\mathrm{E_0}][\mathrm{S}]}{\dfrac{k_{-1} + k_2}{k_1} + [\mathrm{S}]} \tag{6-58}$$

Two further substitutions simplify the form of Equation 6-58. To make the first substitution, we note that the maximum possible enzyme velocity will occur when all the enzyme is in the form of the ES complex,

$$V_{\mathrm{max}} = k_2[\mathrm{E_0}] \tag{6-59}$$

To make the second, we define a constant K_{M}, called the *Michaelis constant,* as

$$K_{\mathrm{M}} = \frac{k_{-1} + k_2}{k_1} \tag{6-60}$$ ◀

Making these substitutions into Equation 6-58, we get an expression for the reaction velocity in terms of substrate concentrations and two constants, V_{max} and K_M:

$$v = \frac{V_{max}[S]}{K_M + [S]} \tag{6-61}$$ ◀

This equation describes the dependence of reaction velocity on substrate concentration shown in Figure 6-12. The curve in which v begins linear in [S] and approaches a maximum is said to be *hyperbolic*. You can see that at very large values of [S], K_M becomes negligible compared to [S], and Equation 6-61 can be written

$$v = \frac{V_{max}[S]}{[S]} = V_{max} \qquad ([S] \text{ large})$$

At very small values of [S], [S] becomes negligible compared to K_M, and Equation 6-61 can be written

$$v = \frac{V_{max}[S]}{K_M} = c[S] \qquad ([S] \text{ small})$$

in which c is a constant. Thus the order (in substrate concentration) of an enzymatic reaction described by Equation 6-61 decreases from 1 at small values of [S] to 0 at large values of [S]. The physical significance of this is that the enzyme velocity increases linearly as substrate is first added because the ES complex is being formed. Once the catalytic sites of all macromolecules become saturated with substrate, further additions create no further ES complexes, and the rate cannot be increased beyond the limit set by the *turnover rate* of the enzyme, equal to k_2.

Some feeling for the physical meaning of K_M can be acquired by solving Equation 6-54 for $(k_{-1} + k_2)/k_1$. This gives

$$\frac{k_{-1} + k_2}{k_1} = \frac{[E][S]}{[ES]} = K_M \tag{6-62}$$

Apparently once [ES] has reached its steady-state value, K_M is analogous to a dissociation constant for the ES complex. Notice that K_M has units of concentration, just as a bimolecular dissociation equilibrium constant would.

EXERCISE 6-13

Show that K_M equals the concentration of substrate for $v = V_{max}/2$.

It is simple to determine both K_M and V_{max} once the initial velocity of an enzymatically catalyzed reaction has been measured as a function of substrate concentration. One of the best ways to plot the data to yield K_M and V_{max} is the *Eadie method*. To do this we rearrange Equation 6-61 to the form

$$\frac{v}{[S]} = \frac{V_{max} - v}{K_M} \tag{6-63}$$

Then a plot of $v/[S]$ against v (Fig. 6-13) gives a straight line with an intercept $v = V_{max}$ at $v/[S] = 0$, and an intercept of V_{max}/K_M at $v = 0$. A second useful method of plotting the kinetic data is the *Lineweaver-Burk double-reciprocal method,* which is based on taking the reciprocal of both sides of Equation 6-61, thereby giving

$$\frac{1}{v} = \frac{1}{V_{max}} + \frac{K_M}{[S]V_{max}} \tag{6-64}$$

If $1/v$ is plotted against $1/[S]$, a straight line is found with intercept $1/V_{max}$ at $1/[S] = 0$, and with slope K_M/V_{max} (see Fig. 6-13).

An Application of Eyring's Theory to the Fumarase Reaction

The hydration of fumaric acid to yield L-malic acid is one of the steps of the citric acid cycle:

$$\begin{matrix} \text{CH—COOH} \\ \| \\ \text{CH—COOH} \end{matrix} + H_2O \underset{}{\overset{\text{Fumarase}}{\rightleftarrows}} \begin{matrix} \text{CH}_2\text{—COOH} \\ | \\ \text{CHOH—COOH} \end{matrix}$$

Massey (1953) interpreted the temperature dependencies of the rate of this and related enzymatic reactions in terms of Eyring's theory; his results are summarized in the reaction profile shown in Figure 6-14.

Although this diagram must be regarded as being highly schematic, it illustrates several important points. The first is that the activated complex, which exists at the highest maximum of the profile, is not the enzyme-

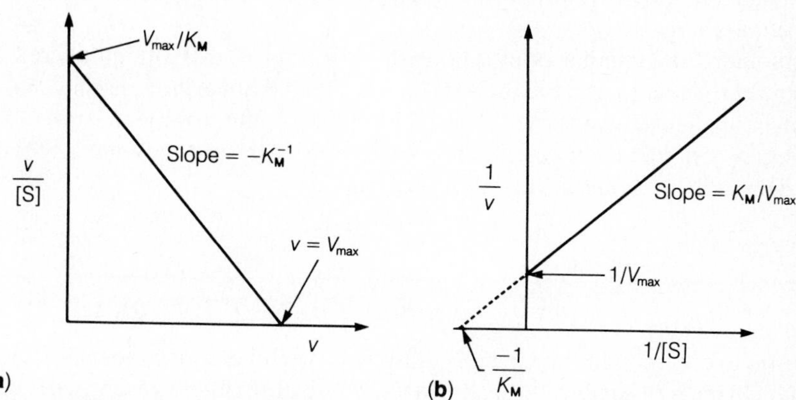

FIGURE 6-13
Two methods of plotting enzyme kinetic data to obtain K_M and V_{max}. **(a)** Eadie method (Equation 6-63), with slope $-K_M^{-1}$ and intercepts V_{max}/K_M and V_{max}. **(b)** Lineweaver-Burk, or double-reciprocal plot, with slope K_M/V_{max} and intercepts $1/V_{max}$ and $-1/K_M$ (Equation 6-64).

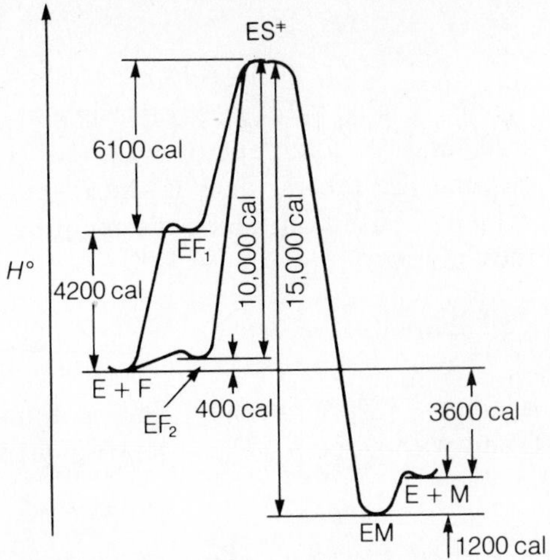

FIGURE 6-14
The reaction profile for the fumarase reaction. E represents enzyme, F represents fumaric acid, and M represents malic acid. The two EF complexes are those detected in the Arrhenius plot of Figure 6-6 for the forward direction. $H°$ is the standard-state enthalpy of the system. (From Dixon and Webb, 1964, based on work by Massey, 1953.)

substrate complex. There are ES complexes for both the forward and backward reactions, and neither is the activated complex. The formation of the ES complex itself is a chemical reaction with a finite rate. Consequently the reactants and the complex are separated by a maximum along the reaction coordinate. This maximum corresponds to a secondary activated complex for the formation of the ES complex. The maxima corresponding to these secondary activated complexes are lower than the main maximum because the chemical conversion is the step with the lowest rate. In other words, the chemical reaction is rate-limiting; the formation and breakdown of the complexes are not.

Another fundamental point about this diagram is that the catalyzed and uncatalyzed reactions have the same $\Delta H°$, which is a thermodynamic quantity that is unaffected by the enzyme. The enzyme alters only the rate of the reaction, and this is reflected in the lower barrier for the enzymatic reaction, as discussed in detail in the next section.

6–17 *The Special Catalytic Power of Enzymes*

Efforts to understand the special catalytic power of enzymes form one of the main currents of modern biochemistry. A line of thought presently accepted by many was proposed by Pauling in 1948. It is that enzymes have evolved to bind to and stabilize the activated complexes of the reactions they catalyze. This binding lowers the transition-state energy, and consequently accelerates the reaction. Pauling expressed this as follows:

> . . . I believe that . . . the surface configuration of the enzyme is . . . complementary to an unstable molecule with only transient existence—namely, the "activated complex" for the reaction that is catalyzed by the enzyme. The mode of action of an enzyme would then be the following: the enzyme would show a small power of attraction for the substrate molecule or molecules, which would become attached to it in its active surface region. This substrate molecule, or these molecules, would then be strained by the forces of attraction to the enzyme, which would tend to deform it into the configuration of the activated complex, for which the power of attraction by the enzyme is the greatest. The activated complex would then, under the influence of ordinary thermal agitation, either reassume the configuration corresponding to the reactants, or assume the configuration corresponding to the products. The assumption made above that the enzyme has a configuration complementary to the activated complex, and accordingly has the strongest power of attraction for the activated complex, means that the activation energy for the reaction is less in the presence of the enzyme than in its absence, and accordingly, that the reaction would be speeded up by the enzyme . . .

The Factor by which an Enzyme Accelerates a Reaction Depends on its Affinity for the Activated Complex

Building on Pauling's idea, Wolfenden (1972) and others have developed the quantitative relationships that follow from it. In doing so, they considered the alternative pathways of uncatalyzed and an enzyme-catalyzed reaction:

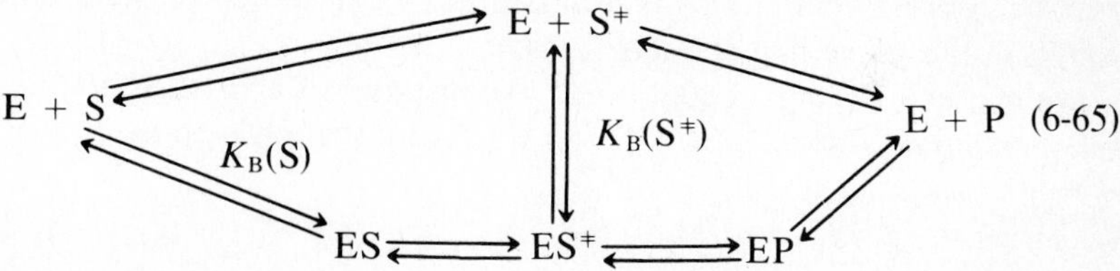

(6-65)

The upper pathway represents the uncatalyzed reaction, in which the substrate S undergoes an activation to the transition state S^+ and then conversion to product P.

The free energy of the intermediates in Reaction 6-65 is represented in Figure 6-15. Notice that there are two paths for converting ES to E + S^+. Since the free energy is a property its change must be the same for both. Therefore, adding the free energies defined in Figure 6-15,

$$\Delta G^{+\circ}(ES^+) - \Delta G_B^\circ(S^+) = -\Delta G_B^\circ(S) + \Delta G^{+\circ}(S^+)$$

Rearranging this equation we get

$$\Delta G^{+\circ}(S^+) - \Delta G^{+\circ}(ES^+) = \Delta G_B^\circ(S) - \Delta G_B^\circ(S^+) \tag{6-66}$$

The left side of Equation 6-66 is the difference in the standard free energies of activation for the uncatalyzed and catalyzed reactions, and the right side

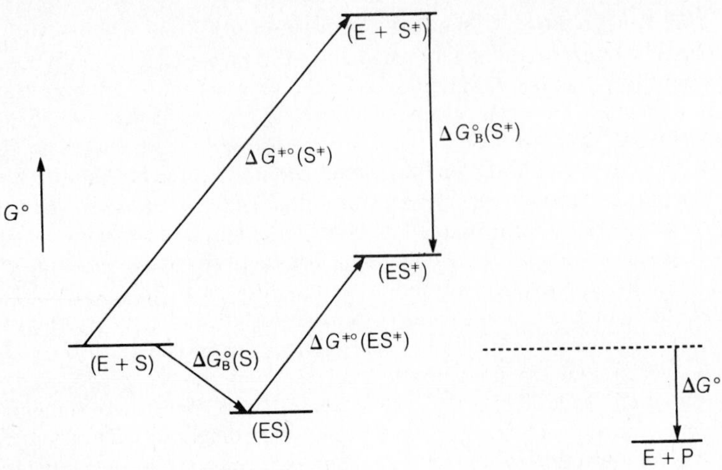

FIGURE 6-15
Standard free-energy changes in Reaction 6-65. The free energies of activation for catalyzed and uncatalyzed reactions are, respectively, $\Delta G^{\ddagger\circ}(ES^{\ddagger})$ and $\Delta G^{\ddagger\circ}(S^{\ddagger})$, while the free energies of binding are $\Delta G^{\circ}_{B}(S)$ for binding S to E and $\Delta G^{\circ}_{B}(S^{\ddagger})$ for binding $S^{\ddagger}$ to E. Additivity of the free energy changes gives the equation $\Delta G^{\ddagger\circ}(ES^{\ddagger}) - \Delta G^{\circ}_{B}(S^{\ddagger}) = -\Delta G^{\circ}_{B}(S) + \Delta G^{\ddagger\circ}(S^{\ddagger})$ for conversion of ES to E + $S^{\ddagger}$.

is the difference in standard free energies of enzyme binding of the substrate (S) and the activated complex ($S^{\ddagger}$). In other words, the greater the free-energy preference of the enzyme for $S^{\ddagger}$ over S, the greater the reduction in the activation free energy and hence increase in the rate that the enzyme can bring about. This is a quantitative expression of the Pauling idea stated in the preceding section.

We can use Eyring's equation, k = constant × exp $(-\Delta G^{\ddagger\circ}/RT)$, and the relationship $K = \exp(-\Delta G^{\circ}/RT)$ to rewrite Equation 6-66 as

$$\frac{k_C}{k_U} = \frac{K_B(S^{\ddagger})}{K_B(S)} \tag{6-67}$$

in which k_C and k_U are respectively the rate constants for the catalyzed and uncatalyzed reactions, and K_B is the binding constant for the enzyme for either $S^{\ddagger}$ or S. For example,

$$K_B(S^{\ddagger}) = \frac{[ES^{\ddagger}]}{[E][S^{\ddagger}]}$$

According to Equation 6-67, *the factor by which an enzyme increases the rate of a reaction is given by the ratio of binding constants of the enzyme for* $S^{\ddagger}$ *and S*. In cases for which the k_C/k_U ratios are known, they fall in the range 10^8 to 10^{14}. Typical values for $K_B(S)$ are in the range 10^3 to 10^5 M^{-1}. Thus the association constants $K_B(S^{\ddagger})$ are likely to be of the order of 10^{15} M^{-1}, which represents an extremely high affinity of the enzyme for the activated-complex form of the substrate.

Enzymes can Accelerate Reactions by Decreasing the Enthalpy of the Activated Complex, by Decreasing its Entropy, or Both

The essence of Equation 6-67 is that enzymes accelerate reactions by binding more tightly to the activated complex than to the unactivated substrate, thus lowering the free energy of the transition state by an amount $\Delta G_B^\circ(S) - \Delta G_B^\circ(S^\ddagger)$, as given by Equation 6-66. Because free energy has contributions from both enthalpy and entropy, $\Delta G = \Delta H - T\Delta S$, it is clear that the enzyme functions either by lowering the enthalpy of the transition state, or by increasing its entropy, or both. Let us consider how the enzyme might accomplish either of these functions.

The activated complex, being a transitory intermediate between two stable compounds, almost certainly is a strained species with a greater enthalpy than either substrate or product. In order to stabilize this intermediate, the enzyme must provide a structure with groups ideally positioned to interact with the substrate in its strained form. In this way binding of the activated complex to the enzyme would be energetically favored, and the enthalpy of the transition state would be lowered by interacting with the enzyme.

Enzymes also can reduce the activation entropy of biochemical reactions. This factor is most apparent when there are two or more substrates in the reaction. All of the substrates can be expected to bind to the enzyme with a favorable binding free energy; once on the enzyme surface they are held in a favorable geometry for the reaction that ensues. Whereas the uncatalyzed reaction requires that the substrates be brought together from free solution to form the activated complex—thus involving substantial entropy loss—the substrates in the enzyme-catalyzed reaction are already held together by their interaction with the enzyme. The *proximity effect* is the name given to this ability of the enzyme to hold its substrates close together in preparation for reaction, thereby making the entropy of activation more favorable.

Transition-State Analogues should be Powerful Enzyme Inhibitors

A corollary to Pauling's hypothesis about the special catalytic power of enzymes is that synthetic compounds which resemble the activated complex for a reaction should bind strongly to the enzyme and thus be potent inhibitors. Work by Wolfenden (1972), Lienhard (1973), and others has shown that this is indeed the case. This discovery opens the way to the synthesis of powerful inhibitors that may prove to be of enormous importance in medicine. As an example, the enzyme aldolase in the glycolytic pathway of any organism is one of two types, depending on the species. Bacterial and fungal aldolases are metalloenzymes that proceed by entirely different mechanisms than do the Schiff-base aldolases of animals. A specific inhibitor of the metalloaldolases would be a powerful antibacterial and antifungal agent.

QUESTIONS FOR REVIEW

1. Define what is meant by reaction mechanism.
2. How is a chemical reaction rate described mathematically?
3. What is a rate law?
4. Define the order of a reaction.
5. Define the reaction rate constant.
6. What are seven important characteristics of the set of elementary steps that constitute a reaction mechanism?
7. Define unimolecular and bimolecular reactions.
8. Define a reversible reaction.
9. What is the principle of microscopic reversibility?
10. What is meant by the principle of detailed balancing?
11. What three lines of evidence are used to infer a reaction mechanism?
12. Describe the steady-state assumption.
13. Define the half-life of a first-order reaction.
14. What is a rad?
15. What is a C_0t curve?
16. What is a reactive encounter?
17. Define the reaction coordinate and the reaction profile.
18. Define the Arrhenius activation energy and the frequency factor.

PROBLEMS

1. Are the following statements true or false?

   (a) The rate of a process is always proportional to its free-energy change.

   (b) The order of a chemical reaction can be determined by examining its stoichiometry.

   (c) The order of an elementary chemical reaction step can be determined by examining its stoichiometry.

   (d) A bimolecular elementary reaction step has a second-order rate law.

   (e) A reaction mechanism can be proved to be correct.

   (f) First-order reactions follow an exponential time course.

   (g) The renaturation of DNA is a second-order reaction.

   (h) The renaturation of DNA is a diffusion-limited reaction.

Answers: (a) F. (b) F. (c) T. (d) T. (e) F. (f) T. (g) T. (h) F.

2. Add the appropriate word or words:

   (a) The ______ describes in detail the transient intermediates that intervene between reactants and products in a chemical reaction.

   (b) The sum of exponents of the concentrations in the rate law is equal to the ______ .

   (c) The rate constant for a bimolecular reaction has units ______ .

   (d) The slowest step in a reaction mechanism is called the ______ .

   (e) Radiation dosage, given as energy absorbed per unit mass, is expressed in ______ .

   (f) According to the Eyring theory, the rate of a reaction depends on the free energy of the ______ .

Answers: (a) Reaction mechanism. (b) (Total) reaction order. (c) Concentration^{-1} time^{-1}. (d) Rate-limiting step. (e) Rads. (f) Activated complex (or transition state).

PROBLEMS RELATED TO EXERCISES

3. (Exercise 6-1) Express the rate of the reaction $H_2O_2 \rightarrow H_2O + \frac{1}{2}O_2$ in terms of the time rate of change of the concentrations, assuming that products appear at the same rate as the reactant disappears.

4. (Exercise 6-2) The rate law for the reaction $H_2 + D_2 \rightarrow 2HD$ is found to be rate $= k[H_2]^{0.38}[D_2]^{0.66}[Ar]$. What is the reaction order with respect to each reactant, and what is the total reaction order?

5. (Exercise 6-3) Explain why the value of the rate constant for a first-order reaction $A \rightarrow B$ is independent of the concentration units chosen (i.e., molar, molal, mole fraction, etc.).

6. (Exercise 6-4) Write the rate law for the elementary reaction step in which NH_4^+ is neutralized by acetate, Ac^-.

7. (Exercise 6-7) Calculate the number of ^{32}P atoms in 1 mCi of ^{32}P.

8. (Exercise 6-8) Calculate the average dose in rads when all the radiation emitted by 1 Ci of ^{3}H is absorbed by 1 g of matter for 10 min.

OTHER PROBLEMS

9. Derive an expression for the reaction half-time of the irreversible second-order reaction $2A \xrightarrow{k_1} B$ in terms of k_1 and the starting concentration A_0.

10. Show that the rate predicted by the reaction mechanism 6-12*a–c*, with the second step assumed to be rate-limiting and the first step assumed to be at equilibrium, is rate $= k_2K_1^{1/2}[Cl_2]^{1/2}[CO]$.

11. Show that the initial rate law predicted by the reaction mechanism 6-12*a–c*, with the first step rate-limiting, is rate $= 2k_1[Cl_2]$. Assume that the $Cl\cdot$ produced in step (3) can be neglected initially.

12. Show that the mechanism for DNA renaturation given in Figure 6-5 implies an apparent Arrhenius activation energy E_A given by $E_A = \Delta H_1^\circ + E_{A2}$ in which ΔH_1° is the enthalpy change of the first step and E_{A2} is the activation energy of the second step. Explain the observation that E_A is found to be negative (about -5 to -10 kcal mol^{-1}).

13. A proposed mechanism for decomposition of ozone (O_3) to oxygen is

$$O_3 \xrightarrow{k_1} O_2 + O\cdot$$

$$O\cdot + O_3 \xrightarrow{k_2} 2O_2$$

Use the steady-state approximation on the concentration of O · atoms to derive the rate law for the process, assuming that the second step is rate limiting and that both steps are irreversible.

14. Calculate the factor by which a reaction rate is increased by an enzyme at 37°C if it lowers the reaction activation energy from 15 kcal mol^{-1} to 10 kcal mol^{-1}.

15. Suppose that the reaction

$$2A \underset{k_{-1}}{\overset{k_1}{\rightleftarrows}} 3B$$

is second-order in the forward direction and third-order in reverse. Derive an expression for the equilibrium constant in terms of the concentrations of A and B by equating the forward and reverse rates.

16. Consider the reversible reaction

$$2A \underset{k_{-1}}{\overset{k_1}{\rightleftarrows}} B$$

which is second-order in the forward direction and first-order in the reverse direction. Write the rate law for the reaction, and integrate it to obtain an expression for the concentration of A as a function of time. The concentrations at $t = 0$ are $[A] = A_0$, $[B] = 0$.

17. How would you explain a biphasic DNA renaturation curve of the following kind?

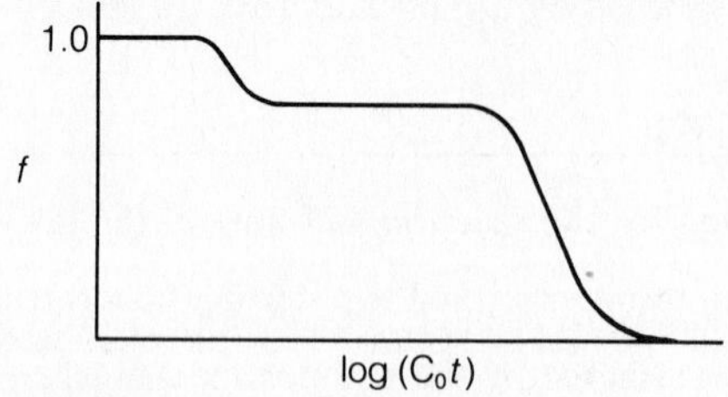

18. When the reciprocal of the free single-stranded DNA concentration [A] is plotted as a function of time, a curve of the following kind is observed:

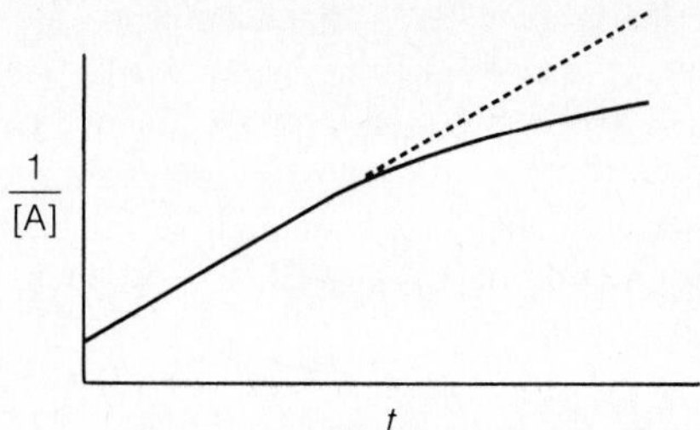

What does the deviation from linearity mean, and how might you explain it?

19. When the drug ethidium binds to DNA, two different complexes are formed. However, unlike the case of proflavine, in which the initial rate of conversion between the two complex forms becomes first order at high concentration, the conversion rate continues to increase quadratically with concentration even at very high concentrations. Propose a reaction mechanism consistent with this observation.

20. A reaction rate increases by a factor of 10 when the temperature is increased from 30°C to 37°C. Estimate the Arrhenius activation energy.

21. The dissociation of the double helix d(AACAA) · d(TTGTT) has an activation energy of 35 kcal mol^{-1} and a rate constant of 10^4 sec^{-1} at 35°C. Calculate the entropy of activation. How might you explain the positive sign of $\Delta S^{\ddagger}$?

22. Given that H_3O^+ reacts with an amine, whose pK is 9.25, with a diffusion-limited rate constant of 4.3×10^{10} M^{-1} sec^{-1}, calculate the rate of reaction of H_2O with R—NH_3^+.

23. Calculate the half-time of the reaction of *lac* repressor with operator, both present at an initial concentration of 10^{-11}M. Assume that the rate constant is $5 \times 10^9 M^{-1}$ sec^{-1}, and that the reverse reaction can be neglected.

24. Derive an expression for the relaxation time of the third-order reaction

$$A + B + C \underset{k_{-1}}{\overset{k_1}{\rightleftarrows}} D$$

25. Use the steady-state approximation on the intermediate $(PD)_{out}$ to calculate the rate of the reaction of P with D to form $(PD)_{in}$:

$$P + D \underset{k_{-1}}{\overset{k_1}{\rightleftarrows}} (PD)_{out} \underset{k_{-2}}{\overset{k_2}{\rightleftarrows}} (PD)_{in}$$

26. Let an enzyme inhibitor I be present in the reaction mixture so that to the reactions

$$E + S \underset{k_{-1}}{\overset{k_1}{\rightleftarrows}} ES \overset{k_2}{\rightarrow} P$$

there is added the reaction

$$\mathrm{E} + \mathrm{I} \underset{k_{-3}}{\overset{k_3}{\rightleftarrows}} \mathrm{EI}$$

in which EI cannot react with substrate (competitive inhibition).
Use the steady-state assumption on ES and EI to show that the reaction velocity is given by

$$v = \frac{k_2[\mathrm{E_0}][\mathrm{S}]}{\dfrac{k_{-1} + k_2}{k_1}\left(1 + \dfrac{k_3}{k_{-3}}[\mathrm{I}]\right) + [\mathrm{S}]}$$

FURTHER READING

Dixon, E. C., and Webb, E. C. 1964. *Enzymes,* 2nd ed. New York: Academic Press. A thorough treatment of enzyme kinetics.

Edwards, J. O., Greene, E. F., and Ross, J. 1968. From Stoichiometry and Rate Law to Mechanism. *J. Chem. Ed. 45,* 381. A systematic introduction to the use of rate laws in finding mechanisms.

Eigen, M. 1968. *Q. Rev. Biophys. 1,* 3. A summary of relaxation methods for measuring rate constants.

Gardiner, W. C. 1969. *Rates and Mechanisms of Chemical Reactions.* Menlo Park, Calif: W. A. Benjamin. A detailed but general introduction to chemical kinetics.

Hendee, W. R. 1973. *Radioactive Isotopes in Biological Research.* New York: John Wiley & Sons. Contains extensive details on the use, detection, and safe handling of radioactive materials.

Jencks, W. P. 1969. *Catalysis in Chemistry and Enzymology.* New York: McGraw-Hill. A book-length essay by one of the greatest authorities on catalysis.

Moore, W. J. 1972. *Physical Chemistry,* 4th ed. Englewood Cliffs, N.J.: Prentice-Hall. A clear summary of chemical kinetics.

Wold, F. 1971. *Macromolecules: Structure and Function.* Englewood Cliffs, N.J.: Prentice-Hall. A readable introduction to enzyme kinetics.

Wolfenden, R. 1972. *Accounts Chem. Res. 5,* 10. A recent summary of ideas about the transition state in enzyme reactions.

OTHER REFERENCES

Atkinson, D. E., Hathaway, J. A., and Smith, E. C. 1965. *J. Biol. Chem. 240,* 3682.

Britten, R. J., and Kohne, D. E. 1968. *Science 161,* 529.

Eigen, M. 1963. Proton Transfer, Acid-Base Catalysis and Enzymatic Hydrolysis. I: Elementary Steps (in German). *Angew. Chem. 75,* 489.

Frost, A. A., and Pearson, R. G. 1953. *Kinetics and Mechanism*. New York: John Wiley & Sons.

Lienhard, G. E. 1973. *Science 180,* 149.

Massey, V. 1953. *Biochem. J. 53,* 72.

Richter, P. H., and Eigen, M. 1974. *Biophys. Chem. 2,* 255–263.

Riggs, A. D., Bourgeois, S., and Cohn, M. 1970. *J. Mol. Biol. 53,* 401–417.

Schechter, A. N. 1970. *Science 170,* 273.

PART TWO

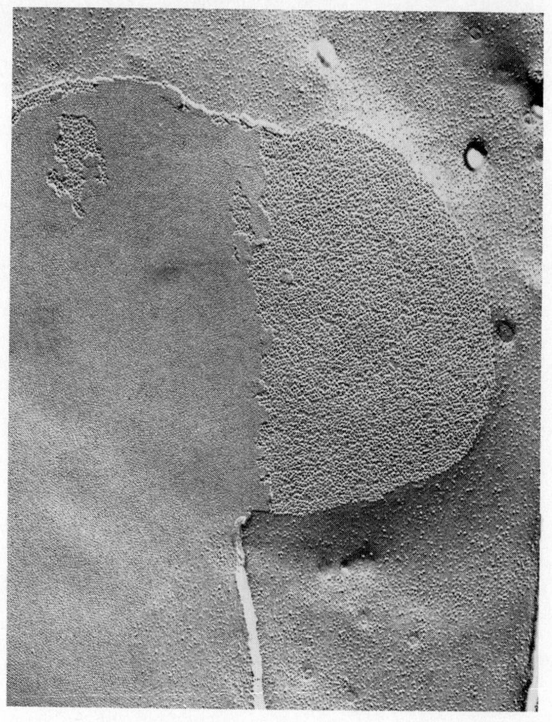

Solutions and Electrochemistry

CHAPTER 7

Solutions

7-1 Macromolecules Form Genuine Solutions

In Chapter 4 we pointed out that it is sometimes advantageous for analysis of macromolecular structural changes to regard the macromolecule as a separate phase. Of far greater general significance for the development of biochemistry and molecular biology was the recognition that macromolecules form genuine solutions. Until the 1930s, macromolecules were generally thought of as *colloids,* an aggregated state of matter containing particles much larger than the constituent molecules. The word *colloid,* from the Greek word for glue, was introduced by Thomas Graham, whose experiments reported in 1861 disclosed the existence of materials such as gelatin that have properties quite different from ordinary salts, sugars, or amino acids. The gelation, or aggregation, of gelatin can be reversed by heating, which indicates a reversible association. It was long believed that all macromolecular substances were noncovalent aggregates of smaller molecules. Well into the twentieth century proteins were regarded as aggregates of polypeptides about 20 amino acids long, nucleic acids were thought of as colloids of a tetranucleotide, and cellulose was considered to contain a noncovalent association of cyclic tetrasaccharides.

In the 1920s there were only a few individuals who did not accept the majority view of colloidal association. The leader of the opposition was Staudinger in Germany, who studied polymeric substances in organic solvents and showed that the same high molecular weight was maintained in all solvents. Colloidal association might have been expected to proceed to different extents when the solvent was changed. The term *macromolecule* was first used by Staudinger as a result of his experiments. More direct evidence on macromolecules came in the 1930s from Svedberg's measurements on the sedimentation of protein in the ultracentrifuge. He found that protein solutions contained components which sedimented at the same rate, implying that the macromolecules were of discrete size; the colloidal association hypothesis had predicted a broad range of particle sizes. At about the same time the crystallization of several enzymes by Sumner and Northrop provided strong support for the macromolecule hypothesis. Only uniform molecules can form regular crystals, so news of the crystallization was surprising to scientists who were astute enough to grasp the implications, as illustrated by the following anecdote. In the late 1920s when Sumner appeared at the door of Svedberg's office in Uppsala, saying, "My name is James B. Sumner; I have crystallized an enzyme," Svedberg concluded that Sumner must be mentally ill and said, "Yes, yes, one moment," as he shut and locked the door for self-protection. But he soon came to believe Sumner's results and their significance: any substance regular enough to crystallize must be discrete in size and shape.

We know now that biological macromolecules are substances of defined size and chemical structure. They usually also have a definite physical structure, as disclosed by x-ray diffraction measurements (see Chap. 17), although there are some flexible macromolecules that have less structural order (see Sec. 14-13). A large number of macromolecules are soluble in aqueous solutions, and much of the biochemical business of the cell is transacted in the solution phase. This makes the solution state of matter of prime importance for understanding biochemical phenomena. Our objective in this chapter is to develop the thermodynamic description of solutions, especially dilute and moderately concentrated solutions of nonvolatile (nonevaporating) solutes such as salts, sugars, amino acids, nucleotides, and the macromolecules that constitute the soluble part of a living cell.

○ 7–2 Solution Concentrations

A *solution* is a phase containing more than one component. The phase rule tells us that we must specify one additional intensive variable for each of these components. The most convenient variable is usually the concentration. Given the tremendous variety of experimental studies of solutions, it is not surprising that there are several different ways of specifying the composition or concentration of the solution. From the point of view of thermodynamics, the most convenient measures are the *mole fraction* and the *molal* concentration. The mole fraction of substance i, X_i, already introduced in Chapter 4, is defined by

$$X_i = \frac{n_i}{\sum_i n_i} \tag{7-1}$$

in which n_i is the number of moles of component i in the solution. Therefore X_i is the fraction contributed by substance i to the total number of moles. The *molal concentration,* or *molality,* given the symbol m, is the number of moles of a component dissolved in one kilogram of solvent. Suppose there are two components, Solvent A and Solute B; m_B is related to the mole fraction X_B by

$$X_B = \frac{m_B}{(1000/M_A + m_B)} \tag{7-2}$$

The quantity $1000/M_A$ is the number of moles of solvent of molecular weight M_A contained in 1000 grams. Mole fraction and molality are preferred by thermodynamicists because the chemical potential and vapor pressure of ideal solutions can be simply expressed in those concentration units, which do not depend on T or P.

The *molar concentration* or *molarity* C_i of substance i is the number of moles of i contained in a liter of solution. This measure is preferred by many chemists and biochemists because a solution volume is easily measured and contains a known number of moles of the solute. For example, 1 ml of 1M NaOH will exactly neutralize 1 ml of 1M HCl. M is the symbol for molar concentration units. Equation 4-37 gives the relationship between molarity and mole fraction for dilute solutions.

Biochemists often must work with materials of unknown molecular weight, so none of the preceding measures of concentration can be used. In this case one choice is percent concentration. A 1% solution contains 1 g of solute in 100 ml of solution. We will denote this concentration by c. Concentrations also can be given in terms of the volume of solute per unit volume of solution; this is called *volume percent.*

7–3 *Partial Molar Quantities Allow One to Represent the Properties of a Solution in Terms of Properties of its Components*

Partial molar quantities express the change in an extensive thermodynamic property per mole of a component added. We already have made extensive use of one partial molar quantity, the chemical potential, which was defined as the change in the free energy G per mole n_i of component i added, with the temperature, pressure, and all other composition variables n_j held constant.

$$\mu_i = \left(\frac{\partial G}{\partial n_i}\right)_{T,P,n_{j\neq i}}$$

This same definition can be generalized to all other extensive properties of the system. For example, the *partial molar volume* $\overline{V}_i$ is

$$\overline{V}_i = \left(\frac{\partial V}{\partial n_i}\right)_{T,P,n_{j\neq i}} \tag{7-3}$$

The meaning of $\overline{V}_i$ is shown graphically in Figure 7-1. The solution volume V changes as component i is added, by an amount ΔV per increment Δn_i, with all other variables held constant. The partial molar volume depends on

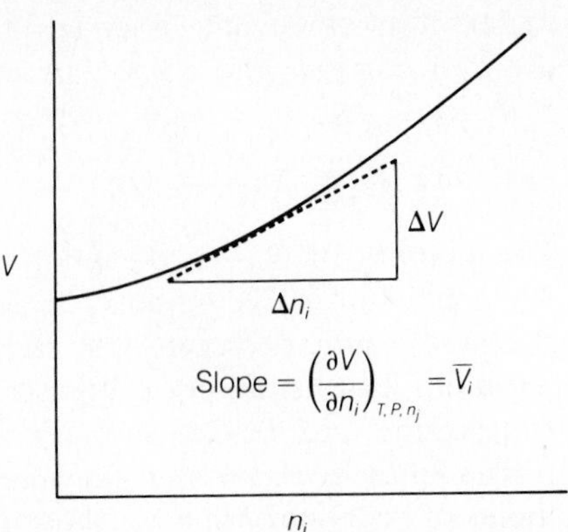

FIGURE 7-1
Graphical determination of the partial molar volume. The solution volume V is measured as a function of the number of moles n_i of component i, holding constant the temperature, pressure, and number of moles n_j of all other components. The slope $(\partial V/\partial n_i)_{T,P,n_j}$ is the partial molar volume $\overline{V}_i$.

the composition of the solution. As another example, the partial molar entropy is

$$\overline{S}_i = \left(\frac{\partial S}{\partial n_i}\right)_{T,P,n_{j\neq i}}$$

All the thermodynamic relationships derived in Chapters 2 through 4 for the extensive quantities also are valid for the partial molar quantities. For example, the relationship $(\partial G/\partial T)_P = -S$ becomes $(\partial \mu_i/\partial T)_P = -\overline{S}_i$. (We use the bar over the symbol for all partial molar quantities except the special case of the chemical potential.)

EXERCISE 7-1

Define the partial molar heat capacity.

ANSWER

$$\overline{C}_{P_i} = \left(\frac{\partial \overline{H}_i}{\partial T}\right)_P \qquad \overline{C}_{V_i} = \left(\frac{\partial \overline{E}_i}{\partial T}\right)_V$$

It also is possible to define the change of an extensive property per unit mass of component i added; these are called *partial specific* quantities. For example, the partial specific volume $\bar{v}_i$ is

$$\bar{v}_i = \left(\frac{\partial V}{\partial w_i}\right)_{T,P,w_{j\neq i}}$$

in which w_i is the mass of component i.

EXERCISE 7-2

Derive a relationship between the partial molar volume and the partial specific volume.

ANSWER

$$\bar{V}_i = M_i \bar{v}_i$$

in which M_i is the molecular weight of component i.

The Sum Rule for Partial Molar Quantities

The partial molar or partial specific quantities are related to the total extensive quantity by an important relationship called the *sum rule*. Suppose we have two components A and B in solution. We can consider the volume to be a function of the number of moles of each component. The total differential of the volume at constant temperature and pressure is

$$(dV)_{T,P} = \left(\frac{\partial V}{\partial n_A}\right)_{T,P,n_B} dn_A + \left(\frac{\partial V}{\partial n_B}\right)_{T,P,n_A} dn_B$$

$$= \bar{V}_A\, dn_A + \bar{V}_B\, dn_B \qquad (7\text{-}4)$$

Equation 7-4 can be integrated directly if we keep the ratio n_A/n_B constant so that $\bar{V}_A$ and $\bar{V}_B$ are constant. (This amounts physically to building up the total volume V by simultaneously adding very small molar increments of A and B, dn_A and dn_B, in the ratio n_A/n_B in the final solution.) Hence

$$V = \bar{V}_A n_A + \bar{V}_B n_B \qquad (7\text{-}5)$$ ◀

This equation shows that we can calculate the volume of the solution by adding partial molar volumes multiplied by the number of moles of each component. The same derivation applies to any partial molar quantity, or to any partial specific quantity. The general sum rules, where $\bar{F}$ is the partial molar quantity and $\bar{f}$ the partial specific quantity, are:

$$F = \Sigma \bar{F}_i n_i \qquad (7\text{-}6)$$ ◀

$$F = \Sigma \bar{f}_i w_i \qquad (7\text{-}7)$$

As an example of the latter equation, the volume V of a two-component solution is given by

$$V = \bar{v}_A w_A + \bar{v}_B w_B \qquad (7\text{-}8)$$

Examples of Partial Molar Volumes

The partial molar volume of organic and biochemical compounds in dilute aqueous solution can be estimated from the sum of contributions by individual atoms and groups. Table 7-1 lists molar volumes for different atoms in a molecule, to which one should add 13 cm^3 per mole for the whole molecule.

The partial molar volumes of ionic solutes usually are smaller than expected; in some cases, such as $MgSO_4$, $\bar{V}$ is actually negative in dilute solution. This means that when a small amount of solid $MgSO_4$ is added to water, the volume of the solution is smaller than the volume of the solvent. The reason is the phenomenon of *electrostriction* in which the small Mg^{2+} ion, with its

TABLE 7-1 *Some Molar Volumes of Atoms*

ATOM	$\overline{V}$, cm³ mol⁻¹	
C	9.9	Add 13 cm³ mol⁻¹ for the whole molecule.
H	3.1	
N	1.5	
O	2.3	
S	15.5	

strong electric field, packs polar water molecules around itself in a smaller volume than they occupy in the bulk solvent.

Dipolar ions such as amino acids also show electrostriction. The partial molar volume of alanine, $^{+}H_3N\underset{\displaystyle CH_3}{\underset{|}{C}}HCOO^{-}$ is 60.6 cm³ mol⁻¹ in dilute solution, whereas $\overline{V}$ for the lactamide $CH_3CHOHCONH_2$, which has exactly the same atomic composition as alanine, is 73.2 cm³ mol⁻¹.

EXERCISE 7-3

Estimate $\overline{V}$ for $CH_3CHOHCONH_2$ from the data in Table 7-1.

ANSWER

70.5 cm³ mol⁻¹. Notice that this is closer to the experimental value for the lactamide than to the value for alanine, which supports the view that $\overline{V}$ for alanine is anomalously small because of electrostriction.

EXERCISE 7-4

From the following data, determine the total volume change for dissolving glycine in 1 kg water to 1 m concentration (1 kg H_2O = 55.51 mol).

GLYCINE CONCENTRATION	GLYCINE	WATER
m	$\overline{V}_2$, cm³ mol⁻¹	$\overline{V}_1$, cm³ mol⁻¹
0	43.20	18.07
1	44.88	18.05
Pure glycine	46.71	—

ANSWER

$$\Delta V = V_{\text{final}} - V_{\text{initial}} = n_1\overline{V}_1 + n_2\overline{V}_2 - (n_1V_1^{\bullet} + n_2V_2^{\bullet})$$
$$= 55.51\,(18.05) + 1(44.88) - [55.51(18.07) + 1(46.71)] = -2.94 \text{ cm}^3$$

This means that the volume of the system H_2O + glycine decreases by 2.94 cm³ when the two components are mixed. Notice that the partial molar volume of glycine is *not* negative, because the volume of the *solution* increases when glycine is added. However, the increase in solution volume is smaller than the volume of glycine added, so the *total* volume change on mixing glycine and H_2O is negative. In measuring the partial molar volume one determines the change in the volume of the *solution* per mole of glycine added.

TABLE 7-2 *Estimated Contribution of Amino-Acid Residues to the Partial Molar Volume of Proteins**

AMINO ACID RESIDUE	VOLUME/cm^3 mol^{-1}
Glycine	36.3
Alanine	52.6
Serine	54.9
Threonine	71.2
Proline	73.6
Phenylalanine	113.9
Methionine	97.7
Cysteine	70.8
Tryptophan	136.7
Tyrosine	116.2
Histidine	91.9
Arginine	109.1
Lysine	105.1
Aspartic acid	68.4
Glutamic acid	84.7
Asparagine	72.6
Glutamine	88.9
Valine	85.3
Leucine	101.8
Isoleucine	101.8

* Adapted from Cohn, E. J., and Edsall, J. T. *Proteins, Amino Acids, and Peptides as Ions and Dipolar Ions*. New York: Reinhold Publishing Co., 1943, p. 372.

The partial molar volume of proteins can be estimated from the amino acid composition by adding the contributions from each amino acid residue, as shown in Table 7-2. The partial specific volume is the total partial molar volume divided by the molecular weight.

7–4 The Gibbs-Duhem Equation Relates the Chemical Potentials of Solute and Solvent

The phase rule states that in a one-phase, two-component system, there are three degrees of freedom ($f = 2 - 1 + 2 = 3$). The variables may be chosen as temperature, pressure, and the chemical potential of one of the components, A. Therefore, since there are only three degrees of freedom, *the chemical potential of the other component B is not an independent variable.*

There must be a relationship between μ_A and μ_B. This is provided in differential form by the *Gibbs-Duhem equation.* We begin with the sum rule (Equation 7-6) for the free energy G, written for simplicity for a two-component system:

$$G = \mu_A n_A + \mu_B n_B \tag{7-9}$$

Using the chain rule for differentiation, we find that the differential dG is

$$dG = \mu_A \, dn_A + \mu_B \, dn_B + n_A \, d\mu_A + n_B \, d\mu_B \tag{7-10}$$

At constant temperature and pressure, the differential dG taken from Equation 4-47, $dG = V\,dP - S\,dT + \Sigma\mu_i\,dn_i$, becomes

$$(dG)_{T,P} = \mu_A \, dn_A + \mu_B \, dn_B \tag{7-11}$$

We can now equate Equations 7-10 and 7-11, with the restriction that T and P are constant, to obtain

$$(n_A \, d\mu_A + n_B \, d\mu_B)_{T,P} = 0 \tag{7-12}$$

The general form of Equation 7-12 is the *Gibbs-Duhem equation:*

$$\left(\sum_i n_i \, d\mu_i\right)_{T,P} = 0 \tag{7-13}$$ ◀

in which the subscripts T and P specify that the differentials are at constant temperature and pressure.

The Gibbs-Duhem equation is extremely important in the thermodynamics of solutions. As we will see, it allows us to determine the chemical potential of nonvolatile solutes from the more easily measured chemical potential of the volatile solvent, water. It applies as well to other partial molar quantities. For example, for the partial molar volume,

$$\left(\sum_i n_i \, d\bar{V}_i\right)_{T,P} = 0$$

and in general, with $\bar{F}_i$ a partial molar quantity,

$$\left(\sum_i n_i \, d\bar{F}_i\right)_{T,P} = 0$$

● *Meaning of the Gibbs-Duhem Equation*

As is apparent from Equations 7-11 or 7-12, the Gibbs-Duhem equation is a *differential* expression, relating *changes* in μ_A, $d\mu_A$, to *changes* in μ_B, $d\mu_B$. As a consequence, it prescribes values only for the *slope* of μ_B as a function of solution concentration, if the slope of μ_A is known as a function of concentration. We will see that any arbitrary constant can be added to μ_A or μ_B without affecting agreement with the Gibbs-Duhem equation, because added constants do not affect the slopes.

To show that the Gibbs-Duhem Equation relates the slopes of chemical potential curves, we first express Equation 7-12 for a two-component solution in terms of the mole fraction of components A and B:

$$X_A\, d\mu_A + X_B\, d\mu_B = 0 \tag{7-14}$$

Dividing through by dX_A gives

$$X_A \frac{d\mu_A}{dX_A} + X_B \frac{d\mu_B}{dX_A} = 0 \tag{7-15}$$

Therefore the ratio of the slopes is

$$\frac{d\mu_A/dX_A}{d\mu_B/dX_A} = -\frac{X_B}{X_A} \tag{7-16}$$

This equation relates the two slopes $d\mu_A/dX_A$ and $d\mu_B/dX_A$ through a factor equal to the ratio of mole fractions. *All two-component solutions, ideal or real, must obey Equation 7-16,* because it is another form of the Gibbs-Duhem equation, and no assumptions have been made about the nature of the system.

The next point we want to demonstrate is that *if one component of a solution follows an ideal expression for the chemical potential, the choice of an ideal expression for the chemical potential of the other component satisfies the Gibbs-Duhem equation*. The chemical potential of component A in an ideal solution (see Equation 4-43*b*) is:

$$\mu_A = \mu_A^\bullet + RT \ln X_A$$

We want to test whether an ideal expression for μ_B,

$$\mu_B = \mu_B^\bullet + RT \ln X_B$$

satisfies the Gibbs-Duhem equation. To do so we calculate the slopes, remembering that T and P are constant, and that $\mu^\bullet$ does not depend on mole fraction. The results are

$$\frac{d\mu_A}{dX_A} = RT \frac{d \ln X_A}{dX_A} = \frac{RT}{X_A}$$

and, similarly, with $dX_B/dX_A = -1$ because $X_B = 1 - X_A$,

$$\frac{d\mu_B}{dX_A} = -\frac{RT}{X_B}$$

Comparing these two equations, we see that

$$\frac{d\mu_A/dX_A}{d\mu_B/dX_A} = -\frac{X_B}{X_A} \tag{7-17}$$

exactly as required by Equation 7-16. Because Equation 7-16 is another form of the Gibbs-Duhem equation, we have proved that our choices for μ_A and μ_B are consistent with the Gibbs-Duhem equation. Figure 7-2(*a*) shows the variation of μ_A and μ_B with X_A for an ideal solution, with ideal chemical potentials. The ratio of the slopes must obey Equation 7-16.

So far we have seen that the value of the slope $d\mu_A/dX_A$ dictates the value of the slope $d\mu_B/dX_A$ through Equation 7-17, and that the choice $\mu_B = \mu_B^\bullet + RT \ln X_B$ satisfies Equation 7-17 when $\mu_A = \mu_A^\bullet + RT \ln X_A$.

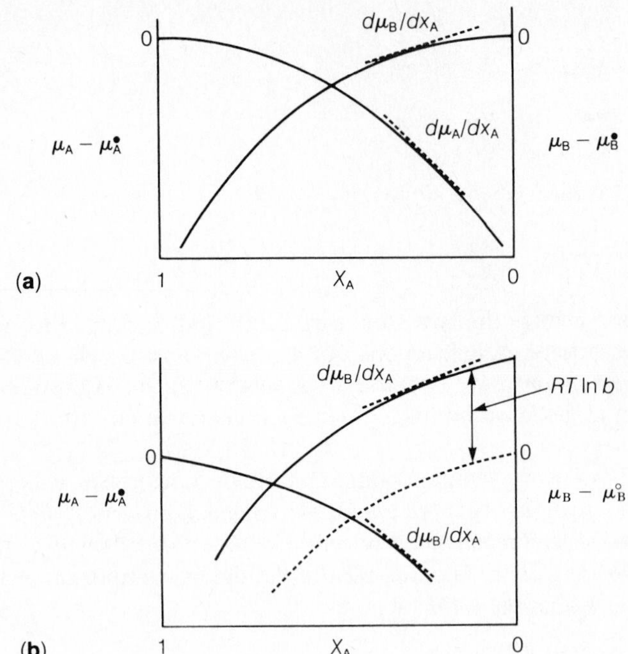

FIGURE 7-2
(a) Schematic variation of the chemical potentials μ_A and μ_B relative to the standard-state values $\mu_A^\bullet$ and $\mu_B^\bullet$ with mole fraction X_A.[1] The Gibbs-Duhem equation leads to Equation 7-16, which states that the slopes of the two curves are related by $(d\mu_A/dX_A)/(d\mu_B/dX_A) = -X_B/X_A$. The choice $\mu_B = \mu_B^\bullet + RT \ln X_B$ satisfies this relationship when $\mu_A = \mu_A^\bullet + RT \ln X_A$. (b) Demonstration that the function $\mu_B = \mu_B^\circ + RT \ln bX_B$ also satisfies the Gibbs-Duhem equation when $\mu_A = \mu_A^\bullet + RT \ln X_A$. The curve for $\mu_B - \mu_B^\circ$ is displaced by $RT \ln b$, but the slope is unchanged. The Gibbs-Duhem equation relates the chemical potentials μ_A and μ_B only through the value of the slopes $d\mu_A/dX_A$ and $d\mu_B/dX_A$. Hence any constant may be added to μ_A or μ_B without affecting the ratio of slopes.

[1] Remember that $\mu_A^\bullet$ is the chemical potential of pure A, whereas μ_A° is the chemical potential of A in its standard state, which need not be the pure substance.

However, we might ask whether some other functions would work equally well. Let us try the choice

$$\mu_B = \mu_B^\circ + RT \ln (b\, X_B) \tag{7-18}$$

in which b is a constant. This is equivalent to

$$\mu_B = \mu_B^\circ + RT \ln b + RT \ln X_B$$

Even without differentiating we can see that the slope of this curve will be the same as $\mu_B = \mu_B^\bullet + RT \ln X_B$, because $RT \ln b$ is the only new term, and it, like μ_B°, is a constant.

Figure 7-2(b) shows why the chemical potential function of Equation 7-18 works just as well as that in Figure 7-2(a). The curve is simply displaced by an amount $RT \ln b$, but the slope is unchanged. Hence *any* value of b can be chosen for the chemical potential function in Equation 7-18. Whatever the choice, μ_B will be compatible with $\mu_A = \mu_A^\circ + RT \ln b'X_A$, in which b' can also have any value. Remember that chemical potentials, like potential energy, are always defined relative to some reference or standard state. An additive constant simply changes the chemical potential in the standard state and does not influence $d\mu/dX$.

7–5 *Ideal Solutions and Raoult's Law*

We have made frequent use of the concept of an ideal solution, with the basic definition that in an ideal solution the chemical potential of component i is $\mu_i = \mu_i^\bullet + RT \ln X_i$. There is another way to define an ideal solution, which turns out to be equivalent, and which provides more insight into the physical nature of the ideal solution state.

Consider the diagram of a two-component solution in Figure 7-3. Suppose that Molecules A and B are identical in size, and that the interactions between A and B are identical to the self-interactions of A with A and B with B. The fact that vapor is in equilibrium with the liquid implies that there is some tendency for molecules to escape from the intermolecular forces into the gas phase. A dynamic equilibrium implies that the rate of escape is equal to the rate of capture of vapor molecules at the liquid surface:

$$\text{Escape rate} = \text{capture rate}$$

The molecules are indistinguishable in size and interactions, so the rate of escape of component A is proportional to the fraction of type A molecule in the liquid,

$$\text{Escape rate} \propto X_A$$

The capture rate will depend on the number of A molecules per unit volume in the vapor, which is proportional to the vapor pressure P_A of component A:

$$\text{Capture rate} \propto P_A$$

Vapor pressure is defined as the pressure of A over the solution when the system is at equilibrium. As we will see, vapor pressure increases exponentially with temperature.

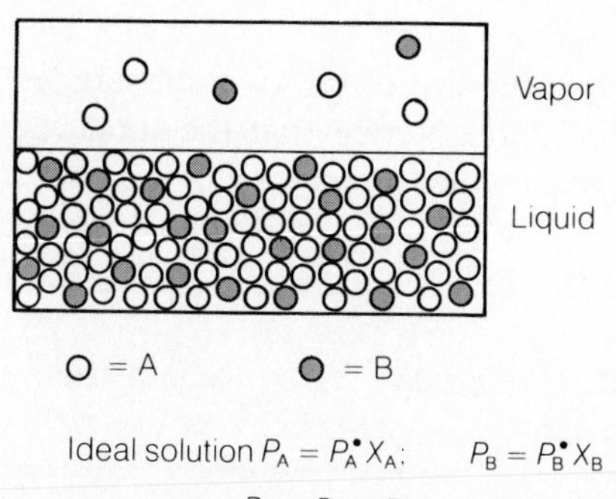

FIGURE 7-3
Physical model of an ideal solution. Molecules A and B are equal in size, and interact with each other exactly as they do with themselves. Hence the rate of escape of a molecule from the solution does not depend on what its neighbors are, and the rate of escape of Component A is proportional to X_A. This leads to Raoult's law, $P_A = P_A^\bullet X_A$ and $P_B = P_B^\bullet X_B$.

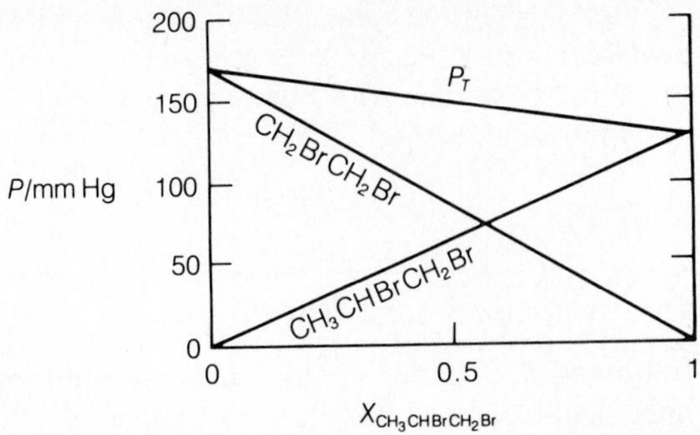

FIGURE 7-4
Experimental verification of Raoult's law for a mixture of ethylene bromide and propylene bromide. The vapor pressure of each component is linear in its mole fraction throughout the entire composition range. Hence the two components form an ideal solution.

Equating the escape and capture rates, we see that the vapor pressure should be proportional to the mole fraction

$$P_A \propto X_A$$

When $X_A = 1$, the vapor pressure must be that of the pure solvent, $P_A^{\bullet}$, so the final equation is

$$P_A = P_A^{\bullet} X_A \qquad \text{(Raoult's law)} \tag{7-19}$$

This is Raoult's (rhymes with *growls*) law for the vapor pressure of an ideal solution. A similar equation holds for Component B, $P_B = P_B^{\circ} X_B$. In 1886, Raoult reported vapor pressure measurements for a number of solutions that closely approximated Equation 7-19.

Equation 7-19 is found to be accurate for materials that approach the conditions required by Figure 7-3 and the accompanying derivation. When the molecules are very similar chemically, their interactions also are similar and the mixtures form nearly ideal solutions. An example is a mixture of ethylene bromide and propylene bromide:

CH_2BrCH_2Br	$CH_3CHBrCH_2Br$
Ethylene bromide	Propylene bromide

Figure 7-4 shows the variation of the vapor pressure of the two substances with the mole fraction X_A of propylene bromide. Notice that the total vapor pressure, the sum of the partial pressures, also changes linearly with X_A.

EXERCISE 7-5

Prove that the total vapor pressure is linearly related to X_A by the equation $P_T = P_B^{\bullet} + X_A(P_A^{\bullet} - P_B^{\bullet})$.

Raoult's Law Describes the Vapor over an Ideal Solution

Our next objective is to show that Raoult's law predicts that $\mu_A = \mu_A^\bullet + RT \ln X_A$, as defined for an ideal solution. To do this we make explicit use of the assumption that the vapor of the components of an ideal solution behaves as an ideal gas. Then the chemical potential of the vapor of A is given by Equation 4-43

$$\mu_A(g) = \mu_A^\circ(g) + RT \ln P_A$$

in which (g) reminds us that this equation gives the chemical potential in the gas phase. However, since solution and vapor phases are in equilibrium, the chemical potential of Component A must be the same in both phases. Therefore

$$\mu_A(\text{sol}) = \mu_A(g) = \mu_A^\circ(g) + RT \ln P_A \tag{7-20}$$

We now can make use of Raoult's law, substituting $X_A P_A^\bullet$ for P_A in Equation 7-20, which becomes

$$\mu_A(\text{sol}) = \mu_A^\circ(g) + RT \ln X_A P_A^\bullet$$

Using the additive property of logarithms, we can rephrase this expression as

$$\mu_A(\text{sol}) = \mu_A^\circ(g) + RT \ln P_A^\bullet + RT \ln X_A \tag{7-21}$$

This equation shows that when $X_A = 1$, the chemical potential $\mu_A(\text{sol}) = \mu_A^\circ(g) + RT \ln P_A^\bullet$. This we define to be the standard state for the solution, so the standard chemical potential of Component A is

$$\mu_A^\bullet(\text{sol}) = \mu_A^\circ(g) + RT \ln P_A^\bullet \tag{7-22}$$ ◀

By inserting this definition into Equation 7-21, we arrive at the expected equation for the chemical potential of A in the ideal solution:

$$\mu_A(\text{sol}) = \mu_A^\bullet(\text{sol}) + RT \ln X_A \tag{7-23}$$ ◀

Derivation of Properties of an Ideal Solution

The simple physical model of an ideal solution shown in Figure 7-3 has important implications for the other thermodynamic properties of an ideal solution. We would expect that when spherical particles A and B of the same size interact in a mixture exactly as they do separately, then there should be no enthalpy or volume change on mixing the two components in any ratio. This result can be derived formally by starting with $\mu = \mu_A^\bullet + RT \ln X_A$, in which $\mu_A^\bullet$ is the chemical potential of pure A, defined to be the standard state. Remembering that $(\partial G/\partial P)_T = V$ (Equation 3-7) and substituting in place of G and V the partial molar quantities μ and $\overline{V}$, we obtain

$$\left(\frac{\partial \mu_A}{\partial P}\right)_T = \overline{V}_A \tag{7-24}$$

However, at constant T the term $RT \ln X_A$ is independent of P so we have

$$\overline{V}_A = \left(\frac{\partial \mu_A}{\partial P}\right)_T = \left(\frac{\partial \mu_A^\bullet}{\partial P}\right)_T = \overline{V}_A^\bullet \qquad (7\text{-}25)$$

Equation 7-25 reveals that the partial molar volume $\overline{V}_A$ of component A in an ideal solution equals the molar volume of the pure component $\overline{V}_A^\bullet$. (Notice that we obtained $\overline{V}_A^\bullet$ because we differentiated the molar free energy of pure A, $\mu_A^\bullet$, with respect to P in Equation 7-25.)

The volume change on mixing the components of a solution is, by definition,

$$\Delta V_{mix} = V(\text{solution}) - \Sigma V(\text{pure components}) \qquad (7\text{-}26)$$

By the sum rule, $V(\text{solution}) = n_A \overline{V}_A + n_B \overline{V}_B$, and $\Sigma V(\text{pure components}) = n_A \overline{V}_A^\bullet + n_B \overline{V}_B^\bullet$. From Equation 7-26, along with $\overline{V}_i = \overline{V}_i^\bullet$, we see that

$$\Delta V_{mix} = n_A \overline{V}_A + n_B \overline{V}_B - (n_A \overline{V}_A^\bullet + n_B \overline{V}_B^\bullet) = 0 \qquad (7\text{-}27)$$

so the volume of change on mixing is zero (at constant T and P).

EXERCISE 7-6

Show that the enthalpy of mixing the components in an ideal solution is zero.

ANSWER

Use $(\partial[\mu_A/T]/\partial T)_P = -\overline{H}_A/T^2$ (see Exercise 4-20).

$$\partial[\mu_A/T]/\partial T = \frac{\partial}{\partial T}[\mu_A^\bullet/T + R \ln X_A] = \partial(\mu_A^\bullet/T)/\partial T = -\overline{H}_A^\bullet/T^2$$

Hence the partial molar enthalpy in the solution ($\overline{H}_A$) equals the molar enthalpy of the pure compound ($\overline{H}_A^\bullet$). Use of the sum rule as before for the volume of mixing yields

$$\Delta H_{mix} = 0 \qquad (7\text{-}28)$$

Hence, as expected, no enthalpy or energy change accompanies the mixing of components in an ideal solution.

However, the free energy of mixing is not zero, because, as we discussed in Section 4-5, the entropy increases on mixing. As we showed in Equation 4-32, the free energy of mixing is $nRT \Sigma X_i \ln X_i$, with $-nR \Sigma X_i \ln X_i$ for the entropy of mixing. Table 7-3 summarizes the thermodynamic parameters of an ideal solution.

TABLE 7-3 *Thermodynamic Properties of an Ideal Solution*

$\mu_i = \mu_i^\bullet + RT \ln X_i$	(Chemical potential)
$\Delta H_{mix} = 0$	(Enthalpy of mixing)
$\Delta V_{mix} = 0$	(Volume of mixing)
$\Delta G_{mix} = nRT \Sigma X_i \ln X_i$	(Free energy of mixing)
$\Delta S_{mix} = -nR \Sigma X_i \ln X_i$	(Entropy of mixing)

7–6 Henry's Law

The physical model of an ideal solution (Fig. 7-3) is a long way from reality. For example, an aqueous solution of macromolecules contains many different kinds of interactions, and the hydrogen bonding between water molecules cannot be equivalent to all the different water-macromolecule interactions, nor can these be equivalent to macromolecule-macromolecule interactions. If solution thermodynamics applied only to ideal solutions, it would be of little use for biochemical systems.

Fortunately there is a more realistic model that we can use as a starting point for considering real solutions. A schematic picture is shown in Figure 7-5, which is different from the ideal solution diagram in Figure 7-3 because the solute is much more dilute and the molecules of solute and solvent are allowed to be physically different in Figure 7-5. The model implies that all the solute molecules are surrounded by solvent molecules, so solute-solute interactions are unimportant. When this is so, the escaping tendency of the solute molecules, B, is proportional to their mole fraction

$$\text{Escape rate} \propto X_B$$

and the capture rate is again proportional to the pressure P_B

$$\text{Capture rate} \propto P_B$$

Hence with escape and capture rates equal

$$P_B \propto X_B$$

In this case, unlike the ideal solution, we do not set the proportionality constant equal to $P_B^\bullet$, because there is no necessary relationship between the escape and capture rates in pure solute ($X_B = 1$) and the escape and capture rates in dilute solution where the solute molecules are surrounded by solvent instead of solute. Thus we write

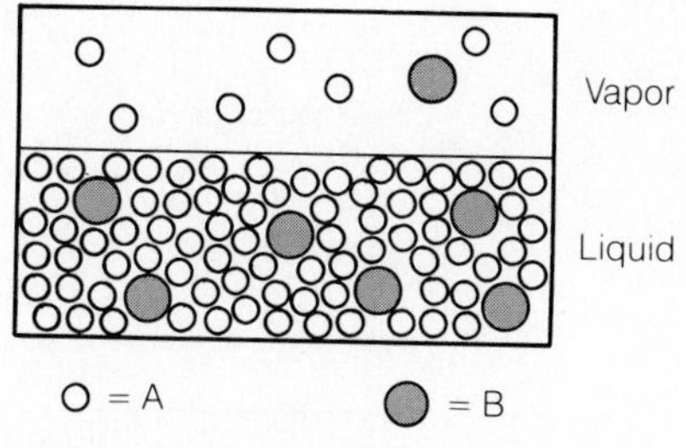

FIGURE 7-5
Physical model of a Henry's law solution. The solute particles B are sufficiently dilute that all are surrounded by solvent molecules A. In this case the escape tendency is proportional to X_B, which leads to Henry's law, $P_B = kX_B$. However, unlike an ideal solution, the environment of the solute changes when X_B approaches 1, so the proportionality constant changes. In general, $k \neq P_B^\bullet$, so Raoult's law is not obeyed for the dilute solute.

$$P_B = k\,X_B \qquad \text{(Solute in Henry's law solution, } X_B \text{ small)} \qquad (7\text{-}29a)$$

in which k is an experimentally determined constant that depends on the interactions between solute and solvent. Equation 7-29*a* is *Henry's law,* obtained originally by William Henry in 1803 for the solubility of gases in liquids. Notice that Equation 7-29 *cannot* be correct when X_B approaches 1 (unless $k = P_B^\bullet$), because it predicts that $P_B = k$ when $X_B = 1$, instead of the observed value, $P_B = P_B^\bullet$.

The solute in a Henry's law solution is dilute, so the solvent is very concentrated, and its vapor pressure will be dominated by the properties of the pure solvent. Therefore for component A, the solvent, we expect

$$P_A = P_A^\bullet\,X_A \qquad \text{(Solvent in Henry's law solution, } X_A \approx 1) \qquad (7\text{-}29b)$$

as prescribed by Raoult's law. Figure 7-6 shows the vapor pressure of a Henry's law solution as a function of the mole fraction of solute, X_B.

Chemical Potential of a Henry's Law Solution

So far we have seen that the chemical potential in an ideal solution varies with $RT \ln X_B$. We shall see here that the chemical potential in a Henry's law solution also varies with $RT \ln X_B$ (or $RT \ln C_B$). However, there is an important conceptual difference between the two solutions. Because the properties of an ideal solution are valid for all compositions, it is convenient to choose the *pure* material as the standard state for each component in an ideal solution. On the other hand, Henry's law is valid only for dilute solu-

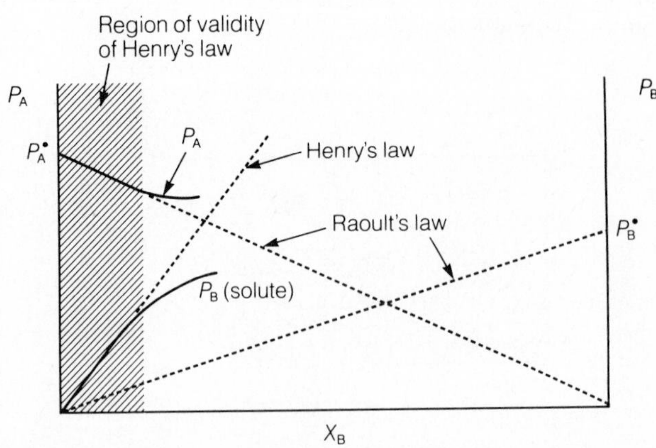

FIGURE 7-6
Henry's and Raoult's laws compared for a dilute solution (small X_B). In sufficiently dilute solution, all solutes obey Henry's law, and the solvent obeys Raoult's law. In an ideal solution, the Henry's law constant k is $P_B^\bullet$. Henry's law *cannot* be obeyed for all X_B unless $k = P_B^\circ$, because the vapor pressure must be $P_B^\bullet$ when $X_B = 1$.

tions, so one can conveniently choose the pure material as the standard state only for the solvent. The properties of the standard state of the solute must be related to the properties of the dilute solution, as we will show in this section.

We obtain the chemical potential of solute in a Henry's law solution by considering the solution to be in equilibrium with the vapor, just as we did for the ideal solution. Rewriting Equation 7-20 to describe the equilibrium of component B in the solution with its vapor.

$$\mu_B(\text{sol}) = \mu_B(\text{g}) = \mu_B^\circ(\text{g}) + RT \ln P_B$$

Substituting Henry's law, $P_B = kX_B$, into this relationship, we obtain

$$\mu_B(\text{sol}) = \mu_B^\circ(\text{g}) + RT \ln k + RT \ln X_B$$

On the basis of this equation, we can choose a standard state for the solute in a Henry's law solution in which the mole fraction $X_B = 1$, which allows us to write for the solution chemical potential

$$\mu_B(\text{sol}) = \mu_B^\circ(\text{sol}) + RT \ln X_B \qquad \text{(Henry's law)} \qquad (7\text{-}30)$$ ◀

in which the Henry's law standard chemical potential of the solute is

$$\mu_B^\circ(\text{sol}) = \mu_B^\circ(\text{g}) + RT \ln k \qquad \text{(Henry's law)} \qquad (7\text{-}31a)$$ ◀

For comparison, the standard chemical potential of Component B in a Raoult's law solution is

$$\mu_B^\bullet(\text{sol}) = \mu_B^\circ(\text{g}) + RT \ln P_B^\bullet \qquad \text{(Raoult's law)} \qquad (7\text{-}31b)$$

Inspection of Equations 7-31*a* and 7-31*b* allows us to see how the standard states are different for Henry's law and Raoult's law (ideal) solutions. Both equations depend on $\mu_B^\circ(\text{g})$, which is the standard chemical potential of solute molecules isolated from each other in the vapor. However, Equation 7-31*a* depends on $RT \ln k$, in which k is the slope of the vapor pressure curve in *dilute* solution only (Fig. 7-6). In contrast, Equation 7-31*b* depends on $RT \ln P_B^\bullet$, in which $P_B^\bullet$ is a property of the *pure* component B. In summary, we conclude that $\mu_B^\bullet$ for a Raoult's law solution reflects the properties of the pure component, whereas μ_B° for a Henry's law solution reflects the properties of a dilute solution.

EXERCISE 7-7

Show that Equation 7-30 for μ_B and Equation 7-23 for μ_A satisfy the Gibbs-Duhem equation.

ANSWER

Equation 7-30 is identical in form to Equation 7-18, which was shown in Figure 7-2(*b*) to be consistent with the Gibbs-Duhem equation when μ_A is given by Equation 7-23.

Because the Gibbs-Duhem equation allows us to add any constant to Equation 7-30, *for dilute solutions in the region of validity of Henry's law,* we can write

$$\mu_B = \mu_B^\circ + RT \ln C_B \qquad (7\text{-}32)$$ ◀

in which C_B *is any measure of concentration that is proportional to the mole fraction in dilute solution*. Equation 7-32 can be written for any solute that does not dissociate into smaller units. The modifications necessary for species such as NaCl that dissociate in solution will be considered in the next chapter.

The Standard State in a Henry's Law Solution is a Hypothetical State

Equation 7-32 expresses the chemical potential of the solute in a dilute Henry's law solution by its value relative to the chemical potential μ_B° in the standard state. In this standard state the concentration is unity, but all solute molecules remain surrounded by solvent molecules. With most choices of concentration units, for example mole fraction or molar concentration, Henry's law is no longer experimentally obeyed when the concentration is unity. Therefore the standard state is a *hypothetical* state, which cannot be achieved in reality. However, its properties can readily be calculated from measurements on dilute solutions. Because the solute particles have the same environment in the standard state and dilute solution, *all partial molar properties which do not depend on the entropy have the same value in the standard state as they do in dilute solution*. This includes, for example, the partial molar volume, enthalpy, and heat capacity. The partial molar entropy varies linearly with the ideal expression $-R \ln C_B$; the partial molar Gibbs free energy (Equation 7-32) varies with $RT \ln C_B$.

Properties of a Henry's Law Solution

The heat of dilution of a Henry's law solution is zero. The conceptual basis for this result can be seen from Figure 7-5. If more solvent is added to the dilute solution, the number of solute-solvent interactions is not changed, because each solute molecule remains surrounded by solvent. The added solvent enters a solvent environment identical to that in the pure material. Let us define the process of dilution as adding $(n'_A - n_A)$ moles of solvent to n_A moles of solvent containing n_B moles of solute,

$$(n'_A - n_A) \text{ solvent} + (n_A + n_B) \text{ solution} \rightarrow (n'_A + n_B) \text{ solution} \qquad (7\text{-}33)$$

Then the heat of dilution can be defined as the difference between the enthalpies of products and reactants,

$$\Delta H_{\text{dil}} = n'_A\overline{H}_A + n_B\overline{H}_B - (n'_A - n_A)\overline{H}_A - n_A\overline{H}_A - n_B\overline{H}_B$$

$\overline{H}$ depends on concentration, so that $\overline{H}_A^\bullet$ and $\overline{H}_A$ are not generally equal. However, they are equal for a Henry's law solution, as can be shown with the aid of chemical potentials. As long as Henry's law is obeyed ($X_B \to 0$, $X_A \to 1$, k a constant),

$$\mu_B = \mu_B^\circ + RT \ln X_B$$
$$\mu_A = \mu_A^\bullet + RT \ln X_A$$

and we find, using the Gibbs-Helmholtz equation (Exercise 4-20), that

$$-\bar{H}_B/T^2 = [\partial(\mu_B/T)/\partial T]_P = [\partial(\mu_B^\circ/T)/\partial T]_P$$

which is independent of X_B, so $\bar{H}_B$ is independent of X_B. For $\bar{H}_A$, we have

$$-\bar{H}_A/T^2 = [\partial(\mu_A/T)/\partial T]_P = [\partial(\mu_A^\bullet/T)/\partial T]_P = -\bar{H}_A^\bullet/T^2$$

(see Exercise 7-6). Therefore

$$\Delta H_{dil} = (n'_A + n_A - n'_A - n_A)\bar{H}_A + (n_B - n_B)\bar{H}_B = 0$$

In a similar manner you can show that

$$\Delta V_{dil} = 0$$

for the process in Equation 7-33, as long as Henry's law is obeyed.

EXERCISE 7-8

A solution of ethanol in water follows Henry's law when X_B is less than 0.1. It is found that mixing ethanol and water to make a dilute solution produces 10.6 kJ of heat per mole of ethanol. How can this be, when ΔH_{dil} is zero?

ANSWER

The result $\Delta H_{dil} = 0$ applies only to the heat of diluting a solution that already is dilute enough to obey Henry's law. When we start with pure ethanol, X_B is outside the range of applicability of Henry's law. A substance can have a large heat of solution when mixed with water, but ΔH_{dil} will still be zero when Henry's law is obeyed. The enthalpy changes on dissolution arise from breaking solvent-solvent and solute-solute interactions. *Dilution* of a Henry's law solution by adding solvent produces no such effects.

These results are valid when X_i is small for the solute and $X_A \approx 1$ for solvent. Any concentration unit proportional to the mole fraction can replace X_i in Table 7-4, but the choice of units affects μ_i°. The Helmholtz free energy of dilution is equal to the Gibbs free energy of dilution because $\Delta H_{dil} = \Delta E_{dil} = 0$. (See the historical sketch of Helmholtz on pp. 291–292.)

Real Solutions Obey Henry's Law when they are Sufficiently Dilute

We have dwelt extensively on the properties of Henry's law solutions because it has been observed experimentally that all real solutions obey Henry's law when they are very dilute. Many years of experiments have shown

TABLE 7-4 *Properties of a Henry's Law Solution*

$\mu_i = \mu_i^\circ + RT \ln X_i$	(Solute chemical potential)
$\mu_A = \mu_A^\bullet + RT \ln X_A$	(Solvent chemical potential)
$\Delta H_{dil} = 0$	(Enthalpy of dilution)
$\Delta V_{dil} = 0$	(Volume of dilution)
$\Delta G_{dil} = \Sigma n_i RT \ln \left(\frac{X_i'}{X_i}\right)$	(Free energy of diluting n_i moles from X_i to X_i', Eq. 7-33)
$\Delta S_{dil} = -\Sigma n_i R \ln \left(\frac{X_i'}{X_i}\right)$	(Entropy of dilution)
$P_A = P_A^\bullet X_A$	(Solvent vapor pressure)
$P_i = k_i X_i$	(Solute vapor pressure)

that the *solvent* vapor pressure in dilute solutions follows Raoult's law, and the *solute* vapor pressure, when it can be measured, follows Henry's law. Therefore, *the properties of a Henry's law solution* (*Table 7–4*) *describe real solutions in the dilute state*. However, a complete description of real solutions requires that we take account of the deviations from these ideal properties that appear when the solutions are made more concentrated. The sections following the historical sketch show how this is done.

○ 7–7 *Real Solutions: A Summary*

In the following sections we will develop quantitative expressions for the chemical potential of real gases and solutions. For readers who do not wish to follow the detailed development, we provide here a brief summary of the results. The chemical potential of solvent (A) and solute (B) in *all* solutions which are extremely dilute can be expressed by the equations for a Henry's law solution,

$$\mu_A = \mu_A^\bullet + RT \ln X_A$$
$$\mu_B = \mu_B^\circ + RT \ln C_B$$

These equations are usually accurate only in the limit of infinite dilution. The equivalent expressions for the chemical potential of more concentrated solutions are

$$\mu_A = \mu_A^\bullet + RT \ln a_A$$
$$\mu_B = \mu_B^\circ + RT \ln a_B$$

in which a_A and a_B are called the *activities* of solvent and solute respectively, and have values such that the equations given for the chemical potential are correct. The activity is always defined so that it approaches a real measure of concentration in the ideal limit. For example, a_A approaches X_A and a_B approaches C_B when $X_A \rightarrow 1$ and $C_B \rightarrow 0$, which are the conditions for a dilute, Henry's law solution.

HERMANN VON HELMHOLTZ

1821–1894

Hermann Helmholtz during his last lecture at the University of Berlin shortly before his death in 1894. (The Bettmann Archive.)

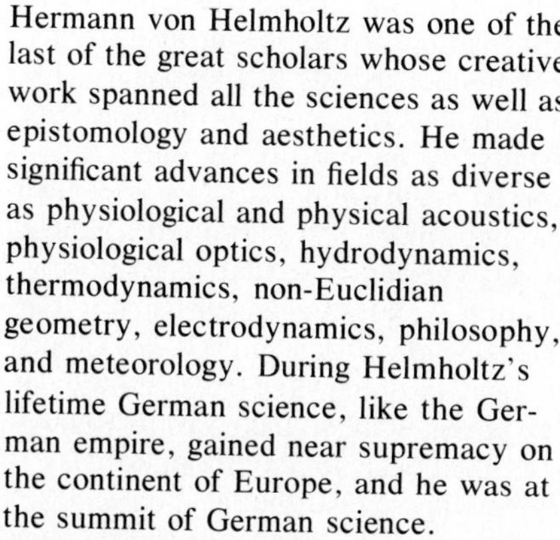

Hermann von Helmholtz was one of the last of the great scholars whose creative work spanned all the sciences as well as epistomology and aesthetics. He made significant advances in fields as diverse as physiological and physical acoustics, physiological optics, hydrodynamics, thermodynamics, non-Euclidian geometry, electrodynamics, philosophy, and meteorology. During Helmholtz's lifetime German science, like the German empire, gained near supremacy on the continent of Europe, and he was at the summit of German science.

Helmholtz was born in Potsdam near Berlin, the son of a teacher of philosophy and literature at the local Gymnasium, or academic high school. The son's poor health confined him to home during his early years, and the father taught him Latin, Greek, Hebrew, French, Italian, and Arabic, as well as much philosophy. The father admired the ideas of Kant, Fichte, and Hegel, and was a close friend of Fichte's son. These philosophers held that notions of space, time, and causation are not based on experience, but rather are innate mental attributes by which we perceive the world around us. The human mind, they believed, does not merely see the order that exists in nature, but rather, organizes the world of perceptions. Increasingly throughout his career Helmholtz came to oppose this view and to believe instead that all knowledge comes through the senses. He also disagreed with the nineteenth-century physiologists that an organism is more than the sum of its physiological parts, or in other words, that there is some vital force distinct from the forces of physics. He believed that all sciences, including biology and meteorology, could and should be reduced to the laws of classical physics.

Helmholtz initially wanted to study physics, but because he could not afford the fees, he enrolled in a medical program that provided free education in return for eight years' service as an army doctor. During his medical studies he also attended lectures on physics, carried out physiological research, worked through the standard texts in higher mathematics on his own, and learned to play the piano—a skill which later helped in his studies of perception of tone. He graduated at age 21 and was assigned to a regiment of the royal guards at Potsdam. His army duties were not demanding and he set up a laboratory in the barracks. He studied the conservation of energy in muscular contraction, and a paper on conservation of energy written when he was 26 so enhanced his reputation as a scientist that he was released from his military obligation. The following year he was appointed associate professor of physiology at Königsberg in East Prussia (now Kaliningrad in the Soviet Union).

In Königsberg, Helmholtz began his work on sense perception, particularly sight. His invention in 1850 of the ophthalmoscope, the now familiar device for illuminating the retina and observing its condition, won him fame throughout Europe. This was followed by studies of color vision, studies on accommodation (focusing) of the lens, and compilation of a three-volume treatise on physiological optics. Helmholtz's fascination with the sense organs was related to his interest in philosophy: he wanted to explore the role of these organs as mediators of external experience in the synthesis of knowledge.

By 1855 Helmholtz was professor of anatomy and physiology in Heidelberg. His work on acoustics culminated in 1862 in his masterful book, *On the Sensations of Tone as Physiological Foundation for the Theory of Music*. This work, still fascinating reading, is a combination of anatomy of the ear, physiology of hearing, physics of vibration, harmonic analysis, and aesthetics of music. Among the topics treated are the basis of hearing in resonance of vibrators in the ear, and the basis for the differences of the tone quality of different musical instruments.

Some feeling for Helmholtz the person comes from the observations of Lord Rayleigh (see Chap. 12) during a visit of the German professor to England. "I had to do the honors at Cambridge for Helmholtz, who came to get an honorary degree," wrote Rayleigh. "He stayed two nights and brought his wife with him. There is not very much to be got out of him in conversation, but he has a very fine head." Rayleigh's son recalled years later being "very much impressed when I was told that the visitor was an even cleverer man than my father. I had a small horse-shoe magnet, as a childish treasure, and he suggested that I should fetch it: but it had been confiscated for some offence in the nursery. It was eventually released in honor of the occasion, and he showed me how the attractive power was concentrated at the ends."

In 1871 Helmholtz was lured to Berlin by a high salary and the construction of a new Institute of Physics for him. He had become increasingly the dominant figure in German science, and in 1882 was even elevated to the nobility. In Berlin he worked mainly in physics. He explained that physiology had grown too complex for any one person to span in its entirety. His many contributions consisted in the extension of classical physics in a variety of areas, including developing the idea of free energy (which, unknown to Helmholtz, had been already formulated by Gibbs). These contributions, however, were eclipsed in significance shortly after his death in 1894 by the discoveries of x-rays, radioactivity, and the quantum, and by the revolution in physics that these findings brought.

The activity is frequently expressed as the product of the concentration times a number called the *activity coefficient*, γ, or

$$a_A = \gamma_A X_A$$
$$a_B = \gamma_B C_B$$

Evidently, γ_A and γ_B approach 1 in the dilute solution limit. The deviation of γ from one is a measure of the nonideality of the solution. When $\gamma < 1$, a high concentration of solute causes its chemical potential to be less than it would be in an ideal solution. This arises from favorable solute-solute and solvent-solvent interactions compared to solute-solvent interactions. When $\gamma > 1$ the chemical potential of the solute is greater than it would be in an ideal solution. Determination of γ is equivalent to determining the chemical potentials μ_A and μ_B. Usually, the chemical potential of the solvent, μ_A, is easily measured, for example from the vapor pressure. The chemical potential of a nonvolatile solute can be calculated from vapor pressure measurements on the solvent by using the Gibbs-Duhem equation. Writing that equation in the form,

$$X_B d\mu_B = -X_A d\mu_A$$

substituting $1 - X_A$ for X_B and dividing by $1 - X_A$, we get

$$d\mu_B = \frac{-X_A}{(1 - X_A)} d\mu_A$$

Determination of μ_B then involves integration of this equation from the dilute Henry's law solution range to a concentrated real solution:

$$\int_{\text{dilute}}^{\text{concentrated}} d\mu_B = -\int_{\text{dilute}}^{\text{concentrated}} \frac{X_A}{(1 - X_A)} d\mu_A$$ ◀

The right side of this equation can be evaluated (after a little rearranging) by graphical integration of data on the solvent vapor pressure. The left side of the equation is the change in the solute chemical potential from the dilute solution limit to a concentrated solution. Since μ_B in the dilute solution limit is known ($\mu_B = \mu_B^\circ + RT \ln C_B$), adding on the value of the integral over $[X_A/(1 - X_A)]d\mu_A$ gives the value of μ_B in the concentrated state. The following four sections show in detail how this can be accomplished.

7–8 Fugacity and Real Gases

Thus far we have considered only idealized states of matter that obey simple equations of state (such as the ideal gas law), or that have simple expressions for the chemical potential, such as Henry's law solutions. Our ultimate objective is to be able to measure, for a real solute dissolved in water, the chemical potential μ relative to μ° in some standard state. The path to this goal is necessarily circuitous; it begins in this section with consideration of deviations of the solvent vapor from ideality. With the equations and concepts developed here we will see in the following section how data on the vapor pressure of the solvent can be used to determine the concentration variation of the chemical potential of the solute, using the Gibbs-Duhem equation.

The equation for the chemical potential of an ideal gas (Eq. 4-43)

$$\mu_i = \mu_i^\circ + RT \ln P_i$$

depends on strict adherence to the ideal gas law, $PV = nRT$. If, as is the case for real gases, this equation is not obeyed except when $P \rightarrow 0$, then Equation 4-43 is also not correct except when $P \rightarrow 0$. At this stage we either could abandon Equation 4-43 and derive a new equation for the chemical potential in terms of P, or we could abandon P_i in Equation 4-43, and replace it with a variable f that makes the relationship correct. The latter course proves to be more convenient. We write

$$\mu_i = \mu_i^\circ + RT \ln f_i \tag{7-34}$$ ◄

in which f, called the *fugacity* by the great American thermodynamicist G. N. Lewis, is the quantity such that Equation 7-34 is correct, with the additional condition that the fugacity approaches the pressure when the gas is very dilute ($P \rightarrow 0$) and the ideal gas law is obeyed:

$$\lim_{P \rightarrow 0} \frac{f_i}{P_i} = 1$$

The substitution of fugacity for partial pressure in Equation 7-34 seems odd on first acquaintance, but its logic grows clearer with exposure. We have replaced the real measure of gas concentration, the partial pressure, by a hypothetical one, the fugacity, which has the virtue of yielding the variation of the chemical potential correctly in Equation 7-34. Furthermore, since the fugacity approaches the partial pressure at low total pressure, we can assign it a value in that limit.

Fugacity can be Measured from Pressure-Volume Data

The fugacity is of no use to us if we cannot calculate it under all conditions. This proves to be possible from experimental data on the pressure dependence of the volume of a real gas. From the differential $dG = n\,d\mu = VdP - SdT$, we can set, at constant temperature and number of moles of gas,

$$(n\,d\mu)_{T,n} = VdP$$

However, according to Equation 7-34 (remember that μ_i° is independent of P but dependent on temperature)

$$(n\,d\mu)_{T,n} = n\,RT\,d \ln f$$

Equating the two expressions for $n\, d\mu$ yields

$$RT\, d \ln f = \frac{V}{n}\, dP \tag{7-35}$$

We now can introduce the symbol α as a measure of the deviation of the molar volume V/n from the ideal gas value $V_{\text{ideal}}/n = RT/P$:

$$\alpha = \frac{V_{(\text{real})} - V_{(\text{ideal})}}{n} = \frac{V}{n} - \frac{RT}{P} \tag{7-36}$$

V in Equation 7-35 is the real volume. From Equation 7-36, $V/n = \alpha + RT/P$. Hence equation 7-35 becomes

$$RT\, d \ln f = RT\, d \ln P + \alpha\, dP$$

Since f and P both approach zero at low pressure, we can integrate this expression to obtain

$$RT \int_0^f d \ln f' = RT \int_0^P d \ln P' + \int_0^P \alpha\, dP'$$

The lower limits $f' = 0$ and $P' = 0$ in these integrals are infinite, $[\lim_{(x)\to 0} (\ln x) = -\infty]$ but they are equal ($f = P$ as $P \to 0$) so they cancel. Therefore we obtain

$$RT \ln f = RT \ln P + \int_0^P \alpha\, dP' \tag{7-37}$$

The integral in Equation 7-37 can be evaluated from a plot of the experimental values of α (see Eq. 7-36) versus P'.

TABLE 7-5 *Fugacity Coefficient of Water Vapor in Equilibrium with Liquid*[a]

T/°C	P/bars[b]	γ_f
0.01	0.00611	0.9995
10	0.01226	0.9992
20	0.02334	0.9988
30	0.04235	0.9982
40	0.07357	0.9974
50	0.12291	0.9964
60	0.19821	0.9950
70	0.30955	0.9933
80	0.46945	0.9912
90	0.69315	0.9886
100	0.99856	0.9855

[a] From Hass, J. L. 1970. *Geochim. Cosmochin. Acta. 34,* 929–932.
[b] 1 bar = 10^6 dyn cm^{-2} = 0.9869 atm ≅ 1 atm.

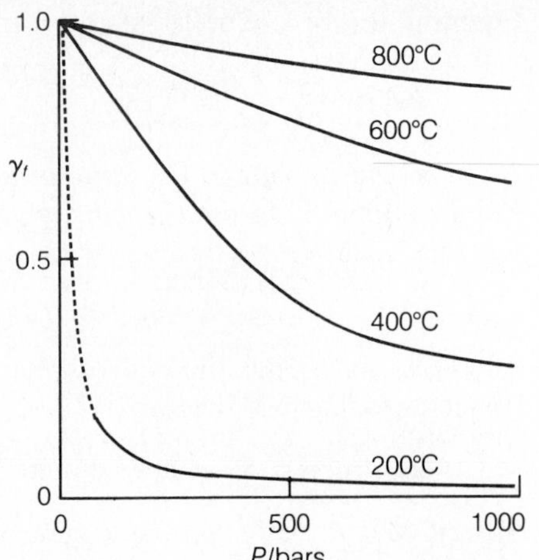

FIGURE 7-7
Fugacity coefficient of water vapor at high pressures, shown for several temperatures. The deviations from $\gamma_f = 1$ indicate departure from ideal gas behavior. [Data from W. T. Holsen, *J. Phys. Chem. 58,* 316–17 (1954).]

A convenient parameter for expressing the fugacity as a function of pressure is the *fugacity coefficient* γ_f, which is the ratio of fugacity to pressure,

$$\gamma_f = \frac{f}{P}$$

Table 7-5 lists the fugacity coefficient of water vapor as a function of pressure for several temperatures. This table indicates that water vapor in the temperature and pressure range considered shows negligible deviation from ideality, since $\gamma_f \approx 1$ in all cases. Only at high pressures, as shown in Figure 7-7, do the deviations from ideality need to be taken into account.

7–9 Real Solutions, Activity, and the Activity Coefficient

Figure 7-8 shows the general thermodynamic picture of a real solution. Since no ideal physical model is used, it is not necessary to specify the molecular details. A two-component solution is considered for simplicity, with Solvent A and dilute Solute B in equilibrium with the vapor. The fugacity of A in the vapor is f_A. Therefore the chemical potential of the solvent in the gas phase is

$$\mu_A(g) = \mu_A^\circ(g) + RT \ln f_A$$

We next introduce into the equation for the chemical potential of the solvent a quantity similar to the fugacity, replacing the ideal expression $\mu_A = \mu_A^\bullet + RT \ln X_A$ by

$$\mu_A = \mu_A^\bullet + RT \ln a_A \qquad (7\text{-}38)$$

Vapor: fugacity f_A

Solution:

Solvent A, activity $a = \frac{f_A}{f_A^\bullet}$

Solute B

Solvent chemical potential:

$\mu_A = \mu_A^\bullet + RT \ln a_A$

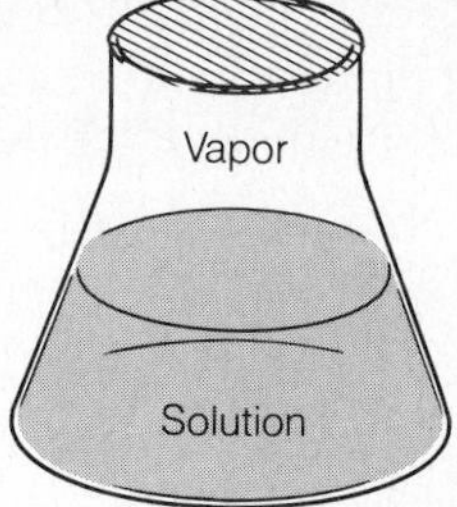

FIGURE 7-8
General thermodynamic description of a real solution. No specific molecular model is required, since deviations from ideal models are allowed.

in which a_A, called the *activity* of A, is the quantity whose value is such that Equation 7-38 is correct for all concentrations.

Just as we did for the fugacity, we are free to set a equal to some particular value in an ideal limit. Because all solvents must follow Raoult's law when the solution is very dilute, it is convenient to adopt the limiting value $a_A = X_A$ in the pure solvent,

$$\lim_{X_A \to 1} \left(\frac{a_A}{X_A} \right) = 1 \tag{7-39}$$

The vapor and solution phases of the solvent are in equilibrium, so we can equate their chemical potentials, given by Equations 7-34 and 7-38

$$\mu_A^\circ(\text{gas}) + RT \ln f_A = \mu_A^\bullet(\text{sol}) + RT \ln a_A \tag{7-40}$$

When $a_A = 1$, Equation 7-39 states that $X_A = 1$ and the liquid is pure solvent. We call $f_A^\bullet$ the fugacity of the vapor A in equilibrium with pure solvent:

$$\mu_A^\circ(\text{gas}) + RT \ln f_A^\bullet = \mu_A^\bullet(\text{sol})$$

Subtracting this equation from Equation 7-40 yields

$$RT \ln a_A = RT \ln (f_A/f_A^\bullet)$$

or

$$a_A = \frac{f_A}{f_A^\bullet} \tag{7-41}$$

Activity can be Deduced from Vapor Pressure Measurements

Equation 7-41 is of great importance for the thermodynamics of solutions because it shows us how to measure the activity of the solvent, and hence its chemical potential relative to the pure solvent which has activity $a = 1$ (Eq. 7-38). If the fugacity of the vapor is known as a function of pressure for the temperature of interest, as shown for water vapor in Figure 7-7, then measurement of the vapor pressure of solvent and solution allows us to calculate the fugacity ratio $f_A/f_A^\bullet$. Equation 7-41 states that this gives the activity a_A.

EXERCISE 7-9

Why are the standard chemical potentials μ_A°(gas) and $\mu_A^\bullet$(sol) different?

ANSWER

Because the gas and solution phases have, by our definition, different standard states. For the solution, $a = 1$ when the solvent is pure, so $\mu_A^\bullet$(sol) is the chemical potential of pure solvent, the solution standard state. For the vapor, the standard state is that with $f = 1$. There is no reason why f should be 1 for vapor in equilibrium with the pure solvent. Hence in general $\mu_A^\circ(\text{gas}) \neq \mu_A^\bullet(\text{sol})$. The exception, of course, is when the pure solvent happens to have unit fugacity, $f_A^\bullet = 1$. It is always true, however, that $\mu_A(\text{gas}) = \mu_A(\text{sol})$ when the two phases are in equilibrium.

EXERCISE 7-10

Dissolution of sucrose in water at 20°C to a mole fraction of 0.0671 reduces the vapor pressure from 17.54 to 15.89 mm Hg. Calculate the activity of water in the solution, and X_w. Assume $\gamma_f = 1$.

ANSWER

$a_w = f_{H_2O}/f_{H_2O}^\bullet = P_{H_2O}/P_{H_2O}^\bullet = 15.89/17.54 = 0.906$
$X_w = 1 - 0.0671 = 0.9329$. Notice that $P_{H_2O} \neq X_w P_{H_2O}^\bullet$, and $a_w \neq X_w$.

The example in Exercise 7-10 reveals that aqueous solutions deviate from $a_A = X_A$ even at fairly low values of X_B. Figure 7-9 shows the variation of the activity of water, a_w, in sucrose solutions; a_w is proportional to X_w in dilute solution, but deviates at higher solute concentration.

Water Activity can be Determined by Isopiestic Transfer

Once water activity as a function of concentration has been established for a solute, such as for sucrose in Figure 7-9, it is no longer necessary to do vapor pressure measurements to determine the activity of water in other solutions. The *isopiestic* method takes advantage of the equilibration of the water activities in two solutions through the vapor in a closed system (Figure 7-10). Solutions (a) and (b) are both in equilibrium with the same vapor, therefore $a_w(\text{a}) = a_w(\text{b})$. If the two solutions have different water activity initially, water will evaporate from one and condense in the other until the two activities are equal. Analysis of the concentration of the reference solution (1) at equilibrium allows determination of a_w in both solutions from the calibration curve, Figure 7-9.

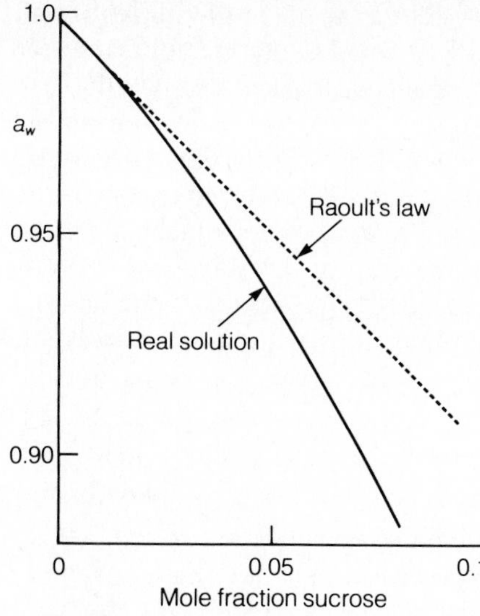

FIGURE 7-9
Activity of water, a_w, in aqueous solutions of sucrose, determined from vapor pressure measurements and the equation $a = P_{H_2O}/P^{\bullet}_{H_2O}$. The straight line for Raoult's law expresses the condition that $P_{H_2O} = P^{\bullet}_{H_2O}X_{H_2O}$, or $a_w = X_w = 1 - X_{sucrose}$. Notice that Raoult's law is obeyed at sufficiently low dilution ($X_{sucrose} \leq 0.01$).

In the Reference State, Activity Equals Concentration

Our treatment of the chemical potential of solute B is similar to that of the solvent, except that the limiting ideal case is a solution that obeys Henry's law instead of Raoult's law. From Section 7-5 we know that the chemical potential must vary linearly with $RT \ln C_B$ at low concentration. Hence we take

$$\mu_B = \mu_B^\circ + RT \ln a_B \tag{7-42a}$$

with the limiting condition

$$\lim_{C_B \to 0} \left(\frac{a_B}{C_B} \right) = 1 \tag{7-42b}$$

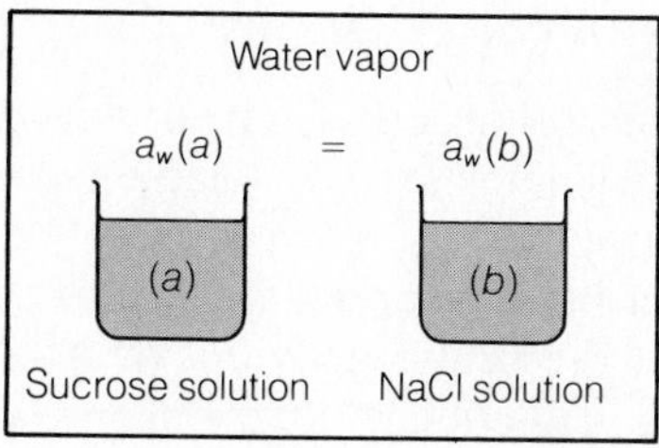

FIGURE 7-10
The isopiestic method for determining the activity of water in a NaCl solution, **(b)**, by equilibrating it with vapor in equilibrium with a reference sucrose solution, **(a)**, for which the relationship between concentration and activity has been established by vapor pressure measurements (Fig. 7-9). Measurement of the concentration of sucrose in solution **(a)** provides a value for a_w, which must be the same in both solutions. Because sucrose and NaCl are nonvolatile, their chemical potentials do not equilibrate in the two solutions.

These equations state that the activity of solute, a_B (a_B is the quantity that makes 7-42*a* correct at all concentrations), is defined so that it approaches the limit C_B in dilute solution. C_B can be any convenient concentration unit, such as molal, molar, or mole fraction.

We will call the dilute solution limit where $a_B = C_B$ the *reference state*, which is not to be confused with the standard state. The reference state is the dilute Henry's law solution diagramed in Figure 7-5, in which all solute molecules are surrounded by solvent. In the case of the solvent, the reference state happens to be the same as the standard state, because the limiting value $a_A = X_A = 1$ in the reference state coincides with the pure solvent, which also is the standard state.

Activity Coefficients

The ratio of activity to concentration, which approaches one as $C \to 0$ according to Equation 7-42*b*, is a useful experimental quantity. It is given the symbol γ and is called, by analogy with the fugacity coefficient for gases, γ_f, the *activity coefficient:*

$$\gamma_B = \frac{a_B}{C_B} \qquad (7\text{-}43) \blacktriangleleft$$

In this equation C_B can be any convenient concentration unit, such as molar, molal, or mole fraction, but the value of γ depends on the choice of concentration units. The equation for the chemical potential of the solute now reads

$$\mu_B = \mu_B^\circ + RT \ln \gamma_B C_B \qquad (7\text{-}44) \blacktriangleleft$$

which is a particularly useful form. The limiting value of γ is always 1 in dilute solution—as long as Henry's law is obeyed, $\gamma = 1$. Deviations from that ideal limit are expressed in tables of γ as a function of concentration.

7–10 *Measurement of Activity Coefficients*

The main problem in measuring activity coefficients is how to determine the solute activity a_B. Many solutes are not volatile enough to allow us to measure their vapor pressure, so we are forced to rely on determination of the solvent activity through its vapor pressure, followed by use of the Gibbs-Duhem equation to get the solute activity. We have already outlined the procedure for doing this in Section 7-7, but will give an explicit example here for a system using molal concentration units for the solute. The overall procedure is an integration of the Gibbs-Duhem equation from low concentration at which $a_B = C_B$ to higher, nonideal, concentrations.

Suppose we have a solution of sucrose in water of molal concentration m_B. It contains m_B moles of sucrose in 1000 g = 55.51 moles of H_2O. The Gibbs-Duhem equation

$$n_A \, d\mu_A + n_B \, d\mu_B = 0$$

therefore becomes

$$55.51\, d\mu_A + m_B\, d\mu_B = 0$$

We replace $d\mu_A$ by using the differential of Equation 7-38 at constant T,

$$d\mu_A = RT\, d \ln a_A$$

in which a_A is known from the vapor pressure. We also replace $d\mu_B$ by using the constant-temperature differential of Equation 7-44,

$$d\mu_B = RT\, d \ln a_B = RT\, d \ln \gamma_B + RT\, d \ln m_B$$

Substituting these expressions into the Gibbs-Duhem equation, we get

$$m_B\, d \ln m_B + m_B\, d \ln \gamma_B = -55.51\, d \ln a_A \tag{7-45}$$

We want to arrange Equation 7-45 into a form that allows integration of $d \ln \gamma_B$ from $m_B = 0$ ($\gamma_B = 1$) to concentration m_B. This is not straightforward, because division of Equation 7-45 by m_B gives two infinities as $m_B \rightarrow 0$, namely $1/m_B$ and $\ln m_B$. The difficulty can be circumvented by defining a function ϕ called the *molal osmotic coefficient*

$$\phi = \frac{-55.51 \ln a_A}{m_B} \tag{7-46}$$

From Equation 7-46 and the chain rule for differentiation, we have

$$d(m_B\phi) = \phi\, dm_B + m_B\, d\phi = -55.51\, d \ln a_A \tag{7-47}$$

Because the right-hand sides of Equations 7-45 and 7-47 are equal, we can equate the left-hand sides, thereby giving

$$m_B\, d \ln m_B + m_B\, d \ln \gamma_B = \phi\, dm_B + m_B\, d\phi$$

Hence, dividing by m_B and rearranging,

$$d \ln \gamma_B = (\phi - 1)\, d \ln m_B + d\phi$$

or, in integral form

$$\int_1^{\gamma_B} d \ln \gamma'_B = \int_0^{m_B} (\phi - 1)\, d \ln m'_B + \int_1^{\phi} d\phi' \tag{7-48}$$

[The lower limit of integration corresponds to $X_B \rightarrow 0$, for which limit $\gamma_B = 1$, $m_B = 0$, and $\phi = 1$. One can see that $\phi \rightarrow 1$ as $m_B \rightarrow 0$ by setting $\ln a_A = \ln X_A = \ln(1 - X_B) = -X_B$ in Equation 7-46. Then $\phi = 55.51\ X_B/m_B = n_A X_B/n_B = n_A n_B/n_A n_B = 1$]. Finally, Equation 7-48 becomes

$$\ln \gamma_B = \int_0^{m_B} \frac{(\phi - 1)}{m'_B}\, dm'_B + \phi - 1 \tag{7-49}$$

The area under a plot of $(\phi - 1)/m_B$ versus m_B is the integral needed to evaluate $\ln \gamma_B$. An example is shown in Figure 7-11 for an aqueous solution of urea. By graphical integration we can determine the activity coefficient of urea from data on the vapor pressure of urea solutions and the fugacity of water vapor.

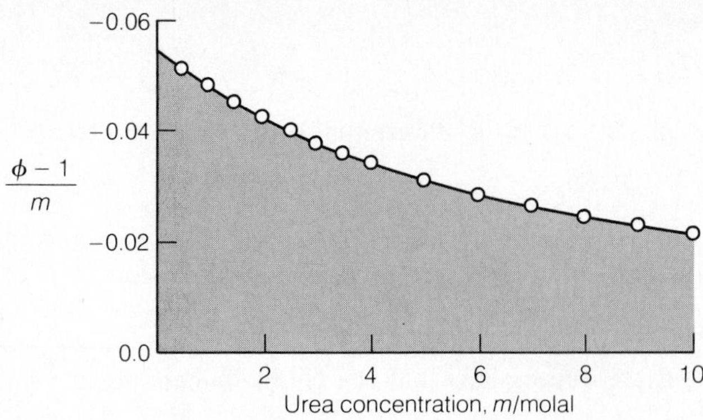

FIGURE 7-11
Graphical determination of the activity coefficient of urea from the osmotic coefficient ϕ. The shaded area under the plot of $(\phi - 1)/m$ is -0.336, and the term $\phi - 1$ in Equation 7-49 is -0.216 at $10m$. Therefore $\ln \gamma = -0.552$ and $\gamma = 0.576$ ($T = 5°C$; Data from Stokes, 1966).

7–11 Physical Interpretation of Activity Coefficients

The activity coefficient is a convenient measure of the deviation of real solutions from the ideal Henry's law limit. Table 7-6 lists activity coefficients for some compounds of biochemical interest. Most of the values of γ either increase from 1 as concentration is increased, or else they decrease from 1. Only occasionally does γ increase above 1 and then decrease below 1 for the same compound.

When $\gamma < 1$ the solute chemical potential is smaller than it would be if there were no nonideality present. Reducing the chemical potential implies that solute-solute interactions decrease the solute free energy; we conclude

TABLE 7-6 *Activity coefficients for some compounds of biochemical interest*[a]

MOLAL CONCEN-TRATION:	0.05	0.2	0.5	1.0
Alanine	—	1.005	1.012	1.023
Glycine	—	0.962	0.912	0.861
Alanylglycine	—	0.931	0.869	0.855
Purine	0.844	0.575	0.374	0.247
6-methylpurine	0.626	0.329	0.185	—
Uridine	0.939	0.808	0.641	—
Cytidine	0.936	0.776	0.580	—

[a] Sober, H. A., ed. *Handbook of Biochemistry*, The Chemical Rubber Co. (1968).; Ts'o, P. O. P., ed. *Basic Principles in Nucleic Acid Biochemistry*, vol. 1, p. 540, (1974).

that *solute-solute interactions are favorable when* $\gamma < 1$. The converse of this statement is that *solute-solute interactions are unfavorable when* $\gamma > 1$.

EXERCISE 7-11

Show that the vapor pressure of component B is proportional to $\gamma_B X_B$ where γ_B is the mole fraction activity coefficient of component B. Assume the vapor is ideal.

ANSWER

Equate the expressions for the chemical potential of B in vapor and solution.

$$\mu_B^\circ(g) + RT \ln P_B = \mu_B^\circ(\text{sol}) + RT \ln \gamma_B X_B$$

and rearrange,

$$RT \ln \frac{P_B}{\gamma_B X_B} = \mu_B^\circ(\text{sol}) - \mu_B^\circ(g)$$

Divide by RT and take the exponential,

$$\frac{P_B}{\gamma_B X_B} = \exp\left\{\frac{\mu_B^\circ(\text{sol}) - \mu_B^\circ(g)}{RT}\right\}$$

Since the right side is a constant at constant T, $P_B = (\text{const})\ \gamma_B X_B$. This exercise shows that if γ_B increases as X_B increases, the slope of P_B versus X_B increases, and if γ_B decreases, the slope of P_B versus C_B decreases. Furthermore, the Henry's law constant, k, is $\exp\{[\mu_B^\circ(\text{sol}) - \mu_B^\circ(g)]/RT\}$.

When the value of γ becomes less than 1 at a relatively low concentration, this is evidence for strong interaction between the solute molecules. For example, Table 7-6 shows that γ for 6-methylpurine is only 0.185 at $0.5m$ concentration. This indicates aggregation of 6-methylpurine in solution, which has been found to derive from stacking of the heterocyclic 6-methylpurine rings together.

The Effect of Dimerization on γ

You can gain further insight into the effect of association reactions on γ by considering the influence of dimerization on the measured value of γ. Suppose that monomeric reactant M dimerizes to D:

$$2\text{M} \rightleftarrows \text{D}$$

The equilibrium constant is $K = C_D/C_M^2$. If we dissolve M in water and measure the chemical potential, we get

$$\mu_M = \mu_M^\circ + RT \ln \gamma\, C_T \tag{7-50}$$

in which C_T is the total concentration of dissolved M. Suppose now that both M and D actually form Henry's law solutions with

$$\mu_M = \mu_M^\circ + RT \ln C_M$$
$$\mu_D = \mu_D^\circ + RT \ln C_D \tag{7-51}$$

or, in other words, the only nonideal process in solution is the dimerization reaction that makes $C_M < C_T$. Comparison of Equations 7-50 and 7-51 shows that $\gamma C_T = C_M$, or

$$\gamma = \frac{C_M}{C_T} \tag{7-52}$$

Hence we only need to calculate C_M from K and C_T to get γ. Alternatively, a measurement of γ can be used to determine K.

EXERCISE 7-12

The activity coefficient of the drug actinomycin at $C_T = 10^{-3}M$ concentration in water at 5°C is 0.379. Assume the deviation from $\gamma = 1$ is due solely to dimerization and calculate the dimerization equilibrium constant.

ANSWER

$C_M = \gamma C_T$; $K = C_D/C_M^2 = \frac{1}{2}(C_T - C_M)/C_M^2 = 3.6 \times 10^3 M^{-1}$. (The structure of the drug includes cyclic peptide side chains attached to a three-ring heterocyclic chromophore. Dimerization occurs by stacking the heterocyclic ring systems together.)

EXERCISE 7-13

ϵ-caprolactam,

```
 O  H
 ‖  |
┌C—N(CH2)5┐
└─────────┘
```

dimerizes in organic solvents by hydrogen bonding between NH·····O═C. In CCl_4 at 22°C the dimerization constant is $1.2 \times 10^2 M^{-1}$. Assume dimerization is the only source of nonideality and derive an expression for the activity coefficient as a function of total concentration and the dimerization equilibrium constant.

ANSWER

Since $K = C_D/C_M^2 = \frac{1}{2}(C_T - C_M)/C_M^2$, we can solve for C_M:

$$2KC_M^2 + C_M - C_T = 0$$

$$C_M = \frac{-1 + (1 + 8KC_T)^{1/2}}{4K}$$

Therefore

$$\gamma = \frac{C_M}{C_T} = \frac{(1 + 8KC_T)^{1/2}}{4KC_T} - \frac{1}{4KC_T} \tag{7-53}$$

⬡ 7–12 *Activity and Chemical Equilibrium*

Our consideration of nonideal solutions forces us to revise Equation 4-56 for the equilibrium constant in order to apply it to systems in which $\gamma \neq 1$. Since the activity a_i replaces C_i in the expression $\mu_i = \mu_i^\circ + RT \ln a_i$ for the chemical potential, a_i must replace C_i in Equation 4-56. Therefore

$$K = \frac{\prod_i a_i^{\nu_i}(\text{products})}{\prod_i a_i^{\nu_i}(\text{reactants})} \qquad (7\text{-}54)$$

in which, as you recall, the exponents ν_i are the stoichiometric coefficients in the reaction

$$\nu_1 A_1 + \nu_2 A_2 \cdots \rightleftarrows \nu_n B_1 + \nu_{n+1} B_2 + \cdots$$

With substitution of $a_i = \gamma_i C_i$, Equation 7-54 becomes

$$K = \frac{\prod_i \gamma_i^{\nu_i} \prod_i C_i^{\nu_i}(\text{products})}{\prod_i \gamma_i^{\nu_i} \prod_i C_i^{\nu_i}(\text{reactants})} \qquad (7\text{-}55)$$

This equation tells us that the equilibrium constant measured at higher concentration will differ from the ideal expression $\prod_i C_i^{\nu_i}$ (products)/$\prod_i C_i^{\nu_i}$ (reactants) by the factor $\prod_i \gamma_i^{\nu_i}$ (products)/$\prod_i \gamma_i^{\nu_i}$ (reactants).

All careful work on the equilibrium constants in solution includes consideration of activity coefficients. As we will see in the next chapter, this is particularly true of reactions involving ionic species. The equations involving K derived in Chapter 4 remain valid for the equilibrium constant defined by Equation 7-55.

Some care is necessary in interpreting quantities such as ΔH° that appear in the Gibbs-Helmholtz (4-59) and van't Hoff (4-61) equations when applied to the standard state. Remember that the standard state partial molar quantities, except the entropy and free energy, are those of the dilute reference state.

EXERCISE 7-14

A zealous chemist decides to test the van't Hoff equation. He measures K for a reaction in very dilute solution at different temperatures, and determines ΔH° from $[\partial \ln K/\partial(1/T)]_P = -\Delta H^\circ/R$. He then makes up solutions of the products and reactants at $a = 1$, taking proper account of activity coefficients, and measures the heat per mole (ΔH) of converting a small amount of reactant to product. He finds that ΔH measured this way for the reaction (when all substances are at unit activity) is very different from the van't Hoff ΔH°, and concludes that van't Hoff was wrong. What is your conclusion?

ANSWER

The overzealous chemist does not understand the nature of the standard state. A real solution at $a = 1$ is *not* the standard state (see Sec. 7-6). ΔH° should be the reaction heat in the dilute-solution limit. It is intuitively clear that one cannot determine the heat of a reaction in a concentrated real solution ($a = 1$) from measurements of K in dilute solution, because solute-solute interactions present in a concentrated solution are absent in the dilute solution.

7-13 The Virial Expansion

Physical chemistry sometimes is described as the science that reduces nature to straight-line plots. Data certainly are easier to analyze if they can be plotted linearly, with the important parameters determined from the slope and intercept of the line. An example is the van't Hoff equation, $\partial \ln K/\partial(1/T) = -\Delta H^\circ/R$; the slope of a plot of $\ln K$ against $1/T$ gives $-\Delta H^\circ/R$. Nature, however, sometimes yields to linearization only grudgingly.

An important function that is naturally nonlinear is the solvent chemical potential, which we write here as the difference between μ_A in the solution and $\mu_A^\bullet$ for the pure solvent.

$$\mu_A - \mu_A^\bullet = RT \ln \gamma_A X_A \tag{7-56}$$

The *virial expansion* is the next best thing to a linearization of this equation. It is a power-series expansion in the solute concentration, a variable that is small in dilute solutions. As you will see, the initial part of the curve is linear, and this result is the basis for useful equations describing the behavior of macromolecules. The virial expansion is a power series in the concentration of the *solute* that expresses the chemical potential of the *solvent*. Equilibration of the solvent chemical potential between different phases or solutions is the starting point for analysis of many macromolecular solution phenomena.

To derive the virial expansion, we begin with the ideal ($\gamma = 1$) form of Equation 7-56, $\mu_A - \mu_A^\bullet = RT \ln X_A$, and substitute $X_A = 1 - X_B$, giving

$$\mu_A - \mu_A^\bullet = RT \ln (1 - X_B) \tag{7-57}$$

There is a convenient power series expression for $\ln (1 - x)$ when x is small compared to 1,

$$-\ln (1 - x) = x + \frac{x^2}{2} + \frac{x^3}{3} \cdots$$

Substituting the series expansion for $\ln (1 - X_B)$ into Equation 7-57, we obtain

$$\mu_A - \mu_A^\bullet = -RT \left[X_B + \frac{X_B^2}{2} + \frac{X_B^3}{3} + \cdots \right] \tag{7-58}$$

If we neglect all powers of X_B higher than one, which is an acceptable procedure when X_B is much smaller than one, we get

$$\mu_A - \mu_A^\bullet \cong -RT\,X_B \tag{7-59}$$

which is a pleasingly linear form of Equation 7-57.

The Virial Expansion and Molecular Weight

The power series expansion of the chemical potential is especially useful in molecular weight determination. Expressing the solute concentration c_B, which we consider to be dilute, in terms of grams per milliliter of solution,

$$X_B \cong \frac{\text{moles B}}{\text{moles A}} = \left(\frac{c_B,\ \text{g/ml}}{M_B,\ \text{g/mole}}\right)(\overline{V}_A^\bullet,\ \text{ml/mole})$$

or

$$X_B = \frac{c_B \overline{V}_A^\bullet}{M_B}$$

Hence, when c_B is small, Equation 7-59 becomes

$$\mu_A - \mu_A^\bullet \cong -RT\,\overline{V}_A^\bullet\,\frac{c_B}{M_B} \tag{7-60}$$

Equation 7-60 is the simplest approximation for the solvent chemical potential, expressed as a linear function of solute concentration c_B.

If X_B^2 is not negligible with respect to X_B, then more terms analogous to those in Equation 7-58 must be included. Furthermore, when the solution is not within the ideal limit ($\gamma = 1$), the coefficients of the power series expansion depend on γ, so we write them as unknown quantities:

$$\mu_A - \mu_A^\bullet = -RT\,\overline{V}_A^\bullet\left(\frac{c_B}{M_B} + Bc_B^2 + Cc_B^3 + \cdots\right) \tag{7-61}$$ ◀

Equation 7-61 is the virial expansion of the solvent chemical potential, and B, the coefficient of c_B^2, is called the *second virial coefficient*. You should notice particularly that the first term in the series, $RT\,\overline{V}_A^\bullet c_B/M_B$, is the same as the first term in the ideal solution expansion (Equation 7-60). In other words, the value of γ has no effect on the first virial coefficient. This turns out to be a consequence of the validity of Raoult's law for the solvent in all solutions at sufficient dilution, because the coefficient of the linear term ($-RT\,\overline{V}_A^\bullet/M_B$) determines the proportionality constant between P_A and X_A (see Problem 7-20). Since the proportionality constant *must* be $P_A^\bullet$ in all cases in the limit as $X_B \to 0$, the coefficient of the first term c_B is always ($-RT\,\overline{V}_A^\bullet/M_B$), even for nonideal solutions. Equation (7-61) may be applied to all solutes except for those that dissociate into smaller units on dissolution (see Chap. 8).

The virial expansion, Equation 7-61, will be left for the moment as a curiosity, but we soon will find it useful in our consideration of colligative properties and membrane equilibria. There are a number of physical techniques that allow determination of the left side of Equation 7-61, which is the chemical potential difference between pure solvent and solvent in solution. Among these are the vapor pressure, freezing point depression, boiling point elevation, and osmotic pressure. Since the limiting slope of the right side depends on $1/M_B$, all of these methods can be used to determine molecular weights M_B. However, only the last one, osmotic pressure, is suitable for study of macromolecules. Before turning to these specific methods, we must first consider the meaning of the average molecular weight of a sample.

7–14 Polydispersity in Macromolecular Solutions

One of the serious problems that plagued early studies of macromolecules was the lack of uniformity among the dissolved large molecules. In general, a polymer is made up of a large number of monomer units. If it has not been fractionated or purified carefully, the sample will contain molecules of different chain lengths. This heterogeneity is given the name *polydispersity,* as opposed to a *monodisperse* solution in which all chains are identical both chemically and in size. The problem of polydispersity is of much less significance for studies of biological macromolecules than it used to be, because modern methods of separation allow preparation of pure, monodisperse materials. Polydisperse samples can and should be avoided in most cases.

The equilibrium thermodynamic methods for measuring the molecular weight of polymers provide us with a value of M that is averaged over all the molecules in the solution. However, there are different ways of taking averages. For example, suppose our sample consists of two molecules, one twice as large as the other. Do we consider the short molecule to be equally as important as the long one, or give it less weight in taking the average because it contains less mass? It turns out that the nature of the average depends on the particular physical technique used for the measurement. Different techniques form the average in different ways.

Molecular Weight Averages

The two main molecular weight averages are called the *number-average* (M_n) and the *weight-average* (M_w) molecular weights. In the number average, each molecule counts equally, but in the weight average molecules contribute according to how much they weigh. Suppose there are N_i molecules in the sample. Figure 7-12 shows a typical molecular weight *distribution* in a polydisperse sample. The distribution is the curve or set of numbers N_i that describe how many (N_i) molecules there are of size M_i. The number-average molecular weight is calculated from the expression

$$M_n = \frac{\sum_i N_i M_i}{\Sigma_i N_i} \tag{7-62}$$

in which M_i is multiplied by the *number* of molecules of that size. According to Equation 7-62, the number average weight of two molecules of mass 1 and $\frac{1}{2}$ is $(1 + \frac{1}{2})/2 = \frac{3}{4}$. The number in each size-class is multiplied by the size and divided by the total number.

In contrast, the weight average is calculated from an expression in which the molecular weight M_i is multiplied by the total *weight* of molecules of that size:

$$M_w = \frac{\Sigma_i W_i M_i}{\Sigma_i W_i} \tag{7-63}$$

in which W_i is the weight of molecules of size M_i. This equation states that the weight-average molecular weight of two molecules of mass 1 and $\frac{1}{2}$ is $(1 \times 1 + \frac{1}{2} \times \frac{1}{2})/(1\frac{1}{2}) = 0.833$. The weight average is always larger than the number average unless the sample is monodisperse, in which case $M_n = W_w$.

Since the weight W_i of molecules of size M_i is $N_i M_i$, Equation 7-63 can be rewritten

$$M_w = \frac{\sum_i N_i M_i^2}{\sum_i N_i M_i} \tag{7-64}$$

Additional molecular weight averages sometimes are defined by noting the progression from Equation 7-63 to 7-64 and continuing it. For example the so-called *z-average* molecular weight is

$$M_z = \frac{\sum_i N_i M_i^3}{\sum_i N_i M_i^2} \tag{7-65}$$

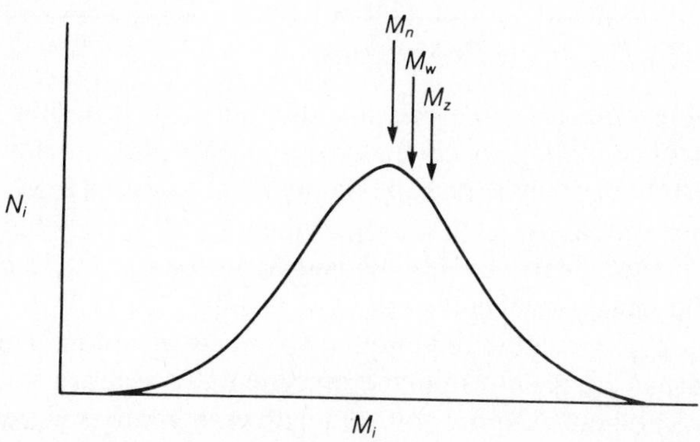

FIGURE 7-12
Schematic illustration of the distribution of molecular weights in a polydisperse sample. The distribution is the set of numbers, N_i, that describes the number of molecules of each size, M_i, in the sample. Hypothetical values for M_n, M_w, and M_z are indicated.

In a polydisperse sample, the z-average is still larger than the weight-average molecular weight, because it pays even more attention to the largest molecules in the sample.

EXERCISE 7-15

Suppose that the change Δx in a measured physical property of a solution depends on the molar concentration C of a solute, $\Delta x = kC$. Show that the ratio $c/\Delta x$, in which c is the weight concentration of a polydisperse sample, depends on the number-average molecular weight.

ANSWER

Let Δx_i be the effect due to molecules of size i. Then

$$\Delta x_T = \sum_i \Delta x_i = k\Sigma C_i = k\Sigma N_i$$

in which $N_i = C_i$ is the number of moles of molecules of size i per liter. Furthermore, c_i = weight of molecules of size i per ml, so $c_i = M_iC_i/1000 = M_iN_i/1000$. The ratio $c/\Delta x = \Sigma c_i/\Sigma \Delta x_i = (\Sigma N_iM_i)/(1000\, k\Sigma N_i) = M_n/(1000\, k)$.

This exercise shows that all physical techniques that depend on the molar concentration of solute, or the number of solute particles per unit volume, will provide us with the number-average molecular weight of a polydisperse sample. The *colligative properties* of a solution are defined as those which depend on the molar concentration or number of dissolved molecules, not their size. Therefore, molecular weight determinations based on the colligative properties always give the number average. In the following sections we consider vapor-pressure lowering, freezing and boiling point changes, and osmotic pressure—all examples of colligative properties.

7–15 Colligative Properties

The colligative properties arise because dissolution of a solute reduces the solvent chemical potential, in effect by diluting the solvent by solute. The basic equation that describes the solvent chemical potential is $\mu_A - \mu_A^\bullet = RT \ln (\gamma_A X_A)$, from which you can see that when $X_A < 1$, $\mu_A < \mu_A^\bullet$. The virial expansion (Equation 7-61) is a convenient form for $\mu_A - \mu_A^\bullet$, because it is linear when the weight concentration c_B is small.

The various experimental techniques to measure colligative properties differ in how $\mu_A - \mu_A^\bullet$ is determined experimentally. All procedures rely on equilibration of Solvent A in the solution with A in another phase, called the reference phase (see Fig. 7-13). The most direct method is measurement of the vapor pressure (or, more precisely, the vapor fugacity; we will assume that the vapor is an ideal gas). In this case the solvent vapor is the reference phase, and equilibration of the chemical potentials requires

$$\mu_A(\text{ref}) = \mu_A(\text{sol})$$

For the vapor phase, $\mu_A(\text{ref}) = \mu_A^\circ(g) + RT \ln P_A$; for the solution phase we use the virial expansion (Equation 7-61),

$$\mu_A(\text{sol}) = \mu_A^\bullet(\text{sol}) - RT\,\overline{V}_A^\bullet \left(\frac{c_B}{M_B} + Bc_B^2 + \cdots\right)$$

and set the two chemical potentials equal to each other,

$$\mu_A^\circ(g) + RT \ln P_A = \mu_A^\bullet(\text{sol}) - RT\,\overline{V}_A^\bullet \left(\frac{c_B}{M_B} + Bc_B^2 + \cdots\right)$$

We subtract from this the equation representing equilibrium of pure solvent and vapor,

$$\mu_A^\circ(g) + RT \ln P_A^\bullet = \mu_A^\bullet(\text{sol})$$

and divide the result by RT to get

$$\ln \frac{P_A}{P_A^\bullet} = -\overline{V}_A^\bullet \left(\frac{c_B}{M_B} + Bc_B^2 + \cdots\right) \tag{7-66}$$

This equation states that a plot of $\ln (P_A/P_A^\bullet)$ versus c_B should have initial slope $-\overline{V}_A^\bullet/M_B$ when c_B is small.

Reference phase	Solution Phase
$\mu_A(\text{ref}) =$	$\mu_A(\text{sol})$

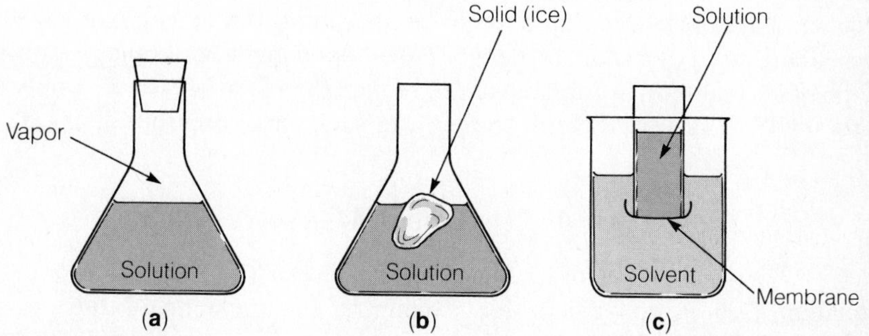

FIGURE 7-13
Equilibration of the solvent chemical potential μ_A in a solution with μ_A in some other phase is the basis for the colligative methods for determining polymer molecular weight. Three examples are shown: **(a)** A reference vapor phase in equilibrium with the solution. The change in the vapor pressure is a measure of $\Delta\mu_A$. **(b)** A reference solid phase in equilibrium with solution. The reduction of the solid melting temperature measures $\Delta\mu_A$. **(c)** A reference liquid phase in equilibrium with the solution through a membrane which allows passage of all but the dissolved macromolecule. $\Delta\mu_A$ is determined from the increased pressure on the solution required for equilibrium.

EXERCISE 7-16

What are the intercept and slope of a plot of $-\ln (P_A/P_A^\bullet)/(c_B \bar{V}_A^\bullet)$ versus c_B?

ANSWER

$1/M_B$ and B.

EXERCISE 7-17

Show that the plot in Exercise 7-16 will yield the reciprocal of the number-average molecular weight as the intercept if the solution is polydisperse.

ANSWER

c_B/M_B is the number of moles per ml. Therefore, using Equation 7-66,

$$-\ln (P_A/P_A^\bullet) = \bar{V}_A^\bullet \left(\frac{c_B}{M_B}\right) = \bar{V}_A^\bullet \sum_i N_i$$

in which N_i is the number of moles of size i per ml. By definition, $c_B = \sum_i w_i$, in which w_i is the weight of molecules of size i per ml, or $c_B = \Sigma N_i M_i$. Hence dividing the equation for $\ln (P_A/P_A^\bullet)$ by $\bar{V}_A^\bullet c_B = \bar{V}_A^\bullet \Sigma N_i M_i$, we get

$$\frac{-\ln (P_A/P_A^\bullet)}{c_B \bar{V}_A^\bullet} = \frac{\Sigma N_i}{\Sigma N_i M_i} = \frac{1}{M_n}$$

This exercise shows that, like all the colligative property methods, the vapor pressure lowering yields the number-average molecular weight.

In practical terms, the vapor pressure lowering is not a feasible technique for measurement of molecular weights of macromolecules, because solutions dilute enough to be dominated by the first virial coefficient do not produce a sufficient vapor pressure reduction to be measured. For example, a 1% solution (10 mg ml^{-1}) of a small protein of molecular weight 10^4 would, according to the first virial term in Equation 7-66, reduce the vapor pressure by only 1.8 parts in 10^5. It is very difficult to measure such small pressure changes.

The Second Virial Coefficient Measures the Effectiveness of a Solvent

Equation 7-66 is convenient for examining the meaning of the second virial coefficient B. When the solution is ideal ($\gamma = 1$), you can show from Equation 7-58 that

$$B_{ideal} = \frac{\bar{V}_A^\bullet}{2M_B^2} \qquad (7\text{-}67)$$

When a small macromolecule of molecular weight $M_B = 10^4$ is dissolved in water ($\bar{V}_A^\bullet = 18$ cm^3 mol^{-1}), $B_{ideal} = 9 \times 10^{-8}$ cm^3 mol g^{-2}. Figure 7-14

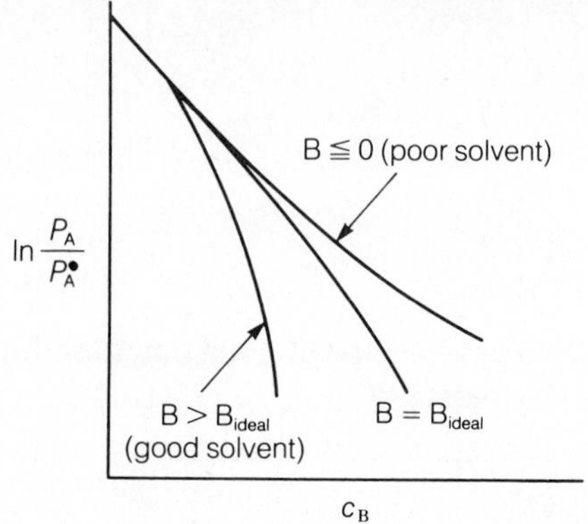

FIGURE 7-14
Dependence of lowering solvent vapor pressure on the value of the virial coefficient B. When $B > B_{ideal}$ the vapor pressure decreases more rapidly with increasing c_B than it does for $B = B_{ideal}$. This means that solute-solvent interactions are more favorable than solute-solute and solvent-solvent interactions. This case is described by the phrase "good solvent." The opposite situation, in which solute-solute and solvent-solvent interactions are preferred, is the "poor solvent" case, in which $B \leq 0$.

shows a plot of Equation 7-66 using the ideal value of B, along with the behavior when B is greater and less than B_{ideal}. When $B > B_{ideal}$, the first and second terms in the virial expansion have the same sign, and $\ln (P_A/P_A^\bullet)$ curves downward. This means that the vapor pressure is *less* than it would be in the ideal solution, and hence that the added solute is "holding on" to the solvent molecules. The implication is that the interactions between solute and solvent are favorable, more so than solute-solute or solvent-solvent interactions. Solvents for which $B > 0$ for a particular macromolecule are *good solvents* for that substance. The value of B for good solvents is in the range of 10^{-5} to 10^{-2}, which is much larger than the ideal value (Equation 7-67).

The other case is a value of $B \leq 0$, for which the solvent is said to be a *poor solvent*. Precipitation of the polymer occurs relatively easily from poor solvents.

Freezing-Point Depression

A dissolved solute also lowers the freezing point of the solvent. In this case the reference phase in Figure 7-13 is the solid solvent. Equating chemical potentials at $T = T_m$, we have

$$\mu_A^\bullet(\text{solid}) = \mu_A^\bullet(\text{solution}) - RT_m\overline{V}_A^\bullet\left(\frac{c_B}{M_B} + Bc_B^2 + \cdots\right)$$

Because the temperature changes, we must take the derivative of this equation with respect to temperature. Keeping only the first term in the virial expansion, dividing by T, recalling that $[\partial(\mu/T)/\partial T]_P = -\overline{H}/T^2$, and neglecting the temperature variation of $\overline{V}_A^\bullet$, the result is

$$\frac{\overline{H}_A^\bullet(\text{solid})}{T_m^{\,2}} = \frac{\overline{H}_A^\bullet(\text{solvent})}{T_m^{\,2}} + \frac{R\overline{V}_A^\bullet}{M_B}\left(\frac{\partial c_B}{\partial T_m}\right)_P$$

The difference $\overline{H}_A^\bullet(\text{solvent}) - \overline{H}_A^\bullet(\text{solid})$ is ΔH_{fus}, so we obtain, after inverting,

$$\left(\frac{\partial T_m}{\partial c_B}\right)_P = \frac{-RT_m^{\,2}\overline{V}_A^\bullet}{\Delta\overline{H}_{\text{fus}}M_B} \tag{7-68}$$

You should be able to derive the analogous equation for the increase of the boiling temperature (Problem 7-23).

EXERCISE 7-18

Calculate the change in the freezing point of water for a 1% solution (10 mg ml) of a protein with molecular weight 10^4. ($\Delta\overline{H}_{\text{fus}} = 5980$ J mole^{-1}.)

ANSWER

$$\Delta T_m = \frac{0.01\ \text{g}\cdot\text{cm}^{-3} \times 8.314\ \text{J K}^{-1}\cdot\text{mole}^{-1} \times (273\ \text{K})^2 \times 18\ \text{cm}^3\ \text{mole}^{-1}}{5980\ \text{J mole}^{-1} \times 10^4\ \text{g mole}^{-1}}$$

$$= 1.87 \times 10^{-3}\ \text{K}$$

This exercise makes it clear that the freezing-point depression method would induce depression in the chemist who selected it for the determination of macromolecular size, because the T_m changes are too small to measure accurately. In fact, only one of the colligative methods has found extensive application in macromolecular physical chemistry: osmotic pressure, which we examine next.

7–16 *Membrane Equilibria and Osmotic Pressure*

All living organisms are separated from their surroundings by a barrier that is permeable to some substances and impermeable to others. It is essential that a cell be able to retain its nucleic acids and the enzymes essential for metabolism. Many intermediate products of metabolism also need to be conserved, but basic raw materials must be taken in and waste products excreted. Biological membranes are exquisitely adapted for these functions, with highly selective permeabilities that depend on the chemical structure of the molecules transported. Some substances are transported *passively*, which means that no work is required for their transfer across the membrane, whereas others are carried across the membrane by *active transport*, which means that work is required. The necessary free energy is supplied by cellular biochemical metabolism (Chapter 5), but the molecular mecha-

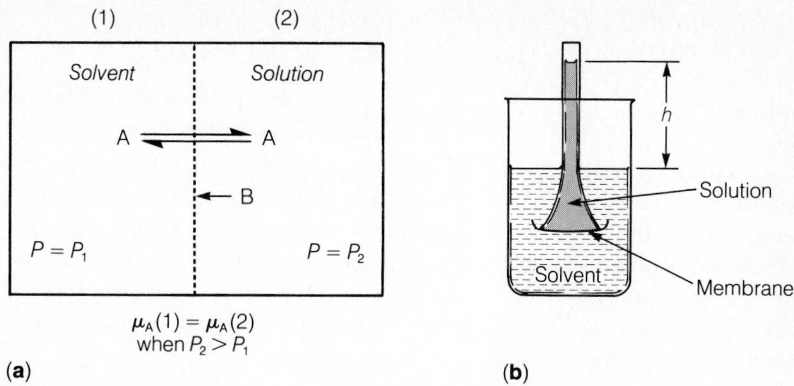

FIGURE 7-15
(a) A membrane permeable to solvent A but not solute B separates solution and solvent. Equilibrium requires that the chemical potential of the solvent be the same on both sides of the membrane. This can be brought about by applying an excess, or *osmotic pressure*, $\pi = P_2 - P_1$, to the solution side of the membrane. The osmotic pressure method is frequently used to determine the molecular weights of macromolecules. **(b)** Schematic diagram of an apparatus to determine osmotic pressure. The excess pressure on the solution phase can be determined from the height, *h*, of the solution column above the surface of the solvent.

nism by which chemical free energy is coupled to transport is at present unknown.

The subject of this section—the relationship between membrane equilibrium and osmotic pressure—is far removed from the nonequilibrium, active-transport characteristics of membranes *in vivo*. However, understanding the equilibrium thermodynamics of membrane phenomena is a necessary prelude to subsequent topics. Furthermore, important use is made of osmotic pressure and dialysis in a number of biochemical and medical applications. We will return to membrane phenomena in Chapter 9 in connection with special effects due to ionic dissociation.

The *semipermeable* membrane (Figure 7-15) is the basis for analysis of the thermodynamics of membrane equilibria. Semipermeable membranes permit the passage of some substances, but not others. The distinction can be based on either chemical or physical properties. A common example is the *dialysis* membrane, which contains pores that allow transport of small molecules but not large ones. Dialysis of a solution of macromolecules allows one to exchange the solvent with a reservoir, while retaining large molecules, such as proteins, inside the membrane. For example, the removal of body wastes, normally accomplished by the kidney, can be performed artificially by dialysis of blood against a solution (the dialysate) that contains small molecules at the desired final blood concentration. Waste products, such as urea and salts, that are present in higher concentrations in blood than in the dialysate will be removed until their blood chemical potential matches that in the dialysate. Larger molecules, such as proteins or cells, do not cross the dialysis membrane because its pores are too small.

Chemical Potentials of Permeable Solutes on Both Sides are Equal

The condition for equilibrium across semipermeable membranes is the same as that for phase equilibrium, with one important exception: *only species that permeate the membrane* must have equal chemical potentials on both sides of the membrane. This equality does not apply to species that do not permeate because the system is not in equilibrium with respect to their distribution. However, no problems in the analysis result from this restriction.

Examining Figure 7-15, we see that the equilibrium condition requires that the chemical potential of the solvent A be equal on the two sides of the membrane. Since Phase 1 is pure solvent, we have

$$\mu_A(1) = \mu_A^\bullet$$

The chemical potential of the solvent on the solution side (Phase 2) is given by the virial expansion,

$$\mu_A(2) = \mu_A^\bullet - RT\bar{V}_A^\bullet \left(\frac{c_B}{M_B} + Bc_B^2 + \cdots\right) \tag{7-69}$$

Equilibrium requires that $\mu_A(1) = \mu_A(2)$, or

$$\mu_A^\bullet = \mu_A^\bullet - RT\bar{V}_A^\bullet \left(\frac{c_B}{M_B} + Bc_B^2 + \cdots\right)$$

If both phases are at the same temperature and pressure, $\mu_A^\bullet(1) = \mu_A^\bullet(2)$ because the standard chemical potential of the solvent ($\mu_A^\bullet$) is independent of solute concentration. Therefore, *equilibrium cannot be established with both phases at the same temperature and pressure, unless $c_B = 0$.* Setting up an experiment like the one diagrammed in Figure 7-15, and allowing both phases to expand and contract at constant temperature and pressure, would result in transport of all of the solvent into the solution compartment until no pure solvent phase remained. This would happen because the solvent chemical potential would be smaller in the solution than in the solvent phase, and the free energy could be reduced by transferring the solvent to the solution phase.

Osmotic Pressure is the Pressure Required to Produce Equilibrium

Suppose, however, that we restrict the volume of the solution phase. As solvent flows in, pressure develops because of the increased amount of matter per unit volume. The increased pressure on the solution phase will increase the solvent chemical potential $\mu_A(2)$ according to the equation (the temperature is held constant),

$$\left[\frac{\partial \mu_A(2)}{\partial P}\right]_T = \bar{V}_A$$

in which $\overline{V}_A$ is the partial molar volume of solvent in the solution. (Recall the equation $dG = V\,dP - S\,dT$.) Consequently,

$$d\mu_A(2) = \overline{V}_A\,dP$$

The partial molar volume of solvent in dilute solution is virtually the same as the molar volume of the solvent $\overline{V}_A^\bullet$, so we replace $\overline{V}_A$ by the constant value $\overline{V}_A^\bullet$ and get

$$d\mu_A(2) = \overline{V}_A^\bullet\,dP \tag{7-70}$$

Allowing for increase of the pressure from P_1 to P_2, the new value of the solvent chemical potential $[\mu_A(2, P_2)]$ is related to the value at pressure P_1 $[\mu_A(2, P_1)]$ by the equation

$$\mu_A(2, P_2) = \mu_A(2, P_1) + \int_{P_1}^{P_2} d\mu_A(2)$$

Substituting Equation 7-69 for $\mu_A(2, P_1)$ and Equation 7-70 for $d\mu_A(2)$ and integrating, we obtain

$$\mu_A(2, P_2) = \mu_A^\bullet - RT\overline{V}_A^\bullet\left(\frac{c_B}{M_B} + Bc_B^2 + \cdot\cdot\cdot\right) + \overline{V}_A^\bullet(P_2 - P_1) \tag{7-71}$$

The new term which makes the solvent chemical potential larger at increased pressure, P_2, is $\overline{V}_A^\bullet(P_2 - P_1)$. We define the pressure increase as the *osmotic pressure* π:

$$\pi = P_2 - P_1$$

Equilibrium with the pure solvent phase now can be established by adjusting the osmotic pressure on the solution. Equating the expression for $\mu_A(2, P_2)$ in Equation 7-71 to $\mu_A(1) = \mu_A^\bullet$, the chemical potential of the pure solvent, leads us to the equation

$$\mu_A^\bullet = \mu_A^\bullet - RT\overline{V}_A^\bullet\left(\frac{c_B}{M_B} + Bc_B{}^2 + \cdot\cdot\cdot\right) + \overline{V}_A^\bullet\,\pi$$

which can be rearranged and solved for the osmotic pressure π,

$$\pi = RT\left(\frac{c_B}{M_B} + Bc_B{}^2 + \cdot\cdot\cdot\right) \tag{7-72}$$ ◀

in which c_B is the concentration in weight per unit volume. Because c_B/M_B is the number of moles per unit volume, it can be replaced by the molar concentration C_B to generate an equivalent equation

$$\pi = RT(C_B + B'C_B^2 + \cdot\cdot\cdot) \tag{7-73}$$

Notice that in using Equations 7-72 and 7-73, the choice of volume units for expressing the concentration dictates the choice of units for R (see Table 2-1); the osmotic pressure π usually is expressed in atmospheres.

EXERCISE 7-19

Estimate the osmotic pressure of a 1% solution (10 mg ml^{-1}) of a protein with molecular weight 10^4 at 25°C.

ANSWER

$$\pi \cong RT\,\frac{c_B}{M_B} = 82.06\ \text{cm}^3\ \text{atm K}^{-1}\ \text{mole}^{-1} \times 298\ \text{K} \times \frac{0.01\ \text{g cm}^{-3}}{10^4\ \text{g mole}^{-1}}$$
$$= 0.0244\ \text{atm, or } 18.6\ \text{mm Hg}$$

This is a readily measurable quantity.

Determination of Molecular Weight

Osmotic pressure is a useful technique for determining macromolecular weights. A substantial pressure must be applied to a solution to increase the solvent chemical potential to that of the pure solvent on the other (lower pressure) side of the membrane. Figure 7-16 shows examples of the use of osmotic pressure to determine the molecular weight of several proteins.

The Biological Significance of Osmotic Pressure

One of the principal characteristics of living organisms is that the concentration of substances inside the membrane that separates them from their environment is different from the outside concentration. This can give rise to a substantial osmotic pressure. To understand the phenonemon, it is essential

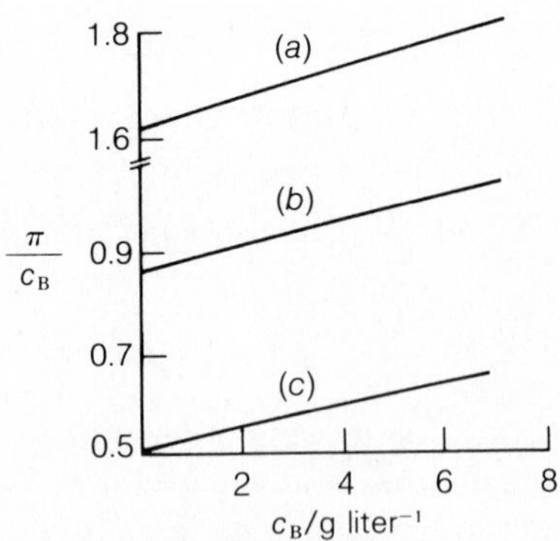

FIGURE 7-16
Osmotic pressure determination of protein molecular weights in a 6 M guanidine hydrochloride solution. The proteins are **(a)** ribonuclease, M = 13.6 × 10^3, **(b)** chymotrypsin, M = 25.6 × 10^3, and **(c)** aldolase, M = 41.9 × 10^3. The osmotic pressure is expressed in cm of solvent; the value of *RT* in the units employed (at 25°C) is 2.2 × 10^4. [From S. Lapanje and C. Tanford, *J. Am. Chem. Soc. 89,* 5030–33 (1967). Reprinted with permission from Journal of the American Chemical Society. Copyright by the American Chemical Society.]

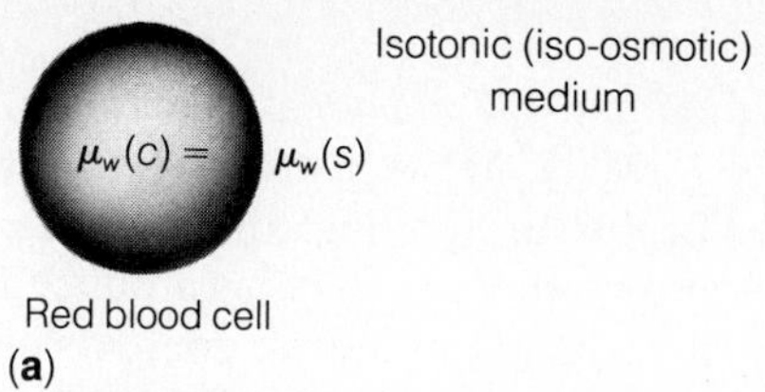

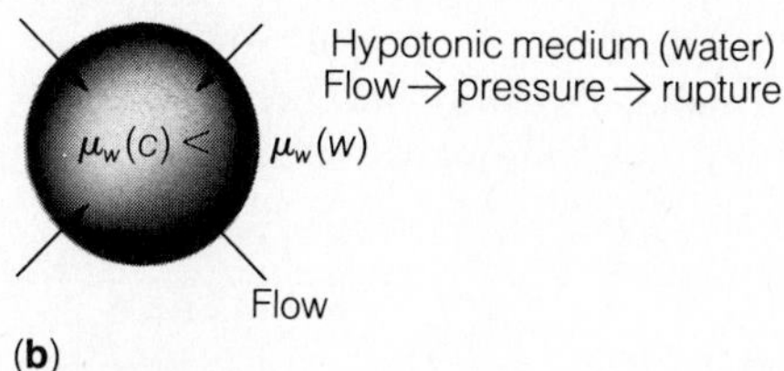

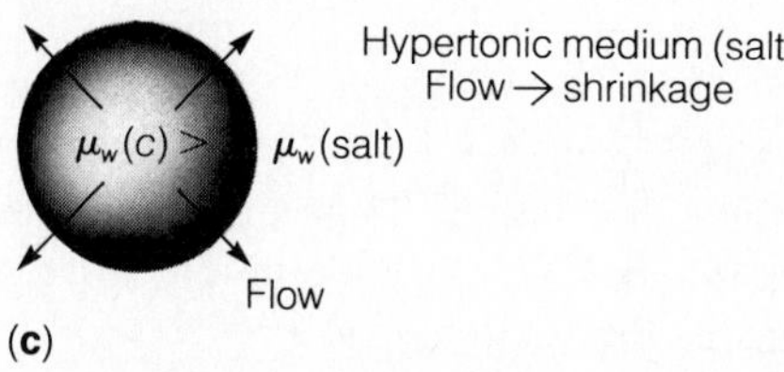

FIGURE 7-17
Red blood cell placed in **(a)** isotonic, **(b)** hypotonic, and **(c)** hypertonic media. In **(a)**, the chemical potential of water is the same inside and outside the cell, so no net flow results. In **(b)**, the higher chemical potential of water outside the cell causes water to flow into the cell, thereby creating an osmotic pressure that ruptures the cell. In **(c)**, water flows out of the cell.

that we examine the origin of the pressure itself. There is some danger that the form of Equation 7-73, which states that the force per unit area is proportional to the solute concentration, could lead you to think that the pressure arises from the unequal impact of solute molecules on the membrane in Figure 7-15. This is not correct. If the membrane were replaced by an impermeable partition, no pressure difference would result. Similarly, when the solute and solvent initially are added to the system, there is no pressure difference between the bulk solutions. The pressure increase accompanies an excess of solvent flow into the solution compartment over the outflow. Development of pressure requires both a semipermeable membrane and an enclosed space that acts to prevent further solvent flow.

We can illustrate these concepts by examining what happens when a red blood cell is transferred from serum to other media. The red cell membrane is permeated more rapidly by water than by other substances. The chemical potential of water inside the red cell is equal to that in serum,[1] so there is no osmotic pressure. If the cell is transferred to 0.31 *M* sucrose, or 0.155 *M* NaCl, the activity of water is still the same inside and outside, because the red cell's contents are osmotically equivalent to 0.31 *M* sucrose. (Only half the molar concentration of NaCl is required because it dissociates into two ions, and the number of particles per unit volume determines the solvent activity.) When no net flow of water results, the medium is said to be *isotonic* [(Figure 7-17(*a*)].

[1] Metabolism in the cell is required to keep it that way, however.

When the cell is placed in a *hypotonic* medium, the chemical potential of water in the cell is less than outside [Figure 7-17(*b*)]. At the instant of mixing, there is no pressure difference, but water begins to flow into the cell. The pressure builds up until the red cell membrane ruptures. The holes produced are large enough to allow escape of some of the macromolecular contents of the cell, primarily hemoglobin. This technique of rupture sometimes is called *osmotic shock*. Many bacteria and algae cannot be ruptured this way because their cell membranes are reinforced by cell walls able to withstand the osmotic pressure generated in hypotonic media. In *hypertonic* media the flow goes in the other direction [(Figure 7-17(*c*)], with shrinkage of the cell. Rupture is less likely in these circumstances.

EXERCISE 7-20

Calculate the osmotic pressure of a solution 0.31 M in sucrose at 37°C, assuming the membrane is permeable to water but not to sucrose.

ANSWER

From Equation 7-78, with only the first virial term, $\pi = RT\,C_B = 0.082$ liter atm K^{-1} $mol^{-1} \times 310$ K $\times$ 0.31 mole liter^{-1} = 7.9 atm. This is the pressure that would develop in a red cell at osmotic equilibrium with water, since 0.31 M sucrose is osmotically equivalent to the contents of a red cell. However, rupture occurs before equilibrium is reached. Only cells with walls to reinforce the membrane are able to withstand such pressures.

EXERCISE 7-21

The red cell wall is permeated rapidly by urea. What would you expect to happen when 0.31 M urea in water is mixed with red cells?

ANSWER

Rupture should occur. The relevant solvent chemical potential includes all components that permeate the membrane rapidly. In this case, the rates of water entry and exit are equal, but the inrush of urea, followed by an inrush of water to equalize the water chemical potential, results in a rupturing pressure.

7–17 *Equilibrium Sedimentation*

One of the important applications of thermodynamics to the study of macromolecules in solution is equilibrium sedimentation. Macromolecules in solution are rotated in an *ultracentrifuge* at speeds up to 100,000 rpm, thereby subjecting them to forces hundreds of thousands of times greater than the force of gravity. If the centrifugal force is large enough, the molecules can be sedimented to the bottom of the tube containing the solution. However, if the rotation is slower, a balance can be set up between the centrifugal force and the randomizing forces that tend to make molecules move from a region of high chemical potential (high concentration) to a region of low chemical potential (low concentration).

The Total Chemical Potential Includes Concentration and Centrifugal Effects

Figure 7-18 shows a schematic diagram of the concentration of a macromolecule at equilibrium in a centrifugal field. The usual condition for equilibrium requires that the chemical potential be constant throughout a system at equilibrium, but since μ depends on C, Figure 7-18 seems to imply that μ is not constant. However, there is now an additional source of potential energy, namely work against the centrifugal force, that must be included in μ before the condition of constancy can be applied. In general, if the potential energy $U(r)$ per mole depends on distance r, then the total chemical potential, μ_T, is

$$\mu_T = \mu_{ch} + U(r) \tag{7-74}$$

in which μ_{ch} is the chemical potential we have used to this point, neglecting all work terms except expansion-compression.

The centrifugal force exerted on a particle of mass m is

$$F = m\omega^2 r$$

in which ω is the speed of rotation (in radians per second) and r is the distance from the axis of rotation. With Equation 2-6 we can calculate the potential energy relative to the energy at $r = 0$. The work required to move the particle from $r = 0$ to r is

$$\begin{aligned} U(r) &= -\int_0^r F(r')\,dr' \\ &= -m\omega^2 \int_0^r r'\,dr' = \frac{-m\omega^2 r^2}{2} \end{aligned} \tag{7-75}$$

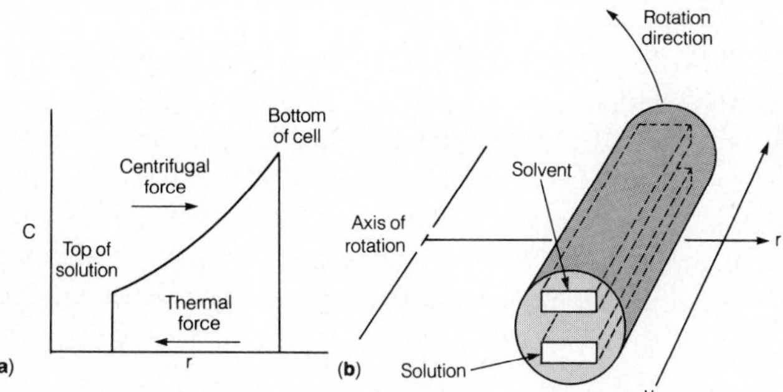

FIGURE 7-18

(a) Schematic diagram of the concentration gradient in a centrifugal field, and the balance of forces produced. The centrifugal field produces a force that moves the particles to the right. The concentration gradient, however, produces an effective randomizing force that tends to make the particles move from high to low concentration. These two forces balance at equilibrium. **(b)** Drawing of an ultracentrifuge cell in which the solution for equilibrium sedimentation is placed. The cell has two sectors, which contain solvent and solution, respectively. The concentration measurement is made by a beam of light that traverses the cell along its y axis; light absorbance (Chapter 12) by the solution, compared to the solvent, is proportional to solution concentration.

The particle of mass m is supported by the mass of liquid it displaces, the so-called buoyancy correction. Let $\bar{v}_B$ be the specific volume ($cm^3\ g^{-1}$) of the particle, and ρ_A the density of the solvent ($g\ cm^{-3}$). Then $m\bar{v}_B\rho_A$ is the mass of liquid displaced by the particle, and $m(1 - \bar{v}_B\rho_A)$ is the excess of mass of the particle over the liquid displaced. Because we want the potential per mole in Equation 7-75, we insert the mass per mole M_B in place of m. Therefore

$$U(r) = \frac{-M_B(1 - \bar{v}_B\rho_A)r^2\omega^2}{2}$$

We now can substitute this expression for $U(r)$, and also $\mu_{ch} = \mu_B^\circ + RT \ln C_B$ for the chemical part of the potential in Equation 7-74 (assuming that the activity coefficient is unity), giving

$$\mu_T = \mu_B^\circ + RT \ln C_B - M_B \frac{(1 - \bar{v}_B\rho_A)}{2} r^2\omega^2 \tag{7-76}$$

At equilibrium this total potential must be constant everywhere in the solution so

$$\frac{d\mu_T}{dr} = 0$$

Differentiation of Equation 7-76 with respect to r, with μ_B° taken to be constant (this is not quite true because the pressure changes through the cell), yields

$$0 = RT \frac{d \ln C_B}{dr} - M_B(1 - \bar{v}_B\rho_A)\omega^2 r$$

or

$$RT \frac{d \ln C_B}{dr} = M_B(1 - \bar{v}_B\rho_A)r\omega^2 \tag{7-77}$$ ◀

This equation is worth considering for a moment. The right side is the centrifugal force acting on one mole of particle B. At equilibrium this force must balance another force, given by the left side of the equation, which is the derivative with respect to distance of the chemical part of the potential. $RT\, d \ln C_B/dr$ is the effective randomizing force that drives particles from high to low concentration. We will make extensive use of this force in Chapter 15 when we consider nonequilibrium processes. Rearranging Equation 7-77, we get

$$\frac{d \ln C_B}{r\, dr} = \frac{M_B\omega^2}{RT}(1 - \bar{v}_B\rho_A)$$

With the substitution of $\frac{1}{2}d(r^2)$ for $r\, dr$, this equation becomes

$$\frac{d \ln C_B}{d(r^2)} = \frac{M_B\omega^2}{2RT}(1 - \bar{v}_B\rho_A) \tag{7-78}$$ ◀

Equation 7-78 states that the slope of a plot of ln C_B versus r^2 should be $M_B\omega^2(1 - \bar{v}_B\rho_A)/(2RT)$. If $\bar{v}_B$ and ρ_A also are measured, the molecular weight is readily calculated. Sedimentation equilibrium has been an extensively used technique for determination of macromolecular weight. More details are given in the books by Tanford (1961) and Van Holde (1971).

7–18 Sedimentation Equilibrium in Multicomponent Systems

The derivation of the sedimentation equilibrium condition given in the preceding section assumed a two-component solution. The problem becomes more complicated if the solvent contains more than one component. For example, suppose that the solution contains water (Component 1), macromolecule (Component 2), and salt (Component 3). Furthermore, suppose that the macromolecule is preferentially *hydrated,* meaning that it interacts more strongly with water than with salt. The water carried along with the macromolecule gives it increased buoyancy, assuming that the solvent (water plus salt) is more dense than the water of hydration. In this case the buoyancy correction in Equation 7-78 is no longer given by $(1 - \bar{v}_B\rho_A)$.

This problem is an important example of the systems that can be treated rigorously by applying thermodynamics to multicomponent systems. Consider the volume element V in Figure 7-19. It contains macromolecules and solvent components 1 and 3 at chemical potentials μ_1 and μ_3. The force on the volume element due to the centrifugal field is the excess of the mass, m, of the volume element over the mass of an equal volume of solvent, m_0, times $\omega^2 r$. We are interested in the force per mole, which we can calculate

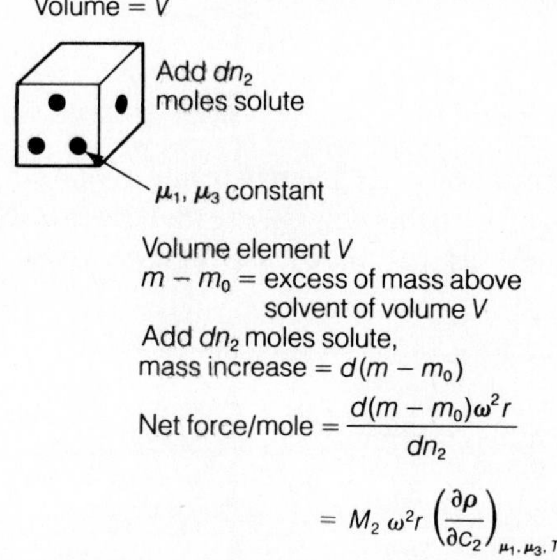

FIGURE 7-19
Derivation of the general expression $(\partial\rho/\partial c_2)_{\mu_1,\mu_3,T}$ for the buoyancy correction for a hydrated macromolecule. To a volume element V we add dn_2 molecules of Component 2, and calculate the additional centrifugal force per mole of solute added. The term $(\partial\rho/\partial c_2)_{\mu_1,\mu_3,T}$ replaces the term $(1 - \bar{v}_B\rho_A)$ found for a two-component system.

from the incremental force due to the addition of dn_2 moles in the volume V. The increase in the force is $d(m - m_0)\omega^2 r$, so the force per mole (F) is

$$F = \frac{d(m - m_0)\,\omega^2 r}{dn_2}$$

Because m_0 is a constant, this can be rewritten

$$F = \frac{dm/V}{dn_2/V}\,\omega^2 r$$

The quantity dm/V is the increase in the density ρ due to addition of dn_2 moles of Component 2, and dn_2/V is the increase in the concentration of Component 2, dC_2 in moles per unit volume, so

$$F = \frac{d\rho}{dC_2}\,\omega^2 r$$

We now replace dC_2 by dc_2/M_2 where dc_2 is the concentration in weight per unit volume. Therefore, in partial derivative notation, the force is

$$F = M_2\omega^2 r\left(\frac{\partial\rho}{\partial c_2}\right)_{\mu_1,\mu_3,T} \tag{7-79}$$

The quantity $(\partial\rho/\partial c_2)_{\mu_1,\mu_3,T}$ is measured easily in a dialysis equilibrium experiment. The polymer at various concentrations is dialyzed to equilibrium against a large excess of solvent having chemical potentials μ_1 and μ_3. The density of the polymer solution is measured as a function of c_2, and the slope is $(\partial\rho/\partial c_2)_{\mu_1,\mu_3,T}$. We call this the *density increment* and give it the symbol φ_2

$$\varphi_2 = \left(\frac{\partial\rho}{\partial c_2}\right)_{\mu_1,\mu_3,T} \tag{7-80}$$

Hence for a multicomponent solution, Equation 7-78 for sedimentation equilibrium becomes

$$\frac{d\ln C_2}{d(r^2)} = \frac{M_2\varphi_2\omega^2}{2RT} \tag{7-81}$$

If the weight concentration c_2, is given in terms of dry polymer, then M_2 is also in terms of dry polymer. In this way we see that the *state of solvation of the polymer has no influence on the molecular weight determined by equilibrium sedimentation.*

EXERCISE 7-22

Show that for a two-component solution, $d\rho/dc = (1 - \bar{v}_B\rho_A)$.

ANSWER

$$\frac{\partial\rho}{dc} = \frac{dm/V}{M_B dn/V} = \frac{dm}{M_B dn}$$

Let $dn = 1$, then $dm = M_B - \overline{V}_B\rho_A$. Hence

$$\frac{dm}{M_B dn} = \frac{M_B - \overline{V}_B\rho_A}{M_B} = 1 - \bar{v}_B\rho_A.$$

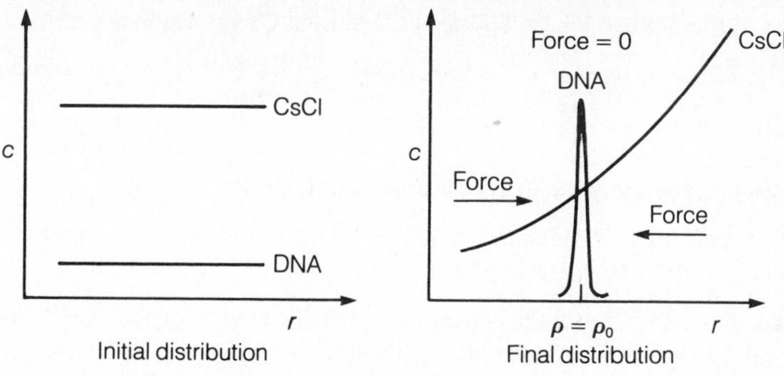

FIGURE 7-20
Illustration of the technique of equilibrium density gradient centrifugation. The application of a centrifugal field to a concentrated CsCl solution converts the uniform initial distribution to the final distribution. A gradient of solution density accompanies the concentration gradient, and DNA molecules collect at the density $\rho = \rho_0$ at which they are neutrally buoyant, characterized by $\phi_2 = \left(\frac{\partial \rho}{\partial c_2}\right)_{\mu_1,\mu_3,T} = 0$. DNA molecules at higher or lower densities than ρ_0 are subject to a force which drives them toward ρ_0. The buoyant density of DNA depends on chemical composition and physical state (native or denatured).

Gradient Sedimentation Measures the Buoyant Density of Molecules

If the density increment $(\partial \rho/\partial c_2) = \varphi_2$ is zero, then there is no centrifugal force on the macromolecule. The solution density required to bring φ_2 to zero is called the *buoyant density* of the molecule. It can be measured by the technique of equilibrium *density gradient* sedimentation. A solution of a heavy salt, such as CsCl, distributes in a centrifugal field according to Equation 7-78. There is a concentration gradient of CsCl at equilibrium, and therefore a density gradient (see Fig. 7-20). DNA molecules, for example, will collect at the density at which $\varphi_2 = 0$, forming a sharp band whose width depends on $1/\sqrt{M_2}$. Given the definition of $\partial \rho/\partial c$, you can see that the banding density ρ_0 depends both on the hydration and the partial specific volume of the macromolecule. The technique has been particularly useful for nucleic acids, since ρ_0 has been found to depend on the base composition and physical state of DNA and RNA. The book by Bloomfield et al. (1974) contains an extensive discussion of equilibrium density gradient sedimentation.

QUESTIONS FOR REVIEW

1. Define the following measures of concentration: mole fraction, molality, molarity, weight percent, and volume percent.
2. Define chemical potential. What is its physical meaning?
3. Define partial molar volume with an equation. What is the physical significance of a partial molar volume? What is the partial specific volume? What is the sum rule for partial molar volumes?

4. What is the Gibbs-Duhem equation? What is its physical meaning?
5. Define the terms *ideal solution* and *vapor pressure*. Which thermodynamic quantities of ideal solutions must be equal to zero?
6. State Raoult's law. State Henry's law. Describe the physical model for each, and the concentration range in which each is applicable.
7. Define enthalpy and volume of dilution. What values do these quantities have for a solution that obeys Henry's law?
8. Define fugacity. Why is the concept necessary? How is the fugacity of a real gas measured? What is the fugacity coefficient?
9. Define activity. Why is this concept introduced in addition to fugacity? What is the reference state?
10. Define the activity coefficient γ. How is γ measured?
11. How does nonideality affect chemical equilibrium?
12. What is the virial expansion and why is it useful?
13. In sedimentation equilibrium, what forces are balanced?

PROBLEMS

1. Are the following statements true or false?
   (a) The chemical potential is the partial molar free energy.
   (b) The chemical potential is defined only for ideal solutions and gases.
   (c) The Gibbs free energy of mixing of an ideal solution must be negative.
   (d) Sufficiently dilute solutions always obey Raoult's law.
   (e) Sufficiently dilute solutions always obey Henry's law.
   (f) For an ideal solution μ_A must be smaller than $\mu_A^\bullet$.
   (g) Pure solvent has unit activity.
   (h) Activity coefficients in biochemical thermodynamics can be neglected for solutions more dilute than 1 *M*.
   (i) Colligative properties yield a number-average molecular weight.
   (j) Freezing point depression is useful for determining protein molecular weights.
   (k) For a system to be at equilibrium, the pressure must be uniform.

Answers: (a) T by definition. (b) F. (c) T. (d) F for the solute. (e) T. (f) T since $X_{\text{solute}} < 1$ in a solution. (g) T. (h) F. (i) T. (j) F. (k) F for osmotic systems.

2. Add the appropriate word:
   (a) Raoult's law states that the vapor pressure of a component over a solution is proportional to its ________ .

(b) The ______ of mixing of an ideal solution must be positive.

(c) In the expression for chemical potential, molarity can replace mole fraction when the concentration is ______ .

(d) The fugacity of a gas approaches the pressure when the gas is ______ .

(e) In the virial expansion the coefficient of the concentration squared is the ______ virial coefficient.

(f) A good solvent is one for which B is ______ than zero.

(g) Water flows into a cell placed in a ______ medium.

(h) A solution containing macromolecular solutes with a spread of molecular weights is called ______ .

Answers: (a) Mole fraction. (b) Entropy. (c) Small. (d) Dilute. (e) Second. (f) Greater. (g) Hypotonic. (h) Polydisperse.

PROBLEMS RELATED TO EXERCISES

3. (Exercise 7-1) Define the partial molar enthalpy.

4. (Exercise 7-2) Calculate the partial specific volume of glycine (CH_2NH_2COOH) from its partial molar volume, 43.20 cm^3mol^{-1}.

5. (Exercise 7-3) Use the data in Table 7-1 to estimate the partial molar volume of glycine (CH_2NH_2COOH).

6. (Exercise 7-5) Show that the total vapor pressure in an ideal Raoult's law solution is linearly related to X_B by $P_T = P_A^\bullet + X_B(P_B^\circ - P_A^\bullet)$.

7. (Exercise 7-6) Show that the heat capacity change is zero when the components of an ideal solution are mixed.

8. (Exercise 7-8) Would you expect to find $\Delta H = 0$ when ethanol is added to a dilute (Henry's law) solution of ethanol in water?

9. (Exercise 7-10) Calculate the activity of water in a sucrose solution when the water vapor pressure is reduced from 20.21 mm Hg to 16.34 mm Hg.

10. (Exercise 7-11) Show that the vapor pressure of component B is given by

$$P_B = \gamma_B k X_B$$

in which k is the Henry's law constant, and γ_B is the mole fraction activity coefficient.

11. (Exercise 7-18) Calculate the change in freezing point of water for a 1% solution of a protein with molecular weight 10^5.

12. (Exercise 7-20) Calculate the osmotic pressure of a solution 0.155 M in NaCl at 37°C.

OTHER PROBLEMS

13. Estimate the partial *specific* volume of a protein molecule containing two glycine, three serine, three phenylalanine, two tyrosine, one aspartic acid, and two lysine residues.

14. Suppose that

$$\alpha = \frac{V(\text{real}) - V(\text{ideal})}{n}$$

is equal to bP, where b is a constant. Derive an equation for the fugacity coefficient of the gas in terms of P, T, and b.

15. The following data on the vapor pressure of aqueous urea solutions can be found in the International Critical Tables.

MOLAL CONCENTRATION, m	$\frac{100(P_0 - P)}{mP_0}$ $(T = 0°C)$
1	1.52
2	1.49
4	1.46
6	1.45
10	1.43

where P = vapor pressure of H_2O in solution, P_0 = vapor pressure of pure H_2O, m = molal concentration. Make a graph showing the vapor pressure ratio P/P_0 as a function of mole fraction of urea, and compare the result with Raoult's law.

16. 1-methylcytosine and 9-ethylguanine interact in nonaqueous solvents to form a hydrogen-bonded complex analogous to a Watson-Crick base pair. Assume that this complex is the only source of solution nonideality and derive an equation expressing γ for 1-methylcytosine in terms of the total concentrations C_1 and C_2 of the two solutes and the association equilibrium constant.

17. A polymer sample contains one molecule of mass $\frac{1}{4}$, one molecule of $\frac{1}{2}$ and one molecule of mass 1. Calculate

    (a) The number-average molecular weight.

    (b) The weight-average molecular weight.

    (c) The z-average molecular weight.

18. Dissolution of sucrose in water at 0°C to a mole fraction of 0.0671 reduces the vapor pressure to $P_A = 0.9059\, P_A^{\bullet}$. Calculate the reduction in the freezing point.

19. Use the vapor pressure lowering data in Problem 18 to calculate the osmotic pressure of the sucrose across a membrane impermeable to sucrose.

20. Begin with the generalized form of Equation 7-59, $\mu_A - \mu_A^{\bullet} = -RT\,kX_B$, and show that when Raoult's law is obeyed, $\lim_{X_B \to 0} k = 1$. (This result proves that the coefficient of the first term in the virial expansion must always be as given in Equations 7-59 or 7-60, since Raoult's law is always followed by the solvent in sufficiently dilute solution.) (*Hint:* Equate $\mu_A - \mu_A^{\bullet} = -RT\,kX_B$ to the equivalent expression for the gas phase, and differentiate with respect to X_B.)

21. Show that for the solvent of a dilute solution, $(d\gamma_A/dX_A)_{X_A=1} = 0$. (*Hint:* Begin with $\mu_A^{\bullet} + RT \ln(\gamma_A X_A) = \mu_A^{\bullet} - RTX_B + \cdots$ and differentiate with respect to X_A.)

22. Show that for a substance that obeys Henry's law, $(\partial \gamma_B / \partial X_B)_{X_B=0}$ is indeterminate.

(*Hint:* Use the Gibbs-Duhem equation and the results of Problem 21.)

23. Start with an expression equating the chemical potential of vapor (1 atm pressure, boiling temperature) and solution, and derive the equation analogous to Equation 7-68 for the change of T_b with concentration c_B. Be careful to get the sign right.

24. Derive an equation describing sedimentation equilibrium in which it is not assumed that $\gamma_B = 1$ but that the dependence of the solute activity coefficient γ_B on c, $\partial \ln \gamma_B / \partial \ln c_B$, is known.

25. Show that the buoyant density ρ_0 of a macromolecule (2) is given by

$$\frac{1}{\rho_0} = \frac{\bar{v}_2 + \Gamma' \bar{v}_1}{1 + \Gamma'} \qquad \text{where } \Gamma' = \left(\frac{\partial m_1}{\partial m_2}\right)_{\mu_1, \mu_3, T}$$

and $\bar{v}_2$ and $\bar{v}_1$ are partial specific volumes of solute and water respectively. Assume v_1 is the same for bulk water and water of hydration. (*Hint:* Add together the volumes of solute and the water of hydration and divide by the mass.)

26. The following data of high precision were obtained by R. H. Stokes [*Australian J. Chem. 20,* 2087 (1967)] for the molal osmotic coefficient of aqueous solutions of urea at 25°C. Use the data and a graphical integration procedure to calculate the activity coefficient at 3.0m concentration. (*Answer:* $\gamma = 0.821$).

M	ϕ
0.5	0.9800
1.0	0.9624
1.5	0.9469
2.0	0.9331
2.5	0.9208
3.0	0.9096

27. The following data on the osmotic pressure of solutions of bovine serum albumin are adapted from the work of Scatchard et al., *J. Am. Chem. Soc. 68,* 2320 (1946). The isoionic protein was dissolved in 0.15 M NaCl to give a solution whose pH was 5.37 at 25°C.

CONCENTRATION IN g/LITER	OSMOTIC PRESSURE IN mmHg
27.28	8.35
56.20	19.33
8.95	2.51
17.69	5.07

Compute the molecular weight of the protein from these data. (*Answer:* about 70,000.)

QUESTIONS FOR DISCUSSION

28. (a) In the late 1960s scientists in the U.S.S.R., England, and the U.S. reported the discovery of "polywater," supposedly a polymerized water that forms in thin glass capillaries. The central observation—and one of the few agreed upon by these scientists—was that "polywater" exhibits a vapor pressure lower than that of ordinary water. What is the thermodynamic implication of this observation? Why should these scientists not have been afraid that "polywater" might be the "Ice IX" of Vonnegut (Problem 4-20)?
    (b) What is a more likely explanation for the lowering of the vapor pressure of water in a thin capillary? (It should be noted that claims of polymerized water died away in the early 1970s.)

29. Some salts have a negative partial molar heat capacity when dissolved in water. What conclusions might you draw about their effect on the structure of water?

FURTHER READING

Lewis, G. N., and Randall, M. (revised by K. S. Pitzer and L. Brewer). 1961. *Thermodynamics,* New York: McGraw-Hill. An excellent thermodynamics text for more advanced students.

Tanford, C. 1961. *Physical Chemistry of Macromolecules.* New York: John Wiley & Sons, Inc. A comprehensive monograph on the solution chemistry of macromolecules, as developed up to 1960.

Van Holde, K. E. 1971. *Physical Biochemistry.* Englewood Cliffs, N.J.: Prentice-Hall. A useful text in biophysical chemistry, especially clear on sedimentation.

Cohn, E. J., and Edsall, J. T. 1943. *Proteins, Amino Acids and Peptides as Ions and Dipolar Ions.* New York: Reinhold Publishing Co. A classic monograph on protein chemistry, and still an indispensable book for those working in biophysical chemistry.

Bloomfield, V., Crothers, D. M., and Tinoco, I. 1974. *Physical Chemistry of Nucleic Acids.* New York: Harper & Row. Further detail on solution chemistry, including hydration and sedimentation.

Prausnitz, J. M. 1969. *Molecular Thermodynamics of Fluid-Phase Equilibria.* Englewood Cliffs, N.J.: Prentice-Hall. A comprehensive, detailed treatment of the fluid phase.

Eisenberg, H. 1976. *Biological Macromolecules and Polyelectrolytes in Solution.* Oxford: Clarendon Press. An excellent account of the thermodynamics of multicomponent solutions of macromolecules.

OTHER REFERENCES

Hearst, J. F., and Vinograd, J. 1961. *Proc. Natl. Acad. Sci. U.S.A. 47,* 1005.

Stokes, R. H. 1966. *Australian J. Chem. 20,* 2087.

CHAPTER 8

Electrolyte Solutions

Electrochemistry is the branch of chemistry that deals with solutions of charged atoms or molecules called *ions,* and with electron transfer reactions that alter oxidation states and produce ions. Electrochemistry cannot be avoided in the study of biochemistry. For example, all living organisms contain solutions of *electrolytes,* such as KCl and $MgSO_4$, which dissociate into ions when they are dissolved in water. The transport of ions across biological membranes is vital to the function of the nervous system. Many crucial biological macromolecules, such as proteins and nucleic acids, are electrically charged in solution. Electron transfer reactions through a series of reversible oxidation-reduction intermediates in mitochondria are essential for oxidative metabolism and the production of ATP. The list of important electrochemical phenomena in biochemistry could be prolonged indefinitely. Our main focus in this chapter will be on the nature and properties of electrolyte solutions. In the next chapter we consider electron transfer reactions, along with the dependence of electrostatic potential on the equilibrium distribution of ions across membranes.

○ 8–1 Coulomb's Law Determines Electrostatic Forces

Coulomb discovered that the electrostatic force acting between two charges q_A and q_B is proportional to the charge on each, and inversely proportional to the square of their separation, r,

$$F \propto \frac{q_A q_B}{r^2}$$

The force is positive (particles repelled) if the charges have the same sign and negative (particles attracted) if they have opposite signs. The constant of proportionality that one uses in this equation depends on the units of charge with which one works.

In *cgs units,* still widely used by chemists and biochemists, the unit of charge is the *electrostatic unit* or *esu*. The esu is defined as the charge that, when placed 1 cm from an equal charge, is repelled with a force of 1 dyne. The charge on the electron is 4.80325×10^{-10} esu. In cgs units, the constant of proportionality is unity, so Coulomb's law is written

$$F = \frac{q_A q_B}{r^2} \qquad \text{(8-1-cgs)} \blacktriangleleft$$

In the *SI* or *international system* of units, now official but often less convenient for electrostatic calculations, the unit of charge is the coulomb (C) = 1 ampere second (A s). The charge on the electron is 1.602×10^{-19} C. The constant of proportionality in Coulomb's law is $1/(4\pi\epsilon_0)$, ϵ_0 being the *permittivity* of a vacuum (8.854×10^{-12} kg^{-1} m^{-3} s^4 A^2 = 8.854×10^{-12} J^{-1} C^2 m^{-1} in SI units). Therefore, Coulomb's law is written

$$F = \frac{q_A q_B}{r^2 4\pi\epsilon_0} \qquad \text{(8-1-SI)} \blacktriangleleft$$

In both systems the force is reduced by a dimensionless factor, $1/\epsilon$, if the charged particles are immersed in any medium other than a vacuum. The factor ϵ is called the *dielectric constant* of the medium.

We can obtain the potential function U that corresponds to Coulomb's law by integration of Equation 8-1 (see Sec. 2-10):

$$U = -\int F\,dr \qquad \text{(8-2)}$$

Substitution of Equation 8-1-cgs for F gives

$$U = -\int_{\infty}^{r} \frac{q_A q_B}{r'^2}\,dr' = -q_A q_B \int_{\infty}^{r} \frac{dr'}{r'^2} = \frac{q_A q_B}{r} \qquad \text{(8-2-cgs)} \blacktriangleleft$$

or, in SI units

$$U = \frac{q_A q_B}{4\pi\epsilon_0 r} \qquad \text{(8-2-SI)} \blacktriangleleft$$

For charges immersed in a medium of dielectric constant ϵ, both equations have an additional factor ϵ in the denominator.

Electrostatic Field and Potential

F and U are respectively the force and potential energy of interaction between two charges; they are related by the equation (see Eq. 2-7):

$$F = -\frac{dU}{dr} \qquad \text{(8-3)}$$

We will make frequent use of two quantities closely analogous to F and U, the electric field and the electrostatic potential. The *electric field* E^* is the *electrostatic force on a unit positive charge;* it is the force F on a charge of magnitude q_A, divided by q_A. (Actually, F and E^* are *vectors,* that is, they have a directional sense that specifies the spatial orientation of the force, but we will not use extensively the vector properties of E^* and F until later in this book.

The *electrostatic potential* Φ *is the electrostatic potential energy of a unit positive charge*. The value of Φ is the potential energy U of a charge q_A, divided by q_A. The electrostatic field and the electrostatic potential are useful because the force on a charge q_A, or the potential energy of the charge, can be expressed in terms of E^* and Φ:

$$F = q_A E^* \tag{8-4}$$

$$U = q_A \Phi \tag{8-5}$$

The SI unit for Φ is volts; Φ in the cgs system is expressed in statvolts, with 1 statvolt = 299.8 volts. Finally, Φ and E^* are related by an equation analogous to Equation 8-3:

$$E^* = -\frac{d\Phi}{dr} \tag{8-6a}$$ ◀

(More precisely, this should be a vector equation; then E^* is the component of the electric field in the direction of the distance change dr.)

EXERCISE 8-1

Calculate the electric field and the electrostatic potential at a distance r from a charge q_B. What is the direction of E^*?

ANSWER

$$F = \frac{q_A q_B}{4\pi\epsilon_0 r^2}$$

so

$$E^* = \frac{F}{q_A} = \frac{q_B}{4\pi\epsilon_0 r^2} \tag{8-6b}$$ ◀

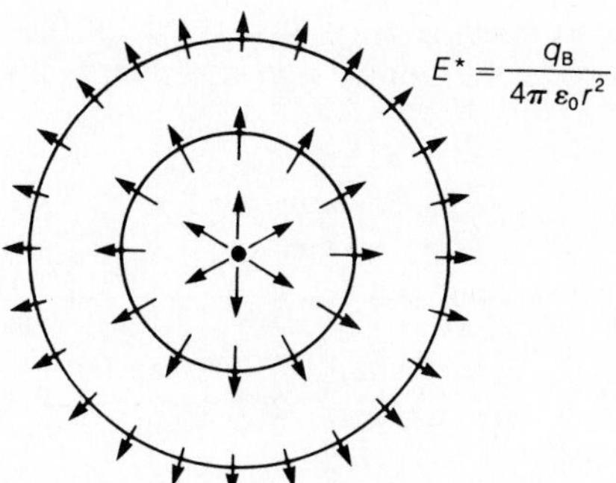

FIGURE 8-1

The electric field E^*, the force experienced by a unit test charge q_A, is directed away from a unit positive charge q_B as indicated by the arrows, which specify the direction of force. (These arrows are perpendicular to the surface of a sphere centered at q_B.) The magnitude of the force depends on the distance according to the equation $E^* = q_B/(4\pi\epsilon_0 r^2)$. The electrostatic potential Φ *is* $q_B/(4\pi\epsilon_0 r)$.

similarly

$$\Phi = \frac{U}{q_A} = \frac{q_B}{4\pi\epsilon_0 r} \qquad (8\text{-}6c)$$

E^* is directed along a line joining q_B and the unit charge or point in space to which E^* refers, as shown in Figure 8-1.

8–2 Ohm's Law States that the Electric Current is Proportional to the Potential Difference, or to the Electric Field

In electrochemistry we frequently have to deal with simple electrical circuits in which there is a flow of electrical charge from one point to another. This flow is called the *current, i,* and is measured in coulomb sec^{-1} (amperes). Figure 8-2 shows an electrical conductor with electrostatic potentials Φ_1 at one end and Φ_2 at the other. The current i is the positive charge transferred from Φ_1 to Φ_2 per unit time; i is directed from Φ_1 to Φ_2 if the potential Φ_1 is greater than Φ_2, because a positive charge q_B will have a lower potential energy ($q_B\Phi$) at the lower potential Φ_2. Ohm discovered that the current i is, in most materials, proportional to the potential difference:

$$i \propto (\Phi_2 - \Phi_1)$$

The proportionality constant is called the *resistance, R,* giving us *Ohm's law:*

$$\Delta\Phi = -iR \qquad (8\text{-}7)$$

in which the potential difference, $\Delta\Phi$, is the final potential minus the initial potential

$$\Delta\Phi = \Phi_2 - \Phi_1$$

In SI units $\Delta\Phi$ is measured in volts, i in coulomb sec^{-1} (amperes) and R in ohms. (It is common to call the voltage drop across the conductor, V, $V = \Phi_1 - \Phi_2$. In this case the minus sign disappears and $V = iR$.)

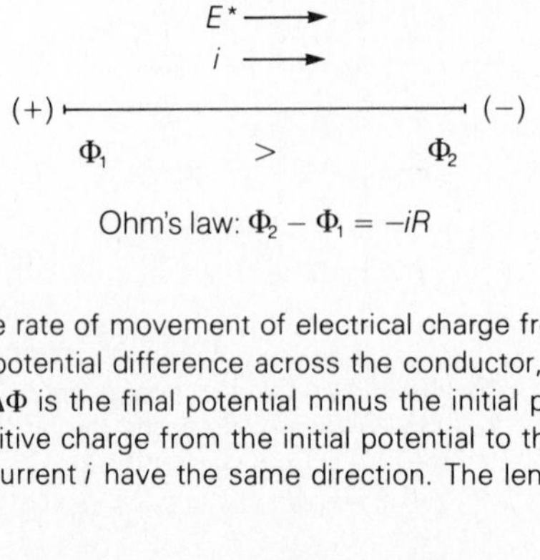

FIGURE 8-2
Ohm's law states that the rate of movement of electrical charge from potential Φ_1 to potential Φ_2 is proportional to the potential difference across the conductor, or $\Delta\Phi = \Phi_2 - \Phi_1 = -iR$. By our sign convention, $\Delta\Phi$ is the final potential minus the initial potential and i is positive for the movement of positive charge from the initial potential to the final potential. The electric field E^* and the current i have the same direction. The length of the conductor is l, and $E^* = -\Delta\Phi/l$.

Equation 8-6, $E^* = -d\Phi/dr$, can be used to relate the current to the electric field. When the resistance per unit length of the conductor is constant, we can replace this equation by the ratio of finite quantities $E^* = -\Delta\Phi/\Delta r$. Therefore, Ohm's law can be written

$$E^* = \frac{iR}{l} \tag{8-8}$$

in which $l = \Delta r$ is the length of the conductor. Equation 8-8 states that the current is proportional to the electric field. Ohm's law (Equations 8-7 and 8-8) applies to most materials, with the exception of those containing special devices, such as semiconductor junctions, diodes, vacuum tubes, and so on.

○ 8–3 *Why Do Electrolytes Dissolve in Water?*

Some substances have the special property that they dissociate into ions when dissolved in water. The salts that have this property are characterized by a strong ionic contribution to the bonding between atoms in the compound. For example, NaCl in the solid state is a regular array of Na^+ and Cl^- ions held together by the attractive electrostatic force between the ions. When placed in water, the interaction of the ions with water is evidently sufficiently strong to overcome the ion-ion bonding, and dissociation occurs. Acids and bases such as HCl and NH_3 also are electrolytes because in one case a proton is given up to H_2O to form H_3O^+ and Cl^-, and in the other, H_2O donates a proton to form OH^- and NH_4^+.

One of the simple questions we can ask about ionic solutions is why the compounds dissolve at all. Na^+ and Cl^- ions separated by a distance r in a vacuum interact with a potential energy given by Coulomb's law (considering the ions to be charges localized at a point):

$$U = \frac{q_{Na^+} q_{Cl^-}}{4\pi\epsilon_0 r} \tag{8-9}$$

The charge on a Na^+ ion is $1.602 \times 10^{-19} C$ (4.803×10^{-10} esu in cgs units), with an equal charge of opposite sign on the Cl^- ion. When two charges interact through a medium other than vacuum, the force and potential energy are reduced by a factor $1/\epsilon$:

$$U = \frac{(q_{Na^+})(q_{Cl^-})}{4\pi\epsilon_0 r\epsilon} \tag{8-10}$$

The dielectric constant ϵ is a property of matter that depends on molecular structure and molecular interactions; its interpretation will be discussed further in Chapter 11. For the present we will be content with the observation that highly polar molecules, such as water, have large dielectric constants. *Polar* molecules (also called *dipolar* molecules) by definition have a separation of the centers of the negative and positive charge distribution, so that they have a negative and a positive end. Figure 8-3 gives a qualitative

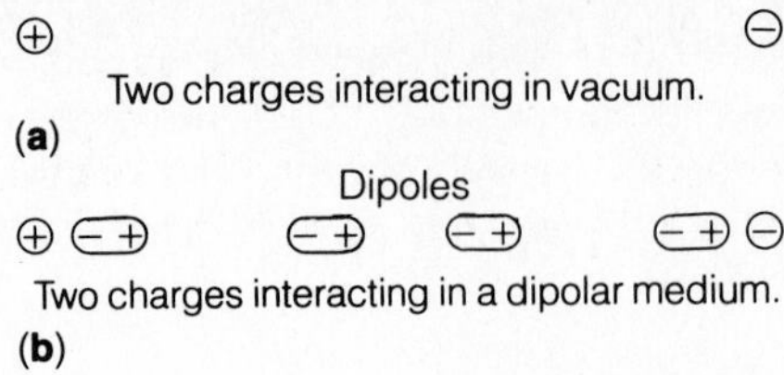
Two charges interacting in vacuum.
(a)
Dipoles
Two charges interacting in a dipolar medium.
(b)

FIGURE 8-3
Effect of a dipolar medium on charge-charge interaction. The dipolar molecules tend to align with their negative ends facing the positive charge and positive ends facing the negative charge. Placing a negative dipole adjacent to the positive charge and a positive dipole near the negative charge reduces the effective charge at those centers and results in a reduction of the interaction energy by a factor $1/\epsilon$ from the value in a vacuum.

interpretation of the dielectric constant. The dipolar molecules tend to line up with their negative ends toward the positive charge and, conversely, with their positive ends toward the negative charge. The positive dipolar charges which accumulate at the negative charge center (and conversely for the positive charge center) reduce the effective size of the charge at the center. Therefore, the ion-ion interaction force and potential energy are reduced; the amount of the reduction ($1/\epsilon$) depends on the material and must be measured experimentally.

EXERCISE 8-2

Calculate the potential energy of interaction between Na^+ and Cl^- separated by a distance of 10Å (1 nm) in a vacuum. Do the calculation in both SI and cgs units.

ANSWER

(a) SI units:

$$U = \frac{-(1.602 \times 10^{-19}\ \mathrm{C})^2}{4\pi \times 8.854 \times 10^{-12}\ \mathrm{kg^{-1}\ m^{-3}\ s^4\ A^2} \times 10^{-9}\ \mathrm{m}}$$
$$= -2.30 \times 10^{-19}\ \mathrm{kg\ m^2\ s^{-2}\ molecule^{-1}}$$
$$= -2.30 \times 10^{-19}\ \mathrm{J\ molecule^{-1}}$$

To get the energy per mole, we multiply by 6.022×10^{23} molecules mole^{-1}, obtaining

$$U = -1.39 \times 10^5\ \mathrm{J\ mol^{-1}} = -139\ \mathrm{kJ\ mol^{-1}}$$

(b) cgs units:

$$U = \frac{-(4.80 \times 10^{-10}\ \mathrm{esu})^2}{10^{-7}\ \mathrm{cm}} = -2.30 \times 10^{-12}\ \mathrm{erg\ molecule^{-1}}$$

Again, the energy per mole is obtained by multiplying by 6.022×10^{23}, or $U = -1.39 \times 10^{12}\ \mathrm{erg\ mol^{-1}} = -139\ \mathrm{kJ\ mol^{-1}}$.

EXERCISE 8-3

Calculate the energy of interaction between Na^+ and Cl^- at a separation of 1 nm in water at 37°C (dielectric constant $\epsilon = 74.2$).

ANSWER

$$U = \frac{-(1.602 \times 10^{-19}\ \text{C})^2}{4\pi \times 8.854 \times 10^{-12}\ \text{kg}^{-1}\ \text{m}^{-3}\ \text{s}^4\ \text{A}^2 \times 10^{-9}\ \text{m} \times 74.2}$$
$$= -3.11 \times 10^{-21}\ \text{J molecule}^{-1}$$

The energy per mole is $U = -1.87\ \text{kJ mol}^{-1}$.

The difference between the answers to Exercises 8-2 and 8-3 is 137 kJ mole^{-1}. This is the amount by which the potential energy of interaction between Na^+ and Cl^-, 10A apart, is reduced when they are immersed in water. *The large dielectric constant of water is a major factor in reducing the energy of interaction between ions and allowing them to separate in solution.*

Another reason that electrolytes dissolve in water is the entropy gained because the ions can move freely in solution. However, this term would apply to all solvents, so it does not explain the special properties of water. Furthermore, ordering of water molecules around the ions can be so extensive that the total entropy sometimes *decreases* on dissolution.

8-4 *Thermodynamics of Ion Solvation*

The reduction in the energy of interaction between two ions in water is accomplished because of the strong interaction called *hydration* or *solvation* between water or another solvent and the dissolved ions. For example, a positive ion will attract the negative end of the water dipole (the oxygen atom; see Fig. 8-4). The negatively charged Cl^- ion will correspondingly attract the positive end of the water dipole. The experimentally determined enthalpy of these solvation interactions is nearly the same as the enthalpy of interaction between Na^+ and Cl^- in the crystal. You can see this by comparing the heat of solution of crystalline NaCl with the heat, $\Delta H_{\text{crystal}}$, of converting the ions in the crystal to gaseous ions at low concentrations:

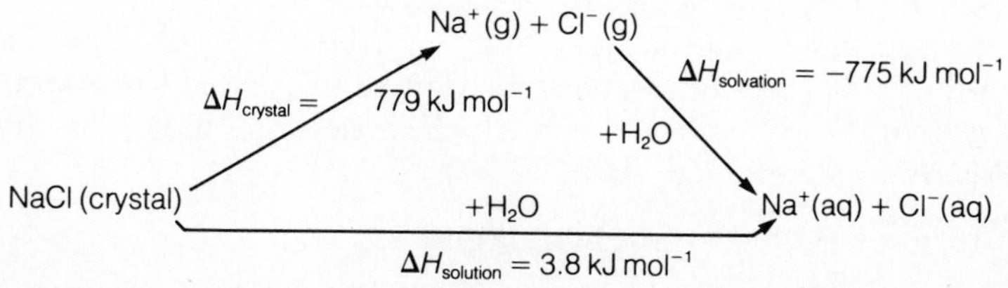

$Na^+(g) + Cl^-(g)$

$\Delta H_{\text{crystal}} = 779\ \text{kJ mol}^{-1}$ $\Delta H_{\text{solvation}} = -775\ \text{kJ mol}^{-1}$ ($+H_2O$)

$\text{NaCl (crystal)} \xrightarrow{+H_2O} Na^+(aq) + Cl^-(aq)$ (8-11)

$\Delta H_{\text{solution}} = 3.8\ \text{kJ mol}^{-1}$

Hydrated Na^+ ion

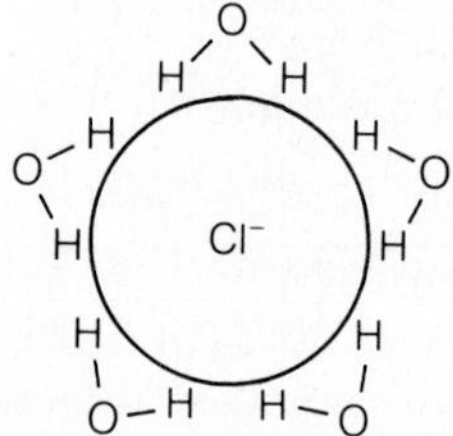

Hydrated Cl^- ion

FIGURE 8-4
Hydrated ions of Na^+ and Cl^- in solution. Water molecules around the ions tend to orient so that the negative end (oxygen atom) interacts with a positive charge and the positive end (hydrogen atom) interacts with a negative charge. Notice that the Cl^- ion is substantially larger than Na^+. The structures shown here are for schematic representation only; the number of water molecules shown has no significance. The enthalpy of interaction of water with ions is favorable ($\Delta H_{\text{hydration}}$ is negative), but the ordering of water molecules provides an unfavorable entropic contribution to the free energy of hydration.

By Hess' law of heat summation (ΔH must be the same for all paths between the same reactants and products),

$$\Delta H_{\text{solution}} = \Delta H_{\text{crystal}} + \Delta H_{\text{solvation}} \tag{8-12}$$

The crystal energy of NaCl, $\Delta H_{\text{crystal}}$, is 779 kJ mole^{-1} (obtained from thermochemical data and ionization energies), and $\Delta H_{\text{solution}} = 3.8$ kJ mole^{-1}. Therefore, $\Delta H_{\text{solvation}} = -775$ kJ mole^{-1}. Heat is released (775 kJ mole^{-1}) on solvating Na^+ (g) and Cl^- (g) ions with water to form a dilute aqueous solution. This is very nearly the same as the heat released (779 kJ mole^{-1}) when Na^+ and Cl^- gaseous ions recombine to form a NaCl crystal.

The heat of solvation is the total enthalpy of interaction between the ions and the surrounding water molecules. The value given is a sum for Na^+ and Cl^- ions; it is not possible to measure separate ionic solvation enthalpies or free energies because one cannot study a solution of Na^+ which does not at the same time contain an equivalent concentration of Cl^- or some other negative ion. (Overall electrical neutrality of the solution must be preserved.)

However, it is possible to measure energy or free-energy changes for adding single water molecules to *gaseous* ions. For example, the free energy change of the general hydration reaction in which single water molecules are added successively to a metal ion, M^+,

$$M(H_2O)_n^+\,(g) + H_2O(g) \rightarrow M(H_2O)_{n+1}^+(g) \tag{8-13}$$

is summarized for $M^+ = Na^+$ and K^+ in Table 8-1.

TABLE 8-1. *Free Energy of Hydration of Gaseous Ions*[a]

BOUND H_2O		ΔG/kJ mol^{-1}	
n	$n+1$	Na^+	K^+
0	1	−70.2	−45.4
1	2	−52.6	−35.5
2	3	−37.1	−25.1
3	4	−25.1	−17.5

[a] From I. Dzidic and P. Kebarle (1970).

These results emphasize an important point that should be kept in mind when considering ionic solutions and ion transport: *if two ions have the same charge, the smaller of the two interacts more strongly with water.* For example, from Table 8-1, we see that the first water molecule bound to gaseous Na^+ releases 70.2 kJ mol^{-1} of free energy, compared to only 45.4 kJ mol^{-1} for the larger K^+ ion. This occurs because the smaller ion has a higher charge density, giving a stronger interaction with the dipolar water molecule.

Another important point to recognize is the entropy decrease associated with orientation of water molecules around an ion. The hydration sphere shown in Figure 8-4 requires loss of freedom of rotational motion for the bound water molecule, providing an unfavorable entropic component. However, the enthalpic component is favorable and larger in magnitude, so the free energy of hydration is large and negative.

The large free energy of hydration is important for ion transport through relatively nonpolar materials such as lipid membranes ($\epsilon \approx 3$), because it implies that *the ion cannot exist bare (unsolvated) in the nonpolar phase.* The free energy cost of complete dehydration cannot be recovered by interaction with the nonpolar material. This leaves two avenues for transport (see Fig. 8-5): either the ion can remain hydrated, or polar groups (B) for solvating the ion can be provided by the membrane during transport. The groups B would replace H_2O molecules while the ion crosses the membrane.

Heats of solution and crystal energies can be used to demonstrate another important general principle: *the absolute value of the solvation enthalpy is significantly larger for small cations than for larger cations of the same charge.*

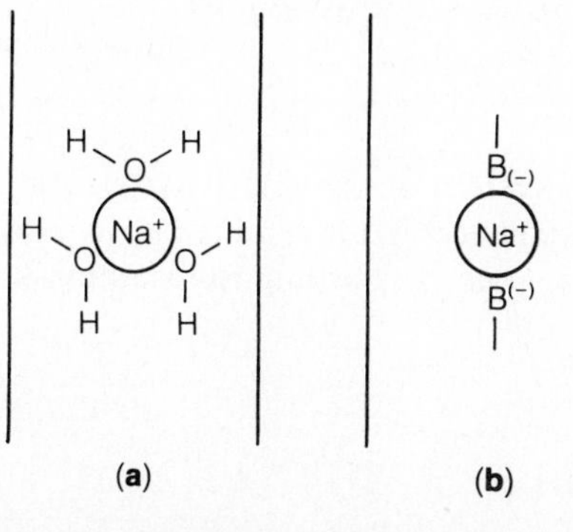

FIGURE 8-5
Two alternative models of the transport of ions across biological membranes. **(a)** Transport of a hydrated ion, and **(b)** solvating groups with electron-rich ends $B^{(-)}$ provided within the membrane. Transport of the bare ion, without solvating groups, is very unlikely because of the large free energy required to remove all water of hydration without alternative solvation interactions.

The crystal energy of KCl is 674 kJ mole^{-1} and the heat of solution is 17.2 kJ mole^{-1}. Therefore the enthalpy of solvation of K^+ and Cl^- is $\Delta H_{\text{solvation}} = \Delta H_{\text{solution}} - \Delta H_{\text{crystal}} = -657$ kJ mole^{-1} (Equation 8-12). The difference between this and -775 kJ mole^{-1} for the solvation enthalpy of Na^+ and Cl^- must be due to the difference between Na^+ and K^+. Therefore $\Delta H_{\text{solvation}}(Na^+) - \Delta H_{\text{solvation}}(K^+) = -775 + 657 = -118$ kJ mole^{-1}. The same qualitative conclusion applies to the difference between Mg^{++} and Ca^{++}. The smaller ion interacts more strongly with the solvent than the larger ion, because smaller ions have higher electric fields near their surfaces (see Eq. 8-6*b*). There are large differences in the extent of hydration of different cations and in the degree of difficulty of removing a water molecule from the first hydration shell (the layer in contact with the ion).

EXERCISE 8-4

The crystal energy of NaBr is 754 kJ mol^{-1}, and the heat of solution is 3 kJ mol^{-1}. Calculate the difference in solvation energy of Cl^- and Br^-.

ANSWER

$\Delta H_{\text{solvation}}(Na^+ + Br^-) = -754 + 3 = -751$ kJ mol^{-1}. $\Delta H_{\text{solv}}(Cl^-) - \Delta H_{\text{solv}}(Br^-) = -775 + 751 = -24$ kJ mol^{-1}. The result of this exercise can be generalized: the solvation energies of anions do not depend as strongly on size as do those of cations, nor is their solvation energy as large, mainly because of the larger size of anions compared to cations. (The ionic radius of Na^+ is 0.96 Å and the ionic radius of Cl^- is 1.81 Å—see Chapter 11.)

8–5 *Coulomb's Law and the Dielectric Constant Give the Free Energy, Entropy, and Enthalpy of Interaction between a Pair of Ions in Solution*

Up to this point we have been referring to the potential energy calculated by Coulomb's law as an energy; more precisely, it is a *free energy* when the ions are immersed in a dielectric medium. Recall from Section 4-3 that the reversible nonexpansion work at constant temperature and pressure is

$$dG = dW_{\text{rev,other}}$$

Therefore the electrostatic free energy of a system containing two ions immersed in a dielectric medium is the nonexpansion work required to move them from infinite separation to a distance r apart. According to Coulomb's law (Equations 8-1-SI and 8-2),

$$dW_{\text{el}} = dW_{\text{rev, other}} = \frac{-q_A q_B}{4\pi\epsilon_0 \epsilon r^2}\, dr$$

Hence

$$G_{el} = -\int_{\infty}^{r} \frac{q_A q_B}{4\pi\epsilon_0 \epsilon r^2}\, dr$$

$$= \frac{q_A q_B}{4\pi\epsilon_0 \epsilon r} \quad (8\text{-}14)$$

The electrostatic entropy (remember that $dG = V\,dP - S\,dT$) is

$$S_{el} = -\left(\frac{\partial G_{el}}{\partial T}\right)_P \quad (8\text{-}15)$$

Because q_A, q_B, ϵ_0, and r are independent of temperature, differentiation of G_{el} with respect to T in Equation 8-14 yields

$$S_{el} = -\frac{q_A q_B}{4\pi\epsilon_0 r}\left[\frac{\partial(1/\epsilon)}{\partial T}\right]_P \quad (8\text{-}16)$$

Using the relation $d(1/x) = -dx/x^2 = -(d \ln x)/x$, we can convert Equation 8-16 to

$$S_{el} = \frac{q_A q_B}{4\pi\epsilon_0 \epsilon r}\left(\frac{\partial \ln \epsilon}{\partial T}\right)_P$$

$$= G_{el}\left(\frac{\partial \ln \epsilon}{\partial T}\right)_P \quad (8\text{-}17)$$

Table 8-2 shows that the dielectric constant of water decreases with increasing temperature. Therefore, with $(\partial \ln \epsilon/\partial T)_P$ negative, two oppositely charged ions which have a *negative* G_{el} according to Equation 8-14 have a *positive* electrostatic entropy of interaction in Equation 8-17. Figure 8-6 shows the qualitative source of the positive electrostatic entropy. Two ions separated by a very large distance ($r \to \infty$) have separate hydration spheres. When the ions approach each other, the hydration spheres overlap, and water molecules can be shared between the two ions. Therefore some water molecules are set free to the less-ordered state of bulk water. This reduction in the number of electrostatically ordered molecules accounts for the entropy increase.

TABLE 8-2 *Dielectric Constant of Water at Several Temperatures*[a]

T, °C	ϵ	T, °C	ϵ
0	88.00	40	73.28
5	86.04	45	71.59
10	84.11	50	69.94
15	82.22	60	66.74
20	80.36	70	63.68
25	78.54	80	60.76
30	76.75	90	57.98
35	75.00	100	55.33

[a] The Chemical Rubber Co. (1966), *Handbook of Chemistry and Physics*, 47th ed., Cleveland.

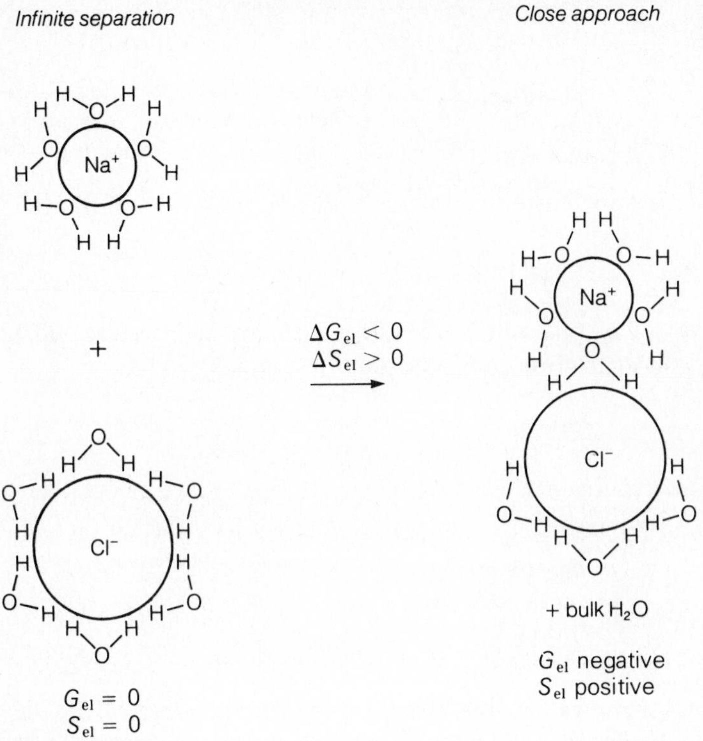

Infinite separation

Close approach

Na^+

Cl^-

$\Delta G_{el} < 0$
$\Delta S_{el} > 0$

Na^+

Cl^-

+ bulk H_2O

G_{el} negative
S_{el} positive

$G_{el} = 0$
$S_{el} = 0$

FIGURE 8-6
The release of ordered water by sharing hydration spheres accounts for the entropy increase when two ions approach each other in solution.

EXERCISE 8-5

Calculate the electrostatic free energy, entropy, and enthalpy of interaction between a positive and a negative ion, both singly charged, separated by a distance of 10 Å (1 nm) in a medium of dielectric constant 74.2 (water at 37°C). Use $(\partial \ln \epsilon/\partial T)_P = -4.64 \times 10^{-3}\ \mathrm{K^{-1}}$.

ANSWER

(See Ex. 8-3.)

$$G_{el} = \frac{-(1.602 \times 10^{-19}\ \mathrm{C})^2}{4\pi \times 8.854 \times 10^{-12}\ \mathrm{kg^{-1}\ m^{-3}\ s^4\ A^2} \times 74.2 \times 10^{-9}\ \mathrm{m}} \times 6.023 \times 10^{23}\ \mathrm{mol^{-1}}$$

$$= -1.87\ \mathrm{kJ\ mol^{-1}}$$

$$S_{el} = G_{el}\left(\frac{\partial \ln \epsilon}{\partial T}\right) = 8.69 \times 10^{-3}\ \mathrm{kJ\ K^{-1}\ mol^{-1}}$$

$$H_{el} = G_{el} + TS_{el} = -1.87 + 2.69 = 0.82\ \mathrm{kJ\ mol^{-1}}$$

H_{el} is *positive* because heat is absorbed when ordered water is released by sharing of hydration spheres between the two ions at a distance of 10 Å.

○ 8–6 *The Conductivity of Aqueous Solutions*

In 1887 Arrhenius concluded that salts dissociate into ions when dissolved in water. His reasoning was based largely on the conduction of electric current by aqueous salt solutions. This result can be understood readily if the solid salt dissociates into oppositely charged ions:

$$\text{NaCl (s)} \rightarrow \text{Na}^+ \text{(aq)} + \text{Cl}^- \text{(aq)}$$

The ions move under the force of the electric field, and in opposite directions because their charges are opposite (Figure 8-7).

According to Ohm's law (Equation 8-8), the current that flows in the conductivity cell in Figure 8-7 should be proportional to the electric field, $E^* = -\Delta\Phi/l$ or

$$i \propto E^* \tag{8-18}$$

The number of ions available as current carriers in a cross-section of the solution is proportional to the cross-sectional area A, so we also expect the current to be proportional to A

$$i \propto A \tag{8-19}$$

Combining Equations 8-18 and 8-19 and inserting a proportionality constant K' gives us

$$i = K' A E^* \tag{8-20}$$

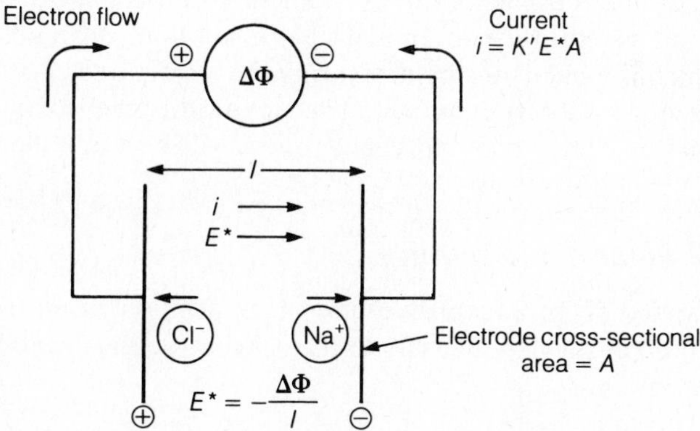

FIGURE 8-7
A conductivity cell. A potential difference, $\Delta\Phi$, is applied to the electrodes, thereby causing ions in the solution to migrate toward the electrode of opposite charge. Electron flow is responsible for the current in the external circuit. The electric field E^* (volts cm^{-1}) in the solution is $-\Delta\Phi/l$, in which l is the length of the cell. Notice that the electric field is constant through the cell, so all ions are subjected to the same force qE^*. The current i is proportional to the area of the electrodes A, to the field E^*, and to the conductivity K' of the material between the electrodes. The electric field E^* is directed from the left (high-potential) electrode toward the right (low-potential) electrode, which causes Na^+ to move to the right and Cl^- to move to the left.

The quantity K' is called the conductivity of the solution. It is given by the ratio

$$K' = \frac{i}{AE^*} \qquad (8\text{-}21)$$

and is equal to the current passed through a solution of unit cross-sectional area, A, under a unit electric field, E^*. When the field E^* is in volts cm^{-1}, i in amperes, and A in cm^2, then K' is measured in $ohm^{-1}\ cm^{-1}$ (or amperes $volt^{-1}\ cm^{-1}$). Conductivities of aqueous solutions are measured with an alternating electric field, usually in the frequency range 1–5 kHz, because a steady voltage and direct (one-directional) current allow the build-up of electrolysis products at the electrodes that alter the current and cause apparent divergence from Ohm's law. The chemical reactions that occur at electrodes are discussed in Section 9-1.

Conductivities of Solutions

The conductivity of pure water (sometimes called Kohlrausch water after the German scientist who distilled the same sample forty-eight times to remove all impurities) is about $0.043 \times 10^{-6}\ ohm^{-1}\ cm^{-1}$. Addition of impurities or salts increases the conductivity of water; ordinary distilled or deionized water containing dissolved CO_2 from the air has a conductivity of approximately $10^{-6}\ ohm^{-1}\ cm^{-1}$. ($CO_2$ reacts with water to form H_2CO_3, which dissociates to form the current-carrying ions H^+ and HCO_3^-.) A 1 M solution of NaOH has a conductivity of about $0.180\ ohm^{-1}\ cm^{-1}$ at 25°C.

Conductivity measurements are convenient to biochemists who wish to determine the concentration of an electrolyte solution. A standard curve is required, which is established by measuring the conductivity as a function of the concentration of a particular salt. The standard curve is used to convert measurements of conductivity to concentration of the same electrolyte.

Equivalent Conductance

The conductivity of an aqueous solution increases when an electrolyte is added. The ratio of the conductivity to the concentration is called the *equivalent conductance*, Λ,

$$\Lambda = \frac{K'}{C'} \qquad (8\text{-}22)$$

By convention, the concentration C' is expressed as equivalents per cm^3. (One mole of $Na^+\ Cl^-$ contains one equivalent; one mole of $Cu^{2+}\ SO_4^{2-}$ contains two equivalents.) The units of Λ are $ohm^{-1}\ cm^2\ equiv^{-1}$.

Because of interaction between the ions, Λ is concentration-dependent. Extensive conductivity measurements in the nineteenth century established that Λ varies linearly with the square root of the concentration for all strong

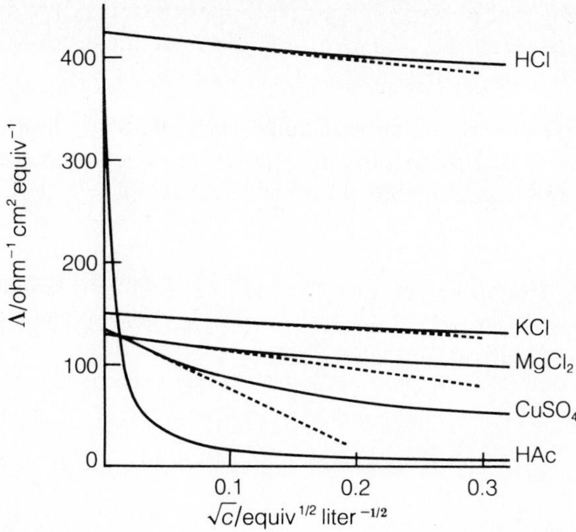

FIGURE 8-8
Equivalent conductances at 25°C as a function of the square root of the concentration in equivalents/liter. For the strong electrolytes (HCl, KCl, $MgCl_2$, $CuSO_4$) Λ varies linearly with $\sqrt{c'}$ at low concentration, with a slope that depends on ion charge. This behavior is explained by the Debye-Hückel theory, Sections 8-10 and 8-11. Conductance of the weak electrolyte acetic acid (HAc) decreases strongly as concentration increases because of association of H^+ and Ac^- to form the nonconducting HAc. The equivalent conductance of substances that produce H^+ or OH^- in aqueous solution is anomalously large because of the special mechanism for migration of these ions, described in Chapter 16. (Data from Harned and Owen, 1958, and Dole, 1935.)

electrolytes in dilute solution; Figure 8-8 shows some examples. The equivalent conductance of *weak electrolytes,* including weak acids such as acetic acid and weak bases such as ammonia, which do not dissociate completely in solution, drops off sharply as the concentrations increase (Fig. 8-8). This occurs because H^+ and CH_3COO^-, for example, combine at higher concentrations, causing a decrease in the number of charge carriers.

The value of Λ extrapolated to zero concentration by plotting the data as in Figure 8-8 is called the equivalent conductance at infinite dilution, Λ_0:

$$\Lambda_0 = \lim_{C' \to 0} \Lambda$$

Λ_0 reflects the properties of isolated ions and can be separated experimentally into contributions from the positively and negatively charged ions:

$$\Lambda_0 = \Lambda_0(+) + \Lambda_0(-) \tag{8-23}$$

where $\Lambda_0(+)$ and $\Lambda_0(-)$ are the *equivalent ionic conductances* at infinite dilution. Experimental data on the conductivity of ionic solutions are frequently tabulated as equivalent ionic conductances (see, for example, Harned and Owen, 1958).

8–7 *The Mobility of an Ion Depends on its Transport Velocity*

The current carried by an ion depends on its rate of steady motion, called the *transport velocity* to distinguish the directed motion between electrodes from the random thermal motion that all particles have at $T > 0$ K. Hence

$$i \propto v$$

in which v is the transport velocity. According to Ohm's law, the current is proportional to the electric field E^*, $i \propto E^*$, so we can set the transport velocity proportional to the electric field:

$$v = u\,E^* \tag{8-24}$$

in which the proportionality constant u is called the *mobility* of the ion. It is given by

$$u = \frac{v}{E^*} \tag{8-25}$$ ◀

which means that *the mobility is the transport velocity of the ion when moving under a unit electric field.* Each electrolyte contains positive and negative ions, characterized by mobilities $u(+)$ and $u(-)$. The usual units for mobility are $v/E^* =$ cm sec^{-1}/(volt cm^{-1}) = cm^2 volt^{-1} sec^{-1}.

It can be shown (see Problem 8-19) that the mobility of an individual ion is related to the equivalent ionic conductance by the equation

$$\Lambda(+) = u(+)F \tag{8-26}$$

in which F is the number of coulombs in one mole of protons, $F = 96{,}487$ C mol^{-1}. F is called a *Faraday,* after Michael Faraday who in 1833 established the relationship between current passed and amount of a chemical reaction (see the historical sketch beginning on page 348). An equation analogous to Equation 8-26 also applies to negative ions. Ionic mobility depends on concentration; the values extrapolated to zero concentration are given subscripts 0, so that, for example,

$$\Lambda_0(+) = u_0(+)F$$

EXERCISE 8-6

The equivalent ionic conductance of NO_3^- at 25°C is 71.4 ohm^{-1} cm^2 equiv^{-1}. Calculate the mobility and the migration velocity under a field of 100 volts cm^{-1}.

ANSWER

$u_0 = \Lambda_0/F = \dfrac{71.4\text{ ohm}^{-1}\text{ cm}^2\text{ mol}^{-1}}{96{,}487\text{ coulomb mol}^{-1}} = 7.4 \times 10^{-4}$ cm^2 volt^{-1} sec^{-1}. (Notice that Ohm's law, voltage = current × resistance, gives us volts = coulomb sec^{-1} ohm.) The mobility, $u = 7.40 \times 10^{-4}$ cm^2 volt^{-1} sec^{-1}, times the electric field, $E^* = 100$ volts cm^{-1}, gives us $v = 7.4 \times 10^{-2}$ cm sec^{-1}.

Interpretation of Mobilities

The mobilities of individual ions vary considerably, as shown in Table 8-3.

TABLE 8-3 *Mobilities of Representative Ions in Water at 25°C*[a]

CATION	$u_0/10^{-4}$ cm^2 volt^{-1} sec^{-1}	ANION	$u_0/10^{-4}$ cm^2 volt^{-1} sec^{-1}
H^+	36.25	OH^-	20.50
Li^+	4.01	Cl^-	7.91
Na^+	5.19	Br^-	8.10
K^+	7.62	I^-	7.97
Rb^+	8.06	NO_3^-	7.40
Cs^+	8.01	ClO_4^-	6.98
$CH_3NH_3^+$	6.08	CH_3COO^-	4.24
$(CH_3)_4N^+$	4.66	$CH_3(CH_2)_2COO^-$	3.38
Mg^{++}	5.50	$Fe(CN)_6^{4-}$	11.5
Ca^{++}	6.17		
Ba^{++}	6.60		
La^{++}	7.20		

[a] Data from Harned and Owen (1958).

Notice that the mobilities of H^+ and OH^- are anomalously large. The reason for this is the special mechanism for migration of these ions that applies when water is the solvent, as considered in Chapter 6. Another anomaly in Table 8-3 is the lower mobility of small ions compared to large ones, for example, Na^+ compared to K^+, or Mg^{++} compared to Ca^{++}. Because Na^+ and K^+ have the same charge, one would expect the smaller ion to experience the same force from the electric field but encounter less frictional resistance because of its smaller size and hence move faster. (The frictional resistance of a sphere is proportional to its radius according to Stokes' law, which will be considered in detail in Chapter 15.) The disagreement with this simple expectation arises from the interactions between water and the ion. The stronger hydration of smaller cations discussed in Section 8-4 causes the effective "Stokes' radius" (the radius that determines the resistance to flow) of the hydrated Na^+ cation to be larger than that of K^+, even though the ionic radius of K^+ is greater than that of Na^+. Equivalent anomalies are not seen for anions. In considering transport of cations across biological membranes, you should keep in mind that the effective radius of hydrated Na^+ is larger than that of hydrated K^+.

MICHAEL FARADAY

1791–1867

Michael Faraday, from a daguerreotype. (The Bettmann Archive, Inc.)

Late starters in science can take heart from the example of Michael Faraday: at age 21 Faraday was a bookbinder's apprentice, with no future other than to be a bookbinder. His father was a blacksmith whose ill health prevented him from providing more than the barest essentials for his family. Faraday recalled later that his mother had once given him a loaf of bread which was supposed to feed him for a week. He also recalled, "My education was of the most ordinary description, consisting of little more than the rudiments of reading, writing, and arithmetic at a common day school. My hours out of school were passed at home and in the streets."

At age thirteen he delivered newspapers for a bookseller and bookbinder, and because of his exemplary conduct, he was accepted as an apprentice without fee. The seven-year apprenticeship that followed had two great benefits for Faraday: he developed an extraordinary manual dexterity that later distinguished his experiments, and he was brought into contact with books, which he read eagerly after working hours. Two books were especially important to his development. The first was *The Improvement of the Mind,* written by an eighteenth-century clergyman, Isaac Watts. Faraday followed Watts' suggestions for self-improvement, including keeping a commonplace book in which he recorded ideas and observations, attending lectures and taking notes, corresponding with others, and founding a discussion group for exchange of ideas. Watts also prescribed accurate observations of facts and precision in language to prevent one from premature generalization. These were traits that Faraday later brought to his science.

The second book of importance in Faraday's development was the *Encyclopedia Britannica,* in which he read the article "Electricity" while rebinding the volume. This article aroused his interest in the subject to which he became the greatest experimental contributor of all time.

Faraday's life was transformed through his contact with Humphrey Davy. Davy was Professor at the Royal Institution (which had been founded by Count Rumford; see Chap. 2). One of the bookbinder's customers offered Faraday tickets to a set of Davy's lectures. Faraday went and took careful notes. He bound these and sent them to Davy, who was flattered but could offer nothing in the way of a job. However, the next year when Davy's assistant was found brawling, he was fired and Faraday got his place.

Several months later Davy decided to make an extended tour of French and Italian laboratories, and asked Faraday to accompany him. This eighteen-month trip was Faraday's introduction to research of that day. Upon returning, he threw himself into chemistry, working first as Davy's assistant and then with increasing independence. His first contributions were in analytical chemistry (for example, his discovery of benzene and chlorides of carbon, such as C_2Cl_6) and industrial chemistry (new alloys of steel). Then he turned to electricity and electrochemistry, and showed how electrical energy can be converted to mechanical energy, and how mechanical energy can be converted to electricity. This second type of conversion was called electromagnetic induction, and is virtually the foundation of our industrial society. By 1823 Faraday's work had received sufficient recognition for him to be elected to the Royal Society. Davy objected strongly to the election over a question of priority, but Faraday refused to permit this to interfere with their friendship.

As his fame increased from his discoveries and from his inspiring lectures at the Royal Institution, Faraday's services as an industrial consultant were in growing demand. In 1830 he received more than a thousand pounds for such work (probably more than $10,000 in current buying power). It seems certain that he could have become immensely rich in time. His discovery in 1831 of electromagnetic induction is, after all, the basis of nearly every practical electrical machine. But he felt, as he later told his colleague John Tyndall, that he had to decide whether to make wealth or science the pursuit of his life, and he chose science. Faraday also rejected the honorary and administrative scientific positions that came his way. For example, both the Royal Society and the Royal Institution offered him their presidencies, but he refused both. In rejecting the offer of the Royal Society, he told his friend Tyndall, "I must remain plain Michael Faraday to the last."

What was the source of inner security that permitted this remarkable self-made man to seek the laws of nature, while rejecting wealth, comfort, and the usual perquisites of success? Certainly one source was religion. Faraday was a lifelong adherent of a fundamentalist Protestant sect known as the Sandemanians. Some of the importance of Faraday's membership in this group is revealed by his associate at the Royal Institution, John Tyndall, an agnostic who was puzzled over Faraday's tie to religion. In his journal Tyndall wrote, "I think that a good deal of Faraday's week-day strength and persistency might be referred to his Sunday Exercises. He drinks from a fount on Sunday which refreshes his soul for a week." Later commentators have noted that Faraday may have gained more than just persistence from his religion. It gave him an unquestioning faith in the unity of the universe and in the interconnections of all phenomena. It also gave him the conviction that he could reveal these interconnections as an instrument of God's will.

8-8 Chemical Potential of Electrolytes

At the same time that Arrhenius was developing his theory of dissociation of electrolytes in solution, van't Hoff was performing osmotic pressure experiments that provided decisive evidence favoring the dissociation hypothesis. He found that dissolution of a salt such as NaCl at a small mole fraction X_B in water produced an osmotic pressure twice as large as an equal mole fraction of a nondissociating substance such as sucrose. He knew that osmotic pressure depends on the number of dissolved particles, and concluded that dissolution of one mole of NaCl produces two moles of particles in the solution, which he identified as Na^+ and Cl^- ions. Our objective in this section is to see how this result affects the expression for the chemical potentials of the solvent (water) and the solute (NaCl).

Chemical Potential for an Ideal Electrolyte Solution

According to Raoult's law, the vapor pressure of the solvent containing a nondissociating solute B is given by $P_A = P_A^\bullet X_A = P_A^\bullet(1 - X_B)$. If the solute dissociates into two particles, the vapor pressure will be lowered twice as much for a given mole fraction, X_B, so the vapor pressure is

$$P_A = P_A^\bullet (1 - 2X_B) \tag{8-27}$$

Remembering that the chemical potential of the solvent in an ideal solution is $\mu_A = \mu_A^\bullet + RT \ln X_A$, and setting $X_A = 1 - 2X_B$ we get

$$\mu_A = \mu_A^\bullet + RT \ln (1 - 2X_B) \tag{8-28}$$

The chemical potential of the solute is obtained from μ_A by using the Gibbs-Duhem equation (Sec. 8-4). First, we take the differential of Equation 8-28 [remember that $d \ln (1 - ax) = -a\,dx/(1 - ax)$]

$$d\mu_A = \frac{-RT\,2\,dX_B}{1 - 2X_B} \tag{8-29}$$

According to the Gibbs-Duhem equation,

$$X_A\,d\mu_A + X_B\,d\mu_B = 0$$

so

$$d\mu_B = -\frac{X_A}{X_B}\,d\mu_A$$

Substituting $X_A = 1 - 2X_B$ and $d\mu_A$ from Equation 8-29, we get

$$d\mu_B = RT\,\frac{2dX_B}{X_B} \tag{8-30}$$

Equation 8-30 implies that

$$\mu_B = \mu_B^\circ + RT \ln (X_B^2) \tag{8-31}$$ ◀

as you can demonstrate by taking the differential of Equation 8-31, obtaining $d\mu_B$ as specified by Equation 8-30.

In summary, the factor 2 in the vapor-pressure lowering requires that we take the square of the mole fraction or concentration in the expression for the chemical potential of the solute.

The reason for the dependence of μ_B on $\ln x_B^2$, as expressed by Equation 8-31, can better be understood by considering that the chemical potential of an ionic compound such as NaCl is the sum of the chemical potentials of the two component ions:

$$\mu_{NaCl} = \mu_{Na^+} + \mu_{Cl^-} \tag{8-32}$$

It is not possible to measure separate chemical potentials for Na^+ and Cl^-, but the theoretical form for μ is the same as for any other solute:

$$\mu_{Na^+} = \mu^\circ_{Na^+} + RT \ln C_{Na^+}$$
$$\mu_{Cl^-} = \mu^\circ_{Cl^-} + RT \ln C_{Cl^-}$$

Substituting these equations into Equation 8-32 we obtain [remember that $\ln x + \ln y = \ln (xy)$]

$$\mu_{NaCl} = \mu^\circ_{NaCl} + RT \ln (C_{Na^+} C_{Cl^-})$$

in which $\mu^\circ_{Na^+} + \mu^\circ_{Cl^-}$ has been replaced by μ°_{NaCl}. Because the solution must be electrically neutral, $C_{Na^+} = C_{Cl^-} = C_{NaCl}$; therefore

$$\mu_{NaCl} = \mu^\circ_{NaCl} + RT \ln C_{NaCl}^2 \tag{8-33}$$ ◄

which is an equation of exactly the same form as Equation 8-31, except that any concentration unit C can replace the mole fraction, as is usual for dilute Henry's law solutions. By this argument we see that *the chemical potential μ_B depends on $\ln X_B^2$ (or $\ln C_B^2$) because the ionic compound consists of two species which can move independently in the solution.*

Chemical Potential of Nonideal Electrolyte Solutions

When nonideal effects influence the chemical potential, Equation 8-33 can be replaced by

$$\mu_B = \mu^\circ_B + RT \ln \gamma_B^2 C_B^2 \tag{8-34}$$

in which γ_B is the activity coefficient. Comparing this equation with $\mu_B = \mu^\circ_B + RT \ln a_B$, which defines the activity, we see that

$$a_B = \gamma_B^2 C_B^2 \tag{8-35a}$$

Equations for the activities of the separate ions also can be written:

$$a_+ = \gamma_+ C_+ \qquad a_- = \gamma_- C_- \tag{8-35b}$$

Because the chemical potential of an ionic compound is the sum of the separate ionic chemical potentials,

$$\mu_B = \mu_+ + \mu_-$$

we can set

$$\mu_B^\circ + RT \ln a_B = \mu_+^\circ + \mu_-^\circ + RT \ln (a_+ a_-)$$

Consequently, with $\mu_B^\circ = \mu_+^\circ + \mu_-^\circ$, we conclude that

$$a_B = a_+ a_- \tag{8-36}$$

which states that the activity of an ionic compound such as NaCl is the product of the individual ion activities. A slightly more complicated equation applies (see below) when the two ions have different charges.

With Equations 8-35*a* and 8-35*b* we can convert Equation 8-36 to

$$\gamma_B{}^2 C_B{}^2 = \gamma_+ \gamma_- C_+ C_- = \gamma_+ \gamma_- C_B{}^2$$

in which $C_+ = C_- = C_B$ because the number of positive and negative ions must be equal to assure electroneutrality of the solution. Defining

$$\gamma_\pm{}^2 = \gamma_+ \gamma_- = \gamma_B{}^2 \tag{8-37}$$

we obtain

$$\mu_B = \mu_B^\circ + RT \ln (\gamma_\pm{}^2 C_B{}^2)$$

which is identical to Equation 8-34, with $\gamma_B = \gamma_\pm$. The geometric mean of the individual ion activity coefficients is an important quantity called the *mean activity coefficient* $\gamma_\pm$. For NaCl it is

$$\gamma_\pm = (\gamma_+ \gamma_-)^{1/2} \tag{8-38}$$

In the more general case of an electrolyte $M_{\nu+}N_{\nu-}$ that dissociates into ν_- negative ions and ν_+ positive ions (of charge 1 or greater)

$$M_{\nu+}N_{\nu-} \rightarrow \nu_+ M^{(+)} + \nu_- N^{(-)}$$

The mean activity coefficient is defined as

$$\gamma_\pm = (\gamma_+{}^{\nu_+} \gamma_-{}^{\nu_-})^{1/\nu} \tag{8-39}$$

where ν is the total number of ions produced,

$$\nu = \nu_+ + \nu_-$$

For example, for ferric sulfate, $Fe_2(SO_4)_3$,

$$\nu_+ = 2,\ \nu_- = 3,\ \nu = 5, \text{ and } \gamma_\pm = (\gamma_+{}^2 \gamma_-{}^3)^{1/5}$$

In this case, if C_B is the molar concentration of the salt, then the concentration C_M of the positive ion is $\nu_+ C_B$, and, for the negative ion, $C_N = \nu_- C_B$. The chemical potential of $M_{\nu+}N_{\nu-}$ is therefore

$$\mu = \mu^\circ + RT \ln (\gamma_+ C_M)^{\nu_+} + RT \ln (\gamma_- C_N)^{\nu_-} \tag{8-40a}$$
$$= \mu^\circ + RT \ln (\gamma_\pm{}^{\nu} C_M{}^{\nu_+} C_N{}^{\nu_-}) \tag{8-40b}$$
$$= \mu^\circ + RT \ln (\gamma_\pm{}^{\nu}\, \nu_+{}^{\nu_+} \nu_-{}^{\nu_-} C_B^{\nu}) \tag{8-40c}$$

EXERCISE 8-7

Write an expression for the chemical potential for $Fe_2(SO_4)_3$ in terms of the mean (molar) activity coefficient $\gamma_\pm$ and the molar concentration C_B of $Fe_2(SO_4)_3$.

ANSWER

$$\mu_{Fe_2(SO_4)_3} = \mu^\circ_{Fe_2(SO_4)_3} + RT \ln (\gamma_\pm^5\, 2^2\, 3^3\, C_B^5)$$
$$= \mu^\circ_{Fe_2(SO_4)_3} + RT \ln (108\, \gamma_\pm^5\, C_B^5)$$

8–9 *Ionic Strength*

From the Debye-Hückel theory, which is considered in the next section, we find that the mean activity coefficient of electrolytes is a function of a quantity called the *ionic strength, I,* defined by

$$I = \tfrac{1}{2} \Sigma\, m_i z_i^2 \qquad \text{(8-41)}$$

in which m_i is the *molal* concentration of ion i, and z_i its charge. When $z = 1$, as for NaCl, I is equal to the molal concentration of NaCl:

$$I = \tfrac{1}{2}(m_{Na^+} + m_{Cl^-}) = m_{NaCl}$$

but when multivalent ions are involved, the ionic strength is greater than the molality.

EXERCISE 8-8

Show that the ionic strength of $CuSO_4$ is four times the molal concentration of $CuSO_4$.

In dilute solutions, the molality m is approximately

$$m_B \approx C_B/\rho_A$$

where ρ_A is the solvent density and C_B is the molar concentration. Therefore the ionic strength is approximately

$$I = \frac{1}{2\rho_A} \Sigma\, C_i z_i^2 \qquad \text{(8-42)}$$

For water, ρ_A is very close to 1, so I can be estimated from $\frac{1}{2} \Sigma\, C_i z_i^2$.

8–10 *A Summary of the Debye-Hückel Theory*

One of the triumphs in the study of electrolyte solutions was the demonstration by Debye and Hückel that theoretical analysis could predict the activity coefficient of dilute electrolytes. The Debye-Hückel theory is of special interest because it begins with *molecular* properties and allows us to calculate a *macroscopic* property, the activity coefficient. The calculation uses the methods of statistical mechanics, a subject we take up in Chapter 14, so in this section we will only summarize the main results of the theory.

The deviations from ideality of an electrolyte solution at low and moderate concentrations (up to about 0.05 M) can be calculated by the Debye-Hückel theory. The physical basis for the calculation is shown in Figure 8-9. Ions of one charge, called counterions, tend to cluster around a central ion of opposite charge. They are attracted there by the favorable electrostatic interaction energy, but their accumulation is opposed by the randomizing thermal forces that cause particles to move away from regions of high concentration. (The thermal energy is proportional to $k_B T$ per molecule, or RT per mole.) According to the Debye-Hückel theory, the free energy of interaction between a central ion and its surrounding counterions or *ion atmosphere* is responsible for the deviations from ideality in dilute solutions of electrolytes.

The main objective of the theory is to calculate the free energy contributed by the formation of an ion atmosphere and by interaction of the central ion with the counterions that make up the ion atmosphere. Two quantities are required for that calculation. First, the average excess concentration of counterions at each point in the ion atmosphere, and second, the electrostatic potental at each point in the atmosphere. The electrostatic free energy is the sum of all the free-energy terms that result from multiplying the charge on each particle by the electrostatic potential. (Remember that potential energy is charge q_B times the electrostatic potential Φ.)

The two main variables that describe the problem are therefore the electrostatic potential $\Phi(r)$ around the ion (r is the distance from the central ion), and the average charge concentration due to the local excess of counterions, given by $\rho(r)$, which is the net amount of charge per unit volume. In cgs units, $\rho(r)$ is measured in esu per cm^3. Because counterions are attracted to

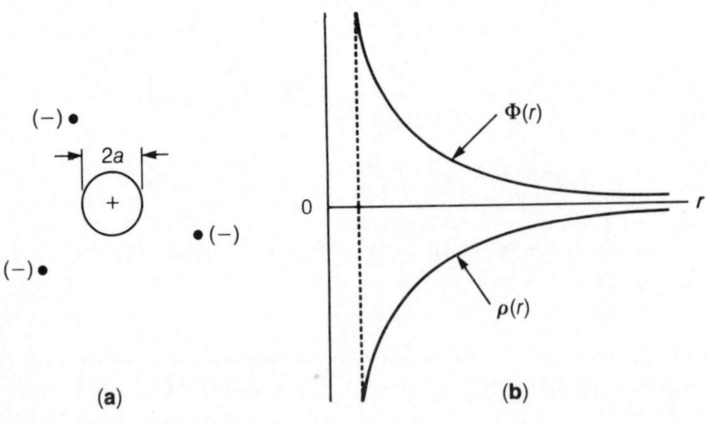

FIGURE 8-9
(a) An instantaneous distribution of negative counterions (considered to be point charges) around a central positive ion of radius a. $\Phi(r)$ and $\rho(r)$ are respectively the average electrostatic potential and average local excess charge density of counterions over positive ions, evaluated at a distance r from the central ion and averaged over all the possible instantaneous charge distributions. **(b)** Schematic variation of $\Phi(r)$ and $\rho(r)$ with r. $\Phi(r)$ increases as $r \to 0$ because of the electrostatic field of the central ion. $\rho(r)$ is negative because the counterions are negative. Both $\rho(r)$ and $\Phi(r)$ approach zero as $r \to \infty$ because the influence of the central ion disappears in that limit, and the average local concentrations of positive and negative ions become equal.

the central ion, we expect $\rho(r)$ to increase in absolute value as r decreases, up to the distance of closest approach between the ions, which we call a, determined by the ionic radii. When r becomes very large, $\rho(r)$ should approach zero because positive and negative charges will, on the average, be present in equal concentrations. In essence, $\rho(r)$ describes the influence of *one* ion, the central ion, on the spatial distribution of all other ions in the solution. Therefore, $\rho(r)$ is called a *distribution function*.

The problem of finding Φ and ρ is difficult because they depend on each other: the tendency of charges to accumulate depends on the potential (Φ), but the potential in turn depends on the density of charge accumulated (ρ). The mathematical approach involves substitution of an equation from statistical mechanics, the Boltzmann distribution, into a second-order differential equation from electrostatics, called the Poisson equation, which relates the second derivative of the potential to the charge density. The mathematical details are worked out in a number of textbooks (see Moore (1972), for example). However, the main conclusions can be understood without the intermediate mathematics.

The Debye Length is the Most Probable Distance from an Ion to Find a Counterion

As expected, the Debye-Hückel theory tells us that ions tend to cluster around the central ion. A fundamental property of the counterion distribution is the thickness of the ion atmosphere, determined by the quantity ($1/\kappa$) called the *Debye length* or *Debye radius*. $1/\kappa$ has the cgs dimensions of cm, so κ has dimensions cm^{-1}; κ is given by:

$$\kappa = \left(\frac{8\pi N_A e^2 \rho_A}{1000\epsilon k_B T}\right)^{1/2} I^{1/2} \tag{8-43}$$

in which ρ_A is the solvent density, e is the protonic charge (4.803×10^{-10} esu), and ϵ is the solvent dielectric constant.

The meaning of the Debye radius can be appreciated better by examining the expression for the counterion distribution (ρ) found from the theory:

$$\rho = B_1 \frac{e^{-\kappa r}}{r} \qquad r \geq a \tag{8-44}$$

in which B_1, a constant independent of r, is

$$B_1 = -\frac{\kappa^2 ze \exp(\kappa a)}{4\pi(1 + \kappa a)}$$

The charge on the central ion is its valence z (± 1, ± 2, etc.) times e. In cgs units, Equation 8-44 gives the charge density in esu cm^{-3}.

There are two ways to use the distribution function ρ to provide a physical interpretation of the Debye radius, $1/\kappa$; both of these are illustrated in Figure 8-10. First, we plot the ratio of ρ to its maximum (absolute) value

$$\rho_{rel} = \rho/\rho_{max}$$

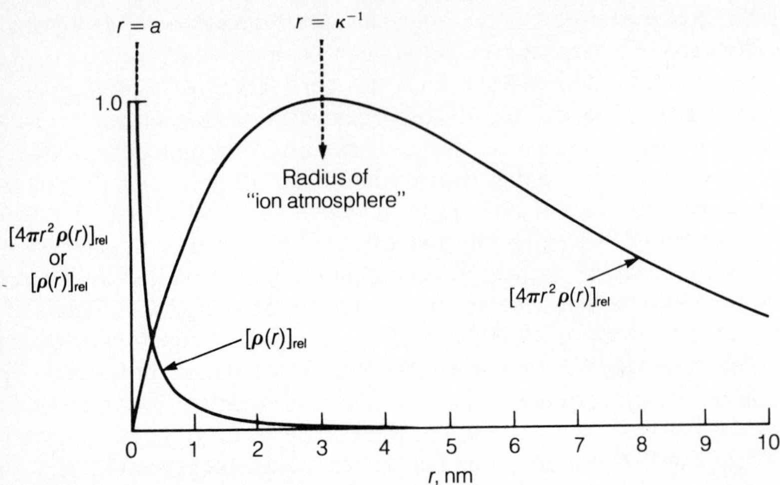

FIGURE 8-10
Two functions that measure the distribution of counterions at a distance r from a central ion. $\rho(r)$ is the time-averaged excess number of ions of one charge over those of the other charge per unit volume. $[\rho(r)]_{rel}$ is the value of this charge density relative to its maximum (absolute) value at the distance $r = a$ of closest approach between ion and counterion. The function $4\pi r^2\rho(r)$ is called the *radial distribution function,* and is shown relative to its maximum (absolute) value, which occurs at $r = \kappa^{-1}$, the Debye radius, or radius of the ion atmosphere. The ionic strength used for the calculation is 0.01, so $\kappa^{-1} = 3.04$ nm at 25°C. The ion radius a was assumed to be 0.1 nm (= 1Å).

as a function of r. According to Equation 8-44, the maximum absolute value of ρ occurs when $r = a$, so $\rho_{rel} = 1$ when $r = a$. The counterion charge density decreases rapidly as r increases from $r = a$; the decay for larger values of r is dominated by the exponential term $exp\ (-\kappa r)$. This term is equal to $1/e$ when $r = 1/\kappa$.

A more direct physical interpretation of $1/\kappa$ results from using ρ to construct the *radial distribution function,* which we define as the excess counterion charge contained in a spherical shell around the central ion, divided by the thickness of the shell. The shell, a distance r from the central ion, is of thickness dr, and its volume is its area ($4\pi r^2$) times its thickness (dr), or $4\pi r^2 dr$. The charge contained in this shell is ρ times the volume, so the radial distribution function is $4\pi r^2\rho\ dr/dr = 4\pi r^2\rho$. Figure 8-10 shows the radial distribution function relative to its maximum (absolute) value,

$$(4\pi r^2\rho)_{rel} = \frac{4\pi r^2\rho}{(4\pi r^2\rho)_{max}}$$

as a function of r. As shown in the figure, the *maximum of the relative radial distribution function occurs when r is equal to the Debye radius* $1/\kappa$ (see Problem 8-16). Because the radial distribution function measures the probability of finding a counterion at a distance r from the central ion, a simple physical meaning of the Debye radius is *the most probable distance from the central ion for finding a counterion.*

TABLE 8-4 *Debye Length*[a] *in Aqueous Solutions at 25°C*

SALT CONCENTRATION MOLAL	TYPE OF ELECTROLYTE[b] 1:1	1:2	2:2
0.0001	30.4	17.6	15.2
0.001	9.6	5.55	4.81
0.01	3.04	1.76	1.52
0.1	0.96	0.55	0.48

[a] Debye lengths κ^{-1} are given in nm: 1 nm = 10 Å.
[b] The type of electrolyte specifies the number of charges on the two ions in the electrolyte. For example, NaCl is 1:1, $CuCl_2$ is 1:2, and $CuSO_4$ is 2:2.

EXERCISE 8-9

Calculate the Debye radius in a 0.01 *M* solution of $MgCl_2$ at 37°C.

ANSWER

$I \cong \frac{1}{2}(0.01 \times 4 + 0.02 \times 1) = 0.03$. From Equation 8-43,

$$\kappa = \left[\frac{8\pi \times 6.02 \times 10^{23} \times (4.8032 \times 10^{-10}\ \text{esu})^2}{1000 \times 74.2 \times 1.381 \times 10^{-16}\ \text{ergs K}^{-1} \times 310\ \text{K}}\right]^{1/2} (I)^{1/2}$$
$$= 3.316 \times 10^{7}(I)^{1/2} = 5.74 \times 10^{6}\ \text{cm}^{-1}$$

Therefore the Debye radius, $1/\kappa$, is 1.74×10^{-7} cm or 1.74 nm. (In cgs units the elementary charge is 4.8032×10^{-10} esu, and Coulomb's law gives the energy in ergs per molecule; hence $k_B T$ also must be in ergs.)

Table 8-4 shows values of the Debye radius in various salt concentrations up to the limit of applicability of the Debye-Hückel theory (concentrations of univalent electrolytes below 0.1 *M*, less for multivalent electrolytes). Notice that the Debye radius depends only on ionic strength, not explicitly on ionic charge. Hence in a solution of $MgCl_2$, the ionic atmosphere is of equal thickness around Mg^{2+} and Cl^-.

The Electrostatic Potential around an Ion is Screened by Counterions

The second important result of the Debye-Hückel theory is the influence of the ionic atmosphere on the electrostatic potential Φ around the central ion. The mathematical solution is

$$\Phi = B_2 \frac{ze}{\epsilon r} \exp(-\kappa r) \tag{8-45}$$

in which B_2 is another constant independent of r,

$$B_2 = \frac{\exp(\kappa a)}{1 + \kappa a} \tag{8-46}$$

and ze is the charge on the central ion. In the simple case of a very dilute solution for which $\kappa a \ll 1$, B_2 is approximately 1. Remembering that the potential Φ_{ion} from an isolated central ion (see Exercise 8-1) is

$$\Phi_{\text{ion}} = \frac{ze}{\epsilon r} \tag{8-47}$$

we find that the potential in the presence of the ion atmosphere (Equation 8-45), with $B_2 = 1$, is

$$\Phi = \frac{ze}{\epsilon r} e^{-\kappa r} = \Phi_{\text{ion}} e^{-\kappa r} \tag{8-48}$$

Equation 8-48 is an example of a *screened Coulomb potential,* in which the ordinary Coulomb's law potential is reduced by the *screening factor* $e^{-\kappa r}$. In essence, the accumulation of counterions around the central ion screens nearby charges from the electrostatic influence of the central ion. Notice that the Debye radius is the distance at which the screening factor is reduced to $1/e$ of its value at $r = 0$.

Screening of the charges on macromolecules by the counterions in aqueous solution is of great importance for the stability of biologically significant structures. For example, the negatively charged phosphate residues on different strands of DNA are separated from each other by about 10 Å. In the absence of positive counterions to screen these charges from each other, the DNA double helix is unstable. Even a small amount of added salt ($I \approx 10^{-4}$) provides sufficient screening of the repulsive electrostatic forces to stabilize the helix. However, quantitative treatment of the electrostatic potential of a *polyelectrolyte* such as DNA is beyond the scope of the simple Debye-Hückel theory.

8–11 Calculation of Ionic Activity Coefficients by the Debye-Hückel Theory

An important result of the Debye-Hückel theory is the calculation of the electrostatic free energy of the solution. In order to perform the calculation we first need to break down the electrostatic potential into the components contributed by the central ion, Φ_{ion}, and by the ion atmosphere, $\Phi_{\text{atmosphere}}$. Recall that the electrostatic potential is the energy of a unit test charge at a point in space. Because energy is an extensive quantity, Φ is the sum of the two components:

$$\Phi = \Phi_{\text{ion}} + \Phi_{\text{atmosphere}} \tag{8-49}$$

The central ion of charge q_{ion} interacts with the test charge by Coulomb's law, or

$$\Phi_{\text{ion}} = \frac{q_{\text{ion}}}{\epsilon r}$$

Hence, with Φ given by Equation 8-48, we get from Equation 8-49

$$\begin{aligned}\Phi_{\text{atmosphere}} &= \Phi - \Phi_{\text{ion}} \\ &= \Phi_{\text{ion}} e^{-\kappa r} - \Phi_{\text{ion}} \\ &= \Phi_{\text{ion}}(e^{-\kappa r} - 1) \\ &= \frac{q_{\text{ion}}}{\epsilon r}(e^{-\kappa r} - 1)\end{aligned} \tag{8-50}$$

This expression gives the electrostatic potential due to the ion atmosphere around a central ion of charge q_{ion}.

Calculation of the Electrostatic Free Energy of the Solution

Our objective now is to calculate the free-energy difference between a solution of ions and a solution of uncharged particles at the same concentration. Imagine that we first place the central ion in the solution, but that it is *uncharged* and therefore has no electrostatic influence on the solution. The electrostatic free energy due to the charge on the central ion is the integral of the electrostatic work dW_{el} ($= dW_{\text{other, rev}}$) required to add the charge gradually to the central ion:

$$G_{\text{el}} = \int dW_{\text{el}} \tag{8-51}$$

Since Coulomb's law does not consider that the interaction of a charge with itself contributes to the electrostatic potential, the electrostatic work is the result of the interaction of the added charge (dq_{ion}) with the electrostatic potential contributed by the ion atmosphere, evaluated at the distance $r = a$ for which the central ion experiences the potential of its atmosphere:

$$dW_{\text{el}} = \Phi_{\text{atmosphere}}(r = a)\, dq_{\text{ion}} \tag{8-52}$$

Therefore the electrostatic free energy is, by Equation 8-51

$$G_{\text{el}} = \int_0^{ze} \Phi_{\text{atmosphere}}\,(r = a)\, dq_{\text{ion}} \tag{8-53}$$

The upper limit on this integral is the final charge on the central ion.

The integral in Equation 8-53 is easy to evaluate if we linearize the expression for $\Phi_{\text{atmosphere}}$ in Equation 8-50. At $r = a$, Equation 8-50 gives

$$\Phi_{\text{atmosphere}} = \frac{q_{\text{ion}}}{\epsilon a}(e^{-\kappa a} - 1) \tag{8-54a}$$

Assuming that ionic strength is low enough so that $\kappa a \ll 1$, we can use the expansion

$$e^{-x} = 1 - x + \frac{x^2}{2} - \cdots$$

Neglecting terms in $(\kappa a)^2$ and higher powers, we get $e^{-\kappa a} = 1 - \kappa a$.

Therefore

$$\Phi_{\text{atmosphere}} = \frac{q_{\text{ion}}}{\epsilon a}(1 - \kappa a - 1)$$
$$= -\frac{q_{\text{ion}}\kappa}{\epsilon} \qquad (8\text{-}54b)$$

Substitution of this expression into Equation 8-53 yields (with κ and ϵ constants)

$$G_{\text{el}} = -\frac{\kappa}{\epsilon}\int_0^{ze} q_{\text{ion}}\, dq_{\text{ion}}$$
$$= -\frac{\kappa(ze)^2}{2\epsilon} \qquad (8\text{-}55)$$

Equation 8-55 gives the free energy contributed to the solution because of the charge on a single ion. Each ion in the solution makes a similar contribution to the total electrostatic free energy. Notice that there is no ion atmosphere around the particle when it is imagined to be present uncharged in the solution. Hence the electrostatic free energy (G_{el}) due to each ion includes the work necessary to form the ion atmosphere. Work is required for this process because of the repulsion between the similarly charged counterions. In addition, there is a work term due to interaction of the ion with its atmosphere. Hence G_{el} can be expressed as the sum of the free energy of formation of the ion atmosphere ($G_{\text{atmosphere}}$) plus the free energy due to interaction of the central ion with a preformed ion atmosphere ($G_{\text{ion-atmosphere}}$):

$$G_{\text{el}} = G_{\text{atmosphere}} + G_{\text{ion-atmosphere}} \qquad (8\text{-}56a)$$

The free energy due to interaction of a central ion of charge ze with a preformed ion atmosphere is, by the definition of the electrostatic potential,

$$G_{\text{ion-atmosphere}} = ze\, \Phi_{\text{atmosphere}}(r = a)$$

which, with the help of Equation 8-54b for $\Phi_{\text{atmosphere}}$ is

$$G_{\text{ion-atmosphere}} = -\frac{\kappa(ze)^2}{\epsilon} \qquad (8\text{-}56b)$$

Combining this result with Equation 8-56a and Equation 8-55 for G_{el} yields

$$G_{\text{atmosphere}} = G_{\text{el}} - G_{\text{ion-atmosphere}} = \frac{-\kappa(ze)^2}{2\epsilon} + \frac{\kappa(ze)^2}{\epsilon}$$
$$= \frac{\kappa(ze)^2}{2\epsilon} \qquad (8\text{-}56c)$$

Hence, we see that G_{el} is made up of a *favorable* (negative) contribution, $-\kappa(ze)^2/\epsilon$, due to interaction of the central ion with a preformed ion atmosphere, and an *unfavorable* (positive) contribution, $\kappa(ze)^2/2\,\epsilon$, which represents the electrostatic work required to form the ion atmosphere. Since the negative term is twice as large in absolute value as the positive term, G_{el} is negative. Because G_{el} is negative, the net interaction between ions is favorable, and we expect the solution activity coefficient to become smaller than 1 as the solution concentration increases.

The Activity Coefficient of an Ion is Related to G_{el}

Now we are able to calculate the activity coefficient. The partial molar free energy of the positive ion is given by

$$\mu = \mu^\circ + RT \ln \gamma_+ + RT \ln C$$

with a corresponding equation for the negative partner. The free energy *per ion* is μ/N_A, or

$$\frac{\mu}{N_A} = \frac{\mu^\circ}{N_A} + k_B T \ln \gamma_+ + k_B T \ln C$$

The nonideality effects are contained in the term $k_B T \ln \gamma_+$. If we assume that electrostatic free energy is the only source of nonideality, then we can equate the electrostatic free energy per ion (G_{el}) with $k_B T \ln \gamma_+$.

$$G_{el} = k_B T \ln \gamma_+$$

from which we conclude, by substituting Equation 8-55 for G_{el} and rearranging, that

$$\ln \gamma_+ = \frac{-\kappa(z_+ e)^2}{2\epsilon k_B T} \qquad \text{(8-57}a\text{)}$$

in which z_+ is the valence of the (positive) central ion. A similar equation applies to the negative ion.

Because we cannot study ions of a single charge we cannot determine γ_+. The only measurable quantity is the mean activity coefficient $\gamma_\pm$. Equation 8-39 reveals that $\ln \gamma_\pm$ is an average of $\ln \gamma_+$ and $\ln \gamma_-$:

$$\ln \gamma_\pm = \frac{\nu_+}{\nu} \ln \gamma_+ + \frac{\nu_-}{\nu} \ln \gamma_-$$

or, substituting Equation 8-57*a* for $\ln \gamma_+$ and $\ln \gamma_-$,

$$\ln \gamma_\pm = -\left(\frac{\nu_+ z_+^2 + \nu_- z_-^2}{\nu_+ + \nu_-} \right) \frac{e^2 \kappa}{2\epsilon k_B T} \qquad \text{(8-57}b\text{)}$$

The quantity z_+ is positive, and z_- is negative. Because the salt is electrically neutral, the number of ions times the charge must be equal for positive and negative ions:

$$z_+ \nu_+ = -z_- \nu_-$$

Therefore the numerator in Equation 8-57*b* can be rewritten

$$\begin{aligned} \nu_+ z_+ z_+ + \nu_- z_- z_- &= -(\nu_- z_- z_+ + \nu_+ z_+ z_-) \\ &= -z_- z_+ (\nu_+ + \nu_-) \end{aligned}$$

With this substitution for $\nu_+ z_+^2 + \nu_- z_-^2$ in Equation 8-57*b*, we obtain the final result

$$\ln \gamma_\pm = -|z_+ z_-| \frac{e^2 \kappa}{2\epsilon k_B T} \qquad \text{(8-58)} ◀$$

in which we place absolute-value bars around z_+z_- to avoid confusion over the sign. This is an important equation because it allows us to calculate a thermodynamic property, the *mean ionic activity coefficient,* from molecular properties. Notice that $\ln \gamma_\pm \leqslant 0$, so $\gamma_\pm \leqslant 1$.

EXERCISE 8-10

Use the Debye-Hückel theory to calculate the mean activity coefficient of a 0.01 *M* solution of $MgCl_2$ at 37°C.

ANSWER

From Exercise 8-9, $\kappa = 5.74 \times 10^6\ \text{cm}^{-1}$. By Equation 8-58,

$$\ln \gamma_\pm = \frac{-(2 \times 1)(4.8032 \times 10^{-10}\ \text{esu})^2 \times 5.74 \times 10^6\ \text{cm}^{-1}}{2 \times 74.2 \times 1.381 \times 10^{-16}\ \text{ergs K}^{-1} \times 310\ \text{K}} = -0.417$$

Therefore, $\gamma_\pm = 0.659$. Clearly, a solution of $MgCl_2$ in 0.03 ionic strength is significantly removed from ideality. The ionic strength in living organisms generally is even higher than 0.03.

Validity of the Limiting Law

We note that the Debye-Hückel theory always predicts $\gamma_\pm \leq 1$. This occurs because the interaction between an ion and its ionic atmosphere is always favorable, so the solute-solute interactions that appear as concentration (or ionic strength) increases cause association of the solute particles. (As discussed in Section 7-11, association implies that $\gamma < 1$.)

The numerical constants in Equation 8-58 can be combined to yield the *Debye-Hückel limiting law* at 25°C:

$$\log_{10}\gamma_\pm = -0.51\left|z_+z_-\right|I^{1/2} \qquad (8\text{-}59)$$

The constant 0.51, which is valid at 25°C, varies with $(\epsilon T)^{-3/2}$. In a mixed solution of electrolytes, z_+ and z_- refer to the component of interest, say $MgCl_2$, but I is the sum of the contributions of all components. Equation 8-59 has been tested against experimental activity coefficients (obtained, for example, from freezing-point depression) and found to be correct in the limit of low ionic strength. Figure 8-11 shows the experimental variation of the activity coefficient of NaCl and KCl, plotted as a function of the square root of concentration. The variation of the equivalent conductance with the square root of concentration (Figure 8-8) is also predicted by the Debye-Hückel theory.

The properties of the solvent enter the Debye-Hückel theory primarily through the dielectric constant. Examination of Equations 8-43 and 8-58 shows that $\ln \gamma_\pm$ is proportional to $(I/\epsilon^3)^{1/2}$. Consequently, the value of I required to produce a given extent of nonideality as measured by γ varies with ϵ^3. We conclude that solvents with dielectric constants only a little lower than water will show significantly greater ion association because of the third

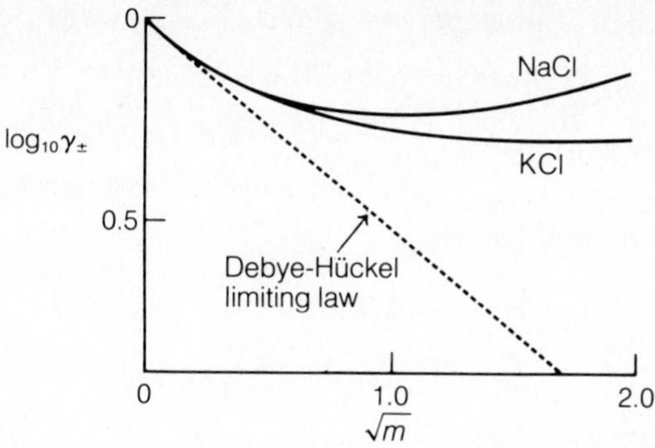

FIGURE 8-11

Experimental activity coefficients of NaCl and KCl at 25°C, plotted as $\log_{10}\gamma_\pm$ versus $\sqrt{m}$ (m is the molal concentration). The dashed line is the Debye-Hückel limiting law, Equation 8-59. Notice that the experimental values begin to deviate significantly from the limiting law around $0.05m$ ($\sqrt{0.05} = 0.22$). Significant deviations begin at even lower concentrations for multivalent electrolytes. (Data from H. S. Harned and B. B. Owen, 1958.)

power dependence on ϵ. Materials with low dielectric constants, such as the interior of a lipid bilayer, yield ion association even at very low ion concentrations.

QUESTIONS FOR REVIEW

1. State Coulomb's law for both cgs and SI units.
2. Define the following: *electrostatic potential, electrostatic potential energy,* and *electric field.* Give the units of each. How are these quantities interrelated?
3. Define *current* and *resistance,* and give their units. State Ohm's law.
4. Define *conductivity, equivalent conductance,* and *mobility* of an ion.
5. How do expressions for the chemical potentials of ionic substances differ from those for nonionic substances?
6. ● Explain the significance of *Debye length, ionic strength,* and *ion atmosphere.* Outline the steps in deriving the Debye-Hückel limiting law. What are the limits of validity of the law?

PROBLEMS

1. Are the following statements true or false?
   (a) The electric field of a charge diminishes by the square of the distance from the charge.
   (b) The electrostatic potential of a charge is independent of distance from the charge.
   (c) The permittivity of a vacuum is unity.

(d) Conductivity is current per unit area per unit potential.

(e) The screening factor in the Debye-Hückel theory comes from counter-ions.

(f) The contribution of an ion to the ionic strength is proportional to its molarity.

(g) If a free-energy change depends on temperature, it must include a contribution from entropy change.

Answers: (a) T. (b) F. (c) T in cgs units, F in SI units. (d) F. (e) T. (f) F, to its molality. (g) T.

2. Add the appropriate word:

(a) The electrostatic potential Φ is the electrostatic potential energy experienced by a unit ________ charge.

(b) The dielectric constant of water ________ with increasing temperature.

(c) The mobility of an ion is its ________ in a unit electric field.

(d) The ionic strength of a solution is one-half the sum over ions of the product of ________ and the charge squared.

(e) The Debye-Hückel limiting law relates $\gamma_\pm$ to the________ of charging an ion and forming its atmosphere.

(f) Ions in media of lower dielectric constant will be ________(more, less) associated than in water.

Answers: (a) Positive. (b) Decreases. (c) Velocity. (d) Molality. (e) Free energy. (f) More.

PROBLEMS RELATED TO EXERCISES

3. (Exercise 8-1) Express Equations 8-6*b* and 8-6*c* in cgs units.

4. (Exercise 8-2) Calculate the energy of interaction between Mg^{2+} and SO_4^{2-} separated by 1 nm in vacuum.

5. (Exercise 8-3) Calculate the energy of interaction between Mg^{2+} and SO_4^{2-} separated by 1 nm in water at 37°C ($\epsilon = 74.2$).

6. (Exercise 8-5) Calculate the electrostatic free energy, entropy, and enthalpy of interaction between Mg^{2+} and SO_4^{2-}, separated by a distance of 1 nm in a medium of dielectric constant 74.2 (water at 37°C). Use $(\partial \ln \epsilon/\partial T)_p = -4.64 \times 10^{-3}\ K^{-1}$.

7. (Exercise 8-6) Calculate the equivalent ionic conductance Λ_0 of H^+ from its mobility at 25°C, $u_0 = 36.25 \times 10^{-4}$ cm^2 volt^{-1} sec^{-1}.

8. (Exercise 8-7) Write an expression for the chemical potential of $AlCl_3$ in terms of the mean (molar) activity coefficient $\gamma_\pm$ and the molar concentration C_B of $AlCl_3$.

9. (Exercise 8-8) Calculate the ionic strength of a 0.01 *M* solution of $Fe_2(SO_4)_3$.

10. (Exercise 8-9) Calculate the Debye radius in a solution of 0.01 *M* $Fe_2(SO_4)_3$ at 37°C.

● 11. (Exercise 8-10) Use the Debye-Hückel theory to calculate the mean activity coefficient of a 0.01 *M* solution of $Fe_2(SO_4)_3$ at 37°C.

OTHER PROBLEMS

● 12. Show that the electrostatic enthalpy of two ions interacting in solution is

$$H_{el} = G_{el}\left(1 + \frac{\partial \ln \epsilon}{\partial \ln T}\right)_P$$

Calculate H_{el} for Mg^{++} and $SO_4^{=}$ ions at 3Å distance in water at 37°C, using $\partial \ln \epsilon/\partial \ln T = -1.44$ and $\epsilon = 74.2$. What *qualitative* contribution does ordering of the solvent make to H_{el} and S_{el}?

13. Calculate the equivalent conductance of HCl at infinite dilution, using $\Lambda_+^\circ\,(H^+) = 349.8$ and $\Lambda_-^\circ\,(Cl^-) = 76.35$ ohm^{-1} cm^2 equivalent^{-1}. Calculate the mobility of H^+ and the average velocity under a field of 100 volts/cm.

● 14. Write an expression for the chemical potential of $MgCl_2$ in nonideal solution. Define the mean activity coefficient for $MgCl_2$ in terms of the separate ion activity coefficients.

15. Calculate the ionic strength of a solution containing 0.1 *M* NaCl and 0.01 *M* $MgCl_2$.

16. Given the result from the Debye-Hückel theory that

$$\rho = B_1 \frac{e^{-\kappa r}}{r},$$

prove that the maximum of the radial distribution function $4\pi r^2 \rho$ occurs when $\kappa^{-1} = r$.

17. (a) Calculate the Debye radius in the solution of Problem 15 at $T = 37°C$.
(b) Assume a screened Coulomb potential and calculate the screening factor by which Φ_{ion} is reduced in the equation $\Phi = \Phi_{ion} e^{-\kappa r}$, using the solution conditions of Problem 15 and $r = 10Å$.

● 18. Calculate the electrostatic free energy, entropy, and enthalpy *per mole* due to the electrostatic nonideality in a solution of 0.05 *M* NaCl at 37°C. (*Hint:* Use the theoretical expression for $\ln \gamma_\pm$ and the approach of Problem 12.)

19. Show that the equivalent ionic conductance of an ion is related to its mobility by $\Lambda_{(+)} = u(+)F$. (*Hint:* Consider a negative electrode of cross-sectional area *A*. All the positive ions within the distance $\Delta x = v\,\Delta t$ of the electrode will strike it in time interval Δt if they move at velocity v. Calculate the current in moles of electrons per second due to motion of the positive ions whose equivalent concentration is C' in the volume element $A\ \Delta x$. Divide this current by $A\ E^*\ C'$ to get the equivalent ionic conductance.)

20. Explain why the ionic mobility of $CH_3(CH_2)_2COO^-$ is smaller than that of Cl^-.

21. Explain why the ionic mobility of $Fe(CN)_6^{4-}$ is larger than that of Cl^-.

22. Discuss the factors that need to be considered in comparing the mobilities of Na^+ and Mg^{2+}.

FURTHER READING

Kirkwood, J. G., and Oppenheim, I. 1961. *Chemical Thermodynamics*. New York: McGraw-Hill. An excellent general treatment of the thermodynamics of electrolyte solutions. For more advanced students.

Robinson, R. A., and Stokes, R. H. 1965. *Electrolyte Solutions,* 2nd ed. London: Butterworths. The general properties of electrolyte solutions.

Harned, H. S., and Owen, B. B. 1958. *The Physical Chemistry of Electrolyte Solutions,* 3rd ed. New York: Reinhold. A detailed and authoritative account of the properties of electrolyte solutions.

OTHER REFERENCES

Dole, M. 1935. *Principles of Experimental and Theoretical Electrochemistry*. New York and London: McGraw-Hill, p. 70.

Dzidic, I., and Kebarle, P. 1970. *J. Phys. Chem. 74,* 1466.

CHAPTER 9

Electrochemical Equilibria

Differences in electrostatic potential are of enormous importance in chemistry and biology. An energy change results when any charged particle is transferred from one electrostatic potential to another, because the potential energy is given by the product of charge and electrostatic potential (see Equation 8-5). In this chapter we will consider the free-energy consequences of transferring electrons and ions from one potential to another.

Electron transfer results from oxidation-reduction reactions. The strength of a reductant can be measured by the relative electrostatic potential of the electrons in equilibrium with the other reactants and products. The important biochemical application of this subject is to the oxidation-reduction reactions that lie at the heart of biochemical metabolism.

Change in the electrostatic potential contributes to the free-energy change when an ion is transferred from one compartment to another. Variation in the electrostatic potential within organisms is of enormous importance to biology. For example, the function of the human brain is totally dependent on both transient and sustained electrostatic potential differences. In the later sections of this chapter we will focus on the fundamental equilibrium thermodynamic description of systems in which an electrostatic potential difference accompanies unequal ion concentrations on opposite sides of a membrane which is permeable to the ion.

○ 9–1 *Oxidation-Reduction Reactions and Electrochemical Cells*

Oxidation-reduction reactions are abundant in nature, both in biological and nonbiological systems. By definition, an oxidation-reduction reaction involves a change in oxidation state, and therefore may be regarded as a transfer of electrons from one compound to another. A simple example is the reduction of Fe^{3+} in cytochrome *c* (see Fig. 9-1) by Fe^{2+} in cytochrome *b* in the mitochondrial electron transport chain:

$$Fe^{3+} \text{ (cyt } c) + Fe^{2+} \text{ (cyt } b) \rightarrow Fe^{2+} \text{ (cyt } c) + Fe^{3+} \text{ (cyt } b) \qquad (9\text{-}1)$$

In this reaction, Fe^{2+} (cyt *b*) is the electron donor, reductant, or reducing agent:

$$Fe^{2+}(\text{cyt } b) \xrightarrow{\text{oxidation}} Fe^{3+}(\text{cyt } b) + e^- \qquad (9\text{-}2)$$

and Fe^{3+}(cyt *c*) is the electron acceptor, oxidant, or oxidizing agent:

$$Fe^{3+}(\text{cyt } c) + e^- \xrightarrow{\text{reduction}} Fe^{2+}(\text{cyt } c) \qquad (9\text{-}3)$$

By definition, we say that the reductant becomes oxidized (loses electrons) and the oxidant becomes reduced (gains electrons) in the oxidation-reduction reaction. The sum of the two half-reactions 9-2 and 9-3 is the complete reaction, 9-1.

Many oxidation-reduction reactions occur by simple electron transfer from one reactant to another, such as from cytochrome *b* to cytochrome *c* in solution. In other cases the electron may be removed in the company of another group, such as in a hydrogen atom. A number of biochemical oxidations proceed in this way, with the extracted hydrogen transferred to the coenzyme NAD^+, thereby reducing it to NADH (see Sec. 9-5). Of special interest for general consideration are the cases in which the electron is transferred along a path which permits work to be done as a result of its passage.

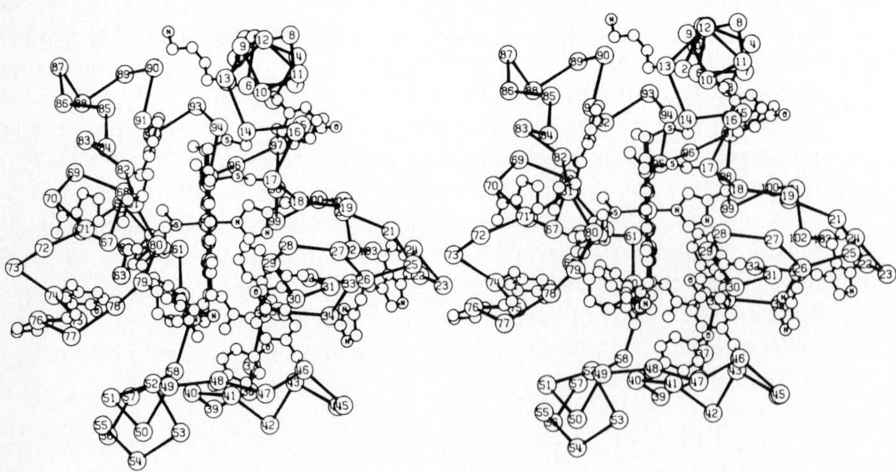

FIGURE 9-1
The structure of cytochrome *c*, shown by a stereopair drawing. Many people can view the molecule in stereo if they hold the page level at about 18 inches away and allow their eyes to drift apart. They see four images, and find they can superimpose the central two.

The iron porphyrin ring is seen edge on at the molecular center. Each amino acid is represented by a numbered sphere placed at the α-carbon position. Only a few amino acid side chains are shown, such as those that coordinate to the heme. (From Takano et al., 1977. *J. Biol. Chem. 252*, 776–785.)

An example is an *electrochemical cell* (a type of battery) in which electrons produced by a chemical reaction at the negative electrode can do electrical work during their transfer through an external circuit to the positive terminal. In the mitochondrial electron transport chain, the passage of electrons (for example from cytochrome *b* to cytochrome *c*) is used to do chemical work, namely the synthesis of ATP from ADP and inorganic phosphate. The analogy between these processes indicates that we can begin with a consideration of electrochemical cells to lay the foundation for understanding the thermodynamics of mitochondrial electron transport.

Electrochemical Cells

A simple electrochemical cell is diagramed in Figure 9-2. A zinc rod in contact with zinc sulfate solution is connected through a sensitive voltmeter to a copper rod immersed in copper sulfate solution. A salt bridge, for example a concentrated solution of KCl in an agar gel, completes the circuit. At the electrodes there is a tendency for electron release and capture to occur:

$$Zn^{2+} + 2e^- \rightleftarrows Zn^\circ$$
$$Cu^{2+} + 2e^- \rightleftarrows Cu^\circ$$

However, since the two electrodes are different, we expect that there will be a difference $\Delta\Phi$ in the potential energy of electrons in equilibrium with the oxidized metal ion and the reduced metal in the electrodes. We let $\Delta\Phi$,

$$\Delta\Phi = \Phi_{Cu} - \Phi_{Zn}$$

be the difference between the electrostatic potentials of the Cu and Zn electrodes. Neither of the potentials Φ_{Cu} and Φ_{Zn} can be measured absolutely,

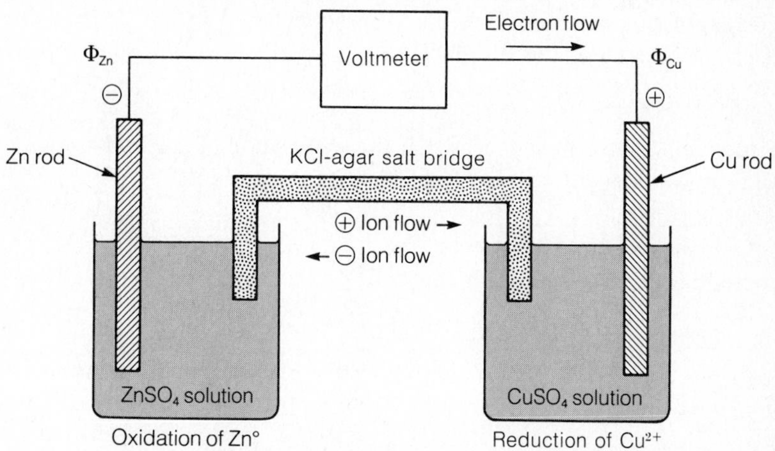

FIGURE 9-2
An electrochemical cell with shorthand notation and the corresponding cell reaction.

but their difference can be. In the example shown, the Cu electrode is more positive, so electrons flow from the negative Zn electrode (the anode) through the external circuit to the positive Cu electrode (the cathode). The cathode always is defined as the electrode to which electrons flow in the external circuit.

Conventions and Notation of Electrochemistry

Conventions are important in electrochemistry to avoid confusion over the sign of the potential difference or the spontaneous direction of the chemical reaction. There is no problem determining experimentally which electrode is positive—the voltmeter does this for us. Furthermore, it is not too difficult to remember the convention that negatively charged electrons flow in the external circuit toward the positive electrode. However, a convention is necessary to represent the cell in shorthand and to indicate by the sign of the potential the spontaneous direction of electron flow. The cell in Figure 9-2 is summarized by the notation

$$\mathrm{Zn}|\mathrm{Zn}^{++}||\mathrm{Cu}^{++}|\mathrm{Cu}$$

The two metal electrodes are at the extreme left and right of the diagram, separated by a vertical line (the phase boundary) from the solutions. The double line represents the salt bridge. By convention, the potential difference $\Delta\Phi$ is the potential of the right electrode minus that of the left electrode:

$$\Delta\Phi = \Phi_R - \Phi_L$$

As you can see by examining Figure 9-2, a *positive difference* $\Delta\Phi$ *means that electrons flow from the left to the right and therefore that oxidation occurs on the left and reduction on the right.* Hence we can specify that if $\Delta\Phi$ is positive, the corresponding reaction with electrons consumed at the Cu (right) electrode and produced at the Zn electrode

$$\mathrm{Zn} + \mathrm{Cu}^{2+}\ (1m) \rightarrow \mathrm{Cu} + \mathrm{Zn}^{2+}\ (1m)$$

is spontaneous. We will see later that the magnitude of $\Delta\Phi$ and even its sign depend on the concentrations of reactants and products.

EXERCISE 9-1

The potential difference $\Delta\Phi$ for the cell

$$\mathrm{Pt},\ \mathrm{H_2}(g,\ 1\ \mathrm{atm})|\mathrm{H}^+\ (1m)||\mathrm{Zn}^{2+}\ (1m)|\mathrm{Zn}$$

is measured and found to be negative. Write the chemical reaction that occurs *spontaneously*.

ANSWER

In this reaction Pt is an inert electrode, as usual. Since $\Delta\Phi$ is negative, electrons flow through the external circuit from right to left, or from Zn to Pt. Hence we want a reduction reaction to occur on the left, with an oxidation reaction on the right. Therefore

$$\begin{array}{ll} 2H^+ \,(1m) + 2e^- \rightarrow H_2 \,(g,\, 1 \text{ atm}) & \text{(reduction)} \\ Zn \rightarrow Zn^{2+} \,(1m) + 2e^- & \text{(oxidation)} \\ \hline 2H^+ \,(1m) + Zn \rightarrow H_2 \,(g,\, 1 \text{ atm}) + Zn^{2+} \,(1m) & \text{(spontaneous reaction)} \end{array}$$

9–2 *The Electromotive Force and Free-Energy Change of Electrochemical Cells*

The potential difference between two electrodes depends on how much current is flowing. The maximum potential occurs when the process is carried out reversibly, which means that the current is negligible. In the limit as current i goes to zero, $\Delta\Phi$ is called the *electromotive force* (emf), $\mathscr{E}$, of the cell:

$$\lim_{i \to 0} \Delta\Phi = \mathscr{E} \tag{9-4}$$

The free-energy change of the electrochemical cell reaction is related to the electrical work done *on* the system (the chemical reaction constitutes the system) by the equation $\Delta G = W_{\text{rev, other}} = W_{\text{el}}$. The electrochemical cell does electrical work $-W_{\text{el}}$ which is equal to the product of the charge transferred reversibly (nF coulombs) times the potential difference ($\mathscr{E}$). (n is the stoichiometric coefficient of the electrons in the oxidation reduction reaction. For example, $n = 2$ in the Zn-Cu cell of Figure 9-2.) Hence, with $-W_{\text{el}} = nF\mathscr{E}$, we can write

$$\Delta G = -nF\mathscr{E}$$

Since ΔG is negative for a spontaneous process, and nF is positive, it is clear that our convention must be that *$\mathscr{E}$ is positive for a spontaneous process.*

EXERCISE 9-2

Is $\mathscr{E}$ positive or negative for the reaction

$$H_2 \,(g,\, 1 \text{ atm}) + Zn^{2+} \,(1m) \rightarrow Zn + 2H^+(1m)$$

corresponding to the cell diagram in Exercise 9-1?

ANSWER

Negative. Notice that if the cell and its reaction are written in the opposite (spontaneous) direction, $\mathscr{E}$ is positive.

The Nernst Equation Relates emf to Concentrations in the Electrochemical Cell

$\mathscr{E}$ for an electrochemical cell is a readily measurable quantity which is related to the free-energy change by a simple equation. Therefore, it is not surprising that many reaction free energies have been determined by electrochemical methods. We begin with Equation 4-55, which states that the free-energy change ΔG is given by

$$\Delta G = \Delta G^\circ + RT \ln \left[\frac{\prod_i C_i^{\nu_i} \text{ (products)}}{\prod_i C_i^{\nu_i} \text{ (reactants)}} \right] \tag{9-5}$$

(When the solutions are not dilute, activities a_i should replace concentrations in the ideal expression, Equation 9-5.) Equating ΔG to $-n F \mathscr{E}$ and rearranging, we get

$$\mathscr{E} = \frac{-\Delta G^\circ}{nF} - \frac{RT}{nF} \ln \left[\frac{\prod_i C_i^{\nu_i} \text{ (products)}}{\prod_i C_i^{\nu_i} \text{ (reactants)}} \right] \tag{9-6}$$

In the standard state the concentrations are unity, and we have for the emf in the standard state ($\mathscr{E}^\circ$)

$$\mathscr{E}^\circ = \frac{-\Delta G^\circ}{nF} \tag{9-7}$$

Substituting this relationship into Equation 9-6, we obtain

$$\mathscr{E} = \mathscr{E}^\circ - \frac{RT}{nF} \ln \left[\frac{\prod_i C_i^{\nu_i} \text{ (products)}}{\prod_i C_i^{\nu_i} \text{ (reactants)}} \right] \tag{9-8a}$$ ◀

or

$$\mathscr{E} = \mathscr{E}^\circ - \frac{2.303RT}{nF} \log_{10} \left[\frac{\prod_i C_i^{\nu_i} \text{ (products)}}{\prod_i C_i^{\nu_i} \text{ (reactants)}} \right] \tag{9-8b}$$ ◀

which is called the *Nernst equation.* It tells us how the emf varies with concentration of products and reactants. By convention, the concentrations in this equation are expressed in molal units, so the hypothetical standard state for measuring and tabulating $\mathscr{E}^\circ$ is 1 molal concentration of reactants and products. At 25°C the Nernst equation constant $2.303RT/F$ is 0.059 volts. When the solution is not dilute, activities a_i should replace the concentrations C_i.

EXERCISE 9-3

Calculate $\mathscr{E}$ and $\mathscr{E}^\circ$ for the following cell at 25°C, assuming ideal solutions. The double line represents a KCl salt bridge.

$$\text{Zn} \mid \text{Zn}^{++}_{(\text{ZnSO}_4)}\ (10^{-5}M) \parallel \text{Zn}^{++}_{(\text{ZnSO}_4)}\ (10^{-4}M) \mid \text{Zn}$$

ANSWER

The reaction is the same on both sides. Only the concentration of Zn^{2+} differs. Therefore $\mathscr{E}^\circ = 0$. According to our convention, we write the cell reaction with oxidation at the left electrode:

$$Zn \rightarrow Zn^{++}\ (10^{-5}M) + 2e^-$$

and reduction at the right:

$$Zn^{2+}\ (10^{-4}M) + 2e^- \rightarrow Zn$$

so the net reaction is

$$Zn^{2+}\ (10^{-4}M) \rightarrow Zn^{++}\ (10^{-5}M)$$

and the Nernst equation gives us

$$\mathscr{E} = \mathscr{E}^\circ - \frac{RT}{nF} \ln \left(\frac{10^{-5}M}{10^{-4}M}\right)$$

or

$$\mathscr{E} = 0 - \frac{8.3143 \text{ J mole}^{-1}\text{K}^{-1} \times 298 \text{ K}}{2 \times 96{,}487 \text{ coulombs mole}^{-1}} \ln (0.1)$$

$$= \frac{-0.059}{2} \log_{10} (0.1) = 0.0296 \text{ volts}$$

The potential difference, 29.6 millivolts, results entirely from the difference in concentration on the two sides of the cell. $\Delta G = -2F\mathscr{E}$ is the free energy of transferring a Zn^{++} from $10^{-4}M$ to $10^{-5}M$ concentration.[1]

A useful relationship that results from the Nernst equation is based on Equation 4-53, $\Delta G^\circ = -RT \ln K$. Substituting $\Delta G^\circ = -nF\mathscr{E}^\circ$, we obtain

$$\mathscr{E}^\circ = \frac{RT}{nF} \ln K \tag{9-9a}$$ ◀

$$K = \exp \left(\frac{nF\mathscr{E}^\circ}{RT}\right) \tag{9-9b}$$ ◀

which are very important equations for the experimental determination of equilibrium constants.

EXERCISE 9-4

$\mathscr{E}^\circ$ for the reaction between cytochromes,

$$Fe^{3+} \text{ (cyt } c) + Fe^{2+} \text{ (cyt } b) \rightarrow Fe^{2+} \text{ (cyt } c) + Fe^{3+} \text{ (cyt } b)$$

is found to be 220 millivolts at 25°C. Calculate K and ΔG° for the reaction.

ANSWER

$K = \exp (nF\mathscr{E}^\circ/RT) = \exp (38.943\, n\mathscr{E}^\circ) = \exp (8.567) = 5.26 \times 10^3$
$\Delta G^\circ = -nF\mathscr{E}^\circ = -1 \times 96{,}487 \text{ coulombs mol}^{-1} \times 0.22 \text{ volts}$
$= -21.1 \text{ kJ mol}^{-1} = -5.1 \text{ kcal mol}^{-1}$

---

[1] No transfer of SO_4^{2-} occurs, so it does not appear in the expression for $\mathscr{E}$ or the free-energy change. Each Zn^{++} that disappears from the left compartment is replaced by $2K^+$ from the KCl salt bridge, and each Zn^{++} that appears in the right compartment is joined by $2Cl^-$ from the salt bridge. No significant electrical work accompanies this process as long as the K^+ and Cl^- fluxes are a negligible perturbation on the spontaneous diffusion of KCl from both ends of the salt bridge. This conclusion depends on the near equality of the mobilities of K^+ and Cl^- ions.

EXERCISE 9-5

The emf of the following cell (with a KCl salt bridge)

$$Cu|CuSO_4(10^{-5}M)||CuSO_4(0.1M)|Cu$$

is found to be 77.2 millivolts at 25°C. Assume the activity coefficient $\gamma = 1$ for $CuSO_4$ at $10^{-5}M$, and calculate the apparent activity coefficient γ for Cu^{++} at $0.01M$.

ANSWER

$$\mathscr{E} = \mathscr{E}^\circ - \frac{RT}{nF} \ln \left(\frac{10^{-5}}{\gamma \times 0.01} \right) \text{ and } \mathscr{E}^\circ = 0$$

$$0.0772 \text{ volt} = \frac{0.02568}{2} \ln (10^3 \gamma) \text{ volts}$$

Therefore $\gamma = \exp(6.012) \times 10^{-3} = 0.41$. (This activity coefficient is an apparent value because the experiment is complicated by the salt bridge, as discussed in the footnote on page 373.)

9–3 Half-Cells and Standard Electrode Potentials

It turns out to be unnecessary to tabulate emf values ($\mathscr{E}^\circ$) for all the electrochemical cells that have been studied, because if the potentials of both electrodes in a cell are known relative to some common reference, we can readily calculate the potential difference between them. Let us return to the Cu-Zn cell in Figure 9-2, calling $\mathscr{E}^\circ_{Zn}$ the difference between the potential of the Zn electrode under standard conditions and some reference potential

$$\mathscr{E}^\circ_{Zn} = \Phi^\circ_{Zn} - \Phi^\circ_{ref}$$

with a similar relationship for the other half of the cell:

$$\mathscr{E}^\circ_{Cu} = \Phi^\circ_{Cu} - \Phi^\circ_{ref}$$

The potential difference $\Delta\Phi$ of the whole cell is

$$\mathscr{E}^\circ = \Delta\Phi^\circ = \Phi^\circ_{Cu} - \Phi^\circ_{Zn} = \mathscr{E}^\circ_{Cu} - \mathscr{E}^\circ_{Zn} \tag{9-10}$$

Therefore we can calculate the potential $\mathscr{E}^\circ = \Delta\Phi^\circ$ of a cell by taking the difference between the standard half-cell potentials relative to some arbitrary reference potential, whose value does not appear in Equation 9-10.

The "arbitrary reference" has been chosen to be the standard hydrogen electrode. This consists of Pt metal in contact with both H_2 gas at one atmosphere pressure and an ideal solution containing H^+ ions at unit concentration. For example, to measure $\mathscr{E}^\circ$ for the Cu-$CuSO_4$ half-cell, we would set up the following cell:

$$\text{Pt, } H_2(g)\,|\,H^+\,||\,Cu^{2+}\,|\,Cu$$

The measured value of $\mathscr{E} = \Phi_{Cu} - \Phi_{H_2/H^+}$ (measured in the ideal dilute state and corrected for the ideal entropy of concentration to the standard state) is $\mathscr{E}^\circ_{Cu}$, the *standard electrode potential* of the Cu-$CuSO_4$ half-cell. The sign convention for standard electrode potentials is that *the sign of $\mathscr{E}^\circ$ is the experimental polarity of the electrode* (*under standard conditions*) *relative to the standard hydrogen electrode.*

The consequence of this convention is that if $\mathscr{E}^\circ$ is positive, the Cu half-cell accepts electrons from the hydrogen electrode:

$$Cu^{2+} + 2e^- \rightarrow Cu^\circ$$

Therefore, $\mathscr{E}^\circ$ values are tabulated as *standard reduction potentials,* with the understanding that *if $\mathscr{E}^\circ$ is positive, reduction is spontaneous under standard conditions when the half-cell is combined with a standard hydrogen electrode.*

EXERCISE 9-6

$\mathscr{E}^\circ$ for the Cu^{2+}/Cu electrode is 0.337 volt and $\mathscr{E}^\circ$ for the Zn^{2+}/Zn electrode is -0.763 volt. Write: (a) the reduction reaction for each of these half-cells, and (b) the spontaneous chemical reaction for each electrode under standard conditions when combined with the standard hydrogen electrode.

ANSWER

(a) $Cu^{++} + 2e^- \rightarrow Cu^\circ$, $\mathscr{E}^\circ = 0.337$ volts
$Zn^{++} + 2e^- \rightarrow Zn^\circ$, $\mathscr{E}^\circ = -0.763$ volts

(b) $Cu^{++} + 2e^- \rightarrow Cu^\circ$
$Zn^\circ \rightarrow Zn^{++} + 2e^-$

This exercise emphasizes the simple consequence of the sign convention: when the standard reduction potential is positive, reduction is spontaneous in combination with the hydrogen electrode, and when the standard reduction potential is negative, oxidation is spontaneous in combination with the standard hydrogen electrode. This rule refers to electrodes in their standard state; the spontaneous direction under other conditions is determined by the sign of $\mathscr{E}$, which is related to $\mathscr{E}^\circ$ by the Nernst equation (9-8).

Calculation of the emf from Standard Electrode Potentials

According to Equation 9-10, we can find $\mathscr{E}^\circ$ for the $Zn\,|\,Zn^{++}\,||\,Cu^{++}\,|\,Cu$ cell by the difference of $\mathscr{E}^\circ$ values for the two half-cells, each relative to the standard hydrogen electrode. Again, it is essential that we give the correct sign to $\mathscr{E}^\circ$ for the chemical reaction as written. The convention that $\mathscr{E}$ is positive for a spontaneous process leads to the following working procedure for calculating $\mathscr{E}^\circ$. Write the reduction reactions with their standard electrode potentials, and subtract one from the other:

$$
\begin{array}{ll}
Cu^{++} + 2e^- \rightarrow Cu^\circ & \mathscr{E}^\circ_{Cu} = 0.337 \text{ volts} \\
-(Zn^{++} + 2e^- \rightarrow Zn^\circ) & -(\mathscr{E}^\circ_{Zn} = -0.763 \text{ volts}) \\
\hline
Cu^{++} + Zn^\circ \rightarrow Cu^\circ + Zn^{++} & \mathscr{E}^\circ = 0.337 + 0.763 = 1.100 \text{ volts}
\end{array} \quad (9\text{-}11)
$$

The standard electrode potential of the reaction that is subtracted is also subtracted. This procedure leads to the correct value of $\mathscr{E}^\circ$ for the chemical reaction as written. You can verify for yourself that if the Cu electrode reaction (and its $\mathscr{E}^\circ$) were subtracted from the corresponding quantities for the Zn electrode, the result would be

$$Cu^\circ + Zn^{++} \rightarrow Cu^{++} + Zn^\circ, \quad \mathscr{E}^\circ = -1.100 \text{ volts}$$

This statement of the cell reaction and its emf is equivalent to that in Equation 9-11, because both the direction of the reaction and the sign of $\mathscr{E}^\circ$ are reversed. The reaction with positive $\mathscr{E}^\circ$ is, of course, the actual spontaneous direction under standard conditions.

Tables 9-1 and 9-2 list standard reduction potentials for a number of substances. Inorganic half-cell potentials appear in Table 9-1, and reactions of biochemical importance in Table 9-2. There is one significant difference in convention between the two tables: they differ in their choice of a standard state for H^+ or OH^-. In Table 9-1, when either of these ions appears as a reactant or product, the standard state is $1m$ concentration of each. In Table 9-2, however, the standard state is a neutral aqueous solution, pH 7, which means that $C_{H^+} = 10^{-7}M$. This standard state is chosen because it is closer to real biochemical conditions than $1m$ H^+ or $1m$ OH^-. Standard emfs based on the pH 7 standard state are given the symbol $\mathscr{E}^*$. All reactants other than H^+ and OH^- are present at unit concentration in the pH 7 standard state. The practical consequence of this is that whenever C_{H^+} or C_{OH^-} appear in the

TABLE 9-1 *Standard Electrode Potentials of Inorganic Substances at 25°C*

REDUCTION REACTION (OXIDIZED → REDUCED)	$\mathscr{E}^\circ$/VOLTS
$Cl_2(g) + 2e^- \rightarrow 2Cl^-$	1.3583
$O_2(g) + 4H^+ + 4e^- \rightarrow 2H_2O$	1.229
$Br_2(aq) + 2e^- \rightarrow 2Br^-$	1.087
$Ag^+ + e^- \rightarrow Ag^\circ$	0.7996
$Fe^{3+} + e^- \rightarrow Fe^{2+}$	0.771
$O_2(g) + 2H^+ + 2e^- \rightarrow H_2O_2$	0.682
$Cu^{2+} + 2e^- \rightarrow Cu^\circ$	0.337
$Cu^{2+} + e^- \rightarrow Cu^+$	0.158
$2H^+ + 2e^- \rightarrow H_2$	0.0000
$Zn^{2+} + 2e^- \rightarrow Zn^\circ$	−0.763
$Na^+ + e^- \rightarrow Na^\circ$	−2.71
$Ca^{2+} + 2e^- \rightarrow Ca^\circ$	−2.76
$K^+ + e^- \rightarrow K^\circ$	−2.92

TABLE 9-2 *Standard Electrode Potential of Biochemical Substances at 25°C (Standard state for H^+ = pH 7)*[a]

OXIDIZED FORM/REDUCED FORM	$\mathscr{E}^{*\prime}$/VOLTS
O_2/H_2O	0.816
Fe^{3+}/Fe^{2+}	0.771
$Fe(CN)_6^{3-}/Fe(CN)_6^{4-}$	0.36
O_2/H_2O_2	0.295
Cytochrome *a* Fe^{3+}/Fe^{2+}	0.29
Cytochrome *c* Fe^{3+}/Fe^{2+}	0.254
Cytochrome c_1 Fe^{3+}/Fe^{2+}	0.22
Cytochrome *b* Fe^{3+}/Fe^{2+}	0.08
Coenzyme Q, quinone/hydroquinone	0.10
Fumarate/succinate	0.03
Methylene blue, leuco/blue	0.011
Pyruvate/lactate	−0.190
Acetaldehyde/ethanol	−0.197
NAD^+/NADH	−0.320
H^+/H_2	−0.42
Ferredoxin Fe^{3+}/Fe^{2+}	−0.43
Acetate + CO_2/pyruvate	−0.70

[a] Sober, H. A., ed. 1968. *Handbook of Biochemistry*. Cleveland, Ohio: The Chemical Rubber Co.

Nernst equation (or in an equilibrium constant based on emf values), they should be replaced by the ratio of the actual concentration (or activity) to that in the standard state, for example, $C_{H^+}/10^{-7}$. (Actually, all activity terms in the Nernst equation, or any equilibrium constant expression, are the ratio of the real activity to that in the standard state, but when the standard-state concentration is unity, this is of no practical consequence.) The following examples illustrate the use of the pH 7 standard state.

EXERCISE 9-7

$\mathscr{E}^*$ (standard state pH = 7) for the electrode reaction acetaldehyde → ethanol,

$$CH_3CHO + 2H^+ + 2e^- \rightarrow CH_3CH_2OH$$

is $\mathscr{E}^* = -0.197$ volts. Calculate $\mathscr{E}$ at pH 6 ($10^{-6}M$ H^+) when ethanol and acetaldehyde are present at $10^{-5}M$, T = 25°C.

ANSWER

$$\mathscr{E} = \mathscr{E}^* - \frac{RT}{nF} \ln \frac{[CH_3CH_2OH]}{[CH_3CHO]([H^+]/10^{-7})^2}$$

where the square brackets indicate concentration. Substituting $10^{-5}M$ for ethanol and acetaldehyde and $10^{-6}M$ for H^+, we get (n = 2)

$$\mathscr{E} = \mathscr{E}^* + \frac{RT}{nF}\ln\left(\frac{10^{-6}}{10^{-7}}\right)^2 = -0.197 + \frac{0.02568}{2}\ln(100)$$
$$= -0.138 \text{ volts}$$

EXERCISE 9-8

Combine the following half-cell reactions and their $\mathscr{E}^*$ values from Table 9-1 to obtain a whole-cell reaction, and calculate the equilibrium constant: acetaldehyde/ethanol; NAD^+/NADH.

ANSWER

$$\begin{array}{ll} CH_3CHO + 2H^+ + 2e^- & \rightarrow CH_3CH_2OH,\ \mathscr{E}^* = -0.197\text{ V} \\ -(NAD^+ \quad + \ H^+ + 2e^- & \rightarrow NADH, \qquad \mathscr{E}^* = -0.320\text{ V}) \\ \hline CH_3CHO + NADH + H^+ & \rightarrow CH_3CH_2OH + NAD^+, \mathscr{E}^* = 0.123\text{ V} \end{array}$$

From Equation 9-10,

$$K^* = \exp\left(\frac{nF\mathscr{E}^*}{RT}\right) = \exp(38.943\ n\mathscr{E}^*)$$
$$= \exp(38.943 \times 2 \times 0.123) = 1.45 \times 10^4$$

The expression for K^* is

$$K^* = \frac{[CH_3CH_2OH][NAD^+]}{[CH_3CHO][NADH]([H^+]/10^{-7})}$$

which is related to the usual equilibrium constant

$$K = \frac{[CH_3CH_2OH][NAD^+]}{[CH_3CHO][NADH][H^+]}$$

by $K^* = 10^{-7}K$.

Prediction of Thermodynamic Spontaneity

Standard reduction potentials are useful for deciding whether a particular oxidation-reduction reaction is thermodynamically favorable. For example, if we mix Ca^{2+} with Cl^- at standard concentrations, do we expect to find Ca° and Cl_2 produced? According to Table 9-1, the standard potential in the reaction $Ca^{2+} + 2\,Cl^- \rightarrow Ca° + Cl_2$ is $-2.76 - 1.36 = -4.12$ volts. This result means that conversion of Ca^{2+} and Cl^- in their standard states to Ca° and Cl_2, also in their standard states, is thermodynamically unfavorable. The Nernst equation can be used to calculate the amount (very small) of Ca° and Cl_2 in equilibrium with Ca^{2+} and Cl^-.

On the other hand, $\mathscr{E}°$ for reaction of K° with Br_2 to form Br^- is found to be positive from the values in Table 9-1, so KBr should be formed spontaneously if K° and Br_2 are mixed in their standard states. Among the biochemical materials, Table 9-2 tells us that cytochrome *c* should oxidize methylene blue, but NAD^+ should not. Instead, the oxidized (leuco or colorless) form of methylene blue should oxidize NADH to NAD^+. Of course, these statements all apply to standard concentration conditions for reactants and products. The Nernst equation also can be used to decide whether a reaction is spontaneous under other concentration conditions.

From the examples presented so far you may have noticed the following general rule: when two half-cells are combined, the one farther up in the table (larger positive $\mathscr{E}°$) tends to go forward, driving the other reaction (lower $\mathscr{E}°$) in the reverse (oxidation) direction. Strong oxidants are materials that are easily reduced, and these, like O_2, have large positive $\mathscr{E}°$ values. Strong reductants, such as H_2, lie at the bottom of the table, with a large negative reduction potential for their conjugate oxidant, such as H^+. Spontaneous biological oxidations, which serve as the primary energy source for nonphotosynthetic organisms, combine two half-cells with sufficiently different reduction potentials to allow formation of ATP by the work term $-nF\mathscr{E}$. Specific examples are considered in Section 9-5.

There is an important qualifying clause attached to these general conclusions: *A reaction that is thermodynamically spontaneous does not always occur at a detectable rate.* A classic example is a mixture of H_2 and O_2, which is thermodynamically unstable with respect to H_2O formation, but actually is stable indefinitely at room temperature in the absence of a catalyst or spark.

9–4 Calculation of a Half-Cell Potential from Two Other Half-Cell Potentials

In the preceding section we saw that the emf of an oxidation-reduction cell could be calculated by subtracting the standard reduction potential of one half-cell from that of the other. The electrode potential which is subtracted is the one associated with the half-cell which occurs as an oxidation in the whole-cell reaction as it is written. It is also possible to combine two half-cell potentials to obtain the potential for another half-cell, but the procedure is a little more complicated and does *not* always involve a simple addition or subtraction of $\mathscr{E}°$ values.

We can provide a general solution for the problem by remembering that the free energy for a process must be the same for all paths, and that $\Delta G = -nF\mathscr{E}$. Suppose that we know $\mathscr{E}°$ for the two reaction steps:

$$\mathrm{A} + n_1e^- \rightarrow \mathrm{B},\ \mathscr{E}° = \mathscr{E}°_{\mathrm{AB}}$$
$$\mathrm{B} + n_2e^- \rightarrow \mathrm{C},\ \mathscr{E}° = \mathscr{E}°_{\mathrm{BC}}$$

What is $\mathscr{E}°$ for the total reaction?

$$\mathrm{A} + (n_1 + n_2)e^- \rightarrow \mathrm{C},\ \mathscr{E}° = \mathscr{E}°_{\mathrm{AC}}$$

The emf is *not* an extensive property, so we *cannot* set $\mathscr{E}°_{\mathrm{AC}}$ equal to $\mathscr{E}°_{\mathrm{AB}} + \mathscr{E}°_{\mathrm{BC}}$. However, we *can* add the separate free-energy changes to get the total free-energy change

$$\Delta G^\circ_{\mathrm{AC}} = \Delta G^\circ_{\mathrm{AB}} + \Delta G^\circ_{\mathrm{BC}}$$

Substituting into this equation the relationship $\Delta G^\circ = -nF\mathscr{E}^\circ$, we get

$$(n_1 + n_2)F\mathscr{E}^\circ_{\mathrm{AC}} = n_1 F\mathscr{E}^\circ_{\mathrm{AB}} + n_2 F\mathscr{E}^\circ_{\mathrm{BC}}$$

Solving for $\mathscr{E}^\circ_{\mathrm{AC}}$ gives us

$$\mathscr{E}^\circ_{\mathrm{AC}} = \frac{n_1\mathscr{E}^\circ_{\mathrm{AB}} + n_2\mathscr{E}^\circ_{\mathrm{BC}}}{n_1 + n_2} \qquad (9\text{-}12)$$

Equation 9-12 reveals that the total emf is an average of the individual $\mathscr{E}^\circ_j$ values, each one weighted by n_j, the number of electrons transferred in step *j*.

In general, problems in which two half-cells must be combined to obtain a half-cell emf should be approached by adding the individual free-energy changes. An illustration is provided by the following exercise.

EXERCISE 9-9

From the following electrode potentials

$$Fe^{3+} + 3e^- \rightarrow Fe^\circ \qquad \mathscr{E}^\circ_1 = -0.036 \text{ V}$$
$$Fe^{3+} + e^- \rightarrow Fe^{2+}) \qquad \mathscr{E}^\circ_2 = 0.771 \text{ V}$$

calculate $\mathscr{E}^\circ$ for

$$Fe^{2+} + 2e^- \rightarrow Fe \qquad \mathscr{E}^\circ_3 = ?$$

ANSWER

$$Fe^{3+} + 3e^- \rightarrow Fe^\circ \qquad \Delta G^\circ_1 = 3F \times 0.036 \text{ V}$$
$$-(Fe^{3+} + e^- \rightarrow Fe^{2+}) \qquad -(\Delta G^\circ_2 = -F \times 0.771) \text{ V}$$
$$Fe^{2+} + 2e^- \rightarrow Fe^\circ \qquad \Delta G^\circ_3 = F(0.771 + 0.108) \text{ V}$$

Therefore $\Delta G^\circ_3 = F \times 0.879$ V, so $\mathscr{E}^\circ_3 = -\Delta G^\circ_3/2F = -0.440$ V.

You should note that the combination of two half-cells to obtain a whole cell can also be treated by adding free-energy changes. Consider the general reactions

$$\Delta G^\circ\text{:}$$
$$\mathrm{A} + n_1e^- \rightarrow \mathrm{B} \qquad -n_1F\mathscr{E}^\circ_1$$
$$-\frac{n_1}{n_2}(\mathrm{C} + n_2e^- \rightarrow \mathrm{D} \qquad -n_2F\mathscr{E}^\circ_2)$$
$$\mathrm{A} + \frac{n_1}{n_2}\mathrm{D} \rightarrow \mathrm{B} + \frac{n_1}{n_2}\mathrm{C} \qquad -n_1F(\mathscr{E}^\circ_1 - \mathscr{E}^\circ_2)$$

in which the second reaction and its accompanying free-energy change are multiplied by n_1/n_2 and subtracted from the first reaction. The resulting free-energy change, $\Delta G^\circ = -n_1F(\mathscr{E}^\circ_1 - \mathscr{E}^\circ_2)$, implies that $\mathscr{E}^\circ = \mathscr{E}^\circ_1 - \mathscr{E}^\circ_2$. Hence our rule, which states that we take the difference of half-cell electrode potentials to obtain $\mathscr{E}°$ for a whole cell, is consistent with the more general principle of addition of free-energy changes.

9–5 A Biochemical Example: The Terminal Oxidation Chain

Mitochondria in aerobic cells carry out a remarkable series of reactions in which the free energy of oxidation of metabolites by molecular oxygen is captured by conversion of ADP and inorganic phosphates to ATP. A number of oxidative steps in metabolism use NAD^+ (nicotinamide adenine dinucleotide) as an electron acceptor or oxidizing agent:

$$\text{Reduced metabolite} + NAD^+ \\ \rightarrow \text{oxidized metabolite} + NADH + H^+$$

The reduction of NAD^+ is a two-electron process, involving transfer of two hydrogen atoms from the metabolite to the nicotinamide ring in NAD^+. One of the hydrogen atoms is released as a proton:

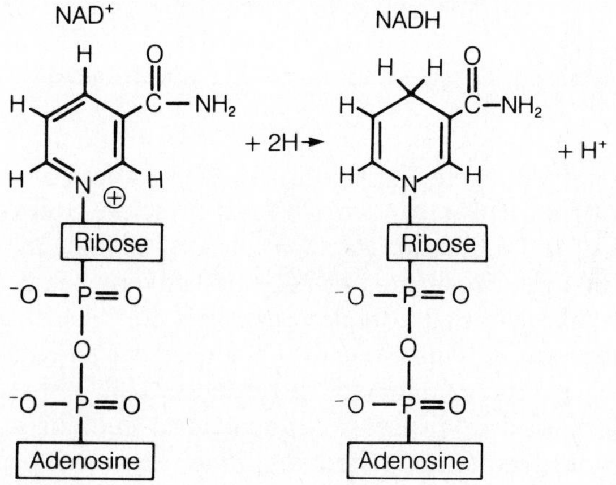

NAD⁺ + 2H → NADH + H⁺

Ribose
Adenosine

Notice that the reaction as written, $NAD^+ + 2H \rightarrow NADH + H^+$, is equivalent to $NAD^+ + H^+ + 2e^- \rightarrow NADH$.

The reduced NADH is oxidized through the terminal electron transport chain, with molecular oxygen as the ultimate electron acceptor. By using standard electrode potentials we can estimate the maximum free energy available to do chemical work (ATP formation) from the overall process. The half-cell reactions are:

$$\begin{array}{lr} \tfrac{1}{2}O_2 + 2H^+ + 2e^- \rightarrow H_2O & \mathscr{E}^* = 0.815\text{ V} \\ -(NAD^+ + H^+ + 2e^- \rightarrow NADH) & -(\mathscr{E}^* = -0.320\text{ V}) \\ \hline NADH + \tfrac{1}{2}O_2 + H^+ \rightarrow H_2O + NAD^+ & \mathscr{E}^* = 1.135\text{ V} \end{array}$$

Therefore the total free-energy change (under standard biochemical conditions) is

$$\Delta G^* = -2 \times \mathrm{F} \times 1.135 \text{ V} = -219 \text{ kJ mole}^{-1}$$

or -52.3 kcal mol^{-1}. Since formation of ATP from inorganic phosphate and ADP under standard biochemical conditions has a positive free-energy change of 7.3 kcal mol^{-1}, the maximum number of molecules of ATP that could be formed per two-electron oxidation of NADH is $52.3/7.3 \cong 7$. The actual number is three, giving the process of biochemical oxidation an efficiency of roughly 40% if carried out under standard conditions. This suggests that biochemical oxidation is not a reversible process, though to be certain we would have to know the concentrations of all metabolites involved. It is characteristic of metabolic reactions that some of the theoretical efficiency of a reversible process is sacrificed for the greater speed of an irreversible process.

EXERCISE 9-10

What happens to the energy of NADH oxidation that is *not* used to do chemical work?

ANSWER

It is released as heat.

Nature has evolved an elegant solution to the problem of utilizing the free energy of NADH oxidation to form ATP. In principle you could imagine a process in which NAD^+ donates electrons directly to molecular oxygen, with the formation of three molecules of ATP. However, it is a general rule of biochemical metabolism that complex reactions are broken down into a sequence of simpler steps. This eliminates the need for a many-body collision of all reactants, permits regulation of the process at several steps, and allows a more nearly reversible process. Mitochondria contain a series of substances of varying reduction potentials. The electrons from NADH are passed through each of these in turn. When a compound receives electrons it is reduced and then reoxidized when the electrons are lost again.

Figure 9-3 shows the cyclic oxidation-reduction reactions that occur between NADH and O_2. This series of reactions is called the *terminal oxidation chain*. The interesting feature of these of reactions from our point of view is the reduction potential of each of the intermediates. These are summarized in Figure 9-4, in which the free-energy changes refer to ΔG^* for transfer of two electrons. We note that there are three steps whose standard free-energy change is comparable to or in excess of the -7.3 kcal mol^{-1} (30.5 kJ mol^{-1}) required for ATP formation: between NADH and coenzyme Q, between cytochromes *b* and *c*, and between cytochrome *a* and O_2. These are the three sites of ATP formation in the electron transport chain. The molecular mechanisms by which ATP generation is coupled to electron transfer remain un-

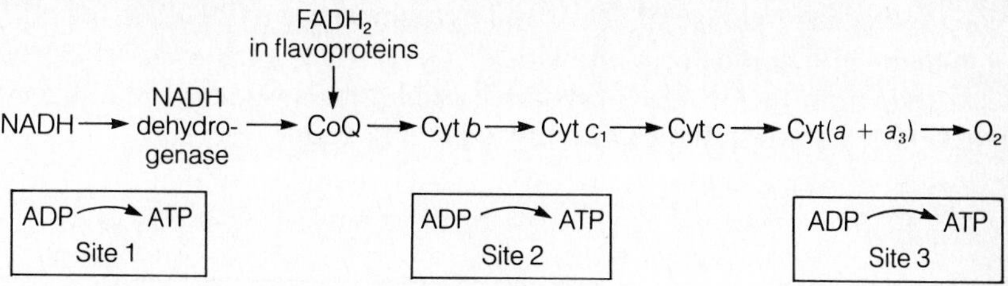

FIGURE 9-3
Sequence of electron carriers in the respiratory chain. Electrons are transferred either from NADH or from the coenzyme $FADH_2$ in flavoproteins through coenzyme Q and the cytochromes to molecular oxygen. NADH is oxidized to NAD^+, and O_2 is reduced to H_2O, while the intermediates undergo cyclic oxidation and reduction. (From *Biochemistry* by Lubert Stryer. W. H. Freeman and Company. Copyright © 1975.)

known. All we can say from thermodynamic studies is that the net reactions, such as

$$
\begin{aligned}
&2\text{Fe}^{++}\ (\text{cyt}\ b) + 2\text{Fe}^{+++}\ (\text{cyt}\ c) + \text{ADP} + P_i \\
&\qquad \rightarrow \text{ATP} + \text{H}_2\text{O} + 2\text{Fe}^{+++}\ (\text{cyt}\ b) + 2\text{Fe}^{++}\ (\text{cyt}\ c)
\end{aligned}
$$

are spontaneous, as judged by $\mathscr{E}^*$ for the cytochrome *b*-cytochrome *c* reaction and the free energy of ATP hydrolysis. Notice that transfer of two electrons from cyt *b* to cyt *c* is required to provide sufficient free energy for ATP formation. The minimum emf required to yield −7.3 kcal in an oxidation-reduction reaction is 0.158 V when two electrons are transferred, and 0.316 V when one electron is transferred.

A remarkable feature of the components of the electron-transport chain, and other iron-containing proteins, is the extent to which the protein environment alters the potential of the $Fe^{3+} + e^- \rightarrow Fe^{2+}$ half-cell reaction. The reduction potential is highest for the free ferric ion (0.771 V) and varies down

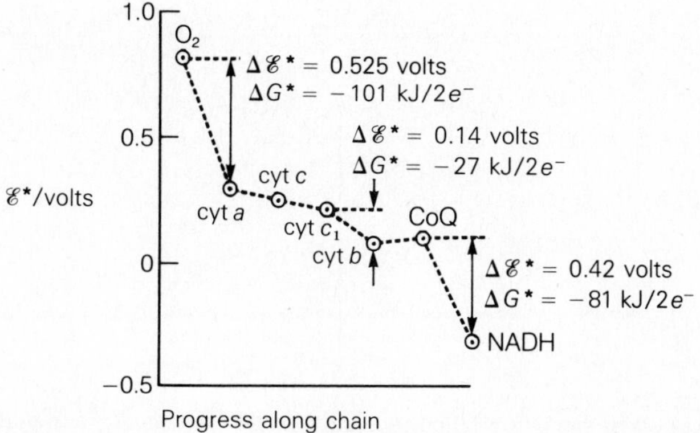

FIGURE 9-4
Variation of standard electrode potentials in the electron transport chain. The standard potential differences and free-energy changes per two electrons transferred are shown for the three sites of ATP formation.

from cytochrome oxidase (0.55 V) and cytochrome *a* (0.29 V) to ferredoxin (a non-heme iron-sulfide protein) with $\mathscr{E}^* = -0.43$ V. In all cases the protein makes reduction to Fe^{2+} less favorable, which means that it stabilizes the ferric (Fe^{3+}) state relative to the ferrous (Fe^{2+}) form.

9–6 *Ionic Effects on Membrane Equilibria*

Ion transport is a common characteristic of biological membranes. In Chapter 7 we saw that semipermeable membranes at equilibrium could yield an osmotic pressure difference across the membrane. If ions are present in a system exhibiting differential ionic permeability, the result can be an analogous difference in electrostatic potential across the membrane. Membrane potentials are extremely important in the function of the nervous system. You should keep in mind, however, that a functioning nerve is not at equilibrium, so even though the equilibrium considerations in this section may provide a logical basis for analyzing membrane potentials, we must not reach too far in extending them to a dynamic system.

The simplest case we can consider is diagramed in Figure 9-5. A membrane permeable only to K^+ ions separates KCl solutions of different concentration. There is a greater tendency of K^+ ions to pass through the membrane from the side of higher concentration to increase the entropy of the whole system. However, only a very small amount (undetectable by chemical analysis) of excess K^+ can build up on the lower concentration side because of the requirement for electrical neutrality of the solution ($C_K = C_{Cl}$). The consequence of a slight excess of positive K^+ ions is an increase of the electrostatic potential Φ on the right (lower concentration) side of the membrane. The electrical free energy per mole of ions (G_{el}) is the potential Φ times the charge zF (Eq. 9-13):

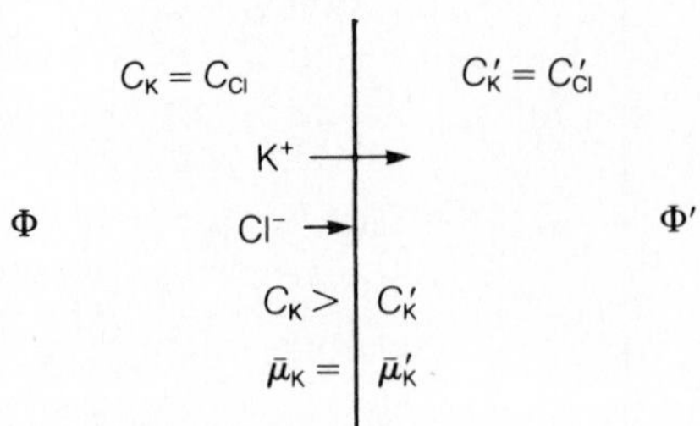

FIGURE 9-5
A membrane permeable only to K^+ ions separates two KCl solutions of different concentration, $C > C'$. A chemically undetectable excess of K^+ over Cl^- builds up on the right side of the membrane because of the transport of K^+ from the higher-concentration (left) side. The result is an increase in electrostatic potential on the right side, so $\Phi' > \Phi$. Equilibrium is characterized by equality of the electrochemical potential $\bar{\mu}_K = \mu^\circ_K + RT \ln C_K + z_K F \Phi$ on both sides of the membrane.

$$G_{el} = zF\,\Phi \tag{9-13}$$ ◄

Here z is the charge of the ion (+1 for K^+, −1 for Cl^-). The electrical free energy increases when Φ increases because of the mutual repulsion of the extra K^+ ions.

The Electrochemical Potential Includes Both Chemical and Electrostatic Free Energies

At this point we find ourselves in the same position as when considering the thermodynamic description of sedimentation equilibrium: the terms so far included in the chemical potential do not fully account for the variation of the partial molar free energy. Just as we added the potential energy due to the centrifugal field to μ, in the present instance we must add the electrostatic free energy to μ to obtain the total chemical potential. We define the *electrochemical potential* $\bar{\mu}$ by

$$\bar{\mu}_i = \mu_i + z_i F\,\Phi \tag{9-14}$$ ◄

in which μ_i is the chemical potential as previously defined. μ_i does *not* include changes in the electrostatic potential of the solution *as a whole.* (Recall from Section 8-11 that μ_i *does* include changes in the *local* electrostatic potential due to interaction of ions and counterions.)

The next step in the solution of the problem set by Figure 9-5 is to equate the electrochemical potentials of K^+ on opposite sides of the membrane, because that is the one ion that is free to move across the membrane and therefore must reach equilibrium:

$$\bar{\mu}_K = \bar{\mu}_K{}'$$

Cl^- is assumed not to permeate the barrier, so a similar equation does not apply to that ion. With the equation for chemical potential

$$\mu_K = \mu_K^\circ + RT\ln C_K$$

and Equation 9-14 we can express the equilibrium of electrochemical potentials as

$$\mu_K^\circ + RT\ln C_K + z_K\,F\,\Phi = \mu_K^\circ + RT\ln C'_K + z_K\,F\,\Phi'$$

which simplifies to

$$RT\ln\left(\frac{C_K}{C'_K}\right) = z_K\,F\,(\Phi' - \Phi)$$

We can solve this for $\Delta\Phi = \Phi' - \Phi$, the difference in potential between right and left sides of the membrane for any ion i:

$$\Phi' - \Phi = \frac{RT}{z_i F}\ln\left(\frac{C_i}{C'_i}\right) \tag{9-15}$$ ◄

The similarity between this equation and the Nernst equation (9-5) is obvious. An electrostatic potential difference can arise in both cases from a concentration difference of ions that can be transferred from one compartment to another. When the solutions are not ideal, activities a_i should replace concentrations C_i in Equation 9-15.

EXERCISE 9-11

A membrane permeable only to Zn^{2+} ions separates $ZnSO_4$ solutions of 10^{-5} m (left) and 10^{-4} m (right). Calculate the transmembrane potential at equilibrium.

ANSWER

$$\Delta \Phi = \Phi_R - \Phi_L = \frac{RT}{zF} \ln \left(\frac{10^{-5}}{10^{-4}}\right) = \frac{0.02568}{2} \ln (0.1) = -0.0296 \text{ V}$$

$$= -29.6 \text{ mV}.$$

The potential is *higher* on the low-concentration side. Notice that this problem, set up exactly the same as Exercise 9-3, has the same potential difference but the opposite sign from that found in 9-3, where the Zn electrode at lower concentration had *lower* potential. You must think carefully about the two problems to understand why the answers have different signs. In Exercise 9-3 the system was *not* at equilibrium; the free-energy change for transferring a Zn^{2+} ion from $10^{-4}m$ to $10^{-5}m$ was found to be $-2F \times 0.0296$ J mol^{-1}. The membrane system is at equilibrium; the free energy of transferring a Zn^{2+} ion from one side to the other is zero. The higher electrostatic potential at low concentration compensates for the greater molar entropy in the less concentrated solution. The two $ZnSO_4$ solutions in Exercise 9-3 have the same electrostatic potentials because they are connected by a salt bridge. The measured potential difference of 29.6 mV is between the two electrodes. To make the free energy of transferring a zinc ion from one electrode compartment to another equal zero, we would have to *increase* the potential of solution and electrode on the low concentration side by 29.6 mV, which would make the potential difference between the electrodes zero. Thus the answer to Exercise 9-3 has the same physical significance as we obtained for the present exercise, when corrected to the condition of $\Delta G = 0$, namely that the lower concentration solution must have Φ greater by 29.6 mV to bring the two solutions into equilibrium.

Active Transport Requires Work

The concept of electrochemical potential allows us to make the important distinction between *active* and *passive* transport of substances across biological membranes. If the electrochemical potential of ion x^+ is greater inside the cell than out (see Fig. 9-6), a positive free-energy change $\Delta\overline{G}$

$$\Delta\overline{G} = \bar{\mu}_x \text{ (in)} - \bar{\mu}_x \text{ (out)} > 0$$

accompanies the transfer of 1 mole of x^+ from outside the cell at electrochemical potential $\bar{\mu}_x$ (out) to the cell interior at electrochemical potential

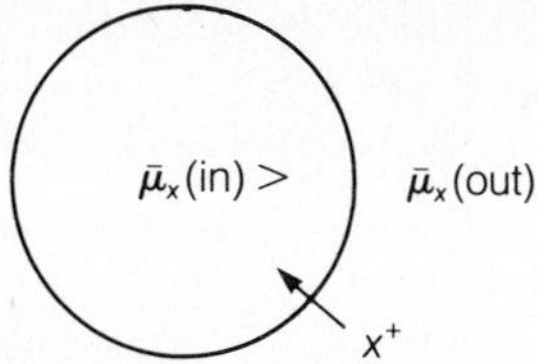

Active transport into cell: $\Delta\bar{G} = \bar{\mu}_x(\text{in}) - \bar{\mu}_x(\text{out}) > 0$

Active transport out of cell: $\Delta\bar{G} = \bar{\mu}_x(\text{out}) - \bar{\mu}_x(\text{in}) > 0$

FIGURE 9-6
The thermodynamic definition of active transport states that if the electrochemical potential inside the cell, $\bar{\mu}_x$(in), is greater than the outside value, $\bar{\mu}_x(\text{in}) > \bar{\mu}_x(\text{out})$, then transport of x^+ into the cell is active. Transport against a gradient in electrochemical potential requires free energy. This required energy comes from a coupled reaction (see Sec. 5-3) in which ATP is cleaved during transport. The coupling agent is a protein called a *pump*.

$\bar{\mu}_x$ (in). Since $\Delta\bar{G} > 0$, *work is required for the process,* and x^+ undergoes *active transport. Passive transport* refers to cases in which *no work is required*—the ion is not transported against a gradient in its electrochemical potential. (When uncharged substances are involved, the chemical potential equals the electrochemical potential.) Note that an ion can in principle undergo active transport either into or out of a cell. Most cell membranes are permeable to water, and the chemical potential of water (which includes the osmotic pressure term) is the same on both sides of the membrane. Hence water crosses the barrier by passive transport.

The important point to look for in considering whether transport is active or passive is the difference in electrochemical potential across the membrane. The following exercise illustrates the procedure involved. (The numbers quoted in the exercise have been slightly altered from the real values to make the results fit a simple idealized model. The conclusions, therefore, are only a first approximation of the truth.)

EXERCISE 9-12

The potassium ion concentration inside a nerve cell at 25°C is measured and found to be 20 times higher than the concentration outside the cell. In contrast, the sodium ion concentration is 20 times higher outside the cell than inside. The potential difference across the cell membrane is 77 millivolts, with the inside negative relative to the outside. Identify the component transported actively. (Assume ideal solutions.)

ANSWER

We must calculate the difference in the electrochemical potentials of K^+ and Na^+ across the membrane. By Equation 9-14, $\bar{\mu}_i = \mu_i + z_iF\,\Phi = \mu_i^\circ + RT \ln C_i + z_iF\,\Phi$. Therefore the difference is

$$\bar{\mu}_i(\text{out}) - \bar{\mu}_i(\text{in}) = RT \ln \left[\frac{C_i(\text{out})}{C_i(\text{in})}\right] + z_iF\,[\Phi(\text{out}) - \Phi(\text{in})]$$

which must be evaluated for each ion. For K^+,

$$\bar{\mu}_i(\text{out}) - \bar{\mu}_i(\text{in}) = RT \ln\left(\frac{1}{20}\right) + 1 \times F \times (0.077)$$

$$= 8.314 \text{ J K}^{-1} \text{ mol}^{-1} \times 298 \text{ K} \times \ln\left(\frac{1}{20}\right) + 96487 \times 0.077 \text{ J mol}^{-1}$$

$$= -7.4 \text{ kJ mol}^{-1} + 7.4 \text{ kJ mol}^{-1} = 0$$

Therefore the distribution of K^+ is at equilibrium, and no active transport is involved. For Na^+, on the other hand, the sign of the first term is changed, since the concentration is higher outside than inside:

$$\bar{\mu}_i(\text{out}) - \bar{\mu}_i(\text{in}) = 8.314 \text{ J K}^{-1} \text{ mol}^{-1} \times 298 \text{ K} \times \ln (20) + 96487 \times 0.077 \text{ J mol}^{-1}$$

$$= 7.4 \text{ kJ mol}^{-1} + 7.4 \text{ kJ mol}^{-1}$$

$$= 14.8 \text{ kJ mol}^{-1}$$

Therefore the electrochemical potential of Na^+ outside the cell is higher by 14.8 kJ mol^{-1} than inside, and active transport is required to pump Na^+ out of the cell. The Na^+ pump is a common characteristic of cells.

EXERCISE 9-13

Calculate the minimum free energy required to operate the Na^+ pump of Exercise 9-12.

ANSWER

The minimum work that must be expended to pump Na^+ is that for the reversible process, and is equal to $\Delta\overline{G}$, with $\Delta\overline{G} = \bar{\mu}(\text{out}) - \bar{\mu}(\text{in})$. Hence $\Delta\overline{G} = 14.8$ kJ mol^{-1} = 3.5 kcal mol^{-1}. Notice that this is less than half the standard free energy of hydrolysis of ATP, so it would be thermodynamically feasible to pump two Na^+ ions per ATP hydrolyzed.

The simple picture of an idealized nerve cell provided by these two exercises includes a nonequilibrium distribution of Na^+ maintained by a Na^+ pump. Extrusion of Na^+ from the cell requires free energy. Steady leakage back in by passive transport means that a constant expenditure of free energy is required to maintain the electrochemical potential gradient. Pumping positive Na^+ ions from the inside leaves it with a negative electrostatic potential relative to the outside. K^+ ions are attracted by the negative potential and accumulate inside the cell by passive transport. Of course, transport in real cells is more complicated and not yet fully understood.

The Donnan Equilibrium is Characteristic of Polyelectrolytes

A more complicated problem is the thermodynamic analysis of membranes permeable to the solvent and small ions, but impermeable to macro-ions such as proteins and DNA. This situation is encountered in dialysis of polyelectrolytes, and gives rise to an asymmetric concentration distribution of the small ions, called the *Donnan effect,* and also to a transmembrane potential called the *Donnan potential.* Real cell membranes also exhibit some of the same characteristics.

Consider the dialysis equilibrium in Figure 9-7. Na^+, Cl^-, and H_2O are free to permeate across the membrane, but the macro-ion of charge z ($z < 0$) cannot. Equilibrium for each of the permeating species requires that the electrochemical potential be the same on both sides of the membrane. Because of the macro-ion we expect an osmotic pressure at equilibrium, so we must add into the chemical potential (Equation 9-14) the corresponding term $\overline{V}_i\pi$ for the solution that contains macro-ions. Hence the electrochemical potential of species i in the presence of the macro-ion is

$$\bar{\mu}_i = \mu_i^\circ + RT \ln a_i + \pi\overline{V}_i + z_iF\,\Phi \tag{9-16}$$

Equating $\bar{\mu}_i$ on one side of the membrane with $\bar{\mu}_i'$ on the other side (μ_i' lacks the osmotic pressure term because π is the excess pressure on the macro-ion solution), we get

$$\mu_i^\circ + RT \ln a_i + \pi\overline{V}_i + z_iF\Phi = \mu_i^\circ + RT \ln a_i' + z_iF\Phi'$$

Solving for $\Phi' - \Phi$, we obtain

$$\Phi' - \Phi = \frac{RT}{z_iF} \ln\left(\frac{a_i}{a_i'}\right) + \frac{\pi\overline{V}_i}{z_iF} \tag{9-17}$$

The potential difference $\Delta\Phi = \Phi' - \Phi$ is called the *Donnan potential*. Our objective here is to solve for it and for the ratio of ion concentrations across the membrane.

To simplify the problem we assume that $\gamma_\pm = 1$ so activities can be replaced by concentrations. Furthermore, we note that the second term on the right in Equation 9-17, $\pi\overline{V}_i/z_iF$, is small compared to the first, so we will

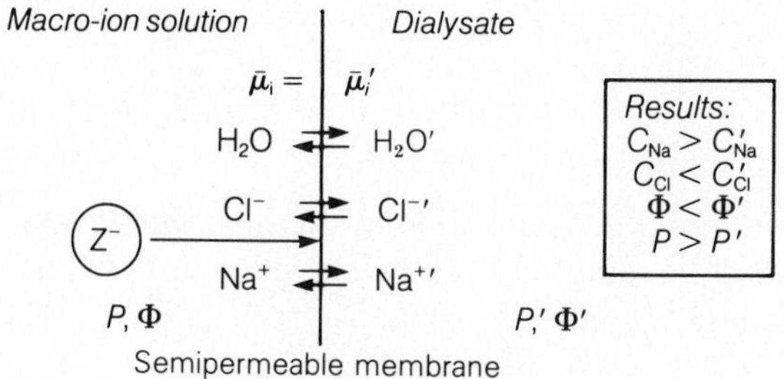

FIGURE 9-7
Dialysis equilibrium of a macro-ion, leading to an asymmetric distribution of ions across the membrane (called the Donnan effect) and a transmembrane Donnan potential. Positive ions accumulate with the negative macro-anion because of the requirement for electrical neutrality. Equality of the electrochemical potential of each ion across the membrane is achieved by a lower electrostatic potential Φ in the solution than in the dialysate. Since the Cl^- ion is negatively charged, its concentration is consequently higher on the dialysate (higher Φ) side than in the solution. Equilibration of the chemical potential of water requires an osmotic pressure $\pi = P - P'$ on the solution side of the membrane.

neglect it here. (Problems 9-13 and 9-14 compare the magnitudes of these terms.) Hence we have

$$\Phi' - \Phi \cong \frac{RT}{z_i F} \ln \left(\frac{C_i}{C_i'} \right) \tag{9-18}$$

Notice the similarity to Equation 9-15. This equation must hold for both Na^+ and Cl^- ions, so we can equate $\Phi' - \Phi$ expressed in terms of the two ions:

$$\frac{RT}{z_{Na} F} \ln \left(\frac{C_{Na}}{C_{Na}'} \right) = \frac{RT}{z_{Cl} F} \ln \left(\frac{C_{Cl}}{C_{Cl}'} \right)$$

With $z_{Na} = 1$ and $z_{Cl} = -1$, this equation reduces to

$$\frac{C_{Na}}{C'_{Na}} = \frac{C'_{Cl}}{C_{Cl}}$$

or

$$C_{Na}\, C_{Cl} = C'_{Na}\, C'_{Cl} \tag{9-19}$$

which states that the product of ion concentrations must be the same on both sides of the membrane. This is equivalent to stating that the activity of the salt NaCl must be the same on both sides of the membrane (see Equation 8-36).

In addition, we must take account of electrical neutrality on both sides of the membrane. This means that the concentration of positive and negative charges must be equal. Hence for the dialysate, containing only NaCl,

$$C'_{Na} = C'_{Cl} = C' \tag{9-20}$$

in which C' is the dialysate concentration of the diffusible salt in equilibrium with the macro-ion. Similarly, for the macro-ion solution, we equate to zero the total charge concentration,

$$zC_B + z_{Na}C_{Na} + z_{Cl}C_{Cl} = 0$$

or, with $z_{Na} = 1$, $z_{Cl} = -1$,

$$zC_B + C_{Na} = C_{Cl} \tag{9-21}$$

in which C_B is the molar concentration of the macromolecular ion of charge z. Equations 9-19, 9-20, and 9-21 can be solved simultaneously by eliminating C_{Cl} and replacing C'_{Na} and C'_{Cl} by C':

$$C_{Na}\,(zC_B + C_{Na}) = C'^2 \tag{9-22}$$

This is a quadratic equation in C. The unknown quantity of interest is the ratio Y of concentrations on the two sides of the membrane:

$$\frac{C_{Na}}{C'} = Y$$

With this substitution, Equation 9-22 becomes

$$Y^2 + \frac{zC_B}{C'}Y - 1 = 0$$

which has the solution

$$Y = \frac{-zC_B}{2C'} + \sqrt{1 + \left(\frac{zC_B}{2C'}\right)^2} \qquad (9\text{-}23)$$

Using Equation 9-23 we can calculate the ratio of concentrations of sodium ions on the two sides of the membrane, and with Equation 9-18 we can calculate the Donnan potential. When the ratio of the macro-ion charge concentration (zC_B) to the electrolyte concentration in the dialysate (C') is small, Equation 9-23 can be approximated by neglecting $\left(\frac{zC_B}{2C'}\right)^2$ compared to 1, yielding,

$$Y = 1 - \frac{zC_B}{2C'} \qquad (9\text{-}24)$$

Significance of the Donnan Equilibrium

We should now pause to examine the physical significance of the results obtained. When the macromolecule is negatively charged, z is negative, so $Y > 1$ (Equation 9-24). This means that the concentration of Na^+ ions, C_{Na}, in the presence of the macro-ion is greater than the concentration in the dialysate, C'. This is because the macro-ion requires counter-ions for electrical neutrality. Furthermore, Equation 9-19 reveals that the negative Cl^- ion will have the reverse distribution,

$$\frac{C_{Cl}}{C'} = Y^{-1} \qquad (9\text{-}25)$$

so the negative ion is less concentrated in the presence of the macro-ion than in the dialysate. The potential difference $\Delta\Phi = \Phi' - \Phi$ is, by Equation 9-18,

$$\Phi' - \Phi = \frac{RT}{z_+F}\ln Y \qquad (9\text{-}26)$$

in which z_+ is the charge on the positive (Na^+) ion. Since $Y > 1$ and $z_+ > 0$, $\Phi' - \Phi$ is positive for a negative macro-ion. This means that *the electrostatic potential Φ in the presence of a negative macro-ion is negative relative to the potential Φ' in the dialysate.* This is the same direction as the potential difference between a nerve cell (containing predominantly negative macro-ions because of its content of nucleic acids) and its surroundings. Hence the Donnan potential is partly responsible for the potential difference ascribed to the Na^+ pump in Exercises 9-12 and 9-13.

EXERCISE 9-14

Consider a solution of DNA molecules at a concentration of 1 mg ml^{-1}, 25°C, dialyzed to equilibrium against 0.001 M NaCl, pH 7. Calculate Y and

$\Phi' - \Phi = \Delta\Phi$, assuming 1 negative charge per nucleotide of molecular weight 340 g mol^{-1}.

ANSWER

The concentration of (negative) macro-ion charge, zC_B, is $zC_B = -1 \times 1$ g liter^{-1}/340 g mol^{-1} $= -2.94 \times 10^{-3}M$. Hence, with $C' = 10^{-3}M$, $zC_B/C' = -2.94$ and $Y = 2.94/2 + \sqrt{1 + (2.94/2)^2} = 3.25$. The concentration $C_{Na} = 3.25 \times 10^{-3}M$, and $C_{Cl} = Y^{-1}C' = 3.08 \times 10^{-4}M$. By Equation 9-26, $\Phi' - \Phi = 0.0303 = 30.3$ mV. The macro-ion solution is negative relative to the dialysate.

9–7 Osmotic Pressure and the Sodium Pump

Cells may expend a sizeable fraction of their metabolic energy on the sodium pump, so one would expect that there may be some universal purpose to this apparently wasteful process. An important factor to keep in mind when considering the problem is the osmotic pressure generated across a membrane permeable to water when the molar concentration of solutes is different on the two sides of the membrane. The presence of macromolecules inside cells produces an internal osmotic pressure, and as we will see here, the pressure is increased by the Donnan effect if the macromolecules are polyelectrolytes. The sodium pump can act to reduce the electrolyte concentration inside the cell and therefore to control the cellular osmotic pressure. The activity of water, which is passively transported across the membrane, can be made equal inside and outside the cell by pumping out electrolyte to compensate for the osmotic effect of intracellular polyelectrolytes and the accompanying accumulation of counter-ions.

First, we calculate the osmotic pressure that accompanies the Donnan effect. To keep our quantitative treatment of this problem as simple as possible, we assume that concentrations are low enough so that only the first virial coefficient need be retained. Therefore Equation 7-73 for the osmotic pressure becomes $\pi = RT\, C_B$. In the present case, where ions are found on both sides of the membrane, this means that

$$\pi = RT\, \Delta\, C_B \tag{9-27}$$

in which ΔC_B is the difference of the molar concentration of all solutes across the membrane. ΔC_B is the sum of concentration differences due to the macromolecule (Component 2) and salt (Component 3):

$$\Delta C_B = C_B(2) + C_B(3) - C'_B(2) - C'_B(3)$$

As usual, C_B is the concentration in the presence of macromolecules inside the cell, and C'_B is the concentration outside the cell. We assume that macromolecules are contained only inside the cell so $C'_B(2) = 0$.

Using the definition $Y = C_{Na}/C'_{Na}$, and $C'_{Na} = C'_{Cl} = C'$, ΔC_B becomes

$$\Delta C_B = C_B(2) + C_{Na} + C_{Cl} - C'_{Na} - C'_{Cl}$$
$$= C_B(2) + C'(Y + Y^{-1} - 2)$$

and hence the osmotic pressure is

$$\pi = RT[C_B(2) + C'(Y + Y^{-1} - 2)] \quad (9\text{-}28)$$

EXERCISE 9-15

Calculate the osmotic pressure for the example of Exercise 9-14, assuming the DNA molecules have a molecular weight of 10^7.

ANSWER

$C_B(2) = 1$ g liter^{-1}/10^7 g mol^{-1} = 10^{-7} mol liter^{-1}. Hence, with $Y = 3.25$ and $C' = 10^{-3}M$, $C_B(2) + C'(Y + Y^{-1} - 2) = 10^{-7} + (3.25 + 1/3.25 - 2)$ $10^{-3} = 1.56 \times 10^{-3}M$. (Notice that the macromolecular concentration, 10^{-7} M, makes a negligible contribution to ΔC_B; the osmotic pressure is due essentially only to the Donnan effect.) Therefore

$$\pi = 0.08206 \text{ liter atm K}^{-1}\text{ mol}^{-1} \times 298 \text{ K} \times 1.56 \times 10^{-3} \text{ mol liter}^{-1}$$
$$= 0.0381 \text{ atm}$$

This example shows that an appreciable osmotic pressure arises from the Donnan effect, even when the polymer molecular weight is so large that it contributes negligible osmotic pressure itself. Much larger contributions to the osmotic pressure by the Donnan distribution can be produced if the salt concentration C' is larger (see Problem 9-13). A corollary of this observation is that it is very difficult to use the osmotic pressure method to determine the molecular weight of polyelectrolytes.

The Na^+ pump can reduce cellular osmotic pressure only if passive transport of K^+ into the cell lags behind the extrusion of Na^+. If passive transport of K^+ establishes equilibrium of the K^+ electrochemical potential, then each Na^+ pumped out of the cell will be replaced by a K^+ ion (see Exercise 9-12), and no reduction in net ion concentration will result. However, it is found that the concentration of Cl^- is about 1.4 times higher outside a red blood cell than inside, which would not arise from simple exchange of Na^+ and K^+. There are two probable contributors to the asymmetric distribution of Cl^-. One is the exclusion of negative ions by negative polyelectrolytes, as predicted for the Donnan equilibrium. The other is a net reduction in internal electrolyte concentration by the pumping out of NaCl. The second effect requires that K^+ entry into the cell must lag behind the pumping of Na^+, meaning that the distribution of K^+ across the membrane would not be at true equilibrium.

The result of a net reduction in internal electrolyte concentration would be an increase in the chemical potential of water inside the cell, and therefore a reduction in the osmotic pressure required to equilibrate the internal water chemical potential with that in the extracellular fluid. In this manner cells could control their osmotic pressure by use of a membrane pump. It is noteworthy that if cellular metabolism is blocked, and hence the sodium pump is inhibited, the rate of red cell lysis is greatly increased, probably at least partly because of increased internal osmotic pressure.

QUESTIONS FOR REVIEW

1. Define an oxidant and a reductant.
2. What is the emf of an electrochemical cell?
3. Outline the steps that lead to the Nernst equation.
4. Explain how to calculate the emf of a cell from standard electrode potentials.
5. Define the electrochemical potential.
6. Define active and passive transport.
7. What is the relationship between the transmembrane potential and the ratio of concentration of permeable ions at equilibrium?
8. What is the origin of the Donnan potential?
9. Explain how the Donnan effect contributes to osmotic pressure.

PROBLEMS

1. Are the following statements true or false?
   (a) When the standard reduction half-cell potential is positive, reduction is spontaneous in all cells.
   (b) The Nernst equation applies only at equilibrium.
   (c) At standard biochemical conditions the concentration of all products and reactants is $10^{-7}M$.
   (d) The electrochemical potential of sucrose in solution is equal to its chemical potential.
   (e) Solutions of an uncharged polymer show no Donnan potential.

Answers: (a) F, it is spontaneous when the half-cell at standard condition is combined with the standard hydrogen electrode. (b) F. (c) F. (d) T. (e) T.

2. Add the appropriate word or words.
   (a) The sign of the electromotive force, $\mathscr{E}$, is ________ for a spontaneous process.
   (b) Oxidation is a ________ of electrons.
   (c) Pt metal in contact with H_2 gas (1 atm) and H^+ (unit concentration) is called the ________ ________ electrode.

(d) The ______ potential is the partial molar free energy including the contribution from the electrostatic potential of the solution as a whole.

(e) ______ transport requires work.

Answers: (a) Positive. (b) Loss. (c) Standard hydrogen. (d) Electrochemical. (e) Active.

PROBLEMS RELATED TO EXERCISES

3. (Exercise 9-3) Calculate the emf of the following cell:

$$\mathrm{Zn}\,|\,\mathrm{Zn^{++}(ZnSO_4)}(10^{-3}m)\,||\,\mathrm{Zn^{++}(ZnSO_4)}(10^{-5}m)\,|\,\mathrm{Zn}$$

4. (Exercise 9-4) $\mathscr{E}°$ for oxidation of cytochrome c_1 by cytochrome a is found to be 70 mvolts. Calculate K and $\Delta G°$ for the reaction.

5. (Exercise 9-6) Write the reduction reactions for the O_2/H_2O and Na^+/Na electrodes, and use the $\mathscr{E}°$ values in Table 9-1 to determine the spontaneous chemical reaction for each electrode under standard conditions when combined with the standard hydrogen electrode.

6. (Exercise 9-7) $\mathscr{E}^*$ for the $NAD^+/NADH$ electrode reaction is -0.320 volt. Calculate $\mathscr{E}$ at pH 8 when $[NAD^+] = 10^{-4}M$, $[NADH] = 10^{-5}M$.

7. (Exercise 9-12) The Na^+ concentration and electrostatic potential are measured on both sides (A and B) of a membrane. Side B has a potential positive relative to A by 50 mv. The Na^+ concentration is higher on side A by a factor 20 over B. Identify the direction of active transport. (T = 25°C.)

OTHER PROBLEMS

8. Use the information

$$Cu^{2+} + 2e^- \rightarrow Cu° \qquad \mathscr{E}° = 0.337 \text{ volt}$$
$$Cu^{2+} + e^- \rightarrow Cu^+ \qquad \mathscr{E}° = 0.158 \text{ volt}$$

to calculate $\mathscr{E}°$ for the reaction

$$Cu^+ + e^- \rightarrow Cu°$$

9. Use the following standard reduction potentials

Pyruvate/lactate	$\mathscr{E}^* = -0.190$ volt
Acetate + CO_2/pyruvate	$\mathscr{E}^* = -0.70$ volt
$NAD^+/NADH$	$\mathscr{E}^* = -0.320$ volt

to calculate ΔG^* for the conversion of one mole of lactate to acetate and CO_2 with coupled reduction of NAD^+ to NADH. Be sure to write balanced equations for all reactions. (Note:

$$\text{Pyruvate} = CH_3\overset{\overset{\displaystyle O}{\|}}{C}COO^-,\ \text{acetate} = CH_3COO^-,\ \text{lactate} = CH_3CHOHCOO^-.)$$

10. The potential difference across a cell membrane is found to be 40 mv, with the inside negative relative to the outside. Assume an equilibrium distribution, and calculate the concentration ratio C_{Mg} (out)/C_{Mg} (in) for Mg^{++} ions; $T = 37°C$.

11. Calculate the free energy required per mole to pump Na^+ out of the cell in Problem 10, assuming equal concentrations inside and outside.

12. A solution of tRNA at a concentration of 100 mg/ml is dialyzed to equilibrium against 0.1 *M* NaCl. Calculate the Donnan potential and the concentration ratio C_{Na} (solution)/C_{Na} (dialysate). Assume there is one charge per nucleotide of $M = 340$; $T = 37°C$.

13. Calculate the osmotic pressure on the solution in Problem 12, with separate calculations for the contribution due to tRNA and that due to the asymmetric ion distribution. (Assume a molecular weight of 25,000 for tRNA.) Calculate the height of a column of water that would be supported by the osmotic pressure on the solution.

Answer: 4.06 atm; 42 *m* H_2O.

14. Use the results of Problem 13 and a partial molar volume of 16.6 $cm^3\ mol^{-1}$ to calculate the ratio of the term $\pi \bar{V}/z_i F$ for Na^+, which was dropped from Equation 9-17, to the term $RT/z_i F \ln (a_i/a_i')$, which was retained.

Answer: 2.25×10^{-3}. Notice that the partial molar volume of a single ion cannot be measured, but the partial molar volume of Na^+ must be smaller than the partial molar volume of NaCl, which is 16.6 $cm^3\ mole^{-1}$.

15. Given the following standard electrode potentials:

	$\mathscr{E}^*$/volts
O_2/H_2O	0.816
O_2/H_2O_2	0.295

Calculate $\mathscr{E}^*$ for the H_2O_2/H_2O electrode.

16. Given the following standard biochemical reduction potentials:

	$\mathscr{E}^*$/volts
Oxygen/water	0.815
Fe^{3+}/Fe^{2+}	0.770

(a) Write a balanced chemical reaction for the cell

$$Pt|Fe^{2+}, Fe^{3+}||H_2O, O_2|Pt$$

(Pt is an inert electrode)

(b) Calculate the standard (biochemical) emf for the cell reaction which you wrote in (a).

(c) Calculate the standard (biochemical) free-energy change for the reaction in (a).

(d) Write an expression for the equilibrium constant for the reaction (a) in terms of the concentrations of the reactants and products. Calculate the value of the equilibrium constant which you wrote.

(e) Calculate the emf of the cell (a) at pH 6, with all other variables in their standard states.

● 17. Given the following standard biochemical reduction potentials:

	$\mathscr{E}^*$/volts
Acetate (CH_3COOH) + CO_2/lactate ($CH_3CHOHCOOH$)	−0.89
Acetate/acetaldehyde (CH_3CHO)	−0.60

(a) Write balanced equations for each half-cell.

(b) Write a balanced equation for the half-cell:

Acetaldehyde + CO_2/lactate

(c) Calculate $\mathscr{E}^*$ for the half-cell in (b).

QUESTIONS FOR DISCUSSION

18. In what ways are electrochemical equilibria similar to chemical equilibria and in what ways are they different?

19. In the medical condition known as *edema,* water accumulates in the interstitial tissue spaces. Explain why edema results from depletion of plasma proteins.

FURTHER READING

Bockris, J. O'M., Bonciocat, N., and Gutman, F. 1974. *An Introduction to Electrochemical Science.* London: Wykeham Publications Ltd. A brief book containing some interesting applications to chemistry, energy storage, biology, and ecology.

OTHER REFERENCES

Takano, T., Trus, B. L., Mandel, N., Mandel, G., Kallai, O. B., Swanson, R., and Dickerson, R. E. 1977. *J. Biol. Chem. 252,* 776–785.

Handbook of Chemistry and Physics, 47th ed. 1968. Cleveland: The Chemical Rubber Co.

Sober, H. A., ed. *Handbook of Biochemistry.* 1968. Cleveland: The Chemical Rubber Co.

Stryer, L. 1975. *Biochemistry.* San Francisco: W. H. Freeman and Co.

PART THREE

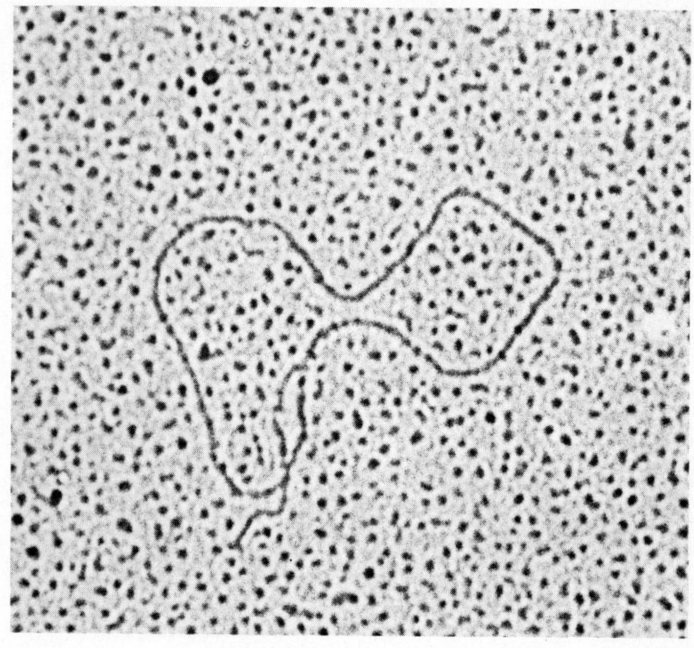

Microscopic Systems

CHAPTER 10

Principles of Quantum Mechanics

Scientists exploring one area sometimes uncover paths that lead to unexpected vistas. Neither Planck's description of the spectrum of radiation emitted by a heated object, Einstein's explanation for the expulsion of electrons from metals by light, nor de Broglie's speculations on waves associated with matter could have been expected to have implications for biological science. Yet these insights about the fundamental nature of matter and radiation laid the groundwork for quantum mechanics, the description of the motions and interactions of particles at the atomic level. Within several years of the formulation of wave mechanics by Werner Heisenberg and Erwin Schrödinger in 1925, the principles of the new mechanics had revealed many rules of chemical bonding and molecular spectroscopy. Today the working chemist or biochemist passes few days without using some spectroscopic technique or interpreting some experiment in terms of chemical bonding. Applied quantum mechanics forms the base for virtually all interpretations of experiments at the molecular level, and success in arriving at correct conclusions depends in part on familiarity with the theory.

To make predictions from quantum theory requires much mathematical sophistication and computational skill, but one can go far in interpreting spectroscopic data and in understanding chemical bonding with only a modest acquaintance with the principles of quantum mechanics. Since most working biochemists use quantum mechanics at the level of *interpretation* rather than of *prediction,* this chapter outlines the theory only to the extent appropriate for understanding elementary material on chemical bonding and molecular spectroscopy, such as the material presented in the following four chapters. Those who desire a more detailed discussion of quantum mechanics may consult the more complete texts listed in the references at the end of this chapter.

10–1 Waves, Particles, and Quanta

By 1900 physicists recognized that electromagnetic radiation has many forms, such as radio waves, infrared radiation, visible and ultraviolet light, and x-rays, differing only in the frequency of oscillation of the electromagnetic wave. The question—argued in the time of Newton—of whether radiation should be thought of as waves or particles seemed to have been answered decisively by the early nineteenth century in favor of waves by the discovery and analysis of *diffraction* phenomena.

You are probably familiar with the basis of diffraction in the interference of light waves whose oscillations are in or out of phase. This topic is treated in Chapter 17 and in basic physics texts; here we will present only a brief review.

A wave of electromagnetic radiation consists of an oscillating electric field E^* and a perpendicular oscillating magnetic field B^*. Figure 10-1(*a*) shows such a wave at an instant in time, with its electric and magnetic field vectors perpendicular to the direction of propagation. The wavelength λ is the distance between successive maxima or minima in the electric or magnetic field vectors. (The wave shown in Fig. 10-1(*a*) has its electric and magnetic fields each oscillating in a single plane, so it is called *plane polarized* radiation.)

An electromagnetic wave moves in the direction of propagation at the velocity of light, $c = 3.00 \times 10^8$ m s^{-1} in vacuum. You can visualize propaga-

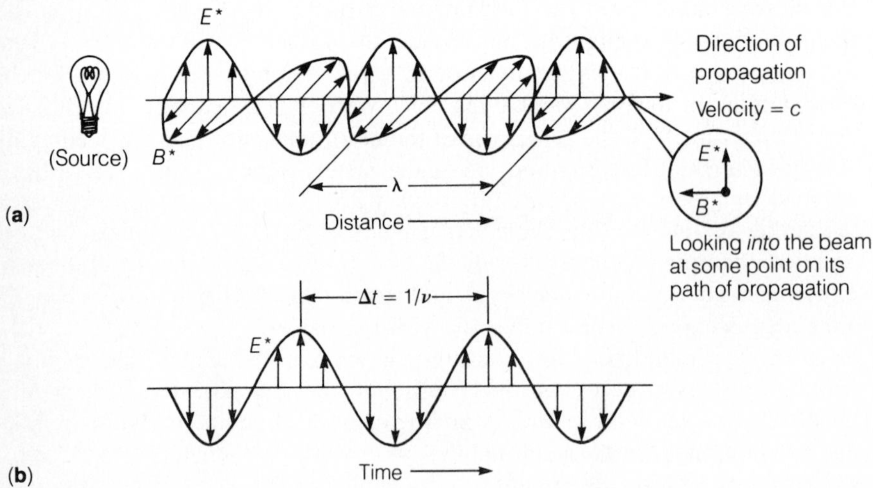

FIGURE 10-1
(a) Oscillation of the electric (E^*) and magnetic (B^*) fields of a plane-polarized electromagnetic wave. The wave is shown at an instant of time as a function of distance along the direction of propagation. The inset shows a view looking into the beam. The wavelength λ is the distance between successive maxima or minima in the field vectors. **(b)** Time-variation of the electric field experienced by a stationary point. The time between successive maxima is $1/\nu$, in which the frequency ν is the velocity of light c divided by the wavelength λ.

tion of the wave by imagining forward motion of the maximum positive electric vector in Figure 10-1(*a*) at the velocity of light. Figure 10-1(*b*) shows the electric field experienced by a *stationary* point as a function of *time*. A similar curve describes the time-variation of the magnetic field.

The time elapsed between successive electric field maxima, Δt, is the distance between the maxima divided by the speed of light, or

$$\Delta t = \frac{\lambda}{c}$$

The reciprocal of Δt is the frequency, ν, with which the electric field oscillates,

$$\nu = \frac{1}{\Delta t}$$

so, substituting $\Delta t = \lambda/c$, we get

$$\nu = \frac{c}{\lambda} \tag{10-1}$$

Frequency has the dimension s^{-1}, a unit called a *Hertz,* Hz.

You will recall from basic physics that two electromagnetic waves interfere *constructively* when they are in phase, as is shown in Figure 10-2(*a*).

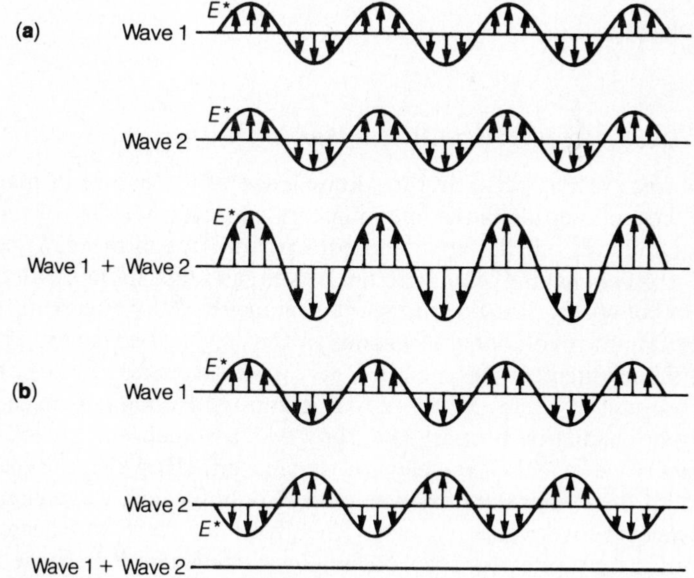

FIGURE 10-2
(a) Constructive interference of two electromagnetic waves. The electric vectors in two light waves in-phase add, so the waves interfere *constructively.* **(b)** Destructive interference of two electromagnetic waves. The electric vectors in two light waves out-of-phase cancel, so the waves interfere *destructively.*

Because they are in phase, their electric vectors add, and the amplitude of the combined wave is the sum of those of the contributing waves. In *destructive* interference, Figure 10-2(*b*), the electric vectors point in opposite directions, so they cancel when the two waves meet.

Diffraction occurs whenever beams of light interfere with each other. Examples are given in Chapter 17. One is the alternate rings of light and dark that are seen when light passes through a pinhole. The diffraction of light was taken in the nineteenth century as proof that light is formed from waves.

Just as diffraction phenomena formed a convincing argument for the wave character of radiation, so too did the laws of motion convince physicists before 1900 that matter consisted of particles. Since the time of Newton it had been accepted that the law of motion

$$\text{Force} = \text{mass} \times \text{acceleration}$$

applied to particles, and that the work done in accelerating a particle of mass m to a velocity v gives a *kinetic energy* (E_K) equal to

$$E_K = \tfrac{1}{2}mv^2 \qquad (10\text{-}2)$$

Another important property of a moving mass is its *momentum* (p)

$$p = mv \qquad (10\text{-}3a)$$

A mass moving in a circular path of radius r has an *angular momentum* L given by

$$L = mvr \qquad (10\text{-}3b)$$

Planck Proposed that Energy is Quantized

The assurance of physicists in their knowledge of the nature of matter and radiation began to erode in the late nineteenth century because observations were accumulating which they could not explain. One of these, a seemingly abstruse problem having to do with the wavelength distribution of light emitted from a hot object, was the origin of a proposal by Max Planck in 1900 that led ultimately to revolutionary changes in the physical sciences. The problem that Planck faced is sketched in Figure 10-3. Classical physics predicted that the amount of energy in the radiation emitted from a hot object (called "blackbody" radiation because the body was assumed not to reflect light) should increase as the wavelength decreased. However, experiments showed that the classical theory was correct only for long wavelengths. The disagreement between theory and experiment at short wavelengths was called, not inappropriately, the "ultraviolet catastrophe," because it was a catastrophe for the predictive powers of Newtonian physics.

Planck's first step in solving the problem was the same as in the classical approach. He assumed that radiation of frequency ν emitted by the blackbody originated from oscillations of matter in the object, also with frequency ν. Planck then found (see historical sketch, p. 409) that he could reproduce the experimental curve by making the radically new assumption that *energy*

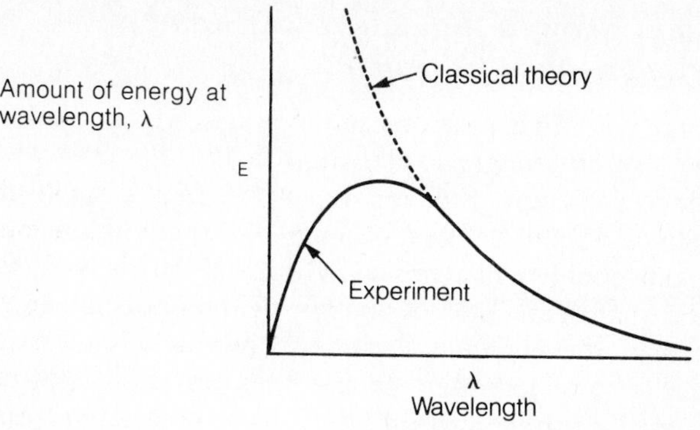

FIGURE 10-3
The blackbody radiation problem faced by Max Planck. Classical physics predicted that increasing amounts of energy should be radiated at shorter and shorter wavelengths by a hot object, but experiments showed that the amount of energy dropped off at short wavelengths.

(E) *could be added or subtracted from the oscillators only in packets or "quanta" of size hν*

$$E = h\nu \qquad (10\text{-}4)$$

in which the quantity h is now called *Planck's constant;* $h = 6.626 \times 10^{-34}$ J s. The consequence of this assumption for the blackbody problem was that high-frequency (short λ) oscillators would have to receive a very large quantum of energy, $E = h\nu$, in order to oscillate at all. It is very improbable (as can be proved by the methods of statistical mechanics, Chap. 14) that the thermal energy provided by heating the blackbody can be sufficiently concentrated at one point to provide such a large energy quantum. Consequently, high-frequency oscillators tend not to oscillate, and therefore little radiation is emitted at high frequency. This explains the experimental energy decrease observed at short wavelengths (Fig. 10-3).

The wider significance of Planck's discovery is the following: quantization of energy means that *any atom or molecule has available to it only a discrete set of energy levels,* as shown in Figure 10-4. Much of quantum mechanics is concerned with calculating the allowed energy levels for particular systems and with characterizing the transitions from one level to another.

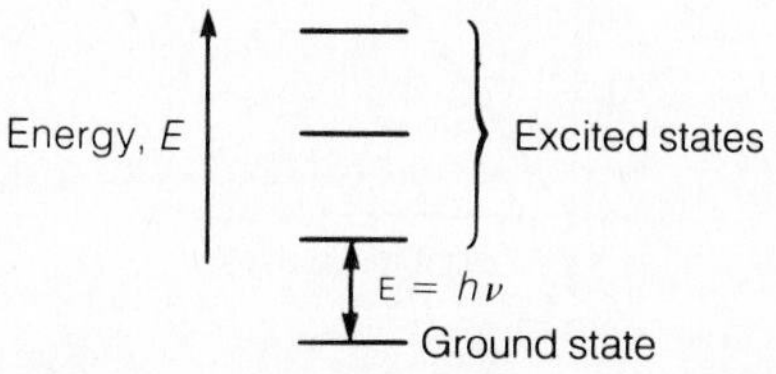

Energy, *E*
Excited states
$E = h\nu$
Ground state

FIGURE 10-4
Quantization of the energy levels in an oscillator (or any atomic or molecular system). The *ground state* is the lowest energy level. Higher energy states are *excited states.* Change from one energy level to another can occur by absorption or emission of a quantum of energy, $E = h\nu$, in which ν is the frequency of the light absorbed or emitted. Energy levels are not necessarily spaced equally.

Einstein Showed that Light has Some Characteristics of Particles

The release of energy in quanta of size $E = h\nu$, in which ν is the frequency of the radiation, has no simple explanation in the classical wave picture of electromagnetic radiation. Einstein suggested that a single light quantum moves in a single direction, like a particle, and that only phenomena involving many quanta could be interpreted in terms of the classical wave theory of radiation. His analysis in 1905 of another of the unexplained phenomena of classical physics, the photoelectric effect, provided solid evidence for the particle-like nature of radiation. It had been observed that shining a light beam on a metallic surface caused electrons to be ejected from the surface. The kinetic energy of the emitted electrons had been measured as a function of the light frequency, with the result [Fig. 10-5(*a*)] that the maximum kinetic energy per electron depended linearly on the frequency of light, but not on the light intensity. Einstein's explanation of this phenomenon was remarkably simple [Fig. 10-5(*b*)]. He proposed that a light quantum strikes an electron in the metal and ejects it from the surface. The maximum kinetic energy the electron can receive is the energy of the quantum, $h\nu$, minus the work, $w = h\nu_{\min}$, required to remove it from the surface

$$E_k(\max) = h\nu - w$$

This equation is in agreement with the experimental results in Figure 10-5(*a*). The classical electromagnetic theory had predicted that the maximum energy should depend on the light intensity.

It frequently is useful to think of a light quantum as a particle, but you must keep in mind that it has zero rest mass and moves at the speed of light. The name *photon* was given to a light quantum by G. N. Lewis, by analogy with names such as electron, proton, and neutron.

Note from Equation 10-4 that the higher the frequency of light, the more energetic is the corresponding photon, so a photon of high-frequency blue

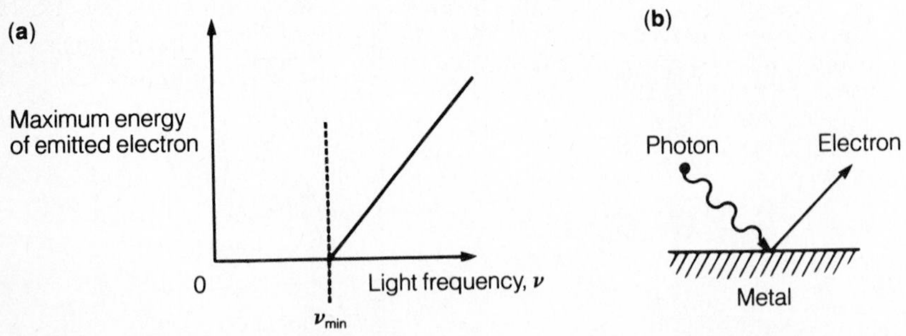

FIGURE 10-5
The photoelectric effect. **(a)** The maximum energy of electrons emitted from a metallic surface illuminated by light depends on the frequency of light, not on the light's intensity. **(b)** Einstein's explanation of the photoelectric effect. A quantum of light (photon) strikes an electron and ejects it from the metallic surface.

light carries more energy than does a photon of low-frequency red light. The energy of a quantum also can be expressed in terms of the wavelength of the radiation. Combining Equation 10-1 ($\nu = c/\lambda$) with Equation 10-4 ($E = h\nu$), we obtain the result

$$E = \frac{hc}{\lambda} \tag{10-5}$$

EXERCISE 10-1

The red color of hemoglobin (and therefore of blood) is a consequence of the strong absorption by this protein of yellow and green light in the wavelength range 500–600 nm, and also of blue light in the wavelength range 400–450 nm, leaving only red light transmitted. What is the energy in ergs of a quantum of yellow light absorbed by hemoglobin at 550 nm?

ANSWER

From Equation 10-5, we have

$$E = \frac{hc}{\lambda} = \frac{(6.63 \times 10^{-34}\ \text{J s} = 6.63 \times 10^{-27}\ \text{erg s})(3.00 \times 10^{10}\ \text{cm s}^{-1})}{(550 \times 10^{-9}\ \text{m})(10^{2}\ \text{cm m}^{-1})}$$
$$= 3.62 \times 10^{-19}\ \text{J} = 3.62 \times 10^{-12}\ \text{erg}$$

EXERCISE 10-2

What is the energy of absorption in units of kJ mol^{-1} and kcal mol^{-1} (using the conversion factors inside the front cover)?

ANSWER

218 kJ mol^{-1} = 52.1 kcal mol^{-1}

Notice that quanta of light are tiny compared to energies we normally experience in our macroscopic world. The energy required to depress a typewriter key is about 10^4 ergs, or about 10^{16} times the size of a quantum of yellow light absorbed by hemoglobin. It is not surprising then that the quantum nature of electromagnetic radiation went undetected until scientific measurements became highly sophisticated.

The Solvay Conference, Brussels, 1911, attended by (1st row from left to right), Nernst, Brillouin, Solvay, Lorentz, Warburg, Perrin, Wien, Madame Curie, Poincaré. (2nd row from left to right), Goldschmidt, Planck, Rubens, Sommerfeld, Lindemann, De Broglie, Knudsen, Hasenohrl, Hostelet, Herzen, Jeans, Rutherford, Kamerlingh, Onnes, Einstein, Langevin. (Courtesy of the Solvay Institute.)

MAX PLANCK

1858–1947

AND

ALBERT EINSTEIN

1878–1955

The personal and professional lives of these two discoverers of the quantum were intertwined, yet their personalities and political philosophies could scarcely have been more different. To understand Planck, said his younger colleague Max Born, one must know about his ancestry of "excellent, reliable, incorruptible, idealistic, and generous men, devoted to the service of Church and (Prussian) State." Disciplined, orderly, formal, and deeply patriotic, he worked at a stand-up desk, yet he was kind to students and friends. Planck had considered careers in classical philology and musical composition but eventually settled on physics. He moved by the conventional steps of a German academic career to become Professor of Physics in Berlin, where, from 1894 on, he concentrated on the problem of the interaction of radiation with matter. In the fall of 1900 he made his great discovery of quantization of energy. As he was walking in the woods with his son, he said, "Today I have made a discovery as important as that of Newton."

Einstein, in contrast, rebelled against authoritarianism in both education and government. He was a citizen of the world but not a patriot of any country; he was fond of mankind in general, but not strongly attached to people around him. Though a prodigy in physics and mathematics, he left his school in Munich without a certificate, apparently because of his resentment of the regimentation of the education. At the age of 16 he passed the entrance exams of the Swiss Federal Polytechnic School (ETH) in Zurich, and embarked upon a four-year college course in physics (1896–1900). That same year he officially renounced his German nationality, thereby becoming a stateless person until he received Swiss citizenship seven years later. He graduated from the ETH, but was the one man in his class to fail to secure an academic post in physics, again apparently because of his independent attitudes. Through the help of one of his teachers he became a patent examiner at the Swiss Patent Office in Bern, work which allowed him to study physics in his spare time over the next five years. He published a few papers on statistical mechanics, and these earned him a Ph.D. from the University of Zurich in 1905. In this same year he wrote three major papers, each of which opened an entire area of physics. One was on the explanation of the photoelectric effect in terms of the quantization of light. A second was his special theory of relativity, with the celebrated equation $E = mc^2$. A third was an explanation of Brownian motion in terms of atomic movements.

Planck and the rest of the physics establishment were quick to recognize Einstein's genius, and soon he was invited to take up professorships at the University of Zurich (1909), Prague (1911), and the ETH in Zurich (1912). Then, in 1913 Planck and Nernst traveled to Zurich to implore Einstein to accept the position as Director of the Kaiser Wilhelm Institute of Physics in Berlin. Though this entailed membership in the Prussian Academy of Sciences, and presumably Prussian citizenship, Einstein accepted.

Einstein's creativity continued in Berlin. He published his general theory of relativity in 1915, and its prediction that starlight is bent as it passes by the sun was dramatically confirmed during the solar eclipse in 1919. His paper on spontaneous emission of radiation in 1917 laid the theoretical groundwork of the laser. By the time he received the Nobel Prize in 1921 he was known beyond scientific circles. Indeed, the Nobel Prize was such a forgone conclusion that his first wife had asked for the prize money as a divorce settlement, even though the divorce occurred several years before the prize was awarded. In Berlin, Einstein met frequently with Planck on scientific matters, and the two also played chamber music with Planck at the piano and Einstein at the violin.

The responses to World War I of Planck and Einstein were entirely different. Planck signed the "Manifesto of the 93," disclaiming Germany's war guilt in the invasion of Belgium and stating that "were it not for German militarism, German culture would have been wiped off the face of the earth." In contrast, Einstein turned increasingly towards pacifism, and he worked for the establishment of a republic in Germany. His request to Planck to sign a statement calling for abdication of the Kaiser was rejected.

As anti-Semitism increased in Germany in the 1920s and early 1930s, Einstein identified himself with his Jewish ancestry, and worked increasingly for Jewish and Zionist causes. With the rise of the Nazis, Einstein became an object for abuse, and Professor Lennard, himself a Nobel Prize winner for his discovery of the photoelectric effect, led the attack. (One example: ". . . the dangerous influence of Jewish circles on the study of nature has been provided by Herr Einstein with his mathematically botched-up theories consisting of some ancient knowledge and a few arbitrary additions.") When Hitler was voted into power in 1933, Einstein's bank account was seized, his apartment and summer house were confiscated, and a book was published with his portrait and the words underneath, "Not yet hanged."

Planck as President of the Kaiser Wilhelm Society attempted to intercede with Hitler on behalf of the great German chemist Fritz Haber, who, like other academics of Jewish ancestry, had been dismissed from his job. Planck reminded Hitler that Haber had saved the German First World War effort with his process for nitrogen fixation from the air. This provided nitrates for fertilizer and explosives when imports were cut off. But Hitler's reaction was a violent outburst against Jews in general. Planck withdrew, probably deciding he must keep peace with the powers to save German science and learning from complete destruction. At the annual meeting of the Kaiser Wilhelm Society in 1933 he read a message sent by the Society to Chancellor Hitler, "The Kaiser Wilhelm Society for the Advancement of the Sciences begs leave to tender reverential greetings to the Chancellor and its solemn pledge that German science is also ready to cooperate joyously in the reconstruction of the new national state."

Einstein and his family took refuge first in Belgium and then the United States, where he worked at the Institute for Advanced Study in Princeton. Undoubtedly his most far-reaching act was to persuade President Roosevelt to initiate an American effort to build an atomic bomb (see the sketches on Oppenheimer in Chapter 11 and Szilard in Chapter 14). Einstein's motivation was the fear that German scientists were on the way to a similar development. Einstein later told Linus Pauling, "I made one great mistake in my life—when I signed the letter to President Roosevelt."

Planck suffered much during the Second World War. His younger son, who held a high government post, was implicated in the July, 1944 plot on Hitler's life and was murdered by the Nazis. His older son had been killed in the First World War. His house in Berlin was destroyed in an air raid on Berlin, and he lost everything, including his library.

When in 1947 Einstein learned of Planck's death, he wrote to Mrs. Planck, "Your husband has come to the end of his days after doing great things and suffering bitterly. It was a beautiful and fruitful period that I was allowed to live through with him. His gaze was directed on eternal truths, yet he played an active part in all that concerned humanity and the world around him. . . . The hours which I was allowed to spend in your house, and the many close conversations which I had with your dear husband, will remain among my happiest memories for the rest of my life. Nothing can alter the fact that a tragic event has affected us both. . . . From this distant place I share your grief and greet you with all the former affection."

Bohr Explained the Spectrum of the Hydrogen Atom

Planck's success with a radical approach to the blackbody radiation problem established the precedent that physicists can allow themselves a limited number of *ad hoc* or unfounded assumptions if these lead to solution of a problem that has not been explained by classical methods. In 1913 Niels Bohr explained the spectrum of light emitted by hydrogen atoms. The problem he faced, and his explanation of it, are sketched in Figure 10-6. Rydberg had

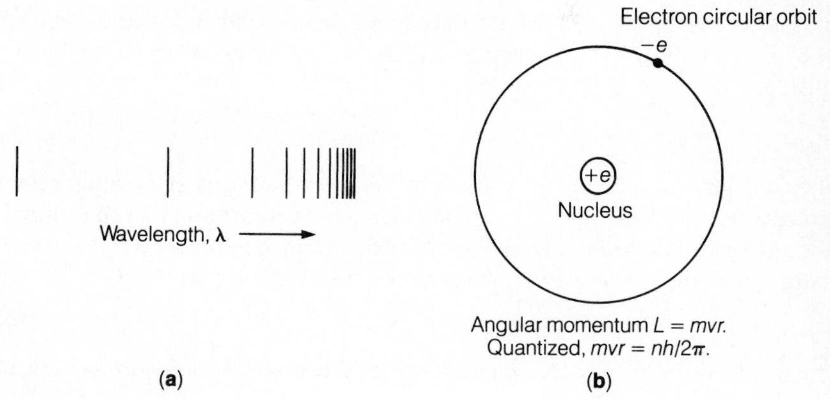

FIGURE 10-6
(a) The light emitted by the hydrogen atom occurs at discrete wavelengths (λ), which can be described by the equation $1/\lambda = R_H[(1/n_2^2) - (1/n_1^2)]$, in which R_H is the Rydberg constant, and n_1 and n_2 are integers (1, 2, 3 . . .). **(b)** Bohr's model for the hydrogen atom. An electron of mass m and charge $-e$ travels in a circular orbit around a heavy nucleus of charge $+e$. Quantization of the angular momentum, $L = mvr = nh/2\pi$, leads to the emission lines observed in **(a)**.

observed that the discrete wavelength (λ) of light emitted by hot hydrogen atoms followed the mathematical relationship

$$\frac{1}{\lambda} = R_H \left(\frac{1}{n_2^2} - \frac{1}{n_1^2}\right) \tag{10-6}$$

in which R_H is a constant that became known as the *Rydberg constant,* $R_H = 109677.59 \text{ cm}^{-1}$, and n_1 and n_2 are integers. Bohr used the result of Rutherford that most of the mass of an atom is concentrated in its nucleus and assumed that the lighter electron of mass m and charge $-e$ travelled in a circular orbit around the hydrogen nucleus of charge $+e$ [Fig. 10-6(*b*)]. As we will see, orbits of different radii (r) become the different allowed energy states in the Bohr theory.

Bohr's radical assumption was that *the angular momentum of the electron is quantized*. The equation he proposed was

$$mvr = \frac{nh}{2\pi} \tag{10-7}$$

in which mvr is the angular momentum, n is an integer called a *quantum number,* $n = 1, 2, 3, \ldots$, and h is Planck's constant. It is an easy matter to show that this *ad hoc* assumption yields for the energy levels of the hydrogen atom

$$E = -\left(\frac{2\pi^2 e^4 m}{h^2}\right)\frac{1}{n^2} \tag{10-8}$$

EXERCISE 10-3

Begin with Newton's equation ($F = ma$) for the motion of the orbiting electron

Coulombic force = mass × centripetal acceleration

and insert the condition for quantization of the angular momentum to obtain Bohr's equation for the energy E (kinetic + potential) of an electron in a hydrogen atom of nuclear charge $+e$.

ANSWER

The centripetal acceleration is given by $\omega^2 r$, in which ω is an angular velocity (radians per second, or 2π × revolutions per second) and r is the radius. According to Coulomb's law (Sec. 8-1) the electrostatic attraction between charges $+e$ and $-e$ is $-e^2/r^2$. Therefore, Newton's equation gives

$$m\omega^2 r = e^2/r^2 \tag{10-9}$$

The linear velocity v is the angular velocity times the radius of the circular orbit, or

$$v = \omega r$$

Substituting v/r for ω in Equation 10-9 yields

$$\frac{mv^2}{r} = \frac{e^2}{r^2} \tag{10-10}$$

or, upon rearranging,

$$r = \frac{e^2}{mv^2} \tag{10-11}$$

Now, use the quantization condition, $mvr = nh/2\pi$, solving it for $v = nh/(2\pi mr)$, and square v to obtain

$$v^2 = \frac{n^2h^2}{4\pi^2m^2r^2}$$

Substitute this expression for v^2 into Equation 10-11 to get

$$r = \frac{4\pi^2m^2r^2e^2}{n^2h^2m}$$

Solve this equation for r:

$$r = \frac{n^2h^2}{4\pi^2me^2} \tag{10-12}$$

These are the *allowed orbits* for the hydrogen atom. The allowed energy levels are

$$\begin{aligned} E &= \text{kinetic energy} + \text{potential energy} \\ &= \frac{mv^2}{2} - \frac{e^2}{r} \end{aligned} \tag{10-13}$$

By using Equation 10-10, we can set

$$\frac{mv^2}{2} = \frac{e^2}{2r}$$

So Equation 10-13 becomes

$$\begin{aligned} E &= \frac{e^2}{2r} - \frac{e^2}{r} \\ &= -\frac{e^2}{2r} \end{aligned} \tag{10-13a}$$

Substituting the allowed radii r of the circular orbits from Equation 10-12 into Equation 10-13a gives the final result

$$E = -\left(\frac{2\pi^2e^4m}{h^2}\right)\frac{1}{n^2}$$

in agreement with Equation 10-8.

As Exercise 10-3 shows, it is only a matter of algebra to derive Equation 10-8 for the energy levels of the hydrogen atom once the acceleration caused by the electrostatic forces and the quantization condition are expressed in mathematical form. An important relationship in the exercise is Equation 10-12 for the allowed radii r of the hydrogen atom orbits. The value of r when $n = 1$ is the smallest allowed radius, a_0,

$$a_0 = \frac{h^2}{4\pi^2 me^2} = 0.05292 \text{ nm} \tag{10-14}$$

which is a measure of length, on the atomic scale, called the *Bohr radius*.

Bohr further assumed that energy emitted in the form of light by a hot hydrogen atom results from switching the electron from one orbit to another, thereby decreasing the energy from E_1 to E_2, with emission of a photon of energy

$$\text{E} = h\nu = E_1 - E_2$$

Changing the energy from level 1 to level 2 means that the quantum number n changes from n_1 to n_2. According to Equation 10-8 the energy released should be

$$E_1 - E_2 = -\left(\frac{2\pi^2 e^4 m}{h^2}\right)\left(\frac{1}{n_1^2} - \frac{1}{n_2^2}\right)$$

Therefore, because $h\nu = E_1 - E_2$, we get

$$h\nu = \left(\frac{2\pi^2 e^4 m}{h^2}\right)\left(\frac{1}{n_2^2} - \frac{1}{n_1^2}\right)$$

and, upon replacing ν by c/λ (Equation 10-1), we find that

$$\frac{1}{\lambda} = \left(\frac{2\pi^2 e^4 m}{h^3 c}\right)\left(\frac{1}{n_2^2} - \frac{1}{n_1^2}\right)$$

Now, if you compare this result with Equation 10-6, you will see that the Rydberg constant is predicted to be

$$R_{\text{H}} = \frac{2\pi^2 e^4 m}{h^3 c} \tag{10-15}$$

Bohr's theory was considered a success because the value of R_{H} it predicted, Equation 10-15, was in close agreement with experiment.

Classical physics demands that an electron rotating about a positive nucleus should emit radiation and fall into the nucleus. Bohr, in contrast, postulated stable angular-momentum states with $n \neq 0$, and thus prohibited the state with $n = 0$, so r cannot be zero. There was no known physical basis for assuming quantization of angular momentum, but the theory was acclaimed because it predicted accurately the energy states and therefore the emission spectrum of the hydrogen atom.

De Broglie Proposed that Matter has Some Characteristics of Waves

In 1924 a French graduate student, Louis de Broglie, made a proposal about the nature of matter so radical that it must have seemed outlandish to many. But to Einstein, de Broglie had "lifted a corner of the great veil." De Broglie's idea was that a fundamental symmetry of nature demands a dual wave-particle character for particles such as electrons, just as for electromagnetic waves. He proposed that there is a characteristic wavelength associated with each particle, given by

$$\lambda = \frac{h}{p} = \frac{h}{mv} \tag{10-16}$$

in which $p = mv$ is the momentum of the particle and m and v are the mass and velocity of the particle. For macroscopic bodies the mass is so large that the de Broglie wavelength is vanishingly small, and the body displays essentially no wave character. But for electrons the de Broglie wavelength can be sufficiently large for them to behave as waves.

EXERCISE 10-4

Compute the de Broglie wavelength associated with a tennis ball of mass 50 g served at 90 miles per hour.

ANSWER

$$\lambda = \frac{h}{mv} = \frac{6.626 \times 10^{-34}\ \text{J sec} \times 3600\ \text{s hr}^{-1}}{0.050\ \text{kg} \times 90\ \text{mi hr}^{-1} \times 1609\ \text{m mi}^{-1}}$$
$$= 3.3 \times 10^{-34}\ \text{m}$$

De Broglie's "matter waves" immediately accounted for the quantization of angular momentum that Bohr had to assume to explain spectral emission lines. A matter wave should exhibit positive and negative interference, just as invoked for radiation to explain diffraction phenomena. In order for an electron to establish a *standing wave* of matter (a stable wave whose maxima do not vary with time), *the circumference must be an integral multiple of the wavelength,* λ (Fig. 10-7). Therefore, we can write

$$2\pi r = n\lambda \tag{10-17}$$

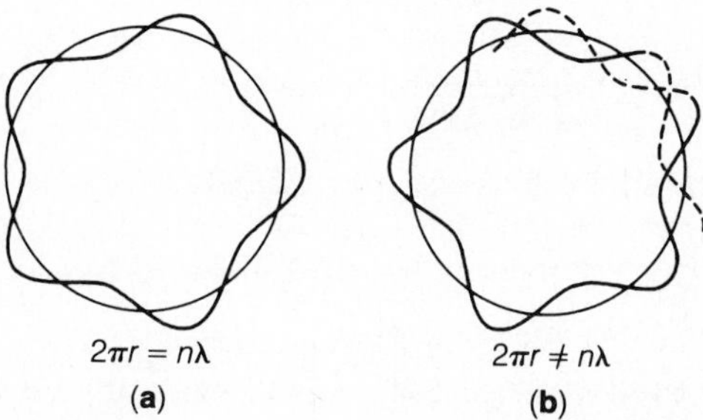

FIGURE 10-7
A de Broglie wave associated with the electron in a hydrogen atom. **(a)** The relationship $2\pi r = n\lambda$ is satisfied so that the state is stable. **(b)** The relationship is not satisfied and the state is not stable. (From Karplus and Porter, 1970.)

Substituting de Broglie's relation for the wavelength, Equation 10-16, we get

$$2\pi r = \frac{nh}{mv}$$

or, upon rearranging

$$mvr = \frac{nh}{2\pi}$$

This is identical with Equation 10-7, assumed by Bohr for quantization of angular momentum. Thus, de Broglie's proposal leads naturally to quantized energy states of atoms.

As radical as de Broglie's proposal was, it was confirmed by experiment within three years when C. Davisson and L. H. Germer in the United States and G. P. Thomson in England observed diffraction of electrons by metals. Diffraction had been considered a property of waves, yet these experiments showed clearly that electrons of the proper velocity experience diffraction, just as though they were light or x-rays. Figure 10-8(*a*) shows the pattern formed on a film by a beam of electrons diffracted by a crystal of the enzyme catalase.

EXERCISE 8-5

A convenient wavelength for electrons in a diffraction experiment is about 0.1 Å = 10^{-9} cm. Electrons from a hot filament are accelerated by a large potential difference $\Delta\Phi$ to a velocity that corresponds with the desired wavelength. What must the potential difference be for a de Broglie wavelength of 0.1 Å?

ANSWER

The final kinetic energy of an electron, $\frac{1}{2}mv^2$, must equal the potential energy it has acquired from acceleration across the potential difference, $e\Delta\Phi$. Thus

$$\tfrac{1}{2}mv^2 = e\Delta\Phi$$

Combining this equation with the de Broglie relationship (Equation 10-16), we find

$$\lambda = \frac{h}{p} = \frac{h}{mv} = \frac{h}{\sqrt{2me\,\Delta\Phi}}$$

Solving for the potential difference $\Delta\Phi$, we get

$$\begin{aligned}\Delta\Phi &= \frac{h^2}{2me\,\lambda^2}\\ &= \frac{(6.63 \times 10^{-27}\text{ erg s})^2}{(2)(9.11 \times 10^{-28}\text{ g})(4.80 \times 10^{-10}\text{ esu})(10^{-9}\text{ cm})^2}\\ &= 50.3\text{ statvolts} = 1.51 \times 10^4\text{ volts}\end{aligned}$$

(Recall that 1 statvolt esu = 1 erg, 1 statvolt = 300 volts, and 1 erg = 1 g cm^2 s^{-2}.)

FIGURE 10-8(a)
The electron diffraction pattern, recorded on film, of a crystal of the enzyme catalase. (Taken by Dr. T. S. Baker.)

The development of one of the most important tools in modern biology, the electron microscope [Figure 10-8(*b*)], grew out of de Broglie's ideas and experimental work on electron lenses. In an electron microscope, electrons are accelerated by a potential difference of about 100 kV and then impinge on the sample. They are diffracted by the sample and then focused by electromagnetic lenses to form an image magnified as much as 800,000 times. This process of diffraction of waves by an object and focusing of the waves to an image by a lens is analogous to the way our eyes focus an image on our retinas. Electromagnetic focusing of electrons is possible because they are charged. Uncharged x-rays cannot be focused this way, nor can they be focused with conventional lenses owing to the very small index of refraction of matter for x-rays. Consequently x-ray diffraction patterns must be interpreted by less direct means (see Chapter 17). The major limitation of electron microscopy in biological work is that biological materials do not have sufficient contrast in scattering power to diffract strongly. Contrast must usually be enhanced by a heavy-metal stain. Unfortunately the heavy-metal coating invariably limits the resolution of the final image.

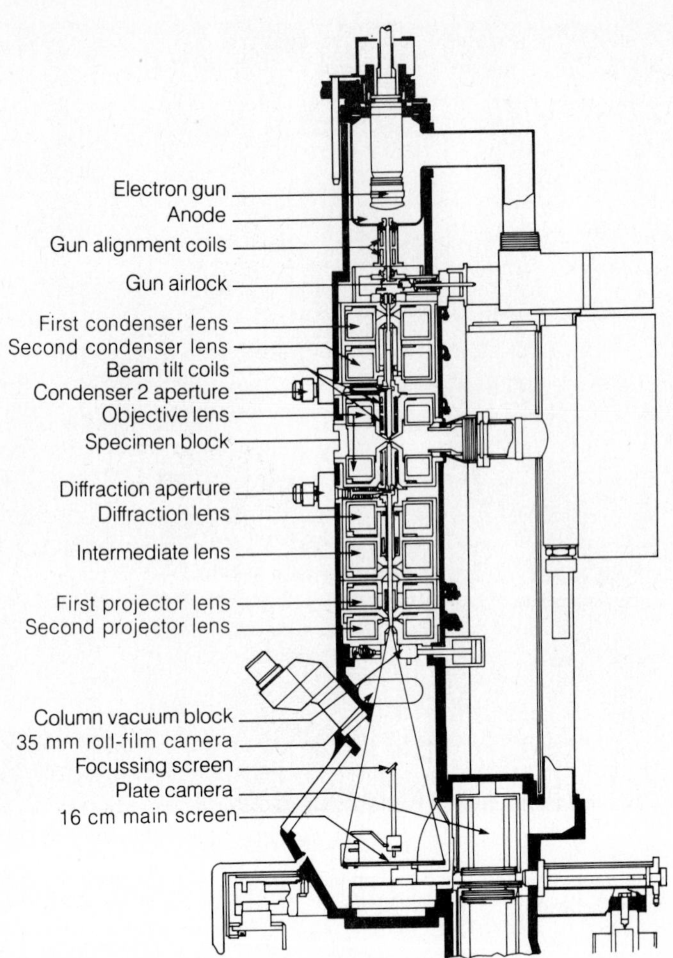

FIGURE 10-8 (b)
A schematic diagram of a modern electron microscope. (Philips Electronic Instruments, Inc.)

The Nature of Light and Matter as Particle-Waves

We have seen in this section that some experiments on particles can be interpreted as though particles have wave-like characteristics, and that some experiments on light waves can be interpreted as though light has particle-like characteristics. This suggests that both matter and light can be viewed in two alternative ways, as either particles or waves. In fact we shall see in the sections that follow that equations which describe the properties of matter accurately use the mathematical language of waves. These equations do not always supply a simple conceptual answer to the question of the nature of matter. Apparently no single concept reflects the present level of human understanding of the nature of the physical world. Physicists often speak of the "particle-wave duality" to describe this situation.

10–2 The Schrödinger Wave Equation Describes de Broglie Waves

In 1926, Erwin Schrödinger formulated quantum mechanics in terms of a differential equation from which, in principle (though not yet in actuality), all of chemistry and much of mechanics can be derived. The equation is a postulate of quantum mechanics, just as Newton's equation $F = ma$ is a postulate of classical mechanics. Schrödinger's equation has the property that the wave character of matter and the quantization of energy emerge as mathematical consequences of solving the equation. These fundamental properties of nature are thus an integral part of the mechanics, rather than additional postulates. Scientists accept the postulate of Schrödinger's equation because its predictions are in accord with experiment.

Wave Functions

The solution of Schrödinger's equation for a particle is a function denoted by ψ, called a *wave function,* which specifies the amplitude of the de Broglie wave of the particle. In the quantum mechanical description of a particle, all possible information about the mechanics of the particle is given by the wave function. Thus knowledge of the wave function is the rough equivalent in quantum mechanics of knowledge of the position and momentum of a particle in classical mechanics. But it is not the exact equivalent, because, as we will see, the wave function yields only the *probable* position of the particle and only average values of other mechanical variables such as velocity and angular momentum.

Though ψ does not have a tangible physical meaning, ψ^2 does: ψ^2 is the *probability density* of the particle described by ψ. This means that

$$|\psi(x,y,z)|^2 = \rho(x,y,z) \tag{10-18}$$

in which $\rho(x,y,z)dx\,dy\,dz$ is the probability that the particle is in the small volume $dx\,dy\,dz$ about the point x,y,z. In other words, if ψ^2 is large at some point in space, there is a high probability of finding the particle at that point, and vice versa. If the wave function describes a system of n particles (say the electrons within a water molecule, or the water molecules within an ice crystal), then

$$|\psi(x_1,y_1,z_1,x_2, \ldots z_n)|^2 dx_1\,dy_1\,dz_1 \ldots dz_n$$

is the joint probability that Particle 1 is in the small volume $dx_1\,dy_1\,dz_1$ around the point x_1,y_1,z_1, Particle 2 is in the small volume $dx_2\,dy_2\,dz_2$ around the point x_2,y_2,z_2, and so forth.

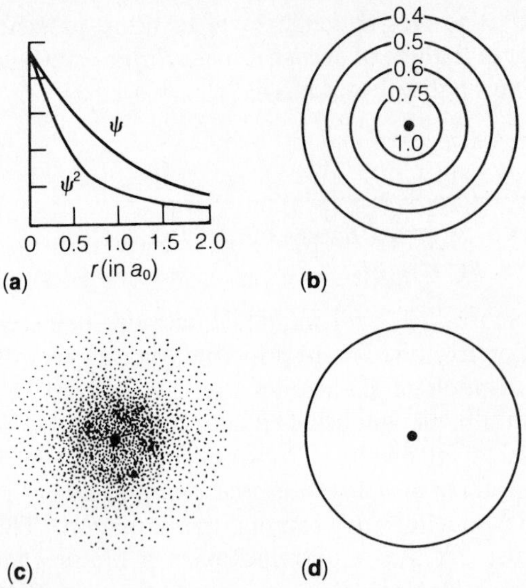

FIGURE 10-9
Four representations of the wave function for the ground state of a hydrogen atom. **(a)** Plots of ψ and ψ^2 against r. **(b)** Contours of ψ in units of $(\pi a_0^3)^{-1/2}$. **(c)** Charge cloud. **(d)** A boundary surface. (From Coulson, 1961.)

As an example of a wave function, consider ψ for the electron in the lowest-energy (or *ground*) state of a hydrogen atom. By methods described later, the function is found to be

$$\psi(r) = \sqrt{\frac{1}{\pi a_0^3}}\, e^{-r/a_0} \tag{10-19}$$

in which r is the distance from the nucleus and a_0 is the Bohr radius. Figure 10-9(a) shows a plot of ψ as a function of r. Note that ψ has its greatest value at the nucleus, and decreases away from the nucleus.

To get ψ^2 we simply square ψ, giving

$$\rho(r) = |\psi(r)|^2 = \frac{1}{\pi a_0^3}\, e^{-2r/a_0} \tag{10-20}$$

Figure 10-9 illustrates the physical meaning of this equation in various ways. In (a), ψ^2 is plotted against r. Clearly, the probability of finding the electron is greatest near the nucleus, and falls to a small value by $2a_0$ (about 0.1 nm) from the nucleus. In (b), contours of constant ψ are drawn about the nucleus; in three dimensions they are concentric spheres with all points on a given sphere having the same ψ value. In (c), the electron is represented as having been drawn out in a cloud, the density of which at any point is proportional to ψ^2 at that point. This shows graphically the fraction of the negative charge of the electron at each point within the hydrogen atom. This representation is the basis of the common description of electron density as a charge cloud. In (d), a particular contour from (b) has been selected such that the total charge outside the contour is a definite fraction of the total electronic charge. The electronic orbitals that we will discuss later often are represented this way.

There are several further points of importance about the wave function:

1. For some systems, ψ is negative in some regions of space; for other systems ψ is a complex number (i.e., it contains $\sqrt{-1}$). If it is complex, then the probability density is given by $\psi\psi^*$, where ψ^* is the complex conjugate of ψ. (See Section 17-4 for a review of complex numbers.) Note that $\psi\psi^*$ is a real number, as a probability density must be.
2. The integral of ψ^2 (or $\psi\psi^*$) over all space must be 1:

$$\int \psi^2\, dV = 1$$

because the probability is unity that the particle will be found somewhere in space. A wave function which satisfies this condition is said to be *normalized*.
3. The wave function ψ must be single-valued and continuous, meaning that it must have a single value at each point in space, and that it cannot change in value by a finite amount for an infinitely small change in distance.

EXERCISE 10-6

Prove that the integral of $\psi^2\, dx\, dy\, dz$ over all space is 1 for the wave function in Equation 10-19, as must be the case if ψ^2 is the probability per unit volume of finding the electron in the volume element $dx\, dy\, dz$.

ANSWER

Integration over all space can be accomplished by noting that the volume of a spherical shell of thickness dr is $4\pi r^2\, dr$. Therefore, the integral sought (I) is

$$I = \int_z \int_y \int_x \psi^2\, dx\, dy\, dz = \int_{r=0}^{\infty} \psi^2 4\pi r^2\, dr$$

Substituting $\psi^2 = e^{-2r/a_0}/(\pi a_0^3)$, we obtain for the integral I

$$I = \frac{1}{\pi a_0^3} \int_{r=0}^{\infty} 4\pi r^2 e^{-2r/a_0}\, dr$$

A table of definite integrals will show you that the integral in this equation equals πa_0^3 so that the integral $I = 1$.

Probability and the Heisenberg Uncertainty Principle

Even though the wave function contains all possible information about a particle, it gives only the *probability* that the particle is at a given point. Thus the exact predictions of position and other mechanical variables made in classical mechanics are not simultaneously possible in quantum mechanics. This is the substance of the *Heisenberg uncertainty principle,* which states that the product of the uncertainty in our knowledge of the momentum of a particle, Δp, times the uncertainty in the position of the particle, Δx, is approximately h:

$$\Delta p \, \Delta x \simeq h \tag{10-21a}$$

Equation 10-21*a* shows that if we know with great precision the position of a particle, there *must* be a great deal of uncertainty about its momentum, and vice versa. Because the wave function ψ gives only the probability of finding a particle at a given position, wave mechanics never permits definitive statements about the position of a particle. This is quite different from the Bohr model, in which the electron has an exact position in its orbit at any instant.

Another useful form of the *Heisenberg uncertainty principle* is

$$\Delta E \, \Delta t \simeq h \tag{10-21b}$$

in which ΔE is the uncertainty in the energy of a state, and Δt is the uncertainty in the lifetime of that state.

Schrödinger's Wave Equation

The Schrödinger equation for the movement of a particle in one dimension may be written

$$\frac{d^2\psi(x)}{dx^2} + \frac{8\pi^2 m}{h^2}[E - U(x)]\psi = 0 \tag{10-22}$$

in which ψ is the wave function for the particle, m is the mass of the particle, $U(x)$ gives the potential energy of the particle as a function of its position, and E is the total energy of the system, a number.

Given an expression for $U(x)$, in principle one can solve this equation for $\psi(x)$. As we shall see in the following sections this is possible for simple potential functions. But when we consider more than one dimension and when $U(x)$ involves the interaction of more than one electron with nuclei, we cannot solve for ψ in closed form, and we must settle for an approximate expression.

The form of $U(x)$ is dictated by the system to be described by Schrödinger's equation. For the hydrogen atom, the electron moves in the electrostatic field of the nucleus. According to Coulomb's law, the potential energy of interaction of two charges q_1 and q_2 separated by r is q_1q_2/r. Since here the nuclear charge is $+e$ and the electronic charge is $-e$, $U = -e^2/r$, in which r is the separation of the nucleus and electron. For the helium atom, U consists of three terms, one for the interaction of each electron with the nucleus, and one for the interaction of the electrons with each other. Clearly, as the number of nuclei and electrons increases, U becomes a function of numerous variables, and solving Equation 10-22 for ψ becomes an increasingly difficult problem.

EXERCISE 10-7

Write the expression for the potential energy U of the hydrogen molecule. Denote the two nuclei by A and B and the two electrons by 1 and 2. Denote the separation between A and 1 by r_{A1}, and so forth for the other separations.

ANSWER

$$U = -e^2\left(\frac{1}{r_{A1}} + \frac{1}{r_{A2}} + \frac{1}{r_{B1}} + \frac{1}{r_{B2}}\right) + e^2\left(\frac{1}{r_{12}} + \frac{1}{r_{AB}}\right)$$

Equations of the form of 10-22 were known before Schrödinger's work. They have the property that *solutions* $\psi(x)$ *exist only for certain values of the energy* E. These values are called characteristic values or, by the German term, *eigenvalues*. They can be numbered in increasing magnitude $E_0, E_1, E_2, \ldots E_n$. To each eigenvalue E_n there corresponds a solution ψ_n, sometimes called an *eigenvector*. In Schrödinger's application, the ψ_n's are the wave functions for the various quantum states of the system, and the E_n's are the corresponding energies of these states. Schrödinger chose an eigenvalue equation to describe quantum mechanics because it builds quantization into the mathematics. The physical system that he set out to describe can exist only in those states characterized by an energy E_n and a de Broglie wave ψ_n. It cannot exist in intermediate states or have an intermediate energy. As always, the physicist selects the mathematical description that predicts the observed physics.

Schrödinger's equation (10-22) was stated above for the motion of a particle in a single dimension. In real applications we will require the equation for three dimensions, so we must add two independent variables and use partial, rather than total, derivatives. The equation now takes the form

$$\frac{\partial^2\psi}{\partial x^2} + \frac{\partial^2\psi}{\partial y^2} + \frac{\partial^2\psi}{\partial z^2} + \frac{8\pi^2 m}{h^2}[E - U]\psi = 0 \qquad (10\text{-}23)$$ ◄

in which ψ and U are functions of x, y, and z.

The Physical Meaning of Schrödinger's Equation: A "Derivation"

It is perfectly reasonable to accept Schrödinger's equation as a postulate, and then to apply it to various problems by inserting the appropriate U and solving for ψ and E. This will give the possible energies of the system and the corresponding wave functions, and from these can be calculated the probability densities for the particle in its various quantum states. Yet Equation 10-22 is not as easy to accept on intuition as is Newton's equation of motion, or even the second law of thermodynamics (if anything can be less intuitive than the second law).

We can acquire some feeling for the physical basis of Schrödinger's equation by first rearranging Equation 10-22:

$$-\frac{h^2}{8\pi^2 m}\frac{d^2\psi}{dx^2} + U\psi = E\psi \tag{10-24}$$

This form has some similarity to the relationship of classical mechanics which states that the sum of the kinetic energy E_k and the potential energy U equals the total energy E:

$$E_k \qquad\qquad + U \quad = E \tag{10-25}$$

It is clear that the right side of Equation 10-24 and the second term on the left correspond respectively to the total and potential energies. This suggests that the first term of Equation 10-24 corresponds to the kinetic energy of the system. We can make this identification stronger by showing that Equation 10-24 can be obtained from Equation 10-25 with some plausible assumptions. This is not a true derivation of Schrödinger's equation (indeed, none is possible) because we will informally associate classical and quantum mechanical concepts.

To obtain Schrödinger's equation from 10-25 we first recall from classical mechanics that

$$E_k = \frac{1}{2}\,mv^2 = \frac{p^2}{2m}$$

in which $p = mv$ is the momentum of a particle of mass m and velocity v. We can eliminate p from this equation by substituting de Broglie's relationship (Equation 10-16):

$$E_k = \frac{1}{2m}\frac{h^2}{\lambda^2} \tag{10-26}$$

Now we introduce a quantity, ψ, which describes wave motion of frequency ν, wavelength λ, and amplitude A, and is a function of time t and distance x.

$$\psi = A\cos 2\pi\left(\frac{x}{\lambda} - \nu t\right) \tag{10-27}$$

You can easily show that ψ is a solution of

$$\frac{d^2\psi}{dx^2} = -\frac{4\pi^2}{\lambda^2}\,\psi \tag{10-28}$$

an equation closely related to the classical equation of wave motion. Solving this equation for $1/\lambda^2$, we get

$$\frac{1}{\lambda^2} = -\frac{1}{4\pi^2\psi}\frac{d^2\psi}{dx^2}$$

and then substituting into Equation 10-26, we find for the kinetic energy

$$E_k = -\frac{h^2}{8\pi^2 m}\frac{d^2\psi}{dx^2}\frac{1}{\psi}$$

Now if we substitute this expression for E_k into Equation 10-25, we obtain Schrödinger's equation in the form of Equation 10-24. This shows that the first term of Schrödinger's equation represents the kinetic energy of the system described by ψ. Since the factor $d^2\psi/dx^2$ of the first term is large when the wave function has large curvature, it seems that the kinetic energy of a system is large when its wave function curves sharply.

EXERCISE 10-8

Prove that Equation 10-27 is a solution of Equation 10-28 by differentiating 10-27 twice and substituting into 10-28.

Quantum Mechanics Permits Prediction of the Value of any Observable Property for any System for Which ψ is Known

We can gain some appreciation for the full power of quantum mechanics by recasting Schrödinger's equation into yet another form. If we factor ψ from the left side of Equation 10-24, we get

$$\left[-\frac{h^2}{8\pi^2 m}\frac{d^2}{dx^2} + U\right]\psi = E\psi \tag{10-29}$$

The expression in brackets is said to be an *operator,* because it defines a mathematical operation to be carried out on ψ. Here the operation is to take the second derivative of ψ, multiply it by a constant, then add to it the product of U and ψ. The first term of the operator is associated with the kinetic energy of the system, and the second term with the potential energy. The full operator is called the *Hamiltonian,* or *energy operator,* of the system, and is denoted $\mathcal{H}$. Thus Schrödinger's equation can be expressed in concise operator form as

$$\mathcal{H}\psi = E\psi \tag{10-30}$$ ◀

This form states that if the Hamiltonian of any system operates on a wave function that describes a state of the system, the result is the same wave function multiplied by the energy of that state.

This third form of Schrödinger's equation is one example of a general postulate of quantum mechanics. This postulate is expressed in terms of an equation of the form of Equation 10-30:

$$Q\psi = q\psi \tag{10-31}$$

This states that if q is a possible value for an observable property of a system described by the wave function ψ, then q must be the eigenvalue in an equation of the form of Equation 10-31, in which Q is the mathematical operator associated with the property q. Quantum mechanics provides a set of rules for expressing the operator Q: the operator for *momentum* is

$(h/2\pi i)d/dx$ (in one dimension), and the operator for *position* is multiplication by the position coordinate x. From this we can see that Schrödinger's equation contains the energy operator $\mathcal{H}$, corresponding to $E = E_k + U$: the kinetic energy is $p^2/2m$ (p is the momentum) so the energy operator is

$$\mathcal{H} = \frac{1}{2m}\left(\frac{h}{2\pi i}\frac{d}{dx}\right)^2 + U$$

or

$$\mathcal{H} = -\frac{h^2}{8\pi^2 m}\frac{d^2}{dx^2} + U$$

(U depends on the coordinate x, so the rules tell us to multiply the expression for U into ψ.) Substitution of this expression for $\mathcal{H}$ into Equation 10-30 yields Equation 10-29.

The physical meaning of Equation 10-31 is that ψ contains all available information on the mechanics of the system it describes. The information about the particular property q can be extracted by operating on ψ with the associated operator Q.

Schrödinger's equation in the form of Equation 10-29 is a special case of Equation 10-31 in which the eigenvalue q is the total energy of the system E, and the operator is the energy operator, the Hamiltonian. In other cases, q could be the position of an electron in a molecule, the dipole moment of a molecule, the average angular momentum of the molecule, and so on. An important example that we will consider later is the case in which q is the angular momentum of the system. For any of these properties, it is possible to compute values for them if the wave function ψ for the system is known, by operating on it with the appropriate operator. Books on quantum chemistry (such as Kauzmann, 1957) discuss the operators for numerous properties. Although for most biochemical applications of quantum mechanics it is unnecessary to be familiar with the applications of these operators, you should be aware that quantum mechanics allows one, in principle, to compute any observable property of any system for which one has an accurate wave function.

Meaning of the Energy E

Students often wonder what is the exact meaning of the energy E of the nth quantum state that is determined by solving Schrödinger's equation. The answer is that this energy is the difference in energy of the system in the nth quantum state, and of the system with zero kinetic energy (all particles at rest) and with zero potential energy. The zero of potential energy is implied by the form of the potential function U. For example, for the hydrogen atom (Sec. 10-6) for which the potential function is $U = -e^2/r$, the zero of potential energy refers to the proton and electron at infinite separation. Thus for the hydrogen atom, the energy E of the ground state yielded by Schrödinger's equation is the difference in energy between the ground state and the state in which the proton and electron are at rest and at infinite separation.

10–3 A Particle Confined to a Box has Quantized Energy Levels

We have not yet seen how Schrödinger's equation leads to quantized energy levels. This emerges from consideration of one of the simplest possible mechanical systems: a single particle confined to move along a line, or in what may be termed a one-dimensional box. Despite its simplicity, this model has much in common with actual quantum chemical systems.

This model is depicted in Figure 10-10. The particle is free to move along the x axis between 0 and a; other regions are forbidden. We specify these restrictions mathematically by equating the potential energy to zero between 0 and a, and to ∞ elsewhere:

$$U(x) = 0, \; 0 < x < a$$
$$U(x) = \infty, \; x \leq 0 \text{ and } x \geq a$$

Now we are ready to solve Schrödinger's equation for the regions where U is infinite and for the region where U is zero. Where $U = \infty$, ψ must be zero. This follows because if the potential energy of the particle is infinite in a region, the particle cannot penetrate the region. If the particle cannot be there, the probability of finding it there is zero. Therefore ψ^2 must be zero in the region, and so must ψ. Thus for $x \leq 0$ and $x \geq a$, $\psi = 0$.

Within the box, $U = 0$, and Schrödinger's equation simplifies to

$$\frac{d^2\psi}{dx^2} + \frac{8\pi^2 m}{h^2} E\psi = 0 \tag{10-32}$$

(a)

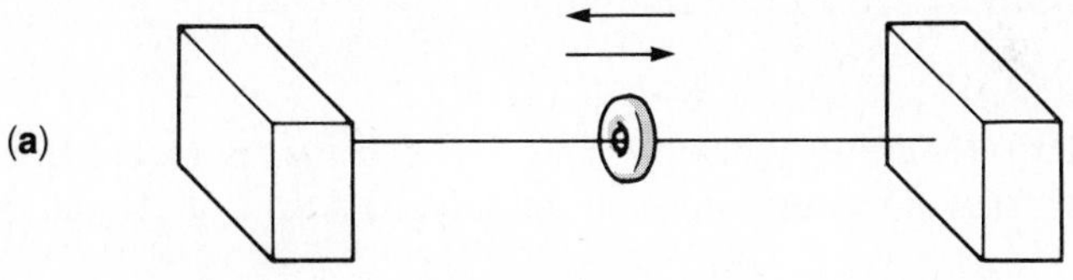

(b)

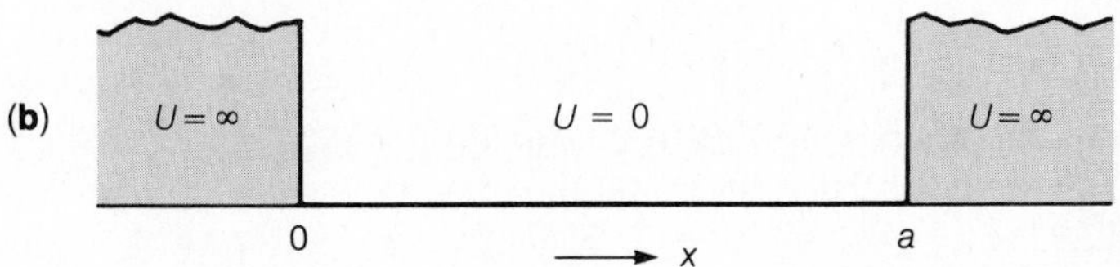

FIGURE 10-10
Particle in a one-dimensional box. **(a)** The example of a bead constrained to move along a frictionless wire, bouncing back and forth between stops at 0 and *a*. **(b)** The potential energy is zero between 0 and *a* and is infinite elsewhere.

We can express this more simply as

$$\frac{d^2\psi}{dx^2} + \beta^2\psi = 0 \qquad (10\text{-}33a)$$

in which

$$\beta = \frac{2\pi}{h}\sqrt{2mE} \qquad (10\text{-}33b)$$

It is straightforward to show that the solution to this equation is

$$\psi = A \sin \beta x + B \cos \beta x \qquad (10\text{-}34)$$

in which A and B are constants.

EXERCISE 10-9

Confirm by differentiation and substitution that Equation 10-34 is a solution to Equation 10-33.

We can evaluate the unknown constant B in Equation 10-34 by utilizing additional information we have about the system. This information is the *boundary condition*. We have determined that ψ must equal zero at $x = 0$ and $x = a$. Since ψ must be continuous, it must approach 0 as x approaches 0 and a. Evaluating Equation 10-34 at $x = 0$, we have

$$0 = A \sin (0) + B \cos (0)$$

Because $\cos x = 1$ when $x = 0$, this equation can be satisfied only if $B = 0$. Therefore the boundary condition at $x = 0$ restricts ψ to the form

$$\psi = A \sin \beta x \qquad (10\text{-}35)$$

The boundary condition that $\psi = 0$ at $x = a$ leads us to the conclusion that the energy of the system must be quantized. At $x = a$, Equation 10-35 becomes

$$0 = A \sin (\beta a) \qquad (10\text{-}36)$$

For this to be true, *the argument of the sine function must be some integral multiple of* π,

$$\beta a = n\pi \qquad (n = \text{integer}) \qquad (10\text{-}37)$$

Substituting our definition of β from Equation 10-33b and rearranging, we find for the energy of the system,

$$E_n = \frac{n^2h^2}{8ma^2} \qquad (n = \text{integer}) \text{ (translation in one dimension)} \qquad (10\text{-}38) \blacktriangleleft$$

Finally, substituting Equation 10-37 for β into Equation 10-35, we get an expression for the wave function

$$\psi_n = A \sin \frac{n\pi x}{a} \qquad (n = \text{integer}) \tag{10-39}$$

Note that because of the boundary conditions, Schrödinger's equation has solutions only for integral values of n. The integer n is the quantum number of the quantum state for which ψ_n is the wave function (or eigenfunction) and E_n is the energy (or eigenvalue).

The wave function ψ and its square, the probability density, are plotted in Figure 10-11. Notice that whereas ψ's for all but the first energy level are negative in some regions of x, ψ^2 is always positive. This is as it must be, because ψ^2 is a probability. As n increases, ψ oscillates more rapidly because the quantum number n is in the argument of the sine function. Also, as n increases, the energy levels spread out because n enters the expression for energy (10-38) as a second power.

The Wave Functions for a Particle in a Box are Orthogonal

An important property of wave functions ψ_n is that they are *orthogonal,* meaning that the product of one wave function times another, integrated over all space, is zero. For the wave functions in Equation 10-39, one can show (by consulting a table of integrals) that ψ_n and ψ_m are orthogonal when $n \neq m$, because

$$\begin{aligned}\int_0^a \psi_n \psi_m \, dx &= A^2 \int_0^a \sin \frac{n\pi x}{a} \sin \frac{m\pi x}{a} \, dx \\ &= 0 \text{ when } n \neq m\end{aligned}$$

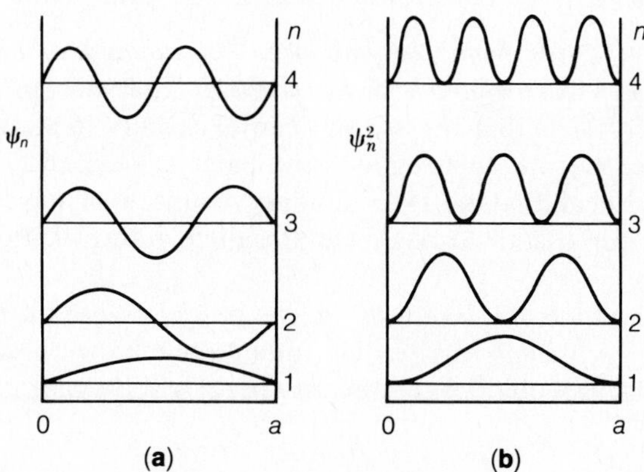

FIGURE 10-11
A particle in a one-dimensional box. **(a)** The wave functions, ψ_n, are shown by the sinusoidal waves. The horizontal lines indicate the relative energies, E_n. Note that E_1 does not equal 0. **(b)** The probability functions, ψ_n^2. (Redrawn from Karplus and Porter, 1970.)

When $n = m$, this expression cannot be zero, of course, because $\int \psi^2\,dx$ must be unity for a normalized wave function.

EXERCISE 10-10

Use the condition that ψ is normalized to determine A.

ANSWER

$$\int \psi_n^{\,2}\,dx = A^2 \int_0^a \sin^2\left(\frac{n\pi x}{a}\right) dx = 1$$

Let $y = n\pi x/a$, so $dy = (n\pi/a)\,dx$ and $dx = a\,dy/(n\pi)$. The integral is transformed by this change of variables to

$$A^2 \int_0^a \sin^2\left(\frac{n\pi x}{a}\right) dx = \frac{A^2 a}{n\pi} \int_0^{n\pi} \sin^2 y\,dy = 1$$

The integral of $\sin^2 y\,dy$ can be found to be $\frac{1}{2}y - \frac{1}{4}\sin 2y$ in a table of integrals, yielding

$$\left(\frac{A^2 a}{n\pi}\right)\left(\frac{n\pi}{2}\right) = 1$$

so we obtain the result that $A = \sqrt{2/a}$.

○ *The Wider Significance of the Particle in a Box*

Though a particle in a one-dimensional box seems too simple and abstract to have much to do with chemistry, the quantum mechanical description of this system illustrates a number of important and general points:

1. *Quantization arises from the imposition of boundary conditions on the de Broglie wave associated with the particle.* The substance of Equations 10-36 and 10-37 is that the system can exist only in states of quantized energy because the de Broglie wave must have a certain value at the boundary. Recall that it was a boundary condition of wave continuity that accounted for quantization in de Broglie's interpretation of the Bohr atom.
2. *In the lowest quantum level, $n = 1$, the particle retains a residual energy, termed the zero-point energy.* The quantum number n cannot be zero, because then ψ would be zero everywhere, and the particle would not be in the box.

EXERCISE 10-11

Place an electron ($m = 9.1 \times 10^{-28}$ g) in a one-dimensional box about the size of an atom (say, 2×10^{-8} cm). Compute the zero-point energy, E_1, and the energy of transition to the next higher energy level.

ANSWER

$E_1 = 1.5 \times 10^{-11}$ erg $= 9.40\,eV = 217$ kcal mol^{-1}. The spacing between level n and $n+1$ is

$$E_{n+1} - E_n = \frac{h^2}{8ma^2}[(n+1)^2 - n^2] = \frac{h^2}{8ma^2}(2n+1)$$

For $n = 1$, $E_2 - E_1 = 3E_1 = 28.2\,eV$.

---

3. For a macroscopic particle at ordinary energies the quantum number is very large. As is illustrated by the following example, the separation between quantum levels is small compared to the energy itself, and the energy appears to be a continuous function. Thus large particles are described adequately by classical mechanics. This is an example of Bohr's *correspondence principle,* which states that in the limit of large quantum numbers, the quantum mechanical result approaches the classical mechanical result.

EXERCISE 10-12

Suppose you are playing a one-dimensional pinball game, in which a 10 g ball is confined to a 100 cm trough. When the ball is rolling at 10 cm s^{-1}, what is the quantum number that describes its motion and what is the energy spacing between quantum levels? Can you observe transitions of the ball among energy states?

ANSWER

Note that when the ball is rolling at 10 cm s^{-1}, its kinetic energy is $mv^2/2 = 500$ ergs. This can be equated to Equation 10-38 to solve for n. Then $n = 3 \times 10^{30}$ and the spacing between levels is about 3.3×10^{-28} erg. This is far too small an energy difference to detect by any means.

---

4. *As the size of the box increases the energy decreases*. The argument here is exactly as in 3. Note that because the potential energy is zero, all energy is kinetic. Thus, as the confinement of a wave-particle is relaxed, its kinetic energy is lowered, or in other words, its stability is increased. Decreased confinement of electrons or *delocalization* leads to increased stability, as happens in aromatic molecules. Another way of viewing this is that decreased confinement permits the spreading of the sinusoidal wave functions of Figure 10-11, so that they are less curved. Since the kinetic energy is proportional to the curvature of the wave function, $d^2\psi/dx^2$, the kinetic energy decreases as the curvature decreases.
5. *The probability functions of Figure 10-11 give only the probability of finding the particle at each point within the box*. It is not possible to say more about where it is. There are *nodes* (points where $\psi^2 = 0$) where the

particle cannot be. For very high quantum numbers, points at which the particle has a high probability of being found will move very close to one another. In other words, the particle can be very many places in the box. This is in accord with our experience at handball, pinball, and other particle-in-a-box activities. It is another example of Bohr's correspondence principle that at high quantum numbers, the quantum mechanical result passes smoothly over to the classical result.

○ 10–4 *The Harmonic Oscillator is a Model for Molecular Vibration*

The harmonic oscillator is a first approximation to a vibrating molecule, and illustrates the principles of quantization in vibrating systems. Vibrating systems at the molecular level include not only the internal vibrations of a single molecule, but also the vibrations of entire molecules about positions of equilibrium in solids, and vibrations of molecules about positions of temporary equilibrium in liquids and solutions. The rates of vibration of substrates on the surfaces of enzymes are thought to be a factor in determining the rates of catalyzed reactions. In this section we review the classical description of harmonic motion, then consider the quantum mechanical description, and finally discuss the analogy of the harmonic oscillator to a vibrating diatomic molecule.

The Classical Harmonic Oscillator Can Have Any Energy

A mass suspended from a spring executes sinusoidal or *harmonic* oscillations [Fig. 10-12(a)] if the force tending to restore the mass to its equilibrium position is proportional to its displacement from equilibrium, x:

$$F = -kx \qquad (10\text{-}40)$$ ◄

in which k is the *force constant*. This way of stating the force law for harmonic motion is exactly equivalent to saying that the potential energy of the oscillator is

$$U = \tfrac{1}{2}kx^2 \qquad (10\text{-}41)$$ ◄

You can prove this equivalence if you recall that the force is the negative derivative of the potential energy with respect to displacement:

$$F = -\frac{d}{dx}\left(\tfrac{1}{2}kx^2\right) = -kx$$

Equation 10-41 tells us that the potential energy has a parabolic dependence on the displacement of the oscillator from equilibrium, rising towards

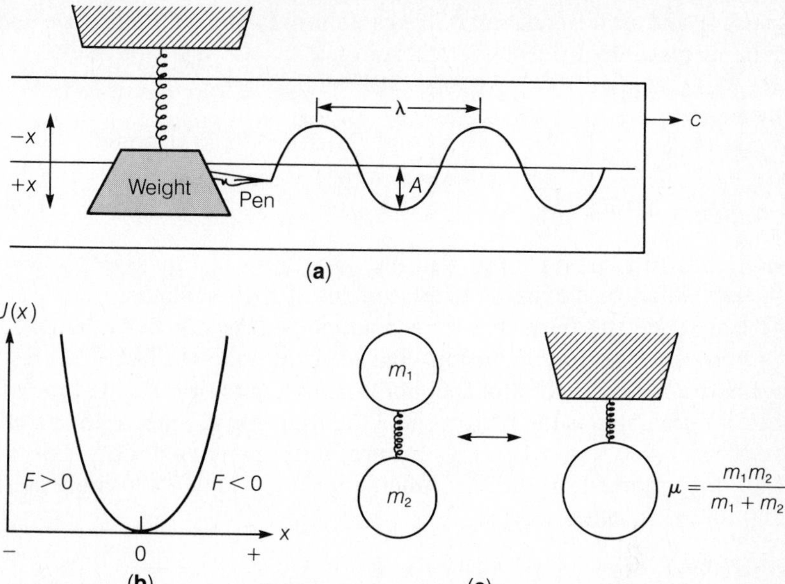

FIGURE 10-12
The harmonic oscillator. **(a)** A mass suspended by a spring executing harmonic vibrations. A pen attached to the mass traces out the displacement x of the mass from its equilibrium position on a piece of paper pulled by the observer at a constant rate of c cm s^{-1}. The wavelength λ and amplitude A of the oscillation are shown on the tracing. The frequency ν s^{-1} of the oscillation is given by c/λ. **(b)** A plot of Equation 10-41 for the potential energy of the oscillator as a function of its displacement from equilibrium. **(c)** The oscillator consisting of masses m_1 and m_2 is mechanically equivalent to an oscillator of a single reduced mass, μ, suspended from a fixed support.

infinity as the displacement increases in either the positive or negative directions [Fig. 10-12(*b*)].

To describe harmonic motion by Newton's equation, $F = ma$, we substitute Equation 10-40 for the force, and represent the acceleration as d^2x/dt^2, thereby giving

$$-kx = m\,\frac{d^2x}{dt^2}$$

or

$$\frac{d^2x}{dt^2} + \frac{k}{m}x = 0 \qquad (10\text{-}42)$$

You can confirm by differentiating twice with respect to t that the general solution to Equation 10-42 is

$$x(t) = A\sin\left(\sqrt{\frac{k}{m}}\,t\right) + B\cos\left(\sqrt{\frac{k}{m}}\,t\right) \qquad (10\text{-}43a)$$

in which A and B are constants. We can simplify the solution if we specify that the displacement x is zero at zero time:

$$0 = A \sin(0) + B \cos(0)$$

This can be true only if $B = 0$, so that Equation 10-28 becomes

$$x(t) = A \sin\left(\sqrt{\frac{k}{m}}\, t\right) \qquad (10\text{-}43b)$$

From Figure 10-12(a) it is clear that the constant A in Equation 10-43b must be the *amplitude* or maximum displacement of the oscillator.

We can relate the frequency of oscillation, ν (the number of oscillations per second), to the force constant and mass as follows. Let τ be the time required for one oscillation of the vibrator; this quantity is called the *period* of the vibration, and is the reciprocal of the frequency. Then as the oscillator moves from $t = 0$ to $t = \tau$, it must return to the same position x. For this to be true, the argument of the sine function in Equation 10-43b must increase from 0 to 2π. In other words,

$$\sqrt{\frac{k}{m}}\,\tau = 2\pi$$

Therefore,

$$\nu = \frac{1}{\tau} = \frac{1}{2\pi}\sqrt{\frac{k}{m}} \qquad (10\text{-}44)$$ ◀

Note that the frequency of a vibration increases with a stronger force constant or a smaller mass.

EXERCISE 10-13

Water absorbs infrared radiation strongly in the wavelength region of 3 μm (3400 cm^{-1}; see Sec. 12-3 on energy units), owing to the O—H stretching vibration of the water molecule. What wavelength would you predict for the corresponding absorption of D_2O?

ANSWER

H_2O and D_2O have the same electronic structure but different masses. The O—H and O—D bonds thus are identical and must have the same stiffness. Therefore you would expect their force constants to be identical. From Equation 10-44 you would predict that their frequencies of vibration (and therefore frequencies of absorption) would differ only because of the different masses. Equation 10-44 suggests that the frequencies of absorption will be inversely proportional to the square roots of the masses of the vibrators. What is vibrating during the O—H stretching motion? Since oxygen is much heavier than hydrogen, you can think of the oxygen nucleus as remaining nearly stationary during the stretching motion, and all motion being confined to the hydrogen. Then you would expect:

$$\frac{\nu_{\text{O-H}}}{\nu_{\text{O-D}}} = \sqrt{\frac{m_{\text{D}}}{m_{\text{H}}}} \simeq \sqrt{\frac{2}{1}}$$

Since the wavelength is inversely proportional to the frequency, you would predict that the O—D absorption would be at wavelength 3 μm $\times$ $\sqrt{2}$ = 4.2 μm. The observed absorption is centered near 4 μm.

This prediction can be refined somewhat by taking into account that both oxygen and hydrogen nuclei move during the vibration. The effective vibrating mass actually is the *reduced mass,* μ, of the oscillator (see Fig. 10-12 and Sec. 10-5) defined as

$$\mu = \frac{m_1 m_2}{m_1 + m_2}$$

in which m_1 and m_2 are the masses of the two objects forming the vibrator. The reduced masses are, in terms of the mass of the proton ($m_H = 1.67 \times 10^{-27}$ kg),

$$\mu_{O-H} = \frac{16 \times 1}{17} m_H = 0.94\, m_H$$

$$\mu_{O-D} = \frac{16 \times 2}{18} m_H = 1.78\, m_H$$

and the square root of their ratio is

$$\sqrt{\frac{\mu_{O-D}}{\mu_{O-H}}} = 1.38$$

This leads to a prediction of 4.1 μm for the wavelength of absorption.

EXERCISE 10-14

Carbon monoxide absorbs infrared radiation strongly at a wavelength of 4.66 μm (2144 cm^{-1}), owing to its internal vibration. (a) What is the frequency of the vibration? (b) What is the period of the vibration? (c) Force constants often are used to characterize the stiffness of a bond. Calculate the force constant for CO, noting that you must use the reduced mass in Equation 10-44. (d) Is the CO bond stiffer or more pliable than the HCl bond? The vibrational absorption for HCl is at 3.46 μm (2890 cm^{-1}).

ANSWER

(a) $\nu = c/\lambda = 3.00 \times 10^{10}$ cm s^{-1} $\times$ 10^{-2} m cm^{-2}/(4.66 $\times$ 10^{-6} m) = 6.44×10^{13} s^{-1}.

(b) $\tau = 1/\nu = 1/(6.44 \times 10^{13}\ \text{s}^{-1}) = 1.55 \times 10^{-14}$ s.

(c) From Equation 10-44, $k = 4\pi^2\nu^2\mu$;

$$\mu = (12.0 \times 16.0/28.0)(1.67 \times 10^{-24}\ \text{g}) = 1.15 \times 10^{-23}\ \text{g};$$

$$k = 4\pi^2(6.44 \times 10^{13}\ \text{s}^{-1})^2(1.15 \times 10^{-23}\ \text{g})$$
$$= 1.87 \times 10^{6}\ \text{g s}^{-2} = 1.87 \times 10^{6}\ \text{dyne cm}^{-1}$$
$$= 1.87 \times 10\ \text{N m}^{-1}$$

(d) $k = 0.438 \times 10^6$ dyne cm^{-1} = 4.38 N m^{-1} for HCl. Therefore the CO bond is considerably stiffer.

EXERCISE 10-15

Show that the total energy of the harmonic oscillator, E, is $kA^2/2$, a constant independent of time and of the position of the oscillator along its path.

ANSWER

$E = E_k + U = mv^2/2 + k\,x^2/2$. Since $x = A \sin(\sqrt{k/m}\,t)$,
$v = dx/dt = A\sqrt{k/m}\cos(\sqrt{k/m}\,t)$, and
$v^2 = A^2(k/m)\cos^2(\sqrt{k/m}\,t)$. Therefore
$E = A^2(k/2)\cos^2(\sqrt{k/m}\,t) + A^2(k/2)\sin^2(\sqrt{k/m}\,t) = kA^2/2$.

This is an important result, because it shows that the energy of a classical harmonic oscillator can have any value. This is not true of the quantum harmonic oscillator.

To compare with our quantum mechanical predictions in the next section, we will want to know the probability of finding the oscillator at any given value of x along its path. It is clear that the faster the oscillator is moving at a given value of x, the less likely we are to find it at that point, and vice versa. Therefore the probability, p_{obs}, of observing the vibrator at x is inversely proportional to the velocity of the vibrator at that x,

$$p_{obs} \propto \frac{1}{\text{velocity}} = \frac{1}{\sqrt{\dfrac{2E}{m} - \dfrac{kx^2}{m}}} \tag{10-45}$$

The last step follows by solving for v from the expression for the total energy, $E = kx^2/2 + mv^2/2$. This probability is shown by the dashed lines in Figure 10-13(b). Note that the oscillator is most likely to be found at the extremes of its path where it moves most slowly, and least likely to be found at the equilibrium position, where it moves most rapidly.

The Quantum Mechanical Harmonic Oscillator has Quantized Energy Levels

To describe the harmonic oscillator by quantum mechanics we must substitute Equation 10-41 for the potential energy into Schrödinger's equation, giving

$$\frac{d^2\psi}{dx^2} + \frac{8\pi^2 m}{h^2}[E - \tfrac{1}{2}kx^2]\psi = 0 \tag{10-46}$$

Solving this equation for ψ is a bit involved, and readers interested in the mathematical procedures are referred to Kauzmann (1957), or other texts on quantum chemistry. The result is that Equation 10-46 has solutions only for certain values of the total energy, given by

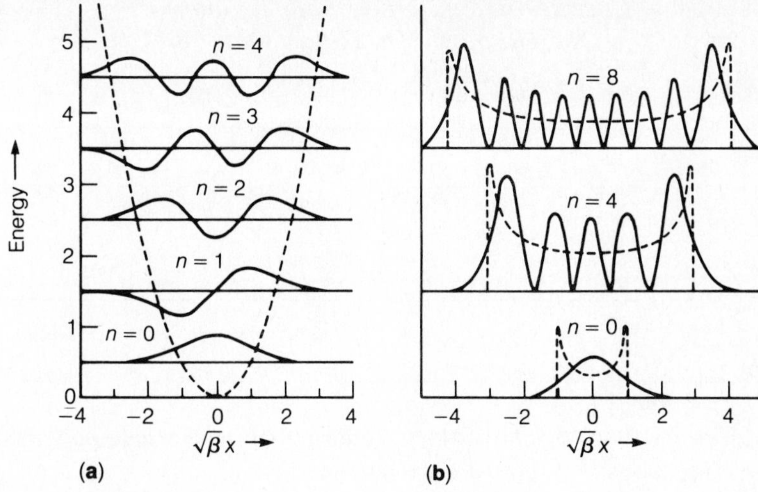

FIGURE 10-13
Wave functions for the harmonic oscillator. **(a)** The horizontal lines show the energy levels in units of $h\nu$. Superimposed on each line is the wave function for that level. The dashed line shows the potential-energy function. Note that the energy for level 0 is not zero. **(b)** Probability functions for levels $n = 0$, 4, and 8. The solid lines show the quantum mechanical probability distributions, ψ_n^2. The dashed lines show the classical probability distributions computed from Equation 10-45, for the same energy E. (Reproduced from Kauzmann, 1957.)

$$E_n = (n + \tfrac{1}{2})h\nu \qquad \text{(vibration in one dimension)} \qquad (10\text{-}47)$$

in which n is an integer called the *vibrational quantum number*. As with the particle in the box, the energy is quantized, but here the energy levels are equally spaced [Figure 10-13(a)]. Notice that in the *ground state* ($n = 0$) the oscillator has a residual vibrational energy, $E = h\nu/2$. This means that even at 0 K all oscillators retain some vibrational energy, which is called the *zero-point energy*.

Wave Functions of the Harmonic Oscillator

The wave function for the nth quantum level is given by

$$\psi_n(x) = H_n(\xi)e^{-\beta x^2/2} \qquad (10\text{-}48a)$$

in which $\beta = 2\pi\sqrt{mk}/h$ and H_n is a polynomial, called *Hermite's polynomial,* given by

$$H_n = (-1)^n e^{\xi^2}\frac{d^n e^{-\xi^2}}{d\xi^n} \qquad (10\text{-}48b)$$

in which $\xi = \sqrt{\beta}\,x$. Values of the Hermite polynomials and wave functions can be derived from Equations 10-48(a) and 10-48(b), and results for the first three levels are summarized in Table 10-1.

TABLE 10-1 *Energies, Hermite Polynomials, and Wave Functions for the Harmonic Oscillator. (Note that* $\xi = \beta^{1/2}x = \left(\frac{2\pi}{h}\right)^{1/2}(mk)^{1/4}x$.)

n	E_n	$H_n(\xi)$	ψ_n
0	$\frac{1}{2}h\nu$	1	$e^{-\beta x^2/2}$
1	$\frac{3}{2}h\nu$	2ξ	$2(\beta)^{1/2}xe^{-\beta x^2/2}$
2	$\frac{5}{2}h\nu$	$4\xi^2 - 2$	$(4\beta x^2 - 2)e^{-\beta x^2/2}$

EXERCISE 10-16

From Equations 10-48(a) and 10-48(b), confirm that the Hermite polynomials and wave functions in Table 10-1 are correct.

EXERCISE 10-17

Confirm by substitution that the wave functions ψ_0 and ψ_1 of Table 10-1 are solutions to Schrödinger's equation for the harmonic oscillator (Equation 10-46).

Several wave functions ψ_n and probability functions ψ_n^2 are plotted in Figure 10-13. As for the particle in the box, the harmonic oscillator wave functions are negative in some regions, but the probability functions are always positive. In the ground state, the probability of finding the oscillator is highest at $x = 0$, exactly where classical mechanics predicts the probability is lowest. Moreover, quantum mechanics shows that there is a finite probability for finding the oscillator at large values of x, beyond the limits of excursion of the classical oscillator. In higher quantum levels, however, the quantum mechanical results begin to approach the classical results, as they must according to the correspondence principle.

EXERCISE 10-18

Suppose that the harmonic oscillator is a valid model for representing vibrations of a diatomic molecule in its ground vibrational state. Compare predictions of the expected distributions of bond lengths from the classical and quantum mechanical models. What meaning is there in citing a bond length for a diatomic molecule?

EXERCISE 10-19

For many diatomic molecules, the force constant for vibration is sufficiently large (the bond is sufficiently stiff) that the size of a vibration quantum ($h\nu$) is large compared to available thermal energy ($k_B T$, where k_B is Boltzmann's constant) at room temperature. Consequently most molecules are in the ground vibrational state. According to Boltzmann's distribution law, the ratio of molecules in the first vibrational level to the number in the ground level, $N_{n=1}/N_{n=0}$, is given by

$$\frac{N_{n=1}}{N_{n=0}} = e^{-\Delta E/k_B T}$$

in which ΔE is the energy separation between the ground and first levels. For HCl, this separation is 2890 cm^{-1} = 8.26 kcal mol^{-1} = 34.6 kJ mol^{-1}. What fraction of molecules are in the $n = 1$ level at 300 K?

ANSWER

9.5×10^{-7}, or about 1 in 10^6.

Relationship of the Harmonic Oscillator Model to the Vibration of a Diatomic Molecule

A little thought tells us that the harmonic oscillator potential function, $U = kx^2/2$, cannot be a complete description of the potential energy of a diatomic molecule as a function of the separation of its nuclei. The reason is that as the internuclear separation, x, is increased, eventually the molecule dissociates into atoms, and the energy of the separated atoms must be independent of their internuclear distance. But the energy of the harmonic oscillator potential function continues to rise indefinitely as x is increased [Fig. 10-13(a)]. Clearly, a more sophisticated function is required.

A number of empirical functions have been devised to represent the potential energy of vibration of a diatomic molecule as a function of the separation of its nuclei. One proposed by P. M. Morse is

$$U(x) = D_e \{1 - \exp[-a(x - x_e)]\}^2 \tag{10-49}$$

This function, plotted in Figure 10-14, displays the correct behavior at large values of x. Its minimum is at the value $x = x_e$, the equilibrium bond length of the diatomic molecule. At smaller values of x the energy rises rapidly, as it must as the bond length is compressed.

EXERCISE 10-20

Show for the Morse potential (Equation 10-49) that $U(\infty) - U(x_e) = D_e$. This quantity is called the *electronic binding energy* of the molecule. Show that near $x = x_e$, the Morse potential has the form $U(x) = D_e a^2 (x - x_e)^2$, which is the equivalent to the harmonic oscillator potential. (*Hint:* Square the bracketed term in Equation 10-49, and then expand the exponential in the series $e^{-x} = 1 - x + x^2/2! - \cdots$.) This shows that the harmonic oscillator is a good representation of a vibrating diatomic molecule for small displacements of the nuclei, or in other words, for the lower vibrational states.

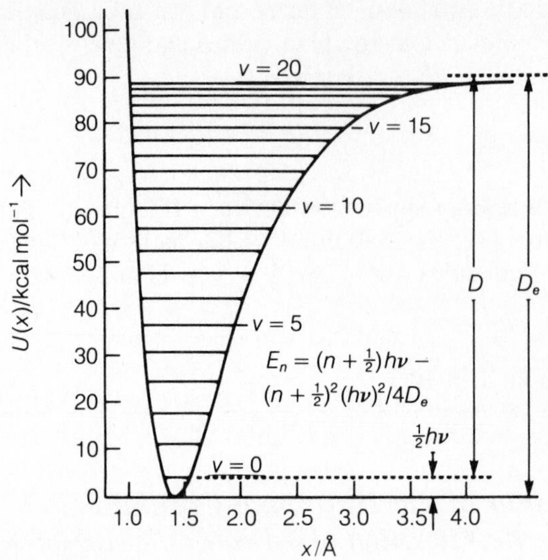

FIGURE 10-14
The Morse potential function (Equation 10-49) for the HBr molecule. The potential function is the continuous curve. The horizontal lines are the energy levels observed in the vibrational spectrum. The parameters for the curve are $D_e = 90.5$ kcal mol^{-1}, $a = 1.814 \times 10^8$ cm^{-1}, and $x_e = 1.41 \times 10^{-8}$ cm. (Redrawn from Moelwyn-Hughes, 1961.)

Schrödinger's equation can be solved when the Morse potential is inserted (Moelwyn-Hughes, 1961) and the result for the possible energies is

$$E_n = (n + \tfrac{1}{2})h\nu - (n + \tfrac{1}{2})^2(h\nu)^2/4D_e \qquad (10\text{-}50)$$

These energy levels are plotted in Figure 10-14, where it can be seen that the levels become more closely spaced at higher quantum numbers. This is a result of the broadening of the potential curve at increased nuclear separation. This closer spacing means that lines in the absorption spectrum corresponding to absorption of light for transitions between adjacent levels move to longer wavelength. This permits determination of the electronic binding energy of a molecule purely from its spectrum of absorbed light.

EXERCISE 10-21

(a) How is the kinetic energy of the vibrating molecule represented in Figure 10-14? (b) What is the physical meaning of the vertical separation of a solid horizontal line and the uppermost dashed horizontal line?

ANSWER

(a) The kinetic energy is the difference between the total energy E_n (horizontal line) and the potential energy curve. (b) The energy required to dissociate the molecule from a particular vibrational state.

10–5 The Rigid Rotor is a Model for Molecular Rotation

Just as the particle in a box is the simplest model for constrained translation, and the harmonic oscillator is the simplest model for vibration, so the rigid rotor is the simplest model for molecular rotations. We review briefly the physics of rotational motion and then sketch in the solution of Schrödinger's equation for the rigid rotor. We find that because rotation is motion in two dimensions, two quantum numbers are necessary for complete description of it, and the wave function contains factors for motion in each of the two dimensions. This foreshadows our discussion of the hydrogen atom in the following section, where we see that the motion of the electron in three dimensions leads to three quantum numbers, and three factors in the wave function. In fact, the wave functions for the rigid rotor make up two of the three factors contained in the wave functions for the hydrogen atom. Beyond the usefulness of the rigid rotor model in describing rotations of molecules, the solution to this model shows us how to measure bond lengths of diatomic molecules from frequencies of spectral lines. This measurement of lengths a few millionths of an inch, sometimes to accuracies of several parts in ten thousand, surely ranks as one of the tangible marvels of quantum mechanics.

Description of the Rigid Rotor

A rigid diatomic rotor consists of two particles of masses m_A and m_B joined rigidly by a connector of length r (Fig. 10-15). We can think of these masses as atoms connected by a chemical bond. The rotor is placed with its center of mass on the origin of the spherical polar coordinate system shown in Figure 10-16. If mass A is distance r_A from the origin and mass B is distance r_B from the origin, then according to the definition of the center of mass

$$r_A m_A = r_B m_B \tag{10-51}$$

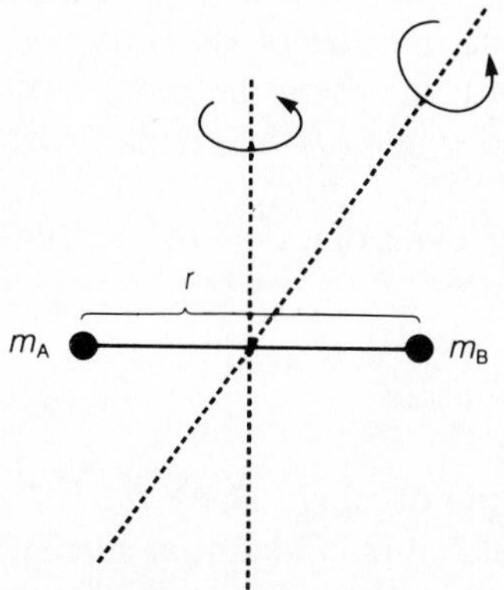

FIGURE 10-15
A rigid rotor consists of two point masses, m_A and m_B, joined by a rigid connector of length r. The motion of the rotator can be thought of as rotation about the two (dashed) axes perpendicular to the bond r. Solution of Schrödinger's equation shows that the energy levels for a rigid rotator are $2J + 1$ degenerate, corresponding to the number of ways that the rotational energy can be partitioned between rotation about the two axes.

Using the relationship

$$r = r_A + r_B$$

and substituting $r_A = r_B m_B/m_A$ obtained by solving Equation 10-51 for r_A, we get

$$r = r_B\left(\frac{m_B}{m_A} + 1\right)$$

This equation can be solved for r_B:

$$r_B = \left(\frac{m_A}{m_A + m_B}\right) r \qquad (10\text{-}52a)$$

with a similar equation for r_A:

$$r_A = \left(\frac{m_B}{m_A + m_B}\right) r \qquad (10\text{-}52b)$$

Now we will see that the rotation of the rigid rotor can be described by the motion of a single object of *reduced mass,* μ. Recall from mechanics that the rotation of a body about an axis through its center of mass is described by a *moment of inertia, I,* defined by

$$I = \sum_i m_i r_i^2 \qquad (10\text{-}53a)$$

in which r_i is the distance of the particle of mass m_i from the axis of rotation. For our diatomic rotor, rotating about an axis perpendicular to the bond, Equation 10-53 becomes

$$I = m_A r_A^2 + m_B r_B^2 \qquad (10\text{-}53b)$$

Substituting Equations 10-52*a* and 10-52*b* into Equation 10-53*a* gives us

$$I = \frac{m_A m_B^2}{(m_A + m_B)^2} r^2 + \frac{m_B m_A^2}{(m_A + m_B)^2} r^2$$

$$= \left(\frac{m_A m_B}{m_A + m_B}\right) r^2 = \mu r^2 \qquad (10\text{-}53c)$$

in which the quantity $\mu = m_A m_B/(m_A + m_B)$ is called the *reduced mass* of the system. This is the analogue in rotational motion of the situation described for vibrational motion in Figure 10-12, in which the motion of two particles can be described by the motion of a single particle with a reduced mass μ.

In the next section we show that the energy levels of a rigid rotor, obtained by solving Schrödinger's equation, are given by

$$E = J(J + 1)\frac{h^2}{8\pi^2 \mu r^2} \qquad \text{(rotational motion)} \qquad (10\text{-}54) ◀$$

in which $J = 0, 1, 2 \ldots$ is an integer quantum number. There are $(2J + 1)$ different wave functions for each energy level J, and in quantum mechanical terminology the energy level E is said to be $(2J + 1)$ *degenerate*. Degeneracy means two or more quantum mechanical states have the same energy. The $(2J + 1)$ ways of obtaining a rotational energy E correspond to the different ways of partitioning the rotational energy between rotation about the two perpendicular axes in Figure 10-15.

Schrödinger's Equation for the Rigid Rotor Separates into Two Equations

The steps for solving Schrödinger's equation for the rigid rotor are the following: (1) The independent variables of Schrödinger's equation are transformed from the Cartesian system x, y, and z to the spherical polar system of Figure 10-16; (2) The moment of inertia is substituted for r^2 times the mass, and the $U(x)$ is set to zero because the energy of the rotor is independent of its orientation (as long as it is not in the presence of an applied electric or magnetic field); (3) The equation is separated into two, each containing only one of the angular variables; (4) Each equation is solved, thereby yielding a quantum number. The expression for the quantized energy contains the moment of inertia, so that an energy measurement can yield the bond length of the diatomic rotor. We only will sketch in these steps here; readers interested in a more detailed description are referred to Kauzmann (1957), Karplus and Porter (1970), and other texts on quantum chemistry.

To express Schrödinger's equation in spherical polar coordinates is a straightforward exercise in transformation of variables; it is described in detail by Karplus and Porter (1970). The result is

$$\frac{1}{r^2}\frac{\partial}{\partial r}\left(r^2\frac{\partial\psi}{\partial r}\right) + \frac{1}{r^2 \sin\theta}\frac{\partial}{\partial\theta}\left(\sin\theta\frac{\partial\psi}{\partial\theta}\right) + \frac{1}{r^2\sin^2\theta}\frac{\partial^2\psi}{\partial\phi^2} + \frac{8\pi^2 m}{h^2}[E - U(r,\theta,\phi)]\psi = 0 \quad (10\text{-}55)$$

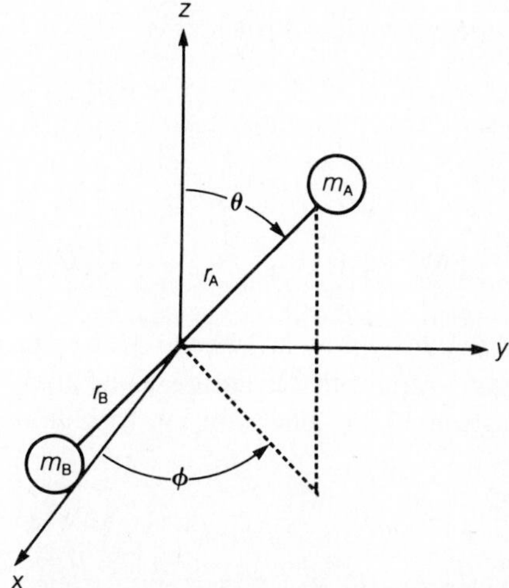

FIGURE 10-16
Spherical polar coordinate system for describing the rigid rotor. The center of mass of the rotor is at the origin of the coordinate system. The masses are separated by a distance $r = r_A + r_B$. Rotation is about an axis perpendicular to the line connecting the masses.

in which θ and ϕ are the angular coordinates of Figure 10-16, and $r = r_A + r_B$ is the effective distance of the particle from the origin. Since r is a constant during movement of the rigid rotor, $\partial\psi/\partial r = 0$, and the first term in Equation 10-55 vanishes. Also for the rigid rotor, the potential energy is independent of position and can be neglected, and r^2 times the mass in Equation 10-55 becomes the moment of inertia. Then Equation 10-55 can be simplified (remember that $d \sin\theta/d\theta = \cos\theta$) to

$$\frac{\partial^2\psi}{\partial\theta^2} + \frac{\cos\theta}{\sin\theta}\frac{\partial\psi}{\partial\theta} + \frac{1}{\sin^2\theta}\frac{\partial^2\psi}{\partial\phi^2} + \frac{8\pi^2 I}{h^2}E\psi = 0 \qquad (10\text{-}56)$$

Because the dependences on θ and ϕ can be segregated into different terms in this partial differential equation, it can be separated into two ordinary differential equations. This is done by assuming a solution of the form

$$\psi = \Theta(\theta)\Phi(\phi) \qquad (10\text{-}57)$$

in which Θ and Φ depend only on the angles θ and ϕ respectively. Substituting Equation 10-57 for ψ in Equation 10-56, using the chain rule of differentiation, multiplying by $\sin^2\theta$ and dividing by $\psi = \Theta\,\Phi$ yields

$$\frac{\sin^2\theta}{\Theta}\frac{d^2\Theta}{d\theta^2} + \frac{\cos\theta\,\sin\theta}{\Theta}\frac{d\Theta}{d\theta} + \frac{1}{\Phi}\frac{d^2\Phi}{d\phi^2} + \frac{8\pi^2 I\,E}{h^2}\sin^2\theta = 0 \qquad (10\text{-}58)$$

Only the third term of this equation depends on ϕ. The only way that the sum of all terms can be zero is if the term that depends on ϕ is a constant $(-m^2)$, and the sum of the terms that depend on θ is minus the same constant (m^2):

$$\frac{1}{\Phi}\frac{d^2\Phi}{d\phi^2} = -m^2 \qquad (10\text{-}59a)$$

$$\frac{\sin^2\theta}{\Theta}\frac{d^2\Theta}{d\theta^2} + \frac{\cos\theta\,\sin\theta}{\Theta}\frac{d\Theta}{d\theta} + \frac{8\pi^2 I\,E}{h^2}\sin^2\theta = m^2 \qquad (10\text{-}59b)$$

These two equations can be rewritten in the more standard forms

$$\frac{d^2\Phi}{d\phi^2} + m^2\,\Phi = 0 \qquad (10\text{-}60a)$$

and

$$\frac{d^2\Theta}{d\theta^2} + \frac{\cos\theta}{\sin\theta}\frac{d\Theta}{d\theta} + \left[\frac{8\pi^2 I}{h^2}E - \frac{m^2}{\sin^2\theta}\right]\Theta = 0 \qquad (10\text{-}60b)$$

Determining the solution to Schrödinger's equation now has been reduced to finding the solution of each of these ordinary differential equations, and then taking the product of functions as in Equation 10-57. The solution to Equation 10-60a is easily found to be

$$\Phi = \exp(im\phi) \qquad (10\text{-}61a)$$

in which $i = \sqrt{-1}$ and m is a constant and an integer, $m = 0, \pm 1, \pm 2 \cdot \cdot \cdot$.

EXERCISE 10-22
Confirm by substitution that Equation 10-61*a* is a solution of Equation 10-60*a*.

The constant m must be restricted to integral values ($m = 0, \pm 1, \pm 2, \ldots$) if the wave function Φ is to have a single value at each point. (To be single valued, Φ must have the same value at $\phi + 2\pi$ as it has at ϕ. Therefore

$$\exp(im\phi) = \exp[im(\phi + 2\pi)] = \exp(2\pi im)\exp(im\phi)$$

Since $\exp(2\pi im) = 1$ only when m is an integer, this equation can be valid only when m is an integer, $m = 0, \pm 1, \pm 2 \ldots$). The integer m is one of the two quantum numbers for rotational motion.

If you compare Equation 10-60*a* to Equation 10-33 you will see that they are essentially the same equation. Yet for Equation 10-33 we write the solution as a sinusoidal function and for Equation 10-60*b* we write the solution as an exponential. This illustrates a general principle of quantum mechanics: a set of wave functions with the same energy can be added or subtracted in *linear combinations* to make a new set of wave functions that are also solutions of Schrödinger's equation. For our example here, we can combine the functions $\exp(i\phi)$ and $\exp(-i\phi)$ for $m = \pm 1$ to yield the alternative solutions

$$\sin\phi = \frac{\exp(i\phi) - \exp(-i\phi)}{2i} \qquad (10\text{-}61b)$$

$$\cos\phi = \frac{\exp(i\phi) + \exp(-i\phi)}{2} \qquad (10\text{-}61c)$$

Which linear combination we use is a matter of convenience.

Equation 10-60*b* is a well-known equation of mathematical physics and is called the *associated Legendre equation*. Solutions of it exist only for certain values of E, given by

$$E = J(J + 1)\frac{h^2}{8\pi^2 I} \qquad (10\text{-}62)$$

in which J is any integer 0, 1, 2, 3, ... For solutions of Legendre's equation the term m is restricted to values from $-J$ to $+J$. Notice that the energy (Equation 10-62) does not depend on the value of m. This means that there are $2J + 1$ quantum states of the same energy corresponding to the different values of m. These states of equal energy are said to be *degenerate*. This degeneracy is removed by an electric or magnetic field if the molecule contains an electric or magnetic dipole moment; then the energy depends on m as well as on J. If we substitute μr^2 for the moment of inertia in Equation 10-62, we get

$$E = J(J + 1)\frac{h^2}{8\pi^2 \mu r^2}$$

as given earlier in Equation 10-54.

TABLE 10-2 *Solutions to Schrödinger's equation for the rigid rotor. The integers J and m are quantum numbers and E is the energy of the corresponding state. The functions $\psi(J,m)$ are the surface spherical harmonics that are solutions to Equation 10-56; they are products of $\Phi(m)$ and $\Theta(J,m)$, which are solutions, respectively, of Equations 10-60a and 10-60b.*

J	m	E (IN UNITS OF $h^2/8\pi^2 I$)	$\Phi(m)$	$\Theta(J,m)$	$\psi(J,m)$
0	0	0	1	1	1
1	0	2	1	$\cos\theta$	$\cos\theta$
1	± 1	2	$\exp(\pm i\phi)$	$\sin\theta$	$\sin\theta \exp(\pm i\phi)$
2	0	6	1	$(3\cos^2\theta - 1)/2$	$(3\cos^2\theta - 1)/2$

The solutions to Equation 10-60b are polynomials that are functions of the integers J and m. Several of them are listed in Table 10-2 where they also are multiplied by the functions $\Phi(\phi)$ to yield the total solutions, ψ, to the rigid rotor equation (10-56).

EXERCISE 10-23

Confirm by substitution from Table 10-2 that $\Theta(1,1)$ is a solution of Equation 10-60b only for $E = 2(h^2/8\pi^2 I)$. This is the condition for quantization of rotational energy stated by Equation 10-62.

Determination of Bond Lengths

Suppose that a rotating diatomic molecule is in a quantum state J and has energy E. Then it absorbs a quantum of energy ΔE as it undergoes a transition to a state of energy E' and quantum number J'. The quantum of energy then is given by

$$\text{E} = E' - E = \frac{h^2}{8\pi^2\mu r^2}[J'(J' + 1) - J(J + 1)] \qquad (10\text{-}63)$$

It can be shown (see, for example, Hanna, 1969) that the only transitions allowed between rotational quantum states are those for which $J' - J = 1$. Therefore Equation 10-63 reduces to

$$\text{E} = \frac{h^2}{8\pi^2\mu r^2} 2(J + 1) \qquad \text{(diatomic rotation)} \qquad (10\text{-}64)$$ ◀

This is an extremely useful equation. It tells us that if we know the energy of a transition to rotational quantum state J' from the neighboring state J for a diatomic molecule with reduced mass μ, we can determine the molecular bond length. Spectral lines associated with transitions between rotational quantum states of molecules are found in the microwave region of the electromagnetic spectrum (see Sec. 12-4).

EXERCISE 10-24

The transition from the $J = 0$ to $J = 1$ rotational states of carbon monoxide causes absorption of microwave radiation at 1.153×10^5 MHz [$= 1.153 \times 10^{11}$ cycles s^{-1} corresponding to an energy of $(1.153 \times 10^{11}\ s^{-1})(6.63 \times 10^{-27}$ erg s$) = 7.64 \times 10^{-16}$ erg]. What is the bond length of CO?

ANSWER

1.13Å.

○ *10–6 The Hydrogen-Atom Wave Functions, Derived from Schrödinger's Equation, are the Basis for Theories of Chemical Bonding*

In January 1926, Schrödinger communicated to *Annalen der Physik* the first of a series of five papers on his new quantum mechanics. This first paper showed that the quantized levels of the hydrogen atom, which Bohr had assumed to explain the atomic spectrum, followed naturally from the new wave equation. Physicists recognized immediately that the new mechanics was capable of describing events on the atomic level. Within a few years, it was realized that the wave functions for the hydrogen atom have a central significance in chemistry. Analysis of these functions or *atomic orbitals* suggested the physical basis for the periodic arrangement of the elements and suggested how atoms bind to one another to form molecules.

In this section we will outline Schrödinger's treatment of the hydrogen atom and discuss the wave functions that he determined and their significance for the periodic table. In the following chapter we will describe how these wave functions lead naturally to theories of the chemical bond.

To apply Schrödinger's equation to the hydrogen atom it is most convenient to use the spherical polar coordinates shown in Figure 10-17(a). The nucleus is placed at the origin, with the electron a distance r away; the position of the electron is described by the three variables r, θ, and ϕ. To generalize the problem beyond the hydrogen atom, the nuclear charge is taken to be Ze^+, so that the solutions apply to H, He^+, Li^{2+}, and any other one-electron ions. These are called *hydrogen-like ions*. The potential energy of the electron interacting with the nucleus is then

$$U = -\frac{Ze^2}{r} \tag{10-65}$$

for which the nuclear charge Z becomes 1 for the case of the hydrogen atom. Figure 10-17(b) shows a plot of this potential energy as a function of r. At infinite separation of the electron from the nucleus, there is no interaction

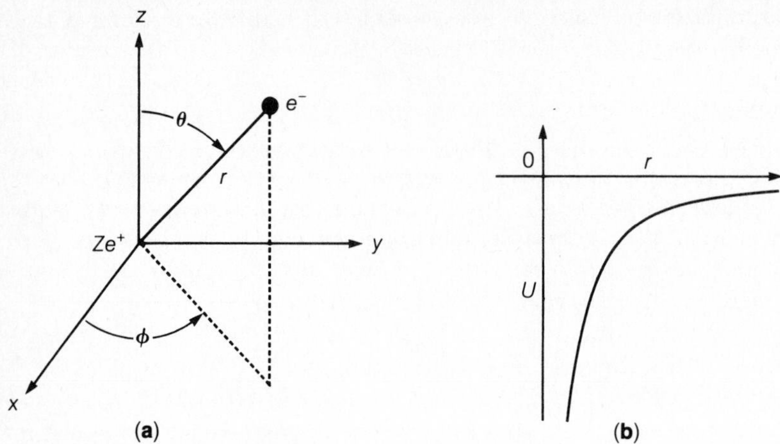

FIGURE 10-17
Mathematical description of the hydrogen atom by Schrödinger's equation. **(a)** Spherical polar coordinate system to describe the position of the electron. The nucleus is at the origin. **(b)** The potential energy as a function of the separation of the electron and nucleus.

and the potential energy is zero. As the electron approaches the nucleus the potential energy decreases (the stability increases).

EXERCISE 10-25

Why doesn't the electron achieve ultimate stability by falling into the nucleus?

ANSWER

The kinetic energy increases as the electron is confined in a smaller space (Sec. 10-3). Thus the electron achieves a balance between lower potential energy and greater kinetic energy.

Solution of Schrödinger's Equation for the Hydrogen Atom

To solve Schrödinger's equation for the hydrogen atom, the potential function (Equation 10-65) is inserted into Equation 10-55, which is Schrödinger's equation expressed in spherical polar coordinates. Since the dependence on r, θ, and ϕ occurs in different terms, and since only the r term involves the potential energy, it is possible to separate the partial differential equation into three ordinary differential equations, one in each of the variables. This is done by inserting into Equation 10-55 a solution of the form

$$\psi(r,\theta,\phi) = R(r)\Theta(\theta)\Phi(\phi) \tag{10-66}$$

in which R, Θ and Φ are functions only of the variables r, Θ and ϕ respectively. The procedure then parallels that of solving the rigid-rotor problem, except that three equations now are obtained, two of them identical to Equations 10-60*a* and 10-60*b* for the rigid rotor, and the third a differential equation for $R(r)$, the factor that describes the radial, or r, dependence of the wave function. This third equation is another eigenvalue equation with solutions that exist only for integral values of a quantum number n.

The wave functions for the hydrogen atom (see Table 10-3) thus are products of three functions. Two of these are the solutions to the rigid rotor, listed in Table 10-2, except that the quantum number l for the hydrogen atom replaces J for the rigid rotor. Also the functions of ϕ are the sinusoidal linear combinations of the form of Equations 10-61*b* and *c*.

One further point about the choice of linear combinations needs to be mentioned. Beginning students of quantum mechanics often are bothered by the fact that the choice of the set of wave functions influences the probability distribution ψ^2 or $\psi\psi^*$ associated with each wave function. For example, $\Phi^2 = \sin^2\phi$ or $\cos^2\phi$ have different angular dependences than do $\Phi\Phi^* = e^{im\phi} \times e^{-im\phi} = 1$. In fact, the choice of linear combinations of the wave functions for the angular dependence of Φ or Θ is entirely arbitrary, and therefore without physical significance, *as long as the potential energy is truly independent of angle*. However, any measurement on an atom requires an electric or magnetic field, and the field causes the potential energy to become angle-dependent. When this happens the wave functions Φ that have different values of m have different energies, and the way they are combined to form the set of wave functions is no longer entirely arbitrary. The wave functions of the form $\sin\phi$ and $\cos\phi$ are used for atomic wave functions because the angular dependence of electron density (Φ^2) that they imply is useful for representing problems like chemical bonding in which there truly is an angular dependence of the electron density.

EXERCISE 10-26

Confirm that the *total* probability density $\Phi^2 = \left(\dfrac{\Phi_1\Phi_1^* + \Phi_2\Phi_2^* + \Phi_3\Phi_3^*}{3}\right)$ is a constant, independent of angle, for the following two choices of wave function sets:

1. $\Phi_1 = \dfrac{\exp(i\phi)}{\sqrt{2\pi}};\ \Phi_2 = \dfrac{\exp(-i\phi)}{\sqrt{2\pi}};\ \Phi_3 = \dfrac{1}{\sqrt{2\pi}}$
2. $\Phi_1 = \dfrac{\sin\phi}{\sqrt{\pi}};\ \Phi_2 = \dfrac{\cos\phi}{\sqrt{\pi}};\ \Phi_3 = \dfrac{1}{\sqrt{2\pi}}$

Hint: $\sin^2\phi + \cos^2\phi = 1$. In the absence of a field that produces angular dependence of the potential energy the total probability distribution for a given energy level is the only physically meaningful quantity (Φ^2), not the contributions due to individual wave functions ($\Phi_i{}^2$) of the same energy. Notice that these functions are angular factors of p orbitals (see Table 10-3).

The solutions to the $R(r)$ equation contribute to the total wave function ψ, a polynomial in r (or, as listed in Table 10-3, a polynomial in $\sigma = Zr/a_0$). There also are additional numerical factors preceding the polynomials in the wave functions in Table 10-3 that normalize the wave function. This means that the probability $\psi^2(n, l, m)$ of finding the electron is unity when integrated over all space.

The physical meaning of the wave functions of Table 10-3 is as follows. Each describes the single electron in one of the energy states of the atom. The lowest energy state, or ground state, is described by $\psi(1,0,0)$. This is the same function we discussed previously in this chapter (Equation 10-19 when $Z = 1$). When the atom absorbs one or more quanta of energy, it is described

TABLE 10-3 *The Hydrogen-like Wave Functions. Z is the nuclear charge, a_0 is the Bohr radius, and $\sigma = Zr/a_0$.*

ELECTRON SHELL	QUANTUM NUMBERS n	l	m	ORBITAL	WAVE FUNCTION $\psi(n,l,m)$
K	1	0	0	$1s$	$\frac{1}{\sqrt{\pi}}\left(\frac{Z}{a_0}\right)^{3/2} e^{-\sigma}$
L	2	0	0	$2s$	$\frac{1}{4\sqrt{2\pi}}\left(\frac{Z}{a_0}\right)^{3/2}(2-\sigma)e^{-\sigma/2}$
	2	1	0	$2p_z$	$\frac{1}{4\sqrt{2\pi}}\left(\frac{Z}{a_0}\right)^{3/2}\sigma e^{-\sigma/2}\cos\theta$
	2	1	±1	* $2p_x$	$\frac{1}{4\sqrt{2\pi}}\left(\frac{Z}{a_0}\right)^{3/2}\sigma e^{-\sigma/2}\sin\theta\cos\phi$
				$2p_y$	$\frac{1}{4\sqrt{2\pi}}\left(\frac{Z}{a_0}\right)^{3/2}\sigma e^{-\sigma/2}\sin\theta\sin\phi$
M	3	0	0	$3s$	$\frac{1}{81\sqrt{3\pi}}\left(\frac{Z}{a_0}\right)^{3/2}(27-18\sigma+2\sigma^2)e^{-\sigma/3}$
	3	1	0	$3p_z$	$\frac{\sqrt{2}}{81\sqrt{\pi}}\left(\frac{Z}{a_0}\right)^{3/2}(6-\sigma)\sigma e^{-\sigma/3}\cos\theta$
	3	1	±1	* $3p_x$	$\frac{\sqrt{2}}{81\sqrt{\pi}}\left(\frac{Z}{a_0}\right)^{3/2}(6-\sigma)\sigma e^{-\sigma/3}\sin\theta\cos\phi$
				$3p_y$	$\frac{\sqrt{2}}{81\sqrt{\pi}}\left(\frac{Z}{a_0}\right)^{3/2}(6-\sigma)\sigma e^{-\sigma/3}\sin\theta\sin\phi$
	3	2	0	$3d_{z^2}$	$\frac{1}{81\sqrt{6\pi}}\left(\frac{Z}{a_0}\right)^{3/2}\sigma^2 e^{-\sigma/3}(3\cos^2\theta-1)$
	3	2	±1	* $3d_{xz}$	$\frac{\sqrt{2}}{81\sqrt{\pi}}\left(\frac{Z}{a_0}\right)^{3/2}\sigma^2 e^{-\sigma/3}\sin\theta\cos\theta\cos\phi$
				$3d_{yz}$	$\frac{\sqrt{2}}{81\sqrt{\pi}}\left(\frac{Z}{a_0}\right)^{3/2}\sigma^2 e^{-\sigma/3}\sin\theta\cos\theta\sin\phi$
	3	2	±2	* $3d_{x^2-y^2}$	$\frac{1}{81\sqrt{2\pi}}\left(\frac{Z}{a_0}\right)^{3/2}\sigma^2 e^{-\sigma/3}\sin^2\theta\cos 2\phi$
				$3d_{xy}$	$\frac{1}{81\sqrt{2\pi}}\left(\frac{Z}{a_0}\right)^{3/2}\sigma^2 e^{-\sigma/3}\sin^2\theta\sin 2\phi$
N	4	0	0	$4s$	$\frac{1}{192\sqrt{\pi}}\left(\frac{Z}{a_0}\right)^{3/2}(24-18\sigma+3\sigma^2-\sigma^3/8)e^{-\sigma/4}$

* For these terms linear combinations of the surface spherical harmonics are taken, rather than the harmonics themselves, in order to give sines and cosines of ϕ, instead of exponentials containing i. (See Exercise 10-26.)

by the appropriate *excited-state* wave functions, such as $\psi(2,1,0)$. The square of the wave function, ψ^2, is the probability density for the electron in the n,l,m quantum state. We will discuss these probability densities below.

Because each of Bohr's electron orbits is replaced in Schrödinger's theory by one of the wave functions of Table 10-3, each wave function is known as an *atomic orbital*. The symbols for the atomic orbitals are given in Table 10-3 and explained in the next section.

○ *Energies and Quantum Numbers for the Hydrogen-like Atoms*

Schrödinger's equation for the hydrogen-like atoms has solutions only for certain values of the energy, given by

$$E_n = -\frac{2\pi^2\mu Z^2 e^4}{n^2 h^2} \tag{10-67}$$ ◀

in which μ is the reduced mass of the electron plus nucleus (virtually equal to the electronic mass) and n is a quantum number.

EXERCISE 10-27

(a) Show that the reduced atomic mass is close to the electronic mass;
(b) Determine whether the energy levels of hydrogen-like atoms are evenly spaced.

ANSWER

(a) $\mu = mM/(m + M)$, where m and M are respectively masses of the electron and of the nucleus, Since $M \gg m$, $\mu \cong mM/M = m$. (b) They are not. The factor n^2 in the denominator means that higher levels are increasingly close together.

The quantum numbers that designate the quantum states of the hydrogen-like atoms are the following:

1. n, the *principal* quantum number, can have values 1, 2, 3, ... The energy of the atom depends only on this quantum number, as shown by Equation 10-67 (except when the atom is in an electric or magnetic field).
2. l, the *azimuthal* (orbital-shape) quantum number, can have values 0, 1, 2, . . . $(n - 1)$, where n is the principal quantum number. In chemical terminology, each hydrogen-like wave function (or atomic orbital) is designated by its principal quantum number n, followed by a letter that gives its azimuthal quantum number. The correspondence between values of l and letters is as follows:

$$\begin{array}{c} l = 0,\ 1,\ 2,\ 3 \\ \updownarrow\ \ \updownarrow\ \ \updownarrow\ \ \updownarrow \\ s,\ p,\ d,\ f \end{array}$$

Thus, for example, chemists usually refer to the $\psi(2,1)$ state as the 2*p* orbital, and the $\psi(3,0)$ state as the 3*s* orbital.

3. *m*, the *magnetic* quantum number, can have values $-l, -(l-1), \ldots, -1, 0, 1, \ldots, (l-1), l$. You can confirm that these are the only values of *m* present in Table 10-3. When the atom is placed in a magnetic field, states with different *m* values have different energies.
4. m_s, the *spin* quantum number, can have the value $+1/2$ or $-1/2$. This fourth quantum number does not follow from Schrödinger's treatment of the hydrogen atom. P. A. M. Dirac showed, however, that it does emerge from an extension that takes account of the relativistic properties of the electron.

○ *Shapes of the Atomic Orbitals*

It is important to acquire a feeling for the spatial properties of the wave functions of Table 10-3—or in other words, a feeling for the shapes of the atomic orbitals. We will do this first by considering the radial dependence of the wave functions (the factors in them that contain $\sigma = Zr/a_0$), and then by considering their angular dependence (the factors that contain θ and ϕ).

Figure 10-18(*a*) contains plots of the radial parts of several atomic orbitals as a function of *r*, the distance from the nucleus. You can confirm that each of these shapes is correct by examining the corresponding function of σ in Table 10-3. For example, $\psi(1,0,0)$, the 1*s* orbital, has radial dependence, $R(1,0,0)$, given by $e^{-\sigma}$, a simple exponential decay. This exponential form is exactly that of the top curve in Figure 10-18(*a*). The $2p_z$ orbital, $\psi(2,1,0)$ has the radial form $\sigma e^{-\sigma/2}$. The pre-exponential factor σ causes ψ to climb nearly linearly from the origin; but then the exponential factor takes over and ψ falls

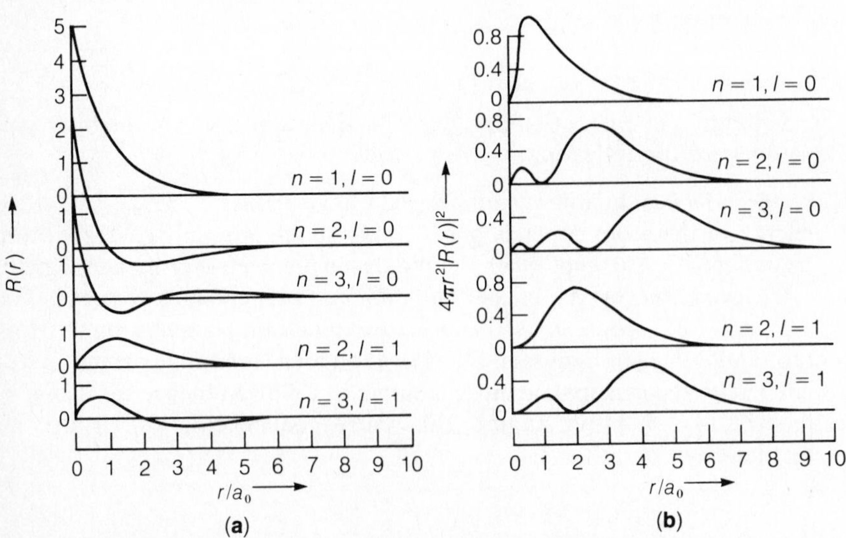

FIGURE 10-18 For legend see opposite page.

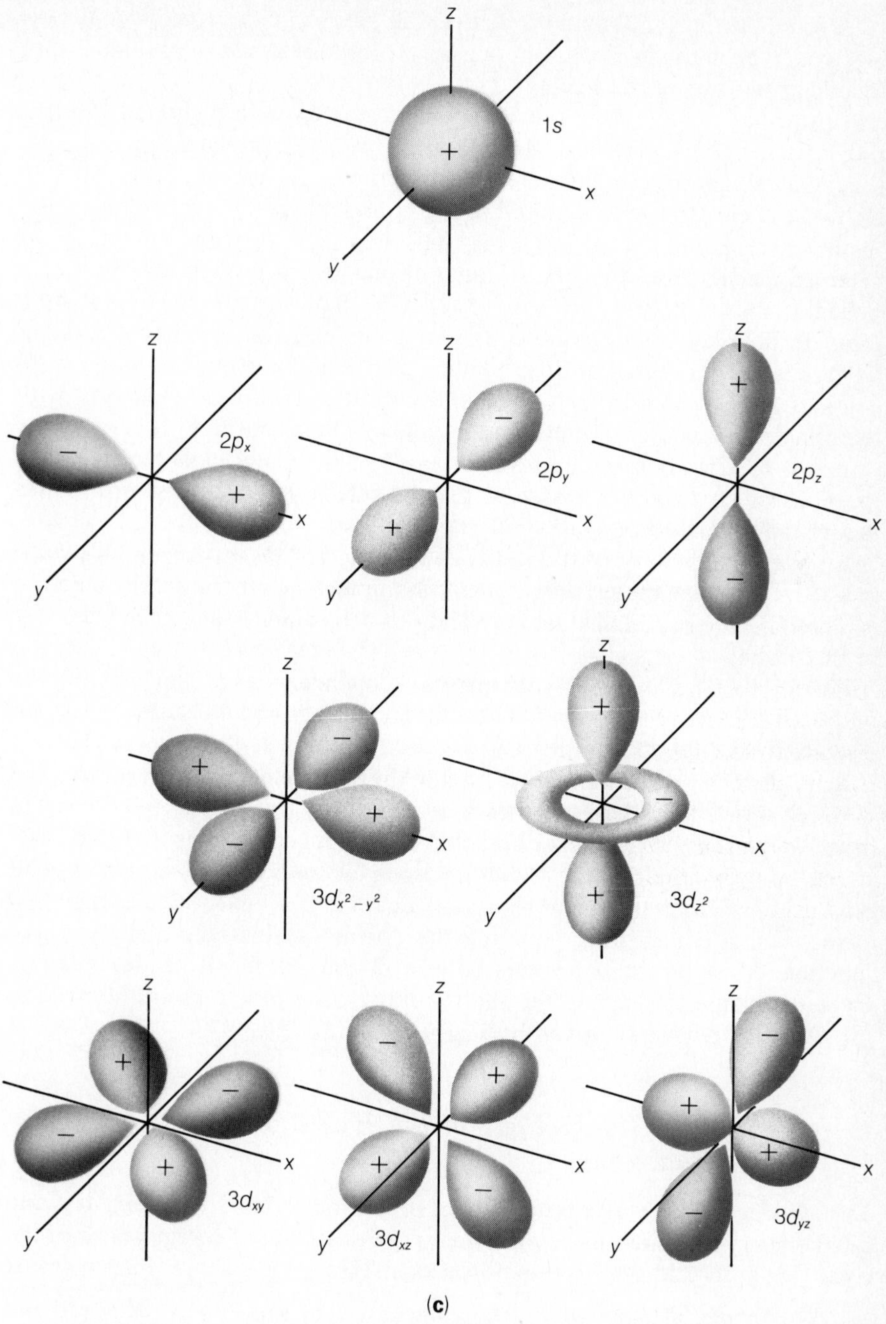

(c)

FIGURE 10-18
Representations of the wave functions for the hydrogen atom. **(a)** The dependence of $R(r)$, the radial factor of the wave function, on r/a_0. **(b)** The radial distribution function for the wave functions of **(a)**. (Redrawn from Daniels and Alberty, 1966.) **(c)** Contour surfaces for the probability densities of several of the hydrogen wave functions. The signs on the lobes show the sign of the wave function (before it was squared to give the probability density). (Redrawn from Daniels and Alberty, 1966.)

off. Note that the $1s$, $2s$, and $3s$ orbitals all have a finite probability for the electron to be at the nucleus, whereas all p orbitals have zero probability for the electron to be at the nucleus.

An alternative plot of the radial dependence is given in Figure 10-18(b) by the *radial distribution function*, $4\pi r^2[R(r)]^2$. This is the probability density for the radial dependence multiplied by the factor $4\pi r^2$, which gives the probability that the electron will be found in the spherical shell of thickness dr at a distance r from the nucleus. (This is so because $4\pi r^2\,dr$ is the volume of the spherical shell.) Note that the distance of maximum probability for the $1s$ orbital ($n = 1$; $l = 0$) is 0.529 Å, exactly the radius of the innermost Bohr orbit. In Schrödinger's picture of the atom the electron also has a substantial probability of being at other values of r. The $2s$ orbital ($n = 2$, $l = 0$) contains a node (point of zero probability) in the radial wave function, with the principal maximum at considerably higher r than for the $1s$ function. The principal maximum of the $3s$ orbital ($n = 3$, $1 = 0$) is at even higher r, and the function contains two nodes. A general rule is that the number of nodes in any radial function is $n - l - 1$ (excluding $r = 0$).

An overall summary of the radial dependence of the hydrogen-like wave functions is that as the principal quantum number (n) increases, the electron is more likely to be found at higher values of r. In short, n determines the size of the orbital.

Figure 10-18(c) depicts the angular dependences of several of the hydrogen-like wave functions. These figures are boundary surfaces for the probability density. They thus correspond to Figure 10-9(d), which shows a surface enclosing a substantial fraction of the entire electronic charge. Figure 10-18(c) shows that the s orbitals are spherically symmetric, whereas the others have some directional character. You can confirm this in Table 10-3, where you will note that the wave functions for the s orbitals do not depend on θ or ϕ, whereas the p and d orbitals do. The p orbitals have a dumbbell shape, which comes from squaring the cosine or sine factor in the wave function. Most of the d orbitals have a double-dumbbell shape. Readers interested in a discussion of the shapes of the wave functions are referred to the excellent derivation given by Karplus and Porter (1970).

The Wider Significance of the Hydrogen-like Wave Functions

The hydrogen-like wave functions have an importance in chemistry and biochemistry that extends beyond their representation of energy states of the hydrogen atom and the hydrogen-like ions. This significance is in two areas:

1. *Basis of the periodic table*. The physicist Wolfgang Pauli proposed the *exclusion principle*, which states that no two electrons in an atom can have all four quantum numbers the same. Thus if we think of building an atom by adding successive electrons to a nucleus, we must assign a new set of quantum numbers to each electron. We can do this by placing two electrons in each of the atomic orbitals of Table 10-3 (each of which is characterized by its own set of the first three quantum numbers) and by assigning different

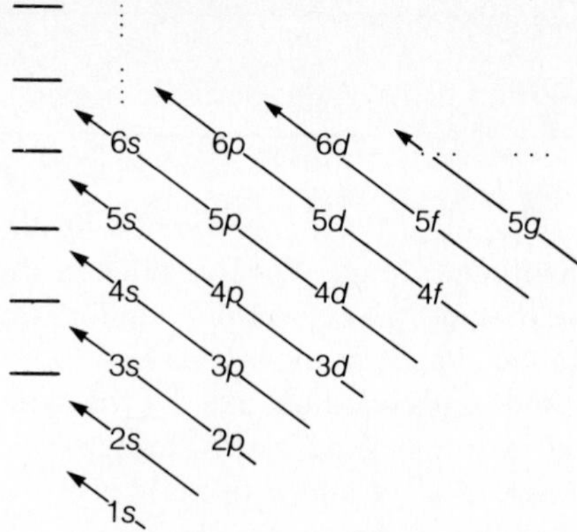

FIGURE 10-19
General order of occupancy of the atomic orbitals. Successive diagonals are drawn from lower right to upper left, starting in the left-hand corner with 1*s*. Orbitals are filled in order of their crossing by the resulting lines: 1*s*, 2*s*, 2*p*, 3*s*, 3*p*, 4*s*, 3*d*, 4*p*, 5*s*, 4*d*, 5*p*, 6*s*, 4*f*, 5*d*, 6*p*, 7*s*, 5*f*.

values of the spin quantum number to each of the two. Thus the Li atom can be thought of as having two electrons with opposite spin in the 1*s* orbital, and one electron in the 2*s* orbital. Proceeding like this, we can rationalize the entire periodic table in terms of atomic structure based on the four quantum numbers.

As you probably know from your earlier study of chemistry, there is a definite order in which the atomic orbitals are filled. For example, after the 1*s* orbital, the 2*s* orbital is filled first, giving Li and Be, before the 2*p* orbitals (B,C,N,O,F, and Ne). Because electrons are placed preferentially in the 2*s* orbitals, they must be lower in energy than the 2*p* orbitals. One of the reasons for this can be seen in Figure 10-18(*b*), in which it is shown that the 2*s* orbital ($n = 2, l = 0$) has higher electron density very close to the nucleus than does a 2*p* orbital ($n = 2, l = 1$). This "penetration" of an outer electron close to the multiply charged nucleus gives the 2*s* orbital an energy advantage over a 2*p* orbital. A more complete explanation of the order of orbital filling requires consideration of repulsions between electrons. The hydrogen-like wave functions are inadequate for this since they include only a single electron. Discussion of the quantum mechanics of many electron atoms is given in Kauzmann (1957) and other references at the end of this chapter.

A convenient diagram for summarizing the general rules for the order of filling the atomic orbitals is given in Figure 10-19. Occasional deviations from the rules occur, such as in the case of Cu, which has a configuration with only one 4*s* electron (instead of two) and ten (instead of nine) 3*d* electrons.

EXERCISE 10-28

What would be the order of filling the atomic orbitals if the energies were unchanged from Equation 10-67 for the hydrogen atom?

ANSWER

The orbitals would be filled only on the basis of the principal quantum number *n*: First the 1*s*, then 2*s* and 2*p* simultaneously, then 3*s*, 3*p*, and 3*d* simultaneously, etc.

EXERCISE 10-29

Show that Figure 10-19 leads to the observed form of the periodic table.

---

2. *Atomic orbitals for chemical bonding*. Chemical bonds are formed by the overlap of atomic orbitals from the constituent atoms. For this reason, the shapes of the atomic orbitals of atoms are of importance in understanding the nature of chemical bonds. While the detailed shapes of the atomic orbitals in heavier atoms such as carbon, nitrogen, and oxygen are not exactly the same as the hydrogen-like orbitals, the general features of the hydrogen-like orbitals are preserved, and the representations of s, p, and d orbitals given in Figure 10-18 are useful in thinking about chemical bonds. We shall make use of these representations in the following chapter.

10–7 Angular Momentum is Quantized

Second in importance only to the principle of quantization of energy in atomic and molecular systems is the quantization of angular momentum. To understand this principle we first must recognize that angular momentum is a vector quantity, as indicated in Figure 10-20. A mass m rotating with velocity v a distance r from the axis of rotation has, as shown by Equation 10-3b, an angular momentum whose magnitude is mvr. However, the rotation occurs in a particular plane, as indicated by the angular momentum vector $\mathbf{L}$, which is drawn *perpendicular to the plane of rotation of m.*

Quantum mechanics shows that two properties of angular momentum are quantized in an atom or linear molecule:

1. The total magnitude of the angular momentum is restricted to discrete values.
2. The orientation of the angular momentum vector relative to some external axis is restricted to discrete values.

The reason these statements can be made is that the atomic or molecular wave function is always found to be an eigenfunction of the operators corre-

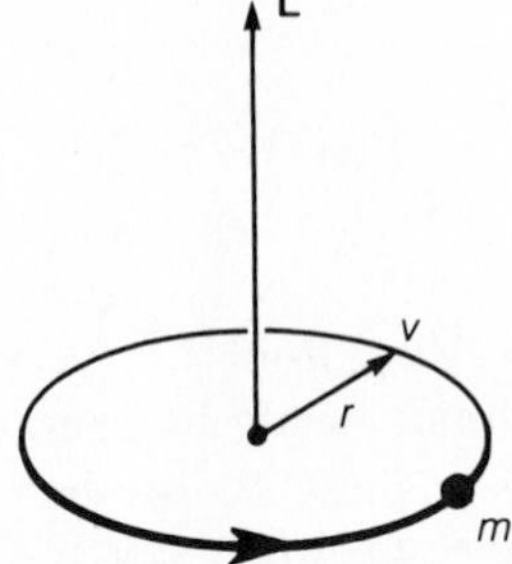

FIGURE 10-20
The angular momentum **(L)** of a mass m rotating in a circular path at velocity v is a vector that is perpendicular to the plane of rotation, and of length or magnitude mvr. (In vector notation, $\mathbf{L} = m\mathbf{r} \times \mathbf{v}$.)

sponding to the square of the angular momentum ($\mathcal{L}^2$) and the z-axis component of the angular momentum (M_z). Just as the relationship $\mathcal{H}\psi = E\psi$ (Equation 10-30) summarizes quantization of energy, the following two equations summarize quantization of angular momentum (of electrons in atoms or of linear molecules):

$$\mathcal{L}^2\psi = l(l + 1)\,\frac{h^2}{4\pi^2}\,\psi \tag{10-68a}$$

$$M_z\psi = \frac{mh}{2\pi}\,\psi \tag{10-68b}$$

More complicated equations are needed to describe the angular momentum of nonlinear molecules, because their symmetry is not as simple as that of atoms or linear molecules.

Equation 10-68*a* states that operation on ψ by the angular momentum operator $\mathcal{L}^2$ (whose nature you can find in the advanced books in the references) gives solutions only for certain values of the angular momentum. Specifically, the magnitude of the total angular momentum ($|\mathbf{L}|$) is restricted to the values

$$|\mathbf{L}| = \frac{\sqrt{l(l + 1)}\,h}{2\pi} \tag{10-69a}$$ ◄

The quantum number l is identical to l for the hydrogen atom, or it can be replaced by J for the rigid rotor

$$|\mathbf{L}| = \frac{\sqrt{J(J + 1)}\,h}{2\pi} \tag{10-69b}$$

Therefore we find that the hydrogen atom quantum number l is associated with angular momentum of the electron due to its orbital motion about the nucleus. In an s orbital $l = 0$, the angular momentum $\sqrt{l(l + 1)}\,h/2\pi$ is also zero. In a p orbital, $l = 1$, and the angular momentum of the electron in its orbital is $\sqrt{2}\,h/2\pi$.

EXERCISE 10-30

What is the angular momentum of the electron due to motion in a d orbital?

ANSWER

$\sqrt{6}\,h/2\pi$.

The Orientation of Angular Momentum is Quantized

The second important principle in the quantization of angular momentum is summarized in Equation 10-68*b*, which states that *the component of the angular momentum along an externally fixed axis can only take on the values $mh/2\pi$*; m is restricted to the values $-l, (-l + 1), \ldots +l$, and is identical with m for the atomic orbitals. Figure 10-21 summarizes the spatial quantization of an

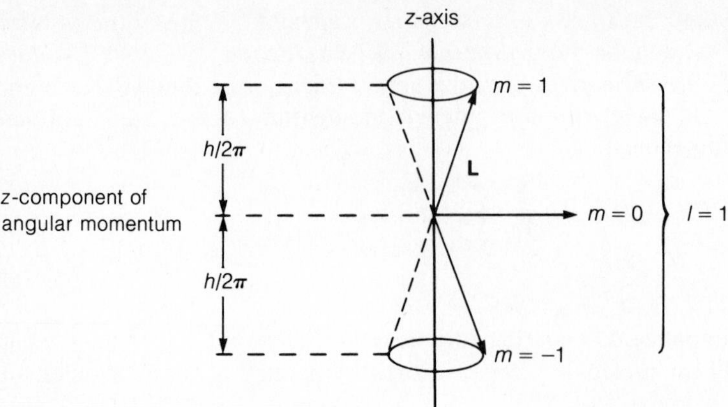

FIGURE 10-21
Possible orientation in space of an angular momentum vector that has $l = 1$ ($|\mathbf{L}| = \sqrt{2}\,h/2\pi$). The component along an axis determined, for example, by an external magnetic field can have the values $h/2\pi$, 0, and $-h/2\pi$ corresponding to $m = 1$, 0, and -1. The vector can rotate about the z-axis, so its component in the x or y direction is not quantized. Notice that there are $2l + 1$ possible values of m.

angular momentum vector that is characterized by $l = 1$ and $m = 1$, 0 and -1. The vector, of length $\sqrt{2}\,h/2\pi$, can have components $h/2\pi$, 0, or $-h/2\pi$ along the z-axis. Notice that the length of the vector is greater than the maximum possible z-component ($h/2\pi$), so **L** never lies exactly along the z-axis. The direction of the vector in the x-y plane is not fixed; instead, it can rotate or *precess* about the z-axis.

The quantization of direction of angular momentum is helpful for understanding the degeneracy of the energy states of the rigid rotor. An angular momentum vector characterized by quantum number J can be oriented so that the z-component is characterized by $m = J, (J - 1), \ldots -J$. Since there are $2J + 1$ values of m, there are $2J + 1$ possible orientations of the angular momentum. These correspond physically to different allowed planes of rotation of the rotor.

Atomic States can be Characterized by their Angular Momentum

As we have just seen, the hydrogen atomic orbitals $s, p, d \ldots$ differ in the magnitude of their angular momentum. In considering these wave functions, one imagines occupancy by a single electron. However, real atoms contain many electrons, so the total angular momentum due to orbital motion, called the *orbital angular momentum,* can have several components. It is not a simple matter to calculate the angular momentum states, because angular momenta are vectors and the rules for addition of vectors must be followed. However, the result is important because atomic states are characterized by their angular momentum. For example, in an atom the total orbital angular momentum is denoted by $S, P, D, \ldots$. By exact analogy with s, p, d for atomic orbitals, the quantum number l_T is 0 for an S state, 1 for a P state, 2 for D, and so forth.

We will give a single example to illustrate how the orbital angular momentum of multi-electron atoms can be calculated. Suppose we have an atomic state with two p electrons with the same spin. Each electron has available the states with quantum numbers $m = 1, 0, -1$. The resultant angular momentum is characterized by a total quantum number m_T,

$$m_T = m_1 + m_2 \tag{10-70}$$

in which m_1 refers to electron 1 and m_2 to electron 2. (The quantum numbers m_1 and m_2 can be added because the z-components of the separate angular momenta can be added to give the total z-component, or

$$\frac{m_T h}{2\pi} = \frac{m_1 h}{2\pi} + \frac{m_2 h}{2\pi}$$

Division by $h/2\pi$ yields Equation 10-70.) Figure 10-22 summarizes the possible values of m_T for the two $2p$ electrons, each having the same spin quantum number. The diagonal values are excluded because they correspond to the same set of all four quantum numbers in the same atom, in violation of the Pauli principle. The shaded values below the diagonal are not counted because they are identical in origin to states with $m_T = 1, 0$, and -1 above the diagonal. (For example, $m_T = 1$ arises from one electron with $m = 1$ and one with $m = 0$. There is no way to tell which electron is which.) The possible values of m_T are therefore 1, 0, and -1. This set requires a quantum number $l_T = 1$ (for $m_T = -l_T, \ldots 0 \ldots l_T$), and we conclude that two $2p$ electrons with the same spin quantum number (spins *unpaired*) should give rise to a P state.

EXERCISE 10-31

Prove that the orbital angular momentum of a filled shell, for example six $2p$ electrons, is zero.

ANSWER

For every electron with $m = +1$ there is an electron with $m = -1$. Therefore, $m_T = 0$. An atom that has only filled shells has an S ground state.

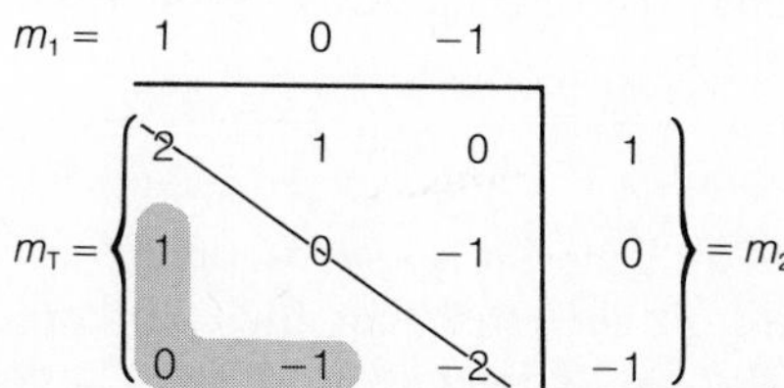

FIGURE 10-22
Possible values of $m_T = m_1 + m_2$ for the total orbital angular momentum of two $2p$ electrons having the same spin. The values on the diagonal are forbidden by the exclusion principle, and those in the shaded area are not counted because they are identical, except for exchanging the number on the electron, with the states 1, 0, and −1 above the diagonal.

Electrons and Nuclei have Spin Angular Momentum

Electrons have angular momentum due to their spin, as do some nuclei. The quantum number that determines the total spin angular momentum, corresponding to the quantum number l for orbitals, is s. The angular momentum of a single electron is

$$|\mathbf{L}| = \sqrt{s(s+1)}\,\frac{h}{2\pi}$$

which is equal to $\sqrt{3}\,h/4\pi$ because the quantum number s that characterizes its spin is $\frac{1}{2}$. The z-component of the electron spin angular momentum can be either one of two values, $\pm h/4\pi$, resulting from substitution of the possible values of $m_s = \pm\frac{1}{2}$ in the expression $m_s h/2\pi$. The two values of m_s correspond to the two possible orientations of electron spin.

If there is more than one electron, the spin angular momenta must be added vectorially. Two electrons in the same atomic orbital *must* have different values of m_s, by the Pauli principle; such electrons are said to be *paired.* (When two electrons are *unpaired,* their m_s values are allowed to be the same, but they may also be different). Since paired electrons have $m_s = +\frac{1}{2}$ and $m_s = -\frac{1}{2}$, the sum of the spin quantum numbers ($m_{sT} = \frac{1}{2} - \frac{1}{2}$) is zero. This requires that the quantum number s_T (analogous to l_T on page 459) is zero. *There is no spin angular momentum for paired electrons.*

Unpaired electrons confer important chemical and physical properties on matter. Two electrons which occupy *separate* atomic orbitals can be unpaired. Because the electrons differ in at least one of the orbital quantum numbers n, l, or m, the Pauli principle does not require that their values of m_s be different. Each electron can have $m_s = \pm\frac{1}{2}$, giving values of $m_{sT} = m_{s1} + m_{s2} = 1, 0,$ and -1. This set requires $s_T = 1$.

The spin quantum number (s_T) of a system of electrons is characterized by specifying the *spin degeneracy.* When s_T is zero, there is only one value of m_{sT}, and the state is called a *singlet.* Two unpaired electrons, with $s_T = 1$ and $m_{sT} = 1, 0, -1$ as deduced above, give a *triplet* state. As you can see, the spin degeneracy is equal to the number of possible values of m_{sT}.

EXERCISE 10-32

What is the degeneracy of the electron spin state if there is one s electron?

ANSWER

$m_s = \pm\frac{1}{2}$, and $s = \frac{1}{2}$. The degeneracy is two, and this would be called a *doublet* state. As usual for angular momenta, the degeneracy is $2s + 1$, analogous to $2J + 1$ or $2l + 1$.

The spin degeneracy of an atomic state is indicated in spectroscopic notation by a number as a superscript preceding the letter that indicates the orbital angular momentum. For example ^3P (read "triplet P") is a state with total orbital angular momentum quantum number $l_T = 1$, and total spin quantum number $s_T = 1$.

EXERCISE 10-33

What are l_T and s_T in the state ^2D?

ANSWER

A doublet D state has $s_T = \frac{1}{2}$ and $l_T = 2$.

In summary, angular momenta are quantized both in magnitude and direction relative to an external field. Coupling of electron spins together, and of nuclear spins, is of importance in such diverse phenomena as chemical bonding, spectroscopy, and nuclear magnetic resonance.

QUESTIONS FOR REVIEW

1. How can the diffraction of radiation be explained?
2. What did Planck assume about the energy of an oscillator?
3. What is meant by the energy levels of a system?
4. Why does the photoelectric effect imply the existence of photons?
5. What assumption did Bohr make to explain the energy levels of the hydrogen atom?
6. How does de Broglie's postulate of matter waves explain Bohr's assumption?
7. State Schrödinger's equation in one dimension.
8. What is the physical meaning of ψ^2?
9. State the Heisenberg uncertainty principle.
10. What is the potential energy function for the particle in a box?
11. Explain how the boundary conditions eliminate most functions A sin $\beta x + \beta$ cos βx as solutions for the particle in the box.
12. Explain what is meant by a normalized wave function and by orthogonal wave functions.
13. What is the zero-point energy?
14. What potential energy function describes a harmonic oscillator?
15. What are the energy levels of a harmonic oscillator?
16. What is the reduced mass of a harmonic oscillator? What is the reduced mass for a rigid diatomic rotor?
17. What is the physical origin of the degeneracy of the energy states of a diatomic rotor?

18. How can rotational spectra be used to determine bond lengths?

19. What quantum numbers characterize the wave function of the hydrogen atom, and what are their allowed values?

20. Describe the angular dependence of an s, a p, and a d atomic orbital.

21. What wider significance can you attribute to the hydrogen atomic orbitals?

PROBLEMS

1. Are the following statements true or false?

   (a) Two light waves will always interfere destructively when they have traveled a different distance from their source.

   (b) Quantization of energy according to frequency, $E = h\nu$, is readily explained by the wave theory of radiation.

   (c) Einstein proposed that light quanta travel in straight lines.

   (d) High energy quanta have short wavelengths.

   (e) The Bohr radius is a reasonable estimate of the size of a hydrogen atom in its ground state.

   (f) The wave function ψ must be a real number.

   (g) ψ must be continuous and single-valued.

   (h) An unconstrained particle would not have quantized energy levels.

   (i) Each separate atomic orbital has a unique set of values of the quantum numbers n, l, m.

   (j) Pauli stated that no two electrons in an atom can have the same values for all three quantum numbers n, l, and m.

Answers: (a) F. (b) F. (c) T (except for the influence of fields in general relativity). (d) T. (e) T. (f) F. (g) T. (h) T. (i) T. (j) F.

2. Add the appropriate words.

   (a) A quantum of light is called a ________ .

   (b) Bohr assumed quantization of ________ ________ in the hydrogen atom.

   (c) ψ^2 is the probability per ________ ________ of finding a particle in space.

   (d) Each wave function or eigenfunction is associated with a particular energy or ________ .

   (e) A harmonic oscillator is a simple physical model for molecular ________ .

   (f) The frequency of a harmonic oscillator ________ (increases, decreases) as the reduced mass increases.

   (g) A rigid rotor is a simple physical model for molecular ________ .

(h) A point at which the wave function $\psi = 0$ is called a ________.

(i) Two wave functions for which the integral over all space, $\int \psi_n \psi_m dv$, is zero are said to be ________.

Answers: (a) Photon. (b) Angular momentum. (c) Unit volume. (d) Eigenvalue. (e) Vibration. (f) Decreases. (g) Rotation. (h) Node. (i) Orthogonal.

PROBLEMS RELATED TO EXERCISES

3. (Exercise 10-1) Calculate the frequency of green light with $\lambda = 550$ nm.
4. (Exercise 10-4) Calculate the de Broglie wavelength of a hemoglobin molecule, molecular weight 67,000, whose kinetic energy is (3/2) $k_B T$ at $T = 310$ K.
5. (Exercise 10-7) Write the expression for the potential energy U of the lithium atom.
6. (Exercise 10-11) Confirm the statement (4) following Exercise 10-12 by repeating Exercise 10-11 for a one-dimensional box of twice the size.
7. (Exercise 10-13) Calculate and compare the reduced masses of

   (a) Cl_2

   (b) HCl

   (c) HI
8. (Exercise 10-19) In Exercise 10-19, what temperature would be necessary for an appreciable population of higher levels, say, $N_{n=1}/N_{n=0} = 0.1$?
9. (Exercise 10-24) The rotation of the molecule $^{14}N^{16}O$ is associated with absorption of microwave radiation of frequency 102.169×10^9 Hz for transition from the state $J = 0$ to the state $J = 1$. Determine the equilibrium NO bond length.

Answer: 1.1511 Å.

10. (Exercise 10-29) Use Figure 10-19 to predict which atomic orbitals are occupied in Fe.
11. (Exercise 10-30) Calculate the magnitude of the orbital angular momentum of a $2p$ electron in a hydrogen atomic orbital.
12. (Exercise 10-31) (a) What are the atomic state or states (in spectroscopic notation) that arise from the electron configuration $(1s)^2(2s)^1(2p)^1$? (b) Give the spectroscopic notation for the atomic state whose electron configuration is $(1s)^2(2s)^2(2p)^6$.

OTHER PROBLEMS

13. The average energy from the sun that strikes the earth's surface is about 1×10^3 J s^{-1} m^{-2}. The average photon energy corresponds to a wavelength of about 550 nm. Calculate the average number of photons striking the earth's surface per square meter per second.

14. Calculate the energy and momentum of photons with the following wavelengths:

    (a) 1.54 Å (x-ray from Cu)

    (b) 280 nm (ultraviolet)

    (c) 550 nm (yellow light)

    (d) 3.1 μm (infrared light)

    (e) 10 m (radio wave)

15. Excitation from the ground state of Li^{2+} causes absorption at wavelengths whose reciprocals are 740, 747 cm^{-1}; 877, 924 cm^{-1} and 925, 933 cm^{-1}. Calculate the minimum energy required to ionize Li^{2+} to Li^{3+}. (*Hint:* see Equation 10-6).

16. Photons absorbed (or reflected) by an object transfer their momentum to the object, producing a force called the *radiation pressure*. Force can be calculated from Δ(momentum)/Δ(time). Use the results of Problems 13 and 14c to calculate the average radiation pressure on the earth exerted by the sun, assuming that all photons are absorbed by the earth. How does the pressure compare in magnitude with the pressure exerted by the earth's atmosphere?

● 17. (a) Prove that the first two wave functions, ψ_1 and ψ_2, for the particle in a box are orthogonal.

    (b) Explain how the solution to the particle-in-a-box problem would be altered if ψ were not required to be continuous.

18. Write Schrödinger's equation for the H_2^+ ion (the one molecule other than H for which a solution in closed form can be found).

19. Calculate the zero point energy per mole of CO due to its vibrational motion. (Absorption wavelength 4.66 μm.)

20. The vibrational frequencies for stretching of single, double, and triple carbon-carbon bonds are as follows:

—C—C— (single bond, with substituent bonds above and below)	900 cm^{-1}
>C=C<	1650 cm^{-1}
—C≡C—	2050 cm^{-1}

    (a) Calculate force constants for these stretching motions.

    (b) How can one represent the increase in energy for a small change in bond length?

21. Identify the quantum number n and l for the following atomic orbitals, and state in each case the possible values of m: 1s, 2p, 4d, 6s, 4f.

22. Use the relationship $\int \psi\psi^* dV = 1$ to show that $E = \int \psi^* \mathcal{H} \psi \, dV$, where the integration is over all space. (*Hint:* Begin with Equation 10-30.)

23. Use Equation 10-31 to show that the average value $\langle q \rangle$ of a variable whose operator is α is given by $\langle q \rangle = \int \psi^* Q \, \psi dV$.

24. Use the result of Problem 23, substituting x for the position operator, to show that $a/2$ is the average position of a particle in the first energy state of a particle in a box.

25. Use the result of Problem 23, substituting $(h/2\pi i)d/dx$ for the momentum operator, to show that the average momentum in ψ_1 for the particle in a box is zero.

26. Use the result of Problem 23, substituting $(h/2\pi i)d/dx$ for the momentum operator to show that the average kinetic energy $(p^2/2m)$ in ψ_1 for the particle in a box is $h^2/8ma^2$.

27. Find the most probable distance from the nucleus of an electron in a 1s orbital, whose wave function is $\psi = \sqrt{z^3/\pi a_0^3}\ \exp(-zr/a_0)$. (The most probable distance is the value of r at which the radial distribution function $4\pi r^2R^2$ in Figure 10-18(b), $n = 1$, $l = 0$, has its maximum value.)

● 28. What are the allowed energy levels for a system analogous to a harmonic oscillator in which $U = kx^2/2$ for $x > 0$, and $U = \infty$ for $x < 0$? (*Hint:* Examine the effect of the new boundary conditions on the choice of allowed solutions ψ found for the standard harmonic oscillator.)

29. Use the method of Problem 23 to find the average kinetic energy of a harmonic oscillator in its ground state (ψ_0). What is the average potential energy in the ground state?

● 30. Are the wave functions for the rigid rotor $\exp(-i\phi)$ and $\exp(i\phi)$ orthogonal?

● 31. Are the alternative rigid rotor wave functions [which are linear combinations of $\exp(i\phi)$ and $\exp(-i\phi)$], given by $\sin\phi$ and $\cos\phi$, orthogonal? In representing atomic orbitals it is useful to use orthogonal wave functions.

● 32. The potential energy (U) for a hydrogen atom approaches infinity when $r \to 0$. What special property does $\partial^2\psi/\partial x^2$ (for $n = 1$, $l = 0$, $m = 0$, for example) have when $r \to 0$ that corresponds to the infinity in the potential energy? Can Schrödinger's equation be satisfied in the limit as $r \to 0$? (*Hint:* The derivative of ψ is discontinuous when $r \to 0$.)

33. Use the definition of the variables θ and ϕ to identify the following atomic orbitals, using Figure 10-18, but without consulting Table 10-3 ($\sigma = Zr/a_0$):

(a) $\psi = \text{const} \times e^{-\sigma}$

(b) $\psi = \text{const} \times (6 - \sigma)\sigma\ e^{-\sigma/3} \cos\theta$

(c) $\psi = \text{const} \times \sigma\ e^{-\sigma/2} \sin\theta \cos\phi$

(d) $\psi = \text{const} \times (24 - 18\sigma + 3\sigma^2 - \sigma^3/8)e^{-\sigma/4}$

(e) $\psi = \text{const} \times \sigma^2 e^{-\sigma/3} \sin\theta \cos\theta \cos\phi$

(*Hint:* The exponential term varies with $e^{-\sigma/n}$, the number of radial nodes is $n - l - 1$, and the angular dependence tells you whether the function is $2p_x$, $2p_y$ or $2p_z$, for example.)

● 34. Calculate the magnitude of the orbital angular momentum in a D atomic state.

● 35. Show that two 2p electrons give rise to the atomic states ^3P, ^1D, and ^1S. *Hint:*

Draw diagrams like Figure 10-22 for the three possible combinations of the two electron spin quantum numbers m_s: $+\frac{1}{2}$ with $+\frac{1}{2}$, $-\frac{1}{2}$ with $-\frac{1}{2}$, and $+\frac{1}{2}$ with $-\frac{1}{2}$. Notice that in this final case the exclusion principle does *not* require elimination of the diagonal terms, and that the states below the diagonal are different from those above the diagonal. When you collect terms you should find a set $m_T = 1, 0, -1$ that can have any of the values $m_{sT} = 1, 0, -1$. These constitute the ^3P state. In addition, there is a set $m_T = 2, 1, 0, -1, -2$ with $m_{sT} = 0$, and one with $m_T = 0$, $m_{sT} = 0$. These last two constitute the ^1D and ^1S states. Hund's rules, by the way, indicate that the state of highest spin multiplicity (^3P in this case) lies lowest in energy, and among states of equal energy, the state of highest angular momentum (^1D in this case) lies lowest. Therefore ^3P is the ground state, followed by ^1D and ^1S.

36. Hydrogen atoms excited to very high energy levels exist in interstellar space, where they radiate low-energy photons in the microwave and radio wave region. An emission that has been detected occurs at the energy expected for transitions from $n = 110$ to $n = 109$. Use the Bohr theory to calculate the wavelength and frequency of the radiation emitted, and to calculate the radius of the orbit with $n = 110$. By what factor is the radius increased over that of the ground state radius, $r = a_0$? Compare the size of such a hydrogen atom with the size of a bacterium ($r \sim 500$ nm).

QUESTION FOR DISCUSSION

37. Quantum mechanics deals with probabilities, not with absolute assertions of fact. For example, a wave function gives only the probability that a particle can be found at such-and-such a coordinate. Given that quantum mechanics provides our most fundamental picture of reality, does this mean that we must abandon the notion of firm causality, in which we can state that event *A* leads to event *B*, and instead that we must be satisfied with stating that event *A* probably leads to event *B*?

FURTHER READING

Coulson, C. A. 1961. *Valence,* 2nd ed. Oxford, England: The Clarendon Press. A lucid and relatively nonmathematical development of quantum mechanics and of applications to chemical bonding by one of the principal contributors to this field. Highly recommended for a deeper excursion into quantum chemistry.

Daniels, F., and Alberty, R. A. 1966. *Physical Chemistry,* 3rd ed. New York: W. J. Wiley. A standard text in physical chemistry with a brief, readable summary of quantum mechanics.

Hanna, M. W. 1969. *Quantum Mechanics in Chemistry,* 2nd ed. Menlo Park, Calif.: W. A. Benjamin, Inc. An excellent, concise exposition of quantum chemistry and spectroscopy.

Karplus, M., and Porter, R. N. 1970. *Atoms and Molecules*. Menlo Park, Calif.: W. A. Benjamin, Inc. A detailed, modern development of quantum chemistry, valence theory, and molecular spectroscopy.

Kauzmann, W. 1957. *Quantum Chemistry*. New York: Academic Press. A thorough exposition of quantum chemistry, with especially helpful material on the physics of wave motion and on the interaction of matter with light.

Moore, W. J. 1972. *Physical Chemistry,* 4th ed. Englewood Cliffs, N.J.: Prentice-Hall. The standard text on physical chemistry. The chapters on quantum chemistry are considerably more detailed and thorough than the material in this text.

Moelwyn-Hughes, E. A. 1961. *Physical Chemistry,* 2nd ed. New York: Macmillan. Another general text, especially detailed on spectroscopy of small molecules.

CHAPTER 11

Chemical Bonds

The stability, variety, and complexity of things in our world come from the properties of chemical bonds. The stability of substances comes from the large energy released when covalent bonds are formed. This energy is far in excess of random thermal energy at ordinary temperatures, so that bonds once formed tend to persist until they are vigorously heated. The variety and complexity come from the property of *multiple valence,* the ability of some atoms, most notably carbon, to form strong bonds simultaneously with two or more other atoms. Carbon frequently bonds to some combination of four carbon, nitrogen, oxygen, sulfur, hydrogen, or halogen atoms, thereby creating an enormous number of possible compounds. This variety is enhanced by the ability of carbon and other atoms to assume various bond-angles and bond-lengths, so that they can participate in linear, ring, and multiple ring structures. The range of possible compounds and properties is increased further by the ability of carbon, oxygen, and nitrogen to form double and sometimes triple bonds, as well as single bonds. Double bonds contain fewer tightly bound electrons which can participate in oxidation-reduction reactions and can be excited by light, leading to the phenomenon of color.

In this chapter we develop some of the principles of the quantum-mechanical theory of chemical bonding. Though this theory is approximate, it is nevertheless of great utility in understanding the properties of substances, and in the synthesis of new substances and prediction of their properties. The precise calculation of energy levels—which is possible only in a few simple cases—is much less important than the broad semi-quantitative understanding that this theory has provided. The most important results of the theory include the following.

1. The basic reason for the formation of covalent bonds between atoms is described briefly in Section 11-1 and is developed more quantitatively in the two following sections.
2. The reason for bond stability is described in Section 11-4. Bond stability is increased when electrons occupy *bonding orbitals,* and instability is increased when electrons occupy *antibonding orbitals*.

3. The reasons for various valences and bond geometries are discussed in Section 11-5. The valence is determined by the number of unpaired electrons ready to couple with unpaired electrons from other atoms, and the geometry is determined by the nature of the atomic orbitals that these electrons occupy.
4. The reason for formation of multiple bonds is discussed in Section 11-4, and has to do with the properties of the atomic orbitals that participate in bonding.
5. The charge-distribution properties of molecules including dipole moment, polarizability, and van der Waals radius are discussed in Sections 11-6 and 11-7. These properties account for intermolecular forces or *noncovalent bonds*.

○ *11–1 Covalent Bonds Form Because of the Decrease in the Electrostatic Potential Energy of Electrons That Accompanies Coupling of Atoms*

Let us begin our consideration of covalent bonds with the simplest of all molecules, H_2^+. It consists of two protons and a single electron. While it is found only in the unbiological environment of a high voltage discharge in hydrogen gas, this molecule is worthy of consideration because it illustrates many of the essential characteristics of chemical bonding that are found in more familiar molecules.

We can describe the electronic states of the H_2^+ molecule by *molecular orbitals*, Ψ, and can think of these as being formed by sums or differences (*linear combinations*) of the atomic orbitals of the hydrogen atom. The lowest energy (ground state) molecular orbital for H_2^+ is

$$\Psi = \psi_1 + \psi_2 \tag{11-1}$$

in which ψ_1 and ψ_2 are 1*s* hydrogen atom orbitals for the two nuclear centers. A plot of the energy of this molecular orbital is shown by the middle curve of Figure 11-1(*a*). There is a minimum in the curve at an internuclear separation of 1.06 Å. This shows that the molecule is stable, or in other words, that a chemical bond can be formed between the nuclei. We will see in Section 11-3 how the actual value of the energy is calculated.

A second possible molecular orbital for the H_2^+ ion is

$$\Psi = \psi_1 - \psi_2 \tag{11-2}$$

The energy of this orbital is shown by the middle curve of Figure 11-1(*b*). There is no minimum, so that the atoms repel each other at all internuclear separations. Thus this orbital is called an *antibonding orbital.* It is denoted σ^*. The first orbital (Eq. 11-1) is called a *bonding orbital,* and is denoted σ.

What accounts for the minimum of the energy curve of the bonding molecular orbital that represents the formation of the covalent bond? It turns out

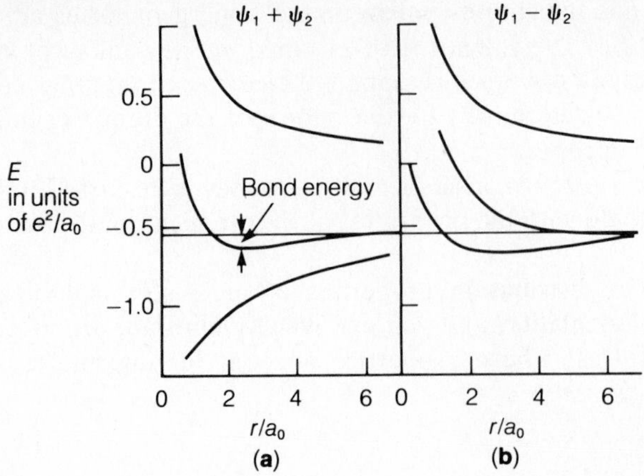

FIGURE 11-1
Energy curves for a H_2^+ molecule. **(a)** For the bonding molecular orbital $\psi_1 + \psi_2$, in which ψ_1 and ψ_2 are hydrogen atom 1*s* orbitals. The top curve gives the energy of nuclear repulsion, the bottom curve gives the electron energy, and the middle curve is the total molecular energy (the sum of top and bottom curves). There is a minimum of 269 kJ mol^{-1} at an internuclear separation of about $2a_0 = 1.06$ Å. **(b)** For the antibonding molecular orbital, $\psi_1 - \psi_2$, the three curves have the same meaning as in **(a)**. (Adapted from Coulson, 1961.)

that both potential and kinetic energy play a role, as can be seen from Figure 11-2. This figure compares the square of the wavefunction for the molecule to the square of the wavefunctions for two 1*s* hydrogen atoms. You will recall that the square of the wavefunctions is the probability density of the electron. For the bonding molecular orbital [Fig. 11-2(*a*)], the squared wavefunction is larger in the region between the nuclei than is the squared wavefunction of the two atoms. Thus there is a greater probability of finding the electron between the nuclei in the molecule. In other words, the charge

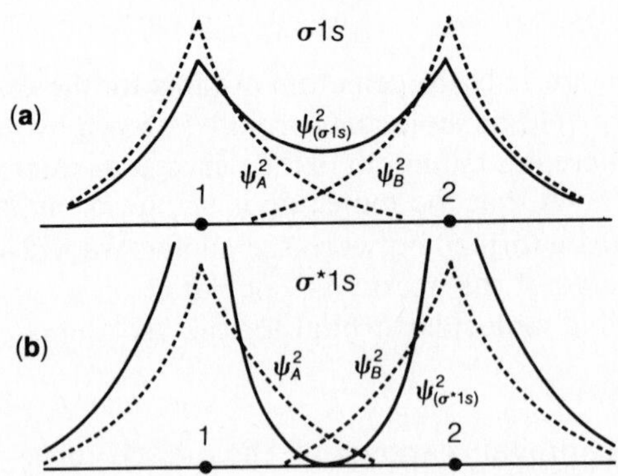

FIGURE 11-2
The probability density, ψ^2, for the **(a)** bonding or σ, and **(b)** antibonding or σ^* molecular orbitals of H_2^+, along the molecular axis. The points 1 and 2 represent the positions of the nuclei. The probabilities for the atomic densities of 1*s* orbitals centered on 1 and 2 are drawn to half scale, to permit comparison of the molecular density with the sum of the atomic densities $\frac{1}{2}(\psi_1^2 + \psi_2^2)$. (Redrawn from Coulson, 1961.)

density shifts to the region between the nuclei as the bond forms. The electron is now shared by both nuclei and the attraction of this negative charge for both positive nuclei lowers the potential energy of the system.

Kinetic Energy Also Plays a Role in Bond Formation

Some workers in the field of quantum chemistry believe that changes in kinetic energy are even more important than changes in potential energy in explaining the formation of chemical bonds (for example, Wilson and Goddard, 1972). This view is based on the fact that the kinetic energy of a particle decreases as it is allowed to occupy a larger region. You will recall that we found this true for the particle in a box (Sec. 10-3). You can see from Figure 11-2(*a*) that the region accessible to the single electron seems greater in H_2^+ than it is in a single 1*s* hydrogen orbital. This is shown by the greater extent of the molecular orbital than that of either atomic orbital. Thus the electron spreads into more space in the molecule, and lowers its kinetic energy. Another way of viewing this is to note that the curvature of the molecular wavefunction is smaller than that of the atomic wavefunction. The kinetic energy of an electron is related to the curvature of its wavefunction, because of the $d^2\psi/dr^2$ term in Schrödinger's equation (see Sec. 10-2).

The Forces in Chemical Bonding Are Electrostatic in Origin

Students sometimes wonder if the energy of chemical bonding might arise from gravitational forces, or some forces other than the Coulombic electrostatic forces that we discussed in Chapters 8 and 9. The answer is that the electrostatic forces are the only significant contribution to binding. The relative strength of the gravitational force is far too small to account for experimental binding energy of molecules. The gravitational energy required to separate two hydrogen atoms, each of mass 1.7×10^{-24} g, from their equilibrium separation of 0.74 Å to infinity is 2.5×10^{-40} J, whereas the bond energy of H_2 is actually 6.7×10^{-19} J. The weak and strong nuclear forces act over too short a range to contribute to chemical binding. They are effectively zero for particles more than 10^{-13} cm apart, whereas atoms in molecules are about 10^{-8} cm apart. Thus the ordinary Coulombic electrostatic force is the only significant force of chemical bonding. Describing this force in terms of quantum mechanics is the subject of the following sections.

○ *11–2 Molecular Orbitals Describe the Motions of Electrons in Molecules*

Molecular orbital theory extends the concept of electron orbitals (Sec. 10-6) from atoms to molecules. The basic assumption is that the motion of each electron in a molecule can be described by a molecular orbital, Ψ.

These molecular orbitals (abbreviated MO) have the following characteristics:

1. The probability that an electron in an orbital is in any given volume $d\tau$ centered on some point in the molecule is given by $\Psi^2 d\tau$ when Ψ is the value of the orbital at the point. For a complex wave function, the probability is $\Psi\Psi^* d\tau$.
2. Each MO is labeled by quantum symbols such as σ, π, and σ^*, to be explained below.
3. The MO's extend over the nuclei of the molecule, rather than being localized about a single nucleus as are atomic orbitals. Figure 11-2 illustrates the difference.
4. No more than two electrons can occupy any MO, and they must be of opposite spin.

What is the mathematical form of the molecular orbitals? Unfortunately, it is impossible to solve Schrödinger's equation for the true Ψ's with systems any more complicated than H_2^+, so we must be satisfied with approximate solutions. A MO can be represented as a linear combination of atomic orbitals (AO's). This is called *the MO LCAO approximation.* The atomic orbitals can be *s*, *p*, *d*, etc., orbitals. For an effective MO to be formed, the contributing atomic orbitals should be about the same energy, and should overlap as much as possible. This representation of molecular orbitals has the form

$$\Psi = c_1\psi_1 + c_2\psi_2 + c_3\psi_3 + \cdots + c_n\psi_n \qquad (11\text{-}3)$$ ◀

in which the ψ's are the atomic orbitals.

The formation of molecular orbitals from atomic orbitals can be written symbolically. An example is the formation of a bond between two lithium atoms:

$$\mathrm{Li}[1s^2 2s] + \mathrm{Li}[1s^2 2s] \rightarrow \mathrm{Li}_2[(\sigma 1s)^2(\sigma^* 1s)^2(\sigma 2s)^2]$$

This equation states that two lithium atoms, each with two electrons in their 1*s* AO's and one electron in their 2*s* AO's, couple to give a Li_2 molecule with two electrons in each of three MO's. Such an explicit listing of the energy states of electrons is called the *electronic configuration.*

The Variation Principle Permits Systematic Selection of the Optimum Constants for the Molecular Orbitals

The usefulness of the MO LCAO method is that there is a systematic way to select the constants in Equation 11-3. This is based on the *variation principle.* To see the meaning of the variation principle, we must recall that Schrödinger's equation can be written in the form

$$\mathcal{H}\Psi = E\Psi \qquad (11\text{-}4)$$

in which $\mathcal{H}$ is the Hamiltonian (energy) operator of the system (see Sec. 9-2), and E is the energy that corresponds to the wavefunction Ψ. If we multiply both sides of Equation 11-4 by Ψ, integrate over all coordinates and remove the constant E from the integral, we obtain (assuming the wavefunction is real)

$$E = \frac{\int \Psi \mathcal{H} \Psi \, d\tau}{\int \Psi^2 \, d\tau} \qquad (11\text{-}5)$$

If Ψ is the wavefunction for the ground (lowest energy) state, then E is the ground state energy, E_g. Equation 11-5 is more useful than 11-4 in molecular orbital theory because it allows us to calculate the energy that corresponds to any wavefunction, not just the true wavefunction Ψ. Let us denote an approximate molecular orbital wavefunction by Ψ' and denote the energy that corresponds to it by E'; then,

$$E' = \frac{\int \Psi' \mathcal{H} \Psi' \, d\tau}{\int \Psi'^2 \, d\tau} \qquad (11\text{-}6)$$

The variation principle is an exact theorem that is proven in books on quantum chemistry (Pilar, 1968, and Prob. 28). It states that the energy E' corresponding to the approximate wavefunction Ψ' is always equal to or *greater* than the true ground-state energy E_g:

$$E' \geq E_g \qquad (11\text{-}7)$$

The equals sign holds only when Ψ' is the true wavefunction Ψ.

The significance of the variation principle is that we can try many different constants in a trial wavefunction, of the form of Equation 11-3. In each trial, we insert the trial wavefunction into Equation 11-6 and compute the energy E'. The constants that yield the lowest energy E' (the energy closest to E_g) are the best set, that is, they produce the approximate wavefunction that is nearest to the true wavefunction.

While it is always possible to try many different constants in the wavefunction, there are more systematic ways of determining the constants. These procedures, as we see below, treat the constants c_i as variable parameters, and determine the values of the c_i that give the lowest energy.

EXERCISE 11-1

Suppose that you have forgotten that the wavefunction for the ground state of the hydrogen atom has the form $\psi = e^{-r/a_0}$, in which a_0 is the Bohr radius ($h^2/4\pi^2me^2$), and corresponds to an energy $E = -2\pi^2me^4/h^2$. You reason, however, that you may be able to determine the energy from the variation principle. You note that the wavefunction most probably has spherical symmetry and falls off exponentially with distance from the nucleus, and therefore it has the form

$$\psi = e^{-cr} \qquad (11\text{-}8)$$

You plan to determine the value of the parameter c that produces the best wavefunction. From the variation principle, you know that this is the value of c that gives the lowest energy E'. Inserting the trial wavefunction (Eq. 11-8) into 11-5, you find after some work that

$$E' = \frac{h^2}{8\pi^2 m} c^2 - e^2 c \tag{11-9}$$

Use this expression to obtain the constant c that produces the lowest energy and best ψ.

ANSWER

When E' is a minimum, $dE'/dc = 0$. Thus

$$\frac{dE'}{dc} = \frac{2ch^2}{8\pi^2 m} - e^2 = 0$$

and

$$c = \frac{4\pi^2 me^2}{h^2} = \frac{1}{a_0}$$

Therefore $\psi = e^{-r/a_0}$ and $E'_{\min} = -2\pi^2 me^4/h^2$. Notice that your answer corresponds to the exact wavefunction and ground-state energy of the hydrogen atom because you chose a form for ψ (Eq. 11-8) that is sufficiently general to include the exact ψ. (To discover what error you would have made by choosing ψ to be of another form, work Problem 11-3.)

11–3 Determination of Constants in a MO LCAO Wavefunction

We can now see how the variation principle can be used to determine the optimum constants in the MO LCAO method. If our wavefunction is of the form

$$\Psi = c_1\psi_1 + c_2\psi_2 \tag{11-10}$$

we can insert it into Equation 11-6 to obtain

$$\begin{aligned} E' &= \frac{\int (c_1\psi_1 + c_2\psi_2)\, \mathcal{H}\, (c_1\psi_1 + c_2\psi_2)\, d\tau}{\int (c_1\psi_1 + c_2\psi_2)(c_1\psi_1 + c_2\psi_2)\, d\tau} \\ &= \frac{c_1^2\int \psi_1\mathcal{H}\psi_1\, d\tau + 2c_1c_2\int \psi_1\mathcal{H}\psi_2\, d\tau + c_2^2\int \psi_2\mathcal{H}\psi_2\, d\tau}{c_1^2\int \psi_1^2\, d\tau + 2c_1c_2\int \psi_1\psi_2\, d\tau + c_2^2\int \psi_2^2\, d\tau} \end{aligned} \tag{11-10a}$$

In the last step it is assumed that $\int \psi_1\mathcal{H}\psi_2\, d\tau = \int \psi_2\mathcal{H}\psi_1\, d\tau$. This relationship is proved in Coulson (1961), and other books on quantum chemistry listed at the end of the chapter. We can simplify Equation 11-10a by introducing the definitions

$$H_{ij} = \int \psi_i\mathcal{H}\psi_j\, d\tau \qquad \text{and} \qquad S_{ij} = \int \psi_i\psi_j\, d\tau \tag{11-11a}$$

Equation 11-10a can be written

$$E' = \frac{c_1^2 H_{11} + 2c_1c_2H_{12} + c_2^2 H_{22}}{c_1^2 S_{11} + 2c_1c_2S_{12} + c_2^2 S_{22}} \tag{11-11b}$$

The variation principle tells us that Ψ is optimized when c_1 and c_2 assume the values that make E' a minimum. This is so when the derivatives of E' with respect to c_1 and c_2 are zero:

$$\frac{\partial E'}{\partial c_1} = 0 = \frac{\partial E'}{\partial c_2} \tag{11-12}$$

Applying these conditions to Equation 11-11*b* yields

$$\begin{aligned} c_1(H_{11} - E'S_{11}) + c_2(H_{12} - E'S_{12}) &= 0 \\ c_1(H_{12} - E'S_{12}) + c_2(H_{22} - E'S_{22}) &= 0 \end{aligned} \tag{11-13}$$

These are called the *secular equations;* they lead to the two unknowns c_1 and c_2 and to expressions for Ψ and E'.

EXERCISE 11-2

Confirm that application of Equations 11-12 to 11-11*b* gives Equations 11-13.

While it is not difficult to solve the secular equations for the general case, it is especially straightforward for the case of a diatomic molecule with both atoms of the same kind (such as H_2^+ or O_2). These molecules are called *homonuclear diatomics*. In this case $H_{11} = H_{22}$ and we shall see that $c_1/c_2 = \pm 1$. Furthermore, for these molecules S_{12} equals S_{21}, and we can drop the subscripts. Also, if ψ_1 and ψ_2 are normalized (Sec. 10-2), then S_{11} and $S_{22} = 1$. With these simplifications, the secular equations become

$$c_1(H_{11} - E') + c_2(H_{12} - E'S) = 0 \tag{11-14a}$$
$$c_1(H_{12} - E'S) + c_2(H_{11} - E') = 0 \tag{11-14b}$$

We can solve each of these equations for the ratio c_2/c_1, giving

$$\frac{c_2}{c_1} = -\frac{H_{11} - E'}{H_{12} - E'S} = -\frac{H_{12} - E'S}{H_{11} - E'} \tag{11-14c}$$

Rearranging, we obtain

$$(H_{11} - E')^2 = (H_{12} - E'S)^2$$

or

$$H_{11} - E' = \pm(H_{12} - E'S)$$

Rearranging again, we find that E' can have two values:

$$E_+ = \frac{H_{11} + H_{12}}{1 + S} \tag{11-15a}$$

and

$$E_- = \frac{H_{11} - H_{12}}{1 - S} \tag{11-15b}$$

EXERCISE 9-3

Show that the wavefunction that corresponds to E_+ is $\Psi = c_1 \psi_1 + c_2 \psi_2$ and the wavefunction that corresponds to E_- is $\Psi = c_1 \psi_1 - c_2 \psi_2$.
Hint: Substitute 11-15*a* and 11-15*b* into 11-14*c*.

If the integrals H_{11}, H_{12}, and S are evaluated for the H_2^+ ion and inserted in these expressions, the lower curves of Figure 11-1*a* and *b* are obtained. By adding the energy of nuclear repulsion ($= e^2/r$), the middle curves of the figure are obtained.

The Physical Significance of S, H_{12}, E_+, and E_-

The quantity S is given by

$$S = \int \psi_1 \psi_2 d\tau \tag{11-16}$$

and is called the *overlap integral,* since it is a measure of the extent of overlapping of the two orbitals. The product $\psi_1 \psi_2$ is very small except in regions where both ψ_1 and ψ_2 have significant values. When the two orbitals overlap completely, the integral (Eq. 11-16) becomes $S = \int \psi_1^2 \, d\tau$. This must be unity if the atomic orbitals are normalized. Thus S is a number that varies from 0 for the two atoms well separated to 1 for the two atoms

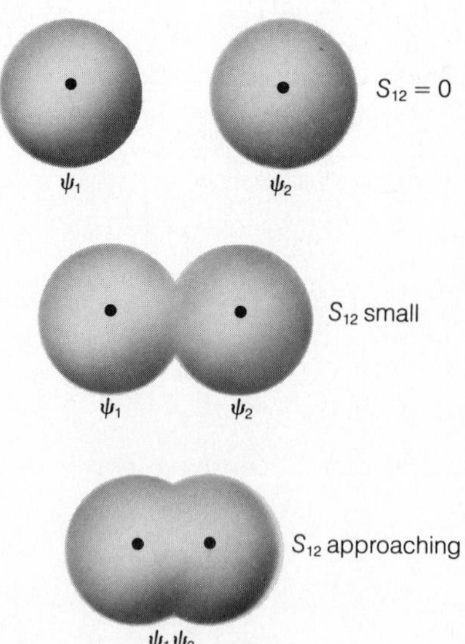

FIGURE 11-3
Increasing overlap of identical atomic orbitals on two atoms.

completely overlapping, and it is proportional to the extent of overlap (see Fig. 11-3).

The quantity H_{12} is called the *resonance integral,* and is a measure of the interaction of the two atomic orbitals ψ_1 and ψ_2. It is always negative, and approaches zero as the atoms are separated.

The quantity H_{11} is called the *Coulomb integral.* It represents the energy of an electron in the atomic orbital ψ_1 perturbed by the electrostatic force of the nucleus of atom 2. It also is negative.

Since both H_{11} and H_{12} are negative, E_+ is the lower energy and E_- is the higher energy. Thus E_+ is the energy for the bonding molecular orbital, $\Psi = c_1 \psi_1 + c_2 \psi_2$, and E_- is the energy for the antibonding orbital, $\Psi = c_1 \psi_1 - c_2 \psi_2$.

The basic methods of applying Schrödinger's equation to the motions of electrons within molecules were pioneered by Born and Oppenheimer (see the historical sketch on Oppenheimer). The MO LCAO method is one technique that followed indirectly from their work.

○ *Summary of MO LCAO Method*

In the MO LCAO method a trial wavefunction is taken to be a linear combination of atomic orbitals

$$\Psi = c_1 \psi_1 + c_2 \psi_2 + c_3 \psi_3 + \cdots + c_n \psi_n$$

The optimum values for the parameters in such a trial function are determined by solving secular equations (of the form of 11-14). The n solutions represent n molecular orbitals. The solutions yield the energies of the n molecular orbitals. The lowest energy is greater than the true ground-state energy of the molecule. The actual evaluation of the energies involves solving integrals, the overlap, Coulomb, and resonance integrals previously

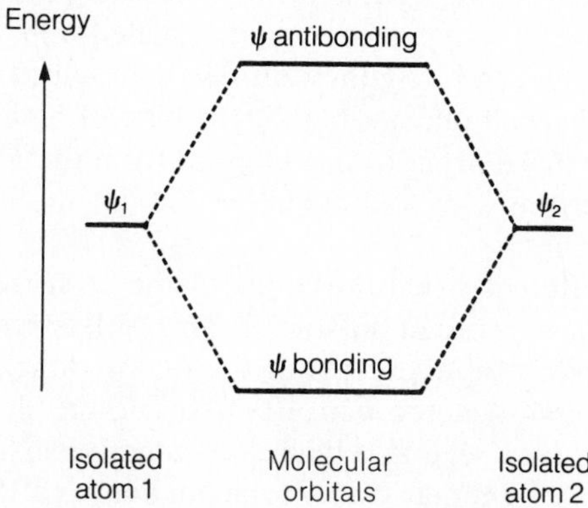

FIGURE 11-4
Formation of molecular orbitals from atomic orbitals contributed by two atoms.

mentioned. Methods for treating these integrals are given in the books on quantum chemistry cited at the end of this chapter.

In the case of two identical 1*s* atomic orbitals coming together to form molecular orbitals, two molecular orbitals are formed, with energies given by Equations 11-15*a* and 11-15*b*. The energy of the bonding molecular orbital E_+ is smaller than that of the contributing 1*s* atomic orbitals, and that of the antibonding molecular orbital is greater (Fig. 11-4). The difference in energy of the molecular orbitals is determined mainly by the resonance integral. The energy spacing between the antibonding orbital and the atomic orbitals exceeds that between the atomic orbitals and the bonding orbital. This arises from the $1 \pm S$ factor in the energy expressions, Equation 11-15. The bonding orbital is denoted $\sigma 1s$. The σ means that the electronic charge is symmetric about the bonding axis, and 1*s* indicates the AO's that have contributed to the MO. The antibonding orbital is denoted $\sigma^* 1s$ (the asterisk denoting its antibonding character). This orbital has a node, or $\Psi = 0$, in the plane between the two nuclei.

○ 11–4 Molecular Orbitals for Diatomic Molecules

So far we have considered only the molecular orbitals formed from the combination of two 1*s* atomic orbitals. These are the bonding MO, $\sigma 1s$ and the antibonding MO denoted $\sigma^* 1s$, shown in column (*a*) of Figure 11-5. Now we must consider the MO's that are formed from other atomic orbitals.

When two 2*s* atomic orbitals come together to form a molecular orbital, the situation is exactly analogous to that with the 1*s* orbitals. A bonding MO denoted $\sigma 2s$ is formed and an antibonding orbital denoted $\sigma^* 2s$ is formed. The shapes of these orbitals are the same as those of the $\sigma 1s$ MO's, except that 2*s* atomic functions replace the 1*s* functions.

The type of MO formed from two atomic *p* orbitals depends on the orientation of the *p* orbitals. If lobes of the *p* orbitals lie along the bonding (*x*) axis, they overlap to form another type of σ orbital, called $\sigma 2p$. This is shown in column (*b*) of Figure 11-5. As with the *s* atomic functions, there is one bonding orbital, called $\sigma 2p$, and one antibonding orbital of higher energy, called $\sigma^* 2p$. The shapes of these orbitals are different from the MO's formed by *s* atomic functions, but they are still symmetric about the *x*-axis, and hence they are termed *σ orbitals*.

The situation is different when the lobes of the combining *p* orbitals lie perpendicular to the axis through the nuclei. This is illustrated in column (*c*). Here the *p* orbitals are symmetric about the *y*-axis rather than the *x*-axis. Now the bonding orbital, denoted $\pi 2p$, has two sausage-shaped regions, one in which Ψ is positive and one in which Ψ is negative. The symbol π indicates that the MO is *not* symmetric about the bonding axis. In fact, the nuclei lie in a nodal plane, in which the value of Ψ is zero as it passes from positive

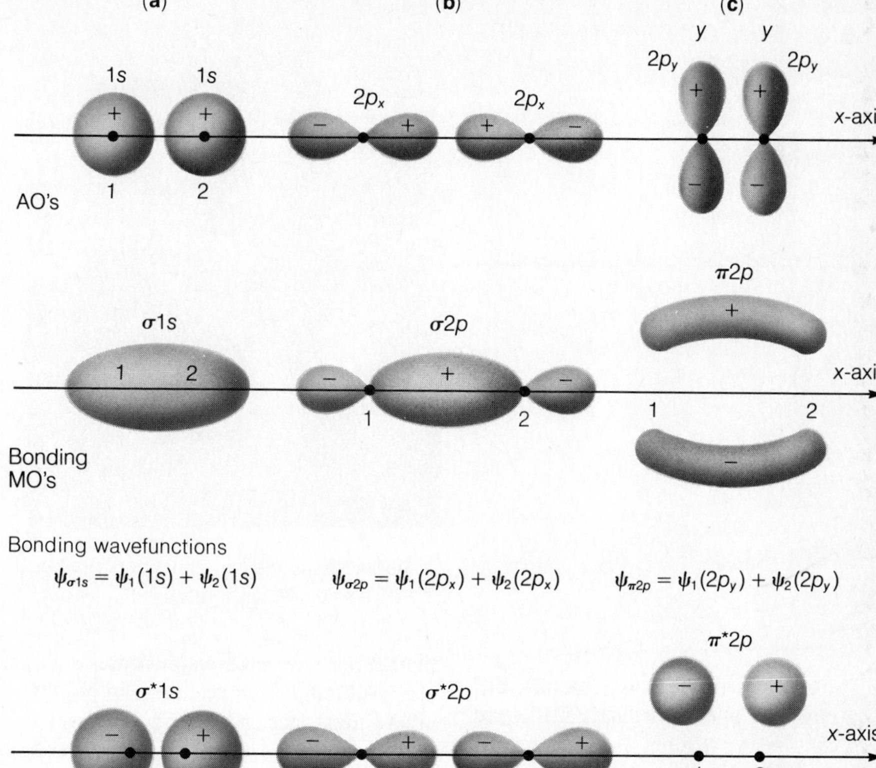

FIGURE 11-5
Formation of molecular orbitals from atomic orbitals.

in one of the sausages to negative in the other. The antibonding orbital denoted π^*2p has four lobes, alternating in the sign of Ψ.

If $2p$ orbitals with lobes along the z-axis are combined, the situation is the same as for those along the y-axis, but both the atomic and molecular orbitals of Figure 11-5(c) are rotated about the x-axis by 90°.

IN THE MATTER OF J. ROBERT OPPENHEIMER

1904–1967

J. Robert Oppenheimer in his pork pie hat. (The Bettman Archive, Inc.)

J. Robert Oppenheimer is remembered as a physicist of extraordinary brilliance, as a charismatic teacher, and as a man of great culture. His long-time associate in physics, Charles Lauritsen, recalled, "This man was unbelievable. He always gave you the answer before you had time to formulate the question." Oppenheimer is also remembered as the founder of a school of American theoretical physics. In fact, the pork pie sombrero he often wore was used at times to symbolize this school. He is remembered too as the man, more than any other, who was responsible for the development of the atomic bomb. Above all, he is remembered for the dramatic events of the following years when, partly on the basis of his participation in a group which had recommended against the crash development of the hydrogen bomb, his loyalty to the United States was questioned and his security clearance revoked. These events were brought on partly by the unique pressures of the Cold War years and partly by Oppenheimer's complex personality. But whatever the cause, Oppenheimer's experience illustrates a general problem which scientists in public affairs must face: whether a gifted scientist must use his talents and discoveries to the ultimate to achieve a technical goal set by others, or whether he has the right to argue against the use of his talents and discoveries on political and moral grounds.

Oppenheimer was a child prodigy of a wealthy commercial family in New York. His father had a good collection of paintings, including three by van Gogh. The son sailed through Harvard in three years, gaining his degree in physics *summa cum laude,* and learning Latin and Greek along the way. Even more remarkable, he took only two years to earn his Ph.D. His thesis, under the supervision of Max Born of the University of Göttingen, laid the basis for the quantum mechanics of molecules by showing that electronic and nuclear motions can be treated separately. After two years of postdoctoral training, Oppenheimer received many offers of academic positions, and he chose to take up a combined post at Berkeley and Caltech. He explained later that he chose Berkeley because of the library's collection of sixteenth and seventeenth century French poetry.

During these years he almost totally ignored politics, economics, and most of the things in daily life that most people care about. He owned neither a telephone nor a radio, never read a newspaper, and learned of the 1929 stock market crash only long after the event. Gradually, however, he became more politically aware, due in part to the influence of his brother and his fiancée, who were sympathetic to left-wing views. By the outbreak of World War II Oppenheimer knew he wanted to contribute to the American war effort, and in 1942 he was appointed leader of a theoretical effort to design the atomic bomb.

When the army sought an isolated location for a secret laboratory for a crash effort to build the bomb, Oppenheimer suggested Los Alamos, New Mexico, located on a remote mesa in country he explored on horseback while vacationing at his ranch not far away. With the construction of the laboratory there in 1943, he became director of the entire effort. Driven by the threat that the Nazis might be developing a similar device, he led every phase of the work with enthusiasm—recruiting the most talented chemists and physicists, overseeing the experimental and theoretical research, and generally managing the operation of the laboratory. Nearly all those who worked at Los Alamos, including the high-ranking army officers, agree that Oppenheimer was the only person who could have organized and led such a successful effort.

The first atomic bomb was exploded at Alamogordo, New Mexico, on July 16, 1945. Later that year Oppenheimer returned to academic posts. In 1947 he became Director of the Institute for Advanced Study at Princeton, which he built into a major center in theoretical physics. He also continued as a government adviser on the use and control of atomic weapons. In 1946, he played a major part in formulating the Acheson-Lilienthal proposal to give control of nuclear arms and energy to an international authority. This proposal, somewhat modified, was presented to the United Nations by Bernard Baruch, but rejected by the Russians. Oppenheimer also served on the general advisory committee of the Atomic Energy Commission (AEC). In 1949 this group voted against a crash program to develop the hydrogen bomb. This recommendation was overruled, and President Truman ordered development to proceed.

By 1953 concern in America about the Cold War, Communist advances, and atomic spies was great. Senator Joseph McCarthy was making wild statements about Communist spies in government. Charges that Oppenheimer was a security risk were made to the AEC. In December of that year, President Eisenhower ordered that a "blank wall" be placed between Oppenheimer and all government secrets. Oppenheimer requested a hearing into all allegations against him. The Personnel Security Board convened, and after extended hearings the proceedings were published under the title, *In the Matter of J. Robert Oppenheimer*. No evidence was found that Oppenheimer betrayed any secrets entrusted to him; indeed he was considered both loyal and exceptionally discreet. His opposition to a crash program to develop the hydrogen bomb in 1949 was not judged to be disloyal, but some of the board viewed this conduct as "disturbing." The major evidence against him was an incident in which Oppenheimer, to protect a former friend from an investigation into an espionage attempt years before, invented a "cock-and-bull story." This story in the end ruined his friend's career. Oppenheimer's only excuse for his baffling and inexcusable behavior was that he

(Oppenheimer) had been an idiot. This whole incident was seen as indicating "defects in his character."

Majorities on both the Personnel Security Board and the Atomic Energy Commission voted to withhold Oppenheimer's security clearance. The decision created a tremendous controversy. Many scientists stated that Oppenheimer's monumental achievement of perfecting the atomic bomb was being rewarded with public humiliation. Some believed that he was being punished because he had opposed the high military officers and their civilian scientific advisors who favored a strategy of "massive retaliation." Others felt that the cock-and-bull story demonstrated a flawed personality and held that, particularly in light of his past left-wing associations, he should not have access to state secrets.

Flawed or not, Oppenheimer's personality was certainly complex. A comment by one of his college friends (Professor John T. Edsall) in a letter to one of the present authors gives some insight into this:

> Oppenheimer had great gifts for making both friends and enemies. His brilliance of mind was of course phenomenal. In our student days he was obviously in a class by himself intellectually, in a group that included several people who became distinguished later. His mind seemed to burn like a flame, and his range of interest was very wide. When we were both students in Cambridge, England (1925–26) he introduced me to the work of several French poets that I never read before; he was devouring Dostoyevsky and other authors, while plunging into the new quantum mechanics—Heisenberg's first papers had just come out, and Schrödinger's were soon to follow. Later his charm, and his powers of persuasion, were famous; indeed people who opposed his policy views sometimes distrusted him for this very reason—they felt he was much too persuasive. Yet also he could be arrogant, and cuttingly sarcastic, with people whom he did not respect intellectually; and he paid a heavy price for this. He was certainly a very complex person; his capacity for friendship was certainly great; he was devoted to his close friends, and they to him.

Looking back over the war years during which his work had changed from contemplation of natural laws to the direction of a vast weapons laboratory, Oppenheimer said, "In some sort of crude sense which no vulgarity, no humor, no overstatement can quite extinguish, the physicists have known sin; and this is a knowledge which they cannot lose."

Oppenheimer died in 1967, four years after receiving the AEC Fermi Award which was viewed widely as an official gesture of restitution.

○ *Each MO Holds Up to Two Electrons With Opposite Spins*

We may think of building up the structure of a diatomic molecule by placing electrons into its MO's, just as building up atoms occurs by placing electrons in AO's (see Sec. 10-6). Each MO holds two electrons of opposite spin. In this way the Pauli principle, that each electron has distinct quantum numbers, is preserved.

Electrons fill the MO's in the order of increasing energy. This order has been determined from molecular spectra. For molecules built from atoms in the first two rows of the periodic table, the order of energies is usually:

$$\sigma 1s < \sigma^* 1s < \sigma 2s < \sigma^* 2s < \sigma 2p < \pi_y 2p = \pi_z 2p < \pi_y{}^* 2p = \pi_z{}^* 2p < \sigma^* 2p$$

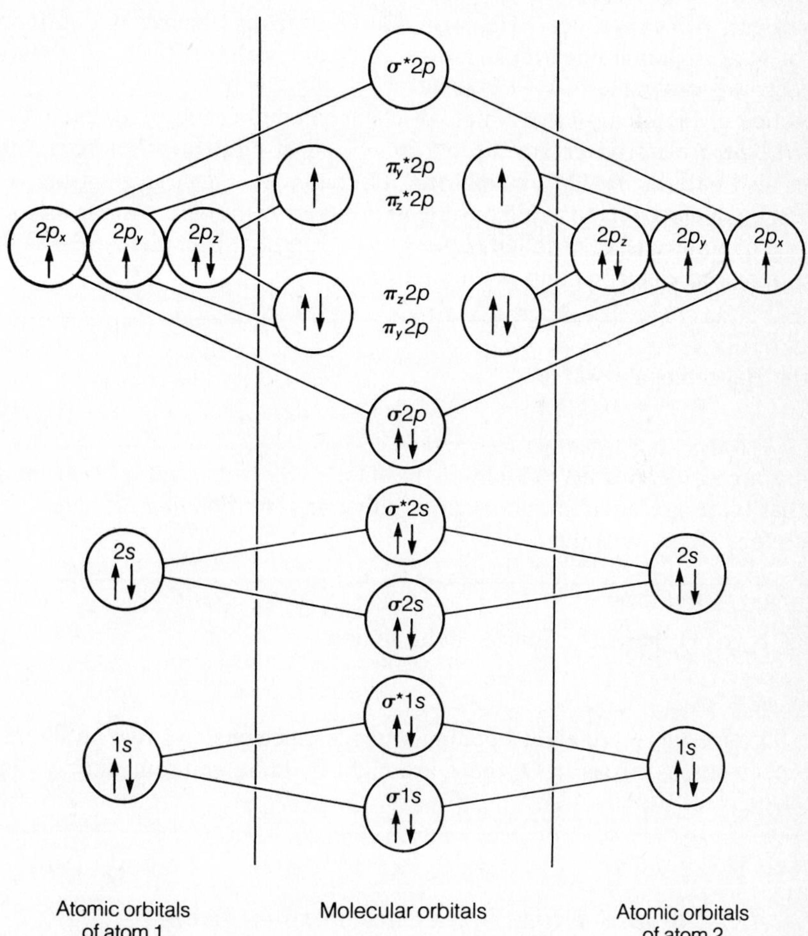

FIGURE 11-6
Molecular orbitals for the O_2 molecule. Electrons are indicated by arrows, the direction of the arrow representing the spin. (Adapted with changes from Coulson, 1961.)

However, since the energies of the $\sigma 2p$ and $\pi 2p$ orbitals are nearly equal, their order of filling is sometimes interchanged, as for the case of C_2 and N_2^+. The precise order is given in Table 11-1.

In thinking of building up diatomic molecules it is useful to keep Figure 11-6 in mind. This figure shows the filling of orbitals for the particular case of O_2, but you can ignore the electrons (arrows), and use it equally well for other molecules. For the case of H_2^+, considered in Section 11-2, the single electron occupies the lowest molecular orbital, $\sigma 1s$. For H_2 a second electron (of opposite spin) enters this orbital, thereby doubling the number of electrons in bonding MO's. Thus one would expect the dissociation energy of H_2 to be greater than that of H_2^+, and inspection of Table 11-1 shows that it is. The molecule He_2^+ contains one electron in the $\sigma^* 1s$ MO. Since this is an antibonding MO, we would expect that the binding energy is decreased compared to H_2. Reference to Table 11-1 again shows that the prediction is consistent with experiment. He_2 would have as many electrons in antibonding orbitals as in bonding MO's, and so we would not expect He_2 to be stable. In fact, this molecule is not observed.

Notice in Figure 11-6 that when an electron enters an orbital, such as π_z^* $2p$, the next electron enters the π_y^* $2p$ orbital of equivalent energy, rather than also entering π_z^* $2p$ and pairing its spin with the first electron. This tendency of electrons to avoid pairing in orbitals when there is an unoccupied orbital of equal energy is called *Hund's rule*. Hund's rule explains why O_2 has two unpaired electrons and is thus paramagnetic.

EXERCISE 11-4

Is the Be_2 molecule stable?

ANSWER

There are two electrons in each of the $\sigma 1s$, $\sigma^* 1s$, $\sigma 2s$, and $\sigma^* 2s$ orbitals, so that there are equal numbers of bonding and antibonding electrons. Therefore Be_2 is unstable.

EXERCISE 11-5

Does N_2 or O_2 have the shorter bond length?

ANSWER

N_2, because there are eight bonding valence electrons and two antibonding electrons, whereas in O_2 there are eight bonding and four antibonding electrons (see Table 11-1).

Double and Triple Bonds are Formed from Both σ and π MO's

In the N_2 molecule, each of the following MO's is filled with two electrons: $\sigma 1s$, $\sigma^* 1s$, $\sigma 2s$, $\sigma^* 2s$, $\sigma 2p$, $\pi_z 2p$, and $\pi_y 2p$. Since there are as many antibonding as bonding $\sigma 1s$ and $\sigma 2s$ electrons, the bonding effectively comes

TABLE 11-1 *Electronic Configurations in Ground States of Homonuclear Diatomic Molecules*[a]

MOLECULE	$\sigma 1s$	$\sigma^* 1s$	$\sigma 2s$	$\sigma^* 2s$	$\sigma 2p$	$\pi 2p$	$\pi^* 2p$	$\sigma^* 2p$	NUMBER OF VALENCE ELECTRONS: BONDING	ANTIBONDING	BOND ORDER	DISSOCIATION ENERGY/eV	SPECTROSCOPIC NOTATION
H_2^+	1								1	0	$\frac{1}{2}$	2.65	$^2\Sigma$
H_2	2								2	0	1	4.48	$^1\Sigma$
He_2^+	2	1							2	1	$\frac{1}{2}$	2.6	$^2\Sigma$
He_2	2	2							2	2	0	0	$^1\Sigma$
Li_2	2	2	2						2	0	1	1.14	$^1\Sigma$
Be_2	2	2	2	2					2	2	0		$^1\Sigma$
B_2	2	2	2	2		2			4	2	1		$^3\Sigma$
C_2	2	2	2	2	1	3			6	2	2		$^3\Pi$
N_2^+	2	2	2	2	1	4			7	2	$2\frac{1}{2}$	6.35	$^2\Sigma$
N_2	2	2	2	2	2	4			8	2	3	7.38	$^1\Sigma$
O_2^+	2	2	2	2	2	4	1		8	3	$2\frac{1}{2}$	6.48	$^2\Pi$
O_2	2	2	2	2	2	4	2		8	4	2	5.08	$^3\Sigma$
F_2	2	2	2	2	2	4	4		8	6	1	1.66	$^1\Sigma$
Ne_2	2	2	2	2	2	4	4	2	8	8	0	0	$^1\Sigma$

[a] From Coulson (1961).

from the two $\sigma 2p$ and four $\pi 2p$ electrons. There is thus a net of six bonding electrons. We therefore call the N_2 bond a *triple bond.* Notice, though, that the bond is not formed from three identical MO's. There is one σ MO that is symmetric about the bonding axis, and there are two π MO's, each with two "sausages." O_2 (Fig. 11-6) has a net of one σ bonding pair and one π bonding pair, and so we say that a *double bond* exists. A single bond can be represented as σ^2, meaning that there the molecule contains a net of two σ bonding electrons. Similarly a double bond can be represented $\sigma^2\pi^2$, and a triple bond as $\sigma^2\pi^4$.

A general measure of bond multiplicity is the *bond order.* It is defined as one-half of the excess of electrons in bonding orbitals over electrons in antibonding orbitals. Bond orders for the homonuclear diatomic molecules are given in Table 11-1.

EXERCISE 11-6

Without consulting Table 11-1, calculate bond orders of He_2^+, Li_2, and F_2.

ANSWER

$\frac{1}{2}$, 1, and 1.

Spectroscopic Notation

Spectroscopists have a special notation for molecules, and this notation contains much information. This notation for homonuclear diatomic molecules is given in final column of Table 11-1. The left superscript is the *spin multiplicity* of the molecule, a measure of the number of unpaired electrons (and hence an indicator of the magnetic moment and ESR spectrum). Each unpaired electron contributes $\frac{1}{2}$ to the resultant spin S, and the spin multiplicity of a molecule is defined as $2S + 1$. Thus O_2 with two unpaired electrons has a multiplicity of 3, as indicated by the superscript 3 in the final column of Table 11-1. A molecule with spin multiplicity of 1 (all electron spins paired) is often called a *singlet,* and a molecule with spin multiplicity of 3 (two unpaired electron spins) is called a *triplet.*

EXERCISE 11-7

By drawing the electronic configuration of N_2^+, confirm that its spin multiplicity is 2.

The capital Greek letter of the spectroscopic symbol gives the quantum number for orbital angular momentum due to electron rotation about the bond axis. If the absolute value of angular momentum is zero, the quantum number is 0 and the symbol is Σ; if the quantum number is 1 (for an odd number of π electrons), the symbol is Π. Electrons in σ orbitals are symmetric about the axis and do not contribute to angular momentum. Electrons in

π orbitals contribute ± 1 and the contributions can cancel if there are two electrons in the MO.

○ *Rules for Combining AO's into MO's*

In forming MO's, we have considered combinations of *s* orbitals with *s* and *p* orbitals, and of *p* orbitals with *p* orbitals. Are other combinations possible? The answer is yes: *d*, *f*, etc., orbitals can be combined with *s*, *p*, *d*, etc., orbitals, but two conditions must be obeyed:

1. The orbitals cannot be of differing symmetries with respect to the bond axis, or else all integrals in the energy expression vanish and there is no interaction. This is illustrated schematically in Figure 11-7.
2. The atomic orbitals cannot differ greatly in energy if they are to combine significantly. This is the reason that valence electrons of one atom tend to combine with valence electrons of another, rather than with the inner electrons of much lower energy.

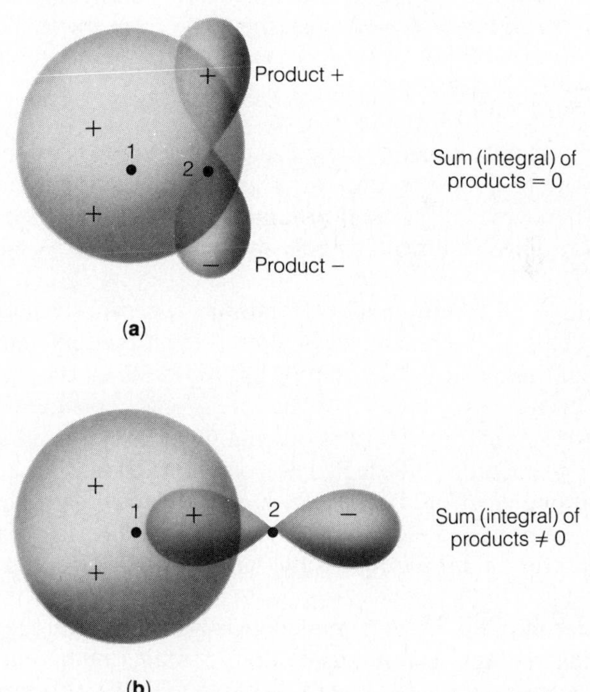

FIGURE 11-7
Atomic orbitals cannot be of different symmetries if they are to combine to form a molecular orbital. **(a)** Different symmetry. **(b)** Same symmetry.

EXERCISE 11-8

Can a p_x orbital from one atom combine with a p_y orbital from another to form a MO?

ANSWER

No, symmetries with respect to the bond axis are different.

EXERCISE 11-9

Can the 1*s* orbital of hydrogen form an effective sigma bond with the 1*s* orbital of F?

ANSWER

No, the 1*s* orbital of F is an inner orbital and of much lower energy than that of valence electrons.

Heteronuclear Diatomic Molecules

Many of the considerations for homonuclear diatomic molecules can be carried over to heteronuclear diatomic molecules (those formed from different atoms). In the latter case, however, the AO's from the two contributing atoms do not enter with equal weights. Thus in the wavefunction,

$$\Psi = c_1\psi_1 + c_2\psi_2$$

the ratio $c_1/c_2 \neq \pm 1$. This means there is a polarity to the bond, because if ψ_1 contributes more to Ψ than ψ_2 does ($c_1 > c_2$), an electron in the MO spends more time around nucleus 1 than around nucleus 2. In other words, the molecule has a dipole moment. Dipole moments are discussed in Section 11-6.

As an example of a heteronuclear diatomic, we can consider NO. The electronic configuration of the nitrogen atom is $(1s)^2(2s)^2(2p)^3$. The electronic configuration of the oxygen atom is $(1s)^2(2s)^2(2p)^4$. When they combine into NO, the molecule has the following electronic configuration: $(\sigma 1s)^2 (\sigma^* 1s)^2(\sigma 2s)^2(\sigma^* 2s)^2(\sigma 2p)^2(\pi 2p)^4(\pi^* 2p)$. The final orbital contains only a single electron identical to the electronic configuration in Figure 11-6 except for removal of one electron, because N has one electron less than O. Thus NO possesses an unpaired electron and is *paramagnetic*. This paramagnetic property is useful in preparing "spin labels" for ESR experiments in biochemistry (Sec. 13-8).

Another heteronuclear diatomic molecule is HCl. The hydrogen atom contributes a single 1*s* AO. The chlorine atom contains numerous AO's, but most of them have energies too low to combine with the 1*s* orbital of hydrogen. (Recall Rule 2 above, which states that the combining orbitals must be of similar energy.) The only suitable AO of chlorine is the $3p_x$ orbital. Thus the HCl bond is a bonding orbital formed from the hydrogen 1*s* orbital and the chlorine $3p_x$ AO.

11–5 Polyatomic Molecules

Bond Properties Make It Possible to Describe Most Polyatomic Molecules in Terms of Localized Molecular Orbitals

You might think that molecular orbitals in polyatomic molecules must contain contributions from atomic orbitals centered on all the atoms. This is true in describing some aromatic molecules and in describing excited states that occur in the presence of light. But for very many molecules, localized molecular orbitals, which contain contributions from AO's on only two atoms, can give a good description of the molecule.

The fact that localized MO's are often sufficient can be deduced from the common values of *bond properties* in different molecules. For example, the O—H bond length is about the same (0.096 nm) whether the bond is in H_2O, H_2O_2, CH_3O—H, or the O—H radical. Similarly, the O—H bond energy is 110 ± 10 kcal mol^{-1} in all these molecules. Moreover, spectroscopic studies show that the frequency of vibration of the O—H group is not changed greatly when the group is incorporated in a variety of larger molecules (such as in the serine residues of a protein). The uniformity of these properties indicates that an O—H group has certain properties that are not appreciably affected by whatever chemical groupings are bonded to the oxygen. This shows that the electrons engaged in the O—H bond are not significantly perturbed by electronic events on the other side of the oxygen atom. A consequence of this situation is that we can describe the bonds of most polyatomic molecules in terms of localized MO's that have contributions only from the AO's of atoms participating in the bond.

Choosing water again as our example, let us write expressions for two localized MO's that describe the two O—H bonds. The two hydrogen atoms have electronic configurations of $(1s)$ and the oxygen atom of $(1s)^2 (2s)^2(2p_z)^2(2p_y)(2p_x)$. Thus by overlapping one of the hydrogen $1s$ AO's with the $2p_x$ AO and the other hydrogen $1s$ AO with the oxygen $2p_y$ AO, two localized MO's are created. These can be written

$$\text{Bond I} = c_1\psi(H\ 1s) + c_2\psi(\text{O}\ 2p_x)$$

and

$$\text{Bond II} = c_1\psi(H\ 1s) + c_2\psi(\text{O}\ 2p_y)$$

The formation of these two MO's is shown schematically in Figure 11-8. If we had chosen ammonia instead of water, the arguments would be similar, but there would be three unpaired *p* electrons from the nitrogen, and thus three AO's that can combine with three hydrogen $1s$ AO's to form three localized MO's.

A little thought tells us, however, that there is an important element missing from this description of H_2O and NH_3 in terms of localized MO's. The

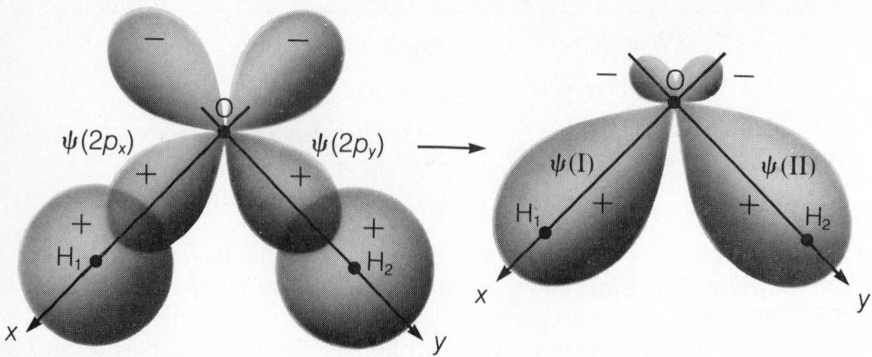

FIGURE 11-8
Formation of two localized MO's in H_2O from $2p$ orbitals of oxygen and $1s$ orbitals of hydrogen. (From Coulson, 1961.)

localized MO's suggest that the O—H bonds in water and the N—H bonds in ammonia are at the angle of the *p* orbitals in atoms, that is, 90°. In fact spectroscopic experiments show that the H—O—H angle in H_2O is 104.5° and the H—N—H angles in NH_3 are 108°. Moreover, when we think about the valence of carbon, the sort of localized MO's we have been considering make little sense. The ground-state carbon atom has the electronic configuration $(1s)^2(2s)^2(2p)^2$, and reasoning such as above would lead us to the conclusion that carbon is usually divalent for hydrogen, with H—C—H angles of 90°. The most common hydrocarbon with a single carbon atom is methane, CH_4, with a bond angle of the tetrahedral angle, 109.47°. The missing element in our description of localized MO's is *hybridization,* which we consider next.

Hybridization is the Mixing of AO's Prior to Their Combining into MO's

The reason that the carbon atom in methane forms four equivalent bonds to hydrogen atoms is that the atomic *s* and *p* AO's in carbon mix or *hybridize* as they combine with the $1s$ AO's of the hydrogen atoms. We can imagine bonding as a two-step process. In the first step, indicated in Figure 11-9, the $2s$ AO is mixed with the three $2p$ AO's to form sp^3 hybrid orbitals of equal energy. The four hybrid orbitals each contain a single electron. In the second step they combine with $1s$ AO's of hydrogen.

The change in orbital shapes during hybridization is depicted in Figure 11-10(*a*). The *s* and *p* AO's add together to form an elongated, directed lobe that can overlap with *s* or *p* orbitals from other atoms. In sp^3 hybridization, as in methane, the four hybrids are at the tetrahedral angle to each other [Fig. 11-10(*b*)]. The mathematical form of the four sp^3 hybrid AO's is a linear combination of atomic orbitals, just as a molecular orbital is a linear combination of atomic orbitals:

$$\chi_1 = 2s + 2p_x + 2p_y + 2p_z$$
$$\chi_2 = 2s - 2p_x - 2p_y + 2p_z$$
$$\chi_3 = 2s + 2p_x - 2p_y - 2p_z$$
$$\chi_4 = 2s - 2p_x + 2p_y - 2p_z$$

and then one of the localized MO's defining one of the C—H bonds can be expressed

$$\Psi_I = c_1(C\ \chi_1) + c_2(H\ 1s)$$

There are of course three other localized bonds, one formed with each of the hybrid AO's.

Carbon is sp^3 hybridized in many compounds other than methane. In the paraffins, for example, butane,

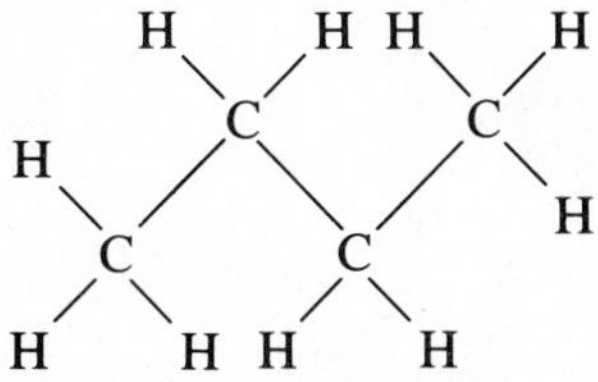

each C—C bond is formed by pairing two tetrahedral hybrids, one from each carbon atom. Each C—H bond is formed by pairing one of the sp^3 hybrids with a $1s$ hydrogen orbital. In diamond, the sp^3 hybrids from neighboring carbon atoms pair to form a tetrahedral structure of great strength and rigidity.

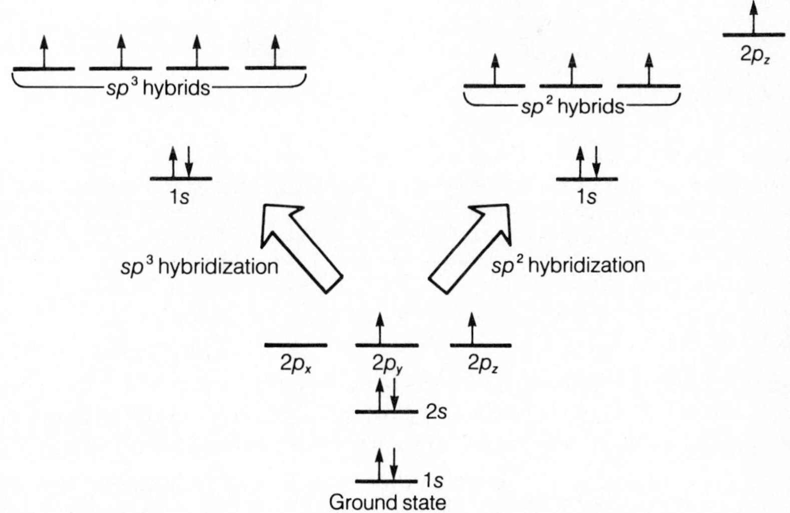

FIGURE 11-9
Mixing of AO's of the ground state of carbon to form hybrid orbitals, ready for bonding. The increase in energy of the $2s$ electrons during hybridization is regained when bonding occurs.

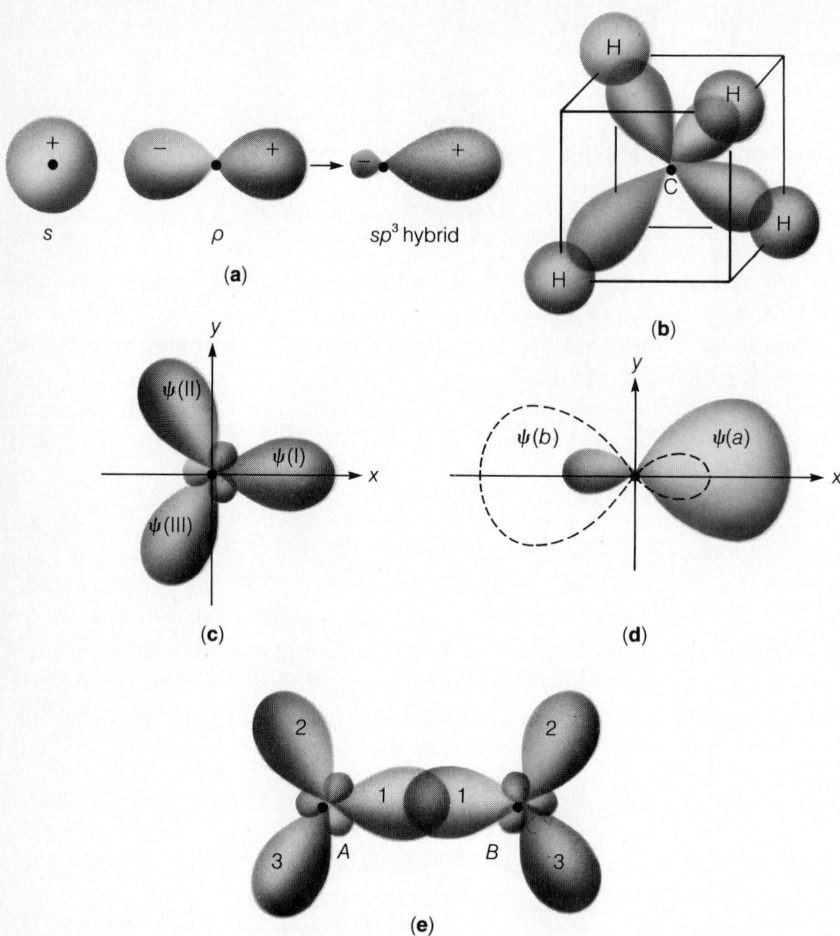

FIGURE 11-10
Hybridization. (**a**) Formation of a hybrid atomic orbital from an s and a p orbital. (**b**) Four tetrahedral sp^3 hybrids of the carbon atom in CH_4, showing the directions of the hybrids and the overlapping with $1s$ orbitals of hydrogen to form four C—H bonds. (**c**) Three trigonal sp^2 hybrids, as in the carbon atoms of ethylene. (**d**) Two sp digonal hybrids as in acetylene. (**e**) Overlap of two sp^2 hybrids to form a σ MO of ethylene. (**a, c, d,** and **e** redrawn from Coulson, 1961; **b** redrawn from Karplus and Porter, 1970.)

A second hybrid of importance is the *trigonal* or sp^2 hybrid. In this hybrid the p_z orbital is left unmixed, and the $2s$ orbital electrons hybridize with only the $2p_x$ and $2p_y$ orbitals (Fig. 11-9). The result is three hybrid AO's that lie in the x-y plane at angles of 120° to each other [Fig. 11-10(c)]. Suppose now that two sp^2 hybridized carbon atoms are positioned such that two of these orbitals overlap [as in Fig. 11-10(e)]. This produces a σ bond between the two carbon atoms. The remaining hybrids can combine with, say, $1s$ orbitals of hydrogen. The p_z electrons of the two carbon atoms are still free to combine with one another, forming a π bond. This is the basis of the double bond in ethylene.

A third important hybrid of carbon is the *digonal* or *sp* hybrid, depicted in Figure 11-10(*d*). Here the 2*s* electrons mix only with the p_x orbital, leaving electrons in the p_y and p_z orbitals. The *sp* hybrids form an angle of 180° with each other.

The bond lengths and bond energies characteristic of the three types of *sp* hybrid are summarized in Table 11-2.

EXERCISE 11-10

Sketch a figure for two overlapping *sp* hybrids that is analogous to Figure 11-10(*e*). How do the $2p_y$ and $2p_z$ electrons of the two carbon atoms pair? What type of bond is formed?

ANSWER

The $2p_y$ and $2p_z$ electrons on the two carbon atoms form two π MO's. The σ MO of the hybrid and the two π MO's constitute a triple bond.

Returning for a moment to the water molecule, we can see how hybridization affects the structure. The 2*s* orbital and three 2*p* orbitals form four sp^3 hybrids as in methane. Two of these contain only a single electron, so that they can each pair with the 1*s* AO's of a hydrogen atom, forming two O—H bonds. Because the oxygen AO's are hybridized the H—O—H bond angle is nearer to 109° than to 90°. The two other hybrid orbitals each contain two electrons, so that they are not free to couple with other atoms. The electrons in them are neither bonding nor antibonding, and are hence referred to as *nonbonding* lone pairs. The two lone pairs form approximately tetrahedral angles with the two O—H bonds (Fig. 11-11). Repulsion between the lone pairs tends to force them out from the tetrahedral angle, and it turns out that this forces the two O—H bonds in to the experimental H—O—H angle

TABLE 11-2 *Properties of Bonds in Hybrid Carbon Compounds* [a]

BOND	HYBRID	LENGTH/Å	BOND ENERGY/ KJ MOL^{-1}	EXAMPLE
C—C	sp^3	1.54	368	Methane
C═C	sp^2	1.34	699	Ethylene
C≡C	sp	1.20	962	Acetylene
C—H	sp^3	1.09	431	Methane
C—H	sp^2	1.07	444	Ethylene
C—H	sp	1.06	506	Acetylene

[a] From Coulson (1961) and Karplus and Porter (1970).

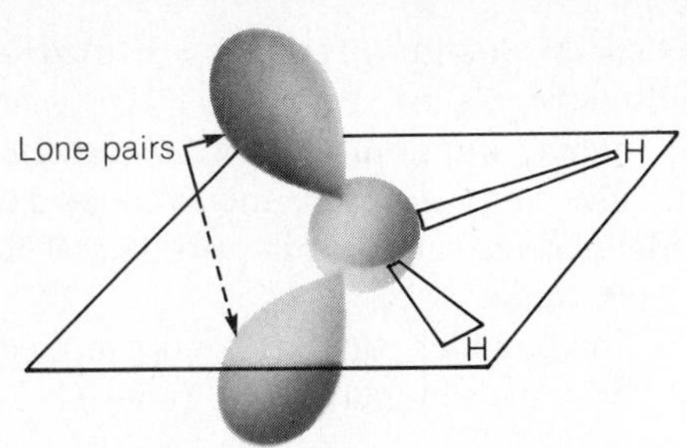

FIGURE 11-11
The tetrahedral character of the water molecule. The three nuclei define a plane from which the lone pairs extend, one up and one down.

of 104.5°. The roughly tetrahedral disposition of the two O—H bonds and the two lone pairs leads to tetrahedral coordination of water molecules, through hydrogen bonding, in ice and liquid water.

EXERCISE 11-11
What type of MO's are formed in NH_3 after sp^3 hybridization of the nitrogen atom? How many lone pairs of electrons are there?

ANSWER
Three σ MO's form, one for each N—H bond. There is one lone pair.

In molecules containing conjugated double bonds, several sp^2 hybrids can overlap to form extended π MO's in which electrons are said to be *delocalized*. In the case of benzene, a planar molecule of formula C_6H_6, each carbon atom is sp^2 hybridized. The hybrids from each carbon atom overlap to form σ bonds with the neighboring carbons and one hydrogen atom. Each carbon also has a $2p$ orbital for π bonding. The six $2p$ orbitals lead to six π MO's. The three MO's of lowest energy are doubly occupied by electrons. The formation of π MO's lowers the energy of the p electrons, and they are said to be stabilized by delocalization. This stabilizing delocalization energy is called *resonance energy* in the valence bond treatment of bonding described in most of the books listed in the references. It is about 37 kcal mol^{-1} for the case of benzene, and is greater for more extended conjugated systems such as naphthalene.

The Peptide Bond has Some Double-Bond Character

Amino acids are linked to form proteins by the *peptide* (sometimes called amide) bond

```
  O  H
  ‖  |
—C—N—
```

X-ray diffraction studies show that all four of these atoms lie in a plane, indicating that a restriction to free rotation about the peptide linkage exists. This restriction has profound implications for the structure of proteins. In this section we apply molecular orbital theory to understand why there is a partial double bond character in the peptide linkage, leading to restriction of rotation.

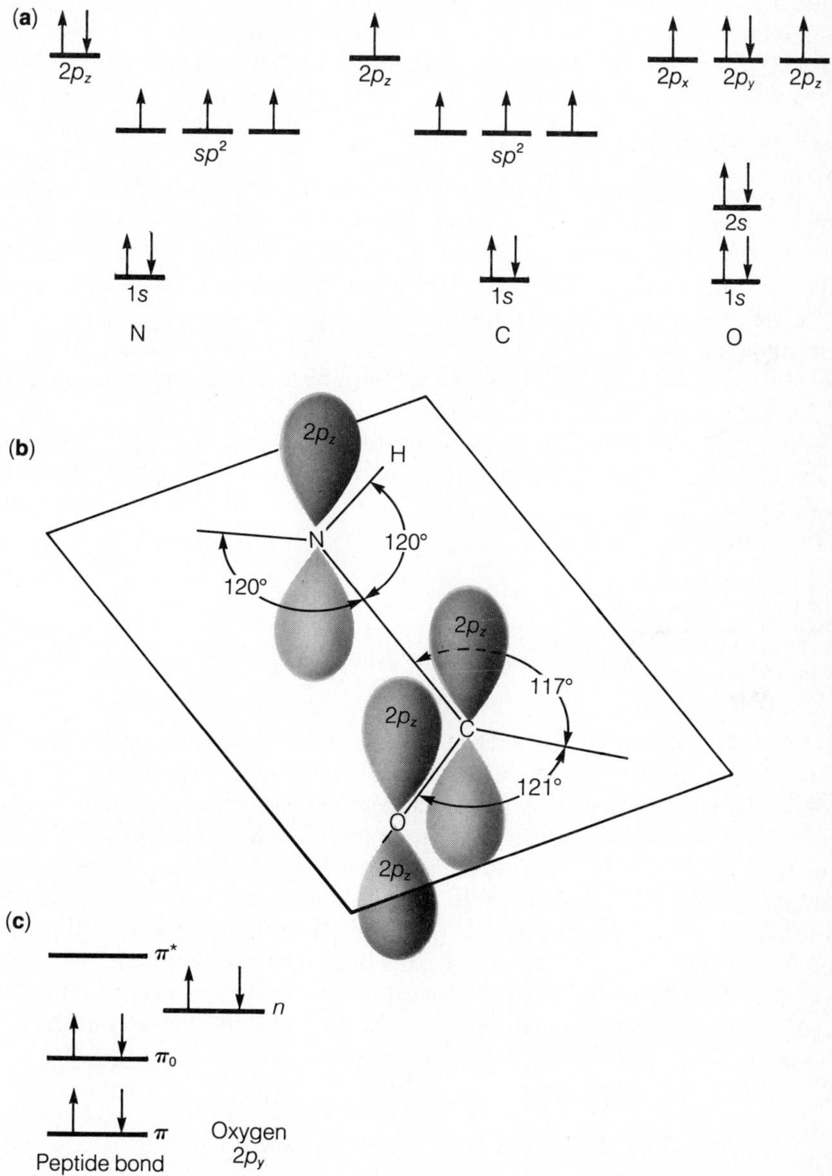

FIGURE 11-12
Covalent bonding in the peptide bond. (**a**) Electronic configuration of the nitrogen, carbon, and oxygen atoms. (**b**) Geometry of the four planar atoms in the peptide group. (**c**) Energy levels of MO's.

The electronic configurations of the nitrogen, carbon, and oxygen atoms involved in the peptide bond are shown in Figure 11-12(*a*). The AO's of carbon and nitrogen atoms are sp^2 hybridized. This is known from the values of about 120° for all angles formed by the bonds from these atoms. Each of these sp^2 hybrids from the nitrogen and carbon atoms overlap with AO's from three neighboring atoms to form σ bonds.

The $2p_z$ orbital of each of the three atoms is perpendicular to the plane containing the atoms [Fig. 11-12(*b*)]. These orbitals can overlap with each other to form MO's of the π type. The wavefunctions for these MO's have the form

$$\Psi = c_1\psi(\text{O } 2p_z) + c_2\psi(\text{C } 2p_z) + c_3\psi(\text{N } 2p_z)$$

Because there are three AO's that contribute, there are three MO's, whose energies are given in Figure 11-12(*c*). Nitrogen contributes two electrons to the system, and oxygen and carbon each contribute one. The lower two MO's are thus each occupied by two electrons, and the upper MO is unoccupied. In the presence of UV light, electrons can be excited from the lower π levels to the upper π^* level. Electrons from the nonbonding $2p_y$ orbital on oxygen can also be excited to higher levels. These excitations are discussed in Section 13-3.

The π type MO's give some double-bond character to the peptide linkage. Experiments confirm this conclusion: not only is free rotation about the N—C bond restricted, but its length (1.32 Å) is somewhat shorter than ordinary C—N single bonds (1.49 Å).

○ 11–6 Charge-Distribution Properties of Molecules

So far in this chapter we have been concerned with covalent, as opposed to noncovalent, bonds. Covalent bonds bind together the atoms of small molecules, and also serve as the linkages of amino acids in proteins and of nucleotides in nucleic acids, and sugars in polysaccharides. But many properties of substances are determined by noncovalent forces. These forces bind polypeptide chains together to form oligomeric enzymes or the coats of viruses, and bind lipids and proteins together to form membranes. Noncovalent forces also determine the properties of liquids and solutions.

Noncovalent forces arise from interactions of charges on neighboring molecules. To gain some appreciation of these forces we must consider the charge-distribution properties of molecules.

The Dipole Moment Describes the Asymmetry of Charge on a Molecule

Consider the particle shown in Figure 11-13(*a*). It consists of a negative charge $-q$ separated by a distance x from a positive charge $+q$. The particle carries no net charge, but its distribution of charges is clearly asymmetric.

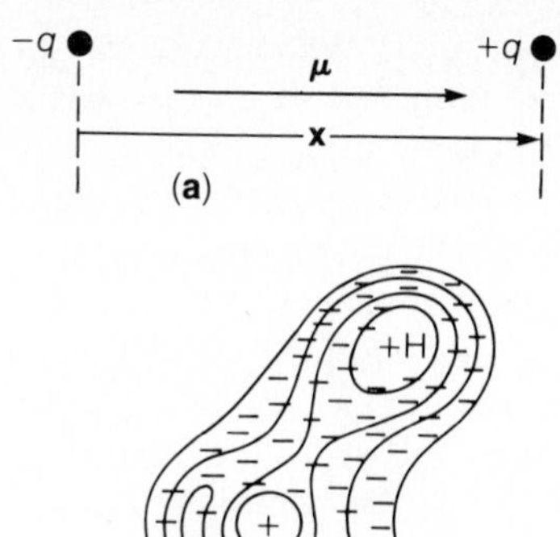

FIGURE 11-13
Two particles with dipole moments. (**a**) A positive charge $+q$ separated by a distance $\mathbf{x}$ from a negative charge $-q$. (**b**) A water molecule, represented by contours of electron density. The dipole moment bisects the H—O—H angle, with its negative end toward the lone pairs of electrons on the oxygen atom (oxygen nucleus represented by +) and its positive end toward the protons (represented by +H). The dipole moment is a vector pointing from the negative to the positive charge.

This asymmetry is described by the *dipole moment*, $\boldsymbol{\mu}$, defined as having a magnitude given by the product of q and x,

$$\boldsymbol{\mu} = q\mathbf{x} \tag{11-17}$$

and pointing from the negative to positive charge. Because it has both magnitude and direction, the dipole moment is a vector.

The definition of the dipole moment can be extended to cover more realistic cases. If a particle consists of a group of charges, the dipole moment is given by

$$\boldsymbol{\mu} = \sum_i q_i \mathbf{x}_i \tag{11-18}$$

in which $\mathbf{x}_i$ is the distance of the ith charge from an arbitrary origin. If the charge distribution is continuous, we can define the dipole moment by replacing the summation by integration:

$$\boldsymbol{\mu} = \int \rho(\mathbf{x})\, \mathbf{x}\, dV \tag{11-19}$$

in which $\rho(\mathbf{x})$ is the charge density and dV is an element of volume. By replacing the charge density with the wave function for the particle, we see how the dipole moment can be calculated from the wave function ψ for the particle:

$$\boldsymbol{\mu} = \int \psi^2 \mathbf{x}\, dV \tag{11-20}$$

The dimensions of a dipole moment are charge times length. Thus in the cgs system its units are esu cm, and in the SI system, C m. Since molecular charges are of the order of 10^{-10} esu (or 10^{-19} C) and molecular distances are of the order of 10^{-8} cm (10^{-10} m), we would expect that molecular dipole moments are of the order of 10^{-18} esu cm (or 10^{-29} C m). This turns out to be so (Table 11-3). Dipole moments frequently are cited in *Debye units*. One Debye, named for the scientist who showed how to measure dipole moments, is 1×10^{-18} esu cm, and is denoted 1 D.

Debye showed that dipole moments can be determined from measurements of the dielectric constant as a function of temperature (see p. 502), and in the past 50 years, dipole moments for many hundreds of molecules have been measured. Values for a few of them are listed in Table 11-3. Molecules sufficiently symmetric to possess a center of symmetry (Sec. 16-1) have zero dipole moment. With increasing asymmetry of charge, dipole moments increase. Water with a moment of 1.83 D is highly polar for a small molecule. This accounts for its high dielectric constant and for its excellent properties as a solvent. The negative end of the H_2O dipole clearly is the oxygen atom with its lone pairs of electrons. The positive end is the two hydrogen atoms [Fig. 11-13(*b*)]. The lone pairs interact with positively charged substances, thereby stabilizing them in solution, and the hydrogen atoms interact with negatively charged substances.

As the charges of a dipole are separated, the moment increases. This is illustrated by the entries in Table 11-3 for glycine and its peptides. Glycine in aqueous solution exists mainly as a *dipolar ion,* or *zwitterion,* in which the proton from the carboxyl group is transferred to the amino group, thereby producing a neutral molecule containing two oppositely charged groups. In glycine dipeptide the negative carboxylate ion is separated from the positive

TABLE 11-3 *Molecular Dipole Moments. (Moments are given in Debye units;* $1\ D = 1 \times 10^{-18}$ *esu cm* $= 3.34 \times 10^{-30}$ *C m.)*

MOLECULE	FORMULA	μ/D
Monatomic ions	H^+, Na^+, K^+	0
Hydrogen	H_2	0
Deuterium	D_2	0
Carbon dioxide	CO_2	0
Ammonia	NH_3	1.48
Water	H_2O	1.83
Methane	CH_4	0
Hydrogen sulfide	H_2S	1.02
Hydrogen chloride	HCl	1.04
Hydrogen bromide	HBr	0.80
Hydrogen iodide	HI	0.38
ortho-Dichlorobenzene	$C_6H_4Cl_2$	2.59
meta-Dichlorobenzene	$C_6H_4Cl_2$	1.67
para-Dichlorobenzene	$C_6H_4Cl_2$	0
Formaldehyde	CH_2O	2.17
Urea	NH_2CONH_2	4.56
Peptide bond	—NH—CO—	~3.7
Glycine	$^+NH_3CH_2COO^-$	~16.7
Glycine dipeptide	$^+NH_3CH_2CONHCH_2COO^-$	~28.6
Glycine tripeptide	$^+NH_3CH_2[CONHCH_2]_2COO^-$	~37.2
Glycine tetrapeptide	$^+NH_3CH_2[CONHCH_2]_3COO^-$	~42.6
Glycine pentapeptide	$^+NH_3CH_2[CONHCH_2]_4COO^-$	~49.5
Glycine hexapeptide	$^+NH_3CH_2[CONHCH_2]_5COO^-$	~52.0
Hemoglobin		Hundreds

ammonium ion by an additional amino acid residue, thereby increasing the dipole moment.

EXERCISE 11-12

In glycine the separation from the positively charged nitrogen group to the point midway between the negatively charged oxygen atoms is just over 3 Å. What magnitude would you expect for the dipole moment of this molecule?

ANSWER

$\mu = 14 \times 10^{-18}$ esu cm $= 14$ D $= 47 \times 10^{-30}$ C m

---

EXERCISE 11-13

Why do the dipole moments of glycine peptides (Table 11-3) not increase linearly with the number of peptide groups?

ANSWER

Because of flexibility in the polypeptide backbone, the separation of + and − charges does not increase linearly with the number of peptide groups.

Dipole moments add like vectors. When several dipolar chemical groups are present in a molecule, it is often possible to estimate the molecular dipole moment by summing the moments of the groups. This principle is illustrated by the dichlorobenzenes in Table 11-3. The largest dipole moment for these compounds is for the *ortho* structure, in which the two polar C—Cl bonds form a small angle with each other. The *meta* structure, in which the C—Cl bonds form an oblique angle, has a smaller dipole moment, and the *para* structure, where they have opposite directions, has zero moment.

The potential energy of a dipolar molecule in an electric field depends on the orientation of the dipole moment with respect to the field. This dependence is given by

$$U = -\mu E^* \cos\theta = -\boldsymbol{\mu} \cdot \mathbf{E}^* \qquad (11\text{-}21)$$

in which θ is the angle between $\boldsymbol{\mu}$ and $\mathbf{E}^*$. When the dipole is aligned in the direction of the field, the potential energy is $-\mu E^*$, and when the dipole opposes the field, its potential energy is $+\mu E^*$.

An Electric Field Induces a Dipole Moment in an Atom or Molecule

Suppose that an atom is subjected to the intense electric field, E^*, of a neighboring ion or dipole. The field will distort the electron distribution of the atom, drawing the electronic charge-cloud of the atom toward the positive pole of the field and the nucleus toward the negative pole of the field (Fig. 11-14). The field thus *induces* a dipole moment into the formerly

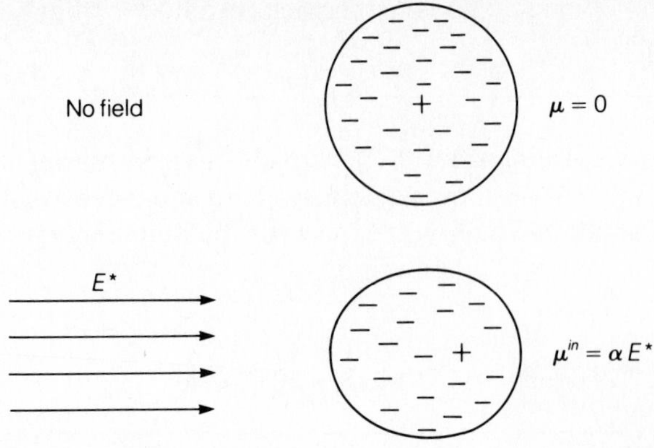

FIGURE 11-14
The induction of a dipole moment, μ^{in}, into an atom of polarizability α by a field of strength E^*.

symmetric atom. At all but very high fields, the magnitude of the induced dipole moment, μ^{in}, is proportional to the field strength:

$$\mu^{in} \propto E^* = \alpha E^* \qquad \text{(11-22 cgs)}$$ ◀

The constant of proportionality, α, in this equation is called the *molecular polarizability*. In the cgs system, the polarizability has units of cm^3, the electric field strength has units of esu cm^{-2}, and their product has the units of a dipole moment, esu cm. In the SI system, the polarizability has units of m^3, and the induced moment is given by

$$\mu^{in} = \alpha\, \epsilon_0\, E^* \qquad \text{(11-22 SI)}$$ ◀

in which ϵ_0 is a constant called the *permittivity of a vacuum* ($\epsilon_0 = 8.8 \times 10^{-12}$ $kg^{-1}\ m^{-3}\ s^4\ A^2$).

EXERCISE 11-14

Show that the dimensions of Equation 11-22 SI are correct using definitions inside the front cover.

ANSWER

$$\begin{aligned}\mu^{in} &= \alpha\, \epsilon_0\, E^* = (m^3)(kg^{-1}\ m^{-3}\ s^4\ A^2)(Vm^{-1}) \\ &= (m^3)\left(\frac{1}{J}\ m^{-1}\ s^2\ A^2\right)(Vm^{-1}) \\ &= (m^3)(m^{-1}\ s\ V^{-1}\ A)(Vm^{-1}) \\ &= Cm\end{aligned}$$

Table 11-4 contains values of the polarizability for several small molecules. Notice that the polarizability has units of volume. Moreover, the values are similar to the actual sizes of molecules, several Å^3. Larger molecules have more electrons and are therefore more polarizable. For some molecules in Table 11-4 the components of the polarizability in each principal direction of the molecule are cited. The mean polarizability is the average of the three components.

EXERCISE 11-15

Explain why Li, which contains only one more electron than He, is nearly 100 times more polarizable than He.

ANSWER

The 2*s* electron is screened from the nuclear charge by the 1*s* electrons, and thus more easily distorted.

Dipole Moments and Polarizabilities can be Determined From Measurements of the Dielectric Constant

The dipole moment and polarizability are molecular quantities, but their determination requires measurement of *macroscopic* properties. The relationship between μ and α and macroscopic properties was derived by Debye and others, and the derivations involve several concepts beyond the scope of

TABLE 11-4 *Some Molecular Polarizabilities. (α is the mean polarizability in all directions; α_1, α_2, and α_3 are components.)*

MOLECULE	$\alpha/10^{-24}$ cm^3	$\alpha_1/10^{-24}$ cm^3	$\alpha_2/10^{-24}$ cm^3	$\alpha_3/10^{-24}$ cm^3	LOCATION OF AXIS
H atom	0.67				
He	0.205				
Ne	0.395				
Ar	1.64				
Kr	2.48	Atoms are isotropic (the same in all directions) and thus have the same polarizability in all directions.			
Xe	4.04				
Li	20.				
Na	20.				
K	36.				
Rb	40.				
H_2	0.79	0.93	0.71	0.71	α_1 = sym. axis
H_2O	1.44				
N_2	1.76	2.38	1.45	1.45	α_1 = sym. axis
O_2	1.60	2.35	1.21	1.21	α_1 = sym. axis
Cl_2	4.61	6.60	3.62	3.62	α_1 = sym. axis
NH_3	2.26	2.42	2.18	2.18	α_1 = sym. axis
C_6H_6	10.3	6.35	12.3	12.3	$\alpha_1 \perp$ ring plane

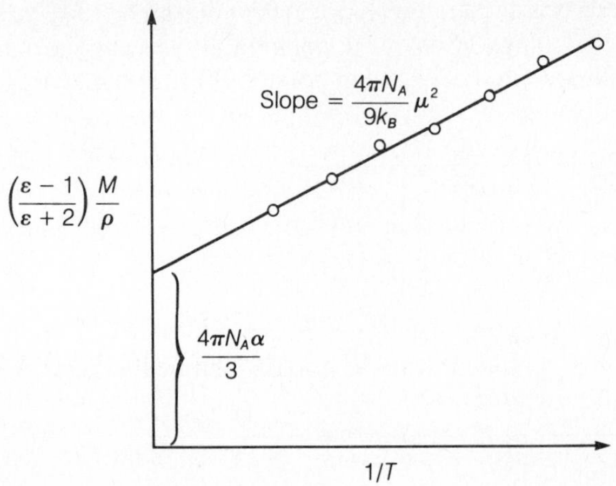

FIGURE 11-15
The determination of the dipole moment and molecular polarizability of a gas by Debye's method.

this book. However, the relationships can be grasped easily without the derivations.

Debye showed that μ and α, for molecules in a gas or in dilute solution in a nonpolar solvent, are related to the dielectric constant ϵ (see Sec. 8-1) by

$$\left(\frac{\epsilon - 1}{\epsilon + 2}\right)\frac{M}{\rho} = \frac{4\pi N_A}{3}\left(\alpha + \frac{\mu^2}{3k_B T}\right) \tag{11-23}$$

in which M is the molecular weight, and ρ is the density. This equation allows you to determine both the dipole moment and polarizability of a molecule if you have measured the dielectric constant of the gas or dilute solution as a function of temperature. You then plot the left side of Equation 11-23 against $1/T$ (Fig. 11-15). The intercept is $4\pi N_A\alpha/3$ and the slope is $4\pi N_A\mu^2/9\,k_B$. Thus you can derive values of both μ and α from the plot.

The physical meaning of the Debye equation is that the dielectric constant arises in part from dipole moments induced in the electron distribution by the applied electric field (the first term on the right). For substances composed of strongly dipolar molecules, a greater contribution comes from alignment of the dipoles in the applied field (the second term on the right). Note in Equation 11-33 that increasing temperature opposes alignment and lowers the dielectric constant.

EXERCISE 11-16

Sketch Debye plots similar to Figure 11-15 for substances 1, 2, and 3, for which $\mu_1 = \mu_2 > \mu_3$ and $\alpha_1 > \alpha_2 = \alpha_3$.

ANSWER

The slopes of 1 and 2 are equal and greater than 3. The intercepts of 2 and 3 are equal and smaller than 1.

Polarizabilities can also be measured from the refractive index of a substance for light (see Sec. 12-4 for definition and further discussion of this property). The reason for this is that the dielectric constant of a substance for electromagnetic fields of high frequency (such as light) is equivalent to the square of the refractive index of the substance. For the dielectric behavior of a substance in the presence of light, Equation 11-23 assumes the form

$$\alpha = \frac{3M}{4\pi N_A\rho}\left(\frac{\epsilon - 1}{\epsilon + 2}\right) = \frac{3M}{4\pi N_A\rho}\left(\frac{n^2 - 1}{n^2 + 2}\right) \tag{11-24}$$

Notice that the last term of Equation 11-23 does not contribute to the polarization by light. The reason is that dipolar molecules cannot move fast enough to align themselves in the direction of the alternating electromagnetic field, when it alternates at the frequency of visible light. When the last term is removed and n^2 is substituted for ϵ, Equation 11-24 is obtained from Equation 11-23.

○ 11–7 Noncovalent Forces are Described by Potential-Energy Functions

You will recall from Section 2-2 that it is convenient to describe the force of interaction of one particle with others by the *potential energy* of interaction. Once we know the potential energy of interaction of two particles, A and B, as a function of their separation, $U_{AB}(r)$, we immediately can determine the force $\mathbf{F}_{AB}$ that one exerts on the other, since it is the negative derivative of the energy,

$$\mathbf{F}_{AB} = -\frac{dU(r)}{dr} \tag{11-25}$$

What are the general features that we might expect for a potential-energy function, U_{AB}, for the noncovalent interaction of two atoms or molecules? For large separations of the two particles we expect no interaction, so it is reasonable to set $U(\infty) = 0$. Since energy always is measured or calculated as a difference from a reference state, we can assign any value to $U(\infty)$, but zero usually is the most convenient. As the two particles approach, we would expect that $U(r)$ becomes negative (an attractive potential energy) for some value of r, because we know that all matter condenses into a solid or liquid under some condition, thereby implying attractive forces between particles. Then as the particles approach even closer we would expect that they repel one another, and that the potential energy becomes positive—that is, $U(r = \text{small}) > 0$. The mutual repulsion of all particles at very small separations is the basis of the finite volume of matter, and arises from the repulsive force that we will consider in the following section.

All these basic features of potential-energy functions are illustrated in the lower curve of Figure 11-16, which plots a potential-energy function for the noncovalent interaction of two water molecules. The functional form of the

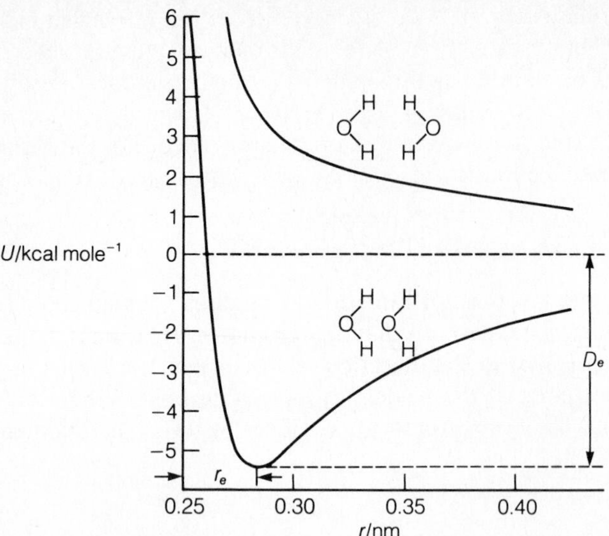

FIGURE 11-16
Potential-energy function for the interaction of two water molecules. Lower curve: Molecules with parallel dipole moments. Upper curve: Molecules with antiparallel dipole moments. The equilibrium separation, r_e, and binding energy, D_e, are indicated.

curve was derived from theory, and the specific constants in it were selected to fit experimental data on the nonideality of water vapor.

The lower curve of Figure 11-16 is for molecules approaching one another with parallel dipole moments. Notice that the equilibrium separation of the two molecules, r_e, is about 0.28 nm, the separation of neighboring H_2O molecules in ice. At this separation the binding energy, D_e, is about 5 kcal/mole-of-dimers (~20 kJ/mole-of-dimers). As r increases, the potential energy of interaction is still negative at 0.4 nm and approaches the axis asymptotically, thus showing that the attractive energy between the dipole moments of two water molecules extends over a relatively long range. As r decreases and the molecular centers come within 0.28 nm, the potential energy rises rapidly, thereby showing that there is a strong repulsive force that operates over a very short range.

EXERCISE 11-17

By applying Equation 11-25 to the potential-energy curve of Figure 11-16, you can determine the range of r for which the *force* between the two molecules is attractive, and the range for which the *force* is repulsive. Does negative $U(r)$ necessarily imply an attractive force?

ANSWER

No. The force is attractive where the *slope* of $U(r)$ is positive, and vice versa. Thus the force can be positive (repulsive) where $U(r)$ is negative, as it is for $0.26 < r < 0.28$ nm.

The Repulsive Force: Noncovalently Bound Atoms Repel Each Other at Close Quarters

Covalent bonds between atoms in the first two rows of the periodic table are all shorter than 0.2 nm (2 Å), but the equilibrium separation of two helium atoms is more than 0.3 nm. Apparently noncovalently bound atoms (or *non-bonded* atoms, for short) repel one another at much greater separations than do atoms that donate electrons of opposite spin to form a covalent bond.

The situation for helium is illustrated by the potential-energy curve in Figure 11-17(*a*). The general shape of the curve resembles those for covalent bonding [see Fig. 11-1(a)], but the equilibrium separation is somewhat greater, and the binding energy is several thousand times smaller. This curve can be represented by a potential function of the form

$$\begin{aligned} U_{AB}(r) &= U_{\text{repul}}(r) + U_{\text{disp}}(r) \\ &= \frac{A}{\exp(ar)} - \frac{C}{r^6} \end{aligned} \tag{11-26}$$

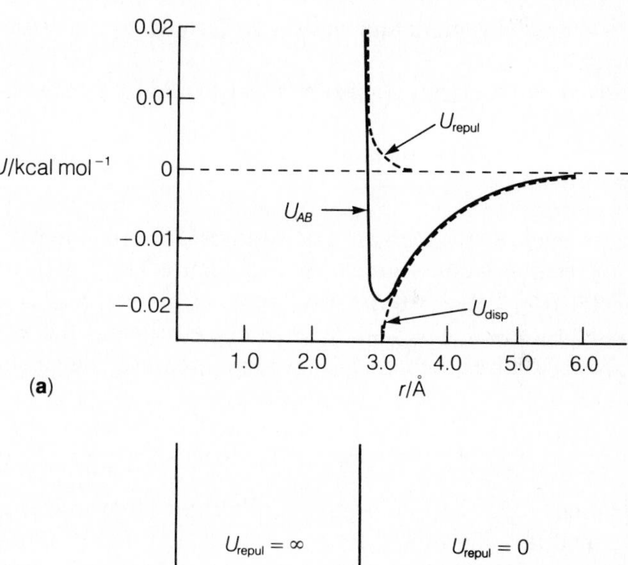

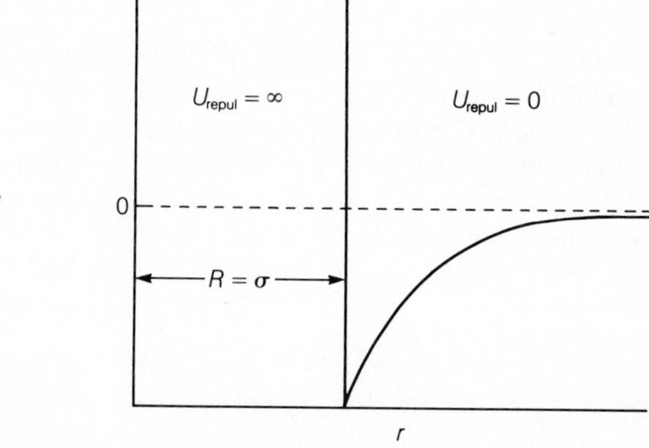

FIGURE 11-17
(a) Solid curve: the potential-energy function for interaction of two helium atoms. Dashed curves: the separate contributions of repulsive energy and attractive dispersion energy. **(b)** Solid curve: the hard sphere approximation to the helium potential-energy function of **(a)**.

in which A, a, and C are the following constants (from Karplus and Porter, 1970):

$$A = 3.82 \times 10^4 \text{ kcal/mole-of-dimers}$$
$$a = 5.04\ \text{Å}^{-1}$$
$$C = 2.03 \times 10 \text{ kcal} \times \text{Å}^6\text{/mole-of-dimers}$$

The second term on the right of Equation 11-26 is negative, and is thus an attractive contribution to the potential energy. It describes the *dispersion force*. This is a weak attractive force operating between all pairs of atoms, arising from the interactions of the instantaneous dipole moments of the atoms. Every atom has an instantaneous dipole moment, because at any instant its electrons are not distributed symmetrically. The motions of electrons in neighboring atoms correlate with each other to produce the weak "induced dipole-induced dipole" of dispersion forces.

The first term describes the *repulsive force*. It increases rapidly over a small range of r because the exponent contains r. You can see from the separate attractive and repulsive contributions in the figure that the repulsive term dominates at small values of r, and the attractive term dominates at large r values. The minimum in U_{AB} is formed by the sum of the two contributions.

The repulsive interaction can also be represented by a high inverse power of r:

$$U_{AB} = \frac{B}{r^n} - \frac{C}{r^6} \qquad (11\text{-}27)$$

in which n is between 12 and 24. For example, Karplus and Porter (1970) represent the helium-helium interaction of Figure 11-17 with this equation where $n = 13$, $B = 1.93 \times 10^4$ kcal Å^{13}/mole-of-dimers, and $C = 2.03 \times 10$ kcal Å^6/mole-of-dimers. The high inverse power of r has the same property as the exponential of Equation 11-26: a rapid increase over a small range of decreasing r.

EXERCISE 11-18

Using Equation 11-27, determine the equilibrium separation, r_e, binding energy, D_e, and the vibrational force constant, k, for the helium dimer. [*Hint:* First determine r_e from the condition that at equilibrium the force between atoms must vanish; that is, $(\partial U/\partial r)_{r_e} = 0$. Then note that $U(r_e) = -D_e$. To determine k, you can suppose that the potential is harmonic near its minimum; then $(\partial^2 U/\partial r^2)_{r_e} = k$.]

ANSWER

$D_e = 1.58 \times 10^{-2}$ kcal/mole-of-dimers (6.60×10^{-2} kJ/mole-of-dimers); $r_e = 2.97$ Å (0.297 nm); $k = 96.6$ erg cm^{-2} (9.64×10^{-2} J m^{-2}).

EXERCISE 11-19

Show that the He_2 molecule does not exist in appreciable concentrations in the gas at room temperature.

ANSWER
RT at room temperature $\cong$ 0.6 kcal mol^{-1}. Therefore, $RT/D_e \cong 40$, and thus the thermal energy greatly exceeds the binding energy.

The Repulsive Force Arises from the Pauli Exclusion Effect

The Pauli exclusion principle (Sec. 10-6) can be expressed as follows: No system can exist in a state where more than one electron has the same set of four quantum numbers.

A system consisting of a single helium atom is in accord with this principle. The 1*s* orbital contains two electrons of opposite spin quantum numbers. But if we attempt to bring two helium atoms sufficiently close to unite them, our system contains two sets of electrons with identical quantum numbers. We must therefore move two electrons into orbitals having higher energy. This required input of energy accounts for the repulsive force.

We also can look at the repulsion from the point of view of molecular orbital theory. If we form a molecule from two helium atoms, there is room in the bonding molecular orbital for only one pair of electrons. The other pair must enter the antibonding orbital. The effect of the antibonding electrons is stronger, and a net repulsion occurs.

The profound effect of the repulsion arising from the Pauli exclusion principle has been stated in graphic terms by Walter Kauzmann (1957):

> Pauli forces are responsible for the apparent 'solidity' of matter. We are told that most of the space in atoms is empty. Why, then, can we not push our hands through a piece of armour plate? The reason is that the electrons in our hands have the same spins and nearly the same velocities as some of the electrons in the armour plate. They therefore repel one another, and the armour plate feels 'hard' to our touch. Clearly, if it were not for the Pauli principle, matter would have very different properties from what we know. In our everyday lives we therefore have a more direct experience of the Pauli principle than of most other laws of nature; a new-born baby becomes conscious of its consequences long before he finds it necessary to take account of the consequences of Newton's laws of motion.

Van der Waals Radii: The Repulsive Potential Rises so Abruptly That Atoms Sometimes can be Represented as Hard Spheres

The repulsive potential rises so sharply with decreasing separation that it can be represented without great loss of reality by a *hard-sphere potential*. This potential depicts an atom as exerting no repulsive force on a second atom, until the second enters within a distance σ from the first. Then the repulsive potential is infinite. This potential-energy curve is shown in Figure 11-17(*b*).

TABLE 11-5 *Van der Waals Radii of Atoms and Atomic Groupings*

ATOM OR GROUP	RADIUS/A[a]	ATOM OR GROUP	RADIUS/A[b]
H	1.2		
N	1.5	$-\overset{\vert}{N}-H$, $-NH_2$, $-NH_3^+$	1.5
O	1.4	$=O$, $-O-$, $-OH$, H_2O	1.4
$-CH_3$	2.0	$-\overset{\vert}{\underset{\vert}{C}}-H$, $-\overset{\vert}{C}H_2$	2.0
Aromatic carbon	1.7	Aromatic carbon	1.85
		Carbonyl carbon	1.5
F	1.35		
P	1.9		
S	1.85	$-S-$, $-SH$	1.85
Cl	1.80		
Br	1.95		
I	2.15		

In atomic groups, the distance r is measured from the nucleus of the heaviest atom.
[a] From Pauling (1960).
[b] From Shrake and Rupley (1973).

Despite the severity of the approximation, you can see that the potential-energy curve is not all that different from the one shown in Figure 11-17(*a*). This is the physical-chemical basis for the usefulness of space-filling models for representing molecules, and for the usefulness of the concept of steric hindrance in organic and biochemistry.

The hard-sphere potential of atoms can be tabulated in terms of *van der Waals radii*. The van der Waals radius is one-half the equilibrium internuclear distance between two nonbonded atoms in a crystal. In other words, the distance of closest approach of two noncovalently bound atoms is expected to be the sum of their van der Waals radii. The values in Table 11-5 were suggested by Pauling (1960), and assigned an uncertainty of about 0.1 Å. Each of these radii is about 0.8 Å greater than the corresponding single-bond radius. Pauling noted that for atoms forming a single covalent bond, the nonbonded radius in directions close to the covalent bond direction is about 0.5 Å less than the van der Waals radius. Thus an atom forming one covalent bond is like a sphere whittled down on the side of the bond.

QUESTIONS FOR REVIEW

1. How do the following quantities change when a H_2^+ molecule is formed from two protons and an electron: (a) Ground-state energy; (b) Charge distribution; (c) Potential energy?
2. What are the characteristics of molecular orbitals?

3. Describe the following: (a) The MO LCAO method; (b) The variation principle, and its use.
4. Summarize the steps involved in determining the constants in a MO LCAO wavefunction by the variation principle.
5. Define the overlap, Coulomb, and resonance integrals. What is the physical meaning of each? How are they related to the ground and excited state energies of H_2^+?
6. What atomic orbitals combine to form σ molecular orbitals? What AO's combine to form π MO's? What are their shapes?
7. What is the usual order of increasing energy of MO's in homonuclear diatomic molecules?
8. Describe the building up of homonuclear diatomic molecules by the filling of electrons into MO's.
9. What are single, double, and triple bonds in terms of MO's? Define bond order.
10. Define *spin multiplicity, singlet,* and *triplet*. Describe the spectroscopist's notation for diatomic molecules.
11. What conditions are required for two AO's to form an effective MO?
12. How do MO's for heteronuclear diatomics differ from those for homonuclear diatomics?
13. What are the characteristics of sp, sp^2, and sp^3 hybridization?
14. Describe the MO's of the water molecule and peptide group.
15. Define dipole moment. Give its dimensions and typical magnitudes in the cgs and SI systems.
16. Define molecular polarizability. Give its dimensions and typical magnitudes in the cgs and SI systems.
17. Explain how dipole moments and polarizabilities are measured.
18. What are the origin and functional form of the following: (a) repulsive forces and (b) dispersion forces? In what substances might dispersion forces be the dominant attractive force?

PROBLEMS

1. Are the following statements true or false?

    (a) Gravitational attraction is a significant contribution to the covalent bond.

    (b) The potential energies of electrons are lowered when a covalent bond is formed.

    (c) s atomic orbitals always form σ molecular orbitals and p atomic orbitals always form π molecular orbitals.

    (d) A molecule with three unpaired electrons is called a triplet.

(e) In a homonuclear diatomic molecule, a π orbital is asymmetric about the bond axis, but its electron density is symmetric by reflection.

(f) There must be two electrons in every MO.

(g) In large molecules, each MO must contain contributions from AO's on every atom.

(h) sp^2 hybrid orbitals are formed from two p AO's and one s AO.

(i) Hund's rule states that electrons tend to pair their spins.

(j) Two noncovalently bound atoms can never approach closer than the sum of their van der Waals radii.

Answers: (a) F. (b) T. (c) F. (d) F. (e) T. (f) F. (g) F. (h) T. (i) F. (j) F.

2. Supply the missing word(s).

(a) The variation principle states that the energy computed from a trial wavefunction is always ______ than the true energy.

(b) Electrons in ______ molecular orbitals contribute to binding energy.

(c) The bond order is given by ______ the excess of bonding over antibonding electrons.

(d) A molecule with net bonding electrons of configuration $\sigma^2\pi^2$ is said to have a ______ bond.

(e) A molecule with no unpaired electrons is said to be in a ______ state.

(f) MO's cylindrically symmetric about the bond axis are called ______ orbitals.

(g) Electrons that neither contribute to nor oppose bonding are said to be in ______ orbitals.

(h) Repulsive forces arise from the ______ principle.

(i) The dispersion energy depends on the inverse ______ power of the separation of the interacting atoms.

(j) The van der Waals radius of an atom is about ______ greater than the corresponding single-bond radius.

Answers: (a) Greater. (b) Bonding. (c) Half. (d) Double. (e) Singlet. (f) σ. (g) Nonbonding. (h) Pauli. (i) Sixth. (j) 0.8 Å.

PROBLEMS RELATED TO EXERCISES

3. (Exercise 11-1) Suppose in Exercise 11-1, you selected function $\psi = e^{-cr^2}$ as a trial wavefunction; this leads to

$$E' = \frac{3h^2}{8\pi^2 m} c - \frac{2e^2\sqrt{2}}{\sqrt{\pi}} \sqrt{c}$$

(a) Show that E' has its minimum value when $c = 8/(9\,\pi\,a_0^2)$.

(b) Find the minimum E'.

(c) By how much does E' differ from the true value of E, $-e^2/2a_0$? Notice that the variation principle produces a good energy even when the wrong functional form is selected for ψ.

4. (Exercise 11-5) Explain the observation that the binding energy of H_2 (108 kcal mol^{-1}) is greater than that of H_2^+ (64.5 kcal mol^{-1}) and that the bond length of H_2 (0.76 Å) is shorter than that of H_2^+ (1.06 Å).

5. (Exercise 11-6) Explain in terms of MO's why among atoms in the first two rows of the periodic table there are single, double, and triple but no quadruple bonds.

6. (Exercise 11-7) The F_2^+ molecule is not listed in Table 11-1. Determine the missing information:

(a) Electronic configuration of MO's.

(b) Bond order.

(c) Stability—do you predict that the F_2^+ molecule exists?

(d) Spectroscopic notation, including spin multiplicity.

7. (Exercise 11-12) Would you expect alanine [$^+NH_3$—$\underset{}{\overset{CH_3}{\overset{|}{C}}}$H—$COO^-$] to have an appreciably different dipole moment than glycine [$^+NH_3$—CH_2—COO^-]?

8. (Exercise 11-16) From the following data (from Böttcher, 1952) for the dielectric constant of gaseous dimethyl ether, determine the molecular polarizability and dipole moment of dimethyl ether. Take the density as 0.001434 g cm^{-3} and the molecular weight as 46.05.

TEMPERATURE/K	DIELECTRIC CONSTANT
298.0	1.004655
338.0	1.004268
378.0	1.003985
418.0	1.003725

9. (Exercise 11-17) Sketch the potential and the force for gravitational attraction of two particles as a function of their separation.

10. (Exercise 11-18) Suppose it is convenient to express the intermolecular potential energy in the form

$$U_{AB}(r) = Ar^{-n} - Br^{-m}$$

(a) Show that $r_e = \left(\dfrac{nA}{Bm}\right)^{1/(n-m)}$

(b) Show that the equilibrium binding energy, $D_e = -U_{AB}(r_e)$ can be expressed

$$D_e = -Br_e^{-m}\left(\frac{m}{n} - 1\right) = -Ar_e^{-n}\left(1 - \frac{n}{m}\right)$$

(c) Show that U_{AB} can be written in the form

$$U_{AB} = \frac{-D_e}{m-n}\left[m\left(\frac{r_e}{r}\right)^n - n\left(\frac{r_e}{r}\right)^m\right]$$

(d) Let σ be the intermolecular separation (other than ∞) at which $U_{AB} = 0$. Show that

$$\sigma = r_e\left(\frac{m}{n}\right)^{\frac{1}{n-m}}$$

(e) Show that U_{AB} can be expressed in terms of σ in the form

$$U_{AB} = \frac{-D_e}{m-n}\left(\frac{n^n}{m^m}\right)^{\frac{1}{n-m}}\left[\left(\frac{\sigma}{r}\right)^n - \left(\frac{\sigma}{r}\right)^m\right]$$

(f) Express U_{AB} in terms of D_e and σ when $n = 12$ and $m = 6$. This is called the *Lennard-Jones potential*.

OTHER PROBLEMS

11. When N_2 loses an electron to form N_2^+, the dissociation energy is decreased. In contrast, when O_2 loses an electron to form O_2^+, the dissociation energy is increased. Account for these observations from the electronic configurations of these four molecules.

● 12. Suppose spectroscopic studies had shown that the ground state of N_2^+ is $^2\Pi$. What would the order of filling MO's for nitrogen then have been?

13. Describe the HF bond in terms of MO's.

14. Write an equation for a molecular orbital in which the electron spends 20% of its time in atomic orbital ψ_A and 80% of its time in atomic orbital ψ_B. Neglect the region of overlap between the orbitals.

15. Write the electronic configurations of the ground states of Na_2 and LiH.

16. Give examples of two types of molecules for which localized MO's are unlikely to give a suitable description.

● 17. What is the spin multiplicity of the NO molecule? What is the spectroscopic notation for it?

18. Write the electronic configurations of Cl and HCl.

19. Draw the electronic configuration of *sp* hybrids in the manner of Figure 11-9.

20. Explain the trend of bond lengths and bond energies of the *sp* hybrids shown in Table 11-2 in terms of their electronic configurations.

21. Prove that a cluster of charges which possess a center of inversion [for each charge at (x, y, z) there is an additional charge at $(-x, -y, -z)$] cannot possess a dipole moment.

22. At 20°C the density of chloroform is 1.490 g cm^{-3} and its index of refraction for the sodium D-line is 1.446. Its molecular weight is 119.4. Compute the molecular polarizability of chloroform.

23. Arrange the following diatomic molecules in order of decreasing dipole moment: Br_2, HBr, HI, HF, CsCl, CO, HCl, KCl.

24. Prove that the dipole moment of a cluster of charges (Equation 11-18*a*) is independent of the choice of origin, provided the cluster has no net charge.

25. (a) Show that the Hamiltonian operator for the H_2^+ molecule, omitting the nuclear repulsion term, is

$$\mathcal{H} = -\frac{h^2}{8\pi^2 m}\left[\frac{\partial^2}{\partial x^2} + \frac{\partial^2}{\partial y^2} + \frac{\partial^2}{\partial z^2}\right] - \frac{e^2}{r_1} - \frac{e^2}{r_2}$$

where r_1 and r_2 are the separations of the electron from the two protons.

(b) Show that the Coulomb integral

$$H_{11} = \int \psi_1 \mathcal{H} \psi_1 \, d\tau$$

can be expressed

$$H_{11} = \int \psi_1 \left(E_0 - \frac{e^2}{r_2}\right) \psi_1 \, d\tau = E_0 - \int \frac{e^2}{r_2} \psi_1^2 \, d\tau$$

in which E_0 is the ground-state energy of the hydrogen atom, that is, energy of the 1*s* orbital.

(c) Show that the resonance integral

$$H_{12} = \int \psi_1 \mathcal{H} \psi_2 \, d\tau = \int \psi_2 \mathcal{H} \psi_1 \, d\tau$$

can be expressed

$$H_{12} = \int \psi_2 \left(E_0 - \frac{e^2}{r_2}\right) \psi_1 \, d\tau = E_0 S - \int \frac{e^2}{r_2} \psi_1 \psi_2 \, d\tau$$

(d) Show that the energy E_+ (Equation 11-15*a*) can be written

$$E_+ = E_0 - \frac{\int \left(\frac{e^2}{r_2}\right) \psi_1^2 \, d\tau + \int \left(\frac{e^2}{r_2}\right) \psi_1 \psi_2 \, d\tau}{1 + \int \psi_1 \psi_2 \, d\tau}$$

(e) Given that ψ_1 and ψ_2 are 1*s* hydrogen orbitals, write out the complete integrals that must be evaluated. Books on quantum chemistry show how these integrations can be performed.

26. $CO_3^=$ is a planar molecule, with C in the middle of a triangle formed by the 3 O atoms.

(a) What is the hybridization of the C atom?

(b) How many σ-bonds does the molecule contain?

(c) Sketch the lowest-energy delocalized π molecular orbital in the molecule.

(d) How many π-molecular orbitals are formed? (include π^* orbitals).

(e) How many electrons occupy the π and π^* molecular orbitals?

(f) How many net π bonds are there in $CO_3^=$?

(g) What structure would you predict for SO_3?

27. Consider the six π-type MO's of benzene. A calculation similar to that in Section 11-3 shows that these six MO's have either 0, 1, 2, or 3 nodal planes of zero electron density perpendicular to the benzene ring. It also shows that the more nodal planes there are in a MO, the higher is its energy. Sketch the six MO's of benzene.

● 28. Let a trial wavefunction, ψ' (Equation 11-6), be normalized ($\int \psi'^* \psi' \, d\tau = 1$), and assume that it can be expressed as an infinite series in the true wavefunctions ψ_i (Equation 11-5):

$$\psi' = \sum_{i=0}^{\infty} c_i \psi_i$$

in which ψ_i are normalized and orthogonal ($\int \psi_i^* \psi_i \, d\tau = 1$; $\int \psi_i^* \psi_j \, d\tau = 0$).

(a) Show that the energy E' (Equation 11-6) is

$$E' = \sum_{i=0}^{\infty} c_i^2 E_i$$

(b) Use the fact that the ψ_i are normalized and orthogonal to show that

$$\sum c_i^2 = 1$$

(c) Let E_g be the lowest-lying energy level, $E_g \leq E_i$. Use the results of (a) and (b) to show that $E' \geq E_g$, as required by the variation principle.

PROBLEMS FOR DISCUSSION

29. (a) How might the greater reactivity of double bonds, as opposed to single bonds, be explained in terms of the overlap of atomic orbitals?

(b) Why is there free rotation about a single bond and highly constrained rotation about double and triple bonds?

30. Covalent bonds are said not to form on the hypothetical planet Nomolcao. Describe the properties of materials there.

FURTHER READING

Coulson, C. A. 1961. *Valence*, 2nd ed. Oxford, England: The Clarendon Press. A lucid, relatively nonmathematical development of the theory of covalent bonding. Highly recommended.

Coulson, C. A. 1973. *The Size and Shape of Molecules*. Oxford, England: The Clarendon Press. A more condensed (only 80 pages) introduction to covalent bonding.

Gray, H. B. 1964. *Electrons and Chemical Bonding*. New York: W. A. Benjamin, Inc. A good introduction to the molecular orbital description of molecules.

Karplus, M., and Porter, R. N. 1970. *Atoms and Molecules*. Menlo Park, Calif.: W. A. Benjamin, Inc. More advanced and more complete than Coulson's book.

Kauzmann, W. 1957. *Quantum Chemistry*. New York: Academic Press. Still more advanced.

Pauling, L. 1960. *The Nature of the Chemical Bond,* 3rd ed. Ithaca, N. Y.: Cornell University Press. An indispensable handbook of information on covalent and noncovalent bonding.

OTHER REFERENCES

Böttcher, C. J. F. 1968. *Theory of Electric Polarization*. New York: Elsevier Publishing Co.

Pilar, F. L. 1968. *Elementary Quantum Chemistry*. New York: McGraw-Hill Book Co.

Wilson, C. W., Jr., and Goddard, W. A., III. 1972. *Theor. Chim. Acta 26,* 195, 211.

Shrake, A., and Rupley, J. A. 1973. *J. Mol. Biol. 79,* 351.

CHAPTER 12

Principles of Spectroscopy

○ *12–1 Biochemists Use Spectroscopy to Monitor Changes at the Atomic Level: An Example*

The current thrust of biochemistry and molecular biology is to understand life processes—replication, movement, communication, and control—at the atomic level. The techniques of physical chemistry that yield the most detailed information on the atomic level are the diffraction and spectroscopic methods. X-ray, neutron, and electron diffraction (Chap. 17 and Sec. 10-1) have the advantage of yielding a direct and often easily interpretable picture of matter, but these methods suffer from severe limitations. The pictures have a limited resolution, and frequently do not reveal as much molecular or atomic detail as one might wish. Moreover, the pictures are at best static snapshots of molecular events. The essential dynamics of enzymatic catalysis or transport—or whatever process is being studied—must be inferred from one or more of these snapshots. In addition, these pictures are most detailed for the solid state (or in the case of electron microscopy, for the dehydrated state) rather than for the solution state that is more characteristic of living systems.

In contrast, spectroscopic techniques are applied readily to solutions, and can be used to follow dynamic processes. Many of them are sufficiently sensitive to detect movements on the order of tiny fractions of an angstrom or a small change in concentration of one component among many. The major limitation of spectroscopic techniques is that the information is in a less easily interpreted form—it consists of a list of the energy levels of the system under study. Although many clever investigators have determined the most intimate details of chemical reactions and structure from spectroscopic measurements, one sometimes has the feeling that a paraphrase of an old Chinese proverb could be applied to biochemistry: one picture is worth a thousand spectra. Thus, to understand any system in much detail, both classes of techniques usually are necessary. The information from each complements the other.

Most spectroscopic studies of biochemical systems are supported by the following chain of logic:

1. *The energy levels of the sample are probed by recording the spectrum of absorbed or scattered light for a spectral region that is likely to be of interest.* The data usually consist of a plot of the intensity of absorption (or scattering) of light as a function of its frequency. The frequency of absorption is related to the energy difference between two levels by the familiar equation $\Delta E = h\nu$. An example is the spectrum of infrared radiation absorbed by the enzyme lysozyme in the neighborhood of 1600 cm^{-1} (6.25 μm). In the following sections we will discuss the various units for measuring intensity and spectral frequency. For the present it is enough to note that lysozyme in D_2O absorbs strongly near 1650 cm^{-1} and 1450 cm^{-1}, and less strongly at about 1565 cm^{-1} and at a shoulder near 1710 cm^{-1}.
2. *Observed spectral bands are assigned to specific atomic processes, such as vibrations, rotations, electronic transitions, and so forth.* This is a critical step in interpretation. Today it is often possible to make correct assignments even in fairly complex systems because there has been much exploratory work on simpler systems. For example, in the lysozyme spectrum we can assign the absorption at 1650 cm^{-1} to the so-called amide I vibration (Susi, 1972), which consists mainly of a C=O stretching motion of the amide group. This assignment is based on the spectra of simpler compounds such as N-methyl acetamide ($H_3C{-}\overset{\overset{\displaystyle O}{\|}}{C}{-}\underset{\underset{\displaystyle H}{|}}{N}{-}CH_3$), which display the same peak. Similarly the peak at 1450 cm^{-1} can be assigned to the amide II band, which arises from a combined C—N stretching and N—D bending. (Note that the protein N—H groups exchange hydrogen nuclei with the D_2O medium to become N—D groups.) The weaker absorptions at 1565 cm^{-1} and 1710 cm^{-1} can be assigned, respectively, to absorption of COO^- and COOD groups, as we shall see presently.

It is relatively difficult to make assignments in biochemical systems for two reasons. First, there are always many types of atomic groupings, each with its own system of energy levels, which gives rise to a complicated superposition of spectra. In fact, it was to eliminate some overlapping spectral bands that D_2O was used in place of H_2O as the solvent in the lysozyme study of Figure 12-1(*a*). The H—O—H bending motion of H_2O gives rise to a strong absorption at 1640 cm^{-1}, and the amide II absorption for amide N—H groups lies at 1550 cm^{-1}. These absorptions would overwhelm the absorptions of interest here, those of COO^- and COOD, if the experiments were carried out in H_2O. Even in D_2O, the strong amide I band at 1650 cm^{-1} is superposed on the bands of interest.

The second reason is that most biochemical work is on dissolved compounds. Spectra of liquids and solutions are harder to interpret owing to the

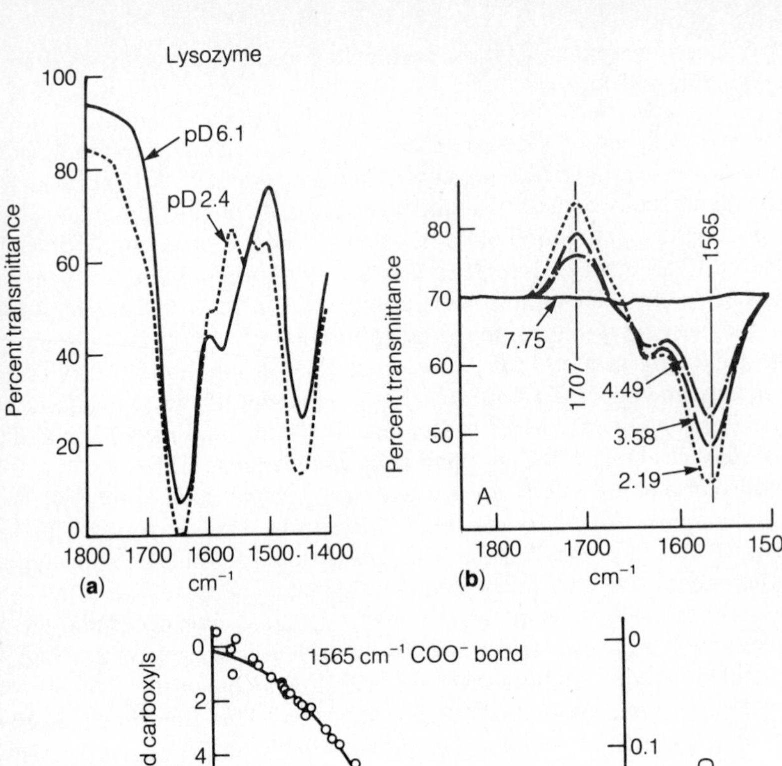

FIGURE 12-1
Spectroscopic titration of lysozyme in D_2O. **(a)** Infrared spectra of lysozyme in D_2O at pD 2.4 and 6.1. The dashed spectrum has been displaced by 10% toward lower transmittance. **(b)** The difference in transmittance of lysozyme at the indicated pD and of lysozyme at pD = 7.75. **(c)** The change in optical density of the 1565 cm^{-1} band as a function of pD. (From Timasheff et al., 1973.)

overlapping of broadened spectral bands arising from the interactions of molecules in these states. This is illustrated in Figure 12-2. Spectra of dilute gases, such as those of atoms, consist of sharply defined lines. Each line corresponds to a transition between two quantum states of the system. As the pressure of the gas is increased, molecules begin to collide and interact with one another, and each molecular quantum state is broadened into a set of states. The spectrum becomes a superposition of lines, the envelope of which is called a *band*. This effect is even more pronounced in a liquid or solution, where the intermolecular forces are stronger. Clearly, assignment is more difficult with broadened bands than with sharply defined lines.

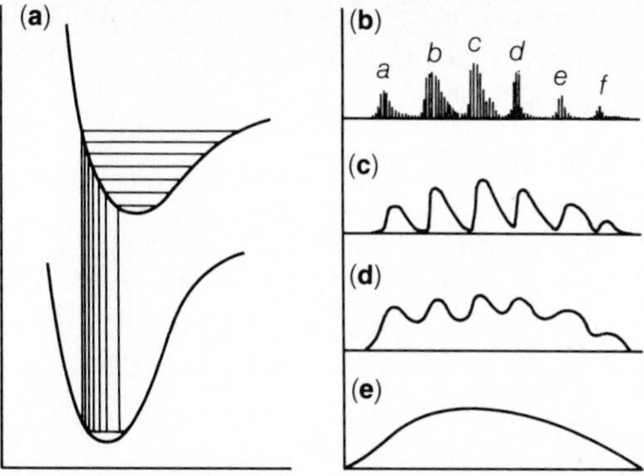

FIGURE 12-2
A schematic illustration of the effect of intermolecular forces on an absorption spectrum. **(a)** An absorption corresponding to the transition of an electron from one orbital to a higher one is represented by the potential-energy curves for the two states. The vertical lines show possible transitions from the ground vibrational state of the lower electronic state. **(b)** The spectrum of the dilute gas with one group of spectral lines for every vertical line in **(a)**. Each of the individual lines corresponds to one rotational state; each group corresponds to one vibrational state. **(c)** In the gas at moderate pressures, intermolecular collisions smear out the individual rotational lines. **(d)** In a liquid with weak intermolecular forces, the vibrational bands are less distinct and shifted slightly in frequency. **(e)** In a liquid with strong intermolecular forces, like water, the vibrational bands are no longer distinguishable. (Redrawn from Kauzmann, 1957.)

3. *The change in the spectrum is monitored as an independent variable of interest is varied.* The variable usually is a concentration of one component; sometimes it is temperature or pressure. In the study of lysozyme, it is the acidity, expressed in this D_2O system as the negative log of deuteron concentration, the pD. Figure 12-1(*a*) shows that the spectrum changes most with pD in the regions of 1565 cm^{-1} and 1710 cm^{-1}. This is confirmed by the *difference spectrum* of Figure 12-1(*b*). Here the spectrum of lysozyme at a given acidity (pD) is compared to that of lysozyme at a pD of 7.75. Note that the peak at 1710 cm^{-1} increases with greater acidity. This confirms its assignment as a titratable group that takes up a deuteron at low pD. The peak at 1565 cm^{-1} decreases with greater acidity, which confirms its assignment as a titratable group that loses a deuteron at lower pD. When the change in absorption is plotted against pD, as in Figure 12-1(*c*), the midpoint is found to be about pD = 4, a value expected for the titration of COO^- groups.
4. *The spectral change is interpreted in terms of a localized molecular change.* In the lysozyme study (Timasheff et al., 1973), the number of titrating COO^- groups was determined, along with the pK of each. This could not have been done by ordinary methods of titration, in which the pH is followed as a function of the amount of acid added, since other protein side chains may titrate in the same pH region and the observed titration curve is a sum for all contributors. In contrast, the spectroscopic study sees only the COO^- and COOD groups because these are the only groups that absorb significantly at the frequencies monitored in the experiment.

Thus, these groups act as probes of the environment of the enzyme molecule, and report to us information about the environment at specific sites of individual molecules.

This is the theme of spectroscopic studies in biochemistry: an experiment is designed around an atomic or molecular probe that will report information of interest about the system. The spectral region that is studied is the one that corresponds to energy levels of the probe. We will see many examples of this in Chapter 13, but first we must discuss the general factors that determine the frequency and the intensity of an absorption.

12–2 The Interaction of Light with Matter: An Overview

Spectroscopy in its broadest sense is the study of the interaction of light, or of some other particle-wave, with matter. When a beam of light impinges on a sample, as in Figure 12-3, some of the photons may have no interaction with the sample. We say that this light is *transmitted*. Other photons may be *absorbed*. These usually produce heat, but sometimes cause a chemical reaction or are re-emitted as light of lower frequency (*fluorescence* or *phosphorescence*). Still other photons, usually a small fraction, may be *scattered* at various angles from the sample.

Absorption of Light

An *absorption spectrum* is a plot of one of the measures of strength of absorption against the frequency of incident light. In Figure 12-1(*a*), the percent transmittance is plotted; in other spectra, the molar absorption coefficient or

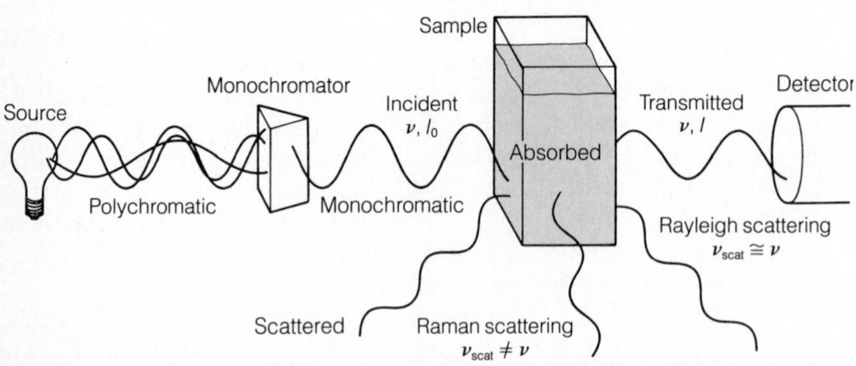

FIGURE 12-3
Various effects that occur when light interacts with matter. The monochromator selects light of a single frequency from the polychromatic source. Some light is absorbed, some is scattered, and some is transmitted. The transmitted intensity I is less than the incident intensity I_0, because of absorption and scattering.

the absorbance is plotted. An instrument that records an absorption spectrum is called a *spectrophotometer*. The reason that light of one frequency is absorbed by a substance and light of other frequencies is transmitted is a major topic of the following four sections. The central idea, however, is simple: light frequency ν is absorbed only if it corresponds by Bohr's principle to the difference in energy of two quantum levels of the sample:

$$\nu = \frac{E_2 - E_1}{h} = \frac{\Delta E}{h}$$

Absorption: $\Delta E = h\nu$ No absorption: $\Delta E \neq h\nu$

in which h is Planck's constant.

The dependence of the amount of light absorption on the amount of absorbing material in the beam is described by the *Beer-Lambert law*, which gives the intensity of transmitted light, I, in terms of the intensity of the incident monochromatic light, I_0, the concentration of absorbers, C, the length of the light path through the absorbers, l, and the *molar absorption coefficient*, ϵ. The number ϵ also is called the *molar extinction coefficient*. It is a measure of the intrinsic absorbing strength of the absorbers, and varies with the frequency of light. The Beer-Lambert law often is expressed as

$$\log_{10}\frac{I_0}{I} = \epsilon lC = A \qquad (12\text{-}1)$$

The quantity $\log_{10}(I_0/I) = A$ is called the *absorbance* or sometimes *optical density* (OD), and is another measure of the strength of absorption of the sample. Biochemists often use the symbol $A^{1\%}_{1\,\text{cm}}$ or $^{1\%}_{1\,\text{cm}}$ to mean the absorbance of a solution with 1 g of absorbers per 100 cm^3 of solution in a cell of 1 cm optical path. A third measure is the *percent transmittance*, defined as 100 (I/I_0).

If there are several absorbing species in solution, and if the presence of each does not affect the absorption of the others, we can extend Equation 12-1 to describe the total absorption:

$$\log\frac{I_0}{I} = l(\epsilon_1 C_1 + \epsilon_2 C_2 + \epsilon_3 C_3 + \cdots) \qquad (12\text{-}2)$$

The meaning of the Beer-Lambert Law is that each layer of solution absorbs the same *fraction* (not the same amount) of the remaining beam of light. We can see this by some rearrangement of Equation 12-1. First we convert to natural logarithms,

$$\ln\frac{I_0}{I} = 2.303\ \epsilon lC$$

then take antilogarithms of both sides and rearrange,

$$I = I_0 e^{-2.303\,\epsilon lC} \qquad (12\text{-}3)$$

then differentiate I with respect to l,

$$dI = I_0 e^{-2.303\,\epsilon lC}(-2.303\ \epsilon C)\, dl \qquad (12\text{-}4)$$

and finally divide Equation 12-4 by 12-3,

$$-\frac{dI}{I} = 2.303\ \epsilon C\ dl \tag{12-5}$$

The quantity $-dI/I$ is the fraction of light absorbed as the beam passes through the small distance dl, and Equation 12-5 shows that it is a constant that is proportional to both the absorption coefficient and the concentration of absorbers.

EXERCISE 12-1

From Figure 12-1(a), estimate the molar extinction coefficient of lysozyme at a pD of 6.1 for an incident frequency of 1645 cm^{-1}. The molecular weight of lysozyme is about 14,600. Assume that the concentration of the enzyme is 80 mg ml^{-1}, and that a cell of length 0.1 mm was used in the experiment.

ANSWER

From Figure 12-1(a), the percent transmittance is about 8.3%. Therefore $I/I_0 = 0.083$, and

$$\epsilon = \frac{\log \dfrac{I_0}{I}}{lC} = \frac{\log\left(\dfrac{1.0}{0.083}\right)}{10^{-2}\text{cm}\left(\dfrac{80\text{ mg ml}^{-1}}{14{,}600\text{ g mol}^{-1}}\right)\left(\dfrac{1.0\text{ g }l^{-1}}{\text{mg ml}^{-1}}\right)}$$

$$= 2.0 \times 10^4\ l\text{ cm}^{-1}\text{ mol}^{-1}$$

EXERCISE 12-2

The absorption of ultraviolet light of wavelength 280 nm by proteins is caused almost entirely by the aromatic amino acids tyrosine and tryptophan. Suppose that a protein of molecular weight 26,000 contains two residues of tryptophan ($\epsilon_{\text{Try}} = 5.0 \times 10^3\ l\text{ cm}^{-1}\text{ mol}^{-1}$) and six residues of tyrosine ($\epsilon_{\text{Tyr}} = 1.1 \times 10^3\ l\text{ cm}^{-1}\text{ mol}^{-1}$). The absorption is recorded from a cell of path length 1 cm, holding the protein at a concentration of 1 mg ml^{-1}. Calculate the absorbance and percent transmission.

ANSWER

$$A = \log\frac{I_0}{I} = l(C_{\text{Try}}\epsilon_{\text{Try}} + C_{\text{Tyr}}\epsilon_{\text{Tyr}})$$

$$= 1.0\text{ cm}\left(\frac{1.0\text{ g }l^{-1}}{2.6 \times 10^4\text{ g mol}^{-1}}\right)(2 \times 5.0 \times 10^3 + 6 \times 1.1 \times 10^3)\ l\text{ cm}^{-1}\text{ mol}^{-1} = 0.64$$

Percent transmission $= 100\ I/I_0 = 100 \times 10^{-0.64} = 23.$

Light Scattering

An isolated atom scatters light because the electric field of the incident light beam (Fig. 12-4) forces the electrons in the atom to oscillate back and forth around their equilibrium position. By the laws of electromagnetism, when a charge changes its velocity, it emits radiation. Light is emitted uniformly in all directions in the plane perpendicular to the oscillation, but decreases in amplitude as the viewing angle shifts away from that plane.

Two major quantities are used to characterize the light scattered from a sample: the intensity and the frequency. The *intensity* of a light beam is a measure of the amount of energy carried in the beam, and depends on the square of the amplitude, the electric field strength E^*. The intensity of light scattered without frequency shift depends on whether the sample contains fluctuations in composition. For example, a crystalline macromolecular sample does not have composition fluctuations because it is well-ordered;

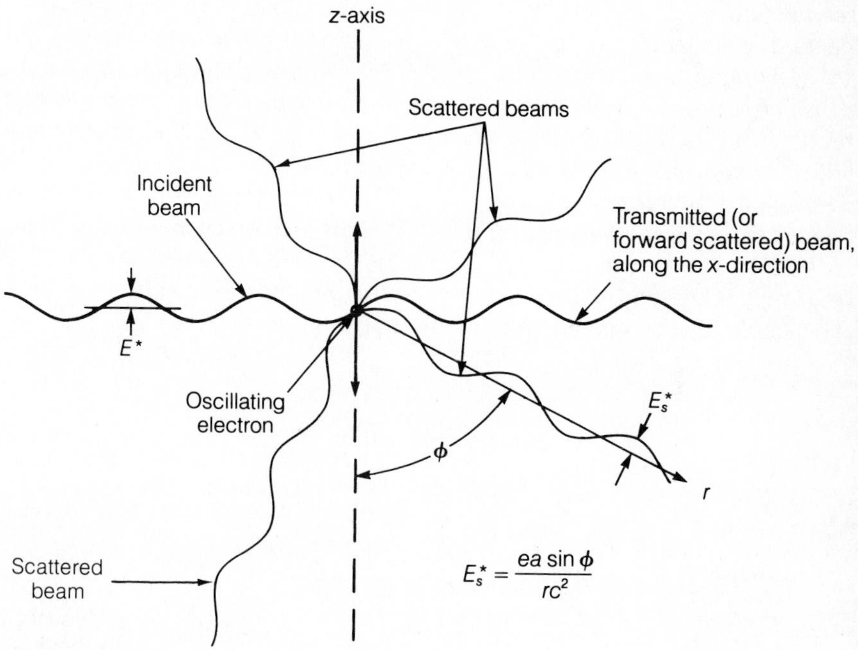

FIGURE 12-4
Radiation of light by an electron forced to oscillate along the z-axis by an incident beam of light. In the plane perpendicular to the direction of oscillation of the electron (the x-y plane), the radiation is emitted uniformly in all directions. However, the amplitude decreases with sin ϕ (where ϕ is measured from the axis of oscillation of the electron) because a smaller component of oscillation of the electron is seen when the viewing angle moves away from the plane perpendicular to the oscillation. The electric field amplitude at a distance r from the oscillator is $E_s^* = ea \sin\phi/rc^2$. $-e$ is the electronic charge, a is the acceleration of the electron, and c is the speed of light.

the crystal appears clear when a nonabsorbed beam of light shines through it. When enough solvent is added to dissolve the macromolecule, it appears cloudy or turbid because it scatters light. All regions of the solution do not contain exactly the same concentration of molecules. As we will see in Section 13-10, light scattering measurements can be used to determine the molecular weight of macromolecules.

Light scattered at essentially the same frequency as the incident beam is called *Rayleigh scattering* (Fig. 12-5), after Lord Rayleigh, who first gave a quantitative expression for light scattering. Two main mechanisms account for shifting the frequency of scattered light away from the monochromatic incident frequency: molecular motion, and change in the internal energy states of the scattering molecule. Molecular motion changes the scattered frequency because the oscillating electron is moving away from or toward the observer. This yields the Doppler shift familiar as the change in pitch (or sound frequency) of a train or automobile horn as it moves past an observer.

Molecular motion leads to broadening of the Rayleigh line because of the *random* molecular movement in the liquid. The extent of the broadening can be used to characterize the rate at which molecules move in fluids (see Chap. 15). Two special scattered lines, the *Brillouin lines* (Fig. 12-5), appear because of *coherent* molecular motion in sound waves that propagate through the fluid. (Two lines are seen because the energy of a sound wave can be added or subtracted from that of the incident light.)

A small fraction of the scattered light differs more substantially in frequency from the incident light (Fig. 12-5). This phenomenon was discovered in 1928 by the Indian scientist C. V. Raman, and is named for him. The frequency is shifted because some of the energy of the photon is absorbed by the sample, or the sample gives off some energy to the photon,

$$\nu_{\text{scat}} = \nu \pm \Delta E/h$$

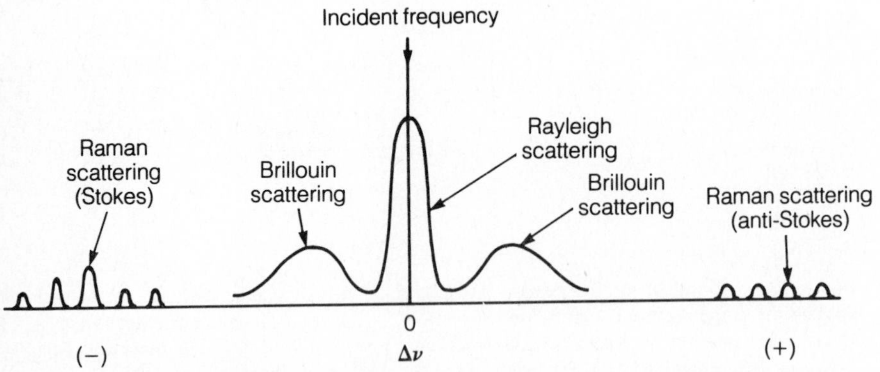

FIGURE 12-5
Frequency shifts in scattered light. The Rayleigh line is broadened and the Brillouin lines are shifted from the incident line by motion of the scattering center. The Raman lines are shifted because the internal energy (vibrational or rotational) of the scatterer is changed by $\Delta E = h\Delta\nu$. *Stokes* and *anti-Stokes* are the names given to Raman-scattered light decreased and increased in frequency, respectively.

in which ΔE is a transition between states of the sample. Therefore, Raman scattering contains information on the energy levels of the sample. This technique, which we discuss in Section 13-1, rivals absorption spectroscopy as a useful tool in chemistry and biochemistry.

Light scattering also gives rise to the phenomena of reflection, refraction, and diffraction. Light scattered in the opposite direction of the incident beam leads to *reflection*. Light scattered in the same direction as the incident beam recombines with the incident beam, which gives rise to the phenomenon of *refraction*. The physical effect of this recombination is to make the transmitted light appear as though it has traveled more slowly through the sample than if it had passed through a vacuum. The *index of refraction, n* (also called the *refractive index*), which describes the magnitude of this effect, is defined as

$$n = \frac{\text{Velocity of light in vacuum}}{\text{Velocity of light in substance}}$$

The index of refraction of a vacuum is 1 by definition; those of other substances depend on the wavelength of light and the state of the substance. For gases and visible light, the index of refraction is only slightly greater than 1. For liquid water and visible light it is 1.33.

The scattered light beams also can recombine with each other, giving rise to the phenomenon of *diffraction*. The intensity of the combined beams depends on the relative positions of the scattering atoms. Thus diffraction can give information on the relative position of atoms in a sample (see Chap. 17).

Fluorescence and Phosphorescence

Frequently the energy absorbed by a molecule from a light beam is converted to molecular motion, such as vibration and rotation, and therefore appears as heat. Sometimes a part of the energy is re-emitted as a photon, usually of lower energy than that originally absorbed. This emitted light is called fluorescence or phosphorescence, distinguished experimentally by the time required for the excited state that gives rise to emission to decay away. *Fluorescence* is the relatively rapidly decaying emission from an excited state that converts by an "allowed" pathway to the ground state, as opposed to the longer-lived *phosphorescence*, which must convert to the ground state by a "forbidden" path. In this chapter we will consider the precise meaning of the terms "allowed" and "forbidden," and in Section 13-4 we will discuss applications of fluorescence in biochemistry.

Optical Rotation and Circular Dichroism

If the incident light is polarized to vibrate in a single plane, the plane of polarization may be rotated as the light passes through the sample. This phenomenon of *optical rotation* also can yield useful information about biochemical systems. A related technique is *circular dichroism* in which the differential absorption of light polarized in different ways is measured. We will discuss these techniques in Section 13-5.

12–3 Energy Changes in Chemical Systems are of Several Types: The Example of Water

In working in chemistry and biochemistry, you will find it is useful to acquire a feeling for the magnitudes of the energy changes that correspond to bond formation, molecular vibration, molecular rotation, and other molecular processes. The energy differences for these processes determine the spectrum of light absorbed and scattered by a substance. Let us first consider the energy differences characteristic of various processes for the water molecule.

Energies are Measured or Calculated as Differences between States

Energies always are measured or calculated as *differences* between the energies of two states. In fact, the only meaning of the absolute energy for a molecule is its Einstein rest-mass energy, $E = mc^2$. This energy is so large (over 10^{10} electron volts for the water molecule) that it would be pointless to discuss other molecular energies as fractions of the rest-mass energy. Instead, we always cite energy changes for a transition between two states. Several energy changes are listed in Table 12-1, in units of electron volts, which is the unit of energy used by many quantum chemists. The electron volt (eV) is the energy acquired by an electron accelerated through a potential difference of 1 volt. An energy difference of 1 eV for a molecular process corresponds to a difference of 23.06 kcal for the same process in a mole of molecules, or 96.48 kJ for a mole of molecules. An erg is a much larger unit of energy: 1 eV = 1.6×10^{-12} erg. These and other conversion factors for units of energy are given inside the front cover. We will discuss their derivation in the following sections.

Energy Changes for Molecular Formation

The standard energy that you select from which to reckon energy differences is a matter of convenience, and depends on your purpose. In quantum chemical calculations, for example, chemists usually choose as the standard energy that of all nuclei and electrons separated and at rest. Then the energy of forming a molecule from these separated particles is called the *total molecular energy*. For the water molecule, its value is −2080.55 eV, the negative sign showing that energy is released during molecular formation. Most of the released energy comes from the attractive potential energy of electrons for

TABLE 12-1 *Comparison of Molecular Energies for Water*

ENERGY DIFFERENCE	eV	kcal mol^{-1}	kJ mol^{-1}
(1) Total molecular energy* at 0 K	−2080.55		
(2) Sum of ground-state energies of separated atoms	−2070.46		
(3) Binding energy = (1) − (2)	−10.086	-2.324×10^2	-9.727×10^2
(4) Zero-point vibrational energy	0.575	13.2	55.4
(5) Energy of formation at 0 K = (3) + (4)	−9.511	−219	−916
Electronic excitation at 1240 Å	10.0	230	964
First ionization energy	12.6	290	1.21×10^3
Lowest vibrational transition	0.198	4.56	19.1
Rotational transition	~0.005	0.1	0.48
Vaporization at 100°C	0.39	8.9	37.6
Fusion of ice I at 0°C	0.06	1.4	5.9
Transition of ice I to ice II at −35°C	7.0×10^{-4}	1.6×10^{-2}	6.7×10^{-2}
Reversal of proton nuclear moment in a field of 10^4 gauss	1.76×10^{-7}	4.06×10^{-6}	1.70×10^{-5}

* For the fictional state of the molecule with nuclei at rest at 0 K.

the nuclei. Positive (repulsive) contributions to the total energy arise from the positive potential energy of interaction of the electrons with each other and of the nuclei with each other. The kinetic energy of the electrons moving in their orbitals is also positive.

A second energy change of chemical interest is that for the formation of a molecule from its atoms. The *binding energy* of a molecule is defined as the difference of the total molecular energy and the sum of the energies of its atoms, all at rest. Table 12-1 shows that the binding energy of the water molecule is −10.086 eV, a small fraction of the total molecular energy. The binding energy is slightly larger in magnitude than the experimental energy of formation, extrapolated to 0 K. The reason is that the actual molecule at 0 K has a residual vibrational energy, the *zero-point energy*. Its value can be calculated from the measured vibrational frequencies of the molecule. When its value (0.575 eV) is added to the binding energy, you obtain the *energy of formation*, −9.511 eV. The negative sign of this energy shows that energy is released during the formation of the molecule from its atoms. If you turn back to Figure 10-14 you will see the graphical representation of the zero-point energy as the difference of the binding energy and the energy of formation.

The total molecular energy (Entry 1 of Table 12-1) is found by summing the binding energy (Entry 3) and the ground-state energies of separated atoms (Entry 2). The binding energy is measured by spectroscopic or thermochemical experiments. The ground-state energies of atoms are determined from the summed ionization potentials of all electrons.

Energy Changes for Molecular Processes

Energy changes for electronic processes such as the excitation of an electron to a higher orbital or the removal of an electron from the molecule (ionization) are similar in size to the binding energy. The energy of electronic excitation corresponding to the absorption band in the water spectrum at 1240Å is about 10.0 eV or 230 kcal mol^{-1}. The energy required to remove the least strongly bound electron from the molecule (the first ionization energy) is 12.6 eV (290 kcal mol^{-1}).

The energy changes for excitation of molecular vibrations or rotations are far smaller than those for electronic processes. Molecules start to rotate and vibrate as they are heated from 0 K, or when they are placed in the presence of electromagnetic radiation of the proper frequency. Because transitions between rotational states require much less energy than transitions between vibrational states, it is the rotations that are excited at lower temperatures or by lower-frequency radiation, and the vibrations that are excited at higher temperatures or by higher-frequency radiation. Table 12-1 shows that the energy for the lowest vibrational transition of the water molecule is 4.56 kcal mol^{-1}, about 2% of the binding energy. A typical rotational transition requires about 0.1 kcal mol^{-1}.

The energy of interaction between molecules is small compared to molecular binding energies. The energy change for vaporization of liquid water during boiling is 8.9 kcal mol^{-1}, which is less than 5% of the binding energy of the molecule. The energy of melting is only 1.4 kcal mol^{-1}, and the energy of hydrogen-bond distortion as ice I is transformed to the high-pressure ice II is only 0.016 kcal mol^{-1}.

As small as these energies are, they are much larger than another class of energy transitions, which are the basis of nuclear magnetic resonance (NMR) measurements. These are the energies of reorientation of atomic nuclei in a magnetic field. Protons and some other nuclei have magnetic moments, and thus tend to align themselves in magnetic fields, just as electric dipolar molecules align themselves in electric fields (Sec. 11-6). The quantized energy of reorientation depends on the magnetic field strength applied to the nucleus. For a proton in a magnetic field of 1 tesla (10^4 gauss), typical of the fields used in NMR experiments, the energy for reorientation is 1.76×10^{-7} eV (4.06×10^{-6} kcal mol^{-1}). Because this is smaller even than the energies of rotational transitions, we would expect NMR experiments to involve lower-frequency radiation.

Other Energy Units

In addition to the electron volt and the calorie, scientists use several other units for energy differences. The official SI unit for energy is the *Joule* (= 1/4.184 cal = 10^7 erg). Spectroscopists frequently use either the wavelength of radiation that corresponds to a transition, or the reciprocal of the wavelength, called the *wave number* and denoted $\bar{\nu}$. These quantities are related to the corresponding energy difference as follows. If we rearrange Planck's equation, we get

$$\nu = \frac{\Delta E}{h} \qquad (12\text{-}6)$$

in which $h = 6.6262 \times 10^{-27}$ erg sec. Since the product of frequency and wavelength is the velocity of light, $\lambda\nu = c$, we can then write

$$\bar{\nu}(\text{cm}^{-1}) = \lambda^{-1} = \frac{\nu}{c} = \frac{\Delta E(\text{erg})}{h(\text{erg s})c(\text{cm s}^{-1})} \qquad (12\text{-}6a)$$

EXERCISE 12-3

Given that 1 eV = 1.602×10^{-12} erg, show that 1 eV = 8065 cm^{-1}.

Another energy equivalent is the temperature at which the thermal energy, k_BT, equals the energy of a transition. This temperature is given by

$$T = \frac{\Delta E}{k_B} \qquad (12\text{-}6b)$$

in which k_B is Boltzmann's constant, 1.38066×10^{-16} erg K^{-1}.

EXERCISE 12-4

From Equation 12-6 and the constants of Exercise 12-3, show that 1 cm^{-1} is equivalent to a thermal energy of 1.439 K.

The cover leaf lists conversion factors among the most commonly used units for energy. Of all these numbers, it is probably most useful to remember that 1 calorie = 4.184 Joule = 4.184×10^{7} erg, and that 1 kcal $\text{mol}^{-1} \simeq$ 350 cm^{-1}. It is also useful to keep in mind the thermal energy that corresponds to room temperature (about 300 K), which is

$$300\text{ K} \simeq 0.6\text{ kcal mol}^{-1} \simeq 2.5\text{ kJ mol}^{-1} \simeq 200\text{ cm}^{-1} \simeq \frac{1}{40}\text{ eV}$$

12-4 *The Electromagnetic Spectrum Detects Energy Changes in Chemical Systems: A Hypothetical Experiment*

Alternating electric currents, radio waves, microwaves, infrared radiation, visible light, ultraviolet light, x-rays, and gamma rays all produce oscillating electric fields, differing in their frequencies and therefore in their properties. We sometimes think of these forms as distinct from one another because the equipment for producing and detecting them varies considerably. Table 12-2 lists these radiations and their corresponding frequencies and wavelengths. Of course, there are no definite limits for the frequency of each type of radiation; each passes smoothly into the next. Table 12-2 also lists the type of atomic or molecular transitions that occur in the presence of radiation of a given frequency. The type of information about molecules that can be derived from spectroscopic experiments with each type of radiation also is

TABLE 12-2 *The Electromagnetic Spectrum and Molecular Spectroscopy*

ν/s^{-1}	λ	SPECTRAL REGION	TYPE OF TRANSITION	SPECTROSCOPIC METHOD	INFORMATION OBTAINED	ΔE/kcal mol^{-1}	$T = \frac{\Delta E}{k_B}$/K
1 K		Alternating electric					
	10^7 cm	current					
10^4							
	10^6 cm	Radio waves:					
10^5		Long wave					
	10^5 cm						
1 M		Broadcast band	Rotation of macromolecules	Dielectric measurements	Dipole moments		
	10^4 cm						
10^7			Nuclear orientations in a magnetic field	NMR	Nuclear environments; molecular interaction; rates of molecular motions		
	10^3 cm						
10^8		FM					
	10^2 cm						
10^9							
	10 cm			ESR	Environments of molecules with an unpaired electron		1
10^{10}		Microwave	Rotations of small molecules	Microwave	Bond lengths	10^{-4}	
	1 cm						10
10^{11}						10^{-2}	
	10^3 μm	Far infrared	Rotation of smallest molecules				10^2

10^{12}		Infrared			Force constants Interatomic and inter-molecular interactions	0.1	
10^{13}	10^2 μm		Vibrations of solids, liquids, and large molecules	Raman and infrared	Identification of functional groups	1.0	10^3
10^{14}	10 μm	Near infrared	Vibrations of small molecules			10	10^4
	10^3 nm				Identification of groups Interactions		
10^{15}		Visible light (Red, Yellow, Blue)	Transitions of outer electrons	Visible		10^2	
10^{16}	10^2 nm	Ultraviolet: Near ultraviolet		Ultraviolet	Ionization energies	10^3	
	10^2 Å	Vacuum ultraviolet	Transitions of inner electrons		Bond dissociation energies		
10^{17}					Energies of inner electrons;		
10^{18}	10 Å	X-rays: Soft x-rays		X-ray	qualitative analysis		
	1 Å	Hard x-rays					
		γ-rays	Transitions within nuclei	Mössbauer	Environment of ^{57}Fe or ^{119}Sn nuclei		

summarized. Much of this chapter and the next deal with how this information can be extracted from the spectrum of absorbed or scattered radiation.

To make the material of Table 12-2 a bit more concrete, let us carry out a thought experiment. We will place a concentrated solution of hemoglobin in a miraculous spectrometer and record its spectrum for the entire range of wavelengths of Table 12-2. Our spectrometer contains a source that emits all wavelengths, a perfect monochromator that allows us to select a narrow range of wavelengths at a time, and a detector that tells us the intensity of absorption at each wavelength. Of course, no such wonderful instrument exists, and the spectrum of hemogloblin has not been recorded in all wavelength regions. Nevertheless a brief discussion of results that might be obtained in such an experiment illustrates a number of points.

Radiowaves and Microwaves

As we sweep through the lowest frequency range in our experiment, we find no significant absorption of energy. The reason is that there are no molecular processes that have transitions in this energy range. As the frequency passes through the range of about 10^7 s^{-1} (10^7 Hz), an absorption band for the solution appears. This absorption is caused by the lagging of hemoglobin dipoles behind the oscillations of the alternating electric field. The dipoles normally rotate about 10^7 times a second, and as the alternating field passes this frequency, they can no longer come into alignment with the field. Let us suppose that our marvelous spectrometer also records the dielectric constant. At frequencies lower than 10^7 s^{-1} the dielectric constant would be greater than that of water, because the hemoglobin dipoles contribute, in addition to the water dipoles. At frequencies above 10^7 s^{-1}, the hemoglobin dipoles do not rotate fast enough to oppose the field, and the dielectric constant falls to that of water. This decline in dielectric constant is called a *dispersion,* and the absorption of energy associated with it is called a *dielectric loss*.

As we continue to increase the frequency of the oscillating field impinging on our hemoglobin solution we find no further absorption, until a frequency of about 10^{11} s^{-1}. This is the rate of rotation of water molecules in the liquid. As the alternating field increases in frequency much beyond 10^{11} s^{-1}, the water molecules start to lag and cause an absorption and a corresponding dispersion. In this dispersion the dielectric constant falls from the normal value for water of about 78 at 25°C to about 5. Now neither the permanent dipoles of water nor the hemoglobin contribute to the dielectric constant, and the only remaining contribution is from polarization of the electron distribution induced by the electric field.

Infrared Radiation and Thermal Energy

As the frequency continues to increase and moves into the far infrared region of the spectrum, the absorption becomes far more complex. These energies of about 0.1–1.0 kcal mol^{-1} are characteristic of vibrations with very weak force constants. These include vibrations of water molecules about positions of temporary equilibrium in the liquid, and loose vibrations of large regions of the hemoglobin molecule. For the infrared region of the spectrum, spectroscopists use the wave number, $\bar{\nu}$, rather than the frequency to characterize transitions. Water absorbs strongly at about 60 cm^{-1} (1.8×10^{12} s^{-1}), 200 cm^{-1}, and from about 300–900 cm^{-1} (9×10^{12} s^{-1} to 2.7×10^{13} s^{-1}). The bands around 60 cm^{-1} and 200 cm^{-1} arise from hindered translations of H_2O molecules, whereas the broad band around 600 cm^{-1} arises from hindered rotations of water molecules.

Increasing the frequency of radiation into the near infrared excites molecular vibrations with stronger force constants. These are the vibrations of covalently bound groups. The amide absorptions mentioned in Section 12-1 are some of these, but there are many others, arising from the many functional groups in the protein. The covalent bond present in highest concentration is the O—H bond of water. The O—H stretching motion absorbs so strongly at about 3400 cm^{-1} (1.0×10^{14} s^{-1}) that water is nearly opaque for radiation of this frequency. The H—O—H bending vibration of the water molecule absorbs at about 1640 cm^{-1} (4.9×10^{13} s^{-1}). Higher vibrational levels for all these motions of water cause absorption in the red visible region, and this absorption in the red end of the visible spectrum accounts in part for the blue color of large volumes of water. Other than this, water does not absorb in the visible region of the spectrum.

An important point to note is that thermal energy at room temperature, about 0.6 kcal mol^{-1}, is sufficient to excite the low-energy vibrations of the far infrared region, but is not large enough to excite vibrations in the stiffer covalent bonds. This is indicated in the last column of Table 12-2. A temperature of 300 K corresponds to a frequency of about 6×10^{12} s^{-1}, or a wave number of about 200 cm^{-1}. Thus thermal energy of 300 K is enough to cause rotations of both water molecules and hemoglobin molecules, and to cause low-frequency vibrations in both, but is insufficient to cause vibrations of covalent bonds.

Visible, Ultraviolet, and X-rays

Hemoglobin absorbs visible light strongly. For the visible and ultraviolet regions, spectroscopists characterize radiation by its wavelength, either in nm (10^{-9} m) or Å. One band is around 500–600 nm (5×10^{14} s^{-1}), the wavelength of yellow light. A second band is in the region 400–450 nm (blue light). Since blue and yellow light are absorbed by the solution, only red light is transmitted, giving hemoglobin its characteristic color. These absorptions of hemoglobin in the visible region arise from the loosely bound π electrons of the conjugated Fe-porphyrin ring.

Increasing the frequency of radiation into the near ultraviolet region excites the somewhat more strongly bound electrons of the aromatic amino acids of hemoglobin. These include tyrosine, phenylalanine, and tryptophan, which absorb around 280 nm and more strongly around 220 nm (1.4×10^{15} s^{-1}). The spectral region beyond 200 nm is termed the *vacuum ultraviolet* because the double bond of oxygen absorbs, and one must record spectra in an evacuated system. Peptide bonds absorb strongly at 190 nm and 150 nm (2×10^{15} s^{-1}). At yet higher frequencies the tightly bound σ electrons of both water and the carbon backbone of the protein are excited to higher orbitals, and many strong absorptions take place. (This frequency region is virtually inaccessible to experimental investigation because, among other factors, no transparent materials exist that can contain samples.)

Absorptions in the x-ray region are characteristic of the inner shell (most tightly bound) electrons. X-ray spectroscopy is mainly useful in qualitative and quantitative analysis of atomic species. Finally, in the γ-ray region, the transitions of energy states within nuclei cause absorption. Mössbauer devised a method for determining γ-ray spectra of certain nuclei such as ^{57}Fe. The method gives information about the distribution of s-electron density around the nucleus.

Spectroscopy in a Strong Magnetic Field

Up to this point we have recorded the spectrum of the hemoglobin solution in the absence of any magnetic field. Now let us throw a switch on our marvelous spectrometer which applies a field of 10^4 gauss to the sample, and rescan the spectrum. We find a whole new set of absorption peaks in the radio wave region of the spectrum, arising from *nuclear magnetic resonance* (NMR). These absorptions are caused by reorientations of atomic nuclei in the applied magnetic field. Several nuclei including the proton, the deuteron, ^{13}C, ^{14}N, ^{19}F, and ^{31}P possess a magnetic dipole moment. The energy of the dipolar nucleus depends on whether it is aligned with the field or against it. Transitions between these states are induced by the radio-frequency field, and absorption of energy takes place. In our sample the protons of water absorb as well as the many protons of the protein. Since the frequency at which a proton absorbs depends on its environment (Sec. 13-6), the absorption peaks are spread out over a range of frequencies. The ^{14}N and ^{13}C nuclei of the protein absorb much less strongly, and make a far smaller contribution to the total spectrum.

With the magnetic field still on, if the frequency of the applied field is increased into the microwave region, still more absorption bands may appear. These bands will be present if the sample contains unpaired electrons. These absorptions are caused by reorientations of the spinning electrons in the applied field. The effect is called *electron spin resonance* (ESR). Hemoglobin contains unpaired electrons when it is in the "deoxy state," with oxygen dissociated from it.

Finally we must return to the dependence on frequency of the dielectric constant (or the refractive index, whose square is the dielectric constant at high frequencies). This dispersion for our hemoglobin solution is shown highly schematically in Figure 12-6. We already have noted the two sharp drops in dielectric constant associated with the rotational frequencies for the hemoglobin and water molecules. At higher frequencies a large number of

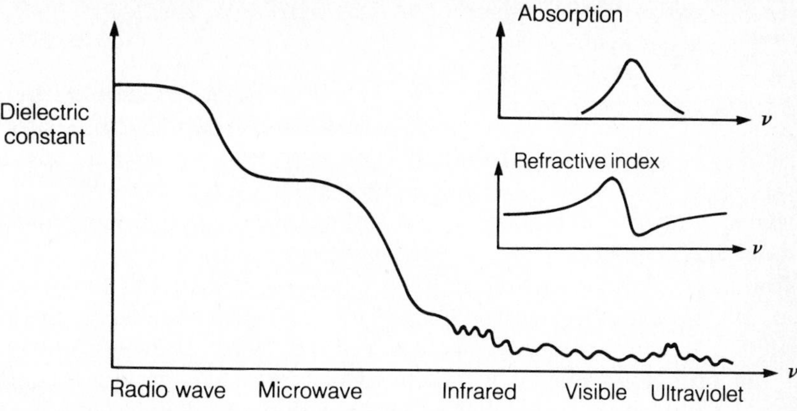

FIGURE 12-6
Schematic representation of the dispersion (dependence on frequency) of the dielectric constant of an aqueous hemoglobin solution. The two large drops in dielectric constant in the radio and microwave regions result from the failure of the hemoglobin and the water dipoles to keep pace with the alternating field. Each of the oscillations of the dielectric constant in the higher-frequency regions is associated with an absorption band. The insets show the characteristic behavior of the refractive index as it passes through a frequency of absorption.

rises and dips occur in the dispersion curve. Each rise-and-dip is associated with a single absorption band, as indicated in the inset to the figure. When the curve is again flat, the dielectric constant is somewhat smaller than it was on the low-frequency side of the rise-and-dip. Thus, the magnitude of the refractive index (or equivalently, the dielectric constant) at a given frequency depends on the number and strength of all absorptions to the high-frequency side of it. In the following section we will see the origin of the rise-and-dip of the refractive index near absorption bands.

○ *12–5 The Classical Mechanical Description of Absorption and Dispersion*

We have seen that every substance absorbs electromagnetic radiation at certain frequencies. These are the frequencies that correspond to energy differences in the substance. We also have seen that the refractive index rises and dips through the absorption band. Why do absorption bands and dispersion curves of the refractive index have the shapes they do? Classical mechanics accounts for these shapes, and although the description is not adequate to explain all observations by physicists, it gives insight into the nature of absorption and dispersion with chemical systems. In the next section we consider the quantum mechanical description of absorption. Just as it is useful to retain the wave-particle dualism for description of matter and radiation, both the classical and quantum mechanical pictures help us understand the interaction of radiation with matter.

Classical Physics Explains Absorption in terms of Resonance

In the classical description, a vibrating body absorbs energy when it is forced to oscillate at its "natural frequency." For the simple harmonic oscillator we considered in Section 10-4, the natural frequency is $\nu_0 = \sqrt{k/m}/2\pi$, in which the subscript 0 indicates that this natural frequency is a property of the oscillator. You can determine ν_0 by setting the oscillator in motion and counting the number of oscillations it makes per second.

Now suppose you force the oscillator to vibrate at some frequency, ν. For the oscillator on a spring diagramed in Figure 12-7, you might apply the force by pulling up and down on the attached string. If you pull very slowly, so that the driving frequency is much smaller than the natural frequency ($\nu \ll \nu_0$), the oscillator will move in phase with the forcing motion—as you pull the string up the oscillator moves up, and vice versa. If you move your hand much faster than the natural frequency, you find that the oscillator moves down as your hand moves up, and vice versa. That is, when $\nu \gg \nu_0$, the oscillator moves 180° out of phase with the driving force. (These relationships can be proved by applying Newton's laws of motion to an oscillating

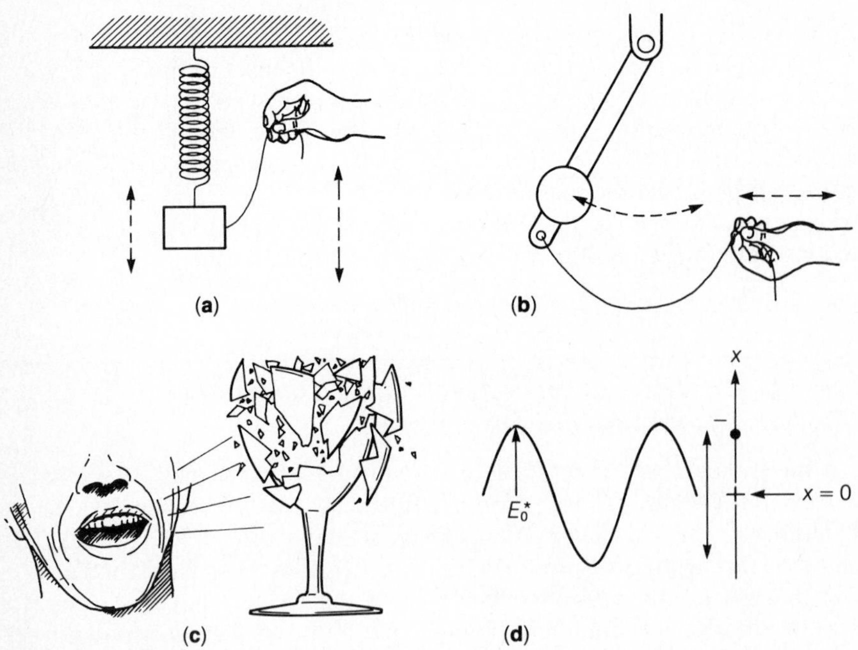

FIGURE 12-7
Examples of resonance produced by forced oscillation of vibrating systems. **(a)** A weight on a string is oscillated by pulling the string. **(b)** A pendulum is swung by pulling a string. (After Kauzmann, 1957.) **(c)** Sound waves from a singer force oscillations of a crystal goblet until it breaks. **(d)** An electromagnetic wave of amplitude E_0^* forces oscillations of an electron bound to a positive charge.

mass.) If you move your hand at just about the natural frequency, the oscillator starts to move with larger amplitude. In other words, when $\nu \simeq \nu_0$, the oscillator absorbs energy from the driving force and, in the terminology of classical physics, the oscillator is in *resonance* with the force. At the resonant frequency ($\nu = \nu_0$) the oscillator is neither exactly in phase, nor exactly out of phase, with the driving force, but is in between. We shall see later that it is 90° out of phase.

You can demonstrate resonance to yourself quite easily by making a pendulum from a yardstick [as in Fig. 12-7(*b*)] and using a string to apply a driving force. First you find the natural frequency ν_0 by letting the pendulum vibrate on its own. Then you can confirm the absorption of energy when you drive it at the natural frequency, and you can observe the motion relative to your hand at higher and lower driving frequencies.

We see examples of resonance in our everyday lives. To increase the amplitude of a child on a swing, one pushes at the natural frequency of the loaded swing. To rock a stuck car out of a ditch, one pushes at the natural frequency of the car in its temporary potential well. The shattering of a crystal goblet by a singer's voice is another example of resonance. The singer changes pitch until the frequency of the sound waves is at the natural frequency of the vibration of the goblet. Resonance occurs. The goblet absorbs energy from the sound waves, and if the amplitude of the goblet's vibrations increases enough, the goblet shatters [Fig. 12-7(*c*)]. In a musical instrument, forced oscillations from a bow or the lips of the musician produce resonance of a string or an air column.

The application of the idea of resonance to describe absorption of electromagnetic waves is depicted in Figure 12-7(*d*). An electron is assumed to be bound to a positive charge in an "atom" by a force $-kx$, where x is the electron's displacement from the positive charge. The electron then oscillates with a natural frequency ν_0. Of course, we know that this model from classical physics is not an accurate description of any real atom, but it is adequate for demonstrating the principles of resonance. The electron's oscillations can be driven by electromagnetic radiation of frequency ν because the electric field of the wave exerts a force on the electron. We shall see that absorption occurs when the frequency of the wave is near to the natural frequency of the electron.

One other concept is needed to understand resonance—*damping*. Every real oscillator eventually comes to rest after the driving force is removed, because there always is some damping force. For the pendulum, the friction of the pivot is the source of damping. For the weight on the spring it is the internal inelasticity of the spring. For the electron in our "atom" it is the dissipation of its energy into re-emission of radiation or into heat. Such damping of electrons always is associated with absorption because the absorbed energy must be converted into light or heat. Damping is described by another frequency, ν', where $2\pi\nu'$ is the reciprocal of the time for the maximum amplitude to be damped to $1/e$ of its value when the driving force is removed. Notice that ν_0 usually is greater than ν', because the time for damping usually is greater than the natural period of an oscillation.

Mathematical Description of Oscillation

The electric field driving the oscillation of an electron varies with time according to the equation

$$E^* = E_0^* \sin 2\pi\nu t = E_0^* \sin \omega t \tag{12-7}$$

in which E_0^* is the amplitude of the wave [see Fig. 12-7(*d*)] and the sine function describes wave motion. On the right side of this equation we have simplified the notation somewhat by replacing the frequency, $2\pi\nu$, with ω, the so-called *circular frequency*. It also simplifies the expressions to work with a circular natural frequency,

$$\omega_0 = 2\pi\nu_0 = \sqrt{k/m} \tag{12-8a}$$

in which m is the mass of the electron. Similarly, we work with a circular damping frequency

$$\omega' = 2\pi\nu' \tag{12-8b}$$

The electron responds to the applied field E^* and the restoring force $-kx$ by periodically changing its displacement from the equilibrium position, $x = 0$. The methods of classical mechanics show that the general time-dependent solution can be written

$$x = x_0 \sin \omega t + x_{90} \cos \omega t \tag{12-9}$$

The component $x_0 \sin \omega t$ is *in phase* with the field $E_0^* \sin \omega t$ because it is zero when the field is zero, and an extreme when the field is at an extreme, but the component $x_{90} \cos \omega t$ is 90° *out of phase* with the force because it is a maximum when the force is zero, and vice versa.

Energy is supplied to an oscillator by doing work on it. At resonance, the energy of the oscillator increases with each cycle if work is done on it.

EXERCISE 12-5

Show that the work done on a charge q in a complete cycle of an electrostatic field, $E^* = E_0^* \sin \omega t = E_0^* \sin \theta$ $(0 \leq \theta \leq 2\pi)$, is zero when the response, $x = x_0 \sin \omega t$, is in phase with the force, $F = qE_0^* \sin \omega t$ [F and x are in phase because both vary with $\sin \theta$—see Fig. 12-8(*a*)].

ANSWER

The work done by the field is $W = \int F\,dx$, in which $F = qE^* = qE_0^* \sin \theta$, and $dx = x_0 \cos \theta\, d\theta$. Therefore

$$W = qE_0^* x_0 \int_0^{2\pi} \sin \theta \cos \theta \, d\theta = \left. \frac{qE_0^* x_0 \sin^2 \theta}{2} \right|_0^{2\pi} = 0$$

The same answer is obtained when the response is 180° out of phase with the force, $x = -x_0 \sin \theta$ [Fig. 12-8(*b*)].

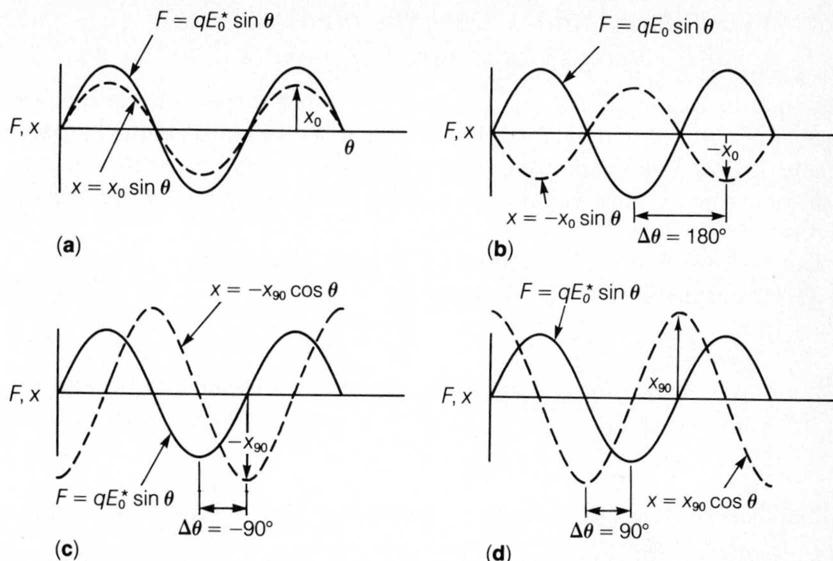

FIGURE 12-8
In- and out-of-phase response of a displacement x to a force F. **(a)** F and x are in phase. **(b)** Successive maxima in F and x are shifted by $\Delta\theta = \pi = 180°$. **(c)** Maxima in the response x lag 90° behind maxima in the force F. **(d)** Maxima in the response x precede maxima in F by 90°. Only in cases **(c)** and **(d)** is there any net energy exchange with the field for a full cycle.

EXERCISE 12-6

Show that the work done on a charge in a complete cycle of an electrostatic field, $E = E_0^* \sin\theta$, is not zero when the response, $x = -x_{90}\cos\theta$, lags 90° behind the force, $F = qE_0^* \sin\theta$ [Fig. 12-8(*c*)].

ANSWER

$W = \int F\,dx;\ dx = x_{90}\sin\theta\,d\theta$. Therefore

$$W = qE_0^* x_{90}\int_0^{2\pi}\sin^2\theta\,d\theta = \pi q E_0^* x_{90}$$

When the response $x = x_{90}\cos\theta$ *leads* the field by 90° [Fig. 12-8(*d*)], the result is

$$W = -\pi q E_0^* x_{90}$$

implying that work is done by the electron on the surroundings.

These exercises show that in a full cycle, *energy is exchanged with the oscillator only through the component of the response that is 90° out of phase with the force.* These phase relationships are summarized in Figure 12-8. Note that any arbitrary harmonic response (x) can be expressed as a sum of sine and cosine components, as in Equation 12-9.

Two Polarizabilities Describe the Resonance of an Electron Driven by a Light Wave

We wish to relate the curves for absorption and dispersion shown in the insets of Figure 12-6 to the forced oscillations of an electron in the "atom" of Figure 12-7(*d*). The main principles behind this are summarized in this section, and some of the mathematical details will be given in the following section.

The driving force, F, exerted on an electron of charge $-e$ and mass m by an electromagnetic wave $E^* = E_0^* \sin \omega t$ is

$$F = -eE^* = -eE_0^* \sin \omega t \tag{12-10}$$

The induced dipole moment, μ^{in}, depends on the polarizability (α) according to the equation

$$\mu^{\text{in}} = \alpha E^*$$

Recall from Section 11-6 that the polarizability is the induced dipole moment per unit field strength,

$$\alpha = \frac{\mu^{\text{in}}}{E^*} = \frac{-ex}{E^*} \tag{12-11}$$

in which the induced dipole moment is the product of the distance (x) by which the negative charge is displaced from the positive center at $x = 0$, times the charge, $-e$.

Both an in-phase and an out-of-phase response are required to describe the motion of the electron, because some of the polarization will not be in phase with the applied field. The general response can be expressed as a sum of sin and cos components:

$$\begin{aligned} x &= x_0 \sin \omega t + x_{90} \cos \omega t \\ &= x_0 \sin \omega t + \beta x_0 \cos \omega t \end{aligned} \tag{12-12a}$$

in which

$$\beta = \frac{x_{90}}{x_0}$$

The induced dipole moment ($\mu^{\text{in}} = -ex$) is

$$\mu^{\text{in}} = -ex_0 \sin \omega t - e\beta x_0 \cos \omega t \tag{12-12b}$$

By analogy with Equation 12-11 we can set

$$\alpha_0 = \frac{-ex_0}{E_0^*} \qquad \alpha_{90} = \frac{e\beta x_0}{E_0^*} \tag{12-13a,b}$$

in which the sign of the second equation is chosen so that α_{90} is a positive quantity, as we will see below. Substituting Equations 12-13 into Equation 12-12*b* yields

$$\mu^{\text{in}} = \alpha_0 E_0^* \sin \omega t - \alpha_{90} E_0^* \cos \omega t \tag{12-14a}$$

$$= \mu_0^{\text{in}} - \mu_{90}^{\text{in}} \tag{12-14b}$$

In these equations, α_0 is the *in-phase polarizability*, because it measures the induction of a dipole moment $\mu_0{}^{\text{in}}$ that varies with sin ωt, just as E^* does. (When α_0 is negative, μ_0 is 180° out of phase with the field.) Similarly, the *out-of-phase polarizability* α_{90} describes an induced dipole moment that varies with cos ωt, and therefore is 90° out of phase with the field. Because $\mu_{90}{}^{\text{in}}$ is found to lag behind the applied field, the sign of the term $-\alpha_{90}E_0^*$ cos (0) is negative. [See Fig. 12-8(c) for an example of a response that lags behind the driving force.]

The two unknown quantities in Equations 12-13 are x_0 (the maximum in-phase electron displacement) and β (the ratio of the maximum out-of-phase displacement x_{90} to the maximum in-phase displacement). As will be shown in the next section, expressions for these quantities can be derived by applying Newton's equations of motion to the oscillating electron. The oscillator is described by its natural frequency $\omega_0 = \sqrt{k/m}$, and its damping frequency ω'; the driving force is applied by radiation of frequency ω. In terms of these quantities it is found that

$$\beta = \frac{-2\omega\omega'}{\omega_0^2 - \omega^2} \tag{12-15}$$

$$x_0 = \frac{-eE_0^*}{m}\left[\frac{\omega_0^2 - \omega^2}{(\omega_0^2 - \omega^2)^2 + 4\omega'^2\omega^2}\right] \tag{12-16}$$

Substituting these quantities into Equations 12-13a and 12-13b, we obtain for the polarizabilities

$$\alpha_0 = \frac{e^2}{m}\left[\frac{\omega_0^2 - \omega^2}{(\omega_0^2 - \omega^2)^2 + 4\omega'^2\omega^2}\right] \tag{12-17}$$ ◀

and

$$\alpha_{90} = \frac{2e^2}{m}\left[\frac{\omega\omega'}{(\omega_0^2 - \omega^2)^2 + 4\omega'^2\omega^2}\right] \tag{12-18}$$ ◀

Because only the term that depends on cos ωt yields absorption of energy, *only* α_{90} *will contribute to absorption*. In the next section we show how Equations 12-17 and 12-18 arise from the classical equations of motion. Readers willing to accept the equations as given can skip to the subsequent discussion of the physical significance of the two polarizabilities.

Classical Mechanics Yields the Values of x_0 and x_{90}

The expressions for the in-phase and out-of-phase polarizabilities result from applying Newton's equation of motion to the oscillating electron. This equation may be written

$$\sum \text{Force} = m\,\frac{d^2x}{dt^2}$$

in which the sum includes all forces acting on the electron, and m and d^2x/dt^2 are respectively the mass and acceleration of the electron. The forces (F) acting on the electron include:

1. The driving force of the electromagnetic wave, $F_1 = -eE_0^* \sin \omega t$.
2. The restoring force between the negative electron and the positive charge, $F_2 = -kx$.
3. The damping force, which is proportional to the velocity of the moving electron and arises from frictional and radiative energy losses, $F_3 = -\eta \, dx/dt$.

Substituting the expression for these three terms into Newton's equation, we obtain

$$-eE_0^* \sin \omega t - kx - \eta \frac{dx}{dt} = m \frac{d^2x}{dt^2} \tag{12-19}$$

Equation 12-19 is a linear second-order differential equation which can be satisfied by the general function

$$x = x_0 \sin \omega t + \beta x_0 \cos \omega t \tag{12-20a}$$

Differentiating Equation 12-20a twice to find dx/dt and d^2x/dt^2 gives

$$\frac{dx}{dt} = \omega x_0 \cos \omega t - \omega \beta x_0 \sin \omega t \tag{12-20b}$$

$$\frac{d^2x}{dt^2} = -\omega^2 x_0 \sin \omega t - \omega^2 \beta x_0 \cos \omega t \tag{12-20c}$$

Substituting Equations 12-20a through c into Equation 12-19 and factoring, we find

$$[-eE_0^* - kx_0 + \eta\omega\beta x_0 + m\omega^2 x_0] \sin \omega t + [-k\beta x_0 - \eta\omega x_0 + m\omega^2 \beta x_0] \cos \omega t = 0 \tag{12-21}$$

If Equation 12-20a is indeed a solution of Equation 12-19, then Equation 12-21 must be true for all values of time t. But sin ωt and cos ωt are oscillating functions of time, and cannot equal zero for all t values. Consequently, for the sum of the two terms to be zero, the factors in the brackets each must equal zero. That is,

$$-eE_0^* - kx_0 + \eta\omega\beta x_0 + m\omega^2 x_0 = 0 \tag{12-22a}$$

$$-k\beta x_0 - \eta\omega x_0 + m\omega^2 \beta x_0 = 0 \tag{12-22b}$$

The second of these equations is readily solved for β:

$$\beta = \frac{\eta\omega}{m\omega^2 - k} \tag{12-23}$$

Similarly, solution of Equation 12-22a for x_0 yields

$$x_0 = \frac{eE_0^*}{m\omega^2 - k + \eta\omega\beta} \tag{12-24}$$

The problem that remains is to express k and η in terms of the characteristic frequencies ω_0 and ω'. When the periodic driving force is removed, Equation 12-19 describes a damped oscillator. This result can be demonstrated mathematically by setting $E_0^* = 0$, in which case the function

$$x = x_0 e^{-\omega' t} \cos [(\omega_0^2 - \omega'^2)^{1/2} t] \tag{12-25a}$$

satisfies the differential Equation 12-19, with

$$\omega' = \eta/2m \tag{12-25b}$$

EXERCISE 12-7

Confirm by differentiation that Equation 12-25*a* is a solution to Equation 12-19 when E_0^* is zero.

Equations 12-25 describe an amplitude x that varies harmonically with time, but is gradually damped by an exponential decay. In classical physics the damped harmonic oscillator is a model for the emission of radiation by an electron which has been set in motion by collisions with molecules or by electrons from a spark.

According to Equation 12-25*b*, the damping constant η can be replaced by $2m\omega'$. Furthermore, the force constant k is related to the oscillator frequency $\omega_0 = \sqrt{k/m}$ by $k = m\omega_0^2$. Entering these substitutions in Equations 12-23 and 12-24 yields the values of β and x_0 given in Equations 12-15 and 12-16.

○ *The Meaning of the Polarizabilities*

At first sight the expressions for the polarizability components α_0 and α_{90} (Equations 12-17 and 12-18) appear to be very complicated, but if each is plotted as a function of the driving frequency, ω, the curves of Figure 12-9

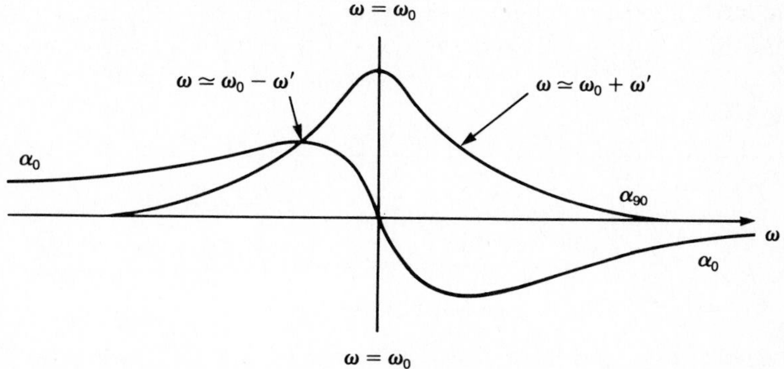

FIGURE 12-9
The in-phase polarizability, α_0, of Equation 12-17, and the out-of-phase polarizability, α_{90}, of Equation 12-18 plotted as functions of the driving frequency, ω.

are obtained. By comparing these curves to the inset in Figure 12-6, you will see that the out-of-phase polarizability, α_{90}, is closely related to the molar absorption (extinction) coefficient, ϵ, and the in-phase polarizability, α_0, is closely related to the refractive index, n. The relationships involve some constants, and are derived in more advanced books (e.g., Kauzmann, 1957). They are:

$$\epsilon = \frac{4\pi N_A}{2303c}\,\omega\alpha_{90} \tag{12-26}$$ ◀

(see Problem 18) and

$$n^2 \simeq 1 + \frac{4\pi N_A \rho}{M}\,\alpha_0 \tag{12-27}$$ ◀

in which c is the speed of light, N_A is Avogadro's number, and ρ and M are the density and molecular weight of the sample, respectively. The intensity of scattered light, I_s, also is related to the in-phase polarizability, α_0, and is given by (see Eq. 13-41*a*)

$$I_s = \frac{I_0 16\pi^4 \alpha_0^2 \nu^4}{r^2 c^4} \tag{12-28}$$ ◀

in which I_0 is the intensity of the beam of light illuminating the sample, and r is the distance to the point of observation, in a plane perpendicular to the direction of polarization of the light beam ($\phi = 90°$ in Fig. 12-4).

The meaning of Equations 12-17 and 12-18 is that the out-of-phase polarizability, α_{90}—and hence absorption— is large only when $\omega \simeq \omega_0$. This is because the term $(\omega_0^2 - \omega^2)^2$ in the denominator of Equation 12-18 approaches zero as ω approaches ω_0. The other term in the denominator is small because it is multiplied by ω', which is a very small number. Hence the fraction is large, and ϵ is large. This is the mathematical description of resonance. The polarizability α_{90} tends toward infinity, but is stopped by the term in the denominator involving damping. When the driving frequency ω is far from the natural frequency, the $(\omega_0^2 - \omega^2)^2$ term in the denominator is large; hence ϵ is small. Thus absorption bands have the shape of the α_{90} curve in Figure 12-9.

EXERCISE 12-8

Show that the maximum of the α_{90} curve is

$$\alpha_{90}^{\max} = \frac{e^2}{2m}\,\frac{1}{\omega'\omega_0}$$

ANSWER

Substitute $\omega = \omega_0$ into Equation 12-18. Notice that the absorption maximum increases for weak damping (small ω').

The meaning of Equation 12-17 is that the in-phase polarizability, α_0, is zero at $\omega = \omega_0$, because $\omega_0^2 - \omega^2$ appears in the numerator. Hence, from Equation 12-27, the refractive index is near unity. When the driving frequency is far from the natural frequency, Equation 12-17 becomes

$$\alpha_0 = \frac{e^2}{m} \frac{1}{\omega_0^2 - \omega^2} \tag{12-29}$$

which is the mathematical description of the "wings" of the α_0 curve in Figure 12-9.

EXERCISE 12-9

Show that Equation 12-17 assumes the form of Equation 12-29 when $\omega \gg \omega_0$ or $\omega \ll \omega_0$.

ANSWER

When $\omega \gg \omega_0$ or $\omega \ll \omega_0$, $(\omega_0^2 - \omega^2)^2 \gg 4\omega'^2\omega^2$ and $4\omega'^2\omega^2$ can be neglected.

Figure 12-10 summarizes the time-dependence of the electric dipole moment induced along the axis of polarization by an electromagnetic wave. At

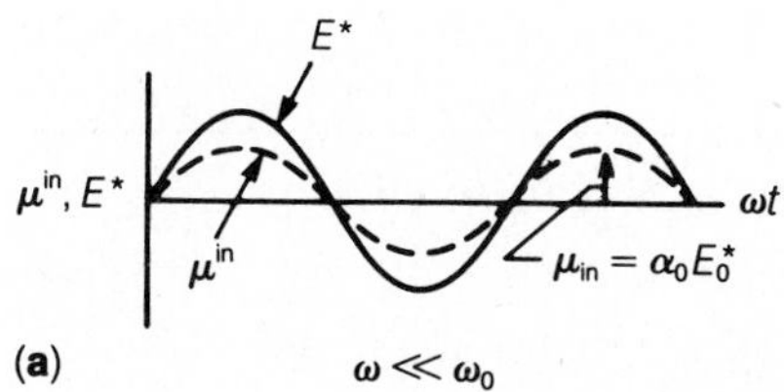

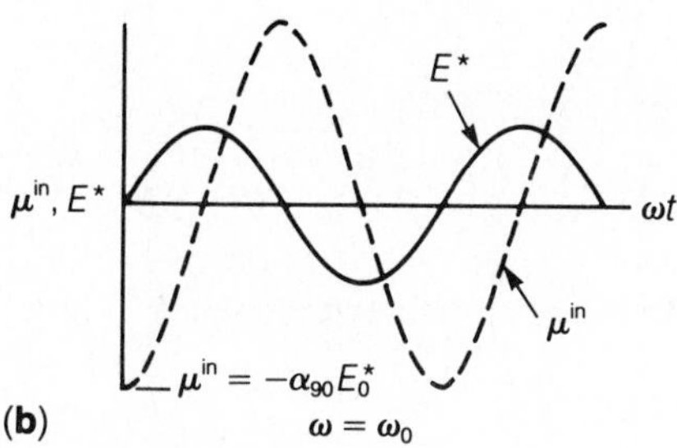

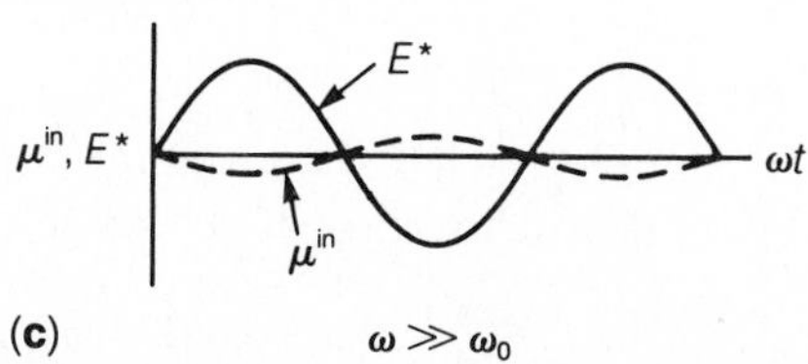

FIGURE 12-10
Summary of the time-dependence of the dipole moment induced along the polarization direction by an oscillating electric field of frequency ω, when the natural oscillation frequency is ω_0. **(a)** At low frequency, μ^{in} is in phase with the applied field. **(b)** When the two frequencies match, $\omega = \omega_0$, the polarizability is a maximum, and so is the maximum induced dipole moment. Notice that μ^{in} when $t = 0$ is $-\alpha_{90}E_0^*$. This can be used as a general definition of α_{90}, which is equal to $-\mu^{in}/E_0^*$ at the time points for which the electric field $E^* = E_0^* \sin \omega t = 0$. **(c)** At high frequency, the lag of the response behind the field has moved from 90° in **(b)** to 180° in **(c)**. The polarizability is very small because electron motion is unable to keep pace with the rapidly oscillating field.

low frequency the induced moment is in the same direction as the field, but as the frequency increases the electron responds more and more sluggishly, until at high frequency the induced moment always opposes the applied field. Notice that the 90° phase shift at $\omega = \omega_0$ applies to the *time* dependence; there is no dipole moment induced at an angle of 90° in *space* relative to E^*. In this simple model, E^* and μ^{in} share a common axis, but can point in opposite directions.

Lord Rayleigh was the first to realize that Equation 12-28 explains the blue color of the sky. Since blue light has nearly twice the frequency of red light, and because the intensity of scattering is proportional to the fourth power of the frequency (Equation 12-28), blue light is scattered about five times more strongly than red light. This is the reason that when we look perpendicularly to the direction of the sun's rays, we see mainly blue scattered light. Planets and moons without atmospheres have black skies because there are no scattering molecules in the sky.

EXERCISE 12-10

Why does light from the sun appear red during sunset and sunrise?

ANSWER

Blue light is scattered away, leaving red transmitted. The sun's rays pass through a greater path length of air at sunset and sunrise than at midday.

○ 12–6 Quantum Mechanical Description of Polarization and Absorption

The derivation of the laws of interaction of light with matter from quantum mechanics is beyond the scope of this book. However, we can summarize the most important results and can emphasize the implications that these results have for understanding chemical and biochemical spectroscopy. These results come ultimately from Schrödinger's equation. They are derived by adding to the Hamiltonian energy operator (Eq. 10-16) the mathematical expression for an electromagnetic wave acting on the system of interest. Then the perturbation of the wave on the system is determined by analyzing how the wave functions are affected by the electromagnetic wave.

JOHN STRUTT, LORD RAYLEIGH

1842–1919

Lord Rayleigh (left) with Lord Kelvin in the laboratory at Terling in 1900. (From Strutt, 1968.)

John Strutt was born on the opposite end of the English economic and social spectrum from Michael Faraday (see Chap. 8). Descended from a long line of members of Parliament and leaders of the country, he married Evelen Balfour, sister of Arthur Balfour (Prime Minister 1902–1905) and niece of Robert, Marquis of Salisbury (Prime Minister in 1885, 1886, and 1895). The stately family home, Terling, was set in formal gardens and surrounded by over 6000 acres of farming land. In a social set accustomed to wealth, privilege, and position, Strutt was considered odd for his determination to pursue a career as a scientist, rather than as an English country gentleman and a member of Parliament.

Strutt's schooling was partly from private tutors at Terling and partly at Eton and Harrow, two of England's most prestigious schools. He entered Trinity College, Cambridge (once Isaac Newton's college) with the special status of "fellow-commoner," which was granted to the eldest sons of peers who paid higher fees. This entitled him to wear a distinctive academic gown and to dine at high table with the fellows (faculty). He studied mathematics during his undergraduate years and emerged from the arduous examinations with top honors. Soon he was elected a fellow of the college.

Shortly after his marriage in 1871, Strutt's life was threatened by an attack of rheumatic fever. To recuperate, he took his bride on an extended houseboat journey up the Nile. In the unlikely setting of the main cabin of the houseboat, Strutt wrote much of his celebrated two-volume treatise, *The Theory of Sound*. Before getting down to writing each morning, he found it necessary to seal the cabin and to kill every fly in the room. He worked seated across the cabin table from his sister-in-law, who was studying mathematics under his direction, and during these hours he was reluctant to allow the boat to land, even for the party to visit the most enchanting temple. The major excitements of the trip were being towed up the First

Cataract by 200 to 300 men, and seeing a caravan bringing slaves from the south. Strutt had taken a pair of derringer pistols with him as a measure of precuation.

Not long after Strutt's return to England, his father died and he succeeded to the title of Third Baron Rayleigh. He took up residence at Terling, now his by inheritance, and had the estate carpenter convert a stable adjacent to the main house to a laboratory, where for most of the remaining 46 years of his life he carried out his research. The quarters consisted of two main rooms above and a small machine shop, a chemical laboratory, and a study below. The light source for spectroscopic experiments was the sun, whose rays entered from south and west windows.

Routine life at Terling began with prayers at 9:00, attended by the family and household staff. While breakfast was being brought in, Lord Rayleigh would open his correspondence in the library. After breakfast he went to his study and began answering letters. At 10:00 his assistant would arrive and Rayleigh would instruct him about work for the day. Rayleigh would then spend an hour or so in his armchair glancing through newly arrived journals, then he would sit at his writing table and work at the numerical reductions of experiments, mathematical analysis, or writing work for publication. The half-hour before lunch was sometimes devoted to a lesson in mathematics for his children or grandchildren.

After lunch he would walk through some of the nearer farms on the estate, and then work for an hour or so in the laboratory with his assistant before tea. At tea he would amuse his children or grandchildren, and when his assistant would return from his own tea at about 6:00, Rayleigh would go back to the laboratory until nearly eight. Rayleigh then put on a black coat for dinner, or if guests were present, dress clothes. After dinner he would study the *Times* and various weeklies. Since the Prime Minister was not infrequently a friend or relative, the papers must have held some interest. He would play a game of whist or bridge when three other players could be mustered, and occasionally he would participate in seances, since for many years he held an open mind to the possibility that spiritualism might be a genuine phenomenon. He sometimes worked in the laboratory or at his writing table for an hour or so before retiring. In the summer there were frequent weekend parties at Terling of a political or scientific nature.

The main exception to this routine was the five-year period after 1879 when Rayleigh was Cavendish professor of experimental physics at Cambridge. On James Clerk Maxwell's death, the chair was expected to go to William Thomson (later Lord Kelvin), but Thomson could not be persuaded to leave the University of Glasgow. Rayleigh was the next choice. The Duke of Devonshire wrote on behalf of the University inviting Rayleigh to take the chair, and suggested that the duties might be onerous to a gentleman of Rayleigh's rank:

> MY DEAR LORD,—
>
> I understand that there is a strong wish among the Cambridge residents originating with those who take a special interest in the experimental along with the mathematical treatment of Physics that your Lordship would consent to accept the chair of Experimental Physics, which has become vacant by the death of the much lamented Professor Clerk Maxwell.
>
> Though it is perhaps somewhat unreasonable to ask you to undertake duties the discharge of which would involve heavy demands on your time, and might very probably be attended with no small personal inconvenience, I feel so strongly the advantage the university would derive from your acceptance of the

office, that I hope you will allow me as Chancellor of the University, and also as taking a special interest in this Professorship, to support the appeal which I am told is about to be made to you, and to express a hope that you will consent to take the proposal into your favorable consideration.

I remain,
My dear Lord,
Yours faithfully,
DEVONSHIRE.

Ordinarily Rayleigh might have preferred to maintain his routine at Terling, but in the face of lower revenues from his estate during a severe agricultural depression, he accepted the Cambridge appointment, along with its stipend and its facilities for research that he could not for the moment support on his own.

Rayleigh's most famous scientific achievement was the discovery and isolation of argon. Its discovery emerged from a puzzle. His high-precision measurements of the density of nitrogen, taken to obtain better values of the atomic weight, showed that the density of nitrogen prepared from ammonia was about one part in two hundred less than the density of nitrogen obtained from air. A lesser scientist might have ignored the discrepancy, but Rayleigh knew it exceeded his precision of measurement and had to be significant. He published a short note calling attention to the dilemma and asking for suggestions as to its origin. Soon after he found a clue in a paper by Henry Cavendish (the discoverer of the composition of water) published in 1795. Cavendish had oxidized nitrogen by sparking and had observed there was always some gaseous residue that was unreactive. Rayleigh repeated these experiments, slowly accumulating enough of the new gas to test its properties. Meanwhile the chemist Sir William Ramsey, having noted Rayleigh's publication, isolated argon by the more efficient method of reacting nitrogen with magnesium. The two scientists presented their discovery of argon in a joint paper. Chemists were skeptical at first that a new element could have remained undetected for so long. For this work Rayleigh and Ramsey received the Nobel Prize in 1904. A more selfish person might not have called attention to the discrepancy, thereby alerting others to the problem, but it was typical of Rayleigh's devotion to the spirit of discovery that he did so.

The isolation of argon was only a tiny fraction of Rayleigh's work, and was not a contribution that he or many others regarded as his most important one. After argon, his best known discoveries were on light scattering and the resolving power of diffraction gratings. During his 55-year career in physics, he published 446 papers, in addition to his treatise on sound. While his work was characterized by great breadth, it was not boldly imaginative like that of Faraday, Maxwell, and Einstein. Rayleigh was the recipient of thirteen honorary degrees and over fifty special awards from learned societies, but like many born to privilege, he prided himself on what he did on his own. In receiving the Order of Merit, one of England's highest accolades, he said, "The only merit of which I personally am conscious is that of having pleased myself by my studies, and any results that may have been due to my researches are owing to the fact that it has been a pleasure to me to become a physicist."

Rules for Absorption of Electromagnetic Energy

Classical mechanics allows absorption of energy in any amount by an oscillating electron, with absorption (extinction) coefficient proportional to a function (α_{90}) called the out-of-phase polarizability. Quantum mechanics restricts the molecule to discrete energy levels, so it is evident that the quantum-mechanical view of energy absorption will differ significantly from the classical picture. We find that the function α_{90} is replaced by a set of *transition probabilities* a_k^2. These tell us the probability that a molecular system (that is, an atom, molecule, or ion) in the quantum state described by the wave function ψ_n, and subjected to an electromagnetic field will be in another state, described by ψ_k, at some later time.

Suppose that we have a set of energy levels, such as the hydrogen atomic orbitals, whose wave functions are given the symbol ψ_k^0. These are called the *unperturbed* wave functions because they are solutions to the Schrödinger equation in the absence of the electromagnetic wave. (The function ψ_k^0 is a little different from those we met earlier for the hydrogen atom, harmonic oscillator, and so forth, because it includes time as a variable.) Under the action of the perturbation supplied by the electromagnetic wave, the solution to Schrödinger's equation is altered because there is a new term in the potential energy. Let the perturbed solution be ψ, which we express as a linear combination of the unperturbed states:

$$\psi = a_1\psi_1^0 + a_2\psi_2^0 + \cdots = \sum_k a_k(t)\psi_k^0 \tag{12-30}$$

Notice that the coefficients $a_k(t)$ are functions of time.

Let us calculate a probability by taking the product $\psi^*\psi$, and integrating over all space

$$\int \psi^*\psi\, d\tau = \int \sum_k a_k^*(t)\psi_k^{0*} \sum_l a_l(t)\psi_l^0\, d\tau \tag{12-31}$$

Remembering that

$$\int \psi_k^{0*}\, \psi_l^0\, d\tau = 0 \text{ when } k \neq l$$

because the wave functions are orthogonal, and

$$\int \psi_k^{0*}\, \psi_k^0\, d\tau = 1$$

because the wave functions are normalized, Equation 12-31 becomes

$$1 = \Sigma a_k^*(t)\, a_k(t) \tag{12-32}$$

On the left side of this equation is the total probability (unity) of finding the system in some state, and on the right side are coefficients $a_k^*(t)\, a_k(t)$ [which we call $a_k^2(t)$ for convenience] that tell us how much each term $\psi_k^{0*}\, \psi_k^0$ contributes to the whole sum. Therefore, when squared, the coefficients $a_k(t)$ in Equation 12-30 give the probability that the system is in the state k at time t.

According to Equation 12-30 the coefficients $a_k(t)$ describe the development of the wave function from some initial state ψ_n^0 [i.e., $a_n(0) = 1$, $a_k(0) = 0$

when $k \neq n$] to a new wave function ψ, which is a mixture of the unperturbed states. The quantity $a_k^2(t)$ gives the probability that the molecule at time t has undergone a transition from the ground state n to an excited state k, as a result of its interaction with the electric field of a photon.

○ The quantum-mechanical theory of spectroscopy is able to predict values of a_k^2 for many simple systems. For a_k^2 to be large it is found that two general conditions must be met:

1. The electromagnetic wave illuminating the sample must have a frequency ν that corresponds to an energy difference $(E_k - E_n)/h$, in which E_k and E_n are the energies of the states ψ_k^0 and ψ_n^0, and h is Planck's constant. We already have discussed at length the implications of this requirement in Section 12-4.
2. The so-called *transition dipole moment,* μ_{kn}, for the transition from n to k must be large. The probability a_k^2 is proportional to the square of this quantity, which is defined as

$$\mu_{kn} = \int \psi_k^0 \, \mu \, \psi_n^0 \, d\tau = \int \psi_k^0 \, ex \, \psi_n^0 \, d\tau \qquad (12\text{-}33)$$ ◄

in which μ is the dipole moment operator, given on the right side of the equation as the charge (e) on the system, multiplied by its distance (x) from the coordinate origin. It can be shown that the transition moment is large only if the dipole moment of the system changes significantly during the transition. In short, for any chemical system to absorb light, its dipole moment must change during the absorption.

For simple systems for which accurate wave functions are known, the transition moments can be calculated exactly. It is found that $\mu_{kn} = 0$ for many possible transitions (that is, for many possible values of $k - n$). These transitions are called *forbidden,* and those for which $\mu_{kn} \neq 0$ are called *allowed.* For the harmonic oscillator (see Sec. 10-4) only transitions with $\Delta n = \pm 1$ are allowed. Thus the quantum number must increase or decrease by 1 during any transition. For the rigid rotor only $\Delta J = \pm 1$ is allowed, and for the hydrogen atom only $\Delta l = \pm 1$ is allowed. Such conditions are called *selection rules.*

● *What is a Transition Dipole Moment?*

Textbooks on quantum mechanics show how Equation 12-33 arises mathematically from Equation 12-30. The new energy term contributed by the electromagnetic wave is (in first approximation) proportional to the field times the instantaneous induced dipole moment, $\mathbf{E}^* \cdot \boldsymbol{\mu}$. It is this new interaction term that causes the unperturbed states ψ_k^0 to mix together to form the new state ψ, and therefore μ_{kn} depends on the integral $E^* \int \psi_k^0 \, \mu \, \psi_n^0 \, d\tau$, which is analogous to the quantity $H_{ij} = \int \psi_i \mathcal{H} \psi_j \, d\tau$ in the theory of chemical bonding (Eq. 11-11*a*).

However, many students who follow the mathematical development find that they still have no physical picture of the nature of a transition dipole

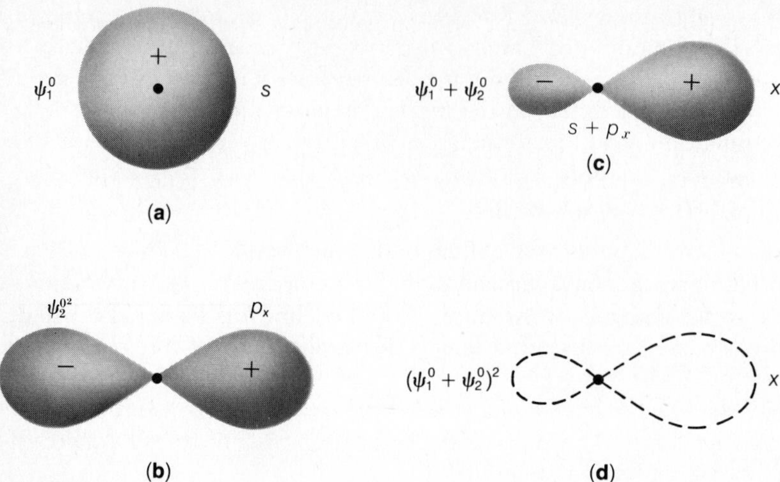

FIGURE 12-11
Dipole moment produced by the asymmetric charge distribution in an orbital that is a mixture of *s* **(a)** and p_x **(b)** orbitals. The *s* + *p* state **(c)** describes the polarization of electron density toward the right along the *x*-axis that would be produced by an electromagnetic wave polarized in that direction. **(d)** shows the electron probability distribution, $(\psi_1^0 + \psi_2^0)^2$, in the mixture of states induced by the radiation. Notice that the electron distribution is polarized along the *x*-axis, creating an induced dipole moment in that direction, which is proportional to the "transition dipole moment." Since the mixture of *s* and *p* states has a dipole moment different from the *s* state, radiation can induce a transition from *s* to *p*.

moment. Combining the classical and quantum mechanical views of the interaction of radiation with matter helps provide such an understanding. For simplicity, we restrict our consideration to the interaction of the electric field of a light beam with electrons in hydrogen-like atomic orbitals. The charge density described by the atomic orbitals has a center of inversion about the nucleus. (For each charge at x, y, z there is an identical charge at $-x$, $-y$, $-z$.) Therefore, there is no dipole moment associated with an atom which has only hydrogen-like atomic orbitals.

Polarization of a hydrogen-like atom by an electric field requires modified wave functions, since the standard atomic orbitals do not permit the asymmetry characteristic of electron polarization. The usual procedure in quantum mechanics is to express the new wave function as a linear sum of the hydrogen-like atomic orbitals. For example, Figure 12-11 illustrates a linear combination of *s* and *p* orbitals to produce a mixture, $s + p$, which has a polarized electron distribution, and therefore possesses an induced dipole moment.

In simple cases, whenever the wave functions ψ_1 and ψ_2 do not have a permanent dipole moment, the transition dipole moment μ_{12} is proportional to the dipole moment of the linear combination $\psi_1 + \psi_2$. You can see this by writing the expression for the dipole moment (μ) of the linear combination (see Eq. 11-18*a* and 12-33):

$$\begin{aligned}\mu &= e \int (\psi_1 + \psi_2)\, x\, (\psi_1 + \psi_2)\, d\tau \\ &= e \int \psi_1\, x\, \psi_1\, d\tau + e \int \psi_2\, x\, \psi_2\, d\tau + 2\, e \int \psi_1\, x\, \psi_2\, d\tau\end{aligned}$$

Since neither ψ_1 nor ψ_2 has a dipole moment, the first two integrals are zero, so

$$\mu = 2e \int \psi_1 x \psi_2 \, d\tau = 2e \int \psi_2 x \psi_1 \, d\tau \tag{12-34}$$

which is proportional to μ_{12} as described by Equation 12-33.

In summary, polarization of electrons by the electric field in a radiation beam mixes the wave functions, as is required in order to obtain a function $\psi = \psi_1 + \psi_2$, which has a dipole moment. The photon combines the states, and there is a finite probability, given by $a_k^2(t)$, that it will be absorbed, leaving the system in the excited state ψ_2, again unpolarized because the photon's field has vanished. The dipole moment of the mixed state $(\psi_1 + \psi_2)$ must be oriented along the direction of polarization of E^* to optimize the interaction with the light beam. The transition dipole moment is equivalent to the dipole moment of the mixed state, assuming no permanent dipole moment of the ground or excited state.

EXERCISE 12-11

Is the transition from an s to a d atomic orbital allowed or forbidden?

ANSWER

Forbidden (consider the charge distribution in a linear combination of s and d orbitals).

A Transition Dipole is Oriented in a Molecule

One of the important principles of spectroscopy is that interaction of the electric field in a light wave with matter is strongest if the molecule contains a changing dipole moment. The maximum strength of interaction occurs when the transition dipole moment is parallel to the electric field vector. This means that the extinction of oriented molecules depends on the spatial relationship between the polarization direction of the light wave and the transition dipole moment.

It is a general rule of dipolar transitions that the transition rate (which determines the extinction coefficient) depends on the *square* of the energy of interaction of the dipole with the field. Thus, for example, the transition probabilities a_k^2 depend on the square of the transition dipole moment, μ_{kn}^2. (Recall from Equation 11-21 that the potential energy of an electric dipole in a field is proportional to the dipole moment μ.) Similarly, if there is an angle θ between the transition dipole moment and the direction of polarization of the electric field in a light beam, the interaction energy depends on the projection of the transition dipole on the field ($U = -E^* \mu \cos \theta$). Therefore, because the transition rate depends on the square of the energy, we find for the extinction coefficient

$$\epsilon = \epsilon_{\parallel} \cos^2 \theta$$

in which $\epsilon_{\parallel}$ is the extinction coefficient for light polarized parallel to the transition moment. The general dependence of the rate of dipolar transitions on the square of the interaction energy is also found useful in considering the rate of nuclear spin transitions in Chapter 15.

It is not difficult to predict some features of the orientation of a transition dipole moment in simple cases. For example, the dipole moment for the $s \rightarrow p_x$ transition in Figure 12-11 is directed along the x-axis; the dipole moment for vibrational transitions of CO is directed along the bond axis, and so forth.

EXERCISE 12-12

Show that the transition moment for a $\pi \rightarrow \pi^*$ transition in O_2 is along the bond axis. The orbital symmetries are given in Figure 11-5.

Absorption Requires Inequality of the Populations of Quantum States

There is an additional requirement for any significant absorption: the initial state must contain a large population of systems compared to the excited state. This is because radiation can stimulate transition between two levels with equal probability in either direction. When the initial state is of higher energy than the ground state, a photon is emitted and the process is called *stimulated emission*. If the two states are equally populated, equal numbers of transitions occur, and there is no net absorption.

To understand the requirement for inequality of population of states, we turn to Figure 12-12. The rate of absorption for a transition from level c to d is the transition probability a_{cd}^2 times the number of systems in state c, N_c, or

$$\text{Absorption rate} = N_c\, a_{cd}^2$$

The probability of emission, a_{dc}^2, is equal to a_{cd}^2, so the rate of emission is

$$\text{Emission rate} = N_d\, a_{cd}^2$$

Therefore the net absorption rate, the difference between absorption and emission, is

$$\text{Net absorption} = a_{cd}^2\,(N_c - N_d)$$

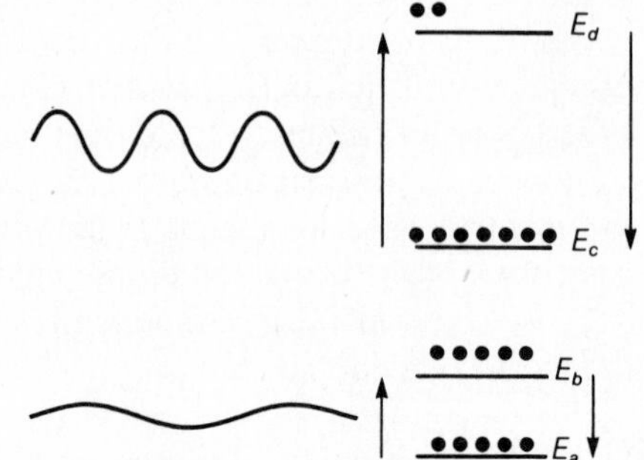

FIGURE 12-12
The rate of absorption equals the rate of emission when the population of the two energy levels is equal, as for E_a and E_b. Only the transition $E_c \rightarrow E_d$ will be observed as an absorption, because the requirement for excess population in the ground state is met.

If the two populations are equal ($N_c = N_d$) then the net absorption is zero. This is the case for the levels *a* and *b* illustrated in Figure 12-12. Consequently, net absorption will be observed only if the population of the ground state (N_c) exceeds that of the excited state (N_d) by an appreciable amount.

Radiationless Transitions

For light to be absorbed, the following conditions must be met: (1) the frequency of the light must correspond to an energy difference in the sample, (2) the absorbing system must experience a change in dipole moment, and (3) there must be a surplus of systems in the initial state. It also is important to note that there can be transitions among energy states that are not caused by absorption or emission of radiation. These *radiationless transitions* are caused by collisions of systems, or by a trading of energy between two systems, and do not need to obey the above conditions. An example of a molecule undergoing radiationless transitions is the H_2 molecule in the gaseous state. An H_2 molecule has vibrational quantum levels of the type described in Section 10-4 and rotational quantum levels of the type described in Section 10-5, and the H_2 gas molecules undergo transitions among these states (because of collisions). However, H_2 does not possess a dipole moment in any of these states or their mixtures. Thus the transition moment is zero, and there is no absorption or emission of radiation. We will encounter other examples of radiationless transitions in the sections on fluorescence and NMR relaxation.

Forbidden Transitions often can be Detected

The transition dipole moment which dictates whether a transition is allowed or forbidden results from the interaction of the *electric field* with only the *electric dipole moment* induced by the field. For that reason it is more properly called an *electric dipole transition moment*. Radiation can interact with matter other than through the electric dipole mechanism, but the other mechanisms are not as important. For example, light contains a magnetic field, and there is a corresponding *magnetic dipole transition moment* that arises from interaction with the induced magnetic moments. In addition, interaction of the electric field with molecular quadrupole moments gives rise to an *electric quadrupole transition moment*. Frequently these other transitions are allowed when the electric dipole transition moment is zero, so a "forbidden" transition may be observed, but with a weak absorption coefficient. If we want to be precise, we state that a transition is "electric-dipole allowed" or "electric-dipole forbidden."

QUESTIONS FOR REVIEW

1. What change might you expect in the spectrum of a substance upon vaporization?

2. Summarize the effects that may be observed when light interacts with matter.
3. What is the distinction among scattering of the Rayleigh, Brillouin, and Raman types?
4. Give the equation for the Beer-Lambert law, and explain its meaning.
5. Describe the types of processes that a molecule can undergo when heated or exposed to electromagnetic radiation, and the relative sizes of the associated energy changes.
6. What wavelengths of electromagnetic radiation are useful for studying each of the processes that you described in Question 5?
7. Define the following terms:
   (a) Dispersion; (b) Damping; (c) Natural frequency; (d) Resonance.
8. Summarize the essential points of the classical description of absorption and dispersion. What is the meaning of the in-phase and out-of-phase polarizabilities? What properties are related to them?
9. Define the following terms:
   (a) Transition probability; (b) Transition dipole moment; (c) Forbidden transition; (d) Selection rule; (e) Radiationless transition.
10. In the quantum mechanical description of absorption, what three conditions must be fulfilled if an absorption spectrum can be observed?

PROBLEMS

1. Are the following statements true or false?

   (a) Intensity of Rayleigh scattered light is proportional to the fourth power of the wavelength of light.

   (b) Transitions among energy levels are impossible unless there is a change in dipole moment.

   (c) The extinction coefficient of a substance in solution is proportional to its concentration.

   (d) Refraction and reflection may be considered consequences of light scattering.

   (e) In the classical description, absorption is proportional to the in-phase polarizability.

   (f) Rayleigh scattering is only in the forward direction.

   (g) Excited states in phosphorescent substances are longer-lived than those in fluorescent substances.

   (h) Water has little absorption in the infrared regions.

   (i) A strong ESR spectrum would be expected from phosphoric acid.

   (j) No absorption occurs if the initial and final quantum states are equally populated.

   (k) All nuclei can experience magnetic resonance.

Answers: (a) F. (b) F. (c) F. (d) T. (e) F. (f) F. (g) T. (h) F. (i) F. (j) T. (k) F.

2. Complete the following statements with the appropriate word.

   (a) $E_{1\,\text{cm}}^{1\%}$ means the ________ of a solution containing 1 g of absorbers per 100 cm^3 of solution in an optical path of 1 cm.

   (b) The ________ ________ of a substance is the ratio of the velocities of light in a vacuum and in that substance.

   (c) The reciprocal of the wavelength of an electromagnetic wave is called the ________ ________.

   (d) ________ spectroscopy utilizes the absorption of γ-rays.

   (e) In the classical description of absorption, greater damping ________ (increases, decreases) the width of an absorption band and ________ (increases, decreases) the height of the absorption band.

   (f) Transitions for which the transition moment $\mu_{kn} = 0$ are called ________.

   (g) Transitions that occur in the absence of an applied electromagnetic field are called ________.

Answers: (a) Absorbance. (b) Refractive index. (c) Wave number. (d) Mössbauer. (e) Increases; decreases. (f) Forbidden. (g) Radiationless.

PROBLEMS RELATED TO EXERCISES

3. (Exercise 12-1) A 100 cm^3 solution contains 50 mg of a colored compound that absorbs half the incident light at 500 nm when the solution is in a 1 cm cell.

   (a) What path length will absorb 75% of the incident light?

   (b) What mass of the compound will have an optical density of 1.5 in solution in the 1 cm cell?

   (c) What concentration of the compound will absorb 90% of the incident light in a cell 2 cm thick?

4. (Exercise 12-2) Derive expressions for determining the concentrations of tyrosine and tryptophan in a protein solution from measurements of its absorbance at 280 nm and 288 nm, given that

$$\epsilon_{280}^{\text{try}} = 5690 \text{ l cm}^{-1}\text{ mol}^{-1} \qquad \epsilon_{280}^{\text{tyr}} = 1280 \text{ l cm}^{-1}\text{ mol}^{-1}$$
$$\epsilon_{288}^{\text{try}} = 4815 \text{ l cm}^{-1}\text{ mol}^{-1} \qquad \epsilon_{288}^{\text{tyr}} = 385 \text{ l cm}^{-1}\text{ mol}^{-1}$$

5. (Exercise 12-3) Express the energy that corresponds to a temperature of 0°C in the following units:
   (a) kcal mol^{-1}. (b) kJ mol^{-1}. (c) kcal molecule^{-1}. (d) erg. (e) cm^{-1}. (f) eV.

6. (Exercise 12-4) What thermal energy corresponds to a vibrational frequency of 3400 cm^{-1}? What wavelength corresponds to this frequency?

7. (Exercises 12-7 and 12-8) Graph α_0 and α_{90} (Equations 12-17 and 12-18) as functions of ω, assuming that $\omega_0/\omega' = 20$.

8. (Exercise 12-9) Why are red lights used to warn pilots in low-flying aircraft of obstacles?

9. (Exercise 12-11) Give an answer in terms of excited states to Exercise 11-15. (Why is Li 100 times more polarizable than He?)

Answer: He has only 1 *s* electron; mixing wave functions to form a polarized state requires mixing 2 *p* with 1 *s*. Because the energy of the 2 *p* state is much higher than 1 *s*, mixing is ineffective. In contrast, the 2 *p* orbitals in Li mix readily with the 2 *s* state because their energies are close together. In general, the more accessible the excited state, the greater the polarizability.

10. (Exercise 12-12) Show that the $\pi \rightarrow \pi^*$ transition moment of benzene is in the plane of the ring.

OTHER PROBLEMS

11. Rank the following energy changes for a protein molecule in expected size from smallest to largest.
    (a) Reorientation of the spin of an unpaired electron of its heme group in a strong magnetic field (say, 10^4 gauss).
    (b) Formation of the molecule from separated nuclei and electrons.
    (c) Reorientation of one of its proton spins in a strong magnetic field (say, 10^4 gauss).
    (d) Formation of the molecule from separated atoms.
    (e) Reorientation of the dipolar protein ($\mu \sim 10^2$ D) in an alternating electric field of 30 volt cm^{-1}.
    (f) Excitation of the vibration of one of its C=O groups.
    (g) Excitation of one of the π electrons in one of its tyrosyl residues.

12. What wavelength of radiation would you use to observe processes (a), (c), (e), (f), and (g) of Problem 11?

13. Calculate in units of kcal mol^{-1} the energies corresponding to light of wavelength 1.5 μm (infrared), 650 nm (visible), 220 nm (UV), 1 Å (x-rays). How do these energies correspond to covalent and H-bond energies?

14. Sketch the solution of the damped oscillator (Eq. 12-25*a*), assuming that $\omega_0/\omega' = 20$. *Hint:* What is the equation of the curve that touches all the maxima of the cosine wave, and what is the equation of the curve that touches all the minima?

15. What is the physical significance of x_0 and ω' in Equation 12-25*a*?

16. Prove that the half-width at half maximum height of α_{90} is approximately ω'. *Hint:* you can equate ω and ω_0 in the numerator of Equation 12-18 since the denominator is dominant in determining the form of α_{90}. Also note that if

$$\omega^2 = \omega_0^2 \pm 2\omega_0\omega'$$

then

$$\omega = \sqrt{\omega_0^2 + \omega'^2} \pm \omega'$$

● 17. For the damped, driven harmonic oscillator, we can define a quantity Φ given by

$$\Phi = \tan^{-1} \frac{\alpha_{90}}{\alpha_0}$$

Sketch the dependence of Φ as a function of the driving frequency ω and explain the physical significance of Φ.

● 18. Begin with the expression

$$\mu^{\text{in}} = \alpha_0 E_0^* \sin \omega t - \alpha_{90} E_0^* \cos \omega t$$

for the response of a bound electron of charge $-e$ to an oscillatory field

$$E^* = E_0^* \sin \omega t$$

(a) Use the relationship between the induced dipole moment, the charge displacement (x) and the charge ($-e$) to write an equation for x as a function of time.

(b) Use the expression $dW = F dx$ to calculate the work done on the bound electron in one complete cycle of the field. (*Note:*

$$\int_0^{2\pi} \sin \theta \cos \theta \, d\theta = 0; \int_0^{2\pi} \sin^2 \theta \, d\theta = \pi.)$$

(c) Use the result of (b) to calculate the work done on the bound electron by the field per unit time. (*Hint:* the frequency gives the number of cycles per second.)

(d) Multiply the result of (c) by the number of particles per cubic centimeter ($N_A C_B/1000$) and the volume $A dl$ to get the energy transferred to the sample per unit time. (A is the cross-sectional area of the beam in cm^2, dl is the path traversed by the beam in cm, and C_B is the molar sample concentration.)

(e) Use the result of (d) to calculate dI, the energy removed from the beam per unit time per unit area.

(f) The intensity of a light beam I is the energy passing through a unit area per unit time, and is given by

$$I = \frac{c E_0^{*2}}{8\pi}$$

in which c is the speed of light. Use the result of (e) to calculate dI/I.

(g) Use the Beer-Lambert law to obtain an expression for dI/I in terms of the molar extinction coefficient ϵ in M^{-1} cm^{-1}.

(h) Combine the results of (f) and (g) to show that

$$\epsilon = \frac{4\pi N_A}{2303 c} \omega \alpha_{90}$$

19. On the morning of June 30, 1908, an impact or explosion of colossal force but unknown origin occurred in remote Siberia. The region was so inaccessible that it was nineteen years before a scientist could reach the site and confirm reports from the region of a cataclysmic event. Estimates of the energy released are in the range of 10^{23} erg, roughly that expected from a 30-megaton hydrogen bomb. The Siberian event in fact had some characteristics of a nuclear detonation, and there is some evidence that the exploding or disrupting object was several thousand feet above the ground. If the energy was indeed released from a nuclear reaction, how much mass was destroyed in the process?

QUESTIONS FOR DISCUSSION

20. What factors contribute to the blue color of a lake?

21. In a laser, a stimulation evokes a large emission of light as particles drop from one quantum state to a lower one. What conditions must prevail prior to stimulation, and how can it be achieved?

22. Suppose the Federal Communications Commission licenses you to operate a radio station of 50,000 watt power, and gives you the choice of broadcasting at 500 kHz or 1100 kHz. Your goal is to broadcast to as many homes as possible. Which frequency do you choose?

23. The "green flash" is a narrow strip of green light sometimes visible at the top of the disc of the setting sun. It can be observed just as the disc drops below a sharp horizon and is best seen over the desert or ocean. Can you give an explanation for the green flash in terms of the combined effects of dispersion, absorption, and scattering?

FURTHER READINGS

Bloomfield, V. A., Crothers, D. M., and Tinoco, I., Jr. 1974. *Physical Chemistry of Nucleic Acids*. New York: Harper & Row. This compendium describes many spectroscopic studies of nucleic acids.

Herzberg, G. 1959. *Molecular Spectra and Molecular Structure*. Princeton, N.J.: D. Van Nostrand Co., Inc. Vol. I. *Spectra of Diatomic Molecules*. Vol. II. *Infrared and Raman Spectra of Polyatomic Molecules*. A monumentally complete treatment of the spectroscopy of small molecules.

Kauzmann, W. 1957. *Quantum Chemistry*. New York: Academic Press. A detailed development of both the classical and quantum mechanical pictures of absorption. Our treatment of classical resonance parallels Kauzmann's, including most notations.

Moore, W. J. 1972. *Physical Chemistry,* 4th ed. Englewood Cliffs, N.J.: Prentice-Hall. A standard text in physical chemistry, with an excellent chapter on molecular spectroscopy.

OTHER REFERENCES

Susi, H. 1972. In *Methods in Enzymology* (eds. C. H. W. Hirs and S. N. Timasheff), Vol. XXVI, 455. New York: Academic Press.

Timasheff, S. N., Susi, H., and Rupley, J. A. 1973. In *Methods in Enzymology* (eds. C. H. W. Hirs and S. N. Timasheff), Vol. XXVII, 548. New York: Academic Press.

CHAPTER 13

Biochemical Spectroscopy

Several spectroscopic techniques have been applied extensively to biochemical problems and offer considerable potential for use in problems in the future. Among these are Raman and infrared spectroscopy, which probe the vibrational energy levels of molecules; visible and ultraviolet spectroscopy, which probe electronic energy levels of molecules; and optical rotatory dispersion and circular dichroism, which probe the asymmetry of molecules. Fluorescence is especially useful for detecting materials at very low concentrations and for measuring distances through the technique of energy transfer. Nuclear magnetic resonance probes the magnetic environment at the nucleus and is of great usefulness in detecting localized changes in macromolecules. Finally, light scattering responds to changes in the degree of order in a solution on a longer distance range than the other techniques and can be used to determine the size and rough shape of macromolecules.

In the preceding chapter we focused on the general principles common to several spectroscopic techniques. In this chapter we consider each technique in turn, with emphasis on the important physical principles at work in each case.

○ *13–1 The Spectrum of Raman Scattered Light Reveals Vibrational Quantum States of the Sample*

In a Raman scattering experiment (Fig. 13-1), an intense beam of monochromatic laser light is directed on the sample. When an incoming photon of frequency ν interacts with the sample, the photon may cause a transition of a system to a higher energy level, thereby losing energy itself. When scattered the photon then will have less energy, and hence a lower frequency, ν_{scat} (Fig. 13-2):

$$\nu_{scat} = \nu - \Delta E/h \quad (13\text{-}1) ◀$$

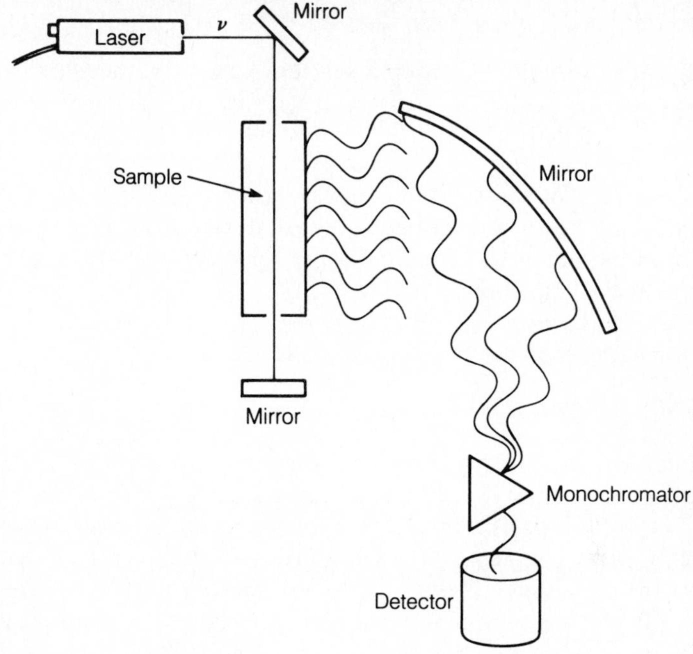

FIGURE 13-1
Schematic drawing of a Raman spectrophotometer.

If the incoming photon instead causes a transition of a system to a lower energy level, then when scattered it will have a greater energy, and hence a higher frequency:

$$\nu_{\text{scat}} = \nu + \Delta E/h \tag{13-2}$$

Thus, by observing the frequency differences, $\nu - \nu_{\text{scat}}$, a spectroscopist probes the spectrum of energy levels of the sample.

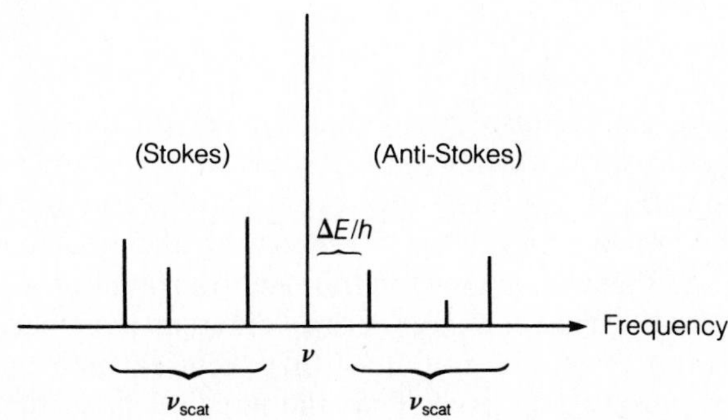

FIGURE 13-2
A Raman spectrum, showing the Stokes and anti-Stokes lines.

Rules for Raman Scattering

What determines whether a sample scatters light of a specific frequency? The probability of scattering light of frequency ν_{scat} is large only if two conditions are fulfilled:

1. The frequency of the scattered photon must equal $\nu \pm (E_k - E_n)/h$, in which E_k and E_n are the energies of the states described by ψ_k and ψ_n. ψ_n describes the initial state of the scattering system and ψ_k describes the state after the transition.
2. The *transition moment,* α_{kn}, for the transition from ψ_n to ψ_k must be large. The transition moment for Raman scattering is

$$\alpha_{kn} = \int \psi_k \, \alpha \, \psi_n \, d\tau \tag{13-3}$$

in which α is the polarizability operator for the scattering system. This integral is large only if the change in the polarizability of the system is large during the transition. Thus, if a molecular vibration changes the polarizability of a molecule, the vibrational transition scatters photons and we say the transition is *Raman active*. Although it often is not easy to see intuitively that α_{kn} is large or small, one general rule is that for molecules with a center of inversion, either α_{kn} or μ_{kn} must equal zero. (A center of inversion exists if for each atom at x, y, z there is an identical atom at $-x$, $-y$, $-z$; see Chap. 16.)

As for absorption, a significant intensity of scattering is observed only if the population of the initial state, ψ_n, is large relative to the final state. If the population is small, the number of transitions is limited and light scattered will be weak, even if the above conditions are fulfilled.

EXERCISE 13-1

Suppose that you are studying the vibrations of molecules of liquid water at 25°C by Raman spectroscopy. Would you expect to observe a greater intensity of scattering at frequencies higher than the illuminating frequency, ν, or at lower frequencies?

ANSWER

At lower frequencies, because of populations. For photons to be scattered at higher frequencies, the vibrational transition must yield energy to the incoming photon (Eq. 13-2). Then the scattering molecule must start in an excited vibrational state when the incoming photon arrives. But from Table 12-1 we see that the smallest vibrational transition of a water molecule is 19.1 kJ mol^{-1}. This is a large energy difference compared to $k_B T$ at room temperature ($\sim$2.5 kJ mol^{-1}), so that very few molecules are likely to be in excited vibrational states. Thus the populations of these states are small, and even though anti-Stokes scattering is allowed by the two rules above, it will not be observed.

Comparison of Raman and Absorption Spectroscopy

In absorption spectroscopy, the frequency of the incident beam of light is varied through a spectral region, and the intensity of the transmitted light at each frequency is observed. The decrease in intensity from the incident to the transmitted beam at each frequency reveals the absorbance for light of that frequency. In Raman spectroscopy, the incident beam is a single frequency. One observes a spectrum of frequencies scattered from the sample. The differences between these scattered frequencies and the incident frequency are proportional to the energies of transition between vibrational or rotational states.

Absorption and Raman spectroscopy often yield complementary information, because some transitions have changing dipole moments and absorb light, whereas others have changing polarizabilities and scatter light. For some transitions both molecular properties change, and these transitions are active both for absorption and Raman scattering. In biochemical work, applications of infrared absorption spectroscopy are somewhat limited because water—the universal biological solvent—absorbs so strongly in some regions of the infrared that it obscures absorption by solutes (see Fig. 13-7).

Resonance Raman spectroscopy is a technique that yields the much stronger scattering needed for biochemical work, where solutes are present in small concentrations, and recently has been applied to proteins by T. G. Spiro and others. This technique is a combination of absorption and Raman scattering. The incident frequency is chosen to be ν_{abs}, a strong visible absorption of the sample. Thus a relatively large number of photons are caused to interact with the sample, and many more photons are scattered than would be without the initial absorption. The frequencies of scattered light then reveal energies of vibrational transitions of the sample, as illustrated in Figure 13-3. Of course, for resonance Raman scattering to be applicable, the sample must include a strongly absorbing chromophore that can absorb the incident beam. Heme proteins, with their strongly absorbing heme groups, are good samples for this technique.

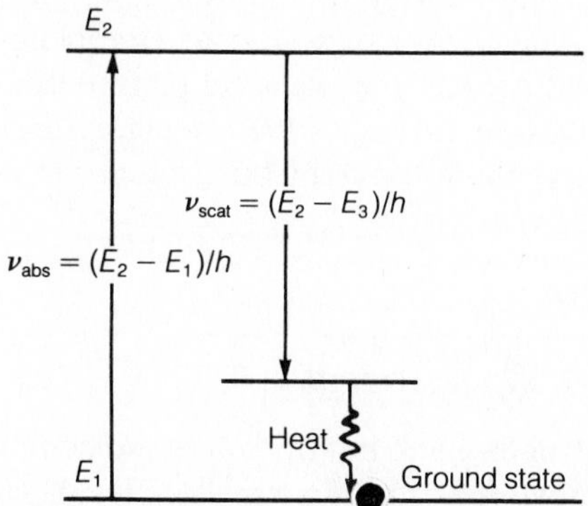

FIGURE 13-3
The energy levels of a molecular system during resonance Raman scattering. A chromophore of the system absorbs a photon of frequency ν_{abs}. The system then scatters a photon of frequency ν_{scat}. The difference between ν_{abs} and ν_{scat} is usually characteristic of a vibrational transition. The system then returns to the ground state by the dissipation of energy into heat.

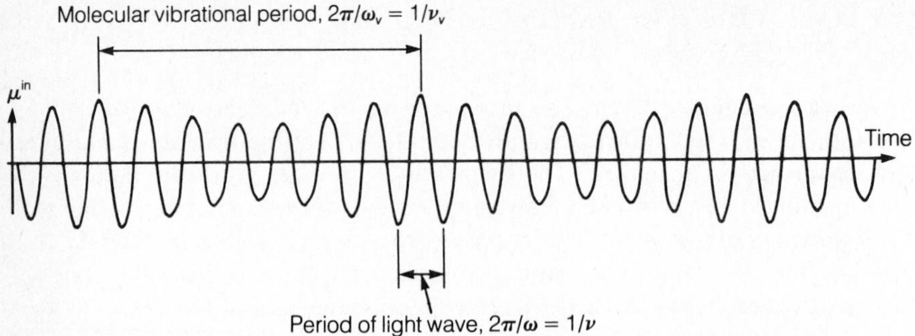

FIGURE 13-4
Schematic diagram of the variation of the induced dipole moment, $\mu^{in} = \alpha E^* = \alpha E_0^* \sin \omega t$, in a vibrating molecule whose polarizability changes as it vibrates. The vibrational frequency ω_v is very much slower than the oscillation of μ^{in} at frequency ω.

Why does Raman Scattering Require a Change in Polarizability?

The classical picture of scattering of radiation helps us understand why the polarizability must change in a vibration (or rotation) for Raman scattering to be observed. Figure 13-4 shows the dipole moment induced by an electromagnetic wave, $E^* = E_0^* \sin \omega t$, in a vibrating molecule, such as O_2. (For simplicity we suppose that the rotation of the molecule is prevented.) Since the polarizability of the molecule changes during the vibration, the maximum induced dipole moment, $\mu_{max}^{in} = \alpha E_0^*$, changes as the molecule vibrates. The molecule scatters light at the frequency of oscillation of the induced dipole moment. The component at frequency ω accounts for Rayleigh scattering (Sec. 12-1).

However, slow alternation of the size of the induced dipole superimposes two new frequencies of oscillation. Decomposition of the curve in Figure 13-4 into its frequency components (the process of Fourier analysis described in Chapter 17) shows that the new oscillations are $\sin [(\omega - \omega_v)t]$ and $\sin [(\omega + \omega_v)t]$, in which ω_v is the frequency of the vibration. These correspond to the Stokes $(\omega - \omega_v)$ and anti-Stokes $(\omega + \omega_v)$ scattered lines. If the polarizability of the molecule did not change during the vibration, no slow change in the induced dipole moment would occur and no Raman (frequency shifted) scattering would be observed. Of course, this classical picture does not predict the higher intensity of Stokes scattering, which depends on a quantum-mechanical effect, namely the predominance of the ground vibrational state.

13–2 *Vibrational (Infrared) Spectroscopy*

Of all techniques for investigating the structure and interaction of molecules, vibrational spectroscopy is one of the most powerful. For small molecules in the gaseous state, this technique yields force constants and bond dissociation energies, and when combined with rotational spectroscopy, it gives accurate values for bond lengths and bond angles (Secs. 10-4 and 10-5). For larger molecules and even complex mixtures, vibrational spectroscopy is an analytical method for determining which chemical groups are present. Moreover, because vibrational frequencies depend somewhat on the environment of the vibrating group, vibrational spectroscopy is one of the best ways for monitoring changes in molecular environments, such as changes in hydrogen bonding or changes in the conformation of a macromolecule.

Transitions between vibrational quantum states are of the size to absorb and emit infrared radiation. For this reason, *infrared spectroscopy* is the term most often used for vibrational absorption spectroscopy. Raman spectroscopy also is very useful for studying vibrational transitions.

The Complexity of a Vibrational Spectrum Depends on the Number of Atoms and the Physical State of the Vibrator

The vibration of any molecule (or indeed of any object) can be described as the superposition of fundamental vibrations called *normal modes.* A normal mode is a vibration in which all nuclei move in straight lines and at the same frequency and in phase. "In phase" means that the nuclei pass through their equilibrium positions at the same time, and reach their turning points at the same time. The normal modes for several molecules are shown in Table 13-1. HBr has a single normal mode, in which the two nuclei move apart and then together. This is called a *bond-stretching mode,* and is denoted $\bar{\nu}_1$.

Any vibrational quantum state of the HBr molecule can be thought of as a state in which the molecule is vibrating as in $\bar{\nu}_1$, with energy $E_n = (n + \frac{1}{2})\, h\nu$. You will recall from Section 10-4 that a vibrational quantum state can be represented as a horizontal line in a potential-energy curve (Fig. 13-5). A vibrational transition occurs as the multiple of the normal mode is increased or decreased by one unit. The increase or decrease must be by only a single unit, because only for $\Delta n = \pm 1$ is the transition moment, μ_{kn}, not equal to zero (Sec. 12-5). Because all transitions are between adjacent quantum states, and because all states are nearly equally spaced, the vibrational spectrum is a single absorption band. The spectrum is a band, rather than a single line, for several reasons. One is that not all the quantum states have exactly equal spacing because the energy curve is not exactly a parabola. A second reason is that each vibrational level contains several rotational quantum levels, and transitions can start or finish in any of them, provided that $\Delta J = \pm 1$. Thus the observed band is an envelope of several lines.

EXERCISE 13-2

Why could CO, HF and other diatomic gases be colored if the selection rule $\Delta n = \pm 1$ did not exist?

ANSWER

Then transitions from the ground state to all higher levels would be possible. Some would have energies large enough to correspond to absorption of visible light.

TABLE 13-1 *Examples of Molecular Vibrations*

MOLECULE	NUMBER OF ATOMS	NUMBER OF NORMAL MODES	NORMAL MODES AND FREQUENCIES OF VIBRATIONS IN WAVE NUMBERS (cm^{-1})
HBr vapor	2	1	← Br H → $\bar{\nu}_1 = 2650$
CO_2 vapor	3	4	←O C O→ $\bar{\nu}_1 = 1340$; O→ ←C O→ $\bar{\nu}_3 = 2349$; O↓ C↑ O↓ $\bar{\nu}_2 = 667$; O up, C down, O up $\bar{\nu}_2 = 667$
H_2O vapor	3	3	O↑ H↙ H↘ $\bar{\nu}_1 = 3652$; O→ H↖ H↖ $\bar{\nu}_3 = 3756$; O↓ H↖ H↗ $\bar{\nu}_2 = 1595$
N-methyl acetamide	6*	12*	I 1650 cm^{-1} (Trans), 1650 cm^{-1} (Cis); II 1550 cm^{-1}, 1450 cm^{-1}; III 1250 cm^{-1}, 1350 cm^{-1} 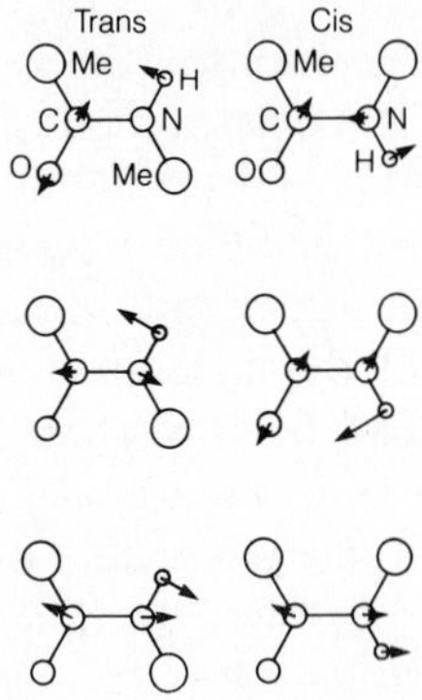
Hemoglobin	~8000	~24,000	

* Considering methyl groups (denoted Me) as single atoms. Counting all atoms there are 30 normal modes. N-methyl acetamide modes reproduced from Fraser and Suzuki (1970).

Looking at the normal modes of the H_2O molecule in Table 13-1, we see that there are three: $\bar{\nu}_1$ and $\bar{\nu}_3$ are essentially bond-stretching modes, and $\bar{\nu}_2$ is a bond-angle bending mode. Any vibration of H_2O can be represented as some combination or multiple of these three motions. And, as with HBr, any vibrational transition is equivalent to the excitation (or de-excitation) of one of the normal modes. But because there are more normal modes for H_2O, there are more bands in the spectrum. There is one band for excitation of each mode, and one for each combination of modes.

How many normal modes does a molecule of n atoms have? To specify the positions of n atoms in three-dimensional space requires $3n$ coordinates. But three of these coordinates can be thought of as defining the center of mass of

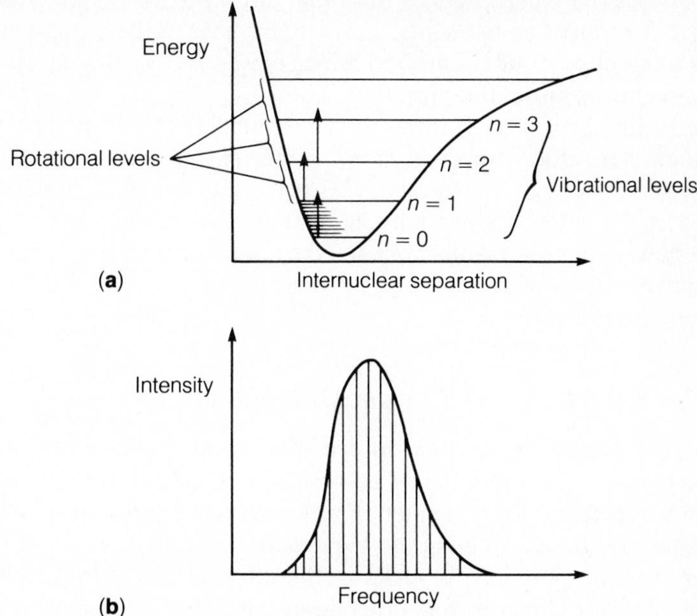

FIGURE 13-5
The basis of vibrational spectroscopy. **(a)** The molecule occupies one of the vibrational quantum states represented by horizontal lines. In each state the molecule is vibrating in one of the normal modes, or with some combination of the normal modes. Transitions between quantum states that differ by $\Delta n = \pm 1$ are allowed. Each vibrational state contains several rotational quantum states. At a given instant, the molecule occupies one rotational state of one vibrational state. **(b)** The spectrum consists of a band composed of several lines. Each line represents the transition from one vibrational and rotational state to another vibrational and rotational state.

the molecule, and three more can be thought of as specifying the orientation in space of the molecule. The remaining $3n - 6$ coordinates describe the vibrational degrees of freedom. These are the $3n - 6$ normal modes. For the special case of a linear molecule, in which all nuclei lie on a line, there are $3n - 5$ normal modes, because there is no rotational motion about the line. In Table 13-1, the number of normal modes for each molecule is determined by the number of atoms, according to these relationships.

EXERCISE 13-3

Consider vibrational transitions that excite each of the normal modes of HBr, CO_2, and H_2O. Which would you expect to be associated with the absorption of infrared radiation, which with Raman scattering, and which with both?

ANSWER

By examining the normal modes shown in Table 13-1, we can see that some of these vibrations are associated with changes of dipole moments, while others are not. For example, for $\bar{\nu}_1$ of HBr, $\bar{\nu}_3$ and $\bar{\nu}_2$ of CO_2, and all three modes of H_2O, there are changes of dipole moments during the vibration. Therefore, the transition moments, $\mu_{kn} \neq 0$ for these modes, and hence they are "infrared-active" and absorb radiation. For $\bar{\nu}_1$ of CO_2, there is no change of dipole moment during the vibration. Thus, $\mu_{kn} = 0$, and this mode is not associated with absorption of infrared. The molecule O═C═O has a center of inversion. Therefore, for each mode, either μ_{kn} or α_{kn} is zero. Since we have just seen that $\mu_{kn} = 0$ for ν_1, ν_1 must be active in Raman scattering. Similarly, the $\bar{\nu}_2$ and $\bar{\nu}_3$ modes must be inactive. All of the H_2O modes involve changes in polarizability, hence they are Raman active (though this is not immediately obvious).

Normal Modes of Larger Molecules

It takes only a little imagination to see that as the number of atoms in a molecule increases, the number of normal modes and the complexity of the vibrational spectrum grow rapidly. Even with N-methyl acetamide (Table 13-1 and Fig. 13-6), determining the movements of the nuclei in each normal mode and assigning observed peaks in the infrared spectrum to particular normal modes is a difficult job. Procedures for doing this are described in advanced books (for example, Brand and Speakman, 1975). In any protein there are thousands of normal modes, and determining the atomic movements in every mode would be impossible.

In addition to the large number of normal modes in biological molecules, there is another factor that complicates interpretation of their vibrational spectra, namely that the spectra are almost always recorded in aqueous solution. Molecules in the dissolved state exert forces on one another, and the solvent also exerts forces on them. These forces slightly change the energy of the molecule, and thus broaden its vibrational energy levels. This

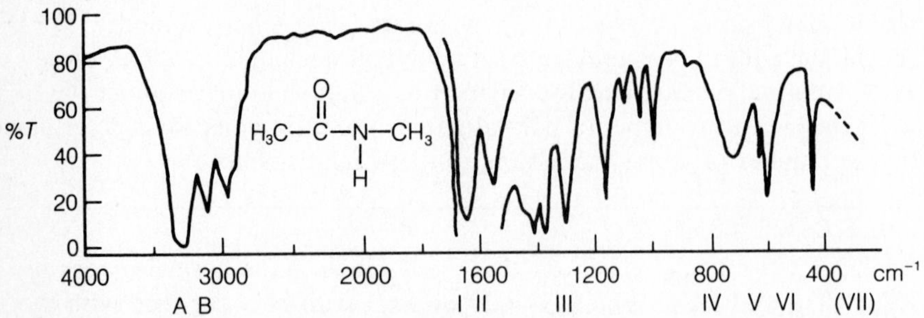

FIGURE 13-6
Infrared spectrum (percent transmission versus wave number) of N-methyl acetamide, a model molecule for the peptide bond of proteins. Regions of strong absorption are assigned to various normal modes, designated by letters and Roman numerals. The modes I, II, and III are shown in Table 13-1. Modes A and B are predominantly N-H stretching motions; modes V, VI, and VII are out-of-plane bending motions. (Redrawn from H. Susi, 1972.)

means that transitions between the energy levels lead to broader bands in the solution spectra than are observed in the vapor-state spectra. This blurring, illustrated in Figure 12-2(*e*), makes interpretation even harder.

Vibrational Spectroscopy of Complex Molecules

Although a complete understanding of a vibrational spectrum requires an analysis of the normal modes of vibration, much can be learned even when an analysis is impossible. The reason is that the frequencies of vibration of functional groups, such as O—H, N—H, and C═O, do not always change greatly as they are incorporated into different molecules. Thus, rigorous analyses of the spectra of small molecules in the vapor phase can be used as guides in interpreting the spectra of larger molecules and of molecules in solution.

For example, in the spectrum of N-methyl acetamide, the broad band with three peaks between 3000 cm^{-1} and 3400 cm^{-1} is close in frequency to the N—H stretching frequency of NH_3 in the vapor phase (3336 cm^{-1}). This suggests that the normal mode that is associated with this band has a large component of N—H stretching motion. Similarly, infrared spectra of proteins, such as that of lysozyme in Figure 12-1(*a*), contain some bands with frequencies close to those of bands in the N-methyl acetamide spectrum. Thus the analysis of the spectrum of this small molecule is a starting point for assignment of bands in the protein spectrum. The basic reason that vibrational frequencies are not shifted greatly from molecule to molecule is that the forces that link the vibrating groups to the rest of the molecule do not necessarily couple the vibrations of the functional group strongly to the rest of the vibrational motions. This is especially true for atoms which are bonded to only one other atom, such as hydrogen in a hydroxyl group or oxygen in a carbonyl and so forth.

Because the vibrational frequencies of a given functional group tend to be roughly the same in many molecules, it is possible to make tables of group absorption frequencies. Figure 13-7 is an example. Tabulations such as this one are helpful in qualitative analyses of complex molecules and mixtures, but it is necessary to keep their limitations in mind. It is always possible that some combination of normal modes gives rise to a spectral band that the table suggests comes from a single normal mode of a completely different functional group.

Shifts in Vibrational Frequencies Yield Information on Molecular Interactions

When hydrogen bonds and other interactions form in solution, the absorption or scattering frequencies of functional groups often change. For example, the O—H stretching modes of water vapor are 3650–3750 cm^{-1} (Table 13-1), whereas the O—H stretching frequencies of hydrogen-bonded H_2O

Wavelength/μm: 1.0 1.2 1.4 1.6 1.8 2.0 2.2 2.4 2.6 2.8 3.0

Group		Values (in order of increasing wavelength)
$-CH_3$		0.02, 0.1, 0.3
$>CH_2$		0.02, 0.1, 0.25
$\geq C-H$		
$-CH$ aromatic		0.1, 0.1
$-CH$ aldehyde		0.5
$-CH$ (formate)		1.0
$-NH_2$ amine	Aromatic	0.04, 0.2, 1.4, 1.5, 30, 30
	Aliphatic	0.5, 0.7, 1.5, 2
$>NH$ amine	Aromatic	0.5, 20
	Aliphatic	0.5, 0.5, 1
$-NH_2$ amide		0.7, 0.7, 3, 0.5, 0.5, 100, 100
$>NH$ amide		1.3, 0.5, 100
$-N<^{H}_{\phi}$ anilide		0.7, 0.4, 0.9, 0.3, 100
$>NH$ imide		
$-NH_2$ hydrazine		0.5, 0.5
$-OH$ alcohol		2, (H), 50
$-OH$ hydroperoxide	Aromatic	1, 1, 1.3, 30, 30
	Aliphatic	2, 0.5, 30
$-OH$ phenol	Free	3, 200
	H-bonded	Variable
$-OH$ carboxylic acid		10-100
$-OH$ glycol	1.2	50, 50
	1.3	20-50, 20-100
	1.4	50-80, 5-40
OH water		0.7, 1.2, 30, 7
$=NOH$ oxime		200
HCHO (possibly hydrate)		
$-SH$		0.05
$>PH$		0.2
$>C=O$		3
$-C\equiv N$		0.1

FIGURE 13-7
Infrared absorption wavelengths of various chemical groups. The bars indicate wavelengths of significant absorption, with the numbers giving the average absorptivity (extinction) in units of l mol^{-1} cm^{-1}. (Modified from Goddu and Delker, 1960.)

molecules in liquid water and ice are around 3400 cm^{-1} (Fig. 13-7). The reason for this reduction in stretching frequency as O—H groups form O—H · · · O hydrogen bonds is that the new neighboring oxygen atom exerts an attractive force on the hydrogen atom, thereby reducing the energy (and thus the frequency) of its stretching motion. A similar reduction in stretching frequency is found for N—H groups as they form hydrogen bonds (from about 3340 cm^{-1} in the vapor phase to 3200 cm^{-1} when in an N—H · · · O hydrogen bond).

An example of how a shift in vibrational stretching frequency can be used to obtain a thermodynamic measure of the strength of hydrogen bonding is a study of the dimerization of δ-valerolactam (Susi, 1972):

$$2\left[\begin{array}{c}\text{C=O}\\ |\\ \text{N—H}\end{array}\right] \rightleftarrows \begin{array}{c}\text{C=O}\cdots\text{H—N}\\ | \qquad\qquad |\\ \text{N—H}\cdots\text{O=C}\end{array}$$

The infrared spectrum of the monomer displays an absorption band at about 6700 cm^{-1}, which shifts toward 5800 cm^{-1} during dimerization. This band is at about twice the stretching frequency of the N—H group, and almost certainly represents the *overtone* frequency $2\bar{\nu}_1$ that results from change in the vibrational quantum number by +2 instead of +1. In water the free N—H band is partly, but not completely, obscured by absorption of the solvent (Fig. 13-8). As the temperature is increased, the absorption of the monomer increases because the equilibrium is shifted toward the monomers. By measuring the absorbance of the solution as a function of concentration, the equilibrium constant can be determined,

$$K_X = \frac{X(\text{dimer})}{X(\text{monomer})^2}$$

in which X denotes mole fraction. From the value of the equilibrium constant (0.75 at 298°K), the standard free energy change of dimerization can be determined:

$$\Delta G^\circ = -RT \ln K_X$$

and from the change of equilibrium constant with temperature the standard enthalpy change can be calculated:

$$\frac{d \ln K}{dT} = \frac{\Delta H^\circ}{RT^2}$$

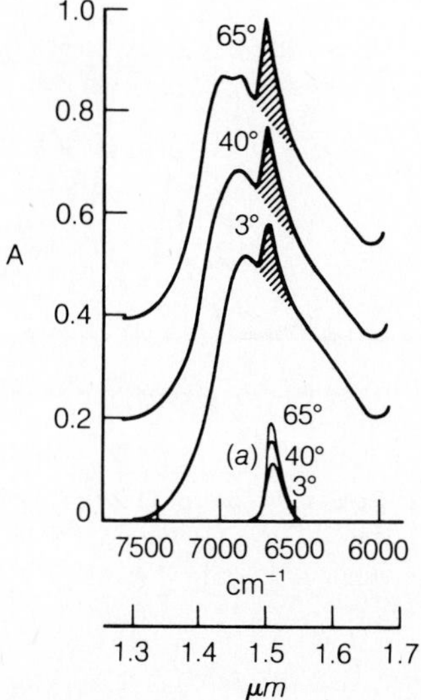

FIGURE 13-8
Infrared absorption spectrum of δ-valerolactam in H_2O as a function of temperature. The inset marked **(a)** is the estimated amide absorption. (From Susi, 1972.)

The value of $\Delta H°$ was found to be -5500 cal per mole of dimer, or about -2.75 kcal per mole of hydrogen-bonds (Susi, 1972). Of course, the standard entropy of dimerization can be determined from $\Delta G° = \Delta H° - T\Delta S°$.

Another example of the usefulness of infrared spectroscopy is the detection of double-stranded nucleic acid in a solution of polynucleotides. Panel (*a*) of Figure 13-9 shows infrared absorption spectra in D_2O of four ribonucleotide monophosphates. In panel (*b*), the dashed lines show the combined absorption of single-stranded poly A and poly U. This absorption is nearly identical to the sum of the absorptions of individual nucleotides. But the absorption of double-helical poly A · poly U complex [shown by the solid line of panel (*b*)] is quite different. Apparently the base-pairing interactions shift the vibrational frequencies. This can be confirmed by studying the absorption of the complex as a function of temperature. The complex is virtually unchanged over the temperature range 0°C–65°C, but if heated above 65°C its spectrum becomes identical to the spectrum of the two single strands. The reason, of course, is that heating the double-stranded structure eventually "melts" it into two single strands.

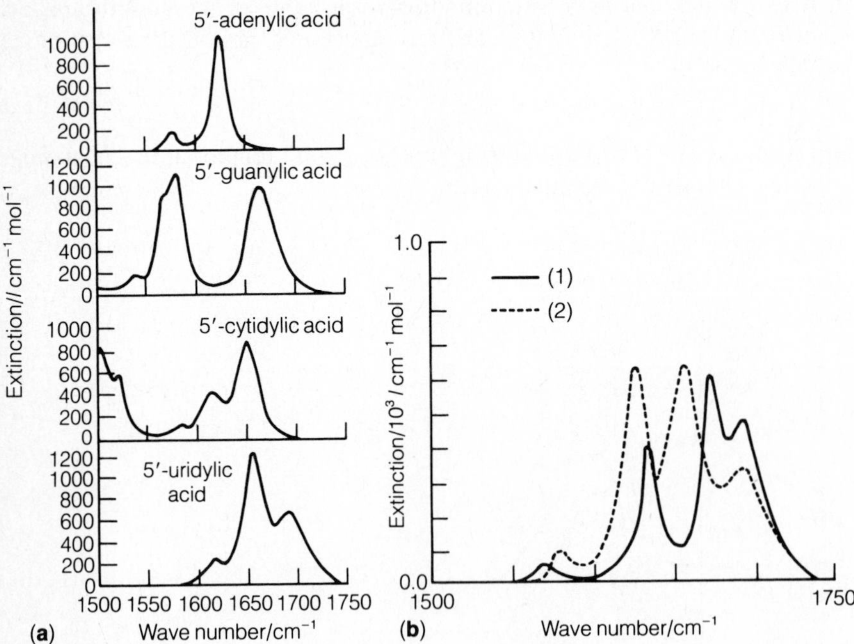

FIGURE 13-9
Infrared spectra of nucleotides and nucleic acids. **(a)** Spectra in D_2O of four ribonucleotide monophosphates. **(b)** Spectra in D_2O of poly A and poly U. Curve (1) is for the double-helical poly A · poly U complex. Curve (2) is for the sum of spectra of single-stranded poly A and single-stranded poly U. (From Bloomfield et al., 1974.)

Symmetry and the Quantum Mechanical Transition Probabilities for Infrared and Raman Spectroscopy

As explained in Section 12-6, the transition probability for infrared absorption depends on the electric dipole transition moment

$$\mu_{kn} = e \int \psi_k \, x \, \psi_n \, d\tau$$

in which the wave functions ψ describe the rotational or vibrational states before (ψ_n) and after (ψ_k) the transition. The transition moment α_{kn} for Raman spectroscopy depends on the polarizability operator α

$$\alpha_{kn} = \int \psi_k \, \alpha \, \psi_n \, d\tau$$

where again the wave functions describe rotational or vibrational states. Examination of the symmetry properties of these expressions will help us understand why for a given normal mode molecules with a center of inversion have either an infrared-active transition or a Raman-active transition, but not both.

We first define an *even* (or *symmetrical*) function as one which does not change sign or value when reflected through the origin, and an *odd* (or *anti-symmetrical*) function as one which does change sign, but not value. For example, the function x changes sign to $-x$ when reflected through the origin ($x = 0$), so x is an odd function.

EXERCISE 13-4

State whether the following atomic orbitals are even or odd: (a) s, (b) p_x, (c) p_y.

ANSWERS

(a) Even. (b) Odd. (c) Odd.

An odd function integrated over all space gives zero, because for each positive contribution there is an equal negative contribution. For example,

$$\int_{-\infty}^{\infty} x \, dx = 0$$

On the other hand, an even function does not yield zero when integrated over all space. Furthermore, you can easily see that the product of two odd or two even functions is even, but the product of an even and an odd function is odd.

Suppose now that the wave functions ψ_n and ψ_k are either odd or even. We now can see that whether the transition probabilities μ_{kn} and α_{kn} are zero depends on whether the products $\psi_k \, x \, \psi_n$ and $\psi_k \, \alpha \, \psi_n$ are even or odd functions.

First, for infrared absorption we know that x is an odd function, so if both ψ_k and ψ_n are even, or both odd, then $\mu_{kn} = e \int \psi_k \, x \, \psi_n \, d\tau$ *must* be zero, but if one of the wave functions is even and the other odd, then μ_{kn} *need not* be zero.

EXERCISE 13-5

Determine on the basis of symmetry whether the harmonic oscillator transitions (a) from $n = 0$ to $n = 1$, and (b) $n = 0$ to $n = 2$ are allowed for infrared absorption.

ANSWER

(a) Allowed; (b) Forbidden.

In the simple case of an isotropic molecule, such as CH_4, the polarizability α is a number, and therefore an even function. (For anisotropic molecules, such as C_6H_6, α is a tensor.) Consequently, α_{kn} will be zero in cases which have one of the two wave functions ψ_k or ψ_n even and one odd, precisely the condition required if μ_{kn} is not to be zero. This implies that molecules which have a center of inversion (so that ψ_k and ψ_n have the even or odd properties we assumed) will have transitions which are active either in the Raman or in the infrared. The simple origin of this effect is that α is an even function, whereas x is odd.

EXERCISE 13-6

Determine on the basis of symmetry whether the harmonic oscillator transition from $n = 0$ to $n = 1$ is allowed or forbidden for Raman scattering.

ANSWER

Forbidden.

This mutual exclusion of Raman and infrared bands applies *only* to molecules with a center of inversion symmetry (so that inversion of every point through the origin, $x \rightarrow -x$, $y \rightarrow -y$, $z \rightarrow -z$, yields the same molecule; see Chap. 16). Less symmetric molecules such as water often show vibrations which are active in both Raman and infrared.

○ 13–3 *Visible and Ultraviolet (UV) Spectroscopy Detect Excitations of Electrons*

Visible or ultraviolet light is absorbed when electrons are excited to higher quantum states. This process is depicted for a diatomic molecule in Figure 13-10. Both the ground and excited electronic states are characterized

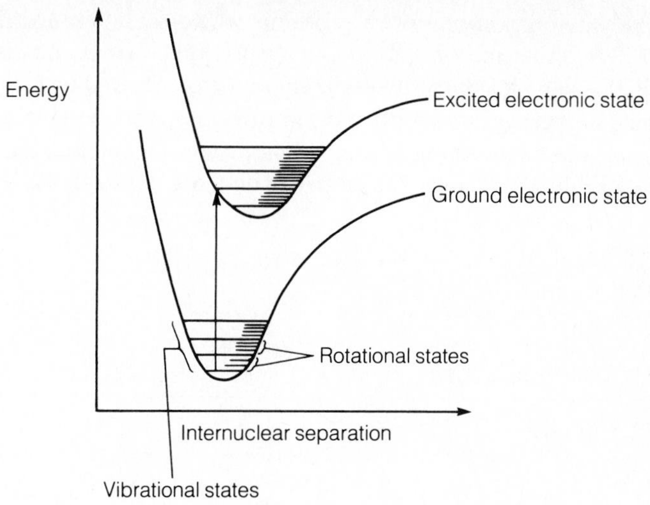

FIGURE 13-10
The basis of electronic spectroscopy. A potential-energy diagram for electronic excitation.

by a plot of potential energy as a function of internuclear separation. The excited state has a greater energy, of course, and may have either a longer or a shorter equilibrium bond length. Often the excited electron enters an antibonding orbital, so that the atoms are less strongly bound; thus the potential-energy curve is displaced toward longer bond length in the excited state. The electronic transition itself is represented by a vertical arrow on the diagram. This is because an electron moves fast compared to the relatively heavy and sluggish nuclei, and thus the internuclear separation does not change significantly during the transition. This separation of nuclear and electronic motion is called the *Bohn-Oppenheimer approximation*.

Visible and UV absorption bands tend to be broad. Notice in Figure 13-10 that each electronic state contains a number of vibrational quantum states and each vibrational state contains numerous rotational states. A transition from the ground electronic and vibrational state may start from any one of a number of rotational states and may end in one of a number of vibrational and rotational states of the excited electronic state, subject to appropriate selection rules. Thus, when many molecules are absorbing, there actually are many vertical arrows, with a distribution of lengths. Consequently, bands in UV and visible spectra often are broad and devoid of sharp features. This is especially true for compounds in solution. Therefore, rigorous assignment of bands to specific transitions is always difficult in biochemical work.

Aromatic molecules, which have loosely bound π electrons, tend to absorb at the relatively low energies of the visible and near ultraviolet (wavelengths longer than 250 nm). Aliphatic molecules, which have more tightly bound σ electrons, tend to absorb UV light in the 100–200 nm region. This is illustrated for the chromophores of proteins in Table 13-2. In examining the table you will find that the only groups to absorb light significantly in the near UV are the aromatic residues: phenylalanine, tyrosine, and tryptophan. The double bonds of the peptide group and the carbonyl group absorb strongly at about 190 nm. Most of these strong absorptions are associated with the excitation of electrons from π molecular orbitals to π^* orbitals.

TABLE 13-2 *UV Absorption Bands of Functional Groups in Proteins*†

CHROMOPHORE	RESIDUES	LOCATION (nm)	log ϵ_{max}	ASSIGNMENT
C—H	All	125[a]	—	$\sigma \rightarrow \sigma^*$
C—C	All	135[a]	—	$\sigma \rightarrow \sigma^*$
C=C	None	175[a]	3.8	$\pi \rightarrow \pi^*$
		200[a]	3	$\pi \rightarrow \pi^*$
O—H	Ser, Thr, water	150[a]	3.2	$\sigma \rightarrow \sigma^*$
		183[a]	2.2	$n \rightarrow \sigma^*$
N—H	Lys, Arg, N-terminal	173[a]	3.4	$\sigma \rightarrow \sigma^*$
		213[a]	2.8	$n \rightarrow \sigma^*$
S—H	CysSH	195[b]	3.3	$n \rightarrow \sigma^*$
S^-	$CysS^-$	235	3.5	$n \rightarrow \sigma^*$
C—S—C	Met	205[b]	3.3	$n \rightarrow \sigma^*$
—S—S—	Cystine	210[b]	3	$n \rightarrow \sigma^*$
		250	2.5	$n \rightarrow \sigma^*$
C=O	None	185[a]	3	$\pi \rightarrow \pi^*$
		260	1.3	$n \rightarrow \pi^*$
COOH	Asp, Glu, C-terminal	175[a]	3.4	$n \rightarrow \pi^*$?
		205	1.6	$n \rightarrow \pi^*$
COO^-	Asp, Glu, C-terminal	200[b]	2	$n \rightarrow \pi^*$
CONH	$AspNH_2$, $GluNH_2$, peptide bond	162[a]	3.8	$\pi^+ \rightarrow \pi^*$
		188	3.9	$\pi^\circ \rightarrow \pi^*$
		225[c]	2.6	$n \rightarrow \pi^*$
Phenyl	Phe	188	4.8	$\pi \rightarrow \pi^*$
		206	3.9	
		261	2.35	
Phenolic	TyrOH	193	4.7	$\pi \rightarrow \pi^*$
		222	3.9	
		270	3.16	
Phenolic	$TyrO^-$	200?	5	$\pi \rightarrow \pi^*$
		235	3.97	
		287	3.41	
Indole	Trp	195	4.3	$\pi \rightarrow \pi^*$
		220	4.53	
		280	3.7	
		286	3.3	
Imidazole	His	211	3.78	$\pi \rightarrow \pi^*$?

† From Donovan, 1969.
[a] Vapor phase spectrum. Those wavelengths not marked by *a* designate positions of absorption bands in water solution.
[b] Shoulder.
[c] Unresolved tail.

Relationship of the UV Absorption of the Peptide Group to its Molecular Orbitals

As an example of the level of analysis of electronic spectra possible with biological systems, let us consider the peptide group. You will recall that the four π electrons occupy two π molecular orbitals (Fig. 13-11). Two other electrons occupy the nonbonding orbital of the oxygen atom at somewhat higher energy. The lowest unoccupied orbital is the π^* orbital. The transition of smallest possible energy is for an electron of the nonbonding (n) orbital to move into the antibonding (π^*) orbital. This so-called n-π^* transition corresponds to absorption at 225 nm. The π-π^* transition is of somewhat greater energy and corresponds to absorption at 188 nm.

The most intense of these absorptions is the one corresponding to the largest transition moment (Sec. 12-6). The transition moments can be estimated from approximate wave functions, and that for the π-π^* transition is found to be larger than that for the n-π^* transition.

EXERCISE 13-7

Show on the basis of symmetry that a $\pi \rightarrow \pi^*$ transition is electric-dipole–allowed.

ANSWER

The π orbital is odd, and the π^* orbital is even, so $\psi_k\, x\, \psi_n$ is even.

EXERCISE 13-8

Show on the basis of symmetry that an $n \rightarrow \pi^*$ transition is allowed when the nonbonded state is s, but forbidden when the nonbonded state is p. You can approximate the π^* orbital in the neighborhood of the nonbonding s orbital or p atomic orbital by an atomic p orbital perpendicular to the bond direction. (Most nonbonded orbitals are a mixture of s and p, so the $n \rightarrow \pi^*$ transition is usually partially forbidden.)

Spectrophotometric Titrations of Proteins

The UV spectra of several amino acids change appreciably as they are titrated. One example is tyrosine, illustrated by the spectrum of the dipeptide glycyl-L-tyrosine in Figure 13-12. Glycyl-L-tyrosine in acid solution has the structure (shown on the following page)

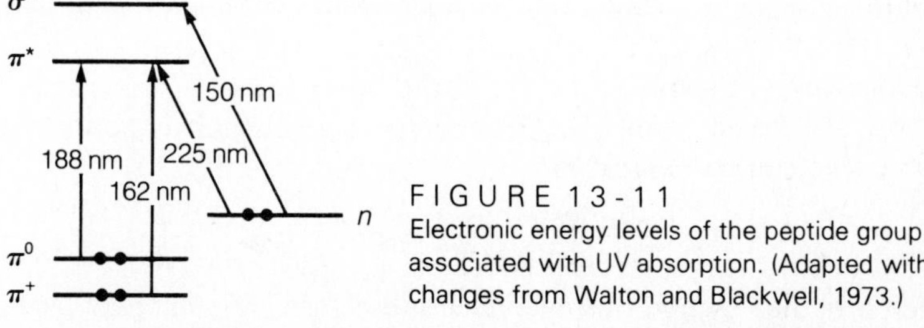

FIGURE 13-11
Electronic energy levels of the peptide group associated with UV absorption. (Adapted with changes from Walton and Blackwell, 1973.)

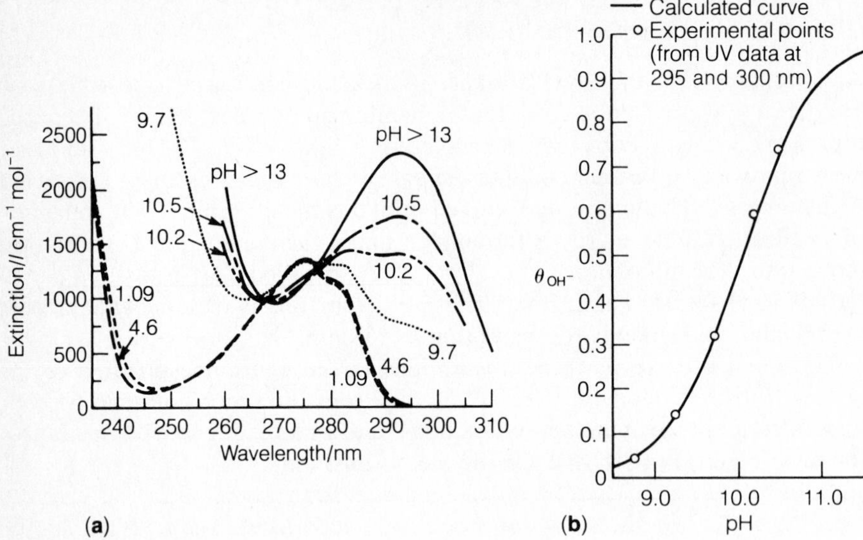

FIGURE 13-12
Spectrophotometric titration of glycyl-L-tyrosine at 25°C, with an ionic strength of 0.16. **(a)** UV absorption spectrum of the dipeptide at indicated pH values. **(b)** The fractional titration α of the phenolic hydroxyl group determined from the data of **(a)**. (From Edsall and Wyman, 1958, based on data of B. H. Gibbons.)

$^{+}H_3N—CH_2—CO—NH—CH(CH_2C_6H_4OH)—COO^{-}$

As base is added, the carboxyl proton is removed first, with a pK of about 3. The ammonium proton is next to titrate, with a pK of slightly greater than 8. Neither of these titrations affects the spectrum appreciably. But the removal of the phenolic proton of tyrosine shifts the absorption maximum from 275 nm to 293 nm and increases the molar absorption (extinction) coefficient from 1375 to 2325. The interest in such a spectrophotometric titration is that it allows us to follow the titration of one type of functional group, without the contributions appearing from other groups (such as the ammonium group).

Let us describe the titration of the phenolic proton in terms of the extinction coefficients ϵ_a for the acid form of the dipeptide (the protonated form) and ϵ_b for the basic form. The Henderson-Hasselbalch equation for the titration curve can be expressed

$$\text{pH} = \text{p}K + \log \frac{[\text{Conjugate base}]}{[\text{Conjugate acid}]} = \text{p}K + \log \frac{\theta}{1 - \theta} \tag{13-4}$$

in which θ is the fraction of molecules in the form of conjugate base. This equation can be derived easily from the expression for ionization of an acid [see Lehninger (1972) for example]. We now must relate θ to the extinction coefficient of the solution. The extinction, ϵ, at any arbitrary pH can be written as the sum of the contributions from the conjugate acid and conjugate base:

$$\epsilon = \epsilon_b\theta + \epsilon_a(1 - \theta) \tag{13-5}$$

Solving for θ, we find

$$\theta = \frac{\epsilon - \epsilon_a}{\epsilon_b - \epsilon_a} \tag{13-6}$$

and then substituting into the Henderson-Hasselbalch equation (13-4), we obtain

$$\text{pH} = \text{p}K + \log \frac{\epsilon - \epsilon_a}{\epsilon_b - \epsilon} \tag{13-7}$$

The value of ϵ_b can be taken as the extinction at pH 13, and the value of ϵ_a can be taken as the extinction at pH 4.6. Then from measurements of ϵ at an appropriate wavelength, θ can be determined from Equation 13-6 at various pH values and plotted. This has been done for glycyl-L-tyrosine in Figure 13-12(*b*). A titration curve can be calculated from Equation 13-7 and compared to the experimental points. Figure 13-12(*b*) shows that the fit is good for glycyl-L-tyrosine. Spectrophotometric titration of histidine and cysteine, as well as tyrosine, are described by Donovan (1969).

EXERCISE 13-9

Prove that at the half-equivalence point of the titration $\epsilon = (\epsilon_a + \epsilon_b)/2$.

ANSWER

At the half-equivalence point, $\theta = \frac{1}{2}$ and pH = pK. Substitution of these relations into Equations 13-6 and 13-7, respectively, yields the expression for ϵ.

EXERCISE 13-10

Spectrophotometric titration can yield an accurate value for the pK of a functional group even when other groups titrate in the same pH range. The pK can be translated into informative thermodynamic quantities. Derive expressions for ΔG° and ΔH° of titration in terms of the pK.

ANSWER

$\Delta G^\circ = 2.303\ RT$ pK and $\Delta H^\circ = 2.303\ R\ d(\text{p}K)/d(1/T)$. These expressions follow from $\Delta G^\circ = -RT \ln K$, and $\Delta H^\circ = -Rd \ln K/d(1/T)$.

Stacked Nucleotides Display Hypochromism for UV Light

The absorption of UV light by nucleic acids displays an interesting deviation from the Beer-Lambert law, called *hypochromism*. This is shown in Figure 13-13, where the UV absorption of double-stranded DNA is compared with

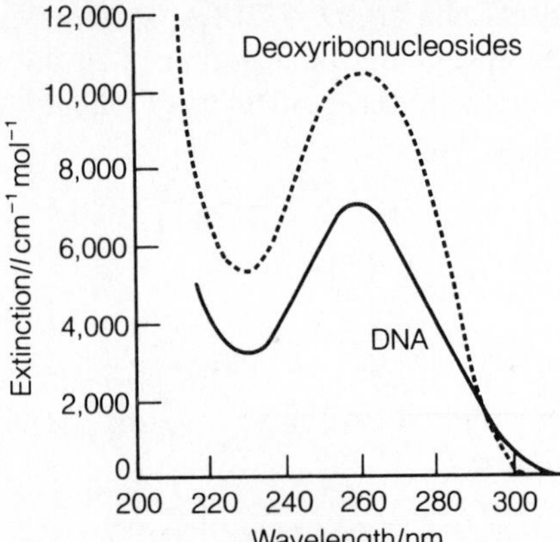

FIGURE 13-13
Hypochromism of double-stranded DNA. The molar absorption (extinction) coefficient of DNA is a maximum at 260 nm, but is substantially smaller than the sum of its component mononucleosides (dashed line). (From Bloomfield et al., 1974.)

that of its component deoxyribonucleosides. This lowering of extinction as bases form a double-stranded helix is hypochromism. The same effect is observed when the bases in a random single-stranded coil stack against one another.

Evidently there is some interaction among stacked nucleotide bases that lowers extinction. We know from Section 12-6 that the intensity of any absorption is governed by the size of the corresponding transition dipole. The absorption of the nucleic acids arises from $\pi - \pi^*$ transitions of the bases. A base in a double helix has a smaller transition dipole than an isolated base, because the dipole moments induced by the light in neighboring bases in turn induce moments in the first base which oppose the moment induced directly by the light. The reduced transition moment lowers the extinction. Bloomfield et al. (1974) discuss absorption by nucleic acids in detail.

○ *13–4 Fluorescence is Rapid Re-emission of Absorbed Radiation*

Fluorescence has become a powerful tool in the biological sciences because of its great sensitivity and because of its capacity to yield information on molecular associations, motions, and separations. Its sensitivity is based on the fact that in a fluorescence experiment one observes only the emitted light, and compares this to darkness. In contrast, all absorption measurements involve the comparison of two beams of light (I_0 and I). Measuring the difference between two nearly equal quantities is inherently less accurate than measuring the absolute value of a single quantity. For this reason, it often is possible to record fluorescence from dilute solutions of macromolecules (rarely as concentrated as 10^{-3} *M*) when it is impossible to determine absorption accurately.

Fluorescence is one mechanism by which a molecule can return to its ground state after it has been excited by absorption of radiation (Fig. 13-14).

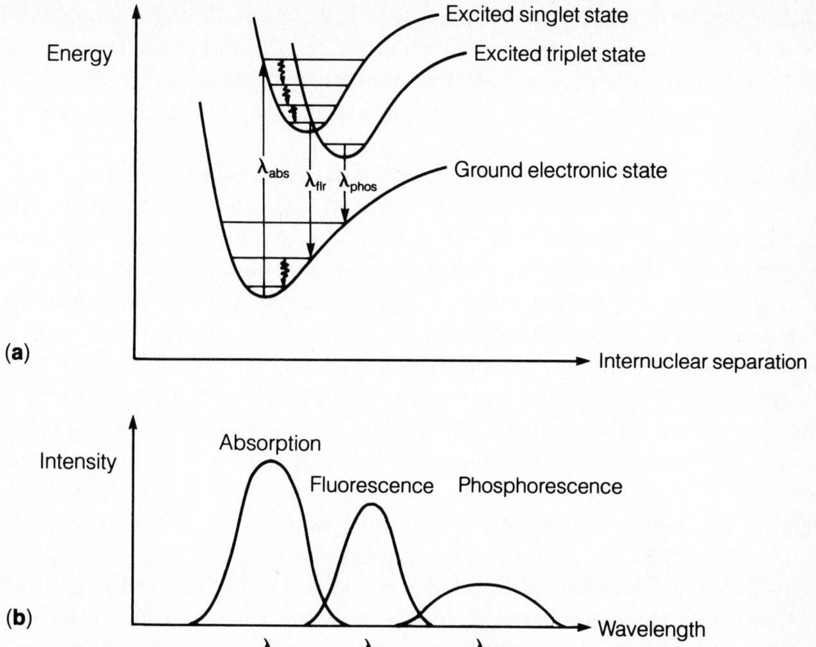

FIGURE 13-14
(a) Quantum states involved in fluorescence and phosphorescence. Straight arrows represent absorption and emission of light; wavy arrows represent radiationless transitions. **(b)** The relative wavelengths of absorption, fluorescence, and phosphorescence indicated in **(a)**.

Another mechanism is *radiationless transition,* in which energy is carried away by collisions with other molecules rather than by re-emission of light. But whatever the detailed path, the first steps usually are radiationless transitions. This is the mechanism by which the molecule passes to the lowest vibrational state of its excited electronic state. In Figure 13-14 these radiationless steps are shown by wavy arrows. These steps also are called *internal conversions*.

Once in the ground vibrational state of the excited electronic state, the molecule can return to the ground electronic state by a further radiationless transition, or by emission of radiation. If the emission is from an excited state in which all electron spins are paired (singlet state), it is termed fluorescence; if it is from an excited state in which a pair of electrons have become unpaired (triplet state) it is termed phosphorescence.

Time Scale of Events in Fluorescence and Phosphorescence

The time scale of these events is important both experimentally and theoretically. The initial absorption of light takes only about 10^{-15} s. Since nearly all ground electronic states are singlet states, and because singlet-triplet transitions are forbidden (that is, they have transition moments near zero), the resulting excited electronic state almost always is another singlet state. Moreover, it is likely that the absorption places the molecule in an excited vibrational state of this excited singlet electronic state (as shown in Fig. 13-14). The time for radiationless transitions to carry the molecule down to the ground vibrational level of this excited state is a few times 10^{-12} s because the interval between molecular collisions in a liquid is of the order of 10^{-12} s. However, once in the ground vibrational level, the molecule must wait the relatively long time of about 10^{-9} s before returning to the ground electronic state. During this "long wait" a macromolecule may experience considerable reorientation or motion. Thus, a comparison of some properties of the absorbed and emitted radiation (for example the polarization) can give information on the motion of macromolecules.

Sometimes it happens that the excited singlet state can undergo transition to an excited triplet state of similar energy (see Fig. 13-14). This is termed *intersystem crossing*. Then the molecule in the electronic triplet state coasts down to its vibrational ground state in a few times 10^{-12} s. But since triplet-singlet transitions are forbidden, it may take as long as a second or more for the molecule to emit radiation (phosphorescence) and return to its ground electronic state. Thus phosphorescence is a slow path of return to the ground state (and thus phosphorescent materials glow for a while in the dark). More often than not, the triplet excited state returns to the ground state by radiationless transitions, and we say that the phosphorescence has been *quenched*.

Wavelengths and Intensity of Fluorescence

What physical parameters characterize fluorescence? First, there are the wavelengths of absorption and emission of the chromophore. The wavelength of emission almost always is longer (frequency shorter) than that of absorption because some energy of the excited state is lost by radiationless transitions (see Fig. 13-14). Several examples of wavelengths of absorption and emission are given in Table 13-3.

A second parameter is the *quantum yield, Q,* which is the measure of efficiency of fluorescence; it is defined by

$$Q = \frac{\text{Number of quanta emitted}}{\text{Number of quanta absorbed}} \qquad (13\text{-}8)$$ ◀

The quantum yield varies between unity (for no pathway of radiationless transitions between excited and ground electronic states) and zero (for no pathway of emission). From the Beer-Lambert law, we know that the intensity of transmitted light, I, is related to the incident intensity I_0 by

$$I = I_0\, 10^{-\epsilon l C} \qquad (13\text{-}9a)$$

The intensity of a light beam is defined as the radiant energy that passes through a unit cross-sectional area per second. For a given frequency, the intensity I is proportional to the number of photons, which we call I'. At other frequencies the proportionality constant between intensity and photons is changed because the energy per photon is different (by Planck's relationship). Since the light absorbed, I_{abs}, has the same frequency as I_0, we can replace Equation 13-9*a* by the equivalent relationship in terms of numbers of photons:

$$I' = I'_0 10^{-\epsilon l C}$$

so that the number of photons absorbed, I'_{abs}, is

$$I'_{abs} = I'_0 - I' = I'_0 (1 - 10^{-\epsilon l C})$$

Therefore, the number of photons emitted in the fluorescent beam is given by

$$I'_{flr} = QI'_{abs} = QI'_0 (1 - 10^{-\epsilon l C}) \quad (13\text{-}9b)$$

When only a small amount of the light is absorbed, ϵlC is near zero, and we can expand the right side of Equation 13-9*b*:

$$\begin{aligned} I'_{flr} &= QI'_0 (1 - e^{-2.303\,\epsilon l C}) \\ &= QI'_0 [1 - (1 - 2.303\epsilon l C + \cdots)] \\ &= 2.303\ QI'_0 \epsilon l C \end{aligned} \quad (13\text{-}10)$$ ◄

Equation 13-10 reveals that the fluorescence intensity of a dilute solution is linearly proportional to the concentration C. Therefore, fluorescence provides a sensitive and accurate analytical tool for detecting small amounts of a fluorescent substance. Of course, a calibration curve is required to determine the quantum yield Q in Equation 13-10.

The change in intensity of fluorescence can give information on the binding of a small molecule to a macromolecule, or on conformational change in a macromolecule. One example is a study of the enzyme lysozyme by Lehrer and Fasman (1967). They found that the intensity of fluorescence at 335 nm is increased by the binding of substrate to the enzyme. Because 335 nm is the wavelength of tryptophan fluorescence, they concluded that tryptophanyl residues participate in binding substrates. From the dependence of pH of the fluorescence, they concluded that carboxyl residues also are involved in binding.

When a macromolecule itself neither contains fluorescent groups nor has fluorescent cofactors, then a clever chemist can bind other fluorescent groups to the macromolecule. These are called *extrinsic* fluorescent probes. An example is the compound called ANS, shown in Table 13-3. Stryer (1965) found that ANS binds to several proteins, and that the increase of its fluorescence and the shift to shorter wavelengths of its fluorescent maximum is a measure of the nonpolarity of the ANS binding site.

Energy Transfer Allows Measurement of Distances

In some cases the intensity of fluorescence is a measure of the separation of two functional groups in a molecule or of groups on neighboring molecules. This is so when there is a donor group and an acceptor group, and when there

TABLE 13-3 *Fluorescence of some Molecules of Biological Significance*

MOLECULE OR GROUP	STRUCTURE	λ_{abs}^{max}/nm	λ_{flr}^{max}/nm	Q	COMMENT
Tryptophan residue	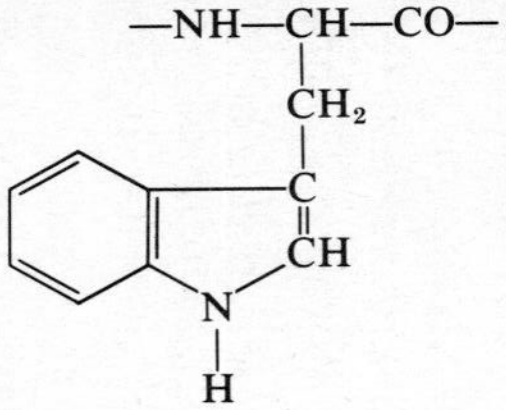 —NH—CH—CO—; CH_2; C=CH; N—H	280	348	0.13	Intrinsic protein probe
1-aniliano-8-naphthalene sulfonate (ANS)	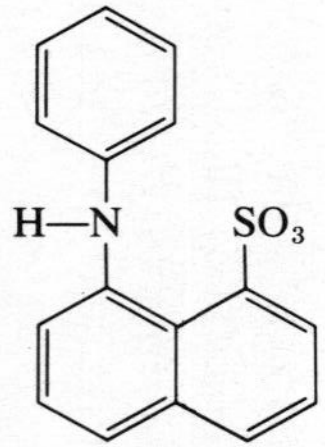 H—N; SO_3		515[a]	[a]	Extrinsic protein probe
Dansyl group	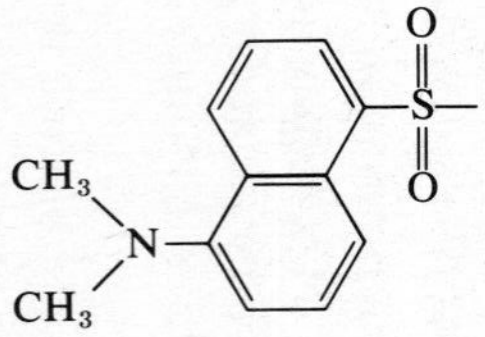 CH_3, CH_3, N; O=S=O	340	510		Linked through poly-L-proline residues in spectroscopic ruler experiment [see Fig. 13-15(*a*)].
Naphthyl group	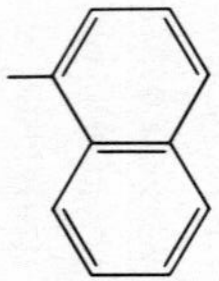	290	350		

[a] Depends strongly on interactions of group: λ_{flr}^{max} shifts from 515 nm toward 454 nm as polarity of solvent decreases. At the same time, the quantum yield increases.

is *resonant energy transfer* between them. Resonant energy transfer is a radiationless form of energy exchange that can take place when the fluorescence spectrum of the donor group overlaps the absorption spectrum of the acceptor. For example, the naphthyl group in Table 13-3 has a fluorescence centered near 350 nm, and the dansyl group has an absorption spectrum centered near 340 nm. Therefore, if light of frequency 290 nm is absorbed by the naphthyl group, the energy can pass by resonant transfer to the dansyl group, and then will be emitted by the dansyl group at its fluorescence wavelength of 510 nm.

What value has all this exchange of light energy? The answer is that intensity of fluorescence of the acceptor (dansyl) group depends on its distance from the donor (naphthyl) group. The absorption of energy by the acceptor is due to its own extinction, ϵ_A, plus the energy transferred from the donor, $E_t\,\epsilon_D$ (E_t is the *efficiency of transfer* from donor to acceptor):

$$E_t = \frac{\text{Quanta transferred to acceptor}}{\text{Quanta absorbed by donor}} \tag{13-11}$$

Therefore ϵ in Equation 13-10 can be replaced by $\epsilon_A + E_t\epsilon_D$, yielding

$$I_{\text{flr}} = 2.303QI_0lC(\epsilon_A + E_t\epsilon_D) \tag{13-12}$$

The important result that Förster calculated is that the efficiency of transfer varies with the inverse sixth power of the distance (just as the dispersion interaction does):

$$E_t = \frac{r^{-6}}{r^{-6} + R_0^{-6}} \tag{13-13}$$

in which r is the separation of the donor and acceptor and R_0 is a constant that combines spectroscopic and geometric quantities. R_0 has the dimensions of length and is the distance of separation between donor and acceptor at which the transfer efficiency is one-half. Typical values of R_0 vary from a few Å to 50 Å. Förster showed that

$$R_0 = 9.79 \times 10^3(Jn^{-4}\kappa^2Q_D)^{1/6} \tag{13-14}$$

in which Q_D is the quantum yield of the donor, n is the refractive index of the medium, κ^2 is a measure of the relative orientation of donor and acceptor groups, and J is the spectral overlap integral between the fluorescence intensity of the donor, F_D, and the extinction coefficient of the acceptor, ϵ_A:

$$J = \frac{\int_0^\infty F_D(\lambda)\epsilon_A(\lambda)\lambda^4\,d\lambda}{\int_0^\infty F_D(\lambda)\,d\lambda} \tag{13-15}$$

The factor κ^2 is one of the main unknowns in the technique of fluorescence energy transfer. It can vary between 0 and 4 depending on mutual orientation. Maximum energy transfer occurs when the electric dipole transition moments for emission by the donor and absorption by the acceptor are parallel. No energy transfer occurs when these two transition moments are

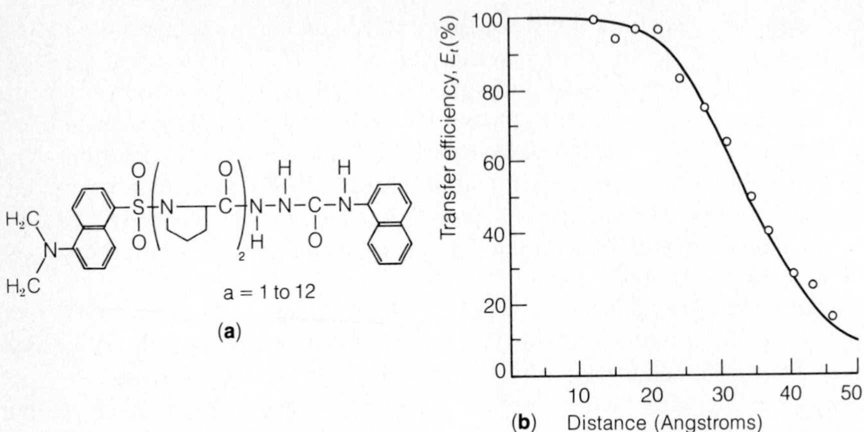

FIGURE 13-15
An example of resonant energy transfer and fluorescence used as a molecular ruler. **(a)** A model system for the study of the dependence of energy transfer on the separation of donor (naphthyl) and acceptor (dansyl) groups. The groups are separated by oligomers of poly-L-proline, having from 1 to 12 prolyl residues. **(b)** The efficiency, E_t, of energy transfer between naphthyl and dansyl groups as a function of their separation (circles). The solid line corresponds to the r^{-6} dependence predicted by Equation 13-13. (From Stryer and Haugland, 1967.)

perpendicular to each other. However, for most orientations κ^2 is reasonably close to the value $\frac{2}{3}$ found if both the donor and acceptor can reorient freely.

The theory of resonant transfer was tested for biochemical systems by Stryer, who showed that the naphthyl-dansyl combination works well for mapping distances in the range of 14–45 Å (see Fig. 13-15). Today this fluorescent ruler is being used to map out relative distances between sites in enzymes, ribosomes, and other sub-cellular assemblies.

EXERCISE 13-11

Sketch the relative positions of electronic energy levels of a pair of functional groups (such as naphthyl-dansyl) to be used for fluorescent energy transfer.

ANSWER

The initial excitation of the donor must require more energy than is released during final fluorescence of the acceptor. The energy difference between the excited state of the donor and the ground state of the acceptor ($\Delta 1$) must equal the difference between the ground and excited states of the acceptor ($\Delta 2$) at the instant that resonant energy transfer occurs (see Fig. 13-16).

EXERCISE 13-12

Estimate R_0 for the naphthyl-dansyl system using the data in Figure 13-15.

ANSWER

About 34 Å.

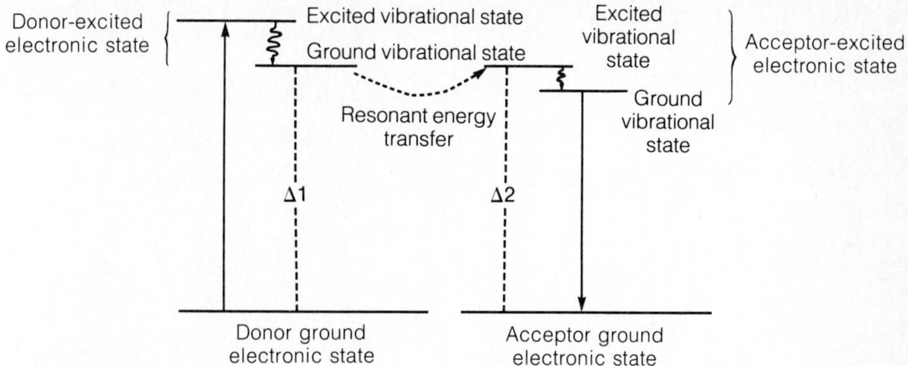

FIGURE 13-16
Energy levels of a pair of functional groups that participate in fluorescent energy transfer.

13–5 *Optical Rotatory Dispersion and Circular Dichroism Reveal Asymmetric Arrangements of Chemical Groups*

Optical rotatory dispersion (ORD) and circular dichroism (CD) are spectroscopic techniques that are based on the dispersion and the absorption of polarized light by matter. Of all spectroscopic techniques, these are most sensitive to the conformations of macromolecules. Therefore ORD and CD are useful for monitoring changes in conformations of proteins, nucleic acids, and polysaccharides. Unfortunately, the theory of these techniques is complicated, so that interpretation of the observed spectra is more difficult than the interpretation of ordinary absorption spectra. In fact, applications of ORD and CD to biochemical systems have been more completely empirical than applications of other spectroscopic techniques. For example, spectra characteristic of the α-helical and β-sheet conformations of proteins can be identified by studying proteins having these structures. Then the spectra of other proteins can be interpreted in terms of those model spectra.

However, it would be wrong to underestimate the power of the optical rotation techniques. It was optical rotation which led Pasteur to postulate that fermentation is a process that requires biological organisms (see historical sketch).

LOUIS PASTEUR

1822–1895

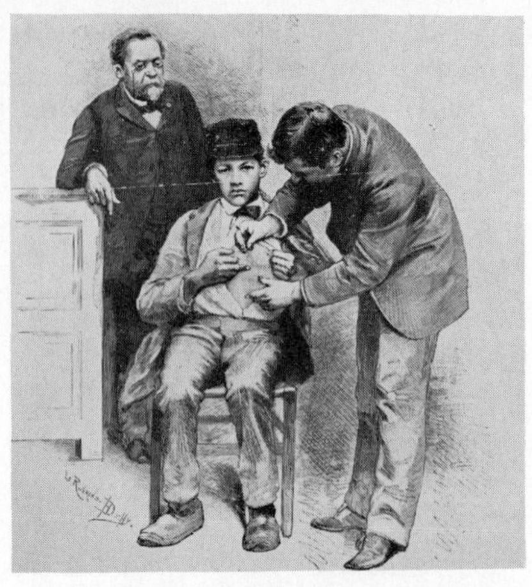

Pasteur (left) supervises the inoculation of Jean Jupille, a 15-year-old shepherd, the second person he saved from rabies. Having seen a dog about to attack a group of children, Jupille diverted and killed the animal but was severely bitten in the struggle. Subsequently the dog was declared rabid, and Jupille was brought to Paris for Pasteur's treatment. (Courtesy of the Pasteur Institute.)

Pasteur wrote that he had been "enchained" by the "almost inflexible logic of my studies." By this he meant that there was a logical thread that led from his first research on crystals and optical rotation through the recognition that fermentation is a biological process; through development of the techniques of microbiology and their application to fermentation of milk, beer, and vinegar; through studies on wine and food preservation; through destruction of the theory of spontaneous generation; through the development of the germ theory of disease and its application to blight in silkworms and anthrax in animals; through the development of ideas of immunity and vaccination; and finally to his vaccine for rabies. Other scientists such as Newton, Faraday, and Einstein have reshaped science and in time transformed civilization, but no other scientist has had the immediate impact of Pasteur. And no living scientist has ever received the veneration accorded to Pasteur. An institute in Paris bearing his name was founded to further his work, and his 70th birthday was celebrated by 2500 students, scholars, and statesmen from world centers of culture. As Pasteur entered on the arm of the President of the French Republic, the audience rose in ovation, and Lord Lister, the father of antiseptic surgery, embraced Pasteur with emotion.

What in the youth of Louis Pasteur foreshadowed such achievement? He was the son of a tanner from eastern France who had served as a sergeant in Napoleon's army. As a boy Louis did reasonably well in school and displayed promise as a painter, but his father steered him toward the more practical career of schoolteacher. His work was of sufficient quality that he won admission to France's most prestigious college, the École Normale Supérieure. In an early example of the perfectionism that would become a trait of his scientific work, Pasteur refused admission because he was only sixteenth in rank in the entrance exam. He returned to his studies and, the next year ranking fifth, accepted his place.

Once in the École he devoted all his energies to mastering science. Willingness to expend total effort was another of his lifelong characteristics. He had the patience and drive to concentrate intensely on one problem, even over periods of years, until he had identified the important variables. In scientific matters he was stubborn, self-confident in the extreme, and combative in controversy. He spent much effort to establish his priority in discovery, and gave credit to others only grudgingly. He was so devastating in debate with one scientific opponent, an 80-year-old surgeon, that the old man challenged him to a duel. Pasteur was devoted to the experimental method and disdainful of speculation and philosophizing in science (though he did his own share), and there can be no doubt that in attention to experimental detail Pasteur has had few equals in the history of science.

Pasteur earned his doctorate for dissertations in both physics and chemistry, focusing on the relationship of crystalline form and optical activity in solution. While waiting for an offer of an academic post, he began a systematic study of the optical activity of tartrate. Tartaric acid was at that time purified from the tartar deposited as a by-product in wine vats. The studies of Biot and Mitscherlich had established that aqueous solutions of tartrate rotate the plane of polarized light to the right. In contrast, racemic acid, which had been claimed by Mitscherlich to be identical to tartrate in atomic arrangement and crystalline form, is optically inactive.

Based on earlier work on quartz crystals, Pasteur was convinced that optical activity is related to asymmetric crystal faces ("hemihedrism"). He soon found that nineteen tartrate compounds which were optically active all had hemihedral facets. In the optically inactive racemic acid, Pasteur expected to find symmetric crystals, and he was disappointed at first to observe hemihedrism. However, some crystals had facets inclined to the right and others to the left. This is in contrast to the tartrates where they are always inclined the same way. Separating the two classes of crystals by hand, and redissolving them separately, Pasteur found that solutions of the left-handed crystals rotated the plane of polarized light in one direction and solutions of the right-handed crystals rotated the plane in the opposite direction, and to about the same extent. Equal weights of the two types of crystals, when dissolved together, produced an optically inactive solution.

Upon observing this, Pasteur's excitement was so great that he rushed into the hallway where he encountered one of the chemistry assistants. Pasteur embraced him saying, "I have just made a great discovery . . . I am so happy that I am shaking all over and am unable to set my eyes again to the polarimeter."

Biot summoned Pasteur to his laboratory and requested that the young scientist repeat the work before his eyes. Pasteur prepared the crystals, and Biot asked him to place on his right hand the crystals that would cause rotation to the right and on his left hand the others. Biot made the solutions, and examined them in his polarimeter. Visibly moved, he seized Pasteur by the hand and said, "My dear son, I have loved science so deeply that this stirs my heart." Pasteur having established himself as a masterful experimenter was appointed professor of chemistry at the University of Strasbourg at age 25.

Pasteur was soon diverted into biological studies, but he did speculate on the spatial arrangements of atoms that might produce molecular asymmetry. In 1860 he suggested that the atoms of a right-handed compound might be "arranged in the form of a right-handed spiral, or . . . situated at the corners of an irregular tetrahedron." With such visionary insight, Pasteur might well have achieved lasting fame in chemistry if his chain of logic had not led him elsewhere. It was some fourteen years after Pasteur's speculation that Le Bel and van't Hoff postulated the tetrahedral carbon atom, and some 90 years later that Pauling, Watson, and

Crick began to consider helical biomolecules.

At Strasbourg Pasteur showed that he could act in his personal life with the same boldness that he did in his scientific life. He met Marie Laurent, daughter of the rector of Strasbourg Academy, and proposed to her within a few weeks. She generally tolerated Pasteur's intense absorption in his work, even serving frequently as his secretary. On one occasion when she expressed feelings of neglect, Pasteur consoled her by saying that he would "lead her to posterity." He was not yet 30.

One of Pasteur's greatest discoveries was that fermentation is a biological process, mediated by microorganisms. One source of his interest in fermentation was his finding that when a solution of racemic acid was infected with a mold, it gradually became more optically active. From this and other experiments, Pasteur became convinced that only living organisms can produce optical activity. Another source of his interest was the request for help by an industrialist. This gentleman produced alcohol by the fermentation of beet juice, and had found that the alcohol was often contaminated with undesirable substances formed during fermentation. Pasteur visited the factory and took samples of the fermenting juice. Under his microscope the juice proved to contain small globules of yeast, and also other structures that were unlike yeast. Examining the juice in his polarimeter, he found it to be optically active. This indicated to Pasteur that the process of fermentation was caused by living "ferments" as he called them (microorganisms as we now call them). This was contrary to the prevailing chemical view, that fermentation was a strictly chemical process. Pasteur believed that the chemicals present acted simply as foods for the ferments, which then produced various end products. Fermentation can be contaminated, he reasoned, if other microorganisms are present which produce undesirable end products. Chemical features of the medium can promote or hinder the growth of any one microorganism.

Once interested in fermentation, Pasteur turned to a simpler system for full exploration, the fermentation of sugar into lactic acid. With this system he developed the basic techniques of bacteriology that were the basis for many of his future discoveries. He then proceeded to study the formation of vinegar and wine. He found that most diseases of wine can be prevented by heating it in closed vessels for an hour or two to 50° or 60°C, thereby killing the contaminating germs. This was the beginning of *Pasteurization,* one of the great advances in preventative medicine.

Next Pasteur turned to diseases of silkworms. A blight of silkworms called *pébrine* had resulted in a revenue loss of over 120,000,000 francs during the 15-year period prior to 1865, and French silk producers were frantic. The minister of agriculture asked Pasteur to study the disease. His work on the problem proceeded with agonizing slowness for over five years, and was accompanied by personal tragedies, including the death of his two-year-old daughter and a partially paralytic stroke in himself. Nevertheless, he discovered in time that *pébrine* was caused by a parasite protozoon. He then worked out effective techniques for breeding uncontaminated silkworms, and organized an educational campaign to transmit his procedures to the silk producers.

From agricultural and industrial processes, Pasteur became increasingly involved in establishing the germ theory of diseases and in finding ways to prevent and to vaccinate against infection. His most widely known work was the development of a treatment for the mysterious and dread disease of rabies. There have been few episodes of science more dramatic than that in which Pasteur saved the first life from rabies. And no example of the admiration Pasteur inspired is more dramatic than the way in which the same life was later sacrificed to his honor.

On July 6, 1885 a nine-year-old boy, Joseph Meister, was brought to Pasteur suffering from fourteen rabid dog bites on his hands, legs, and thighs. Pasteur had been experimenting on animals with a rabies vaccine, but was not yet ready to apply it to humans. When physicians assured him that the boy would die, Pasteur decided to try. He started by injecting virus from the spinal cord of an infected rabbit which had been attenuated by fourteen days' drying. In twelve successive injections, Meister received an increasingly active virus until on July 16 he was injected with a virulent spinal cord that had just been removed from the body of a rabbit that had died from an injection of virus.

Joseph Meister did not contract rabies. He regained his good health and returned to his family. So successful was this first Pasteur treatment that within the next fifteen months, 2490 persons received the vaccine. Meister grew up to become gatekeeper of the Pasteur Institute. Then in 1940, 55 years after he made medical history, he killed himself rather than open Pasteur's burial crypt to German soldiers.*

One Property of a Light Wave is its Plane of Polarization

In Chapter 12 and in this chapter we have considered several properties of a light wave—its wavelength or frequency, its amplitude, and its direction. An additional property is the direction in which its electric field points. The directions of its electric (E^*) and magnetic (B) fields are perpendicular to each other, and also are perpendicular to the direction of propagation of the wave [Fig. 13-17(*a*)]. In an ordinary unpolarized beam of light, the electric-field directions of the various waves point in various directions. In contrast, in a *plane-polarized* beam all electric waves point in the same direction. The plane defined by this direction and the direction of propagation is called the *plane of polarization* of the beam. Thus, looking into a beam of polarized light, we would see the electric fields of all waves lying on a line, whereas looking into a beam of unpolarized light, we would see electric fields oscillating in all directions:

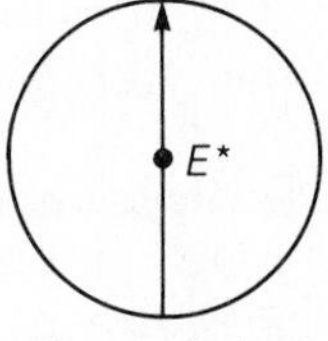

Plane polarized

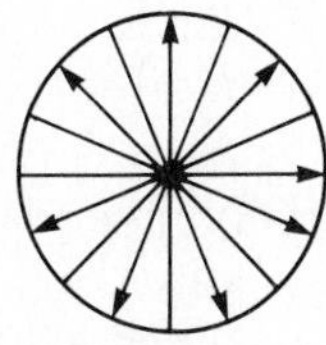
Unpolarized

* According to a private communication to Professor René Dubos from Pasteur's grandson, Professor Vallery-Radot.

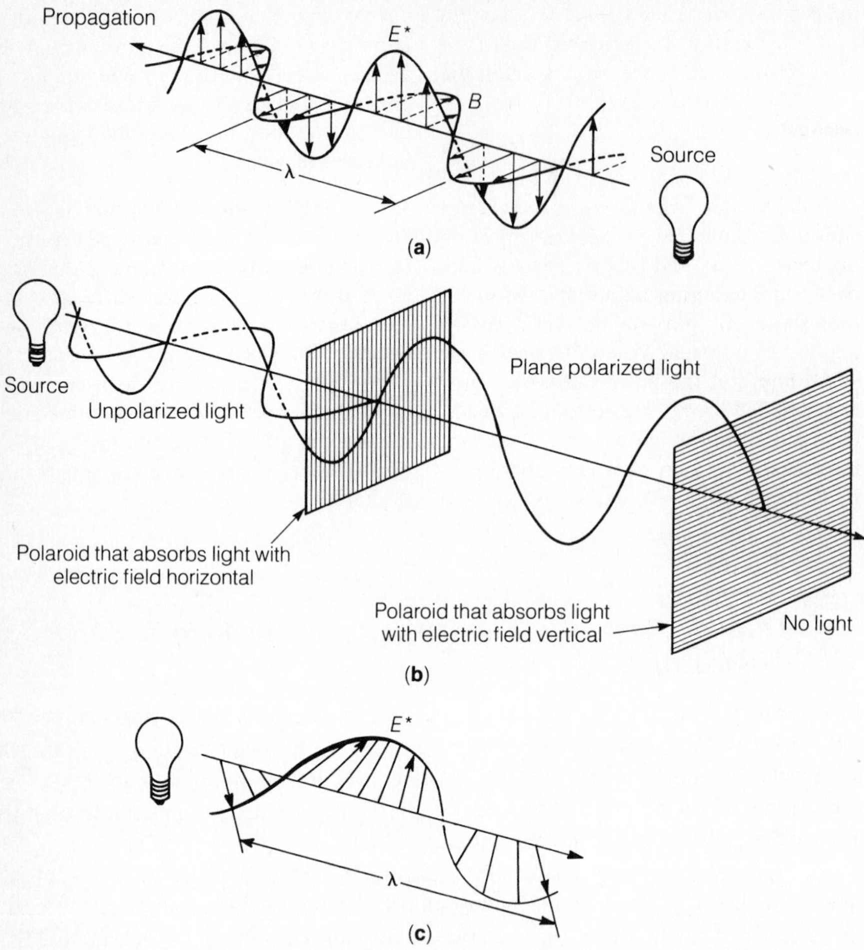

FIGURE 13-17
Polarized light. **(a)** A plane-polarized beam. **(b)** Polarizers rotated at 90° to each other, thereby removing all light from the beam. **(c)** A right-circularly polarized beam. [**(a)** and **(c)** adapted from K. E. Van Holde, 1971, *Physical Biochemistry* (Englewood Cliffs, N.J.: Prentice-Hall, Inc.). Reprinted by permission of Prentice-Hall, Inc.]

Plane-polarized light can be produced by passing unpolarized light through a highly anisotropic (uni-directional) material, such as Polaroid, which absorbs all light with the electric field pointing in one direction (i.e., horizontal), thereby leaving only rays whose electric fields point vertically. Another method is to pass the unpolarized light through an optical device called a *Nicol prism,* which removes light polarized in one direction by reflection. Reflection from almost any surface produces partially polarized light.

EXERCISE 13-13

View the light reflected from an automobile windshield or other surface through a lens from a pair of Polaroid sunglasses. As you rotate the lens about the direction of view, you will observe that the intensity of transmitted light increases and decreases. Explain this observation.

ANSWER

The reflected light is partially polarized, so that its electric field points mainly in one direction (call it vertical). When rotated into one position, the Polaroid lens absorbs light with its electric field pointing vertically; thus it transmits little intensity. When the lens is rotated 90°, it absorbs only the small amount of reflected light with the electric field pointing horizontally. Then most of the reflected light is transmitted [see Fig. 13-17(*b*)].

Light also can be *circularly polarized.* A circularly polarized wave has its electric-field direction rotating about the direction of propagation, such that the electric-field direction makes one complete cycle in one wavelength of the light [Fig. 13-17(*c*)]. If the rotation is clockwise as one views the oncoming beam, the light is said to be *right-circularly polarized,* and if the rotation is counterclockwise, it is said to be *left-circularly polarized.* Circularly polarized light can be produced by passing plane-polarized light through an optical crystal called a *quarter-wave plate,* or through an electro-optical device called a *Pockels cell.*

Optical Rotation

When plane-polarized light passes through any optically active substance, the plane of polarization is rotated. This phenomenon of *optical rotation* can be measured with a polarimeter (Fig. 13-18). This instrument consists of a source, monochromator, and polarizer, which together produce a

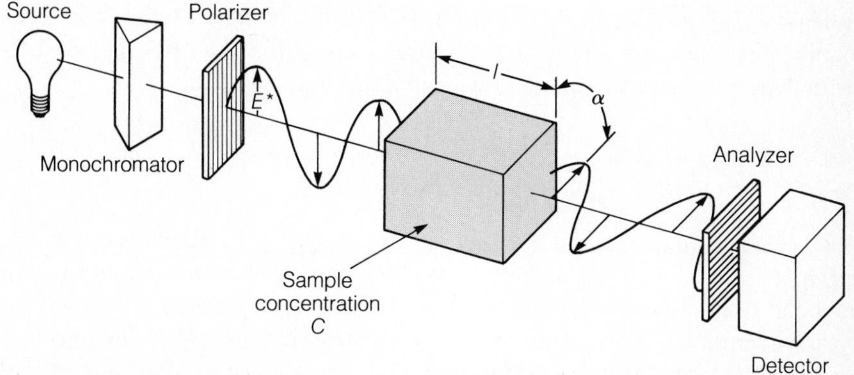

FIGURE 13-18

Schematic representation of a polarimeter. The angle α (here 90°) is the rotation of the plane of polarization.

monochromatic plane-polarized light beam that is directed onto the sample. A second polarizer (called the analyzer) is positioned beyond the sample. Beyond the analyzer is a detector that measures the intensity of transmitted light. In the absence of a sample, and with the polarizer and analyzer rotated about the direction of the light 90° with respect to each other, the transmitted intensity is a minimum. This is because the polarizer removes all light with components of its electric field in one direction (say vertical) and the analyzer removes all light with components of its field in the other (horizontal) direction. If an optically active sample is now inserted, as the light interacts with the molecules, the planes of polarization of individual waves are changed. The net effect is that the plane of polarization of the entire beam is rotated. The angle of rotation, α, is defined as the angle through which the analyzer must be rotated to produce once again a minimum in transmitted intensity.

The intrinsic ability of a substance in solution to rotate the plane of polarization is given by its *specific rotation* in degrees, which is defined by

$$\left[\alpha\right]_{\lambda}^{t} = \frac{100\alpha}{lc} \tag{13-16}$$

in which l is the length of the cell in decimeters (1 decimeter = 10 cm), c is the concentration in grams of solute per 100 ml of solution, the subscript is the wavelength of light, and the superscript is the temperature of the sample in °C. For α-D-glucose in water, $[\alpha]_D^{20}$ is +112.2. The positive sign indicates that the plane is rotated to the right as one looks into the beam, and D indicates that the sodium D line is the light source. For L-serine, $[\alpha]_D^{20}$ is −6.8. For macromolecules, the optical rotation often is given by the reduced residue rotation, defined as

$$[m'] = \frac{3}{(n^2 + 2)} \frac{m}{100} [\alpha]_{\lambda}^{t} \tag{13-17}$$

in which n is the refractive index of the solution and m is the mean residue molecular weight. For amino acids m is about 115. The factor $3/(n^2 + 2)$ roughly describes the effect of the solvent on rotation, and is included to correct the rotations to values they would have in a vacuum.

Optical Rotatory Dispersion

Optical rotatory dispersion is the dependence of optical rotation on the wavelength of incident light. A machine that varies the incident wavelength with a monochromator, and determines the value of α at each wavelength is called a *spectropolarimeter*. The results often are expressed as a plot of $[m']$ against wavelength, as in Figures 13-19(a) and 13-20(a). Figure 13-19(a) shows that the ORD of a single asymmetric chromophore has the same dip-and-rise shape as the dispersion of the ordinary refractive index (Figs. 12-6 and 12-9).

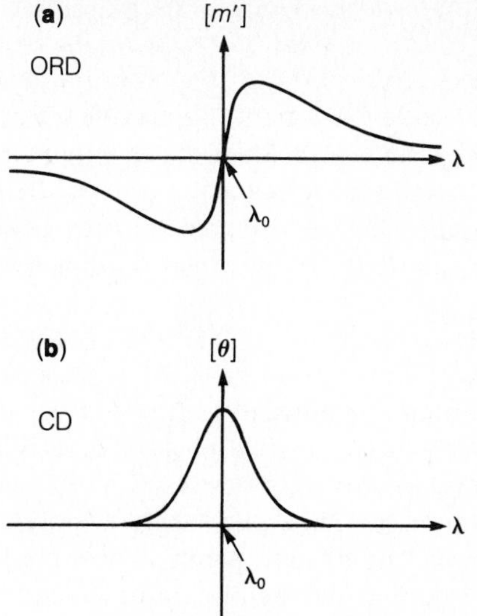

FIGURE 13-19
The ORD (a) and CD (b) spectra of a single optically active chromophore. For the enantiomorph (mirror image) of the chromophore, the ORD spectrum would be the mirror image of the one shown, and the CD spectrum would be a trough rather than a peak.

A single dip-and-rise in the ORD is often called a *Cotton effect*. The ORD of any biological substance has a more complicated shape because the substance contains several chromophores, each of which contributes a dip-and-rise such as that shown in Figure 13-19(*a*). Figure 13-20(*a*) shows the ORD of poly-L-lysine, when the polypeptide is in three different conformations. It is clear that ORD is sensitive to the arrangement in space of the polypeptide backbone.

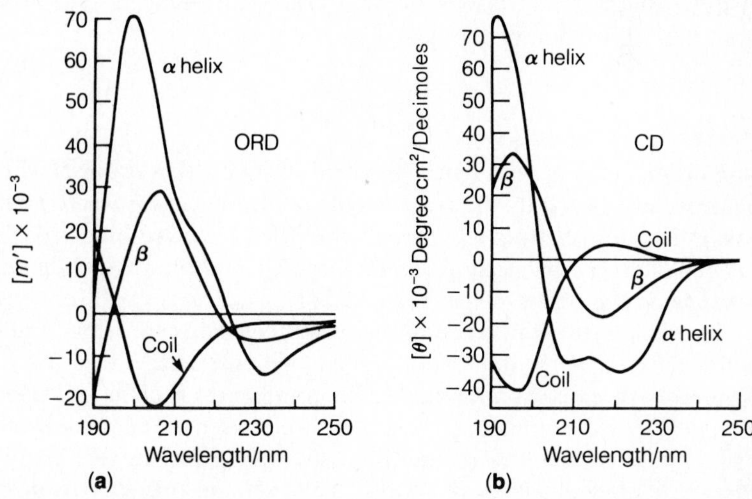

FIGURE 13-20
The ORD spectrum (a), and the CD spectrum (b), of poly-L-lysine in various conformations. (From Adler, Greenfield, and Fasman, 1973.)

Circular Dichroism

Circular dichroism (CD) is the differential absorption of a sample for left-circularly polarized and right-circularly polarized light. Optically active samples absorb the two types of polarized beams to different extents. The extinction coefficients of a sample for left- and right-circularly polarized light are represented ϵ_L and ϵ_R, respectively, and the extent of circular dichroism usually is expressed by a quantity called the *ellipticity,* θ, which is defined as

$$\theta = \frac{(2303)9}{2\pi}(\epsilon_L - \epsilon_R) \cong 3.3 \times 10^3(\epsilon_L - \epsilon_R) \tag{13-18}$$

When the ellipticity of a single chromophore is plotted against the wavelength of the incident light, as in Figure 13-19(*b*), either a peak or trough is observed, depending on whether the left- or the right-circularly polarized light is absorbed more strongly. For a complex substance such as poly-L-lysine [Figure 13-20(*b*)], a superposition of troughs and peaks is observed.

Both ORD and CD are sensitive to the conformations of macromolecules, as Figure 13-20 illustrates and both, in principle, convey the same information. Often the bands in a CD spectrum are more fully resolved, and thus a CD spectrum sometimes is easier to interpret than an ORD spectrum. ORD does have one advantage over CD, however. From Figure 13-19 you can see that the CD falls to near zero as the frequency of the incident radiation moves away from the frequency of absorption of the chromophore. (This, of course, is just the natural resonance absorption discussed in Section 12-5.) In contrast, the ORD approaches zero more slowly as the frequency is changed. This means that the ORD can be detected farther away from the absorption wavelength than can CD. Since many chromophores of biological interest lie in the far UV, this means that ORD can be useful in the more accessible spectral region of the near UV.

● *Relationship of the ORD and CD*

Circular dichroism bears the same relationship to ORD that ordinary absorption spectroscopy bears to the wavelength dependence (*dispersion*) of the refractive index. Recall from Chapter 12 that the refractive index of a substance depends on the in-phase polarizability, α_0, and the extinction coefficient depends on the out-of-phase polarizability, α_{90}. As we will see, optical rotation depends on the difference in the refractive index for right- and left-circularly polarized light, and therefore on the difference in α_0 for right- and left-circularly polarized light, while circular dichroism depends on the difference in extinction coefficients between right- and left-circularly polarized light, and therefore on the difference in α_{90} for the two beams.

To see the relationship between optical rotation and refractive index, we begin by recognizing that a plane polarized wave can be thought of as a combination of right- and left-circularly polarized waves. Figure 13-21(*a*) shows the electric vectors that characterize these waves. Rotation of the two equal electric vectors at the same rate leads to oscillation of the sum of their

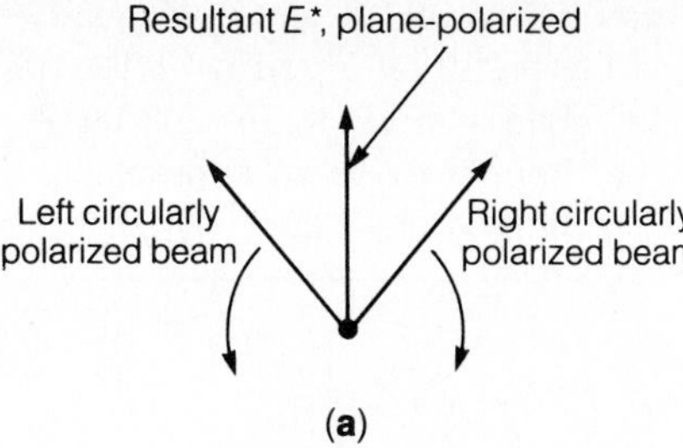

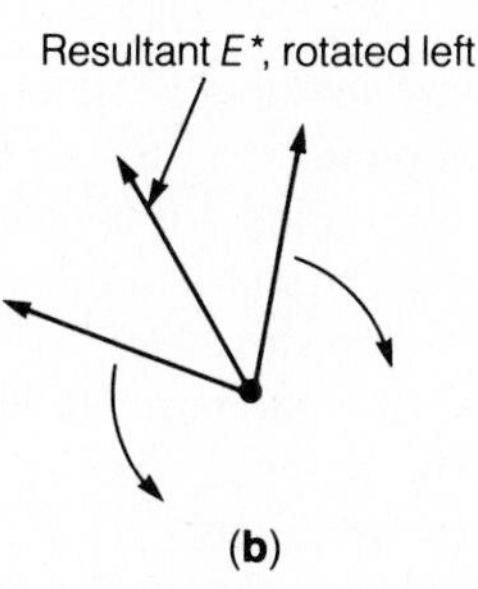

FIGURE 13-21
Electric vectors, as observed looking into a plane-polarized beam. **(a)** A plane-polarized beam is composed of right and left circularly polarized beams. The resultant electric field E^* oscillates in a plane. **(b)** When the right circularly polarized beam is retarded relative to the left circularly polarized beam ($n_R > n_L$), the leftward rotating beam will have rotated farther than the rightward rotating beam. As shown in the figure, the resultant electric vector is rotated left. Similarly, when $n_L > n_R$, the plane-polarized beam is rotated to the right.

electric fields in a plane. In Figure 13-21(*b*) we see the consequence of a difference in velocity for the two beams: if the right-circularly polarized beam moves more slowly, because its refractive index n_R is greater than n_L, its circular rotation falls behind that of the leftward rotating beam. Therefore, the resultant of the two beams is rotated to the left.

In summary, optical rotation requires that α_0 be different for left- and right-circularly polarized light, and circular dichroism requires that α_{90} be different for left- and right-circularly polarized light. In Section 12-5 we saw that α_0 and α_{90} are related to each other. Therefore, it is not surprising that circular dichroism and optical rotation are related; the equation expressing this is the *Kronig-Kramers transformation:*

$$[m']_\lambda = (2.303)\,\frac{9000}{\pi^2}\int_0^\infty (\epsilon_L - \epsilon_R)_{\lambda'}\,\frac{\lambda'}{(\lambda^2 - \lambda'^2)}\,d\lambda' \qquad (13\text{-}19)$$

This equation states that if we know the circular dichroism $\epsilon_L - \epsilon_R$ at every wavelength λ', we can calculate the optical rotation at wavelength λ by integrating the expression on the right side of the equation. This is an example of what is called an *integral transformation,* in which one variable is expressed as an integral over the other. The circular dichroism can also be expressed as an integral transformation of the optical rotation, by an equation similar to Equation 13-19.

EXERCISE 13-14

Convince yourself by drawing diagrams like Figure 13-21 that if ϵ_L is greater than ϵ_R, but the refractive indexes n_L and n_R are equal, then passing a plane polarized beam through the solution will produce an elliptically polarized beam, with the major axis of the ellipse parallel to the original polarization direction. (In an *elliptically polarized* beam the resultant electric vector follows the circumference of an ellipse, rather than a circle as in circularly polarized light. Plane polarized light can be thought of as an eliptically polarized beam in which the minor ellipse axis is zero.)

The ORD and CD of a Substance Depend on Both Electric and Magnetic Transitions

In our discussion in Section 12-6 of the quantum-mechanical basis of absorption, we saw that the strength of absorption depends on the size of the electric transition moment, μ_{kn}, induced in the absorbing substance by the oscillating electric field. Quantum-mechanical analysis of optical rotation shows that the strength of rotation depends not only on the electric transition moment, but also on the transition induced by the magnetic field of the electromagnetic radiation. This is described by the magnetic transition moment, m_{kn}. Each absorption band of the sample contributes to the optical rotation of the sample by an amount R, the *optical rotatory power,* given by

$$R = Im\ (\boldsymbol{\mu}_{kn} \cdot \mathbf{m}_{kn}) \tag{13-20}$$

in which the symbol Im means the imaginary component of the quantity in parentheses, which itself is a scalar product of two vectors. To those not familiar with complex numbers and vectors, this quantity will have little meaning. What is important about it for our purposes is that R equals zero for any group of atoms that is symmetric (that is, has a center of inversion; see Chap. 16), and that the contributions to R from any asymmetric group are equal in magnitude and opposite in sign to the contribution of the enantiomorph (mirror image) of the asymmetric group. Thus symmetric substances, and mixtures of equal amounts of D and L groups, have no optical rotation.

The reduced mean rotation $[m']$ of a substance containing i chromophores can be expressed as a sum of the contributions of each of its i absorption bands,

$$[m'] = \frac{96N\pi}{hc}\ \frac{(n^2 + 2)}{3} \sum_i \frac{\lambda_{0i}^2 R_i}{\lambda^2 - \lambda_{0i}^2} \tag{13-21}$$

in which h is Planck's constant, c is the speed of light, λ_{0i} is the wavelength of absorption of the ith band, and R_i is the optical rotatory power of the ith band. This equation is similar in form to the "in-phase polarizability" of ordinary dispersion.

EXERCISE 13-15

From Equation 12-17, the in-phase polarizability can be expressed as

$$\alpha_0 \propto \frac{1}{\omega_0^2 - \omega^2}$$

Show that it also can be expressed as

$$\alpha_0 \propto \frac{\lambda_0^2 \lambda^2}{\lambda^2 - \lambda_0^2}$$

The Physical Basis for Optical Rotation and Circular Dichroism

The quantum mechanical Equations 13-20 and 13-21 tell us that optical rotation relies on both the electric and magnetic dipole transition moments, but they do little to provide us with insight into the physical origin of ORD and CD. Careful consideration of the classical view of absorption and dispersion leads to greater enlightenment. For this purpose we need some elementary results from electromagnetic theory, sketched in Figure 13-22. As shown there, a current provides a magnetic field, and a changing magnetic field induces a current.

Optical rotation requires an object which is not identical to its mirror image. For this purpose we choose a right-handed helix that conducts a current, as shown in Figure 13-23(a). Also shown are left- and right-circularly polarized beams passing perpendicular to the helix axis. Our objective is to show that *the polarizability of the helix is different for the left- and right-circularly polarized beams.*

The two beams will induce magnetic and electric fields in the helix; the components along the helix axis are shown in Figure 13-23(b). The position

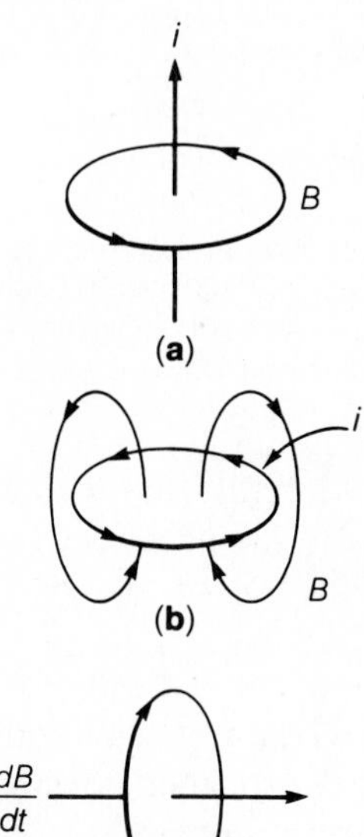

FIGURE 13-22
The magnetic field B of a current i, and the electric field E^* induced by a changing magnetic field. **(a)** The magnetic field of a linear conductor. **(b)** The magnetic field of a circular conductor. **(c)** The electric field E^* induced by a magnetic field increasing in the direction indicated. E^* produces a current, which induces a magnetic field that opposes the increase in B.

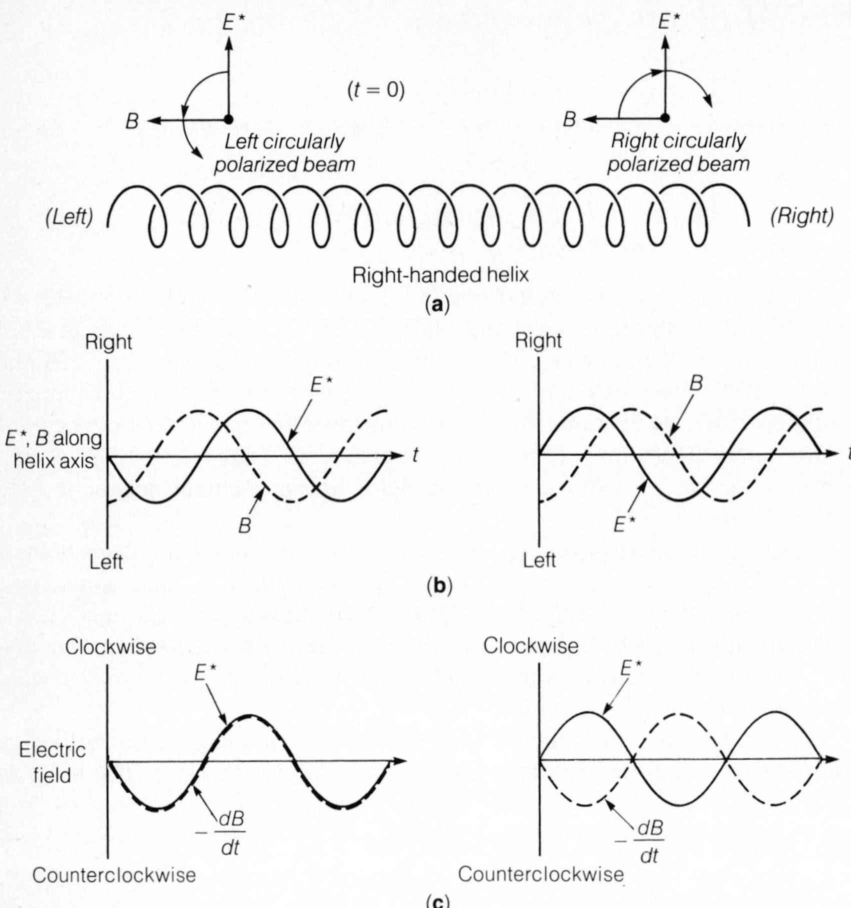

FIGURE 13-23
The origin of optical rotation and circular dichroism by a right-handed helical conductor. **(a)** A right-handed helix is perpendicular to a left and a right circularly polarized beam. **(b)** The magnetic and electric fields along the helix axis. Right is considered positive and left negative. **(c)** The electric field causing movement of charge around the helix axis, as viewed from the left end of the helix. Clockwise is considered positive, and counterclockwise negative.

of the fields at zero time is as shown in Figure 13-23(a), and they vary with time in different ways because the direction of rotation of the field vectors is different for the two beams. Notice that the values of E^* and B at $t = 0$ in Figure 13-23(b) are those given in panel (a).

An electric field along the helix axis causes current to flow. Because the conductor is a helix, current must flow in a helical path. The direction of the circular component (around the helix axis) of the current depends on the handedness of the helix. The circular flow of current implies an electric field that parallels the current flow. Figure 13-23(c) shows the direction of this field (E^*), either clockwise or counterclockwise as viewed along the axis of the helix of panel (a) from the left end. Note, for example, that for right circularly polarized light the flow is initially clockwise as time increases from 0.

The changing magnetic field also induces an electric field around the helix axis, as demanded by the principle of Figure 13-22(*c*). This figure shows us that E^* is counterclockwise when B increases to the right, as viewed along the direction of increase (along the arrow). Therefore, since counterclockwise is negative in Figure 13-23(*c*), we can set the *magnetically induced* electric field equal to $-dB/dt$. This is done in Figure 13-23(*c*).

Figure 13-23(*c*) shows that the right-handed helix responds differently to left- and right-circularly polarized light, because *in one case E^* and $-dB/dt$ reinforce each other, and in the other they oppose each other.* The two effects work together in the case of left-circularly polarized light so a given electric field intensity will produce greater polarization current in this case. Therefore, we conclude that the polarizability will be greater for the left-circularly polarized beam, and the $\epsilon_L - \epsilon_R$ will be positive for a right-handed helix in which the direction of polarization is along the helix backbone.

EXERCISE 13-16

Show that $\epsilon_L - \epsilon_R$ is negative when the helix in Figure 13-23(*a*) is left-handed.

The demonstration contained in Figure 13-23 refers only to a single orientation of the helix, so it is merely a qualitative indication of the origin of circular dichroism. Furthermore, the result depends on the orientation of the transition moment on the helix, so you should not make the mistake of thinking that all right-handed helices will have positive circular dichroism bands.

The larger polarizability of the right-handed helix in Figure 13-23(*a*) to left-circularly polarized light also explains the ORD curve. At low frequency α_0 is positive (Equation 12-17). A greater positive polarizability means a larger refractive index (Equation 12-27). Therefore, the left-circularly polarized beam moves more slowly, and the argument summarized in Figure 13-21 indicates that plane polarized radiation will be rotated to the right, just as shown in Figure 13-19. The dispersion of the refractive index implied by Equations 12-17 and 12-27 indicates that α_0 becomes negative at high frequency (above ω_0), so $n_L - n_R$ changes sign and so does the rotation.

ORD and CD can Reveal Conformations of Proteins in Complex Biological Tissues

An example of the usefulness of CD is in the study of membranes (for example, Holzwarth, 1972). Figure 13-24 shows the CD spectra of red-blood-cell membranes dissolved in phosphate buffer, and also in 2-chloroethanol. The membrane lipids apparently do not contribute to these spectra, because lipid extracted from the membranes displays no substantial

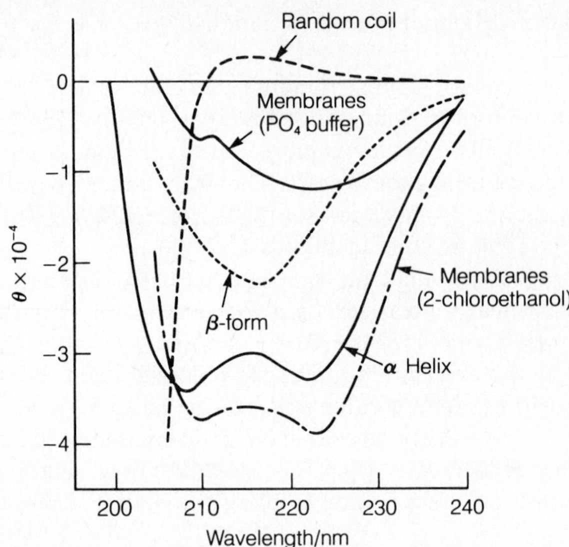

FIGURE 13-24
The CD spectra of red blood cell membranes suspended in 0.008M phosphate buffer, at pH 7.7, and dissolved in 2-chloroethanol. The spectra of poly-L-lysine in the α-helical and β-sheet forms, and of poly-L-glutamic acid in the random-coil form are plotted for comparison. (From Lenard and Singer, 1966.)

CD at wavelengths greater than 215 nm. Moreover, treatment of the membranes with the enzyme phospholipase C, which is known to release 60–70% of membrane phosphorus, has no effect on the CD spectrum. Thus the CD arises mainly from the membrane proteins, which appear to have some α-helical character. Because the CD spectrum does not change as red-blood-cell membranes are solubilized by 0.1% sodium dodecyl sulfate (a detergent), the conformation of the proteins does not change appreciably during detergent treatment. However, when membranes are transferred to 2-chloroethanol, the proteins become substantially more α-helical, thereby showing that this solvent disrupts the integrity of the membranes.

13–6 Nuclear Magnetic Resonance is a Versatile Spectroscopic Tool Characterized by Sensitivity and High Resolution

The enormous usefulness of nuclear magnetic resonance (NMR) stems from the fact that the atomic nuclei are chromophores when placed in an applied magnetic field. The proton, a nucleus present in all biological molecules, is the most strongly absorbing of all. Moreover, the local environment of a nucleus determines its frequency of resonance absorption; thus an NMR spectrum contains much information on the interactions and motions of the molecules containing the nuclei. Also, because the frequencies of resonance absorption of the nuclei can be measured to about 1 part in 10^8, different absorptions can be distinguished from one another. This feature of *high resolution* of NMR spectra is an advantage over other spectroscopic methods, such as UV and CD, where the absorptions of chromophores often overlap, and the assignment of an absorption band to a particular group is difficult.

The Energy of a Spinning Nucleus Depends on its Orientation in a Magnetic Field

Any spinning charge has a magnetic dipole moment. Many atomic nuclei behave as though they were spinning; hence they have magnetic moments. In the presence of a magnetic field, the energy of a magnetic moment depends on its orientation, just as the energy of an electric dipole moment depends on its orientation in an electric field. One example is a compass needle in the earth's magnetic field. If pointed east and then freed to rotate, the compass needle turns north, thereby lowering its energy in the external field. Another example is a spinning atomic nucleus in a sample tube between the poles of a strong electromagnet.

A central difference between the compass needle and the nuclear magnet is that the nuclear magnet, being very small, cannot point in all possible directions, thereby having all possible energies. It has quantized energy and therefore can point only in certain directions. Because the spinning nucleus has quantized energy levels, transitions between these quantum levels can be induced by electromagnetic radiation of the proper frequency. When a field that has the proper frequency to cause resonance is applied, the nuclei absorb energy and are excited to a higher energy level that corresponds to an orientation opposing the magnetic field (Fig. 13-25).

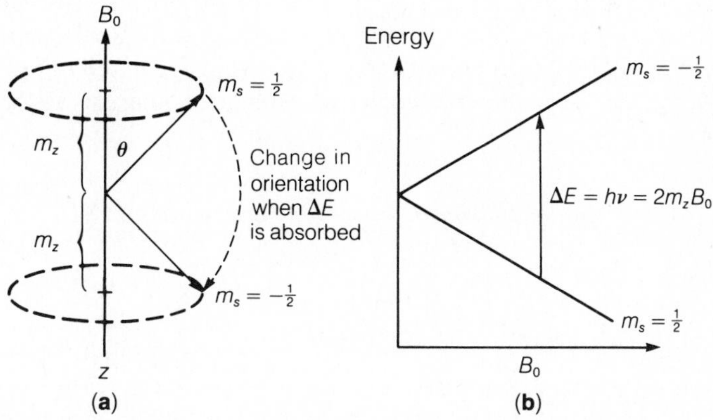

FIGURE 13-25
(a) Possible orientations of a nuclear magnetic dipole of spin $\frac{1}{2}$ in an applied magnetic field B_0. The magnetic dipole must lie at an angle θ to the field. Therefore, the tip of the dipole must be on one of the dashed circles. The magnetic field B_0 is directed along the z-axis. The movement of the magnetic dipole around the dashed circle in the x-y plane is called *precession.* (b) The separation of the energy levels corresponding to the two orientations of (a). The separation is proportional to both m_z and B_0.

Resonance Absorption by Nuclei

It is straightforward to describe the frequency of electromagnetic radiation that causes resonance absorption in terms of the size of the nuclear magnetic moment, **m,** and strength of the applied field, B, but first we must define a number of terms relating to magnetism. For a compass needle in an applied field, the energy E is given by

$$E = -\mathbf{m} \cdot \mathbf{B} = -|\mathbf{m}|B \cos \theta = -m_z B \qquad (13\text{-}22)$$ ◀

in which θ is the angle between the direction of the applied field and the direction of the magnetic moment. The quantity m_z is the component of the moment in the direction of the applied field. For the compass needle m_z can have any value between $|\mathbf{m}|$ and $-|\mathbf{m}|$. For a spinning nucleus, m_z is restricted by quantized energies to certain values [Fig. 13-25(a)]. The values are given in SI units by the relationship

$$m_z = g_n \left(\frac{eh}{4\pi m_p} \right) m_s \qquad \text{(SI)} \qquad (13\text{-}23a)$$ ◀

and in cgs units (which differ in the dimensions of charge) by

$$m_z = g_n \left(\frac{eh}{4\pi m_p c} \right) m_s \qquad \text{(cgs)} \qquad (13\text{-}23b)$$

Here g_n is an empirical constant called the *nuclear g-factor,* which gives the magnitude of the magnetic moment, e and m_p are the charge and mass of the proton, c is the speed of light, and m_s is a quantum number corresponding to the possible orientations of the nuclear spin angular momentum. The value of m_s can be only $I, I - 1, I - 2, \ldots, -I$, where I is called the *spin* of the nucleus. (Quantization of nuclear spin is exactly analogous to the quantization of angular momentum and electron spin discussed in Sec. 10-7.) Values for g_n and I for several nuclei of biological interest are listed in Table 13-4.

TABLE 13-4 *Nuclei of Importance for Biological NMR*

ISOTOPE	SPIN (I)	g_n	NATURAL ABUNDANCE (%)	RELATIVE SENSITIVITY	NMR FREQUENCY (MHz) IN A 1 T FIELD
^{1}H	½	5.585	99.98	1.000	42.57
^{2}H(D)	1	0.857	0.0156	0.0096	6.54
^{13}C	½	1.405	1.108	0.0159	10.70
^{14}N	1	0.403	99.64	0.0010	3.08
^{19}F	½	5.257	100.00	0.834	40.05
^{31}P	½	2.263	100.00	0.0664	17.24

EXERCISE 13-17

Using constants from Table 13-4, calculate the maximum and minimum values of m_z for a proton.

ANSWER

Because the spin, I, for the proton is ½, the quantum number m_s can be only ±½. Thus the maximum and minimum values of m_z are

$$m_{z=} \pm g_n \frac{eh}{4\pi m_p}\left(\frac{1}{2}\right) \tag{13-24a}$$

$$= \frac{(\pm 5.585)(1.602 \times 10^{-19}\ \text{C})(6.626 \times 10^{-34}\ \text{J s})(0.5)}{4\pi(1.6726 \times 10^{-27}\ \text{kg})}$$

$$= \pm 1.410 \times 10^{-26}\ \text{CJ s kg}^{-1} \tag{13-24b}$$

$$= \pm 1.410 \times 10^{-26}\ \text{JT}^{-1} \tag{13-24c}$$

$$= \pm 1.410 \times 10^{-30}\ \text{JG}^{-1} \tag{13-24d}$$

In Equation 13-24c and Table 13-4, the SI symbol T (for *tesla*), the unit of magnetic field, has been introduced; 1 T = 1 kg s^{-2} A^{-1} = 1 kg C^{-1} s^{-1}. In Equation 13-24d, the symbol G (for gauss), the older unit of magnetic field, is used. The relationship between these units is 1 T = 10^4 G.

EXERCISE 13-18

How many allowed orientations in a magnetic field (and therefore how many energy levels) are there for the ^{19}F and ^2H (D) nuclei?

ANSWER

The spin, I, of ^{19}F, like that of the proton, is $\frac{1}{2}$. Therefore $m_s = \pm\frac{1}{2}$, and there are two orientations and two corresponding energy levels. The spin of ^2H is 1, so m_s equals 1, 0, or −1, and there are three orientations and three energy levels. The levels are equally spaced, and the selection rule for NMR absorption is $\Delta m_s = \pm 1$; therefore, only one absorption band occurs. Generally, for a spin of I, there are $2I + 1$ energy levels.

Now we are prepared to calculate the frequency, ν, at which resonance absorption will occur. For the proton there are only two energy levels, one for each of its orientations. Transitions can be only between these levels. From Equations 13-22 and 13-23, the energy separation of these levels is

$$\Delta E = 2m_z B \tag{13-25}$$

and the frequency of radiation-producing absorption is

$$\nu = \frac{\Delta E}{h} = \frac{2m_z B}{h} \tag{13-26a}$$ ◀

For an applied magnetic field of the strength (say, 1 T = 10^4 G) that can be generated by a strong electromagnet,

$$\nu = \frac{(2)(1.410 \times 10^{-26}\ \mathrm{JT^{-1}})(1\ \mathrm{T})}{6.626 \times 10^{-34}\ \mathrm{J\ s}} = 4.256 \times 10^{7}\ \mathrm{s^{-1}}$$
$$= 42.56\ \mathrm{MHz} \tag{13-26b}$$

(Note that 1 MHz equals 10^6 cycles per second.) Thus the energy differences that correspond to the transitions of nuclear spins are of the size to absorb radio waves.

Equation 13-25 shows us that the energy separation of the two quantum levels of a proton in a magnetic field is proportional to the magnetic field strength. This means there are two ways in which an NMR experiment can be carried out. With nuclei in a constant magnetic field, the frequency of radio waves can be varied until the resonance condition of Equation 13-26*a* is satisfied and absorption takes place. Alternatively, the radio frequency field can be fixed (for example, at 100 MHz = 10^8 cycles per second) and the magnetic field increased until the resonance condition is satisfied and absorption occurs. Figure 13-26 illustrates the design of an NMR spectrometer.

NMR Chemical Shifts Reflect the Local Environment of a Nucleus

The frequency, ν, at which resonance absorption by a nucleus occurs (Equation 13-26*a*) depends not only on the strength of the applied magnetic field, $\boldsymbol{B}_0$, but on the strength of the magnetic field, $\boldsymbol{B}$, actually acting on the nucleus.

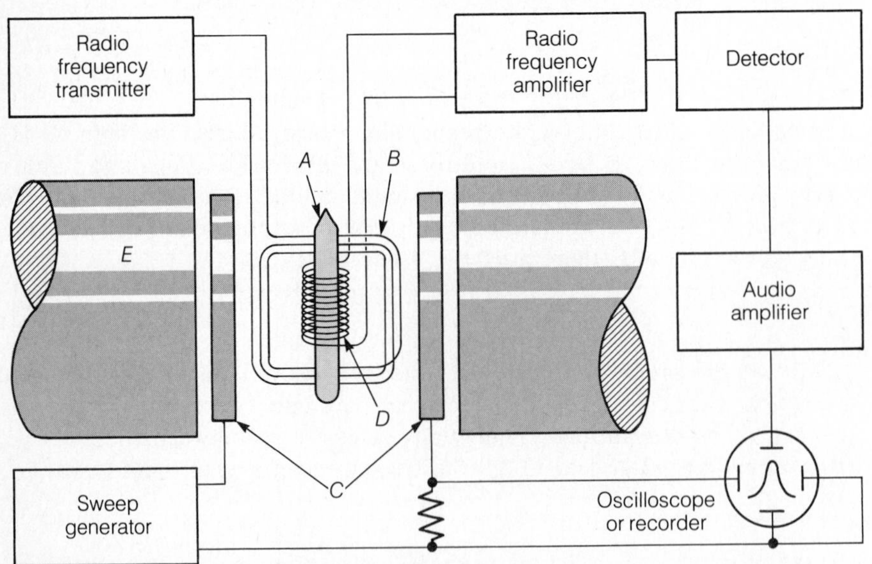

FIGURE 13-26
Schematic diagram of an NMR spectrometer. The sample is contained in tube *A*, which is typically about 5 mm in diameter. The tube is placed between the coils of the magnet *E*, which in present-day instruments may have a field strength of 10–60 kG (1–6 T). The radio frequency transmitter applies an electromagnetic field of frequency ν by means of the coil *B*. Coil *D* detects absorption. *C* is a smaller "sweep" magnet that varies the magnetic field strength until the resonance condition is satisfied. (From Bovey, 1972.)

These strengths differ, in part because the applied magnetic field produces motions in the electrons surrounding the nucleus, and these motions give rise to a magnetic field $\sigma \boldsymbol{B}_0$ that opposes $\boldsymbol{B}_0$. The constant of proportionality, σ, is called the *shielding constant,* and is characteristic of the electronic configuration around the nucleus. The field acting on the nucleus then is given by

$$\boldsymbol{B} = \boldsymbol{B}_0 - \sigma \boldsymbol{B}_0 = \boldsymbol{B}_0(1 - \sigma) \tag{13-27}$$

Thus, the magnetic field at which absorption occurs is shifted somewhat from the strength of the applied field. This is called the *chemical shift,* δ, which is defined as the fractional change in magnetic field strength from the field at which some reference compound absorbs radiation:

$$\delta(\text{parts per million}) = \frac{\boldsymbol{B}_0(\text{reference}) - \boldsymbol{B}_0(\text{sample})}{\boldsymbol{B}_0(\text{reference})} \times 10^6 \tag{13-28}$$

For the proton in various environments, these shifts can be as large as 16 parts per million (16 ppm). Figure 13-27 gives characteristic shifts for the proton, relative to the protons in tetramethylsilane [$Si(CH_3)_4$], which is a common reference. Another chemical-shift scale used by some workers also is shown in Figure 13-27—the τ scale in which $\tau = 10 - \delta$.

What determines the size of the chemical shift of a nucleus in a given environment? From Equation 13-27 you can see that a large shielding constant means that there is a large local magnetic field opposing the

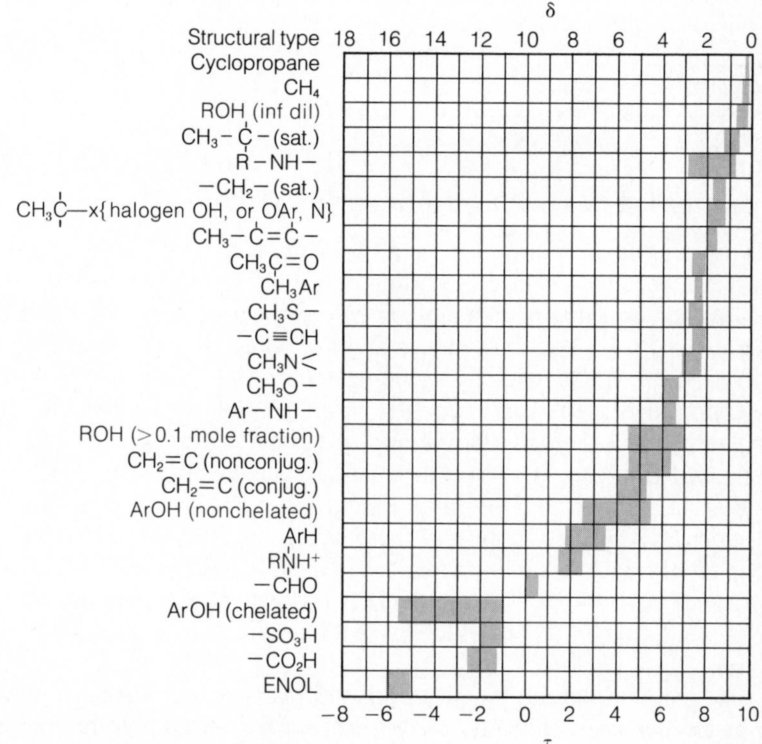

FIGURE 13-27
Proton chemical shifts relative to tetramethylsilane. (From Bovey, 1972.)

applied field, and hence that the applied field must be increased to achieve the resonance condition. The amount of shielding depends on several factors. One is the density of electrons around the resonating nucleus. As the density is increased, the shielding is increased, and resonance occurs at greater values of B_0. For example, in the series of groups containing protons,

$$CH_3{-}O{-} \qquad CH_3{-}N\!\!< \qquad CH_3{-}\overset{|}{\underset{|}{C}}{-}$$

the electronegativity of the atom bonded to the methyl group decreases toward the right. Therefore electrons are shifted from around the methyl carbon decreasingly toward the right of the series, and hence σ increases toward the right. This means that the magnetic field strength for resonance increases toward the right. Other factors that affect shielding include nearby π electrons or nearby paramagnetic ions (see, for example, Bovey, 1972).

EXERCISE 13-19

Figure 13-30 shows the proton NMR spectrum of ethanol. (a) Does the magnetic field strength increase toward the right or toward the left? (b) Which proton has the smallest shielding? (c) Which resonance has the largest chemical shift relative to tetramethylsilane?

ANSWER

(a) Right. (b) O—H. (c) O—H.

The Physical Origin of Magnetic Shielding

The chemical shift (δ) of a proton or other nucleus depends on its chemical environment, because chemical bonding determines electron density. Electrons shield against an applied magnetic field through a general phenomenon called *diamagnetism*. You can understand this effect if you recall a simple result from physics, which is illustrated in Figure 13-28. A charge moving in a magnetic field is turned in a circular path. Electrons in molecules have kinetic energy, and their motion occurs in circular paths when a magnetic field is applied. Figure 13-28(*c*) shows—as also illustrated in Figure 13-22(*b*)—that an electron moving in a circular path produces a magnetic field which opposes B_0 inside the path of the charge, and is in the same direction as B_0 outside the path. Usually, a nucleus is most influenced by the electrons that surround it, so the field at the nucleus is smaller than the applied field, by the amount of the induced field. This is the origin of magnetic shielding by electrons.

The induced field can be thought of as arising from a magnetic moment, whose properties are analogous to those of an electric dipole moment. In diamagnetic materials, the net magnetic moment induced by the applied field opposes the applied field. Just as would be found for an electric dipole

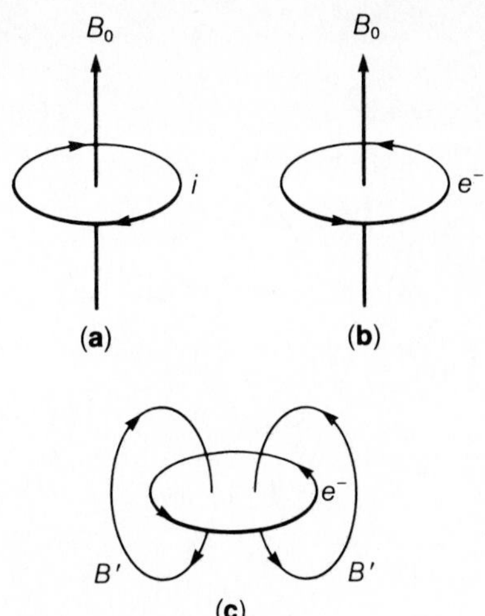

FIGURE 13-28
(a) A magnetic field B_0 causes a current i to follow a circular path; i is the motion of positive charge. **(b)** An electron moving in a molecule is turned into a circular path by an applied magnetic field. **(c)** The magnetic field induced by the circular motion of positive charge, i, or negative charge, e^-, in **(a)** and **(b)**. Notice that the field B' inside the circumference of the electron path opposes the applied field B_0 so the effective field there, $B = B_0 - B'$, is smaller than B_0. Therefore, a nucleus inside the electron path is shielded from the field.

oriented against an applied electric field, there is a force that pushes an induced magnetic moment out of a magnetic field which opposes it. The unfavorable interaction energy between the field and the magnetic moment is smaller when the field is smaller, so diamagnetic materials are forced toward lower field.

Paramagnetic materials contain unpaired electrons, which have a magnetic moment because of their spin. The spin magnetic moment can align with the field, lowering the energy by doing so. In this case the interaction of the magnetic moment and the magnetic field becomes more favorable when the field is stronger, so a paramagnetic material is attracted into a magnetic field. The size of the force can be used to calculate the number of unpaired electrons in the sample.

The Ring Current Shift

An important example of diamagnetic shielding is the *ring current* induced in aromatic molecules. A conjugated system of double bonds that forms a closed path provides a means by which electrons can move in a large circular path. Benzene is a simple example, with the π electrons free to circulate around the molecule.

Motion of the electrons in a magnetic field leads to induction of an opposing field through the center of the ring, as illustrated in Figure 13-28(*c*). Therefore, a nucleus located over the center of the ring will experience a net magnetic field smaller than it would if isolated in space. The result is that a higher magnetic field is required for resonance of a nucleus that is shielded by being next to an aromatic ring. This is called the *ring current shift*.

Figure 13-29 illustrates the influence of the ring current shift on the NMR spectrum of an oligonucleotide. At low temperature (25°C) the bases in the trinucleotide ApApGp tend to stack on top of each other, forming a structure

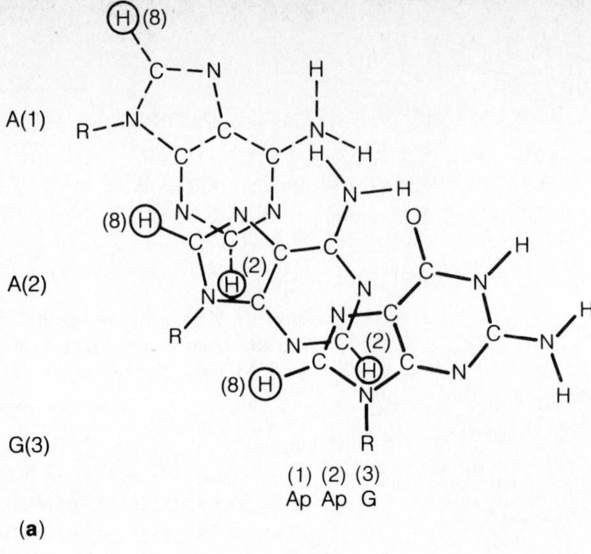

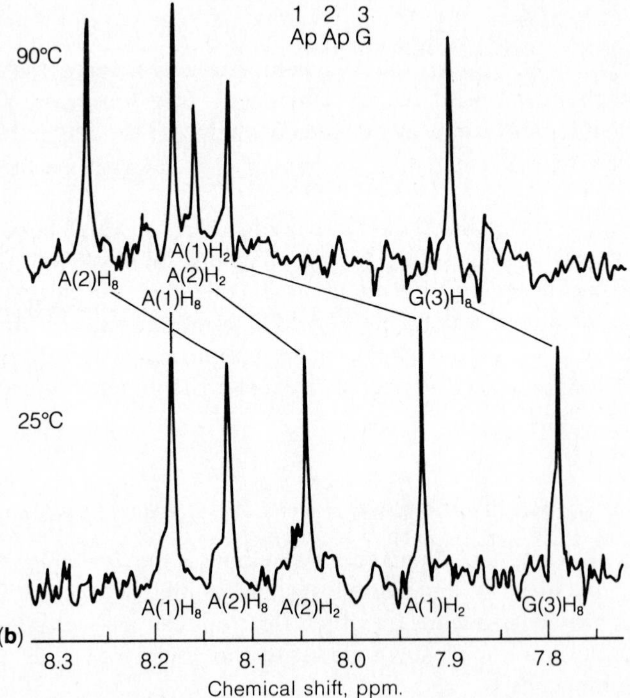

FIGURE 13-29
(a) Stacking of the bases in ApApG due to formation of a right-handed helix. The bases are attached to the ribose rings (R), which form part of the ribose-phosphate backbone that joins the bases together. This view of the structure looks down the helix axis, with the 3′ terminal guanine base closest to the observer. Each successive stacked base is roughly parallel to its neighbors, moving farther from the observer toward the 5′ end of the chain. **(b)** High resolution (270 MHz) photon NMR spectrum of ApApGp at two temperatures, in the region of the spectrum corresponding to the nonexchangeable purine ring protons. All proton resonances except A(1) H8 move downfield as the temperature increases, because the ring current shielding from adjacent bases is lost as the structure becomes disordered. (Data from Shum, 1977.)

similar to the helical ordering of one strand in double helical RNA. A view of this structure, looking down the helix axis, is shown in Figure 13-29(*a*). Notice that the H8 protons of the second adenine, A(2), and guanine, G(3), are close to the adjacent base, as are both adenine H2 protons. Figure 13-29(*b*) shows how temperature increase affects the NMR spectrum. All proton resonances but one move downfield when the temperature rises from 25°C to 90°C, an effect due to loss of the ring current shielding contributed by the adjacent base. The shielding effect is lost because the ordered, stacked structure shown in Figure 13-29(*a*) is disrupted at 90°C. The exceptional proton which shows no downfield shift is H8 from A(1). You can understand the basis for this by observing in Figure 13-29(*a*) that A(1) H8 is too far distant from an adjacent base to experience any appreciable ring current shift.

The ring current shift is frequently used to investigate structural details of proteins and nucleic acids. In addition to the nucleic acid bases (especially the purines such as adenine and guanine), the aromatic side chains in proteins, such as phenylalanine, tyrosine, and tryptophan, can produce substantial ring current shifts. The size of the ring current shift is very sensitive to the distance between the nucleus and the ring, so that movements as small as 0.01 nm can be detected.

EXERCISE 13-20

Use the ring current concept to explain why aromatic protons resonate at relatively low fields.

Hint: What induced field would be experienced by a proton attached to the outside of the ring in Figure 13-28(*c*)?

○ *Spin-Spin Interactions Broaden and Split NMR Absorption Bands*

The interactions of neighboring nuclear spins affect the shape of NMR absorption bands. These effects can help in assigning bands to particular nuclei, and can yield information on chemical rates and structure.

As an example, consider the NMR spectrum of ethanol, which is shown at low resolution in Figure 13-30(*a*). From our discussion of the chemical shift, we know that the hydroxyl, methylene, and methyl protons absorb at different field strengths because they "feel" different local magnetic fields. The ratio of the areas of these three absorption bands is 1 : 2 : 3 because this is the ratio of the numbers of their protons. But why are three *bands* observed, rather than the three sharp lines that one might expect for a transition between two sharp energy levels? The answer is that the methyl protons, for example, also experience small magnetic fields from the neighboring methylene protons. Since each of the two methylene protons can have two orientations, there are four possibilities for the orientations of the two nuclei:

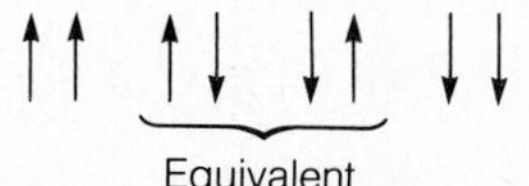

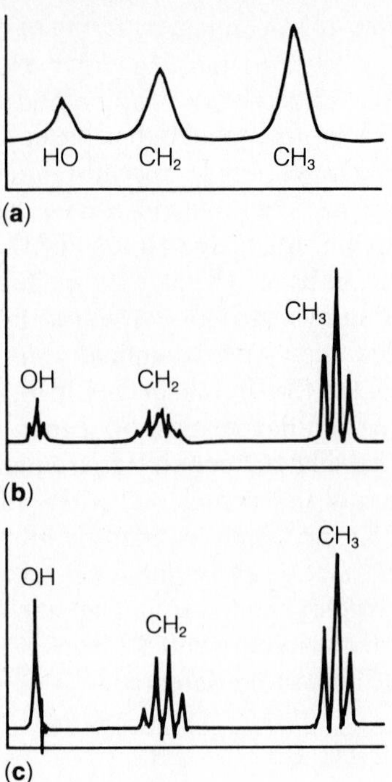

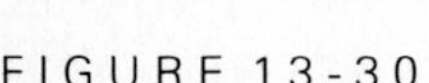
FIGURE 13-30
The proton magnetic resonance spectrum of ethanol. **(a)** Low resolution, showing broad bands for the methyl, methylene, and hydroxyl protons, with band areas in the ratio 3 : 2 : 1. **(b)** At high resolution, the spin-spin splitting is resolved. This spectrum is for pure, dry ethanol. **(c)** Spectrum at high resolution for ethanol with a trace of acid, which catalyzes exchange of the hydroxyl proton. (Reproduced with changes from Moore, 1972.)

where an arrow represents a component of a nuclear magnetic moment in the direction of the applied field. Since each nuclear magnet gives rise to its own small magnetic field, the nearby methyl proton may feel any of three distinct fields. The two other combinations give rise to the same field. Thus, we might expect that the methyl band actually is composed of three peaks, each corresponding to one of the three possible local magnetic fields. In fact, under high resolution the methyl band is seen to correspond to three peaks [Fig. 13-30(*b*)]. The O—H band, which also is near the methylene protons, is split into three bands. The methylene band is split into four peaks because of the neighboring methyl protons. The combinations of methyl proton spins are the following:

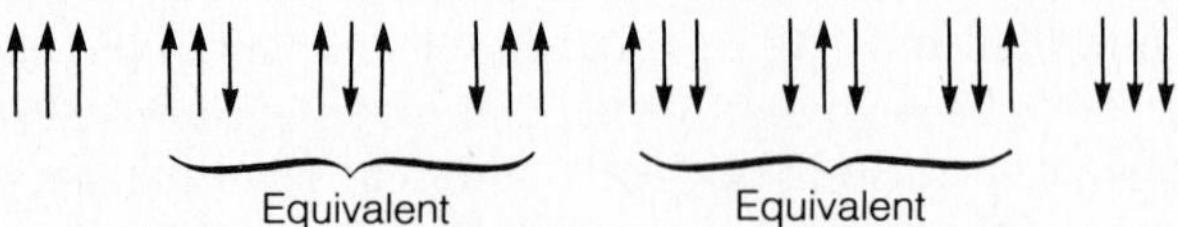

Since all three members of the two center groups give rise to equivalent magnetic fields, there are four distinct arrangements thereby producing the observed splitting.

In general, if a resonating nucleus has n neighboring nuclei of spin I, its NMR absorption band is split into $2nI + 1$ peaks.

EXERCISE 13-21

Sketch the NMR spectrum of CHO—CH_3 at both low and high resolution.

ANSWER

The methyl proton resonance is at higher magnetic field than the aldehyde proton resonance. At low resolution the methyl band has an area three times that of the aldehyde band. At high resolution the methyl band is split into two peaks (a doublet), and the aldehyde band is split into four peaks (a quartet).

NMR band shape also can give some information on rates, as is illustrated in Figure 13-30. Panel (*c*) shows that when a trace of acid is added to pure, dry ethanol, the hydroxyl resonance changes from a triply split band (triplet) to a single band (singlet). The reason is that the acid catalyzes rapid exchange of protons at the hydroxyl group. This exchange is much more rapid (about 10^{11} times per second) than the frequency of the NMR measurement (about 10^8 Hz), so that during a single pass of the alternating electromagnetic field that causes the resonance, many different protons have been part of the O—H group. The observed single band reflects an average of their interactions with neighboring protons. In contrast, in pure, dry ethanol the hydroxyl proton is fixed in place, and its resonance is split by its interactions with the —CH_2—protons. Therefore, the spectra of panels (*b*) and (*c*) show that the rate of proton exchange at the hydroxyl group in acid solution is greater than 10^8 times per second. In this example the precise rate of exchange is too fast to be determined by NMR methods, but in many cases NMR alone can be used to determine a rate (Sec. 13-8).

● *Equivalent Protons show no Spin-Spin Splitting*

You may have wondered why we neglected the other methyl protons in our consideration of spin-spin splitting of the methyl proton resonance by the methylene protons. After all, each methyl proton should experience a variable magnetic field depending on the orientation of the other methyl proton magnetic moments. There is a rule which states that *nuclei whose chemical shifts are equal exhibit no spin-spin splitting due to their interaction*. As we will see here, the physical origin of this effect is that *transitions which change the total spin angular momentum of a set of magnetically equivalent nuclei are forbidden*. There is a close analogy between this phenomenon and the fact that singlet-triplet electronic transitions are forbidden, since the latter require a change in the total electron spin angular momentum.

Consider the problem for the specific case of the NMR spectrum of a hydrogen molecule, H_2. Since the two hydrogens are equivalent, the rule tells us to expect no spin-spin splitting, and indeed a single proton resonance is observed. To begin, we must consider the spin angular momentum of both protons together. There are two possible values of the total nuclear spin angular momentum, which are analogous to the singlet and triplet states possible for a system of two electrons.

In the "singlet" state, the two nuclear spins must point in opposite directions, yielding a total spin angular momentum quantum number (m_{sT}) of zero:

$$m_{sT} = m_s \text{ (nucleus 1)} + m_s \text{ (nucleus 2)}$$
$$= +\tfrac{1}{2} \quad -\tfrac{1}{2} = 0$$

Therefore the total nuclear spin quantum number, s_T, is zero, and the magnetic moment is zero. This form of H_2 is called *para-hydrogen.* In the "triplet" state, the two nuclear spins are always parallel, with three permitted values of $m_{sT} = 1$, 0, and -1. You can visualize these states (see Fig. 10-22) as

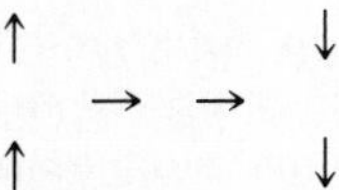

each having the same total angular momentum. Notice that the middle state has quantum number $m_{sT} = 0$, but it is not identical with the singlet state, which also has $m_{sT} = 0$ but would be represented as

because its total angular momentum is zero. The triplet state of H_2 is called *ortho-hydrogen,* which can be obtained free of para-hydrogen because the two interconvert very slowly when the sample is pure. (Impurities, especially paramagnetic materials, can catalyze the conversion.)

Now we are ready to consider the nuclear spin energy levels of H_2 in a magnetic field (Fig. 13-31). The spin-spin interactions within the triplet state are the same because the identical spins are always parallel, but the triplet spin-spin interaction energy is different from the corresponding energy in the

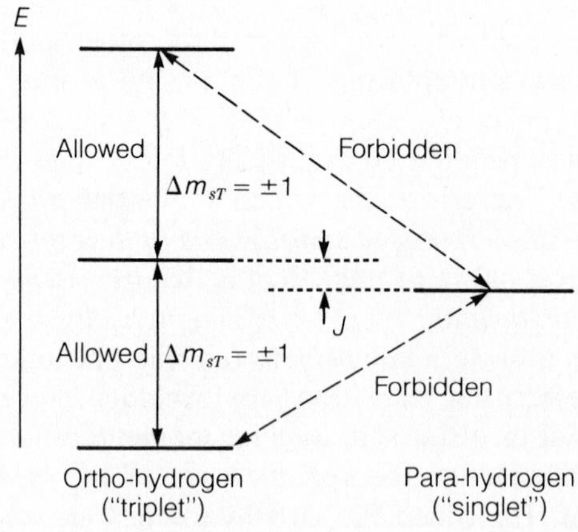

FIGURE 13-31
Energy levels due to orientation of the nuclear magnetic moments of H_2 in a magnetic field. Para-hydrogen, with the two proton spins opposed, has a spin-spin interaction energy (J) different from that of the middle level of the "triplet" state, which also has $m_{sT} = 0$. However, transitions between the singlet and triplet states are forbidden, so only one resonance line appears, corresponding to $\Delta m_{sT} = \pm 1$ within the set of levels in the triplet state.

singlet state. Thus, in principle there are three different transition energies, one between the levels of the triplet state, and two between the triplet and singlet states.

At this point the selection rule intervenes and leaves us with only one allowed transition energy. Transitions between singlet and triplet states are forbidden (Fig. 13-31), so only the $\Delta m_{sT} = \pm 1$ transitions between the levels of the triplet state remain. Therefore, we find that two equivalent protons provide only one resonance line.

Of course, this entire argument relies on the selection rule, and you might wonder whether there is a simple conceptual basis for this rule. Selection rules have their origin in quantum mechanical transition moments (such as the electric dipole transition moment, Sec. 12-6). These transition moments are very dependent on the symmetry properties of the initial and final states, as discussed for infrared and Raman transitions in Section 13-2. The symmetries of singlet and triplet states are such that they are not mixed by the magnetic vector in a beam of radiation. If you want to consider the problem in a more conceptual classical sense, the singlet state has no magnetic moment, and so it does not interact with the magnetic field in the radiation. The triplet state interacts because it has a magnetic moment, but all the radiation can do is turn the magnetic moment in a different direction, the transition described by $\Delta m_{sT} = \pm 1$.

You might also consider what happens when two protons have nearly, but not quite, identical chemical shifts. In this case the selection rules in Figure 13-31 are not absolute. Small resonance absorptions appear at the frequencies which correspond to the forbidden transitions. The intensities of these increase as the protons become less alike in their magnetic environment (see Problem 13-25). When the two protons are sufficiently different, two doublet resonance lines appear, as expected for the spin-spin splitting of each proton resonance because of the two possible orientations of the other proton.

13–7 Nuclear Relaxation and Fourier Transform NMR

Modern high-sensitivity NMR spectrometers work on a slightly different principle than that characteristic of the spectrometer in Figure 13-26, in which the magnetic field is varied slowly until each proton in turn comes into resonance. This technique is intrinsically inefficient because only one small region of the spectrum can be investigated at any instant in time. The technique of *Fourier transform (FT) NMR spectroscopy* is more efficient because all the nuclei in the sample can be investigated simultaneously, with Fourier analysis (see Sec. 17-5) of the system's response used to extract the individual resonance frequencies.

In order to understand FT NMR we first have to learn something about how the population of spin states and the orientation of nuclear magnetic moments change with time. There are two important time constants for this motion, called the *spin-lattice relaxation time*, T_1, and the *spin-spin relaxation time*, T_2. The reason there are two time constants is that the magnetic moment along the z-axis (direction of B) is quantized, and therefore behaves differently from the magnetization in the x-y plane. Precession of the individual magnetic moments about the z-axis (Fig. 13-25) causes a time-dependent magnetic moment in the x and y directions.

The Populations of Different Spin States are Almost Equal

Recall from Section 12-6 that for a sample to absorb electromagnetic radiation, the frequency of radiation must be near the natural resonance frequency of the sample, the transition moment must be nonzero, *and* the population of particles in the ground state must be significantly larger than that in the excited state. In contrast to other types of spectroscopy we have considered in this chapter, with NMR it often can happen that the ground state has no greater population of nuclei than does the excited state; consequently, absorption cannot occur. The reason is that these states are so close in energy that thermal agitation populates both states to almost the same extent. If n_1 is the number of particles excited to an upper quantum state by thermal agitation and n_0 is the number of the ground state, then the ratio n_1/n_0 is given by the Boltzmann distribution,

$$\frac{n_1}{n_0} = e^{-\Delta E/k_B T} \tag{13-29}$$

in which ΔE is the separation of the quantum levels, and $k_B T$ is the product of Boltzmann's constant and the absolute temperature. From Equation 13-25, for proton resonance in a magnetic field of 1 T (10^4 G), we find that

$$\begin{aligned}\Delta E &= 2m_z B \\ &= 2(1.410 \times 10^{-26}\ \text{J T}^{-1})(1\ \text{T}) \\ &= 2.82 \times 10^{-26}\ \text{J}\end{aligned}$$

At room temperature $k_B T$ equals about 4.14×10^{-21} J so,

$$2m_z B/k_B T \cong 6.8 \times 10^{-6}$$

Because the exponent in Equation 13-29 is so small, we can neglect all but the first two terms in its expansion,

$$e^{-x} = 1 - x + \frac{x^2}{2!} - \cdots$$

giving,

$$\frac{n_1}{n_0} = 1 - \frac{2m_z B}{k_B T} = 1 - 6.8 \times 10^{-6} \tag{13-30}$$

This means that the ground state contains only a few nuclei per million more than the excited state.

Spin-Lattice Relaxation is the Radiationless Transition of Nuclei between Excited and Ground States

When an electromagnetic field is applied to a sample in a magnetic field, more nuclei are excited to the upper level than the other way around because radiation has equal probability of inducing transitions in either direction between two states. Therefore, if there were no return of nuclei to the lower state, it would take little time for the populations to become equal and for absorption of radiation to cease. However, some nuclei do change their orientations and return to the ground state, by a radiationless mechanism. This process is called *spin-lattice relaxation,* and occurs with a characteristic time T_1, which can range from 10^{-3} to 10^3 s, depending on the nature and physical state of the sample. A nucleus can change its orientation, and thus return to the ground state, only when it experiences an oscillating magnetic field of frequency ν. Moving electrons and nuclei in neighboring molecules are two sources of such fields.

The spin-lattice relaxation (T_1) time can be given a quantitative definition in terms of the time-dependence of the z-component of the net magnetic moment of the sample. If there are n_0 spins with magnetic moment m_z and n_1 spins with magnetic moment $-m_z$, the net magnetic moment in the z direction, M_z, is

$$M_z = (n_0 - n_1)m_z \tag{13-31}$$

In other words, application of a magnetic field to a sample results in a net orientation of the nuclear magnetic moments such that the resultant magnetic moment is parallel to the applied field. (The diamagnetic response of the electrons in the sample, for which the induced field opposes the applied field—Section 13-6—usually is larger than the nuclear paramagnetism, so materials that do not contain unpaired electrons generally are diamagnetic in their total response to the field.)

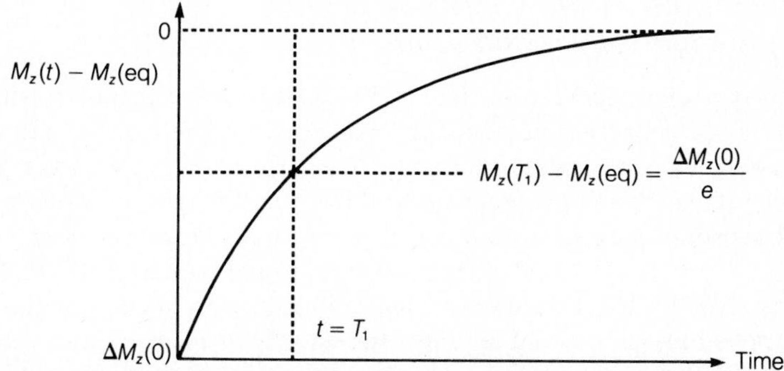

FIGURE 13-32
Exponential recovery of the z-component of the nuclear magnetization towards equilibrium, in the absence of an applied oscillating magnetic field. An electromagnetic field pulse at time zero was used to reduce the net magnetization to zero. Time T_1 is required for the displacement from equilibrium to be reduced to $1/e$ of its initial value $\Delta M_z(0)$ at $t = 0$.

By application of a perturbation, such as an oscillating magnetic field of frequency ν, the magnetization M_z(t), now a function of time, can be caused to take on a value different from the equilibrium value M_z(eq) when the perturbation is removed. M_z(t) returns to its equilibrium value by the radiationless process described by T_1. The time dependence is given by the equation:

$$M_z(t) - M_z(eq) = \Delta M_z(0)e^{-t/T_1} \qquad (13\text{-}32)$$

This equation states that the difference between M_z and its equilibrium value decays to zero as an exponential function of time (Fig. 13-32). T_1 is the time required for M_z(t) − M_z(eq) to be reduced to $1/e$ of its initial value at time $t = 0$ [which is given by $\Delta M_z(0)$].

EXERCISE 13-22

Show that, for a sample containing N protons,

$$n_1 = \frac{N}{2}\left(1 - \frac{|m_z|B}{k_B T}\right)$$

and

$$n_0 = \frac{N}{2}\left(1 + \frac{|m_z|B}{k_B T}\right)$$

are consistent with Equation 13-30.

EXERCISE 13-23

Use the result of Exercise 13-22 and Equation 13-31 to show that

$$M_z(eq) = \frac{N m_z^2 B}{k_B T}$$

The Width of an NMR Band Reflects the Spin-Spin Relaxation Time, T_2

The second nuclear relaxation time is T_2, which is a measure of the time required for spins precessing about the z-axis to lose their phase coherence. The phenomenon is described by Figure 13-33. Panel (a) shows an important result from physics, and gives a way to think about magnetic resonance in a classical sense: a magnetic moment in a magnetic field precesses (moves circularly) about the field, at a frequency (ν) equal to the NMR resonance frequency (Eq. 13-26). Resonance occurs when the frequency of the radiation or applied magnetic field matches the natural or *Larmor* precession frequency. Thus the usual condition for absorption of radiation is met when the radiation frequency equals the Larmor frequency (see Sec. 12-5).

A sample of matter in a magnetic field contains many identical nuclei precessing about the field axis. By techniques to be described, it is possible to obtain a set of identical nuclei that are "in phase" because the y-

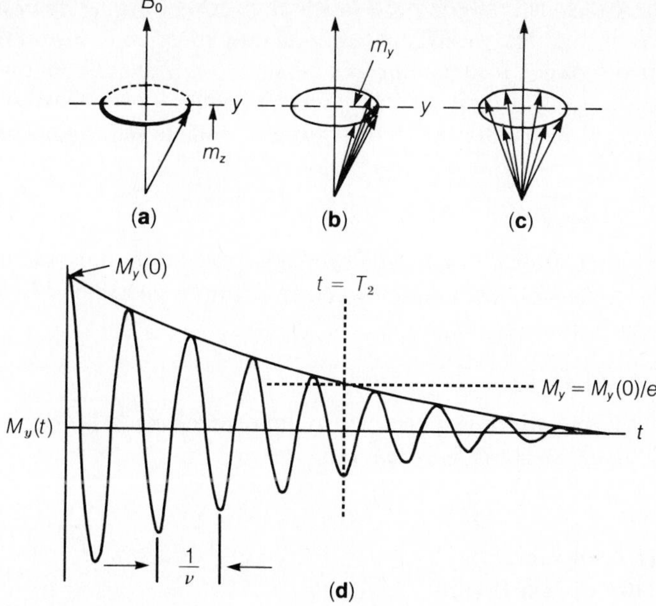

FIGURE 13-33
The spin-spin relaxation process T_2. **(a)** Precession of a magnetic moment (z-component $= m_z$) about the field B at the Larmor frequency $\nu = 2m_zB/h$. **(b)** A set of nuclei initially in phase, so that their y-components of magnetization are equal. **(c)** After a time long compared to T_2, the nuclei have lost phase coherence, and the average x or y component of the magnetization is zero. **(d)** Time course of decay of the y-component of the magnetization, with characteristic time T_2. Oscillations occur at the Larmor frequency ν.

components of their magnetic moments are equal. This means that they will all point at the same spot on the circle which describes their path of precession [Fig. 13-33(b)]. A number of influences can cause the spins to reorient or to precess at different rates, even though the nuclei are identical. For example, the magnetic field may not be exactly the same for all nuclei, either because the applied field is not quite constant over the whole sample, or because the nuclei have different environments due to intermolecular interactions. These and other factors contribute to loss of phase coherence of the nuclei, as indicated in Figures 13-33(c) and (d). Panel (c) shows the final state, in which the magnetic moments are distributed at random around the precession path, and the average magnetization in the x or y direction is zero. Panel (d) shows the time course of decay of the initial magnetization, $M_y(0)$, present in panel (b). Oscillations occur at the Larmor frequency ν as the spins precess about the z-axis. [You can visualize this as rotation of the vector m_y in panel (b).] Each successive maximum is a little smaller than the one before, decaying according to the equation

$$M_y(\text{max}) = M_y(0) \exp(-t/T_2) \tag{13-33}$$ ◀

This equation gives a quantitative definition to the spin-spin relaxation time T_2.

The time T_2 also is related to the width of an NMR resonance line. If nuclei are precessing at slightly different rates, as they must be to lose phase coherence, their resonance frequencies will be slightly different from one another. Therefore, absorption will cover a range of values of ν. The relationship between T_2 and $\Delta\nu_{1/2}$, the width of a band at half its maximum height, is

$$T_2 = \frac{1}{\pi\,\Delta\nu_{1/2}} \qquad (13\text{-}33b)$$

Equation 13-33*b* states that a short spin-spin relaxation time produces broad NMR absorption lines. High-resolution spectra are possible only when T_2 is long.

EXERCISE 13-24

Calculate the line width in parts per million (ppm) of an NMR resonance measured at 100 MHz if T_2 is 0.1 sec.

ANSWER

$\Delta\nu_{1/2} = 1/(T_2\pi)$; therefore $\Delta\nu_{1/2}/\nu = 1/(T_2\pi\nu) = 1/(0.1 \times \pi \times 10^8) = 0.0318 \times 10^{-6} = 0.0318$ ppm.

The main physical basis for nuclear relaxation is a fluctuating magnetic field. When the local field fluctuates at the Larmor frequency, it is very likely to cause reorientation of a nuclear spin, the T_1 process. Such field fluctuations can arise because of movement of one spin magnetic moment relative to another, as can occur when a molecule tumbles in solution, for example. Measurement of T_1 relaxation times is therefore an important method for detecting motion of macromolecules in solution.

Slowly fluctuating magnetic fields contribute strongly to the T_2 relaxation process, because they cause the magnetic environment and therefore the precession rate of one nucleus to be different from that of another. A slowly fluctuating magnetic field arises if a molecule tumbles slowly. In contrast, rapid tumbling results in rapid averaging of the magnetic environment. Large macromolecules tend to tumble slowly, so they have short T_2 values and broad NMR resonances (Eq. 13-33*b*). With present spectrometers, the usual upper limit for high-resolution proton NMR spectoscopy in solution is reached with molecular weights of 50,000–100,000.

FT NMR Measures Precession in the x-y Plane

The simplest form of operation of a Fourier transform NMR spectrometer can be understood by referring to Figure 13-33. A mixture of proton spins lacking phase coherence, as shown in Figure 13-33(*c*), has no net magnetization in the *x-y* plane. However, a briefly pulsed magnetic field along the *x*-axis causes the spins to precess about the *x*-axis, tipping them into the *y* direction if the pulse is timed to produce 90° rotation. After the pulse is terminated, the magnetization in the *x-y* plane oscillates as shown schematically in Figure 13-33(*d*). However, all protons present contribute to the oscillation, and the resulting complex oscillatory curve can be resolved into its Fourier components (Section 17-5), yielding the Larmor frequencies of all protons (or another selected nucleus) in the sample. In this way data can be obtained for all frequencies simultaneously, greatly increasing the efficiency of data collection.

13–8 Nuclear Magnetic Resonance can be used to Measure Kinetics

Nuclear magnetic resonance provides a powerful tool for kinetic measurements and is capable of covering a very wide range of time scales. Consider a chemical reaction such as

$$\mathrm{A} \underset{k_B}{\overset{k_A}{\rightleftarrows}} \mathrm{B}$$

for which a proton or other nucleus has a different chemical shift in states A and B. Let us suppose that by a change of temperature or some other variable we are able to alter the rate at which A and B equilibrate. We will find that the observed NMR spectrum varies greatly, depending on the chemical conversion rate, as shown in Figure 13-34. When the chemical reaction rate is slow (compared to the difference in nuclear resonance frequencies, to be discussed), the NMR experiment "sees" two different nuclei having different chemical environments. In contrast, when the chemical reaction is fast, the nuclear environment is seen only as an average, and a single line

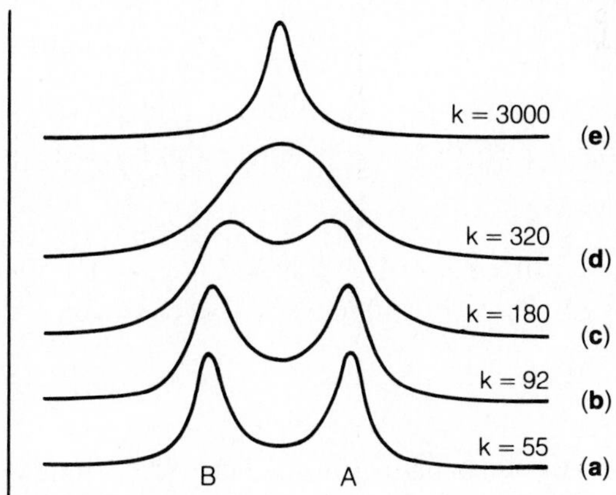

FIGURE 13-34
Calculated variation of the NMR spectrum of two resonance lines A and B separated by $\nu_A - \nu_B = 128\ \mathrm{sec}^{-1}$, as a function of the rate constant $k = k_A = k_B$ (in sec^{-1}) for converting A to B. In **(a)** k is substantially smaller than $\nu_A - \nu_B$, and two separate lines are observed, but in **(e)** k is much greater than $\nu_A - \nu_B$, leaving a single line at the average resonance frequency. Notice that coalescence to a single line occurs between $k = \nu_A - \nu_B$ and $k = \pi(\nu_A - \nu_B)$. (From Klevan, 1978.)

appears. In the intermediate range we see two lines with varying degrees of broadening and partial collapse to a single line at the average chemical shift.

The width of an NMR line is determined by the transverse, or spin-spin relaxation time, T_2. Specifically, the NMR absorption $g(\nu)$ depends on frequency ν according to the relationship

$$g(\nu) = \frac{2T_2}{1 + 4\pi^2 T_2^2(\nu - \nu_0)^2} \qquad (13\text{-}34a)$$

in which ν_0 is the resonance frequency. The maximum value of $g(\nu)$ is $2T_2$ when $\nu = \nu_0$; $g(\nu)$ has half the maximum value when $\nu - \nu_0 = \pm 1/(2\pi T_2)$. Therefore the width of the line at half-height is given by

$$(\Delta\nu)_{1/2} = 2(\nu - \nu_0)_{1/2} = 1/(\pi T_2) \qquad (13\text{-}34b)$$

The widening of the resonances at the limiting conditions (*a*) and (*e*) in Figure 13-34 can be understood in simple physical terms. Consider first the case in which the chemical reaction is slow. The separate lines for A and B are widened slightly by an amount that depends on the lifetime $\tau_A = 1/k_A$ of each state, because one way that the set of nuclei in state A can lose phase coherence is for some of them to be transferred to state B. The rate of this process is added to the intrinsic spin-spin relaxation rate $1/T_{2A}$ for the molecules in state A, or

$$1/T_2 = 1/T_{2A} + 1/\tau_A \qquad (13\text{-}34c)$$

This leads to a wider line, as given by Equation 13-34*b*, with $1/T_{2A} + 1/\tau_A$ replacing $1/T_2$. Comparison of line widths with and without exchange can then be used to calculate τ_A. We note that in contrast to relaxation kinetics, in which the rate always includes a contribution from both forward and reverse rates, the *NMR method can measure the lifetime* τ_A *of a state directly.*

EXERCISE 13-25

A proton resonance is 0.20 Hz wide in absence of exchange. Calculate T_{2A}. Calculate also the line width when exchange occurs, with $\tau_A = 1$ sec.

ANSWER

$T_{2A} = 1/(\pi\Delta\nu_{1/2}) = 1/(0.2\pi \text{ sec}^{-1}) = 1.59$ sec. $1/T_2 = 1/1.59$ sec $+ 1/1$ sec (Eq. 13-34*c*) $= 1.63 \text{ sec}^{-1}$. $T_2 = 0.61$ sec. Therefore $\Delta\nu_{1/2} = 1/(0.61\pi$ sec) $= 0.52 \text{ sec}^{-1}$.

Line-broadening by exchange with another state sometimes is called *uncertainty broadening* because of its relationship to the uncertainty principle, which states that the product of the uncertainty in energy, ΔE, and lifetime, Δt, of a state is given by

$$\Delta E \Delta t = h/2\pi$$

in which h is Planck's constant. Long-lived states have greater uncertainty in their lifetimes than short-lived states, and their energy levels therefore are defined more precisely. Since $\Delta E = h\Delta\nu$, the uncertainty in the frequency of absorption is

$$\Delta\nu = \frac{1}{2\pi\,\Delta t}$$

Anything that reduces the state lifetime, measured by Δt, increases the frequency uncertainty $\Delta\nu$. One contribution to a short lifetime of state A is conversion to state B, at rate $1/\tau_A$. Therefore the resonance line should be broadened in frequency by approximately $1/\tau_A$. In essence, the nucleus does not spend a long enough time in state A to have precisely defined energy levels in the magnetic field.

When the chemical relaxation rate is sufficiently large that the two separate resonances collapse into one [Fig. 13-34(*e*)], the line width is determined by T_2':

$$\frac{1}{T_2'} = \frac{P_A}{T_{2A}} + \frac{P_B}{T_{2B}} + 4\pi^2 P_A{}^2 P_B{}^2(\nu_A - \nu_B)^2(\tau_A + \tau_B) \tag{13-34d}$$

in which P_A and P_B are the probabilities of states A and B, respectively, and ν_A and ν_B are the resonance frequencies of the nucleus in states A and B. The first two terms on the right of Equation 13-34*d* represent the average of $1/T_2$ values in states A and B. To this is added the term that depends on $(\nu_A - \nu_B)^2$ and $(\tau_A + \tau_B)$. This additional broadening arises from inexact averaging over the environments of the proton in states A and B. The observation of an average resonance peak at $\nu = P_A\nu_A + P_B\nu_B$ depends on each nucleus spending a fraction P_A of its time in state A and P_B in state B. However, if the rate of hopping back and forth from A to B is not very fast compared to the NMR sampling time, then those average values will not be strictly observed for each nucleus. This is analogous to the random-sampling error observed with any finite sample. The relevant sampling time is proportional to the reciprocal of the frequency difference, $(\nu_A - \nu_B)^{-1}$, because the time required for nucleus A to move one turn ahead of nucleus B in its precession about the magnetic field is $(\nu_A - \nu_B)^{-1}$. A sampling time of this order is required to measure the frequencies ν_A and ν_B with good accuracy compared to their difference $\nu_A - \nu_B$. Detailed analysis of the problem shows that inexact averaging contributes a term $4\pi^2 P_A{}^2 P_B{}^2(\nu_A - \nu_B)^2(\tau_A + \tau_B)$ to the rate of loss of phase coherence $1/T_2'$.

EXERCISE 13-26

A proton spends half its time in environment A and half in environment B. Given that $T_{2A} = T_{2B} = 2$ sec, $\nu_A - \nu_B = 100$ Hz, and $\tau_A = \tau_B = 1 \times 10^{-3}$ sec, calculate T_2' and the linewidth.

ANSWER

$$\frac{1}{T_2'} = \frac{0.5}{2\text{ sec}} + \frac{0.5}{2\text{ sec}} + \frac{4\pi^2}{16}(100\text{ sec}^{-1})^2(2 \times 10^{-3}\text{ sec})$$

$$= 49.3\text{ sec}^{-1}$$

$$T_2' = 0.020\text{ sec}$$

$$\Delta\nu_{1/2} = \frac{1}{0.02\pi} = 15.9\text{ Hz}$$

An example of the use of proton NMR to measure kinetics is shown in Figure 13-35, which shows the resonances of the exchangeable NH hydrogens of arginine [Fig. 13-35(*a*)] at varying temperature. At low temperature, peak (1) [Fig. 13-35(*b*)] corresponds to the three $—NH_3^+$ protons, peak (2) is the N(1)—H proton, and peaks (3) and (4) are the four guanidinium $—N(2)H_2$ protons. These split into two groups of two protons each because of slow rotation about the N(1)—*C*(ϵ) bond, causing the groups *cis* and *trans* to the N(1)—H proton to have different chemical shifts. Rotation about the C(ϵ)—N(2) bonds is fast at these temperatures, so the two $—N(2)H_2$ protons are equivalent in environment and chemical shift.

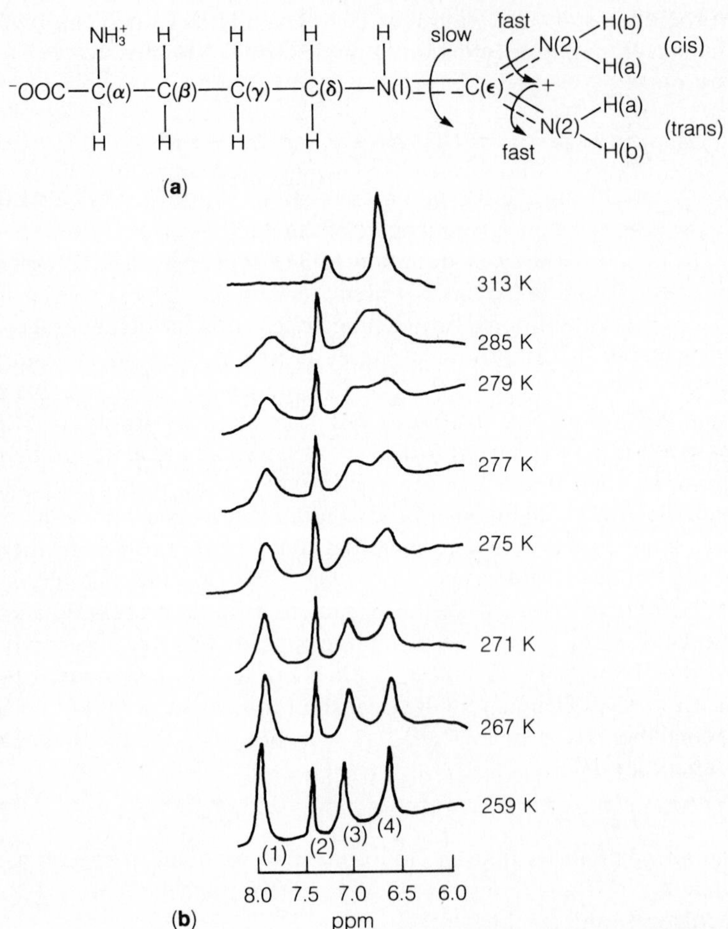

FIGURE 13-35
(a) Structure of arginine. **(b)** NMR spectrum of arginine in the region of the NH proton resonances. Peaks (3) and (4), corresponding to guanidinium $—C(NH_2)_2^{\oplus}$ protons, coalesce at higher temperatures, allowing measurement of the rate of bond rotation. Peak (1), corresponding to $—NH_3^+$, disappears at high temperature because of fast exchange of the protons with solvent water. (From Klevan, 1978.)

13–9 Electron Spin Resonance (ESR) Detects Unpaired Electrons

Just as the energy of a spinning nucleus depends on its orientation in a magnetic field, so the energy of a spinning electron depends on its orientation in a magnetic field. Electrons, like protons, have only two spin quantum states, and thus only two orientations in a magnetic field. Resonance between the two states can be induced by applying an alternating electromagnetic field of the correct frequency. In contrast to NMR, *electron spin resonance* (ESR; also called *electron paramagnetic resonance,* EPR) is not observed with most substances, because the spins of electrons usually are paired, so that an atom has no net magnetic moment. But for paramagnetic substances, including free radicals, triplet ground states such as the O_2 molecule, paramagnetic ions of transition elements, and photochemical intermediates in the triplet state, electron spin resonance is observed.

The principles of ESR are quite similar to those of NMR. The magnetic moment of an unpaired electron is given by

$$m = -g_e\left(\frac{eh}{4\pi m_e}\right) m_s \quad \text{(SI)} \qquad (13\text{-}35)$$

in which g_e is the *electronic g-factor* (a number very nearly equal to 2), $-e$ and m_e are the electronic charge and mass, and m_s is the spin quantum number (equal to $\pm\frac{1}{2}$). The quantity in parentheses in Equation 13-35 is called the *Bohr magneton,* and has the value of 9.2732×10^{-24} J T^{-1} in SI units (9.2732×10^{-21} erg G^{-1}). In an applied magnetic field of strength B, a transition of an electron from ground to excited state requires an energy

$$\Delta E = g_e\left(\frac{eh}{4\pi m_e}\right) B \quad \text{(SI)} \qquad (13\text{-}36)$$

In a magnetic field of 2T (20 kG), this energy corresponds to absorption of radiation of frequency

$$\nu = \frac{\Delta E}{h} = \frac{2(9.273 \times 10^{-24}\ \text{J T}^{-1})(2\ \text{T})}{6.62 \times 10^{-34}\ \text{J s}} = 5.6 \times 10^{10}\ \text{Hz}$$

which is in the microwave region of the spectrum.

Paramagnetic substances are detected readily by ESR. About 10^{-13} mole of a substance gives an observable signal, so this technique is one of the most sensitive of all spectroscopic tools.

ESR Bands are Split by Hyperfine Coupling of Electron and Nuclear Spins

The effect of a neighboring nuclear spin on the resonance of an unpaired electron is called *hyperfine coupling*. It corresponds to spin-spin interaction in NMR. For an electron in a magnetic field there are two orientations and two quantum states [Fig. 13-36(*a*)]. A nearby nucleus with spin $\frac{1}{2}$ also has two

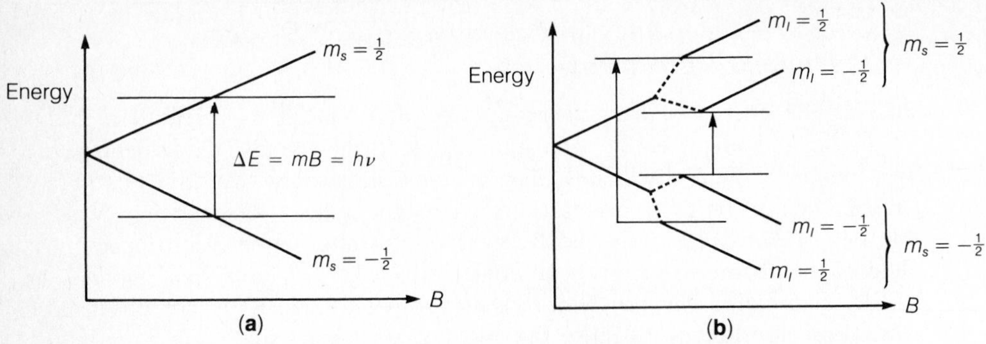

FIGURE 13-36
Energy levels of an unpaired electron with spin quantum $m_s = \pm\frac{1}{2}$ (a) in a magnetic field, and (b) in a magnetic field and coupled to a nuclear spin of $I = \frac{1}{2}$, with nuclear spin quantum number $m_s = \pm\frac{1}{2}$.

orientations and two quantum states. This gives a possible combination of four quantum states for the electron-nucleus pair [Fig. 13-36(*b*)]. It is said that the nuclear spin "splits" each electron quantum state into two states. Because the selection rules for transitions in hyperfine coupling are $\Delta m_s = \pm 1$ and $\Delta m_I = 0$ [see Fig. 13-36(*b*)], there can be only two transitions among these four states for which electromagnetic radiation can be absorbed. These two transitions are shown by vertical arrows in Figure 13-36(*b*).

If the spinning nucleus that couples with the electron has a spin greater than $\frac{1}{2}$, there are more than four hyperfine levels. For the ^{14}N nucleus, for example, $I = 1$, so there are three nuclear spin quantum states. Thus a nearby ^{14}N nucleus splits the electronic levels into six levels. Three transitions are allowed among the six levels; consequently the spectrum consists of three absorption bands.

Spin Labels are Paramagnetic Nitroxide Molecules that Serve as Probes in Macromolecules and in Membranes

McConnell and his colleagues showed that a great deal of information can be derived about macromolecules and membranes from the ESR spectra of bound nitroxide molecules. These are stable molecules that possess an unpaired 2*p* electron. The unpaired electron endows the molecules with strong ESR spectra. A commonly used nitroxide is 2,2,6,6-tetramethylpiperidinol-N-oxyl, abbreviated TEMPOL. Its structure is

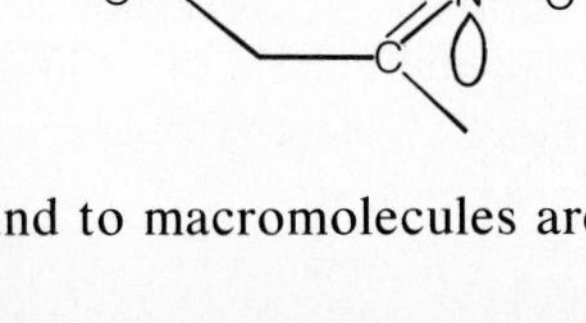

Nitroxide molecules bound to macromolecules are called *spin labels*.

Because the ^{14}N nucleus in a nitroxide is near the unpaired electron, there is an interaction between them, thereby producing hyperfine splittings in the ESR spectrum. The ^{14}N nucleus has a spin of 1, and consequently three absorption bands appear in the ESR spectra. ESR spectra usually are recorded as the first derivative of the absorption spectrum, so instead of three bands there are three rise-and-dip spikes, which are the derivatives of the three bands. These can be seen in Figure 13-37.

Spin labels can give various kinds of information about the molecules to which they are bound. They can report the rate of motion of a macromolecule to which they have been covalently bound, or the amount of thermal motion in a membrane in which they have been inserted. The principle is that the bands of the ESR spectrum are broad when the spin label is immobilized, and narrow when it is tumbling rapidly. The narrowing comes from the more rapid relaxation of the spin when neighboring groups are moving rapidly with respect to the spin label. This is illustrated by the ESR spectra of Figure 13-37 for TEMPOL and another spin label dissolved in glycerol. As the temperature of the solutions is lowered, and the spin labels are increasingly immobilized, the bands are broadened.

A second type of information that can be reported by a spin label is the polarity of its environment. The extent of splitting of the side bands from the central band depends on the dielectric constant of the medium in which the spin label is dissolved. Solvents of high dielectric constant augment the

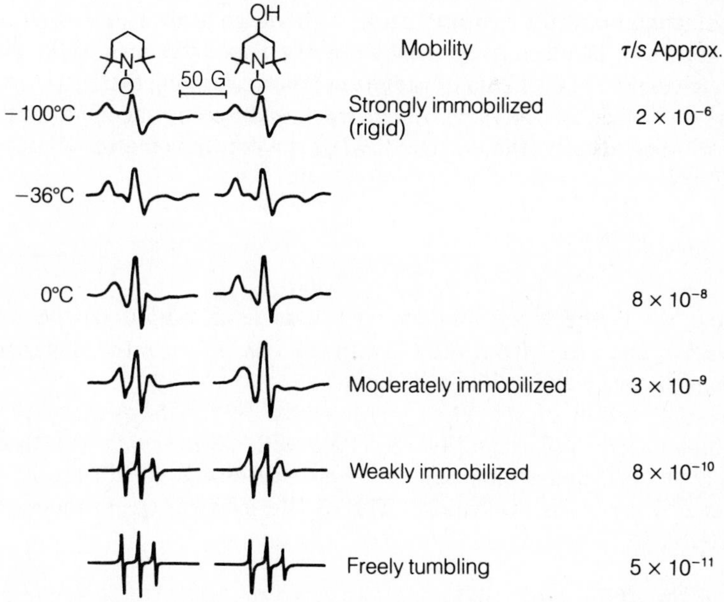

FIGURE 13-37
ESR derivative spectra at various temperatures of two spin labels dissolved in glycerol. The concentration of each solution is 0.5 mM. The frequency of absorption is 9.5 GHz (9.5×10^9 cycles per second). The rotational correlation times for the spin labels are given on the right. (From Dwek, 1973.)

polarity of the N—O bond and increase the splitting. By measuring the splitting an estimate can be made of the polarity of the surroundings of the spin label. This is of interest, for example, when a spin label is bound to a membrane, since it allows one to determine if the label is bound near the polar head groups or near the nonpolar hydrocarbon chains (Mehlhorn and Keith, 1972).

13–10 Light Scattering Allows Determination of Macromolecular Size and Shape

Most of the spectroscopic methods that are useful in biochemistry involve absorption of radiation—in the classical description they reflect the properties of the out-of-phase polarizability, α_{90}. The two major exceptions to this rule are optical rotation (ORD), which we discussed in Section 13-5, and light scattering. Both of these techniques depend on the in-phase polarizability, α_0, and both can, in the right circumstances, detect changes in the shape of macromolecules. However, whereas ORD is sensitive to localized structural changes, light scattering is sensitive to overall macromolecular size and shape. As we will see, the intensity of light scattered by a solution depends on the molecular weight, and the angular dependence of the scattering depends on the shape of the molecule (but only, it turns out, if the molecule is not very small compared to the wavelength of radiation scattered).

Our starting point for consideration of Rayleigh scattering is Figure 12-4, which shows an electron oscillating along the z-axis because of the alternating electric field E^* in a beam of radiation polarized along the z-axis. According to the laws of physics, a charge undergoing acceleration emits radiation—specifically, the electric field of the emitted (scattered) beam (see Fig. 12-4) is

$$E_s^* = \frac{ea \sin \phi}{rc^2} \tag{13-37}$$

in which $-e$ is the electronic charge, a its acceleration, ϕ the angle of observation measured from the z-axis (see Fig. 12-4), r the distance from the electron, and c the velocity of light.

The acceleration of the electron depends on the frequency (ν) of the light, and on the polarizability, α_0. (We will assume that the frequency is far from any absorption bands—light scattering is subject to severe anomalies when α_{90} is important.) The dipole moment (μ^{in}) induced by the field, E^*, is

$$\mu^{\text{in}} = \alpha E^* \qquad (\alpha = \alpha_0)$$

The induced moment can be replaced by

$$\mu^{\text{in}} = -ex$$

in which x is the displacement of the electron from the positive charge center. Therefore,

$$x = \frac{-\alpha E^*}{e} \qquad (13\text{-}38a)$$

To calculate the acceleration of the electron, we must insert explicitly the time dependence of x. With the electric field oscillating according to

$$E^* = E_0^* \sin(2\pi\nu t)$$

x also oscillates harmonically:

$$x = \frac{-\alpha E_0^*}{e} \sin(2\pi\nu t) \qquad (13\text{-}38b)$$

The acceleration of the electron is d^2x/dt^2, so

$$a = \frac{4\pi^2\alpha\nu^2}{e} E_0^* \sin(2\pi\nu t) \qquad (13\text{-}39)$$

Inserting this expression for a in Equation 13-37 gives us

$$E_s^* = \frac{4\pi^2\alpha\nu^2}{rc^2} \sin\phi \, E_0^* \sin(2\pi\nu t) \qquad (13\text{-}40)$$

This equation states that the field E_s^* of the scattered radiation oscillates in phase with the incident beam. Furthermore, the magnitude of the electric field is proportional to the polarizability α, and to the square of the frequency ν. Notice that the ν^2 dependence arises because of the dependence of the electron acceleration on the square of the oscillation frequency, Equation 13-39.

In light-scattering measurements one usually compares the intensity of the scattered beam (I_s) with the intensity (I_0) of the incident beam. The intensity of a light beam depends on the energy flow per unit area (energy time^{-1} length^{-2}), and is proportional to the square of the electric field intensity, $I \propto E^{*2}$. Therefore, the ratio of scattered to incident intensity is

$$\frac{I_s}{I_0} = \left(\frac{E_s^*}{E^*}\right)^2$$

Dividing Equation 13-40 through by $E^* = E_0^* \sin(2\pi\nu t)$ and squaring yields

$$\frac{I_s}{I_0} = \frac{16\pi^4\alpha^2\nu^4 \sin^2\phi}{r^2c^4} \qquad (13\text{-}41a)$$

Use of the relationship $\nu = c/\lambda$ gives the result for scattering from a single particle:

$$\frac{I_s}{I_0} = \frac{16\pi^4\alpha^2 \sin^2\phi}{r^2\lambda^4} \qquad (13\text{-}41b) \blacktriangleleft$$

This equation emphasizes the strong wavelength dependence of the intensity of scattered light. The factor $\sin^2\phi$ applies to polarized light. When

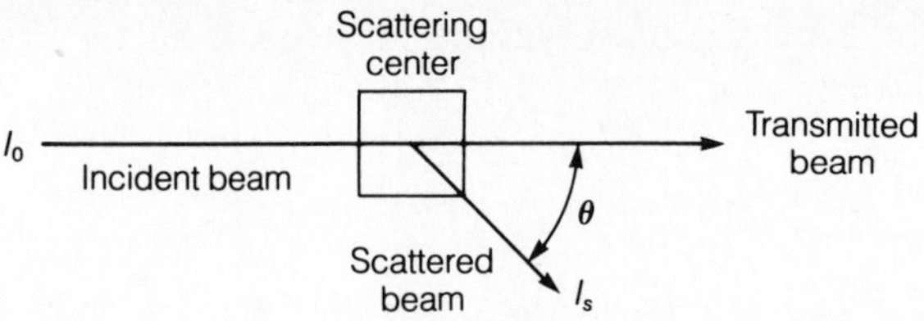

FIGURE 13-38
Definition of the scattering angle θ. Scattering of unpolarized light depends only on θ.

unpolarized light is used $\sin^2\phi$ is replaced by a different angle dependence, $(1 + \cos^2 \theta)/2$, in which θ is the angle between the emerging transmitted beam and the scattered beam (Fig. 13-38) (Tanford, 1961).

Light Scattering Requires Fluctuations in Polarizability or Refractive Index

A fundamental physical fact about light scattering—which you must know to understand how light scattering is used—is that *perfectly ordered materials show negligible light scattering* if the wavelength of light is long compared to the separation of atoms. The physical basis for this is indicated in Figure 13-39(*a*). The scattered beams from a perfectly ordered material can always be combined in pairs that are out of phase and therefore cancel (see Kauzmann, 1957, for a detailed discussion). Only in the forward direction, along $I_{\text{transmitted}}$, is the interference not destructive. In that case there is an alteration in the phase of the light, which is equivalent to an altered velocity in the medium, as given by its refractive index.

On the other hand, if there are fluctuations in the number of particles in each cell, as in Figure 13-39(*b*), the two beams which cancel in (*a*) need no longer do so. Therefore, *scattering occurs from materials containing fluctuations in the scattering power.* Since the scattering power depends on polarizability (or on the refractive index, which is how polarizabilities are frequently measured), we can ascribe light scattering to fluctuation in polarizability or refractive index.

Light Scattering from a Solution Requires that the Refractive Index Change with Concentration

Suppose now that an occasional cell in Figure 13-39(*a*) is occupied by a solute molecule. If the solute has the same polarizability as the solvent, the scattering power will be unaffected, and pairwise cancellation of scattered beams will still occur. However, if the polarizability of the solute and solvent are different, their scattering will no longer cancel. Hence *light scattering depends on having a difference between the refractive index of the solution and the pure solvent.*

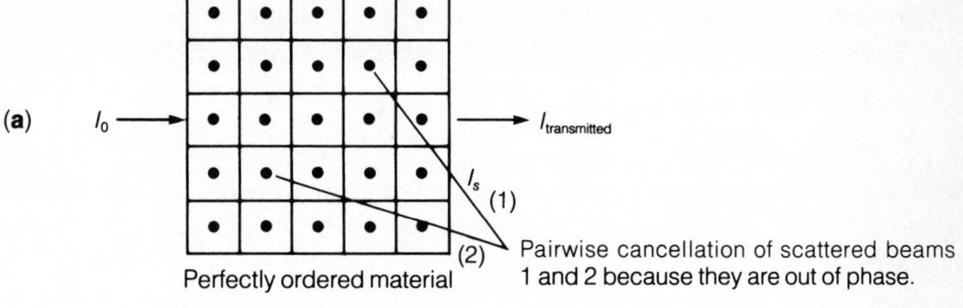
(a)
I_0
$I_{transmitted}$
I_s
(1)
(2)
Perfectly ordered material
Pairwise cancellation of scattered beams 1 and 2 because they are out of phase.

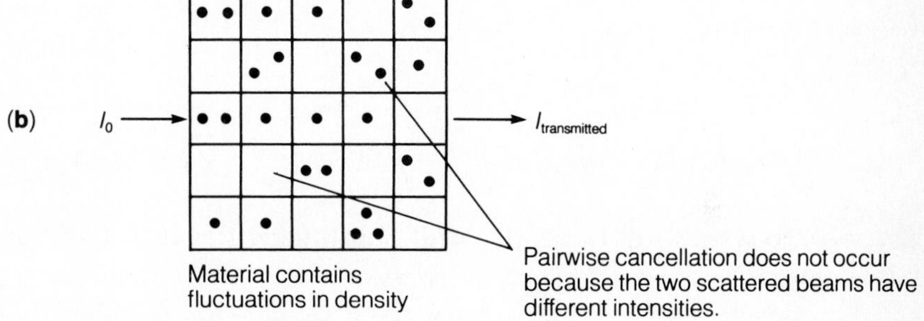
(b)
I_0
$I_{transmitted}$
Material contains fluctuations in density
Pairwise cancellation does not occur because the two scattered beams have different intensities.

FIGURE 13-39
(a) A perfectly ordered material shows negligible light scattering (for wavelengths long compared to the interatomic spacings) because the scattered beams can always be combined in pairs such that they are 180° out of phase and therefore cancel each other. **(b)** When there are fluctuations in the density, the number of scatterers in the two cells which cancelled in **(a)** need not be the same. Therefore, the scattered beams can be of different intensity and need not cancel. For this reason a liquid scatters more light than a crystalline solid.

This conclusion suggests a simple model for light scattering in which molecules of polarizability α in a solvent of polarizability α_0 are replaced by a solution of molecules of polarizability $(\alpha - \alpha_0)$ in a material of negligible polarizability. In other words, the solute molecules are responsible for scattering because they represent fluctuations in the local polarizability, of size $(\alpha - \alpha_0)$. Therefore $(\alpha - \alpha_0)^2$ replaces α^2 in Equation 13-41*b*:

$$\frac{I_s}{I_0} = \frac{16\pi^4(\alpha - \alpha_0)^2 \sin^2 \phi}{r^2\lambda^4} \tag{13-42}$$

Now we need a relationship between $\alpha - \alpha_0$ and the concentration dependence of the solution refractive index. The required equation (see Eq. 12-27) is

$$4\pi N(\alpha - \alpha_0) = n^2 - n_0^2 \tag{13-43}$$

in which n_0 and α_0 are the refractive index and polarizability of the solvent, respectively, n and α are the same quantities for the solution, and N is the number of solute particles per unit volume.

Equation 13-42 gives the scattering from a single particle. When it is multiplied by the number N of particles per unit volume to yield the scattering from a unit volume, and combined with Equation 13-43, the result is

$$\frac{I_s}{I_0} = \frac{\pi^2(n^2 - n_0^2)^2}{Nr^2\lambda^4} \sin^2 \phi \tag{13-44}$$

This equation can be simplified by using a Taylor's series expansion for n about its value n_0 when solute concentration c_B (in weight per unit volume) is zero:

$$n = n_0 + \left(\frac{\partial n}{\partial c_B}\right) c_B + \cdots$$

therefore we find the result

$$n^2 - n_0^2 = \left[n_0 + \left(\frac{\partial n}{\partial c_B}\right) c_B + \cdots\right]^2 - n_0^2$$
$$= 2n_0 \left(\frac{\partial n}{\partial c_B}\right) c_B \tag{13-45}$$

when only terms linear in c_B are retained. Substituting Equation 13-45 and the relationship

$$N\,(\text{particles vol}^{-1}) = \frac{c_B(\text{wt vol}^{-1})N_A(\text{particles mol}^{-1})}{M(\text{wt mol}^{-1})}$$

into Equation 13-44 we obtain the equation (for polarized light)

$$\frac{I_s}{I_0} = \frac{4\pi^2 n_0^2(\sin^2 \phi)(\partial n/\partial c_B)^2 c_B M}{r^2\lambda^4 N_A} \tag{13-46}$$ ◀

Notice that Equation 13-46 predicts that the *intensity of light scattered depends linearly on the molecular weight, M*. This is the basis for the determination of molecular weights by the light scattering method.

The Rayleigh Ratio Corrects for Geometric Factors

The primary experimental variable in light scattering measurements is the ratio of scattered to incident intensities, corrected for the geometric factors $\sin^2 \phi$ (or $1 + \cos^2 \theta$ for unpolarized light) and r^2, called the *Rayleigh ratio*, R_θ. For polarized light

$$R_\theta = \frac{I_s}{I_0}\frac{r^2}{\sin^2 \phi} \tag{13-47}$$

Comparing this expression with Equation 13-46*b* we see that

$$R_\theta = \left[\frac{4\pi^2 n_0^2(\partial n/\partial c_B)^2}{\lambda^4 N_A}\right] c_B M \tag{13-48}$$

This can be shortened to

$$R_\theta = K c_B M \tag{13-49}$$

in which K is a constant equal to the quantity in square brackets in Equation 13-48. Clearly M can be determined from the ratio of R_θ to $K c_B$.

Light Scattering Depends on the Second Virial Coefficient

Equation 13-49 is valid only in the limit as the concentration c_B approaches zero. More advanced textbooks (Tanford, 1961) will show you that scattering from higher concentration solutions is described by the equation

$$\frac{K c_B}{R_\theta} = \frac{1}{M} + 2 B c_B + \cdots \tag{13-50}$$

in which B is the second virial coefficient defined in Section 7-15. Thus a plot of $K c_B / R_\theta$ against c_B has slope $2B$ and intercept $1/M$. This is an important technique for determination of the molecular weight of macromolecules.

The Shape Factor is Important for Molecules Larger than a Tenth of the Radiation Wavelength

When a molecule's size becomes appreciable compared to the wavelength of light, its scattering is reduced in all directions except $\theta = 0$, equivalent to forward scattering. The origin of this reduction is the correlation in position of scattering elements. In essence, a large molecule occupies more than one cell in Figure 13-39. This produces a region which is locally isotropic in scattering power, so that the fluctuations in refractive index are reduced.

The effect of molecular size is accounted for by a shape factor $P(\theta) \leq 1$, which multiplies the Rayleigh ratio. Equation 13-50 is therefore modified to read

$$\frac{K c_B}{R_\theta} = \frac{1}{P(\theta)} \left[\frac{1}{M} + 2 B c_B + \cdots \right] \tag{13-51}$$

The factor $P(\theta)$ depends on the scattering angle θ; it equals 1 for all molecules in the limit as $\theta \to 0$, but is less than 1 for other angles when the size of the molecule is comparable to the wavelength λ. Advanced textbooks (Tanford, 1961) will show you that $P(\theta)$ can be expressed as a power series, the first terms of which are

$$P(\theta) = 1 - \frac{16 \pi^2 R_G^2}{3 \lambda^2} \sin^2 \left(\frac{\theta}{2} \right) + \cdots \tag{13-52}$$

R_G is called the *radius of gyration*, which is a measure of the size of a polymer molecule, defined by the equation

$$R_G = \left(\frac{\Sigma m_i r_i^2}{\Sigma m_i}\right)^{1/2} \tag{13-53}$$ ◀

in which r_i is the distance from the center of mass of each mass element m_i in the polymer molecule. Notice that R_G has the dimension of length, and is roughly the radius of the polymer molecule.

EXERCISE 13-27

Calculate the radius of gyration of a sphere of uniform density and radius a.

ANSWER

$$R_G = \left[\frac{\int_0^a \text{density} \times 4\pi r^2 \cdot r^2\,dr}{\int_0^a \text{density} \times 4\pi r^2\,dr}\right]^{1/2}$$

$$= \left[\frac{3}{5}\frac{a^5}{a^3}\right]^{1/2} = \sqrt{\frac{3}{5}}\,a$$

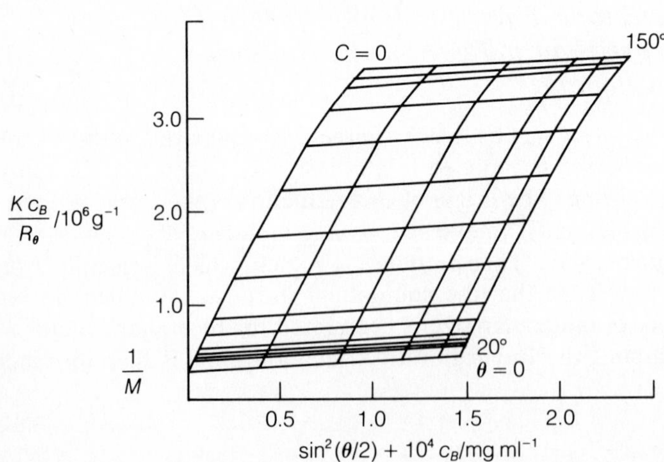

FIGURE 13-40
Light scattering determination of the molecular weight, radius of gyration, and virial coefficient of a DNA sample. According to Equation 13-51, the slope of the line at zero angle (θ) can be used to determine the virial coefficient B, and the slope at zero concentration (c_B) yields the radius of gyration (R_G), when combined with Equation 13-52. The molecular weight is the reciprocal of the intercept of the vertical axis. The results of this experiment give $M = 3.67 \times 10^6$g mol^{-1}, $R_G = 213$ nm, $B = 5.3 \times 10^4$ ml mol g^{-2}. Notice that a molecular weight of 3.67×10^6g mol^{-1} corresponds to a total DNA length of 1882 nm, assuming the B-form structures for DNA. If DNA were a stiff rod, R_G would be the total length divided by $\sqrt{12}$ (Problem 32), or 543 nm. Since the actual value of R_G is smaller than this, one concludes that the molecule is able to bend. The conformation of flexible polymers in solution is considered in detail in the next chapter. [From D. Jolly and H. Eisenberg, *Biopolymers 15,* 61 (1976).]

The utility of Equation 13-52 is that it can be used to determine the radius of gyration of a polymer molecule. For example, when the scattering angle θ is 40° and $R_G = 0.1\lambda$, then the second term in Equation 13-52 is 0.0616. This means that the Rayleigh ratio will be $1 - 0.0616 = 0.9484$ times as large as it would be if the particle were a point scatterer. This difference is detectable, and can be used to determine R_G. In the limit as $P(\theta)$ approaches 1, measurements of light or x-ray scattering can be used to determine R_G. For larger angles θ, other shape factors have a pronounced influence on $P(\theta)$.

An Example: Determination of the Molecular Weight, Radius of Gyration, and Virial Coefficient of a DNA Sample

Figure 13-40 illustrates the use of light scattering to determine the size and extent of interaction of macromolecules in solution. The dependence of the ratio $K\ c_B/R_\theta$ on angle (θ) and concentration c_B (Eq. 13-51) is expressed through a double extrapolation to zero concentration and angle called a *Zimm plot,* explained in detail by Tanford (1961). The slope at zero angle ($\theta = 0$) is $2B$, in which B is the second virial coefficient, and the slope at zero concentration can be used to calculate the radius of gyration, R_G, from Equation 13-52. The double intercept on the vertical axis is $1/M$.

QUESTIONS FOR REVIEW

1. Describe a Raman scattering experiment. What determines the frequency of scattered light? What determines the intensity of scattered light?
2. What is resonance Raman spectroscopy and what is its principal advantage?
3. How many normal modes of vibration does a molecule have? Why are some associated with Raman scattering, some with infrared absorption, and some with both?
4. Sketch an energy-level diagram that shows electronic, vibrational, and rotational transitions, with typical values.
5. How can spectroscopy be used to measure $\Delta G°$ and $\Delta H°$ for an interaction? How can spectroscopy be used to record the titration of a chemical group?
6. Under what conditions is fluorescence observed? When is phosphorescence observed? What physical quantities characterize fluorescence?
7. How can resonant energy transfer be used to determine separations?
8. Define plane polarization, circular polarization, optical rotation, and specific rotation.
9. Compare optical rotatory dispersion and circular dichroism. What are experimental differences in these techniques, and what are the advantages of each? Define ellipticity.
10. What conditions are necessary for nuclear magnetic resonance?

11. What molecular properties are reflected in ORD and CD spectra?
12. Define nuclear g-factor, shielding constant, chemical shift, and chemical exchange.
13. Explain the physical basis of T_1 and T_2, and how they are measured.
14. In what respects does ESR differ from NMR? Explain hyperfine coupling. What is a spin label?
15. What is the physical basis of light scattering? How is light scattering related to polarizability?
16. How is information on molecular weight and molecular shape determined by light scattering?

PROBLEMS

1. Are the following statements true or false?

   (a) Raman scattered light always has frequency smaller than that of the incident light.

   (b) Intense infrared absorptions always are accompanied by a change in the polarizability of the absorbing group.

   (c) All substances display an ORD spectrum.

   (d) The stretching frequency of an N—H group is reduced when it forms a hydrogen bond.

   (e) In UV spectroscopy the light beam passes through a monochromator before the sample, whereas in Raman spectroscopy it passes through the monochromator after the sample.

   (f) Contributions of individual absorption bands are better resolved in CD than in ORD.

   (g) $\Delta G°$ for a process (say ionization) can be measured spectroscopically, but $\Delta H°$ cannot.

   (h) A plane polarized beam of light is composed of right- and left-circularly polarized light.

   (i) Optical rotation depends on both electric and magnetic transitions.

   (j) All nuclei have magnetic moments.

   (k) ORD and light scattering depend on the in-phase polarizability, α_0.

Answers: (a) F. (b) F. (c) F. (d) T. (e) T. (f) T. (g) F. (h) T. (i) T. (j) F. (k) T.

2. Complete the following sentences with the appropriate words.

   (a) A vibrational transition is ________ active if the corresponding vibration scatters photons strongly.

(b) In ________ Raman scattering, the frequency of incident light corresponds to an absorption band of the sample.

(c) The lowering of UV absorption as polynucleotides form a double helix is called ________ .

(d) Intersystem crossing is ________ (faster, slower) than internal conversion.

(e) The ratio of the number of quanta emitted to the number absorbed is called ________ ________ .

(f) Absorption bears the same relationship to refractive index dispersion as CD does to ________ .

(g) The frequency of nuclear resonance increases as the magnetic field strength ________ (increases, decreases).

(h) ESR is observed for systems with ________ electrons.

Answers: (a) Raman. (b) Resonance. (c) Hypochromism. (d) Slower. (e) Quantum yield. (f) ORD. (g) Increases. (h) Unpaired.

PROBLEMS RELATED TO EXERCISES

3. (Exercise 13-1) How might heating affect the Raman spectrum of a substance?

4. (Exercise 13-3) Describe the normal modes of vibration of SO_2 (bent) and CS_2 (linear). Which are Raman active and which are infrared active?

● 5. (Exercise 13-4) State whether the following functions are even (symmetric) or odd (antisymmetric):
(a) $y = x$ (b) $y = |x|$ (c) $y = \sin x$ (d) $y = \cos x$

● 6. (Exercises 13-5 and 13-6) Determine on the basis of symmetry whether the harmonic oscillator transition from $n = 1$ to $n = 2$ is allowed for infrared absorption and Raman scattering.

7. (Exercise 13-9) Derive the Henderson-Hasselbalch equation (13-4).

8. (Exercise 13-10) The phenolic —OH group of tyrosine has a pK of about 10.95 at 25°C. Suppose you determine from spectrophotometry that the pK decreases by 0.29 units as the temperature is increased from 15°C to 35°C. Determine $\Delta G°$, $\Delta H°$, and $\Delta S°$ of titration at 25°C.

9. (Exercise 13-18) The ^{17}O nucleus has a spin $I = 5/2$. How many orientations are allowed for this nucleus in a magnetic field?

● 10. (Exercise 13-25) A proton resonance A is 0.1 Hz wide in the absence of exchange, and widens to 1 Hz when exchange occurs (slow exchange limit). Calculate τ_A.

● 11. (Exercise 13-26) A coalesced proton resonance in the fast exchange limit is 0.1 Hz wide when exchange is infinitely fast, and 1 Hz wide when exchange is slowed. Calculate τ_A, assuming $P_A = P_B$, $\nu_A - \nu_B = 100$ Hz.

OTHER PROBLEMS

Problems on Optical Methods

12. How many normal modes of vibration do each of the following molecules have? (a) CO. (b) Acetylene. (c) Ethanol.

13. The following stretching vibrations occur in motions of proteins: $\overleftarrow{\mathrm{C}}{=}\overrightarrow{\mathrm{O}}$, $\overleftarrow{\mathrm{O}}{-}\overrightarrow{\mathrm{H}}$, $\overleftarrow{\mathrm{C}}{=}\overrightarrow{\mathrm{N}}$, $\overleftarrow{\mathrm{C}}{=}\overrightarrow{\mathrm{C}}$, and $\overleftarrow{\mathrm{N}}{-}\overrightarrow{\mathrm{H}}$. Which do you predict will contribute mainly to infrared absorption and which mainly to Raman scattering?

14. Explain the following observations.

    (a) The formation of a hydrogen bond by an O—H group of H—O—H is accompanied by an increase in the frequency of the H—O—H bending mode.

    (b) The formation of a hydrogen bond by an O—H group is accompanied by a shift of the proton resonance to lower magnetic field strengths.

    (c) Some nucleic acids isolated from cells have ESR spectra but after treatment with the chelating agent EDTA, they do not.

    (d) Polysaccharides do not absorb light in the visible or near UV region.

15. The $^{127}I_2$ molecule displays Raman scattering at frequency 213 cm^{-1}.

    (a) Estimate the force constant of I_2.

    (b) Estimate the energy required to stretch the I_2 bond length by 0.2 Å.

    (c) Estimate the fraction of molecules in the first excited vibrational state at 298 K.

16. (a) Predict whether the following motions are infrared active or inactive. (i) Vibration of O_2. (ii) Rotation of CO. (iii) Rotation of benzene. (iv) Rotation of phenol.

    (b) Predict whether the following motions are Raman active or inactive. (i) Vibration of O_2. (ii) Rotation of O_2. (iii) Rotation of benzene. (iv) Rotation of NH_3.

17. Sketch the energy of a molecule that undergoes absorption and then fluorescence as a function of the log of the time after the absorption.

18. The rotational Raman spectrum of $^{35}Cl_2$ consists of a series of lines separated by 0.9752 cm^{-1}. Calculate the bond length of Cl_2, given that the selection rule is $\Delta J = \pm 2$ for rotational Raman scattering.

19. A unit of light energy used in photochemistry and photobiology is the *einstein*, defined as a mole of photons.

    (a) Write an expression for the energy of an einstein as a function of the wavelength of light.

(b) Our visual pigment is rhodopsin, which has a strong absorption at about 500 nm. Compute the energy of photons of this wavelength in erg photon^{-1} and in kcal einstein^{-1}.

20. In applications of Equation 13-13 to determine the separation of donor and acceptor fluorescent groups, the value of E_t is determined from

$$E_t = \frac{\tau_D - \tau_A}{\tau_D}$$

in which τ_D is the fluorescence lifetime of the donor, and τ_A is the fluorescence lifetime of the donor in the presence of the acceptor. The fluorescence lifetime is the time required for the fluorescent molecule to return to the ground state.

Suppose that naphthyl and dansyl groups can be bound to two distinct sites on a protein. The naphthyl group has a value of τ_D of 23 ns when on the protein. When the dansyl group is added to the second site, the naphthyl fluorescence lifetime drops to 17 ns. What is the separation of the two groups?

21. Calculate the optical rotation α for the Na D line of a 500 mM solution of α-D-glucose in water at 20°C in a 1 cm cell. Use the value of +112.2 for $[\alpha]_D^{20}$.

Problems on Resonance Methods

22. Calculate the magnetic field strength for proton magnetic resonance at a frequency of 220 MHz. What is the energy in kcal mol^{-1} that corresponds to this transition?

23. Calculate the frequency of electromagnetic radiation required for electron spin resonance in a field of 2 T = 20,000 G. What energy difference in kcal mol^{-1} does this correspond to?

● 24. Draw a diagram similar to Figure 13-36(*b*) for the splitting of electron spin states by a ^{14}N nucleus.

● 25. The proton NMR spectrum of 2-bromo-5-chlorothiophene,

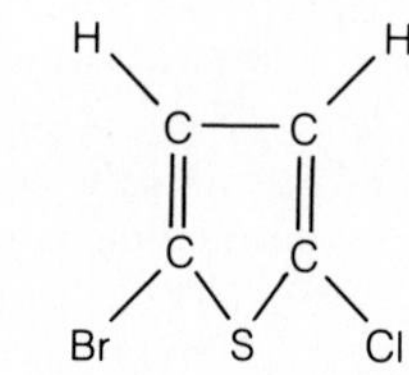

is

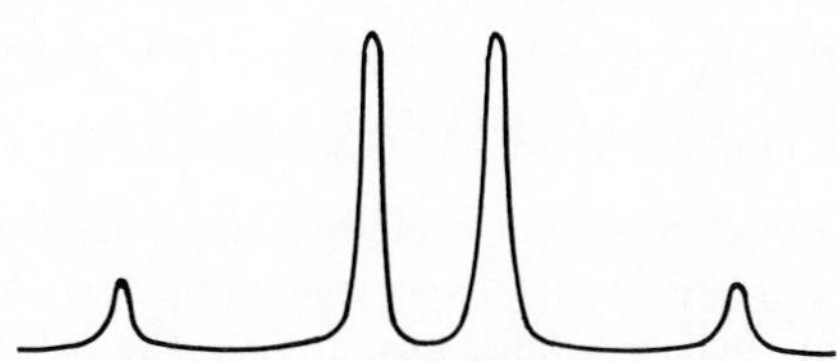

[Anderson, W. A., *Phys. Rev. 102,* 151 (1956)].

Draw an energy-level diagram (see Fig. 13-31) to explain the origins of the resonance lines. Why are two of the lines lower in intensity than the other two? What should be the spectrum of 2,5-dibromothiophene? What would you expect to see for the spectrum of 2-bromo-5-fluorothiophene? Draw schematic energy-level diagrams in each case.

26. Given that OH^- reacts with $R{-}NH_3^+$ with a diffusion-limited rate constant of $3.4 \times 10^{10} M^{-1}\ sec^{-1}$, calculate the pH at which exchange of the $R{-}NH_3^+$ protons with water would contribute a broadening of 50 Hz to the amino proton resonance line at 270 MHz.

27. Suppose that an NMR line in absence of exchange is 1 Hz wide, and that a 10% widening can be detected experimentally. What is the longest lifetime τ that can be measured?

28. Suppose that resonances A and B each are 1 Hz wide in absence of exchange, and that under conditions of rapid conversion $P_A = P_B$. What is the *shortest* lifetime that can be detected if a 10% widening of the average resonance can be measured, and $\nu_A - \nu_B = 3$ ppm at 100 MHz?

29. Electrons have a magnetic moment due to their *orbital* angular momentum. The component (m_z) of the magnetic moment along the z-axis (in SI units) is

$$m_z = \left(\frac{eh}{2\pi m_e}\right) m$$

in which m is the hydrogen atom quantum number. The transition from the electronic state $(1s)^2(2s)^2$ to the singlet state with electron configuration $(1s)^2(2s)(2p)$ shows a single spectral line in the absence of a magnetic field. However, in the presence of a magnetic field, three lines appear instead of one, which is called the *Zeeman effect*.

(a) Explain why three lines are observed.
(b) Calculate the Zeeman splitting in cm^{-1} between two adjacent lines in (a) above when the field is 1 T.

30. The volume magnetic susceptibility (χ_0) is defined as the magnetic moment per unit volume induced in a sample by a field B. Use the results of Exercise 13-23 to show that the nuclear contribution to χ_0 is $\chi_0 = N|m_z|^2/(k_B TV)$, in which N is the number of nuclei in volume V, and calculate the nuclear χ_0 for protons in water at T = 303 K. (The diamagnetic contribution to χ_0 is about -10^{-6}.)

Problems on Light Scattering

31. Show that light scattering measures the weight-average molecular weight.

32. Show that the radius of gyration of a rod of uniform density and length b is $b/\sqrt{12}$.

33. Light scattering by a high-molecular-weight solution leads to an apparent absorbance of light as measured by most spectrophotometers, because their optics are set to collect only light emerging near $\theta = 0$. The total amount of scattered light can be calculated by integrating I_s over the surface of a sphere. For a fixed value of θ rotated about the $\theta = 0$ axis, the area element is $2\pi r^2 \sin\theta\, d\theta$. Replace $\sin^2\phi$ in Equation 13-47 by $1 + \cos^2\theta$ to take account of the use of unpolarized light, resolve the equation for I_s, multiply by the area element and integrate

from $\theta = 0$ to π to show that the total scattered light, $I_0 - I_{\text{transmitted}}$, is $(16\pi/3)R_\theta I_0$. Take the radius of the sphere as r.

34. Use the results of Problem 33, and the assumption that $I_{\text{transmitted}}$ is nearly equal to I_0, to show that the apparent absorbance of a turbid (scattering) solution is $A = (16\pi/3)\ R_\theta/2.303$.

35. A suspension of a bacterial virus of molecular weight 200 million (size ~ 100 Å) and concentration 0.5 mg/ml shows considerable turbidity. When a small amount of phenol is added to rupture the virus and release DNA molecules of molecular weight 100 million, plus smaller protein molecules, no turbidity is visible. Explain this observation. [*Hint:* You will have to consider $P(\theta)$.]

36. The following values of $c_B I_0/r^2 I_s$ were measured by Outer, Carr, and Zimm [*J. Chem. Phys. 18*, 830 (1950)] using polarized light to study the scattering of polystyrene in butanone at 20°C. [$n_0 = 1.378$, $dn/dc_B = 0.214$ ml/g, $\lambda = 546$ nm, $c_B = 0.312$ mg/ml, ϕ (Fig. 12-4) = 90°].

θ/DEG	$[c_B I_0/r^2 I_s]$/g cm^{-2}
26.	1.58
36.9	1.66
66.4	1.82
90.	1.95
113.6	2.17
127.	2.28

Assume that c_B is low enough that the second virial term can be neglected, and extrapolate to zero angle to obtain M. Use the results to calculate R_G.

QUESTIONS FOR DISCUSSION

37. (a) You wish to devise a method for detecting small amounts of a drug in the blood of athletes. What spectroscopic tools might you use and what preliminary experiments might you perform?

(b) What are the essential spectroscopic properties of an effective suntan oil?

(c) Why are many electron-transport proteins colored?

(d) Would it be an advantageous adaptation for humans to have the wavelength range of reception of their retinas extended to longer values (into the infrared)?

FURTHER READING

Bloomfield, V. A., Crothers, D. M., and Tinoco, I., Jr. 1974. *Physical Chemistry of Nucleic Acids*. New York: Harper & Row. This compendium describes many spectroscopic studies of nucleic acids.

Bovey, F. A. 1972. *High Resolution NMR of Macromolecules*. New York: Academic Press. Contains a summary of principles and many applications to polymers, both biological and synthetic.

Dwek, R. A. 1972. *Nuclear Magnetic Resonance in Biochemistry*. Oxford: The Clarendon Press. Another excellent monograph on NMR, but more narrowly focused on application to enzymes.

Hirs, C. H. W., and Timasheff, S. N., ed. 1972 and 1973. *Methods in Enzymology*, vols. 26 and 27. New York: Academic Press. These two volumes contain 66 articles on various aspects of biophysical chemistry, including spectroscopic techniques. Some articles contain only applications to proteins; others include much background information.

Leach, S. J., ed. 1969 and 1970. *Physical Principles and Techniques of Protein Chemistry, Parts A and B*. New York: Academic Press. Several authors have contributed a total of 16 articles on spectroscopic and transport methods in protein chemistry. Each chapter contains background material as well as applications.

Slayter, E. M. 1970. *Optical Methods in Biology*. New York: John Wiley & Sons. Aimed mainly at microscopists, this book contains a great deal of useful information on optics, diffraction, and spectroscopy.

Van Holde, K. E. 1971. *Physical Biochemistry*. Englewood Cliffs, N.J.: Prentice-Hall. A useful introduction to several techniques of biophysical chemistry, including an especially helpful chapter on CD and ORD.

Walton, A. G., and Blackwell, J. 1973. *Biopolymers*. New York: Academic Press. This book is more of a compendium than a text but it does contain many references to papers on biochemical spectroscopy.

OTHER REFERENCES

Adler, A. J., Greenfield, N. J., and Fasman, G. D. 1973. In *Methods in Enzymology*, vol. 27, ed. C. H. W. Hirs and S. N. Timasheff. New York: Academic Press.

Brand, J. C. D., and Speakman, J. C. 1975. *Molecular Structure*. London: Edward Arnold, Ltd.

Donovan, J. W. 1969. In Leach (1969), cited in Further Readings.

Edsall, J. T., and Wyman, J. 1958. *Biophysical Chemistry*. New York: Academic Press.

Frazer, R. D. B., and Suzuki, E. 1970. In Leach (1970), cited in Further Readings.

Goddu, R., and Delker, D. 1960. *Anal. Chem. 32,* 140.

Holzwarth, G. 1972. In *Membrane Molecular Biology*, ed. C. F. Fox and A. D. Keith. Stamford, Conn.: Sinauer Associates, Inc.

Klevan, L. 1978. Thesis, Yale University.

Lehrer, S. S., and Fasman, G. D. 1967. *J. Biol. Chem. 242,* 4644.

Lehninger, A. L. 1972. *Biochemistry*. New York: Worth Publishing Co.

Lenard, J., and Singer, S. J. 1966. *Proc. Natl. Acad. Sci. U.S.A. 56,* 1828.

Mehlhorn, R. J., and Keith, A. D. 1972. In *Membrane Molecular Biology,* ed. C. F. Fox and A. D. Keith. Stamford, Conn.: Sinauer Associates, Inc.

Shum, B. 1977. Thesis, Yale University.

Stryer, L. 1968. *Science 162,* 526.

Stryer, L., and Haugland, R. P. 1967. *Proc. Natl. Acad. Sci. U.S.A. 58,* 719.

Susi, H. 1972. In *Methods in Enzymology,* vol. 26, ed. C. H. W. Hirs and S. N. Timasheff. New York: Academic Press.

Tanford, C. 1961. *Physical Chemistry of Macromolecules.* New York: W. J. Wiley.

Timasheff, S. N., Susi, H., and Rupley, J. A. 1973. In *Methods in Enzymology,* vol. 27, ed. C. H. W. Hirs and S. N. Timasheff. New York: Academic Press.

PART FOUR

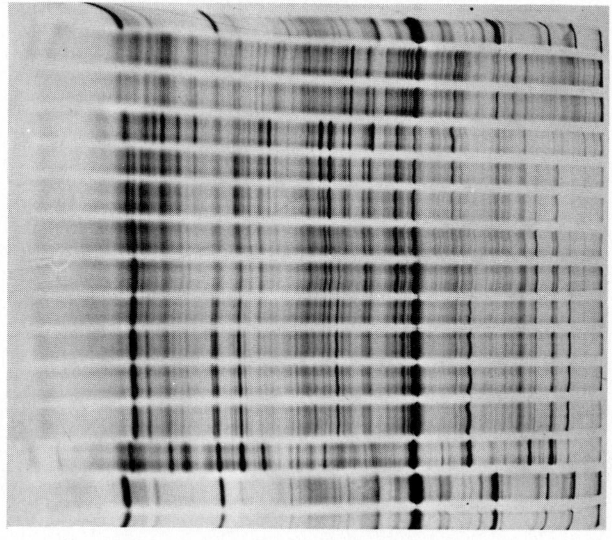

Bridging the Macroscopic and Microscopic

CHAPTER 14

Statistical Mechanics

○ 14–1 Statistical Mechanics Provides the Connection between Molecular and Macroscopic Properties

In Parts One and Two of this book we developed a thermodynamic description of the macroscopic properties of matter. Part Three focused on the behavior of isolated molecules or of small numbers of interacting molecules. Since a macroscopic state of matter consists of a collection of many individual molecules, you might expect that the macroscopic properties can be derived from molecular parameters. However, the derivation does not look easy because of the huge number of molecules (of the order of magnitude $N_A = 6 \times 10^{23}$) that make up a macroscopic system. The branch of physical chemistry that deals with this problem is called *statistical mechanics*.

The term *statistical* appears because the enormous number of molecules in a macroscopic system forces us to rely on statistical averages. It would be impossible to write equations governing the motion of 10^{23} particles and solve these equations simultaneously for macroscopic properties such as pressure, volume, energy, and entropy. However, as you will see in this chapter, it turns out to be unnecessary to keep track of all molecules individually; rather, we need only to calculate the *probability* that the molecules are in particular energy states. For example, for an ideal monatomic gas the only energy contribution considered is the kinetic energy of motion of the particles. Quantum mechanics give us the energy levels, ϵ_n, for motion in one of the three dimensions of ordinary space. In our study of statistical mechanics (Fig. 14-1) we will calculate the probability that a molecule is in each of these energy states; this probability turns out to be proportional to $\exp(-\epsilon_n/k_BT)$. We then use this result to calculate the total energy of a collection of such molecules. As we will see, all of the thermodynamic properties of an ideal gas can be calculated with high accuracy by the methods of statistical mechanics. In this chapter the energy of a macroscopic system (E) will be distinguished from the quantum mechanical energy levels of a molecule by reserving the symbol ϵ_n for the latter.

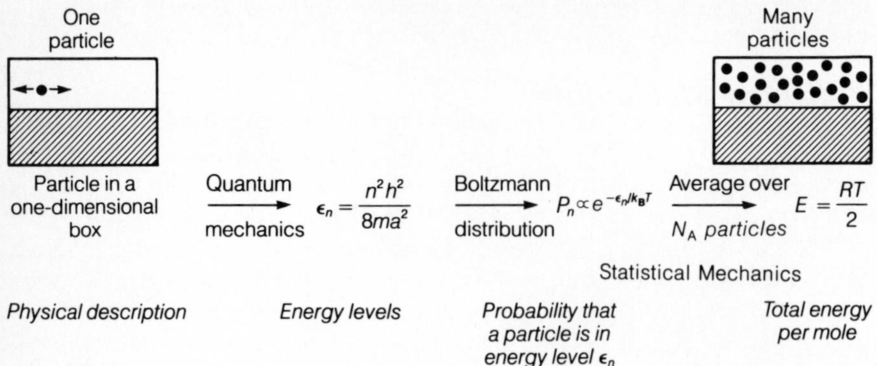

FIGURE 14-1
Schematic outline of the procedure by which statistical mechanics is used to calculate the energy per mole of particles moving in a one-dimensional box. The equation for the energy levels ϵ_n comes from Chapter 10. Equations for the probability of occupancy of each energy level and for the average energy will be developed in this chapter.

For certain ideal states of matter, such as crystals and the ideal gas, statistical mechanics is highly accurate and relatively simple. However, the application of statistical mechanics to disordered, condensed systems, such as liquids and solutions, is a still-developing science. For these complex problems, heavy reliance currently is placed on the calculation power of high-speed computers. In *molecular-dynamics* calculations, for example, the computer keeps track of the position and velocity of all particles in the sample as they collide with each other. Even though the sample size must be restricted to a few hundred molecules (the limit is set by computer memory size and calculation speed), many interesting equilibrium and dynamic properties of condensed states can be investigated.

One of the virtues of statistical mechanics is that it gives a conceptual basis for thermodynamics. The quantities defined in thermodynamics (especially entropy, free energy, and the distinction between heat and work) are abstract and always difficult for students to grasp. Statistical mechanics provides a means of thinking about these quantities in concrete physical terms. The theory originally was developed in the latter part of the nineteenth century by Boltzmann in Austria, Maxwell in England, and Gibbs in the United States, and it formed a logical basis for the science of thermodynamics. The advent of quantum mechanics considerably altered the formalism of statistical mechanics, providing an explanation for some early failures of the theory. We will not develop the formalism of the older "classical" statistical mechanics, in which potential and kinetic energy were considered as continuous rather than quantized variables. Classical and quantum statistical mechanics differ in their prediction only when the temperature is so low that the lowest-lying energy levels are the only ones occupied. Once the temperature is high enough that ϵ_n/k_BT is small, so that a particle is likely to be in a high energy level [$P_n \propto \exp(-\epsilon_n/k_BT) > 0$ for large n], then the special effects of quantizing the energy vanish.

14–2 Entropy is a Measure of Disorder or Degeneracy

It is often said that entropy is a measure of disorder. Statistical mechanics gives quantitative expression to this concept. Suppose that a collection of molecules has a total energy E. Usually the molecules can be arranged in a number of different ways to yield the same total energy E—for example, by having some particles moving faster than others or being oriented in different directions. The number of ways a state with fixed total energy can be achieved is called the *degeneracy* of that state, to which we give the symbol Ω. The entropy of the system is related to the degeneracy by the equation proposed by Boltzmann:

$$S = k_B \ln \Omega \qquad (14\text{-}1)$$

where the proportionality constant k_B is the *Boltzmann constant.*

This relationship is one of the central equations of statistical mechanics. It states that the entropy of a macroscopic system in a given state is proportional to the logarithm of the number of arrangements that lead to that state. Since Ω is a statistical quantity, and S is a thermodynamic property, Equation 14-1 establishes the connection between thermodynamics and statistical mechanics. The derivation of this equation is implicit in the sections that follow. We will see that maximization of the degeneracy Ω for a system at constant volume and energy allows us to predict thermodynamic results such as the dependence of the energy of an ideal gas on $3RT/2$, or the dependence of free energy on the logarithm of partial pressure or concentration. Since the second law of thermodynamics states that entropy also is maximized at equilibrium in a system of fixed volume and energy, we conclude that degeneracy Ω and entropy S are related quantities.

Statistical mechanics can be developed without any reference to thermodynamics, without using any of the names such as entropy and free energy that arise in thermodynamics. However, it is found that statistical mechanics yields equations exactly analogous to those derived from the laws of thermodynamics. From this correspondence it can be concluded that a statistical mechanical variable such as $k_B \ln \Omega$ must be the same as the thermodynamic variable S, with a similar correspondence for other quantities such as the free energy. For the purpose of learning it is simpler if we point out each correspondence as we go along. In that spirit, Equation 14-1 is offered here.

If entropy depends on disorder as measured by degeneracy, the proportionality of S to $\ln \Omega$ is inescapable. For example, consider a nuclear spin ($s = \frac{1}{2}$) in a magnetic field so weak that the two possible orientations of the spin (+ parallel to and − opposed to the field; see Chap. 13) have essentially identical energies. Then the degeneracy due to the nuclear spin state is $\Omega = 2$. If there are two independent nuclei, each spin can be in either of the two states + and −: (++, + −, − +, − −). The degeneracy Ω is the product of the number of ways of orienting each spin, $\Omega = 2^2 = 4$. Similarly, if there are N spins, $\Omega = 2 \times 2 \times 2 \times 2 \times \cdots = 2^N$. In general, if we have two

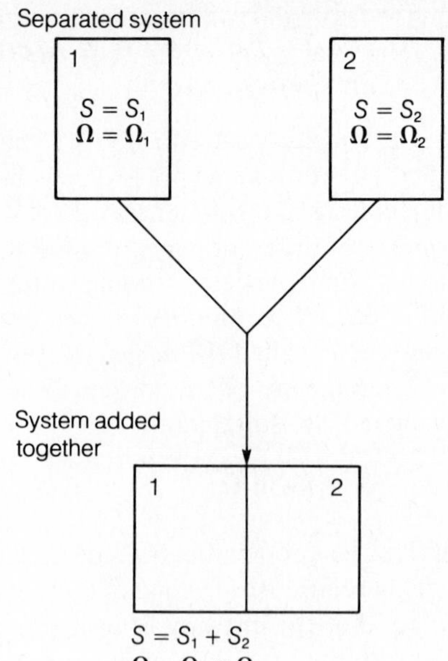

FIGURE 14-2
Entropy is an extensive (additive) property, while degeneracy is multiplicative. Therefore S must be proportional to ln Ω.

samples of matter, one with degeneracy Ω_1, and the other Ω_2, the total degeneracy is the product, $\Omega = \Omega_1\Omega_2$, since for each state of Sample 1 there are Ω_2 states of Sample 2. But because entropy is an extensive property (Sec. 1-2), the total entropy is additive: $S = S_1 + S_2$. The only way to obtain an additive function S from the multiplicative function Ω is to set S proportional to ln Ω, or $S = k_B \ln \Omega$ (Fig. 14-2). In short, because when two systems are combined, their entropies add but the number of arrangements open to the combined system multiply, there must be a logarithmic relationship.

The Boltzmann constant has the value $k_B = 1.38 \times 10^{-16}$ erg K^{-1}. As we will see when we consider the statistical mechanics of an ideal gas, the product of Avogadro's number N_A and k_B is the gas constant R:

$$k_B N_A = R$$

R has the value of 1.99 cal K^{-1} or 8.31 J K^{-1}.

EXERCISE 14-1

Calculate the entropy contributed to a crystal of one mole of carbon monoxide if the molecules can orient with the C═O bond axis in either direction. Assume the energy of the crystal is independent of which way the molecules are oriented. (Which is nearly, but not quite, true).

ANSWER

Each molecule can be in either of two orientations. Therefore, the degeneracy of the crystal is $\Omega = 2^{N_A}$. Hence $S = k_B \ln 2^{N_A} = N_A k_B \ln 2 = R \ln 2$, which is 1.38 cal K^{-1} mol^{-1} = 5.76 J K^{-1} mol^{-1}.

The degeneracy Ω of a macroscopic system is an extraordinarily large number. When $\Omega = 2^{N_A}$, it is approximately $10^{1.8 \times 10^{23}}$, or 1 followed by 1.8×10^{23} zeroes. Furthermore, since the thermodynamic quantity entropy is proportional to the logarithm of Ω, large errors can be made in the estimation of Ω without appreciable effect on the value of S. The following exercise shows the mathematical basis for this.

EXERCISE 14-2

Suppose that our estimate of $\Omega = 2^{N_A}$ is too large by a factor of 10^{1000}. Calculate the resulting error in the entropy.

ANSWER

$\Omega_{true} = 2^{N_A}/10^{1000}$, so $S_{true} = k_B (N_A \ln 2 - 2300)$. Therefore, $S_{true} = k_B (4.17 \times 10^{23} - 2300)$. The correction factor 2300 is negligible compared to 10^{23}, so there is no change in the estimated entropy.

The relationship between entropy and degeneracy provides a simple conceptual basis for the second law of thermodynamics (Chap. 3). One way of stating the law is that in an isolated system a spontaneous process will be accompanied by an increase in entropy. In an isolated system the total energy stays constant, and the degeneracy, Ω, is just the number of ways of achieving that energy state. The system will change from State 1 to State 2 if $\Omega_2 > \Omega_1$, because there are many more ways of achieving State 2 than State 1, and State 2 is therefore much more probable.

As an example, let us compare an unlikely (ordered) state with a much more likely set of disordered states. Suppose you have ten books on your desk and, driven to desperation by studying physical chemistry, you decide to divert your thoughts by setting the books in alphabetical order. There is only one correctly ordered state, so $\Omega_1 = 1$. All the other arrangements of the books constitute the disordered state, State 2. We can calculate the total number of arrangements of the books by noting that the first book, chosen at random, could be any one of ten, leaving nine possibilities for the second book, eight for the third, and so on. Therefore, the total number of arrangements, $\Omega_1 + \Omega_2$ is

$$\begin{aligned} \Omega_1 + \Omega_2 &= 10 \times 9 \times 8 \times 7 \times \cdots \times 2 \times 1 \\ &= 10! = 3{,}628{,}800 \end{aligned}$$

(The product 10! is read "10 factorial.")

The probability that the books would happen by chance to fall into the ordered state is $\Omega_1/(\Omega_1 + \Omega_2) = 1/3{,}628{,}800 = 2.76 \times 10^{-7}$. The disordered state has probability $\Omega_2/(\Omega_1 + \Omega_2)$ or 0.99999972. Even with a sample as small as 10 objects, the probability of spontaneously ordered states is very small. When the sample contains 10^{23} molecules, the probability of suddenly forming an ordered state, such as ice from water at 20°C, is negligible.

The Boltzmann equation allows us to calculate the entropy of the ordered and disordered states of the books on the desk. Using $S = k_B \ln \Omega$, we find

$S_1 = k_B \ln 1 = 0$, and $S_2 = k_B \ln \Omega_2 = 1.38 \times 10^{-16}$ erg $K^{-1} \times \ln(3{,}628{,}799) = 2.08 \times 10^{-15}$ erg K^{-1}.

You should keep in mind that calculation of the relative probabilities of different states using the degeneracy Ω can be done only when the total energy is constant. Systems open to exchange of energy cannot be thought of in this simple way. The results of the next section will permit us to consider the statistical-mechanical interpretation of energy changes.

14–3 The Probability of a Quantum State is Proportional to exp $(-\epsilon_n/kT)$

One of the main uses of statistical mechanics is to calculate the expected relative populations of the quantum states available to a molecule. Consider the problem in the following terms: each particle in a macroscopic sample has quantized energy levels ϵ_n—for example, the one-dimensional translational energy levels are $\epsilon_n = n^2h^2/8ma^2$. Assume that there are N particles and total energy E. Our problem is to calculate the number of molecules that are found in each energy level. Suppose we could examine at any instant the energy state of a particular molecule. The probability, P_n, that it would be in the state with quantum number n is, by definition,

$$P_n = N_n/N \tag{14-2}$$

in which N_n is the number of particles with quantum number n.

The set of numbers P_n describes the distribution of molecules among the energy levels n, so it is called a *distribution function*. There are only two physical restrictions on how the molecules are distributed, arising from constancy of both the total energy and the total number of particles. The energy constraint means that the sum of the individual molecular energies must be E, or

$$\sum_n N_n\epsilon_n = E \tag{14-3}$$

while the total number of particles must be N,

$$\sum_n N_n = N \tag{14-4}$$

Since there are only these two simple constraints on how the molecules can be distributed among the energy levels, you might wonder how it is possible to solve for the function which determines P_n. To appreciate the problem more clearly, consider Figure 14-3, which shows three possible distributions of 10 molecules over equally spaced energy levels. All three distributions have the same total energy, and therefore are permitted by the constraints of Equations 14-3 and 14-4. Since the three distributions are very different, how can we expect to describe the actual distribution by a single equation?

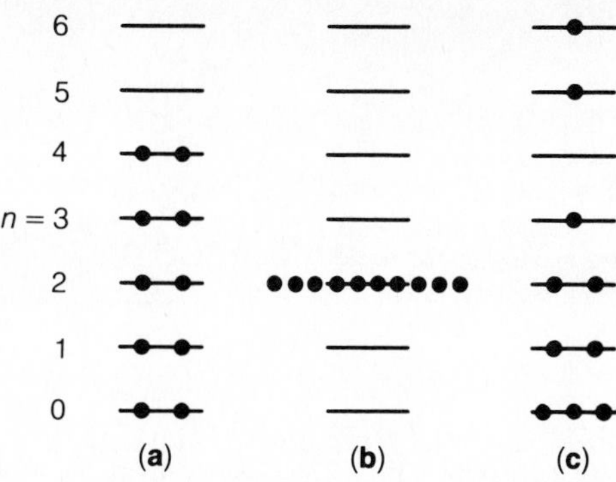

FIGURE 14-3
Three possible distributions of molecules (●) over seven equally spaced energy levels. All distributions have the same total energy, *E*, and the same number of molecules, *N*. All three distributions are equally probable, but there are more distributions like **(c)**—which resembles the Boltzmann distribution—than any other kind, so the system of ten molecules is most likely to be found with a distribution like **(c)**. The distribution is described by the set of numbers N_n. For example, for **(c)**, $N_0 = 3, N_1 = 2, N_2 = 2, N_3 = 1, N_4 = 0, N_5 = N_6 = 1$. The distribution also can be characterized by the set of numbers $P_n = N_n/N$.

The factor which distinguishes one distribution from another is the *degeneracy,* the number of distinguishable substates. For example, there is only one substate that has all molecules in the $n = 2$ energy level, as shown in Figure 14-3(*b*). On the other hand, assume that the molecules are individually identifiable, and imagine interchanging molecules in Figure 14-3(*c*): for example, Molecule 1 in energy level $n = 0$ could be interchanged with Molecule 4 in energy level $n = 1$. This interchange provides a new substate, but leaves the distribution unaltered, because the distribution is characterized only by the *number* of molecules in each level.

The distribution over energy levels can be found by utilizing the following principle: *the greater the number of substates which contribute to a distribution, the greater is the probability of that distribution.* By determining the distribution (P_n) that maximizes the degeneracy, it is found that one distribution has a much greater degeneracy than any other, and that all other distributions can be neglected (assuming a large number of molecules). The result, derived in the following paragraphs, is one of the most important equations in statistical mechanics:

$$P_n = \frac{\exp(-\epsilon_n/k_B T)}{\sum_n \exp(-\epsilon_n/k_B T)} \tag{14-5}$$ ◀

This is called the *Boltzmann distribution* (Fig. 14-4). It gives the probability P_n that a molecule chosen at random will be in the quantum state n, characterized by energy ϵ_n. In what follows, we will abbreviate this phrase by saying that P_n measures the probability of quantum state n. Equation 14-5 reveals that the state of highest probability is the state of lowest energy (the ground state). However, the probability of excited states increases at the

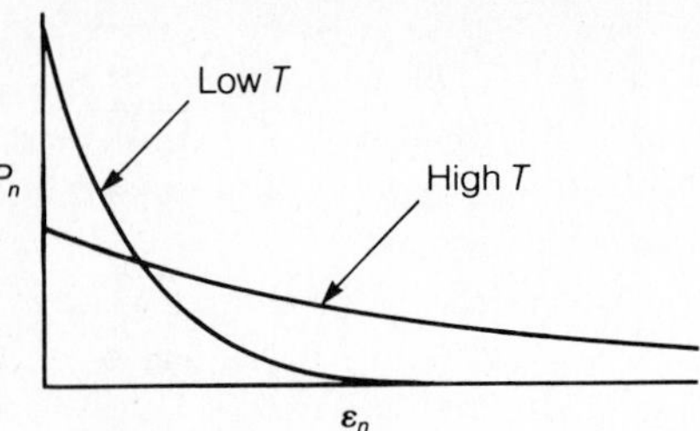

FIGURE 14-4
The Boltzmann distribution P_n as a function of energy ϵ_n at two temperatures. P_n decreases exponentially with ϵ_n in both cases, but high temperature causes the excited states to be relatively more populated than at a low temperature.

expense of the ground states as temperature increases because of the dependence of P_n on $\exp(-\epsilon_n/k_BT)$. This phenomenon is called *thermal excitation* of the system. Notice from Equation 14-5 that the sum of P_n over all values of n is 1 at all temperatures.

Derivation of the Boltzmann Distribution

We will outline here the derivation of Equation 14-5; you can find a full account in the advanced books given in the reading list. Our problem is to find the distribution that maximizes Ω. To do so we need to express Ω mathematically. Consider a crystal containing N particles (atoms or molecules), each of which has available the energy levels ϵ_n. The degeneracy Ω is the total number of ways the energy $E = \sum_n N_n\epsilon_n$ can be distributed among the $N = \Sigma N_n$ particles. By a distribution we mean the set of numbers $N_n = NP_n$ that describes the way the energy is allocated. The only thing that matters for a distribution is the *number* of particles in each energy level, not which particles have which energy. There are many possible distributions, and each distribution generally will have a large degeneracy Ω' because of the different ways of assigning the individual particles so that there are N_n in energy level ϵ_n. The total degeneracy Ω is the sum of Ω' values for all possible distributions:

$$\Omega = \Sigma\Omega'$$

It turns out that one distribution (namely the one with $P_n \propto e^{-\epsilon_n/kT}$) can be achieved in so many more ways than any other that we can approximate Ω adequately by the number of ways (Ω_B) of achieving that particular (Boltzmann) distribution, or

$$\ln \Omega \cong \ln \Omega_B \tag{14-6}$$

(See the discussion of the influence of errors on $\ln \Omega$ in Exercise 14-2.) Hence our problem is to find the *most probable distribution,* i.e., the one which maximizes Ω.

Each particle in the crystal can be distinguished from the others because it occupies a defined position in space. Therefore, we can number the particles 1, 2 . . . N. A *substate* of the crystal will have Particle 1 in energy level ϵ_i, Particle 2 in energy level ϵ_j, and so on. Our initial problem is to find the number Ω' of single substates there are in a distribution that has N_1 particles in energy level ϵ_1, and so on. We can solve this problem by the following logic (Fig. 14-5). We line up a series of boxes corresponding to the energy levels ϵ_n, with each box to accept N_n particles from the crystal. We pick particles at random from the crystal and assign the first N_1 to box ϵ_1, the next N_2 to box ϵ_2, and so on. The number of ways this can be done is equal to the number of different orders in which the particles can be chosen from the crystal. The first particle can be chosen from any of N. Consequently there are N different ways of beginning. There are $N - 1$ particles remaining, so the second step can be taken in $N - 1$ ways. Hence the number of ways of taking the first two steps is $N(N - 1)$. Continuing through all steps yields $N! = N(N - 1)(N - 2)(N - 3) \ldots (3)(2)(1)$.

The problem is not quite solved, however, because we have overcounted the ways of achieving a given distribution. Suppose, for example, that Particles 1 and 2 are both in energy level ϵ_1. It makes no difference whether we first place Particle 1 in Box ϵ_1, then Particle 2, or the other way around. Since these orders of choice are counted as being different in $N!$, we have to correct by dividing by the number of orders that should not be considered distinct. By the previous logic, the N_1 particles in Level 1 can be chosen in $N_1!$ different orders, and similarly for all other energy levels. Hence the degeneracy Ω' of a given distribution is $N!$ divided by the product over all $N_n!$:

$$\Omega' = \frac{N!}{\prod_n N_n!} \tag{14-7}$$

The next problem is to find the distribution that maximizes Ω', which will be the Boltzmann distribution. The solution requires some slightly complicated mathematics. We first note Stirling's approximation for $\ln N!$:

$$\ln N! = N \ln N - N \tag{14-8}$$

This approximation is good to 13% when $N = 10$, and improves for larger N. Then we convert the problem of maximizing Ω' to finding the maximum of $\ln \Omega'$ by combining Equations 14-7 and 14-8:

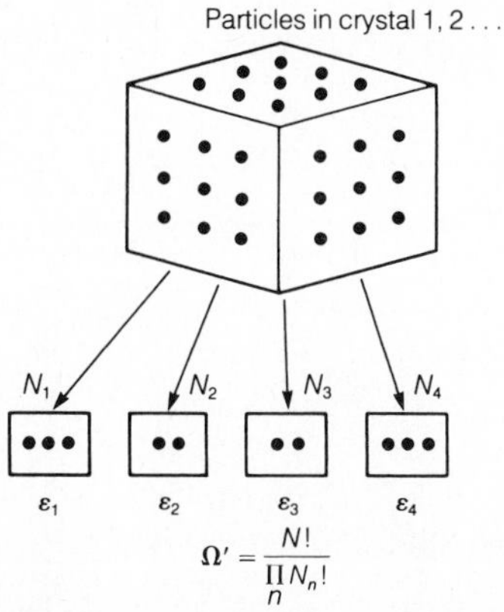

FIGURE 14-5
Determination of the degeneracy Ω' for a given distribution. The particles can be chosen in $N!$ different orders and placed sequentially in the boxes. Substates that differ in the order in which the particles in a given box are inserted are not different, so $N!$ must be divided by $\prod_n N_n!$ to obtain Ω'.

$$\begin{aligned}\ln \Omega' &= \ln N! - \sum_n \ln N_n! \\ &= N \ln N - \sum_n N_n \ln N_n \\ &= \left(\sum_n N_n\right) \ln \sum_n N_n - \sum_n N_n \ln N_n \end{aligned} \tag{14-9}$$

Thus the problem is to find the values of N_n that maximize Equation 14-9, subject to the restrictions

$$\Sigma N_j = N \tag{14-10a}$$

$$\Sigma \epsilon_j N_j = E \tag{14-10b}$$

This problem is solved using Lagrange's method of *undetermined multipliers*. At the maximum of Ω' the differential of $\ln \Omega'$ must be zero:

$$d \ln \Omega' = \sum_j \left(\frac{\partial \ln \Omega'}{\partial N_j}\right)_{N_{n \neq j}} dN_j = 0 \tag{14-11}$$

For generality we multiply Constraint (14-10a) by the undetermined quantity α, and then take the differential, obtaining (since N is constant):

$$\alpha \Sigma \, dN_j = 0$$

Similarly, Equation 14-10b is multiplied by β to make it more general, yielding the differential

$$\beta \sum_j \epsilon_j \, dN_j = 0$$

Subtracting these two equations from Equation 14-11 gives

$$\sum_j \left(\frac{\partial \ln \Omega'}{\partial N_j} - \alpha - \beta\epsilon_j\right) dN_j = 0$$

This expression will be generally equal to zero only if each term is zero, or

$$\frac{\partial \ln \Omega'}{\partial N_j} - \alpha - \beta\epsilon_j = 0 \tag{14-11b}$$

The derivative $\partial \ln \Omega'/\partial N_j$ can be calculated using the chain rule and Equation 14-9 for $\ln \Omega'$. (Because all other values of N_n are held constant in taking the derivative, $\partial N_n/\partial N_j$ is zero unless $n = j$, in which case it is 1.) Thus,

$$\begin{aligned}\frac{\partial \ln \Omega'}{\partial N_j} &= \frac{\partial}{\partial N_j}[(\Sigma N_n) \ln \Sigma N_n - \Sigma N_n \ln N_n] \\ &= \frac{\Sigma N_n}{\Sigma N_n}\left[\Sigma\left(\frac{\partial N_n}{\partial N_j}\right)\right] + \Sigma\left(\frac{\partial N_n}{\partial N_j}\right) \ln \Sigma N_n \\ &\quad - \Sigma\left(\frac{\partial N_n}{\partial N_j}\right) \ln N_n - \Sigma \frac{N_n}{N_n}\left(\frac{\partial N_n}{\partial N_j}\right) \\ &= 1 + \ln \Sigma N_n - \ln N_j - 1 \\ &= \ln N - \ln N_j \end{aligned}$$

Substituting this result into Equation (14-11b) yields

$$\ln N - \ln N_j - \alpha - \beta\epsilon_j = 0$$

or

$$\ln N_j = \ln N - \alpha - \beta\epsilon_j$$

Taking the exponential of this equation, we get

$$N_j = Ne^{-\alpha}e^{-\beta\epsilon_j} \tag{14-12}$$

Equation 14-12 is the result we sought. It states that the occupancy of an energy state ϵ_j is proportional to $\exp(-\beta\epsilon_j)$. Comparison with thermodynamic expressions, which will be implicit in the following sections, shows that

$$\beta = \frac{1}{k_B T} \tag{14-13}$$

Furthermore, for normalization of Equation 14-12 so that $\sum_j N_j/N = 1$, it is apparent that

$$e^{\alpha} = \sum_j \exp(-\epsilon_j/k_B T) \tag{14-14}$$

Combining Equations 14-12, 14-13, and 14-14, and replacing the index j by n, gives the final result:

$$P_n = N_n/N = \frac{\exp(-\epsilon_n/k_B T)}{\sum_n \exp(-\epsilon_n/k_B T)} \tag{14-15}$$

This equation tells us how many particles to put in each quantum state n to maximize the degeneracy Ω' of the distribution over the energy levels.

○ *Physical Consequence of the Boltzmann Distribution*

According to the Boltzmann distribution, when energy is stored in a system many molecules are in their lower energy levels and only a few are excited to high energies. The reason for this is that excitation of one molecule to high energy reduces the number of possible substates of the other particles. It is unlikely that the available energy will be sufficiently localized on one particle to raise it to an energy level far above the average energy of all the molecules. A few energetically excited molecules do exist, however, and they frequently are responsible for important phenomena such as chemical reactions, which may require energy in order to overcome an initial "activation" barrier (see Chap. 6).

○ *14–4 Calculation of the Total Energy*

The problem we consider next is how to calculate the total energy, E, of a collection of particles, using the Boltzmann distribution given in Equation 14-5. This problem can range from rather simple to extremely difficult, depending on the nature of the system under study. The primary factor that determines the difficulty is whether the energy terms are independent of each other. The molecules in an ideal diatomic gas, for example, can have energy due to their translational motion, or due to rotation or vibration. If the quantum states for all three are independent of each other, we can write the energy as a sum of three independent terms:

$$\epsilon = \epsilon_{\text{trans}} + \epsilon_{\text{rot}} + \epsilon_{\text{vib}}$$

The total energy per particle, ϵ, will be a sum of the translational, rotational, and vibrational energies, and each of these can be considered as a separate and independent problem.

This separation of the modes of motion of a single molecule is only an approximation. For example, the rotational motion of a molecule can depend on the vibrational state if the vibration is vigorous enough to alter the bond length and therefore the moment of inertia of the molecule. However, we will not consider complications of this kind.

A very important factor in deciding the ease of calculating the average energy is whether the molecules are independent of each other. If they are, the energy of N particles is N times the average energy, $\langle\epsilon\rangle$, of an isolated particle,

$$\begin{aligned} E &= N\langle\epsilon\rangle \\ &= N(\langle\epsilon_{\text{trans}}\rangle + \langle\epsilon_{\text{rot}}\rangle + \langle\epsilon_{\text{vib}}\rangle) \end{aligned} \tag{14-16}$$

In this equation, $\langle\epsilon_{\text{trans}}\rangle$, $\langle\epsilon_{\text{rot}}\rangle$ and $\langle\epsilon_{\text{vib}}\rangle$ are the average translational, rotational, and vibrational energies.

You can appreciate the advantage of treating noninteracting molecules by considering what happens when the particles are *not* independent, as in a liquid. In that case the particles interact with each other, and the energy depends on how far apart they are, how fast the neighbors are moving, and so on. Then we no longer can obtain the total energy by multiplying the average energy for an isolated particle by N. We will not consider problems of that degree of difficulty here.

Let us continue with the example of an ideal one-dimensional monatomic gas to illustrate how the Boltzmann distribution may be used to calculate macroscopic energies for a collection of independent particles. Equation 14-5 expresses the probability that a particle is in a particular quantum state with energy ϵ_n. The total energy of N particles is

$$E = \sum_n N_n\epsilon_n = N\sum_n P_n\epsilon_n \tag{14-17}$$

Combining this equation with (14-5) gives us

$$E = \frac{N\sum_n \epsilon_n \exp(-\epsilon_n/k_B T)}{\sum_n \exp(-\epsilon_n/k_B T)} \tag{14-18}$$

The Molecular Partition Function

Equation 14-18 is a clumsy expression which can be simplified by using a shorthand notation: let q be

$$q = \sum_n \exp(-\epsilon_n/k_B T) \tag{14-19}$$ ◄

This summation is a very important quantity, called the *partition function*. Since the energies ϵ_n refer to the energy levels of a single molecule, we call q the *molecular* partition function, to distinguish it from the *system* partition function Q to be introduced later, in which macroscopic energies E_n replace ϵ_n (Sec. 14-9).

Now we want to express Equation 14-18 in terms of q. Notice that the expression in the numerator, $\epsilon_n \exp(-\epsilon_n/k_B T)$, can be calculated by differentiating $\exp(-\epsilon_n/k_B T)$, holding ϵ_n (and therefore the length of the box) constant:

$$\frac{\partial}{\partial T} \exp(-\epsilon_n/k_B T) = \exp(-\epsilon_n/k_B T) \frac{\partial}{\partial T}(-\epsilon_n/k_B T)$$

$$= \frac{\epsilon_n}{k_B T^2} \exp(-\epsilon_n/k_B T)$$

Therefore

$$\epsilon_n \exp(-\epsilon_n/k_B T) = k_B T^2 \frac{\partial}{\partial T} \exp(-\epsilon_n/k_B T) \tag{14-20}$$

The same equation holds for the entire summation:

$$\Sigma \epsilon_n \exp(-\epsilon_n/k_B T) = k_B T^2 \frac{\partial}{\partial T} \Sigma \exp(-\epsilon_n/k_B T)$$

Or, in terms of q,

$$\Sigma \epsilon_n \exp(-\epsilon_n/k_B T) = k_B T^2 \frac{\partial q}{\partial T} \tag{14-21}$$

Consequently, Equation 14-18 is

$$E = \frac{N k_B T^2}{q} \frac{\partial q}{\partial T}$$

Substituting $\partial q/q = \partial \ln q$, this equation becomes (with the box size, or more generally, the volume V held constant)

$$E = N k_B T^2 \left(\frac{\partial \ln q}{\partial T} \right)_V \tag{14-22}$$ ◀

Equation 14-22 provides us with an efficient expression for calculating the energy E of independent particles, since all we have to do is find q and its derivative with respect to T.

Degeneracy and the Partition Function

Equation 14-19 expresses q as a summation over the allowed molecular quantum states of energy ϵ_n. Sometimes an energy ϵ_n has more than one quantum state, as do the rotational energy levels of a diatomic molecule. The number of quantum states corresponding to an energy level ϵ_j is called the *degeneracy* of that energy level, and is given the symbol ω_j. If the summation for q is carried out over the energy levels, j, rather than over the quantum states, n, as in Equation 14-19, ω_j must multiply the term $\exp(-\epsilon_j/k_B T)$. Thus,

$$q = \sum_j \omega_j \exp(-\epsilon_j/k_B T) \tag{14-23}$$ ◀

Equation 14-23 is another way of expressing the molecular partition function q.

The Energy of Particles in a Box

As an example of the use of Equation 14-22 to calculate the energy, we consider the particle in a one-dimensional box. The energy levels are given by $\epsilon_n = n^2h^2/8ma^2$, in which $n = 1,2,3 \ldots$. Using the definition of q in Equation 14-19 we get for the molecular partition function for translational motion, q_{tr},

$$q_{tr} = \sum_{n=1}^{n=\infty} \exp(-n^2h^2/8ma^2k_BT) \tag{14-24}$$

In general, many translational levels are populated ($h^2/8ma^2k_BT \ll 1$), and Equation 14-24 can be approximated by an integral over n from zero to infinity because the terms in the summation are so close together in value:

$$q_{tr} \cong \int_0^\infty \exp(-n^2h^2/8ma^2k_BT)\, dn$$

This integral can be solved using the relationship

$$\int_0^\infty e^{-\alpha x^2}\, dx = \frac{1}{2}\sqrt{\frac{\pi}{\alpha}}$$

which you can find from a table of definite integrals. Hence the translational partition function is

$$q_{tr} = \frac{a\sqrt{2\pi mk_BT}}{h} \tag{14-25}$$ ◀

and, from Equation 14-22, the energy is

$$E = Nk_BT^2 \frac{\partial}{\partial T} \ln\left(\frac{a[2\pi mk_BT]^{1/2}}{h}\right)$$

$$= Nk_BT^2 \frac{\partial}{\partial T} \ln T^{1/2} = \frac{Nk_BT}{2}$$

For one mole of particles moving in one dimension, $N = N_A$ and $N_Ak_B = R$, so

$$E = \frac{RT}{2} \tag{14-26}$$ ◀

EXERCISE 14-3

Calculate the energy for translational motion in three dimensions.

ANSWER

Motion in each dimension is independent of motion in the other directions, so the energy will be three times as large as for motion in one dimension. Hence,

$$E = \frac{3}{2} RT$$

Notice that this equation gives the energy of an ideal monatomic gas, as introduced in Chapter 2.

The Energy of Harmonic Oscillators

We also can readily calculate the energy of a collection of independent harmonic oscillators of frequency ν. The energy levels are given by $\epsilon_n = (n + 1/2)\, h\nu$, $n = 0,1,2 \ldots$. Therefore, the molecular partition coefficient for vibrational motion, q_{vib}, is

$$\begin{aligned} q_{\text{vib}} &= \sum_{n=0}^{\infty} \exp\left[-\left(n + \frac{1}{2}\right) h\nu/k_B T\right] \\ &= \exp(-h\nu/2k_B T) \sum_{n=0}^{\infty} \exp(-nh\nu/k_B T) \end{aligned}$$

The sum is a familiar infinite series of the form

$$\sum_{n=0}^{\infty} x^n = \frac{1}{1 - x} \qquad |x| < 1$$

in which $x = \exp(-h\nu/k_B T)$. Therefore q_{vib} is

$$q_{\text{vib}} = \frac{\exp(-h\nu/2k_B T)}{1 - \exp(-h\nu/k_B T)} \tag{14-27}$$

When the temperature is high enough so that $h\nu/k_B T$ is much less than one, Equation 14-27 can be simplified. We use the power series expansion for an exponential, keeping only the linear term,

$$\exp(-h\nu/k_B T) \cong 1 - \frac{h\nu}{k_B T}$$

Inserting this approximation, and a similar one for $\exp(-h\nu/2k_B T)$ in Equation 14-27 we get

$$q_{\text{vib}} = \frac{1 - h\nu/2k_B T}{1 - [1 - h\nu/k_B T]}$$

or, neglecting $h\nu/2k_B T$ relative to 1,

$$q_{\text{vib}} = \frac{k_B T}{h\nu} \qquad (k_B T \gg h\nu) \tag{14-28}$$ ◀

For the energy per mole, Equation 14-22 yields

$$\begin{aligned} E &= N_A k_B T^2 \frac{\partial}{\partial T} \ln (k_B T/h\nu) \\ &= N_A k_B T = RT \qquad (k_B T \gg h\nu) \end{aligned} \tag{14-29}$$ ◀

This is the energy of one mole of one-dimensional harmonic oscillators in the limit of high temperature.

14–5 The Classical Principle of Equipartition of Energy

Since all of the expressions that we have derived for the total energy (such as Equations 14-26 and 14-29) come out in multiples of $RT/2$, you might suspect that there is some underlying general principle. That suspicion is correct. The principle is referred to as *equipartition of energy,* and it states that *at high temperature the energy contains a contribution RT/2 per mole for each degree of freedom in the partition of the energy.* One degree of freedom is supplied by each spatial coordinate, and one by each velocity component, on which the energy depends quadratically. For example, the kinetic energy of an ideal gas particle in three dimensions is calculated from the square of the x, y, and z components of the velocity. The particle energy does not depend on position, so the spatial coordinates provide no degrees of freedom. Hence, there are three degrees of freedom in the energy, and $E = (\frac{3}{2})RT$ (see Ex. 14-3). In contrast, a one-dimensional harmonic oscillator has one degree of freedom due to the kinetic energy of vibration, and another due to the potential energy, which depends on x^2. Hence $E = RT$ (see Eq. 14-29).

The principle of equipartition of energy applies only when the effects of quantizing the energy are unimportant. This happens when the spacing between the energy levels is small compared to $k_B T$.

$$\epsilon_{n+1} - \epsilon_n \ll k_B T$$

which ensures that many energy levels will be populated. Population of many levels resembles closely enough a continuous distribution of energies that classical equations can be substituted for quantum mechanical results.

We can use the equipartition principle to calculate the energy due to rotation of a diatomic molecule. Consider the diatomic molecule in Figure 14-6, which is lined up with the bond direction along the z-axis. The molecule is free to rotate about the x- and y-axes, with moment of inertia I. The atoms are considered to be points on the z-axis, so the moment of inertia for rotation about the z-axis is zero, and we cannot count that motion as a degree of freedom for partitioning the rotational kinetic energy. Furthermore, we are considering a free rotator, so there is no potential energy. Hence, there are two degrees of freedom, and $E = RT$ in the classical limit.

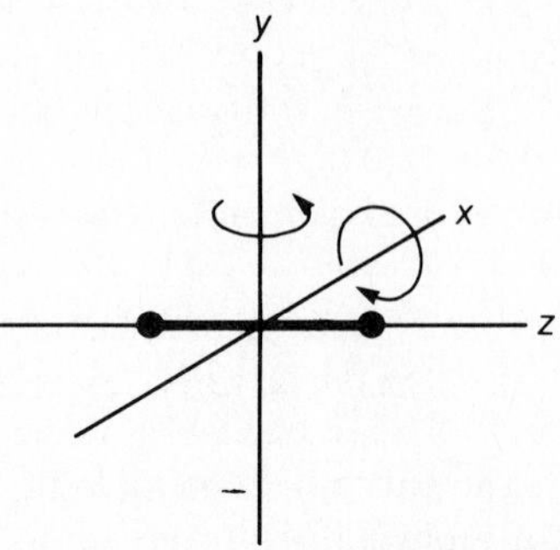

FIGURE 14-6
There are two degrees of freedom for rotation of a diatomic molecule, corresponding to rotation about the x and y axes. Rotation about the z axis does not contribute because the molecule has no moment of inertia about that axis. According to the principle of equipartition of energy, $E = RT$ per mole.

EXERCISE 14-4
Calculate the heat capacity, C_V, in the classical limit for an ideal diatomic gas.

ANSWER
Recall that $C_V = (\partial E/\partial T)_V$. The energy E is given by the sum of the translational $[(\frac{3}{2})RT]$, vibrational (RT), and rotational (RT) degrees of freedom, or

$$E = \frac{7}{2} RT$$

Hence

$$C_V = \frac{7}{2} R$$

Most diatomic gases do *not* obey this relationship, for reasons considered in the following section.

14–6 Quantum-Mechanical Effects Influence the Energy and Heat Capacity

Classical statistical mechanics predicts that the heat capacity of an ideal diatomic gas should contain contributions from translational $[(\frac{3}{2})R]$, rotational (R), and vibrational (R) motion, totaling $(\frac{7}{2})R$ (see Ex. 14-4). The measured heat capacity of most diatomic gases near room temperature is approximately $(\frac{5}{2})R$. This failure of the theory was a source of severe disappointment for Willard Gibbs and his scientific contemporaries, who could see no plausible basis for that result. We now know that the reason is the quantization of energy, but it was five years after Gibbs' death before this became apparent.

The culprit in the failure of the heat-capacity calculation for a diatomic gas is the lack of excitation of vibrational motion. The derivation of Equation 14-28 for the vibrational partition function assumes that $T \gg h\nu/k_B$. However, for most diatomic molecules at room temperature, $T < h\nu/k_B$ (see Sec. 13-2). When this condition applies, the relative probability of the first excited vibrational state, $P_1 \propto \exp[-(1 + \frac{1}{2})h\nu/k_B T]$, is much smaller than the probability of the ground vibrational state, $P_0 \propto \exp(-h\nu/2k_B T)$. Therefore, nearly all of the molecules will be in their ground vibrational states (see Fig. 14-7), and there are essentially no degrees of freedom for partitioning vibrational energy. This means that the vibrational heat capacity contribution is nearly zero. In qualitative terms, as diatomic molecules are heated at room temperature, the thermal energy available is too small to excite vibrations. Thus, no energy goes into vibrations, and the heat capacity is consequently smaller than it is at ~3000 K where the thermal energy ($k_B T$) is large enough to excite vibrations.

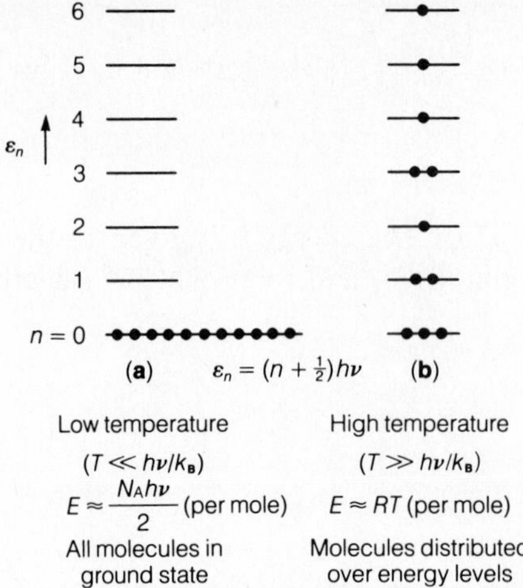

FIGURE 14-7
Comparison of the distribution over vibrational energy levels at low **(a)** and high **(b)** temperatures. All the molecules enter the ground state when T approaches 0 K. Increasing the temperature slightly does not provide enough energy to raise them to the first excited energy level, thus the system absorbs little energy for a given temperature increment. In **(b)** the molecules are distributed over the energy levels, and C_V has its classical value of R for a one-dimensional harmonic oscillator.

EXERCISE 14-5

Use Equation 14-27 to derive an expression for the energy and heat capacity per mole for harmonic oscillators when nothing is assumed about the value of $h\nu/k_B$.

ANSWER

$$E = N_A k_B T^2 \frac{\partial \ln q_{vib}}{\partial T}$$

$$= N_A \left(\frac{h\nu}{2} + \frac{h\nu e^{-h\nu/k_B T}}{1 - e^{-h\nu/k_B T}}\right)$$

The first term in the parentheses is the zero-point energy, which is the only contribution when $T = 0\ K$. Similarly

$$C_V = \left(\frac{\partial E}{\partial T}\right)_V$$

$$= \frac{R\left(\frac{h\nu}{k_B T}\right)^2 e^{-h\nu/k_B T}}{(1 - e^{-h\nu/k_B T})^2}$$

EXERCISE 14-6

Calculate the value of $h\nu/k_B$ for carbon monoxide, which absorbs infrared radiation at 4.66 μm (2144 cm^{-1}) because of its internal vibration.

ANSWER

The frequency $\nu = c/\lambda$. Hence $h\nu/k_B = hc/\lambda k_B = 3088$ K. This is the approximate temperature at which the CO vibrations become appreciably excited.

EXERCISE 14-7

Calculate the absorption wavelength of a vibration for which $h\nu/k_B = 295$ K (room temperature).

ANSWER

$\lambda = 48.8\ \mu\text{m}\ (205\ \text{cm}^{-1})$.

The rotational energy levels of diatomic molecules are more closely spaced than the vibrational levels. Therefore, rotational motion is thermally excited at lower temperature than is characteristic for vibrational motion. Still lower temperatures are sufficient to excite translational motion. Table 14-1 gives characteristic temperatures for excitation of vibrational (Θ_v) and rotational (θ_r) motion of diatomic molecules. The vibrational contribution to the heat capacity reaches half the classical value (R) when $T = \Theta_v/3$. Approximately the same relationship also holds for rotational motion.

Heat Capacity of Crystals and Proteins

Quantization of vibrational energy levels influences the heat capacity of solids. Atoms or molecules in a crystal vibrate through their equilibrium position when the temperature is high enough. For example, vibrational motion of the atoms in metals such as Fe, Al, and Cu becomes appreciably excited in the temperature range 60 K–150 K. Three-dimensional vibration is possible, so the limiting vibrational energy is $3RT$ when the temperature is high. Figure 14-8 shows schematically how the heat capacity of a crystalline

TABLE 14-1 *Characteristic Temperatures for Rotational and Vibrational Motion of Diatomic Molecules*

MOLECULE	Θ_v/K [a]	θ_r/K [b]
H_2	6210	85.4
N_2	3340	2.86
O_2	2230	2.07
CO	3070	2.77
NO	2690	2.42
HCl	4140	15.2
Cl_2	810	0.346

(a) $\Theta_r = h\nu/k_B$
(b) $\theta_r = h^2/(8\pi^2 I k_B)$

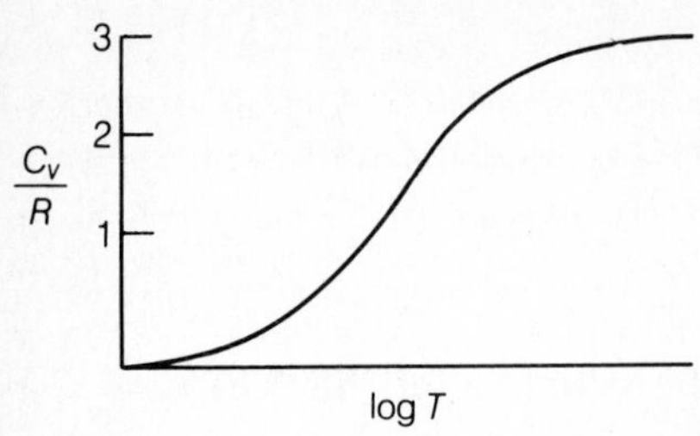

FIGURE 14-8
Schematic variation of the heat capacity of crystalline solids like Al, Cu, Pb, or C (diamond) with the logarithm of temperature. The temperature at the midpoint of the rise of C_V/R is different for different metals. (From Hill, 1960.)

solid increases from zero toward its limiting value of $3R$ as the temperature rises.

Sturtevant (1977) has pointed out that changes in the frequency of vibrational modes in proteins could contribute to the heat capacity changes that accompany conformational alterations. For example, unfolding or denaturation of proteins is characterized by a large increase in the heat capacity. Suppose that the protein has "soft" vibrational modes whose frequency is roughly $k_B T/h$; remember that stretching of chemical bonds generally occurs at much higher frequencies, so that the "soft" vibrations must correspond to distortions of the protein produced by coupled bending and rotation of several bonds. If the frequency of "soft" vibrations is decreased when the protein unfolds, the heat capacity will increase because there will be greater excitation of vibrational motion at the same temperature. Decreasing the vibrational frequency has the same effect as increasing the temperature in Figure 14-8.

The other main contributor to the heat capacity increase when proteins unfold is thought to be the increased ordering of water molecules produced by the enhanced exposure of hydrophobic groups to the solvent after unfolding (see Sec. 4-11). Ordered water molecules increase the heat capacity because heat uptake is required to "melt" such aggregates when the temperature increases.

14–7 A Statistical-Mechanical Interpretation of Heat, Work, Entropy, and Information

According to Equation 14-17, the total energy of a collection of N independent particles is $E = N \sum_n P_n \epsilon_n$, in which the ϵ_n are the energy levels. Our study of thermodynamics told us that the energy E can change either through the heat (Q) absorbed by the system or the work (W) done on the system, with $\Delta E = Q + W$. You can gain insight into the difference between heat and work by considering their statistical-mechanical interpretation. E is a sum of terms that are products of two quantities, P_n and ϵ_n. Either of these quantities can change when the conditions change, so the differential of E (assuming N is constant) is, from Equation 14-17,

$$dE = N \sum_n (P_n \, d\epsilon_n + \epsilon_n \, dP_n) \tag{14-30}$$

[Remember that $d(xy) = x\,dy + y\,dx$.] The two terms $N\Sigma P_n\,d\epsilon_n$ and $N\Sigma\epsilon_n\,dP_n$ correspond, respectively, to the reversible work and heat contributions to dE. You can understand this intuitively by considering an ideal gas. When a sample is heated at constant volume, the energy levels ϵ_n do not change because they depend on the size of the box but not on the temperature. Since $\Delta V = 0$, no work is done, and the energy change is due only to the heat absorbed, or $dE = dQ_{\text{rev}}$. Setting $d\epsilon_n = 0$ in Equation 14-30 yields the expression, at constant volume (where $dQ = dE = dQ_{\text{rev}}$),

$$(dE)_V = dQ_{\text{rev}} = N\sum_n \epsilon_n\,dP_n \tag{14-31}$$

In general, *heat absorbed or given off reversibly is accompanied by a change in the number of molecules in the various energy levels.* If heat is added to the gas, more molecules enter the higher excited states (dP_n positive), with fewer in the lower energy states (dP_n negative). All the heat energy, Q_{rev}, is accounted for by this shift of P_n values.

The work term must account for the remaining energy, or

$$dW_{\text{rev}} = -N\sum_n P_n\,d\epsilon_n \tag{14-32}$$

Thus, *reversible work corresponds to the energy change due to a shift in the energy levels,* with a fixed occupation of each level. In summary, you can think of a reversible heat increment as producing a change in the distribution over energy levels, and reversible work as reflecting a change in the energy levels themselves. Of course, *irreversible* heat and work changes do not follow this principle because their differentials are indeterminant. For example, an ideal gas expanding isothermally against zero external pressure does zero work, $W = 0$, and absorbs no heat, $Q = 0$. However, the changes in P_n and ϵ_n have the same values they would for a reversible isothermal expansion.

Understanding the thermodynamic definition of entropy ($dS = dQ_{\text{rev}}/T$) in terms of statistical mechanics is a little more complicated. We begin by noting that Ω, as given by Equation 14-7, can be written

$$\Omega = N!/\prod_n N_n!$$

$$= N!/\prod_n (P_nN)!$$

Using Stirling's approximation for the factorials gives

$$\ln\Omega = N\ln N - N - \sum_n (P_nN)\ln(P_nN) + \sum_n (P_nN)$$

Remembering that $\sum_n P_n = 1$, we obtain

$$\ln\Omega = -N\sum_n P_n\ln P_n$$

The differential of this equation (notice that $\sum_n dP_n = 0$) is

$$d\ln\Omega = -N\sum_n \ln(P_n)\,dP_n$$

Substituting the Boltzmann distribution gives us $\ln P_n = -(\epsilon_n/k_B T + \ln q)$, so

$$k_B \, d \ln \Omega = N \sum_n \frac{\epsilon_n}{T} dP_n + N k_B \ln q \sum_n dP_n$$

$$= N \sum_n \frac{\epsilon_n}{T} dP_n \qquad (14\text{-}33)$$

The left side of this equation is dS, because $S = k_B \ln \Omega$. According to Equation 14-31 the right side of Equation 14-33 is dQ_{rev}/T, so we arrive at the conclusion that

$$dS = dQ_{rev}/T$$

exactly as prescribed by the second law of thermodynamics. If you examine this derivation carefully you will see that it is the dependence of P_n on $\exp(-\text{energy}/T)$ that leads to the dependence of dS on heat (energy)/T.

Statistical mechanics also can be used to define the relationship between entropy and the amount of information needed to describe a system precisely. For example, in a computer, a single memory element can be in one of the two states, *on* or *off*. One binary digit (0 or 1), or *bit*, is necessary to define the state of the element. A *word* in the computer memory consists of a sequence of n memory elements, so n bits are necessary to define the value of the stored word. The total number of possible values is $\Omega = 2^n$. Thus the information (I) required to specify the value precisely is $I = n = \log_2 \Omega$. Clearly there is a linear relationship between I and entropy, $S = k_B \ln \Omega$. One of the major contributors to establishing this relationship was Leo Szilard, whose life is sketched briefly starting on page 672.

○ 14–8 The System Partition Function is a Product of Molecular Partition Functions

In Section 14-4 we introduced the symbol q as a convenient shorthand for the sum over energy levels $\sum_n \epsilon_n/k_B T$, which we called the molecular partition function. It turned out that the average energy could be expressed conveniently in terms of the derivative of q, or $\langle \epsilon \rangle = E/N = k_B T^2 (\partial \ln q/\partial T)_V$. We will see in the next section that free energy and entropy also are simply related to a partition function. However, since the free energy depends not only on the number of molecules but also on their concentration, we first must develop the idea of a partition function for a collection of particles, which we call the *system partition function*.

The system partition function, Q (which is not the same as heat, although it has the same symbol), is also a sum over energy levels, but this time it is a sum over the total energy of all particles, which can have values E_j,

$$Q = \sum_j \Omega_j \exp(-E_j/k_B T) \qquad (14\text{-}34) \blacktriangleleft$$

A particular value of E_j results from assigning each particle to an energy level ϵ_n, and summing, or

$$E_j = \sum_n N_n \epsilon_n$$

There is a value of E_j for each way of assigning the particles to the energy levels ϵ_n, and Ω_j is the number of ways of assigning the particles to obtain energy E_j.

The system partition function is a product of the molecular partition functions when the particles are independent (meaning that the energy levels of one do not depend on the energy state of its neighbors). Suppose that we have N identical, independent molecules, which are distinguishable because they occupy different positions in a crystal. Then

$$Q = q^N \tag{14-35}$$ ◄

in which q is the molecular partition function.

Equation 14-35 can be derived formally, but it can be readily understood through an example. Suppose we have a system consisting of two particles, and each particle has two energy levels ϵ_1 and ϵ_2 (Fig. 14-9). Then

$$q = \exp(-\epsilon_1/k_B T) + \exp(-\epsilon_2/k_B T)$$

By Equation 14-35,

$$Q = q^2 = \exp(-2\epsilon_1/k_B T) + 2\exp[-(-\epsilon_1 + \epsilon_2)/k_B T)] + \exp(-2\epsilon_2/k_B T) \tag{14-36}$$

Notice that each term in the sum corresponds to $\exp(-E_j/k_B T)$, with the possible total energies E_j given by $2\epsilon_1$ (both molecules in level 1), $\epsilon_1 + \epsilon_2$ (one molecule in level 1 and one in level 2), and $2\epsilon_2$ (both molecules in level 2). The energy level $E = \epsilon_1 + \epsilon_2$ has degeneracy 2, because the first molecule can be in level ϵ_1 or level ϵ_2. Therefore, Equation 14-36 is equivalent to

$$Q = \Sigma \Omega_j \exp(-E_j/k_B T)$$

which is the definition of the system partition function Q. If you work out other examples you will find that Equation 14-35 is always equivalent to Equation 14-34 for independent, distinguishable particles.

A

One molecule
Molecular partitition function = q

A B

Two distinguishable molecules
System partition function $Q = q^2$

FIGURE 14-9
The system partition function of two distinguishable, independent molecules, whose molecular partition function is q, is equal to q^2. Generally, any partition function which combines independent degrees of freedom, such as for two independent molecules, is the product of the partition functions for the separate degrees of freedom.

LEO SZILARD

1898–1964

Leo Szilard, a colleague said, had "a power of foresight, bordering on the visionary." He seemed to understand the long-term consequences of a scientific discovery or of a social development almost immediately. He saw the potential of nuclear energy before fission had been discovered, the possibility of building an atomic bomb before World War II began, the danger of an atomic arms race before the first atom bomb had been exploded, and the excitement of molecular biology before Watson met Crick.

Szilard was the eldest of three children of an architect-engineer who lived in Budapest, Hungary. His studies in electrical engineering were interrupted by World War I, when he was drafted into the Austro-Hungarian Army. After the war he left for Berlin to continue his studies in engineering. Before long, however, he switched into physics because of the excitement created by the presence in that city of Einstein, Planck, von Laue, Schrödinger, Nernst, and Haber. Szilard studied with von Laue, who suggested a problem for his thesis. However, in only a few weeks of working on his own, he developed a proof that the

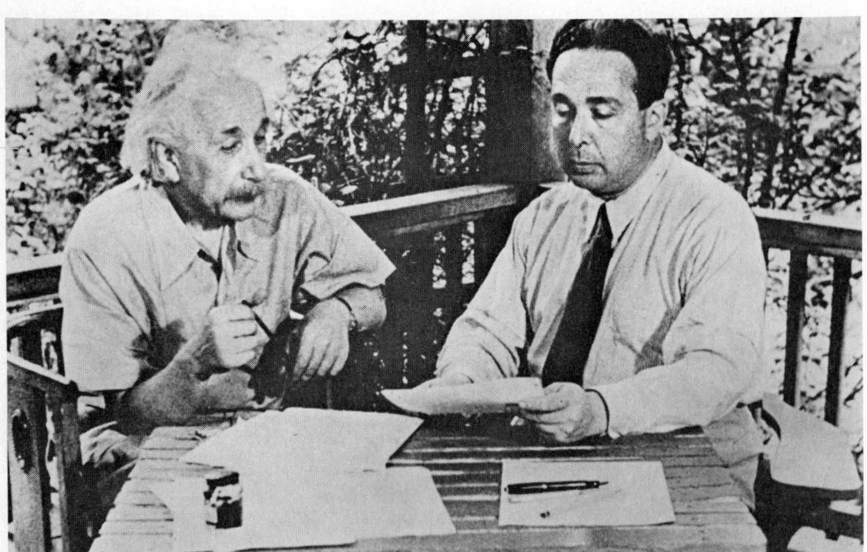

Leo Szilard with Albert Einstein writing the 1939 letter to President Roosevelt. This is actually a photograph taken in the late 1940s of a re-enactment. (Reprinted from *The Collected Works of Leo Szilard: Scientific Papers* by B. T. Feld and G. W. Szilard by permission of The MIT Press, Cambridge, Massachusetts. Copyright © 1972 by The MIT Press.)

second law of thermodynamics can describe fluctuations from thermodynamic equilibrium. The existence of such a proof so surprised Einstein and von Laue that the paper describing it was accepted as his Ph.D. thesis. He was invited to join the staff at the Kaiser Wilhelm Institute in Berlin, and soon he wrote a second landmark paper. This paper established the connection between entropy and information, a link that has become central in the new science of information theory.

When Hitler was voted into office in 1933, Szilard was living in the faculty club of the Institute. With his remarkable foresight, he kept his belongings packed in two suitcases, ready to go. Just after the Reichstag fire in the spring of 1933 (which triggered the start of Nazi terror) he took his suitcases and left on a train for Vienna, and the train, he noted, was practically empty. The same train on the next day was overcrowded and the passengers had to get out to be questioned by the Nazis. The conclusion Szilard drew from this was, ". . . if you want to succeed in this world you don't have to be much cleverer than other people, you just have to be one day earlier than most people."

Szilard moved to London where he helped to set up an organization to find jobs for academics expelled from Germany. He himself, however, was without a job. He lived in hotels and thought about what he should do next. In the newspaper one day he read a speech in which Lord Rutherford said that expectations of power from atomic transmutations were "the merest moonshine." Szilard had never worked in nuclear physics, but he began to ponder ways in which this power might be liberated. While waiting for a red light during a walk, it occurred to him that if he could find an element which absorbs one neutron and then splits and emits two neutrons, a large enough mass of the element might sustain a chain reaction, and hence the liberation of great energy. He mentioned his idea to several distinguished physicists, but it evoked no enthusiasm.

Szilard recalled that at this time he lived in the Strand Palace Hotel. His room did not contain a bath, so he would bathe in a bathroom down the corridor. He would go into the bath around nine o'clock in the morning and soak and think about possible experiments on nuclear energy. Around noon the maid would knock and ask, "Are you all right, sir?" Then he would get out and make a few notes on his thoughts. This procedure worked well; in the spring of 1934 Szilard applied for a British patent which described the laws governing a nuclear chain reaction. Because of its possible use for bombs, he did not want the idea to become public, and consequently had to assign the patent to the British Admiralty. (This was one of many patents filed by Szilard during his lifetime. Another was with Einstein on an electromagnetic pump for liquid refrigerants. This process now serves as the basis for circulation of liquid coolants for metal in nuclear reactors.) Wanting to put his ideas to experimental test, Szilard persuaded the director of physics at a London hospital to let him use the hospital's radium source. He found some novel results, though not the chain reaction, and published two papers in *Nature* that brought him recognition among nuclear physicists and a faculty post at Oxford.

Because of the threat of war, Szilard moved to the United States in 1938 and was again without a job. He heard that Hahn in Germany had discovered uranium fission: an atom breaks in two when it absorbs a neutron. Szilard realized at once that if enough neutrons are emitted in fission, his postulated chain reaction would be possible in uranium. He wanted to keep this idea secret from the Germans, who he feared might construct fission bombs. With this in mind he contacted Fermi and Rabi, who were physicists at Columbia

University and who might think along similar lines. Fermi's first reaction to these possibilities was to say, "Nuts." He explained that there was only a "remote possibility" of perhaps "ten percent" that a chain reaction in uranium is possible. Rabi responded, "Ten percent is not a remote possibility if it means we may die of it. If I have pneumonia and the doctor tells me that there is a remote possibility that I might die, and that it's ten percent, I get excited about it."

Szilard lost no time in showing that neutrons are emitted during uranium fission. He got a temporary appointment at Columbia, where there was equipment suitable for his experiments. To rent a gram of radium needed in the work he borrowed $2,000 from a wealthy friend. After this, teaming up with Fermi, he explored ways of sustaining a chain reaction. It all seemed possible.

As always, Szilard was thinking about developments in international politics. He realized that the Belgian government should be warned not to sell to Germany any of the vast quantities of uranium that they held in their African colony, the Congo. But how could he, a recent immigrant and a relatively unknown physicist, possibly influence the Belgian government? It then occurred to him that Einstein had the necessary prestige and connections. He went out to Long Island to see Einstein, who was staying at a vacation cottage. Einstein saw all the implications when told about the chain reaction and offered to do anything that needed to be done. During a second visit it was decided that Einstein should write to President Roosevelt. Szilard drafted the now-famous letter. It began,

> "Sir: Some recent work by E. Fermi and L. Szilard, which has been communicated to me in manuscript, leads me to expect that the element uranium may be turned into a new and important source of energy . . . This new phenomenon could also lead to the construction of bombs, and it is conceivable—though much less certain—that extremely powerful bombs of a new type may thus be constructed . . ."

The letter went on to urge a permanent contact between the Administration and physicists working on chain reactions. This was the start of the American atomic bomb effort (mentioned in the sketch on Oppenheimer in Chap. 11). Szilard himself, with his background in engineering, contributed significantly to the construction of the atomic pile which permitted a sustained chain reaction.

Towards the end of the war, when it became apparent that a German defeat was near, Szilard began to worry about the consequences of dropping atomic bombs on Japan. He was concerned both with the moral issues and with the likelihood of stimulating an atomic arms race with the Soviet Union. He drafted a petition against dropping the bomb which was signed by many physicists and biologists in the project (but not by many chemists). He pressed for an opportunity to present his ideas to President Truman, and at length obtained an interview with James Byrnes, soon to be appointed Secretary of State by Truman. Szilard found himself unable to convince Byrnes of the danger of an arms race, and left the interview depressed. He later recalled, "I thought to myself how much better off the world might be had I been born in America and become influential in American politics, and had Byrnes been born in Hungary and studied physics. In all probability there would have been no atomic bomb, and no danger of an arms race between America and Russia."

What can account for Szilard's extraordinary insights into human nature, and his seeming ability to foretell political trends? Perhaps one factor was his detachment from the society around him.

Though a participant from time to time in important scientific and political events, he was not as strongly coupled to, nor so blindly influenced by, his immediate surroundings as are most of us. His university positions were of short duration and he never had charge of a laboratory for an extended period. He lived mainly in hotels or faculty clubs and traveled constantly. He married late in life and had no children. Lacking the security of a fixed home and of a permanent job, he lived by his wits. His mind was constantly working, analyzing the possibilities, examining the eventualities.

Calculation of the Energy using the System Partition Function

The energy of a system of independent particles is given by Equation 14-22 as $E = Nk_BT^2 (\partial \ln q/\partial T)$. This is equivalent to

$$E = \frac{k_B T^2 \partial \ln q^N}{\partial T}$$

Substituting $Q = q^N$, we obtain

$$E = \frac{k_B T^2 \partial \ln Q}{\partial T} \tag{14-37}$$

Clearly, the molecular and system partition functions are closely related for independent, distinguishable molecules.

The System Partition Function for Indistinguishable Particles Requires a Factor 1/N!

So far all we have said is that we can construct a system partition function for identical independent particles in a crystal by raising the molecular partition function to the Nth power. This is true when the particles can be distinguished, for example because they occupy different spatial points in a crystal. However, when the particles are in a gas or in solution, they become *indistinguishable,* and an important correction is necessary. This correction amounts to dividing q^N by $N!$,

$$Q = \frac{q^N}{N!} \quad \text{(indistinguishable particles)} \tag{14-38}$$

The reason is that N distinguishable particles can be arranged in $N(N - 1)(N - 2) \ldots (3)(2)(1) = N!$ orders, whereas all these orders are the same for indistinguishable particles. Therefore the degeneracy of every state in Q is smaller by a factor $1/N!$ than it would be if the particles were distinguishable. The term $N!$ is very significant. For example, it gives rise to the dependence of free energy on the logarithm of concentration, as we will see in Section 14-9. (See Problem 14-29 for some complications that result from dividing by $N!$)

○ 14–9 Free Energy and the Partition Function

The Helmholtz free energy, A, can be calculated from the partition function by the relationship

$$A = -k_B T \ln Q \tag{14-39}$$ ◀

which is one of the most important equations of statistical mechanics. We will justify the equation by a simplified argument. First, we approximate Q by the largest term in the sum of Equation 14-34,

$$Q \approx [\Omega \exp(-E/k_B T)]_{\max}$$

Taking the logarithm of this equation and multiplying by $-k_B T$ yields

$$-k_B T \ln Q = E - k_B T \ln \Omega$$

With the Boltzmann equation, $k_B \ln \Omega = S$, we have

$$-k_B T \ln Q = E - TS$$

Using the definition $A = E - TS$, it is clear that $-k_B T \ln Q$ must be A. A rigorous derivation of Equation 14-39 can be found in the more advanced textbooks given in the reading list.

The Free Energy of a Gas Depends on RT Times the Logarithm of Concentration

We are now in a position to show another important correspondence between a thermodynamic result and statistical mechanics. Recall from Chapter 4 that the Gibbs and Helmholtz free energies of a gas vary linearly with $RT \ln P$, or $RT \ln C$, in which C is any measure of the gas concentration. The system partition function for a monatomic gas is, by Equation 14-38

$$Q = \frac{(q^3_{\text{trans}})^N}{N!} \tag{14-40}$$

in which q_{trans} is the molecular translational partition function in one dimension, Equation 14-25. (We raise q to the third power because we want the partition function for three dimensions—motion in each dimension is independent of the other directions.) Substituting

$$q_{\text{trans}} = a\sqrt{2\pi m k_B T}/h$$

into Equation 14-40, we get

$$Q = \frac{1}{N!}\left[\frac{V(2\pi m k_B T)^{3/2}}{h^3}\right]^N \tag{14-41}$$

in which the volume V has replaced a^3. Now, using $A = -k_B T \ln Q$, we find

$$A = -Nk_B T \ln\left[\frac{V(2\pi m k_B T)^{3/2}}{h^3}\right] + k_B T \ln N! \tag{14-42}$$

To proceed we need to evaluate $\ln N!$ Stirling's approximation is

$$\ln N! = N \ln N - N$$

Therefore

$$A = -Nk_BT \ln\left[\frac{V(2\pi mk_BT)^{3/2}}{h^3}\right] + k_BT(N \ln N - N)$$

Next, we let $N = nN_A$ and $R = N_Ak_B$, yielding

$$A = -nRT \ln\left[\frac{V(2\pi mk_BT)^{3/2}}{h^3}\right] + nRT \ln (nN_A) - nRT$$

Using the properties of logarithms (Sec. 2-9) and dividing by n, we can rearrange this equation to

$$\frac{A}{n} = -\frac{3}{2} RT \ln\left(\frac{2\pi mk_BT}{h^2}\right) + RT \ln N_A - RT + RT \ln\left(\frac{n}{V}\right) \quad (14\text{-}43)$$

Equation 14-43 is the result we have been seeking. It shows two important results:

1. The last term on the right side of the equation indicates that the free energy per mole (A/n) depends on RT times the logarithm of the number of moles per unit volume. This is the same result we found in Chapter 4 from thermodynamic principles.
2. The equation shows that the free energy of an ideal gas can be calculated from molecular parameters. Since the energy $E = k_BT^2 (\partial \ln Q/\partial T)$ also is known (see Eq. 14-37), the entropy can be calculated from

$$S = \frac{E - A}{T}$$

Therefore, as we alluded to in Chapter 3 in connection with the third law of thermodynamics, statistical mechanics enables us to calculate the entropy of a gas.

EXERCISE 14-8

Evaluate the Helmholtz free energy of 1 mole of argon gas at 1 atm pressure 298 K.

ANSWER

$V = RT/P = 0.08205$ l atm K^{-1} $\times$ 298 K/1.0 atm = 24.45 l = 2.445×10^4 cm^3. Mass Ar = 39.94 g mol^{-1}/N_A = 6.63×10^{-23} g

$$\frac{A}{n} = -\frac{3}{2} \times 8.314 \times 10^7 \times 298$$

$$\times \ln\left\{\frac{2\pi \times 6.63 \times 10^{-23} \times 1.380 \times 10^{-16} \times 298}{(6.624 \times 10^{-27})^2}\right\}$$

$$+ 8.314 \times 10^7 \times 298 \times \ln (6.023 \times 10^{23}) - 8.314 \times 10^7 \times 298 - 8.314 \times 10^7 \times 298 \times \ln (2.445 \times 10^4)$$

$$= -4.239 \times 10^{11} \text{ erg mol}^{-1} = -42.39 \text{ kJ mol}^{-1}$$

Notice that the energy of the gas is $\frac{3}{2}RT = 3.716$ kJ mol^{-1}. Therefore the entropy is 154.7 J K^{-1} mol^{-1}.

Additional Degrees of Freedom Provide More Energy and Entropy

Suppose that our ideal gas has rotational and vibrational motion. As long as each type of energy remains independent of the others, the calculation remains simple. A gas molecule with independent translational, rotational and vibrational degrees of freedom has a molecular partition function which is a product of the individual partition functions,

$$q = q_{\text{trans}}^3 q_{\text{rot}} q_{\text{vib}} \tag{14-44}$$

According to Equation 14-22,

$$\begin{aligned} E &= Nk_BT^2(\partial \ln q/\partial T) \\ &= Nk_BT^2(\partial \ln q_{\text{trans}}^3/\partial T + \partial \ln q_{\text{rot}}/\partial T + \partial \ln q_{\text{vib}}\ \partial T) \\ &= E_{\text{trans}} + E_{\text{rot}} + E_{\text{vib}} \end{aligned}$$

Because the energy involves the derivative of the logarithm of q, we must take a *product* of partition functions to get a *sum* of energies.

The free energy must always be calculated using the system partition function ($q^N/N!$ for independent particles) which takes account of the fact that the particles are indistinguishable. For a gas with additional degrees of freedom, Equation 14-44 is used for q.

14–10 Calculating the Probability of Complex Molecular States

We are nearly ready to consider applications of statistical mechanics to biochemical problems. To do so it will be convenient to generalize the Boltzmann distribution in Equation 14-5. Molecules in biochemical systems exist in complex quantum states which no one can yet calculate as solutions of the Schrödinger equations. The molecules have many vibrational, rotational, and electronic degrees of freedom, presumably described by many quantum numbers, and they interact strongly with the solvent (water and ions). A large number (ω_i) of the molecular quantum states may have nearly the same energy ϵ_i. The Boltzmann distribution states that the probability of energy level i is

$$P_i = \frac{\omega_i \exp(-\epsilon_i/k_BT)}{\Sigma\omega_i \exp(-\epsilon_i/k_BT)}$$

Using Boltzmann's equation in the form $s_i = k_B \ln \omega_i$, in which s_i is the entropy per molecule in energy level ϵ_i, we find that the probability P_i is proportional to $\exp(s_i/k_B - \epsilon_i/k_BT)$ or $\exp(-a_i/k_BT)$:

$$P_i = \frac{\exp(-a_i/k_B T)}{\Sigma \exp(-a_i/k_B T)}$$

in which a_i is the Helmholtz free energy per molecule in free-energy level i. Multiplying a_i and $k_B T$ by N_A, we get the result

$$P_i = \frac{\exp(-A_i/RT)}{\Sigma \exp(-A_i/RT)} \tag{14-45}$$

in which A_i is now the molar Helmholtz free energy in free-energy level i.

Equation 14-45 involves the Helmholtz free energy and is appropriate for use when temperature and volume are held constant. As usual, the Gibbs free energy is more convenient for biochemical conditions. The analogue to Equation 14-45 when T and P are held constant is

$$P_i = \frac{\exp(-G_i/RT)}{\Sigma \exp(-G_i/RT)} \tag{14-46}$$ ◀

We will use the symbol δ for the (molecular) partition function obtained by summing $\exp(-G_i/RT)$:

$$\delta = \Sigma \exp(-G_i/RT) \tag{14-47}$$

Calculating Averages

Use of Equation 14-46 requires that we know how to calculate the average of a molecular property, since the average is in many cases the only observable quantity. If a quantity m has value m_i in free-energy level i, the average, $\langle m \rangle$, is

$$\langle m \rangle = \sum_i m_i P_i = \frac{\sum_i m_i \exp(-G_i/RT)}{\sum_i \exp(-G_i/RT)} \tag{14-48}$$ ◀

This equation is the usual definition of a linear average: multiply the number of times a given state occurs by the value in that state and divide by the total number in the sample.

Visualizing the Complex States of a DNA Molecule

A specific application will help you understand the meaning of Equation 14-46. When both the strands of a double helical DNA molecule are circular, the molecule can have *superhelical* turns, as illustrated in Figure 14-10. Two different enzymatic mechanisms are known that will introduce a swivel into the molecule by temporarily breaking a bond in one of the chains. The swivel allows superhelical turns to unwind, as shown in the "obligate nicked intermediate" in Figure 14-10. When the superhelical turns are free to come to

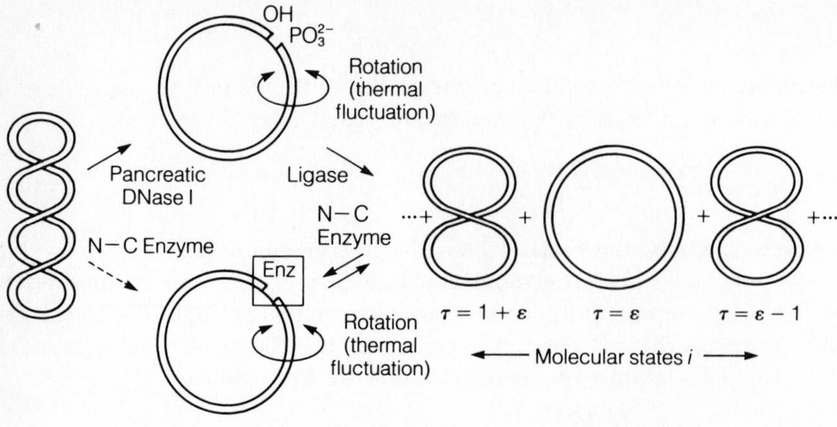

FIGURE 14-10
Conversion of a closed circular DNA molecule (left) through a "nicked" intermediate containing a swivel to a set of molecules containing an equilibrium population of superhelical turns. The conversion can be brought about by deoxyribonuclease I (DNase I) to open the nick and a ligase enzyme to seal it again, or by the nicking-closing (N-C) enzyme. The Boltzmann distribution describes the relative amounts of the molecular states at the right of the figure. ε is the smallest possible number of superhelical turns τ in a closed circular molecule. (From Pulleyblank et al., 1976.)

equilibrium, they will exist with the probability given by Equation 14-46. The instantaneous distribution can be "frozen" by enzymatically resealing the broken bond, producing products at the right of the figure, which differ by steps of one in the number of superhelical turns.

Thus, for our example, the complex states characterized by free energy G_i in Equation 14-46 are the molecules at the right of Figure 14-10, which differ from each other by one superhelical turn. The number of superhelical turns in a molecule is given by the symbol τ (negative for right-handed turns, positive for left-handed turns). The minimum number of superhelical turns in a closed molecule is not necessarily zero, since a small amount of twist, ε (between $-\frac{1}{2}$ and $+\frac{1}{2}$ turn), may be necessary to bring the two broken ends into position to close the strand break.

The complex states i can be separated from each other by gel electrophoresis (see Chap. 15). Figure 14-11(a) shows the profile obtained; each peak corresponds to a different number of superhelical turns. Clearly the equilibrium mixture contains a number of molecular states. The most probable state is that near the center of the distribution with $\tau = \varepsilon$. It must have the minimum free energy, corresponding to minimum stress introduced by superhelical turns. However, because of thermal fluctuations, some molecules will contain superhelical turns at the moment of closure. These are the bands of lower concentration in Figure 14-11(a); they must have a higher free

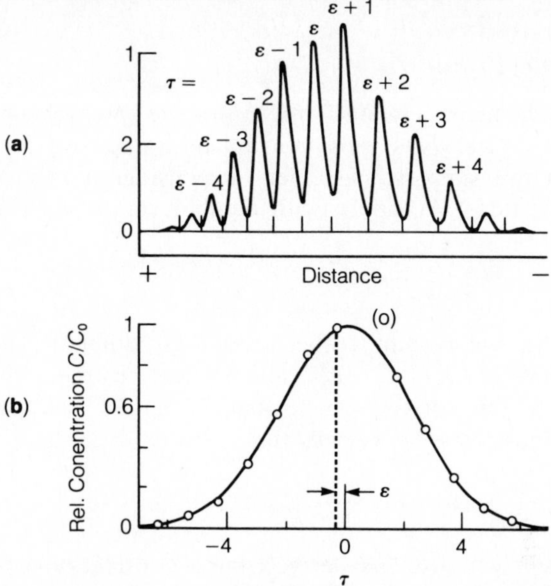

FIGURE 14-11
(a) Relative amounts of the various closed circular DNAs in the equilibrium mixture of Figure 14-10. The highest peak is slightly contaminated by the linear molecule and was therefore not used further in the analysis. The number over each peak indicates the number of superhelical turns τ. **(b)** Fit of the data of **(a)** to the distribution predicted by Equation 14-46. B and ε were adjusted for best fit. (From Pulleyblank et al., 1976.)

energy G_i because they contain more superhelical turns than the minimum (ε).

The free energy G_i of a closed circular molecule is found to vary with the square of τ_i, just as the potential energy of a harmonic oscillator varies with the displacement (x) squared. The equation is

$$G_i = \frac{B\tau_i^2}{2} \tag{14-49}$$

in which B is a constant. Therefore, the probability of complex state i should vary with τ_i according to

$$P_i = \frac{\exp(-B\tau_i^2/2RT)}{\sum_i \exp(-B\tau_i^2/2RT)}$$

Figure 14-11(b) shows this equation [a *Gaussian* distribution because it depends on $\exp(-x^2)$] compared to the relative concentrations of the states. B and ε were adjusted for best fit. Clearly, the Boltzmann distribution in the form of Equation 14-46 accurately describes the relative amounts of the macromolecular states.

14-11 A Biological Application: Multiple Equilibria

Complex biochemical systems often involve multiple equilibria, and statistical mechanics gives us a convenient framework for treating such problems. As the simplest case, let us consider a macromolecule that can be in two states, A and B, with equilibrium constant K between the two states;

$$\mathrm{A} \overset{K}{\rightleftarrows} \mathrm{B}$$

According to thermodynamics, the free energy difference between A and B, $\Delta G^\circ_{\mathrm{AB}}$ equals $-RT \ln K$. We can set the reference free energy of state A equal to zero, or $G_{\mathrm{A}} = 0$, so that $G_{\mathrm{B}} = -RT \ln K$, and $e^{-G_{\mathrm{A}}/RT} = 1$, $e^{-G_{\mathrm{B}}/RT} = K$. Equation 14-46 reveals that

$$P_{\mathrm{A}} = \frac{1}{1+K} \text{ and } P_{\mathrm{B}} = \frac{K}{1+K}$$

which is a result you also can derive from a consideration of the chemical equilibrium expression.

Next, suppose a macromolecule R_0 has a single binding site for a ligand A,

$$R_0 + \mathrm{A} \overset{K}{\rightleftarrows} R_0\mathrm{A}$$

According to thermodynamics, the free-energy change required to convert one mole of R_0 to R_0A each at the total macromolecule concentration C_R depends on the concentration [A] by the relationship

$$\begin{aligned}\Delta G &= \Delta G^\circ - RT \ln ([\mathrm{A}][C_R]/[C_R]) \\ &= -RT \ln K - RT \ln [\mathrm{A}]\end{aligned}$$

(Notice that the concentrations of R_0 and R_0A do not enter into the expression for the free energy because these states are considered explicitly in the probability expressions that follow.) Therefore, the states of the macromolecule R_0 (State 0) and R_0A (State 1), have free energies $G_0 = 0$ (by definition of the reference free-energy state) and $G_1 = \Delta G = -RT \ln K - RT \ln [\mathrm{A}]$. Equation 14-46 gives for probability P_0

$$\begin{aligned}P_0 &= \frac{\exp(-0/RT)}{\exp(-0/RT) + \exp\{(-RT \ln K - RT \ln [\mathrm{A}])/RT\}} \\ &= \frac{1}{1 + K[\mathrm{A}]}\end{aligned}$$

Similarly for P_1,

$$P_1 = \frac{K[\mathrm{A}]}{1 + K[\mathrm{A}]}$$

By Equation 14-48, the average number of A molecules bound per macromolecule—which we will call $\bar{\nu}$—is

$$\bar{\nu} = 0 \times P_0 + 1 \times P$$
$$= \frac{K[\mathrm{A}]}{1 + K[\mathrm{A}]}$$

The same equation can be derived by purely thermodynamic reasoning.

EXERCISE 14-9

Suppose a macromolecule can exist in two states, T_0 and R_0, with equilibrium constant L^{-1} between them, and that the R_0 form has two identical binding sites for ligand A with binding constant K:

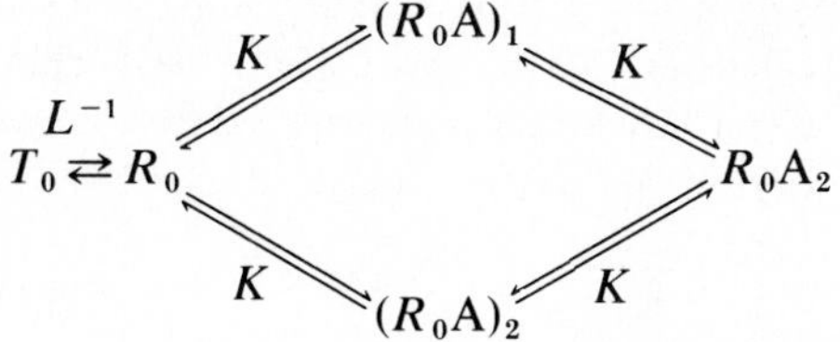

Calculate the average number of ligands bound per macromolecule.

ANSWER

There are five possible states, two of which have identical free energies $(R_0\mathrm{A})_1$ and $(R_0\mathrm{A})_2$, which correspond to one ligand in either of the two identical sites. Let the state R_0 have free energy $G_R = 0$, $\exp(-G_R/RT) = 1$, so that state T_0 has free energy $G_T = -RT \ln L$, and $\exp(-G_T/RT) = L$. The free energy of the states $(R_0\mathrm{A})_1$ and $(R_0\mathrm{A})_2$ is $G_{R\mathrm{A}} = -RT \ln K - RT \ln [\mathrm{A}]$, so $\exp(-G_{R\mathrm{A}}/RT) = K[\mathrm{A}]$ for each. Similarly, the free energy of $R_0\mathrm{A}_2$ is $G_{R\mathrm{A}_2} = -2RT \ln K - 2RT \ln [\mathrm{A}]$, and $\exp(-G_{R\mathrm{A}_2}/RT) = K^2[\mathrm{A}]^2$. Therefore, by Equation 14-48,

$$\bar{\nu} = \frac{2K[\mathrm{A}] + 2K^2[\mathrm{A}]^2}{L + 1 + 2K[\mathrm{A}] + K^2[\mathrm{A}]^2}$$

Titration of Polyprotic Acids

The statistical mechanical equations just derived can help us treat mathematically the complex equilibria that arise when protons can be bound at more than one site on a molecule. As our example we choose glycine, which has two ionizable groups:

$$
\begin{array}{ccccc}
 & & \mathrm{HOOCCH_2NH_2} & & \\
 & \overset{K_1}{\nearrow\!\!\swarrow} & & \overset{K_3}{\searrow\!\!\nwarrow} & \\
{}^{-}\mathrm{OOCCH_2NH_2} & & & & \mathrm{HOOCCH_2NH_3^+} \\
 & \underset{K_2}{\searrow\!\!\nwarrow} & & \underset{K_4}{\nearrow\!\!\swarrow} & \\
 & & {}^{-}\mathrm{OOCCH_2NH_3^+} & &
\end{array}
$$

Unlike the example of Exercise 14-9, the two binding sites for the proton are not equivalent, and for generality we take four different intrinsic association constants. Notice, however, that the equilibrium constant for adding two protons is the product of the equilibrium constants for the two separate steps, and must be independent of the path followed. Therefore $K_1K_3 = K_2K_4$ (see Equation 6-7).

Use of Equation 14-46 to calculate the probability of each possible state of glycine requires that we know the relative free energy of all the states. Taking the form $NH_2CH_2COO^-$ as the reference state with free energy zero gives the free energy of NH_2CH_2COOH, $G_1 = \Delta G° - RT \ln [H^+] = -RT \ln K_1 - RT \ln [H^+]$. Hence the term $\exp(-G_1/RT)$ is $K_1[H^+]$. Similarly the term $\exp(-G_2/RT)$ for the state $NH_3^+CH_2COO^-$ is $K_2[H^+]$ and the term $\exp(-G_3/RT)$ for the state $NH_3^+CH_2COOH$ is $K_2K_4[H^+]^2$. Equation 14-48 for the average number of protons bound, $\bar{\nu}$, gives

$$\bar{\nu} = \frac{(K_1 + K_2)[H^+] + 2K_2K_4[H^+]^2}{1 + (K_1 + K_2)[H^+] + K_2K_4[H^+]^2} \tag{14-50a}$$

At this point it is important to make a distinction between the *intrinsic* equilibrium constants $K_1, K_2 \ldots$ and the *apparent* equilibrium constants K' that would be measured by titrating glycine. When HCl is added to a solution of glycine in alkaline medium, H^+ binds to glycine, but there is no way of knowing from this simple experiment whether binding occurs at $-NH_2$ or $-COO^-$. Therefore the equilibrium can be formulated as

$$\text{Glycine}^- + H^+ \overset{K_1'}{\rightleftarrows} \text{glycine} + H^+ \overset{K_2'}{\leftrightarrows} \text{glycine}^+$$

In other words, the first step of proton binding can occur at either site, and a simple measurement of proton uptake cannot distinguish the two. Only two equilibrium constants can be determined from such an experiment. (Notice, however, that a spectroscopic experiment, such as measurement of the ^{13}C NMR chemical shift of the carboxylate carbon, could be used to monitor directly the proton binding at $—COO^-$.)

Calculation of the relative free energies of glycine$^-$, glycine, and glycine$^+$ and use of Equation 14-48 yield

$$\bar{\nu} = \frac{K_1'[H^+] + 2K_1'K_2'[H^+]^2}{1 + K_1'[H^+] + K_1'K_2'[H^+]^2} \tag{14-50b}$$

Comparison of this result with Equation 14-50*a* gives

$$K_1' = K_1 + K_2 \qquad K_2' = \frac{K_2K_4}{K_1 + K_2} = \frac{K_3K_4}{K_3 + K_4}$$

Measurement of K_1' and K_2' is simple in the cases exemplified by glycine, in which the first step of H^+ binding is much stronger than the second. A *titration* is carried out, in which the degree of proton binding is measured as a function of the pH $= -\log_{10}[H^+]$ in the solution (Fig. 14-12). Upon adding H^+ to $NH_2CH_2COO^-$, the proton binds first primarily to $—NH_2$. Neglecting

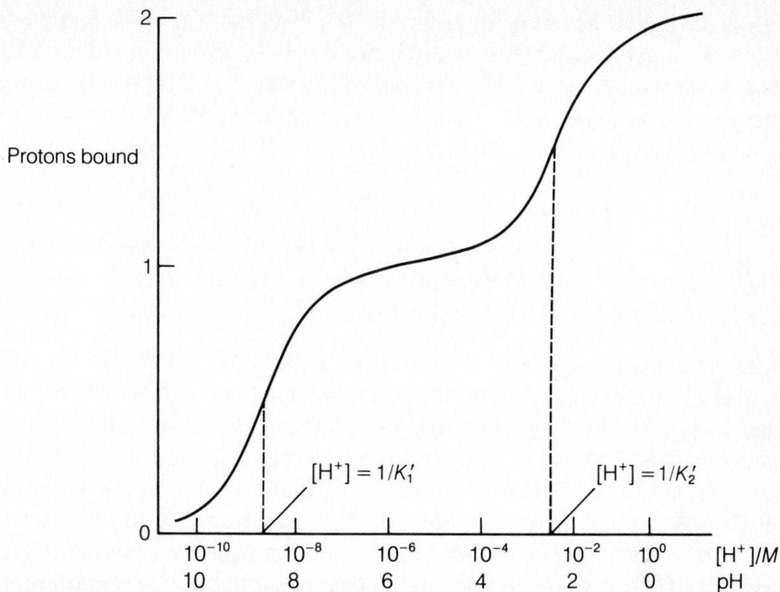

FIGURE 14-12
Schematic diagram of a titration of $NH_2CH_2COO^-$ by addition of H^+. The two equivalence points correspond to successive proton binding to $—NH_2$ and $-COO^-$.

terms in Equation 14-50*b* which represent the binding of two protons, we see that

$$\bar{\nu} = \frac{K_1'[H^+]}{1 + K_1'[H^+]}$$

when $\bar{\nu} = \frac{1}{2}$, this expression yields $K_1' = 1/[H^+]$. Therefore the proton binding constant K_1' is the reciprocal of $[H^+]$ at the first half-equivalence point in Figure 14-12. Notice that K_1' is the reciprocal of the dissociation constant K_1^d usually used to express acid strength, so that $[H^+] = K_1^d$, and the pH_1 at the first equivalence point is given by

$$pH_1 = pK_1^d$$

where $pK_1^d = -\log_{10} K_1^d$.

For the second half of the titration we can neglect 1 compared to $K_1'[H^+]$, allowing us to approximate Equation 14-50*b* by

$$\bar{\nu} = \frac{K_1'[H^+] + 2K_1'K_2'[H^+]^2}{K_1'[H^+] + K_1'K_2'[H^+]^2}$$
$$= \frac{1 + 2K_2'[H^+]}{1 + K_2'[H^+]}$$

At the second equivalence point in the titration $\bar{\nu} = \frac{3}{2}$, which occurs when $K_2' = 1/[H^+]$. Hence pK_2^d is equal to the pH at the second equivalence point.

In the case of glycine, K_1 is much smaller than K_2, so K_1' can be taken equal to K_2, and K_2' equal to K_4. Thus protonation takes primarily the lower of the two paths shown on page 683, through the *zwitterion* or doubly charged species. In more complicated cases, additional information may be necessary to determine the intrinsic binding constants.

14–12 Flexible Polymers have a Statistical Distribution of Conformations

Most macromolecules of biological significance have a definite structure. Their shape is either globular, as are most enzymes and tRNA, or fibrous, as are double helical DNA or triple helical collagen. Most synthetic polymers, on the other hand, do not have a definite structure in solution. Such chains are flexible, and their conformation is a statistical average of the large number of ways the molecule can be folded. A loose general term often used to describe these materials is "random coil," which implies a coiling polymer chain whose conformation changes from one possible folded state to another as a function of time. The globular proteins are not random coils because their three-dimensional interactions favor a single structure. However, in many cases this three-dimensional structure can be unfolded by an increase of temperature or by addition of a "denaturing" solvent. Under appropriate conditions protein and nucleic acid chains can exist as random coils.

Flexible polymer chains contain single bonds, such as C—C, C—O, or P—O, about which rotation is possible. For example, consider the C—C bond in Figure 14-13. Suppose that each carbon atom has an R or R' sub-

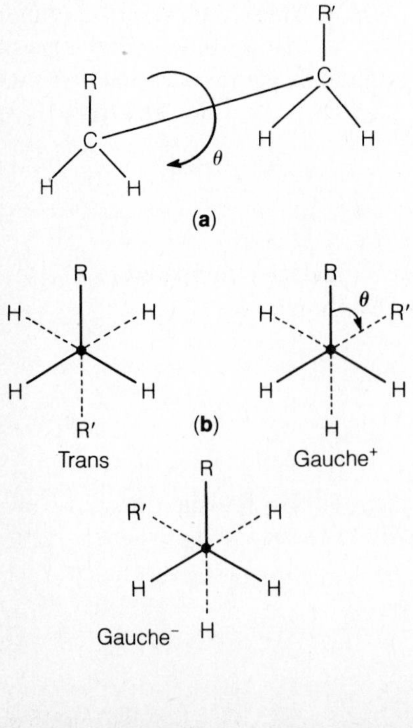

FIGURE 14-13
The preferred dihedral rotational angles of a single bond. The rotation occurs about the C—C bond, as seen in **(a)**. In **(b)**, the line of sight is directed along the C—C bond, with the group R in the front. The *gauche*$^+$ (g^+) state results from clockwise rotation of the group (R') at the far end of the bond, and *gauche*$^-$ (g^-) from counterclockwise rotation of R'. The dihedral angle θ measures the clockwise rotation of R' from R.

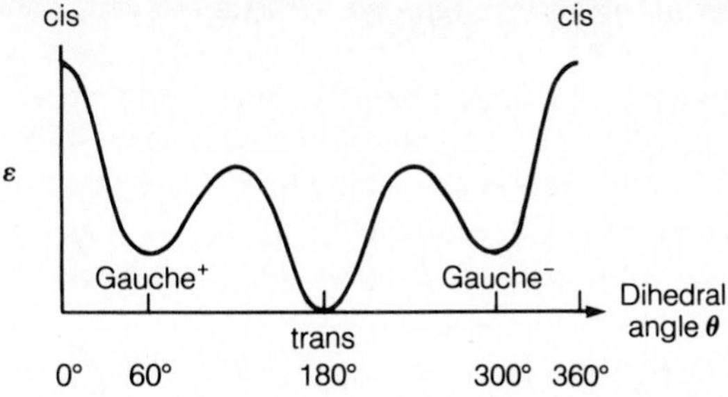

FIGURE 14-14
Schematic drawing of the potential energy ϵ as a function of the dihedral angle θ in Figure 14-12(*a*). When the molecule is in solution, so that the quantum states are not calculable, the free energy per molecule should replace the energy ϵ in the Boltzmann probability distribution, Equation 14-51.

stituent, and two bonded hydrogen atoms. The geometry of bonding to the carbon atom is tetrahedral, so if we direct our line of sight along the C—C bond axis we will see the orientation of atoms shown in Figure 14-13(*b*) as the C—C bond is rotated. During rotation the potential energy varies as indicated schematically in Figure 14-14. When the two *R* groups are in the same plane as the two carbon atoms, the conformation is called *cis,* and the *dihedral angle* of bond rotation is defined as zero. The potential energy is a maximum in the cis conformation because of the repulsive interactions between the *R* groups. These repulsive interactions diminish in the two *gauche* and the *trans* states, thereby producing local potential-energy minima.

The conformation of a flexible molecule in solution is determined by two factors. One of these is minimization of the potential energy. The discipline that deals with this problem is called *conformational analysis*. Its objective is to calculate potential-energy functions of the sort shown in Figure 14-14, and similar energy terms for other nonbonded interactions. The result is a *preferred conformation* with minimum energy for the molecule. This procedure has been reasonably successful for molecules the size of nucleotides or small peptides.

As the molecule becomes larger, however, a second factor becomes so important that it cannot be neglected. The difference between potential energy minima, for example, *gauche*$^+$ compared to *trans,* usually is not very much greater than k_BT. According to the Boltzmann distribution the probability of the *gauche*$^+$ (g^+) state is

$$P_{g^+} = \frac{\exp(-\epsilon_{g^+}/k_BT)}{\exp(-\epsilon_{g^+}/k_BT) + \exp(-\epsilon_{g^-}/k_BT) + \exp(-\epsilon_t/k_BT)} \tag{14-51}$$

with similar expressions for P_{g-} and P_t. When one energy is not much smaller than the others, there is a reasonable probability that the bond will be in one of the higher-energy rotational isomeric states. Instead of a single preferred conformation, the result is a set of conformations whose probabilities depend on their energies. For real molecules in solution, the *free* energy per molecule should replace the energy ϵ in Equation 14-51 because rotational oscillations and interactions with the solvent contribute entropic terms.

Flexible polymeric molecules have a huge number of conformations, and cannot be thought of as existing in a single shape. Each conformation has different geometric properties, including, for example, the separation between the two ends of the chains (Fig. 14-15). If $W(L)$ is the probability that the two ends are a distance L apart, then the mean square end-to-end distance, $\langle L^2 \rangle$, is

$$\langle L^2 \rangle = \int_0^\infty W(L) L^2 \, dL \tag{14-52}$$

Equation 14-52 is the definition of the average of the square of the end-to-end distance in the chain. The square root of $\langle L^2 \rangle$, $\langle L^2 \rangle^{1/2}$, the root-mean-square end-to-end distance, has the dimensions of length and is one measure of the size of the molecule in solution. We use it here because it will later (Chap. 15) be related to experimental measurements of the size of polymer molecules. The theory for calculating measurable quantities such as $\langle L^2 \rangle$ for a polymer from geometric and potential energy information was developed principally by P. J. Flory, who received the Nobel Prize in Chemistry for his many contributions to polymer science.

The simplest theoretical model for a flexible polymer molecule is a *random flights chain,* which is based on what is called a *random walk.* For this model, we assume that the molecule is made up of rigid segments of length b, joined by completely flexible joints. After each length b, the chain can turn in any direction with equal probability, which is why it is called a random walk. The most readily calculated quantity for a random walk is the distance L that the walker has moved away from the starting point after N steps. It turns out

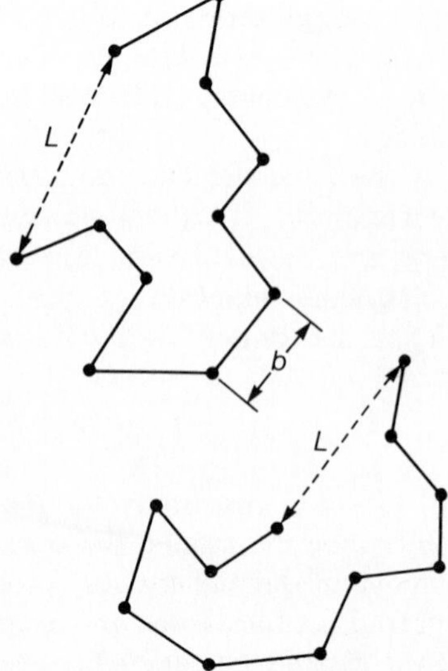

FIGURE 14-15
Schematic drawing of two conformations of a flexible polymer. The polymer is assumed to be a sequence of bonds of length b connecting the residues (●), with flexible joints between the bonds. One measure of the extension in space of the polymer is the distance, (L) from one end of the chain to the other. The mean square end-to-end distance is the average of L^2, $\langle L^2 \rangle$.

that the probability of a given value of L is proportional to the Gaussian or error function $\exp(-3L^2/2Nb)$. This result can be justified by considering a random walk in just one dimension.

The One-Dimensional Random Walk

In a one-dimensional random walk, there is equal probability of moving to the left or the right along the x-axis at each step. The random expectation result is that there will be $N/2$ left $(-)$ steps and $N/2$ right $(+)$ steps, which would bring the walker exactly back to the origin after N steps. However, we must allow for the possibility that there will usually be a few more steps of one kind than the other. Let there be $N/2 + x = N_+$ plus steps, and $N/2 - x = N_-$ minus steps. After N steps, the distance from the origin will be $N_+ - N_- = 2x$. The probability of a given x depends on the number of ways, Ω, that the corresponding combination of N_+ and N_- can be obtained. Following the logic of our derivation of the Boltzmann distribution (Sec. 14-3),

$$\Omega = \frac{N!}{N_+!N_-!}$$

$$= \frac{N!}{\left(\frac{N}{2} + x\right)!\left(\frac{N}{2} - x\right)!} \tag{14-53}$$

Using Stirling's approximation ($\ln N! = \text{constant} + N \ln N - N$), Equation 14-53 can be simplified to

$$\ln \Omega = \text{constant} - \left(\frac{N}{2} + x\right) \ln \left(1 + \frac{2x}{N}\right) - \left(\frac{N}{2} - x\right) \ln \left(1 - \frac{2x}{N}\right)$$

which, with expansion of the logarithm $\ln (1 + y) = y - \frac{y^2}{2} + \cdots$, yields

$$\ln \Omega = \text{constant} - \frac{2x^2}{N}$$

Since the probability depends on Ω, we have

$$P(x) \propto \exp(-2x^2/N)$$

(Notice that we leave this result as a proportionality because the proportionality constant is not obtained correctly with the simplified form of Stirling's approximation that we are using.) The important point is the calculated variation of the end-to-end distance probability with an exponential dependence on $-2x^2/N$.

The Three-Dimensional Random Walk as a Model for a Flexible Polymer

For a three-dimensional random walk, the probability that one end of the chain will be found in a volume element dV at a distance L from the other end is (Tanford, 1961):

$$W'(L,N)\,dV = \left(\frac{3}{2\pi Nb^2}\right)^{3/2} \exp\left(\frac{-3L^2}{2Nb^2}\right) dV \tag{14-55}$$

The *distribution function* $W'(L,N)$ is shown in Figure 14-16(*a*); it is half of a bell-shaped, *normal* or Gaussian distribution. Notice that W' is a maximum when $L = 0$, so if we sample a volume element dV, the most likely place to find the end of the chain is in the vicinity of the other end, where W' is a maximum.

We also can ask for a slightly different kind of distribution function. What is the probability that one end of the chain is within a spherical shell of thickness dL at a distance L from the other end? This function, which we call the *radial distribution function,* $W(L,N)$, can be obtained by expressing the volume of the spherical shell as $dV = 4\pi L^2\,dL$ in Equation 14-55 (see the discussion of the radial distribution function in Sections 8-10 and 10-6). The result is

$$W(N,L)\,dL = 4\pi\left(\frac{3}{2\pi Nb^2}\right)^{3/2} \exp\left(\frac{-3L^2}{2Nb^2}\right) L^2\,dL \tag{14-56}$$

$W(N,L)$ is shown in Figure 14-16(*b*).

The most significant result of this analysis is the calculation of the root-mean-square (rms) end-to-end distance. Using Equation 14-52

$$\langle L^2\rangle = \int_0^\infty W(L,N)\,L^2 dL$$

which, with the help of integral tables, we find to be

$$\langle L^2\rangle = Nb^2 \tag{14-57a}$$ ◀

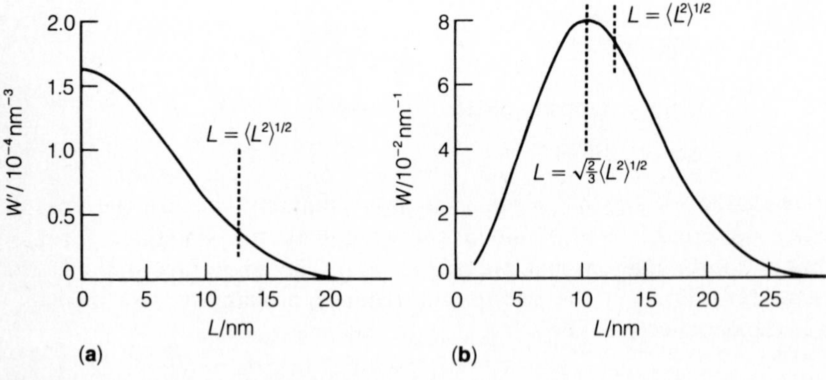

FIGURE 14-16
Distribution functions for a flexible polymer chain containing $N = 1000$ segments, each of length 0.4 nm. **(a)** Volume distribution function W' (Equation 14-55), giving the probability per unit volume that one end of the chain will be found a distance L from the other. The root-mean-square end-to-end distance $\langle L^2\rangle^{1/2}$ is $N^{1/2}b$ (Equation 14-57). **(b)** Radial distribution function W (Equation 14-56), giving the probability per unit length that one end of the chain will be found at any point in a spherical shell of radius L around the beginning of the chain ($L = 0$). As shown in Exercise 14-10, the maximum of W occurs when $L = \sqrt{2/3}\,\langle L^2\rangle^{1/2}$.

or

$$\langle L^2 \rangle^{1/2} = N^{1/2} b \qquad (14\text{-}57b)$$

This simple result says that the *root-mean-square (rms) distance between the ends of the chain varies with the square root of the number of segments in the chain.*

EXERCISE 14-10

Find the value of L at which W has its maximum value.

ANSWER

$$\frac{dW}{dL} = 0 = \frac{d}{dL}\left\{\exp\left(\frac{-3L^2}{2Nb^2}\right) L^2\right\} = -\frac{3L^3}{Nb^2} + 2L$$

Therefore

$$L_{\max} = \sqrt{\frac{2Nb^2}{3}}$$

Since the root-mean-square end-to-end distance is $\langle L^2 \rangle^{1/2} = N^{1/2} b$,

$$L_{\max} = \sqrt{\frac{2}{3}}\langle L^2 \rangle^{1/2}$$

Two Models for Polymers of Limited Flexibility

Solvent conditions for many polymer molecules can be found such that the rms distance, $\langle L^2 \rangle^{1/2}$, does indeed vary with the square root of the number of segments. This condition is called a *theta solvent*. However, some modifications of Equation 14-57 are necessary to describe such polymers accurately. For large N let us take the ratio

$$\frac{\langle L^2 \rangle}{Nb^2} = C_\infty \qquad (14\text{-}58)$$

C_∞ is called the *Flory characteristic ratio*. According to Equation 14-57, C_∞ should be 1. Actual values often are found to be considerably greater than 1; an example is a single-stranded polynucleotide for which $C_\infty = 17$. This result arises from the stiffness or limited flexibility of the segments in the chain. The random-flights model assumes complete flexibility at each joint. It can be shown rigorously using the isomeric state model (g^+, g^-, t rotational states) that limited rotational flexibility yields C_∞ values different from 1, but retains the dependence of $\langle L^2 \rangle$ on N as long as there are no long-range interactions in the polymer.

One way to model a stiff chain is as a collection of N_e effective segments of length b_e, with $N_e b_e^2 = \langle L^2 \rangle$ (Fig. 14-17). Another constraint is that the contour length Nb be constant, so $N_e b_e = Nb$. The quantity b_e is called the *Kuhn statistical segment length.* For stiff molecules such as double helical DNA, b_e can be very large; values for DNA range between 100 nm and 150 nm.

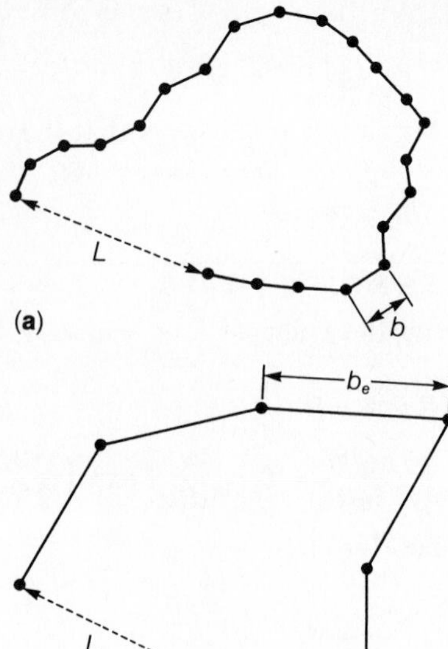

FIGURE 14-17
A polymer chain whose N segments are of length b and which has limited flexibility or range of angles at each residue, as in **(a)**, can be represented, as in **(b)**, by a smaller number N_e of segments of length b_e. The requirements of the model are that the contour lengths and the mean square end-to-end distances of the two molecules **(a)** and **(b)** must be the same. b_e is the Kuhn statistical segment length, related to the Flory characteristic ratio by $C_\infty = b_e/b$.

EXERCISE 14-11

Show that the Kuhn statistical segment length $b_e = C_\infty b$.

ANSWER

Since $\langle L^2 \rangle = C_\infty b^2 N$ and $\langle L^2 \rangle = b_e^2 N_e$, $C_\infty b^2 N = b_e^2 N_e$. Combining with $b_e N_e = bN$ yields $b_e = C_\infty b$.

EXERCISE 14-12

Calculate the root-mean-square end-to-end distance in a bacterial DNA molecule containing 1.5×10^6 base pairs (molecular weight 10^9), using 120 nm for the Kuhn statistical segment length. The distance between base pairs in double helical DNA is 0.34 nm. Assume a random flights chain.

ANSWER

By Exercise 14-11, $C_\infty = b_e/b = 120/0.34$; using Equation 14-58, $\langle L^2 \rangle = C_\infty N b^2 = (120/0.34) \times 1.5 \times 10^6 \times (0.34)^2 = 6.12 \times 10^7 \text{ nm}^2$. Therefore, $\langle L^2 \rangle^{1/2} = 7.8 \times 10^3 \text{ nm} = 7.8 \times 10^{-3}$ mm. Notice that this is larger than the size of a typical bacterium (~1000 nm), so the DNA in a bacterium must be "packaged" or folded in some way, rather than free to take on its equilibrium dimensions.

Another useful model for a stiff chain is the "wormlike" chain, which is slightly flexible at all points. A characteristic quantity for this model is the *persistence length*. Let the z-axis of the molecule be in the direction parallel to

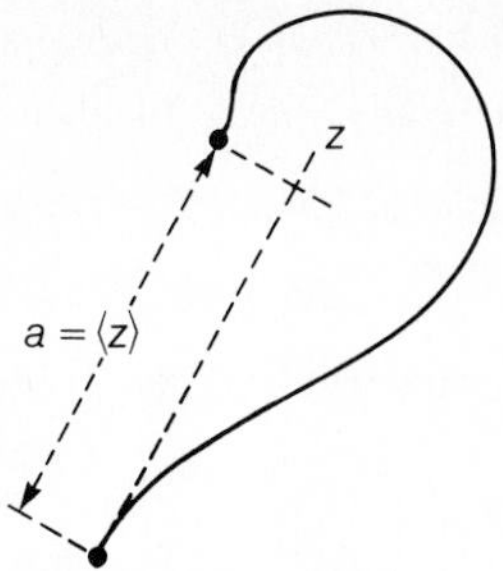

FIGURE 14-18
The wormlike chain, a useful model for a polymer with limited flexibility. This model differs from the Kuhn model (Fig. 14-17) in that curvature is permitted at all points in the chain, rather than only at the junction of the segments. The persistence length a is the average projection of the end-to-end vector $\langle L \rangle$ on the initial direction of the chain at one end (defined to be the z-axis).

the chain at one end (Fig. 14-18). We calculate the projection of the end-to-end vector along the z-axis. If the molecule is stiff, many segments will "persist" in the initial direction; the average projection is called the persistence length a. It can be shown (Bloomfield et al., 1974) that the Kuhn statistical segment model and the wormlike chain, when applied to the same molecule, yield the same $\langle L^2 \rangle$ if

$$b_e = 2a \tag{14-59}$$

Excluded Volume Means That $\langle L^2 \rangle$ No Longer Depends Linearly on N

Many polymers do not obey Equation 14-58, which implies a constant value for C_∞ when N becomes large. Rather, it is found that the exponent is increased by an amount ϵ,

$$\langle L^2 \rangle \propto N^{1+\epsilon} \qquad \epsilon \geq 0 \tag{14-60}$$

In other words, the chain expands more rapidly with increasing N than would be expected if the *rms* end-to-end distance depended on the square root of the number of segments. This arises from long-range interactions in the chain, or *excluded volume effects*. In its simplest form, this effect arises from the inability of two segments of the polymer to occupy the same point in space. More rigorously, the effect depends on the interaction energy between two polymer segments, and the solvent-solvent interaction energy, compared to the interaction between polymer and solvent. With the right choice of solvent the excluded volume effect can be eliminated ($\epsilon = 0$). A solvent in which $\epsilon = 0$ is called a theta solvent. Most biopolymers, such as DNA in aqueous solution, show appreciable excluded volume effects. Experimental methods to detect such influences include light scattering (Sec. 13-9) and sedimentation and viscosity measurements, which are discussed in Chapter 15.

QUESTIONS FOR REVIEW

1. State Boltzmann's equation relating entropy to the degeneracy of a macroscopic state.

2. What is the Boltzmann distribution?
3. How is the average energy related to the molecular partition function?
4. What is the meaning of the degeneracy of an energy level ϵ_n?
5. State the principle of equipartition of energy.
6. Why do quantum mechanical effects influence the heat capacity at low temperatures?
7. Define the system partition function.
8. Why does the system partition function contain the factor $1/N!$ when the particles are indistinguishable?
9. What replaces $\exp(-\epsilon_n/k_BT)$ in calculating the probability of complex molecular states?
10. How does the probability of superhelical DNA molecules in an equilibrium mixture depend on the number of superhelical turns?
11. Define a Kuhn statistical segment.
12. What is meant by the persistence length of a stiff chain?
13. What effect does excluded volume have on the dimensions of a flexible polymer chain?
14. Define the Flory characteristic ratio.

PROBLEMS

1. Are the following statements true or false?
   (a) The degeneracy of an isolated system increases in a spontaneous process.
   (b) According to the Boltzmann distribution, it is always less likely that a molecule is in an excited energy level compared to the ground state.
   (c) If vibrational and rotational motion are independent, their total energy contribution is the sum of the separate vibrational and rotational energies.
   (d) If vibrational and rotational motion are independent, their total molecular partition function is a product of the separate vibrational and rotational molecular partition functions.
   (e) The mean square end-to-end distance in a long random flights chain is proportional to the number of segments in the chain.
   (f) The mean square end-to-end distance in a flexible polymer chain in solution always increases linearly with the number of residues in the chain.

Answers: (a) T. (b) F, the probability also depends on the degeneracy of an energy level. (c) T. (d) T. (e) T. (f) F, not if there are excluded volume effects.

2. Add the appropriate words or symbols.

   (a) Boltzmann's equation states that entropy is a measure of ________.

   (b) The sum over all quantum states, n, of the terms $\exp(-\epsilon_n/kT)$ is called the ________.

   (c) The classical (high-temperature) value of the energy of a molecule free to vibrate in three dimensions is ________.

   (d) The classical (high-temperature) heat capacity of a rigid diatomic rotor is ________.

   (e) The highest-energy rotational isomer is usually the ______ form.

   (f) The stiffness of a wormlike polymer chain is characterized by the ________.

   (g) The ratio of the Kuhn statistical segment length b_e to the real segment length b is equal to the ________.

Answers: (a) Degeneracy. (b) Molecular partition function. (c) 3 RT. (d) R. (e) *cis*. (f) Persistence length. (g) Flory characteristic ratio.

PROBLEMS RELATED TO EXERCISES

3. (Exercise 14-1) Calculate the entropy of a crystal of CH_3D, if it is assumed that the deuterium atom can occupy any of the four equivalent tetrahedral positions with equal probability.

4. (Exercise 14-3) Calculate the classical energy of a particle free to oscillate in three dimensions.

5. (Exercise 14-4) Calculate the classical heat capacity due to the vibrational motion of a diatomic molecule.

6. (Exercises 14-5 and 14-6) Calculate the heat capacity due to vibration of CO at 298 K; CO absorbs infrared radiation at 4.66 μm.

7. (Exercise 14-8) Calculate the Gibbs free energy of one mole of argon gas at 1 atm pressure, 298 K.

8. (Exercise 14-9) Derive an expression for $\bar{\nu}$ if the reaction scheme in Exercise 14-9 is extended so that there are three independent binding sites instead of two.

9. (Exercise 14-10) Find the value of L at which the radial distribution function W has its maximum value.

OTHER PROBLEMS

10. Calculate the entropy change when 20 previously alphabetized books are randomized in their order.

11. Calculate the number of ways of assigning 12 distinguishable particles to 4 different energy levels, with 3 particles in each energy level.

12. Calculate the ratio of the probability of the one-dimensional translational energy level for an electron with $n = 1000$, to the probability of the level with $n = 1$, for a box with side $a = 1$ cm at 37°C.

13. What is the value of the translational quantum number n at which $\epsilon_n = k_B T$ for an electron in a 1 cm one-dimensional box at 37°C?

14. The energy levels of a rigid rotor are $\epsilon_J = J(J + 1)h^2/8\pi^2 I$, in which I is the moment of inertia. Show that the partition function at high temperature is $q_{rot} = 8\pi^2 I k_B T/h^2$. (*Note:* Do not forget that the degeneracy of energy level J is $2J + 1$ because of the $2J + 1$ different directions that the angular momentum vector can point.) (*Hint:* Convert the sum to an integral.)

15. Calculate the molar heat capacity due to vibrational motion of O_2 at 37°C, using the information in Table 14-1. Why is your result small compared to the classical value (R)?

16. Calculate the molar energy, Helmholtz free energy, and entropy of helium gas at 37°C, 1 atm. You may consider that the gas is ideal.

17. Use your knowledge of statistical mechanics, and the accompanying information, to calculate the molar energy, Helmholtz free energy and entropy of HCl at 1 atm, 37°C. (For rotational motion, $h^2/8\pi^2 I k_B = 15.2\ K$; for vibrational motion, $h\nu/k_B = 4140\ K$.) You may consider that the gas is ideal; use the result of Problem 14.

18. The molar entropy calculated from the spectroscopic properties of CH_3D is greater than that measured calorimetrically [by integration of

$$S = \int_0^T (dQ_{rev}/T')\, dT' + \Delta S \text{ (phase changes)}]$$

by 2.8 cal K^{-1}. How would you explain this phenomenon? (*Hint:* Reread Section 3-10.)

19. The molar entropy of O_2 calculated for its vibrational, rotational and translational motion is about 2.2 cal K^{-1} *smaller* than measured by integration of dQ_{rev}/T. How would you explain this discrepancy? (*Hint:* The ground electronic state of O_2 is a triplet.)

20. The vibrational motion of Cl_2 is characterized by $h\nu/k_B = 810$ K. Calculate the ratio of the populations of the first two excited states at 200 K and 800 K.

21. A macromolecule has two conformations T_0 and T_1, which differ by 6 kcal mol^{-1} in free energy. What fraction of the molecules is in the less probable state T_1 at 37°C?

22. A polymer contains N independent binding sites for a small molecule, each with binding constant K. Show that the partition function $\delta = \Sigma \exp(-G_i/RT)$ is $\delta = (1 + KC)^N$, in which C is the molar concentration of the unbound small molecule. (Assume a dilute solution.)

23. Use the results of Problem 22 to show that θ, the fraction of sites occupied, is given by $\theta = KC(1 + KC)$. (*Hint:* Notice that θ can be expressed as $\theta = (1/N)(\partial \ln \delta/\partial \ln C)$; see Exercise 14-9.)

24. Using the bromine atoms to define the dihedral angle, draw the *trans, gauche*$^+$, and *gauche*$^-$ rotational states of $F_2BrCCCl_2Br$. Which pair are optical iso-

mers? Sketch your prediction for the potential energy as a function of dihedral angle.

25. Show that the square of the average end-to-end distance, $\langle L \rangle^2$, in a polymer chain with radial distribution function given by Equation 14-56 is $\langle L \rangle^2 = (8/3\pi) Nb^2$.

26. A flexible polymer chain (such as rubber) exerts a restoring force when its ends are pulled apart. This force arises even if the conformational energy of the chain does not depend on its extension, as assumed for a flexible polymer chain. The force, F, can be calculated (see Section 4-3) from

$$F = -\frac{dA}{dL}$$

in which A is the Helmholtz free energy and L the length. Use the fact that $\Omega(L)$, the number of polymer configurations that end at a given point in space, is proportional to the distribution function $W'(L)$

$$\Omega(L) \propto W'(L)$$

to show that the force exerted by a random flights polymer is

$$F = -\frac{3Lk_BT}{Nb^2}$$

27. Assuming that the Kuhn statistical segment length of DNA is 120 nm, calculate the root-mean-square end-to-end distance in a DNA molecule from T2 bacteriophage, containing 1.7×10^5 base pairs (molecular weight 115 million). The distance between base pairs in DNA is 0.34 nm. Comment on the comparison of $\langle L^2 \rangle^{1/2}$ with the diameter of the virus (~100 nm).

28. What should be the value of the second virial coefficient (Chap. 7) in a theta solvent?

Answer: $B = B_{\text{ideal}} \approx 0$.

29. Our derivation of the Boltzmann distribution assumed distinguishable particles, and we obtained $\Omega' = N!/\Pi N_n!$ for the number of ways of assigning the N particles, with N_n in quantum state n. However, when we considered indistinguishable particles, we divided the partition by $N!$. Therefore, each term Ω' is divided by $N!$, leaving $1/\prod_n N_n!$. This number is less than 1 unless all N_n are either 0 or 1 ($1! = 1$, $0! = 1$). Obviously, the number of ways of achieving a distribution cannot be anything but an integer, so we have a problem. (See Mayer and Mayer, 1940, pp. 63–67 for a careful discussion.)

The resolution of the problem is to recognize that there are two exact statistics for indistinguishable particles, neither identical with Boltzmann statistics corrected by division by $N!$, but both converging to the same results when the temperature is high and many quantum states are available. *Fermi-Dirac statistics* applies to particles such as electrons, protons, neutrons, etc. that follow the Pauli principle, so that no two identical particles can have identical quantum states. *Bose-Einstein statistics* applies to photons and allows any number of identical particles to be in the same quantum state.

Consider the following two cases:

(a) All the particles are in different quantum states. Show that $\Omega' = 1$ for Fermi-Dirac, Bose-Einstein, and corrected Boltzmann statistics.

(b) Each occupied quantum state contains two particles. Show that $\Omega' = 0$ for Fermi-Dirac statistics, $\Omega' = 1$ for Bose-Einstein statistics, and $\Omega' = 2^{-N/2}$ for corrected Boltzmann statistics. Generalize this result to show that Ω' for corrected Boltzmann statistics is always between the limits set by Fermi-Dirac and Bose-Einstein statistics.

30. The antibiotic actinomycin (A) can bind two dinucleotides (D) of sequence d-(GpC):

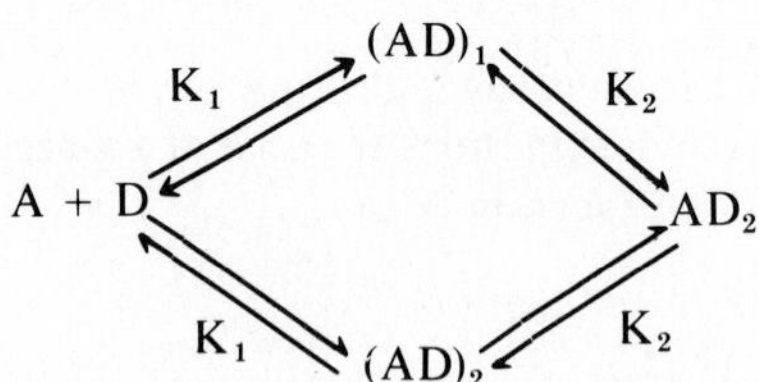

where K_1, K_2 are the association equilibrium constants for each step. (Binding the second dinucleotide is much more favorable than the first because the sequence d-(GpC) is self-complementary and forms a miniature double helix around the intercalated actinomycin chromophore.)

(a) Obtain an expression for the average number of dinucleotides bound ($\bar{\nu}$) per actinomycin as a function of K_1, K_2 and the dinucleotide concentration [D].
(b) Calculate $\bar{\nu}$ when $K_1 = 10^2\ M^{-1}$, $K_2 = 10^5\ M^{-1}$, $[D] = 10^{-4}$ M.

31. A linear DNA molecule with unpaired complementary bases at each end is able to cyclize:

However, there is a competing dimerization reaction

$$2 \; \overline{} \rightleftarrows \overline{}\;\overline{}$$

The objective of this problem is to calculate the molar concentration of linear DNA at which there is an equal probability of dimerization and cyclization. Use your knowledge of polymer statistics to calculate the effective concentration of one end of the linear polymer immediately adjacent to the other end. Take the Kuhn statistical length of DNA = 120 nm, and the total DNA contour length as 1.2×10^4 nm. Convert your result to a molar concentration. Deduce from this the molar concentration at which cyclization and dimerization are of equal probability.

QUESTIONS FOR DISCUSSION

32. Which do you find conceptually easier, thermodynamics or statistical mechanics? Why?

33. Do you think it should be possible, in principle, to use statistical mechanical techniques to calculate the entropy of a living organism?

34. Assuming that muscle contraction results from a conformational change of muscle muscle proteins, what is the physical origin of the force exerted?

REFERENCES

Bloomfield, V. A., Crothers, D. M., and Tinoco, I., Jr. 1974. *Physical Chemistry of Nucleic Acids*. New York: Harper & Row.

Pulleyblank, D. E., Shure, M., Tang, D., Vinograd, J., and Vosberg, H.-P. 1975. *Proc. Natl. Acad. Sci. U.S.A. 72,* 4280.

Sturtevant, J. M. 1977. *Proc. Natl. Acad. Sci. U.S.A. 74,* 2236.

Tanford, C. 1961. *Physical Chemistry of Macromolecules*. New York: John Wiley & Sons.

Walton, A. G., and Blackwell, J. 1973. *Biopolymers*. New York: Academic Press.

FURTHER READING

Besides the standard physical chemistry textbooks listed in Chapters 1–3, there are a number of more advanced textbooks on statistical mechanics which you can read for further information. Among the excellent texts are the following.

Davidson, N. 1962. *Statistical Mechanics*. New York: McGraw-Hill.

Flory, P. J. 1969. *Statistical Mechanics of Chain Molecules*. New York: Interscience.

Hill, T. L. 1960. *Introduction to Statistical Thermodynamics*. Reading, Mass.: Addison-Wesley Publishing Co.

Mayer, J. E., and Mayer, M. G. 1940. *Statistical Mechanics*. New York: John Wiley & Sons.

CHAPTER 15

Transport Processes

○ 15–1 Transport Processes Involve a Flow of Matter or Energy

Chemical kinetics, which was the subject of Chapter 6, deals with one kind of spontaneous process, in which matter is converted from one composition to another at a measurable rate. The reaction is driven by the decrease in free energy that accompanies the conversion. We can think of that driving free energy as providing a "force" that causes matter to "flow" from one form to another. There are many other forces that can be applied to matter which also produce a flow if the system is not at equilibrium. When a force applied to a system causes a flow, we call it a *transport process*.

A conceptually simple example of a transport process is the motion of a macromolecule through a solution when the molecule is subjected to a force. For example, positive and negative electrodes at opposite sides of a solution produce an electric field that forces a charged macromolecule to move toward one electrode. The force (F) acts parallel to the electric field (Fig. 15-1). If there were no solvent around the macromolecule, the force would cause continuing acceleration, a, described by Newton's law, $F = ma$, in which m is the mass of the molecule. Because of continuing acceleration, an unimpeded particle moves faster and faster the longer the field is left on. However, the frictional resistance provided by the solvent prevents the velocity from increasing indefinitely. The net rate of motion in one direction is called the *transport velocity*.

Frictional resistance also is a force, which is proportional to the transport velocity, v, as long as motion is not too fast. The frictional force acts in the direction opposite to the velocity, so it is given by $-fv$. The proportionality constant f is called the *frictional coefficient*. The net force on the molecule is the applied force plus the frictional force, or $F - fv$, which must, by Newton's law, equal mass times acceleration, or

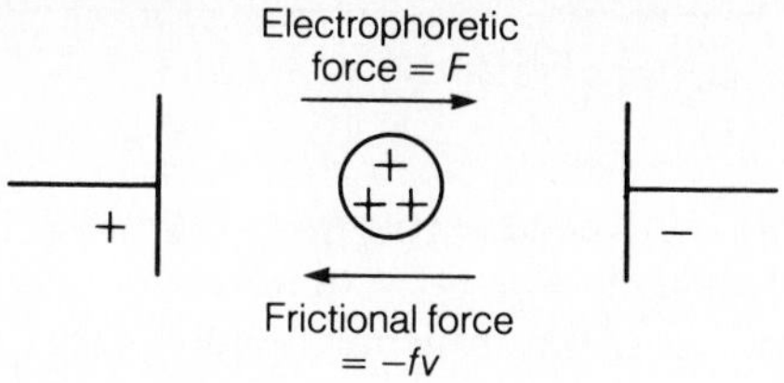

Electrophoretic force = F

+ −

Frictional force = $-fv$

FIGURE 15-1
Electrophoretic force (F) and the oppositely directed frictional force ($-fv$) which act on a charge particle subjected to an electric field. At steady state the two forces balance, $F = fv$, so that the acceleration $a = dv/dt$ is zero and the velocity is unchanging.

$$\frac{mdv}{dt} = F - fv \tag{15-1}$$

In Equation 15-1, the acceleration has been replaced by its equivalent, the derivative of velocity with respect to time, dv/dt.

Equation 15-1 predicts that the transport velocity will vary as shown in Figure 15-2. The equation that describes the curve is

$$v(t) = \frac{F}{f}[1 - \exp(-ft/m)] \tag{15-2}$$

which, as you can verify by differentiation, satisfies Equation 15-1. Immediately after the field is applied ($t = 0$), the velocity rises, approaching F/f exponentially as t becomes large compared to m/f. Since m/f is usually of the order of 10^{-12} sec, the velocity reaches the limiting, *steady state,* value, v:

$$v = F/f \tag{15-3}$$

virtually instantaneously in all transport measurements.

Because the acceleration phase is so exceedingly short-lived, transport equations usually are written for the steady-state condition, neglecting acceleration terms. The equations are all analogous to Equation 15-3, with a *flow* (the velocity or similar term) proportional to a *force, F,*

$$\text{Flow} = \text{coefficient} \times \text{force}$$

Other examples are *heat conduction,* in which energy flows as a result of a "force" provided by a temperature gradient, *sedimentation,* in which molecules move under a centrifugal force, and *diffusion,* in which the force is provided by a concentration gradient.

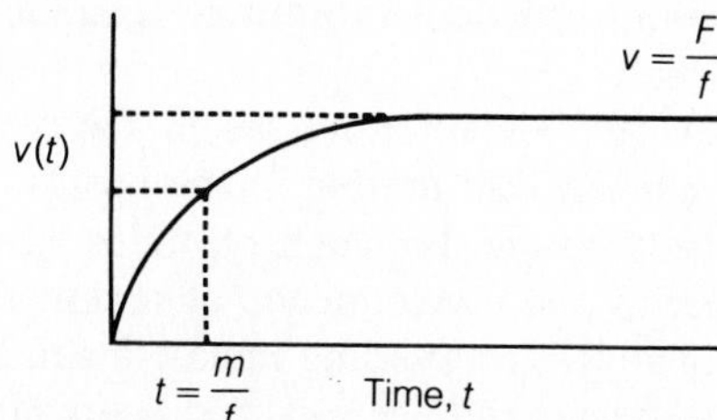

FIGURE 15-2
Increase of the particle velocity toward the steady-state value after application of the electric field in Figure 15-1. The rising curve is exponential, with time constant m/f, which has typical values of 10^{-12} sec.

15–2 How the Transport Velocity Differs from the Molecular Velocity

It is important to understand the distinction between the transport velocity, v, and the average rate of motion of a molecule, v. For the example of an ion moving under the influence of an electric field, the transport velocity measures the net rate of motion toward one electrode, but it would be a mistake to think of the process as a steady motion at the transport velocity. Rather, the molecule is undergoing much more rapid motion in all directions. The probability of motion is nearly the same in all directions, but there is a very slight preference for moving in the direction driven by the force of the electric field. The transport velocity is the result of the slight preference for moving with the field rather than against it.

The Kinetic Theory of Gases Allows Calculation of the Molecular Velocity Distribution

The *kinetic theory of gases* is useful for providing a general view of the distribution of the rate of molecular motion in a sample of matter. We begin with the model of an ideal monatomic gas, in which the particles have only kinetic energy, which is independent of position in space. Furthermore, we assume that the particle's kinetic energy E_x due to motion at velocity v_x in the x direction can be described by the classical expression

$$E_x = \frac{1}{2} m v_x^2$$

with a similar equation for motion in the y and z directions. According to the Boltzmann distribution, the probability of energy E_x, $P(E_x)$, is given by the proportionality

$$P(E_x) \propto \exp(-E_x/k_B T)$$

Integrating the expression $\exp(-E_x/k_B T) = \exp[-mv_x^2/(2k_B T)]$ over all possible values of v_x from $-\infty$ to ∞ yields $(2\pi k_B T/m)^{1/2}$. Therefore the normalized probability is

$$P(v_x) = \left(\frac{m}{2\pi k_B T}\right)^{1/2} \exp[-mv_x^2/(2k_B T)]$$

This equation tells us the probability that a particle in an ideal gas has velocity v_x. Notice that $P(v_x) = P(-v_x)$, which means that a particle chosen at random is equally likely to be moving in the $+x$ direction as in the $-x$ direction. However, the equation does *not* mean that motion in the positive and negative directions will always precisely cancel for each particle. You can understand this distinction by considering the consequence of repeated flipping of a coin. Each flip has equal probability of yielding heads or tails, but twenty trials will not always yield ten heads and ten tails. In a similar manner, an individual molecule may by chance experience more motion in the $+x$ than in the $-x$ direction. The consequence is random, drifting motion of the particle through space. This process is called *self-diffusion*.

The speed, s, of a particle is defined as the magnitude of the vector sum of its individual velocity components v_x, v_y, and v_z. By the Pythagorean theorem we can set

$$s^2 = v_x^2 + v_y^2 + v_z^2$$

The probability of particular values of v_x, v_y and v_z is given by the product of the individual probabilities:

$$\begin{aligned} P(v_x,v_y,v_z) &= \left(\frac{m}{2\pi k_B T}\right)^{3/2} \exp[-m(v_x^2 + v_y^2 + v_z^2)/2k_B T)] \\ &= \left(\frac{m}{2\pi k_B T}\right)^{3/2} \exp[-ms^2/(2k_B T)] \end{aligned} \tag{15-4a}$$

To calculate the probability of a particular value of s, we must recognize that a certain s can result from many combinations of values of v_x, v_y, and v_z. Consider a vector of length s that points outward from the center of the coordinate axes v_x, v_y, and v_z. Any point on the surface of the sphere swept out by the end of s has the same value of s, but different values of v_x, v_y, and v_z. The number of values of v_x, v_y, and v_z that correspond to s is proportional to the area of the sphere, $4\pi s^2$. Therefore

$$\begin{aligned} P(s) &= 4\pi s^2 P(v_x,v_y,v_z) \\ &= 4\pi \left(\frac{m}{2\pi k_B T}\right)^{3/2} s^2 \exp\left(-\frac{ms^2}{2k_B T}\right) \end{aligned} \tag{15-4b}$$ ◀

This important equation, first derived by Maxwell, is called the *Maxwell distribution* of velocity or speed. Notice that because the speed s is the length of the velocity vector, it cannot be negative and can vary only from 0 to ∞, in contrast to v_x in Equation 15-4*a*, which can vary from $-\infty$ to ∞.

The average molecular speed can be calculated in several ways. For example, the mean square speed $\langle s^2 \rangle$ is

$$\langle s^2 \rangle = \int_0^\infty s^2 P(s) ds$$

which can be shown with the help of a table of integrals to be

$$\langle s^2 \rangle = \left(\frac{3k_B T}{m}\right)$$

The root-mean-square speed is the square root of this quantity. Similarly, the mean speed $\langle s \rangle$ is

$$\begin{aligned} \langle s \rangle &= \int_0^\infty sP(s)\, ds \\ &= \left(\frac{8k_B T}{\pi m}\right)^{1/2} \end{aligned}$$

These equations, derived for an ideal gas, can be applied to a macromolecule in solution to estimate its rate of random thermal motion. For example, $\langle s \rangle$ for a molecule of molecular weight 25,000 is 1.6×10^3 cm sec^{-1} at 300 K. In contrast, one has to wait hours for a tRNA molecule of this molecular weight to move several centimeters in a typical electrophoresis experiment. Therefore, the transport velocity is generally much smaller than the average rate of molecular motion.

15–3 The Forces that Drive Transport can be Obtained by Differentiating a Potential Energy or a Free Energy

Recall that the potential energy U of a harmonic oscillator is

$$U = \frac{kx^2}{2}$$

in which x is the displacement from equilibrium and $-kx$ is the restoring force on the particle. The force can be obtained by differentiating U, $F = -dU/dx$, or, in general,

$$\text{Force} = \frac{-d(\text{potential energy})}{d(\text{distance})} \tag{15-5}$$

The effective force that arises due to a concentration gradient can be obtained by differentiating the *chemical potential,* μ, with respect to distance. When the chemical potential of a substance is greater at one point in a solution than another, the molecules tend to move from the high to the low potential. Let the solution be ideal, so that the chemical potential is:

$$\mu = \mu^\circ + RT \ln C$$

If we assume that μ° does not depend on distance x, then (see Figure 15-3)

$$\text{Force} = \frac{-d\mu}{dx} = -RT\frac{d \ln C}{dx} \tag{15-6}$$ ◀

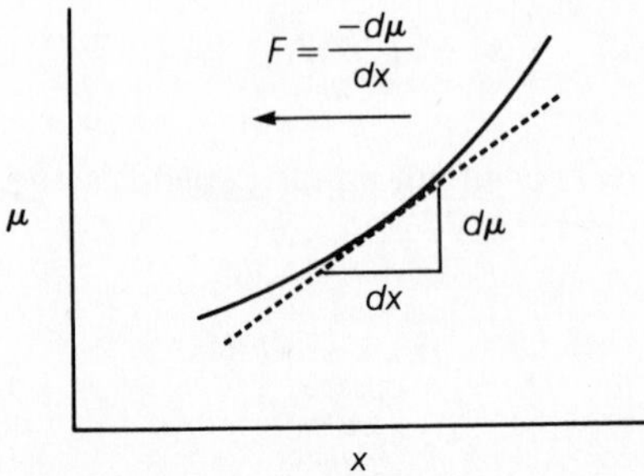

FIGURE 15-3
The average effective force acting on molecules in a concentration gradient is $-d\mu/dx$. The concentration increases with distance (x) from left to right, causing an increase in the chemical potential (μ). Notice that the force, $-d\mu/dx$, acts to move the molecules toward lower concentration.

Since μ refers to one mole, the force does too, so Equation 15-6 gives the force per mole of particles. According to Equation 15-3, this force, when divided by N_A times the frictional coefficient per molecule yields the average transport velocity, v,

$$v = \frac{\text{force}}{N_A f} = \frac{-RT}{N_A f}\frac{d \ln C}{dx} \tag{15-7}$$

Equation 15-7 states that the average transport velocity of particles moving to equalize a concentration gradient is proportional to $d \ln C/dx$.

15–4 *The First Law of Diffusion States that the Flux is Proportional to the Concentration Gradient*

The effective force due to a concentration gradient, given by Equation 15-6, produces transport of matter at an average velocity specified by Equation 15-7. Transport by this mechanism is called *diffusion,* which is described by Fick's first and second laws. The first of these relates the concentration gradient to the amount of matter transported. To derive it, Equation 15-7 is rewritten as

$$v = \frac{-RT}{CN_A f}\frac{dC}{dx} \tag{15-8}$$

and multiplied through by C to obtain Cv, the *flux* of matter, which is the amount of material crossing a unit area in the solution per unit time. To see that Cv is the flux defined in this way, we can first use a dimensional argument. If C is measured in particles/cm^3 (grams or moles can be substituted for particles) and v is in cm/sec, Cv is expressed in particles cm^{-2} sec^{-1}, which has the units of amount of matter per unit area per second.

The relationship between the flux and the transport velocity can be obtained more rigorously using Figure 15-4. A volume element of thickness dx and cross-sectional area A contains concentration C. Since the velocity is dx/dt, all the particles in the volume element will move a distance dx and therefore will cross A during time dt. The amount of matter that crosses A is

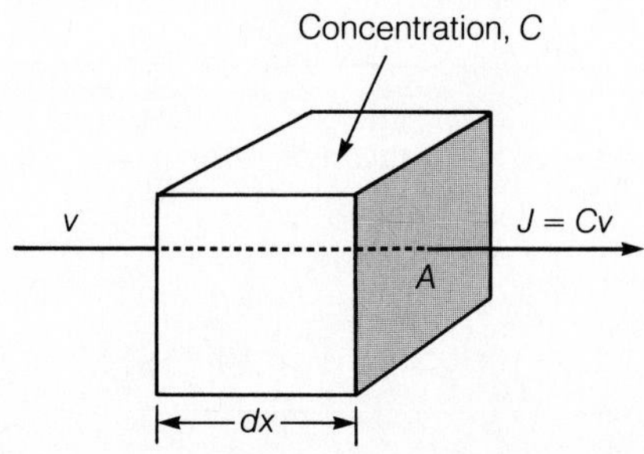

FIGURE 15-4
Molecules move at velocity v through a volume element of thickness dx and cross-sectional area A perpendicular to the velocity vector. The flux, J, is the amount of matter that crosses the face of the volume element per unit area per unit time. It is shown in the text that $J = Cv$.

the concentration times the volume, $C\,dV$, or $CA\,dx$; the amount that crosses per unit area per unit time is $CA\,dx/A\,dt$, or $C\,dx/dt = Cv$. We give the flux Cv the symbol J.

According to Equation 15-8, with $Cv = J$

$$J = -\frac{RT}{N_A f}\frac{dC}{dx} = \frac{-k_B T}{f}\frac{dC}{dx} \tag{15-9}$$

where we have divided R by N_A to obtain k_B. The quantity k_BT/f is called the *diffusion coefficient, D*:

$$D = \frac{k_B T}{f} \tag{15-10}$$ ◄

Equation 15-10 is an important one, since it relates a *macroscopic* quantity, D, to a *molecular* quantity, the frictional coefficient f.

Equation 15-9 now reads

$$J = -D\frac{dC}{dx} \tag{15-11}$$ ◄

This is called *Fick's first law of diffusion,* derived by A. Fick in 1855. It states that the flux of matter is proportional to the concentration gradient in the solution. The proportionality constant, the diffusion constant, has typical values from 10^{-6} cm²/sec to 10^{-8} cm²/sec. The negative sign in the proportionality means that when dC/dx is positive, the flux will be in the direction of decreasing x, or decreasing concentration.

EXERCISE 15-1

Calculate the frictional coefficient of ribonuclease, which has a diffusion constant of 1.1×10^{-6} cm²/sec at 20°C.

ANSWER

$f = k_BT/D = 1.381 \times 10^{-16} \times 293/1.1 \times 10^{-6} = 3.7 \times 10^{-8}$ dyne cm^{-1} sec, or 3.7×10^{-8} g sec^{-1} since 1 dyne = 1 gm cm sec^{-2}.

EXERCISE 15-2

Calculate m/f for ribonuclease (M = 13,683) to estimate the time constant for decay of acceleration terms (see Eq. 15-2).

ANSWER

$m/f = M/N_Af = 13{,}683/(6.022 \times 10^{23} \times 3.7 \times 10^{-8}) = 6.2 \times 10^{-13}$ sec.

15–5 Diffusion can be Thought of as a Random Walk

The particles in a solution that are moving by diffusion are undergoing random motion, sometimes called *Brownian* motion, produced by collisions with other particles in the solution. The force term that appears in Equation 15-6 is only an *effective* force that can be used to calculate the average transport velocity, and should not be thought of as a constant mechanical force acting on each molecule. It may seem to you contradictory that net motion in one direction can result from random motion of all the particles in a solution. The explanation is to be found in the concentration gradient term in Equation 15-9. Consider the diagram in Figure 15-5. There are fewer molecules on the low-concentration side of the imaginary partition, so there will be more motion from high concentration to low than in the reverse direction, even though all molecules have equal *a priori* probability of moving in either direction. The high probability of moving out of high concentration (high chemical potential) is the physical origin of the effective force that appears in Equation 15-6.

The concept that diffusion arises from random motion allows us to focus on the average behavior expected for a single particle. Each molecule follows a random walk of the kind considered for polymer chains in Section 14-12, as shown in Figure 15-6. The mean square distance moved from the origin ($t = 0$) at time t is $\overline{x^2}$. The number of steps N in the random walk is proportional to the time interval Δt. Since $\overline{x^2} \propto N$ (Eq. 14-57*a*, with $\overline{x^2}$

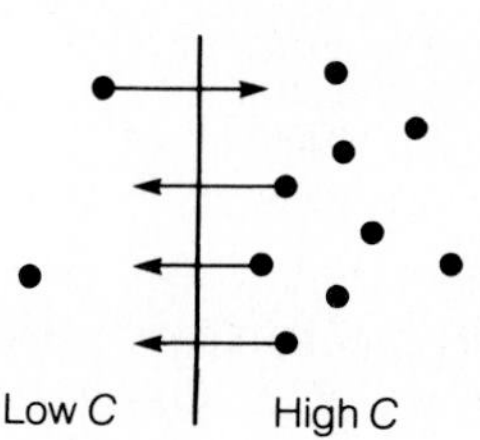

FIGURE 15-5
A high concentration to the right of an imaginary partition produces greater probability for motion to the left than to the right. Each molecule has an equal probability of crossing the partition, but because there are more molecules on the right, the number that move from right to left is greater than the number crossing in the other direction.

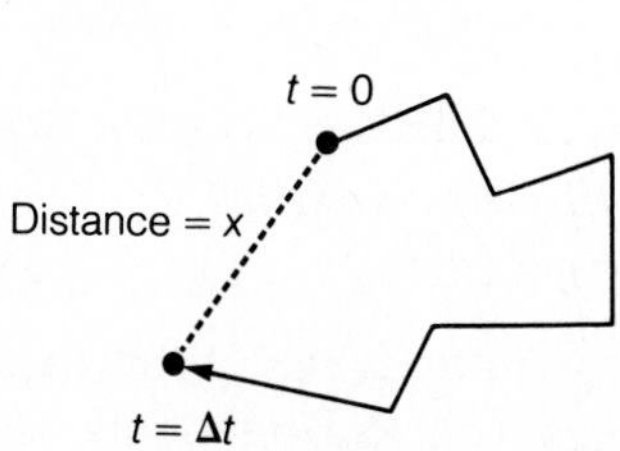

FIGURE 15-6
Brownian motion of a molecule can be described by a random walk, in which x is the net distance moved in time Δt. From the analysis of the random walk in Section 14-12 we expect that the mean square displacement $\overline{x^2}$ (analogous to the mean square end-to-end distance in a polymer chain) should be proportional to Δt. The exact equation, $\overline{x^2} = 2D\Delta t$, was originally derived by Einstein.

replacing L^2), $\overline{x^2}$ must be proportional to the interval Δt. Einstein showed that the proportionality constant is $2D$

$$\overline{x^2} = 2D\,\Delta t \qquad (15\text{-}12)$$

in which D is the diffusion constant. You can see from Equation 15-12 that the dimensions of D must be cm²/sec.

Diffusion is an Effective Transport Mechanism only over Short Distances

Equation 15-12 is extremely important because it allows us to estimate how far and how fast, on the average, a molecule can move when it is free to do so in a cell or in solution. This is illustrated by the following exercises.

EXERCISE 15-3

Calculate the average time required for a protein molecule of diffusion coefficient 10^{-6} cm²/sec to move the length (10^{-4} cm) of a bacterial cell by diffusion.

ANSWER

$\Delta t = x^2/2D = 10^{-8}\ \text{cm}^2/2 \times 10^{-6}\ \text{cm}^2\ \text{sec}^{-1} = 5 \times 10^{-3}$ sec, or 5 msec.

EXERCISE 15-4

Suppose the diffusion constant for a molecule moving across a lipid bilayer is 10^{-8} cm² sec^{-1}. How long would be required to traverse the 100 Å = 10^{-6} cm width?

ANSWER

$\Delta t = 10^{-12}\ \text{cm}^2/2 \times 10^{-8}\ \text{cm}^2\ \text{sec}^{-1} = 5 \times 10^{-5}$ sec, or 50 μsec.

Notice that diffusion is an extremely effective mechanism for moving particles over small distances but the $\overline{x^2}$ dependence of Δt makes the process very slow for long distances. To move the full 1 m length of a nerve cell, for example, would require about 5×10^9 sec (160 years) for a molecule of diffusion constant 10^{-6} cm² sec^{-1}. It is evident that some mechanism other than diffusion is required for communication between the parts of such large cells.

○ *15–6 The Second Law of Diffusion Results when Conservation of Mass is Included*

Equations such as 15-3, which relate velocity to force, can be written for all transport systems. In the cases of electrophoresis, sedimentation, or viscosity, the force is externally applied and no other equation is needed. However, in diffusion the molecules being transported are responsible for the force, so the force changes as transport occurs. Hence the problem cannot be solved without another equation, as you can see if you look at Equation 15-11. We may know the value of dC/dx at the beginning of the experiment, but the flow of matter will change it. In order to calculate how it changes, we must take account of the conservation of matter. (The problem of heat conduction is exactly analogous, except that energy is transported instead of matter.)

Because we will introduce time as a variable, we must replace Equation 15-11 with a partial derivative, which specifies that the derivative with respect to x is taken at a certain time t:

$$J = -D \left(\frac{\partial C}{\partial x} \right)_t \tag{15-13}$$

Figure 15-7 shows the flux into and out of a box with length Δx and cross-sectional area A at the ends. The change in concentration in the box per unit time is the matter that flows into the box per unit time (AJ_{in}) minus the amount that flows out per unit time (AJ_{out}), divided by the volume of the box ΔV. The latter is $\Delta V = A\ \Delta x$. Thus $\Delta C/\Delta t = A(J_{\text{in}} - J_{\text{out}})/\Delta V$, or

$$\frac{\Delta C}{\Delta t} = \frac{-D \left(\frac{\partial C}{\partial x} \right)_t + D \left[\left(\frac{\partial C}{\partial x} \right)_t + \Delta \left(\frac{\partial C}{\partial x} \right)_t \right]}{\Delta x}$$

Converting the limits Δ to the derivative gives, by the definition of first and second derivatives,

$$\left(\frac{\partial C}{\partial t} \right)_x = D \left(\frac{\partial^2 C}{\partial x^2} \right)_t \tag{15-14}$$ ◀

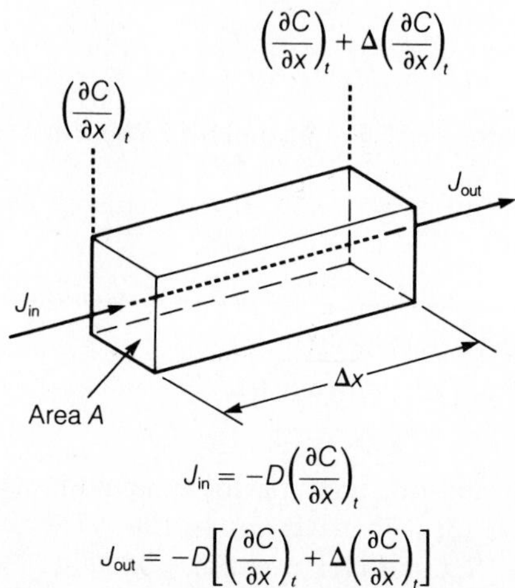

FIGURE 15-7
Flux into and out of a box which has concentration gradient $(\partial C/\partial x)_t$ at one end and $(\partial C/\partial x)_t + \Delta(\partial C/\partial x)_t$ at the other. The concentration change in the box per unit time is proportional to the flux in minus the flux out, a relationship which leads to Fick's second law of diffusion.

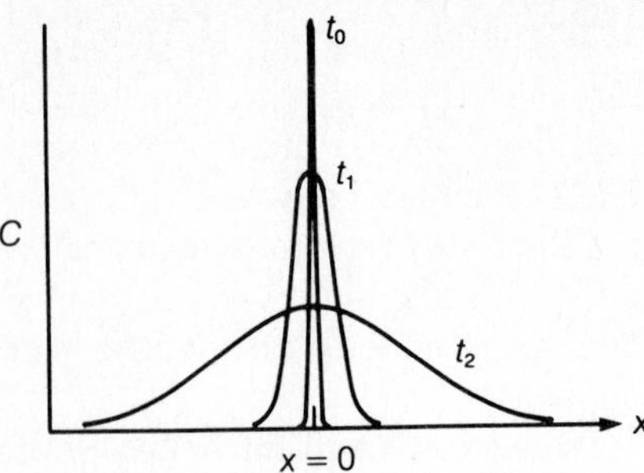

FIGURE 15-8
Schematic diagram of the spreading by diffusion of molecules initially concentrated in the plane at $x = 0$. The curves follow a Gaussian distribution, just as found for the random walk problem in Section 14-12. Exercise 15-5 shows that the mean square distance moved is $\overline{x^2} = 2Dt$, which can be taken as a derivation of the Einstein relationship, Equation 15-12.

This is called *Fick's second law of diffusion*. It is a second-order differential equation whose solution depends on the boundary conditions, just as did the solutions of Schrödinger's equation in Chapter 10. The conditions include the initial concentration profile and the spatial limits on the system. One example is shown in Figure 15-8. All the material is initially concentrated in an infinite plane. It moves a distance x into the solution according to the equation

$$C(x,t) = \frac{C_0}{2(\pi Dt)^{1/2}} \exp\left\{\frac{-x^2}{4Dt}\right\} \tag{15-15}$$

in which C is the relative concentration, and C_0 is a constant. You can see that the Gaussian function $\exp\left\{\frac{-x^2}{4Dt}\right\}$ is analogous to the solution of the random walk problem, Equation 14-55.

EXERCISE 15-5

Calculate the mean square distance moved from the plane in Figure 15-8.

ANSWER

$$\overline{x^2} = \frac{\int_{-\infty}^{\infty} x^2 C(x)\, dx}{\int_{-\infty}^{\infty} C(x)\, dx} = 2\,Dt$$

(Consult a table of integrals.) Compare this with Equation 15-12, the Einstein relationship.

15–7 *Measurement of Light Scattering Fluctuations can be Used to Determine the Diffusion Coefficient*

Light is scattered by a solution when the solution contains local, spontaneous fluctuations in concentration that cause local fluctuations in the refractive index (Sec. 13-10). Of course, these fluctuations are not constant, but instead are continually changing, so the amount of light scattered by the solution also fluctuates with time. The rate at which the molecules can move about by diffusion determines the rate with which the light scattering intensity can fluctuate, since the solute particles must move relative to each other in order to produce a new concentration fluctuation.

The information about molecular motion that is inherent in light scattering is easily mistaken for noise. Figure 15-9(*a*) shows schematically the variation in scattered light intensity that might be observed for a macromolecular solution. Random variations in intensity occur at a rate or frequency that depends in part on how fast the molecules are moving.

The Autocorrelation Function Summarizes the Dynamic Properties of a Fluctuating System

The dynamic information in Figure 15-9 can be extracted from the fluctuation trace by computing what is called the *autocorrelation function* $F(\tau)$, defined by

$$F(\tau) = \langle I(t)I(t + \tau)\rangle \tag{15-16}$$

which has the following meaning: take the value of the intensity I at time t, multiply it by the value at a later time $t + \tau$, and repeat the procedure at

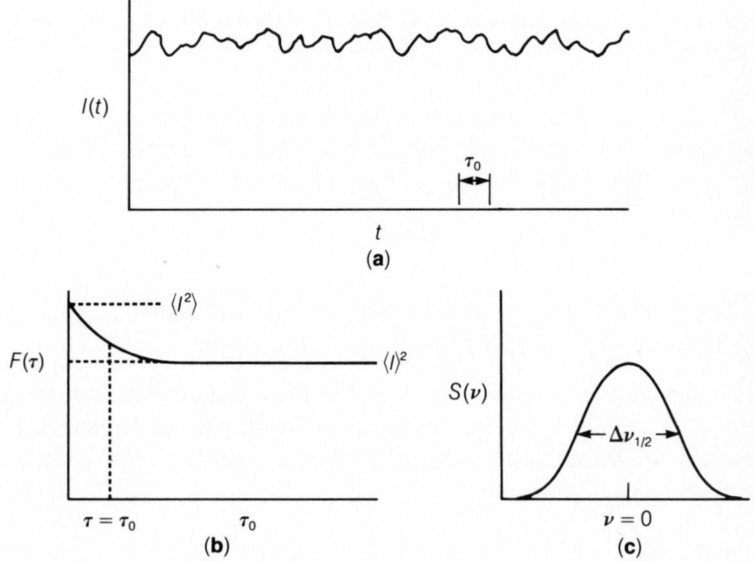

FIGURE 15-9
Three methods of presenting data from intensity fluctuation spectroscopy. **(a)** Time-variation of the intensity of light scattered by a macromolecular solution. The scattering changes with time because the concentration fluctuations change with time. **(b)** Exponentially decaying autocorrelation function for the intensity fluctuations in **(a)**; $F(\tau) = \langle I(t)I(t + \tau)\rangle$ decays from $\langle I^2\rangle$ to $\langle I\rangle^2$. The magnitude of the decay time τ_0 is shown on the axis of **(a)** for comparison. **(c)** $S(\nu)$, the Lorentzian power spectral density curve which corresponds to exponentially decaying autocorrelation spectral function in **(b)**. The width of the peak at half-height is $1/(\pi\tau_0)$.

many different initial times t to obtain the average of $I(t)I(t + \tau)$, as indicated in Equation 15-16 by the brackets. Another value of τ is then chosen, and the process is repeated. Frequently, the result is a curve that decays exponentially with time τ, as shown in Figure 15-9(*b*).

The name of this function—"autocorrelation"—implies that it indicates the extent to which the value of a variable is correlated with itself, or, more specifically, how rapidly the self-correlation dies away. A property that changes slowly has a slowly decaying autocorrelation function, and a rapidly fluctuating property has a rapidly decaying autocorrelation function. Periodic functions have periodic autocorrelation functions, as illustrated by the following exercise.

EXERCISE 15-6

What is the autocorrelation function of the sound intensity produced by a $33\frac{1}{3}$ rpm phonograph record which is blank except for a scratch from the center to the edge?

ANSWER

The sound of the scratch appears $33\frac{1}{3}$ times a minute, or every 0.03 minutes. Therefore there is a strong correlation between the sound at time 0 and time $\tau = 0.03$, 0.06, 0.09, etc., minutes. For all other intervals, either $I(t)$ or $I(t + \tau)$, or both, are zero. Hence the autocorrelation function is zero except for spikes at 0, 0.03, 0.06, 0.09, etc., minutes.

Notice that the exponentially decaying autocorrelation function in Figure 15-9(*b*) decays from the mean square intensity $\langle I^2 \rangle$ to the square of the mean intensity, $\langle I \rangle^2$. The former is always larger than the latter for fluctuating quantities.

The Decay of the Autocorrelation Function is Determined by the Diffusion Coefficient

The time constant for exponential decay of the autocorrelation function for scattered light depends on the diffusion coefficient. More advanced books, such as those by Berne and Pecora (1976) or Chu (1974) demonstrate that the relaxation time τ_0 is given by

$$\tau_0 = \frac{1}{2K^2D} \tag{15-17}$$

in which D is the diffusion coefficient and K is the quantity

$$K = \left(\frac{2\pi n}{\lambda_0}\right) \sin(\theta/2) \tag{15-18}$$

As usual, the refractive index is symbolized by n, λ_0 is the wavelength, and θ is the scattering angle (Fig. 13-38). Notice that rapidly moving molecules have large diffusion constants, D, and, according to Equation 15-17, they have short autocorrelation decay times τ_0.

EXERCISE 15-7

Show that $1/K$ has the dimensions of length, and write Equation 15-17 in a form analogous to the Einstein relationship, Equation 15-12.

ANSWER

n, the refractive index, is dimensionless, as is sin $(\theta/2)$. Therefore K has the dimensions of 1/wavelength, or length $^{-1}$, and K^{-1} has dimensions of length. The Einstein relationship is $\overline{x^2} = 2D\,\Delta t$. Equation 15-17, written analogously, is $(1/K)^2 = 2D\tau_0$. The analogy tells us that τ_0 is equal to the time required for the molecule to diffuse through a distance $1/K$, approximately equal to $\lambda_0/2\pi$ when θ is large. The reason the wavelength of light sets the characteristic distance for diffusion is that the molecule must move through a substantial portion $(1/2\pi)$ of a wavelength in order to change appreciably the phase of its scattered light. Shifting of the relative phases of light scattered by different particles is what gives rise to fluctuations in the total scattered intensity.

The Power Spectral Density is the Fourier Transform of the Autocorrelation Function

Another way of presenting the information present, but hidden, in the trace of time-dependent fluctuations in Figure 15-9(a) is a plot called the *power spectral density,* shown schematically in Figure 15-9(c). You can understand the qualitative meaning of this graph by considering that the fluctuation record in Figure 15-9(a) contains some regions where the scattered light intensity is varying slowly or even is constant and other regions in which it varies at higher frequency. The power spectral density curve gives the relative contribution of each frequency to the total fluctuation trace. Notice that it has a maximum at $\nu = 0$, the zero frequency component, which corresponds to a constant intensity. The curve includes negative frequencies for symmetry reasons. You can think of these as the frequencies that would result if the fluctuation trace ran backwards, on a reversed time axis.

The power spectral density, $S(\nu)$, is shown in more advanced texts (Chu, 1974) to be the Fourier transform (Sec. 17-5) of the autocorrelation function:

$$S(\nu) = \int_0^\infty \langle I(t)I(t+\tau)\rangle \cos(2\pi\nu\tau)\,d\tau \tag{15-19}$$

Therefore $S(\nu)$ *gives the relative weights of the Fourier components at frequency* ν *for the fluctuation curve in Figure 15-9(a).* When the autocorrelation function follows an exponential decay curve, proportional to $\exp(-\tau/\tau_0)$, the power spectral density is

$$S(\nu) = \int_0^\infty \exp(-\tau/\tau_0)\cos(2\pi\nu\tau)\,d\tau \tag{15-20}$$

This integral can be looked up in a table. Its value is

$$S(\nu) = \frac{1/\tau_0}{(1/\tau_0)^2 + (2\pi\nu)^2}$$

$$= \frac{\tau_0}{1 + (2\pi\nu\tau_0)^2} \tag{15-21}$$

An Exponential Autocorrelation Function has a Lorentzian Power Spectral Density

Equation 15-21 is an important functional form called a *Lorentzian,* which has the general form $a/[b^2 + (\omega - \omega_0)^2]$. The quantities a and b are constants, and the angular frequency is $\omega = 2\pi\nu$. A Lorentzian function is symmetric about $\omega = \omega_0$, where it reaches a maximum; such functions always characterize the frequency components of small fluctuations, just as exponential decay always characterizes the relaxation of small fluctuations or perturbations toward equilibrium (see Sec. 6-7).

EXERCISE 15-8

Show that the NMR line shape function $g(\nu)$ (Eq. 13-34a) is Lorentzian.

If fluctuation data are presented as the power spectral density, the exponential decay time for the autocorrelation function is readily determined from the spectral density curve, because the width of the Lorentzian at half-height is

$$\Delta\nu_{1/2} = \frac{1}{\pi\tau_0} \tag{15-22}$$

which can be verified by substituting the values $2\pi\nu = \pm 1/\tau_0$ required to give $S(\nu)$ half its maximum value. The frequency distance between these two half-height points is $1/(\pi\tau_0)$, as shown in Figure 15-9(*c*). Again, notice the relationship to nuclear magnetic resonance, with T_2 equivalent to τ_0.

For completeness, we also observe that the autocorrelation function is the Fourier transform of the spectral power density. When $S(\nu)$ is Lorentzian, evaluation of the Fourier transform of $S(\nu)$ reveals that the autocorrelation function is exponential, as we knew it must be since Equation 15-20 stated that the Fourier transform of an exponential decay is a Lorentzian.

The Width of a Lorentzian Power Spectral Density Curve Determines the Diffusion Coefficient

Figure 15-10 shows an application of intensity fluctuation spectroscopy, giving the power spectral density for light scattered from a suspension of casein micelles. The slight deviation of the observed curve from calculated Lorent-

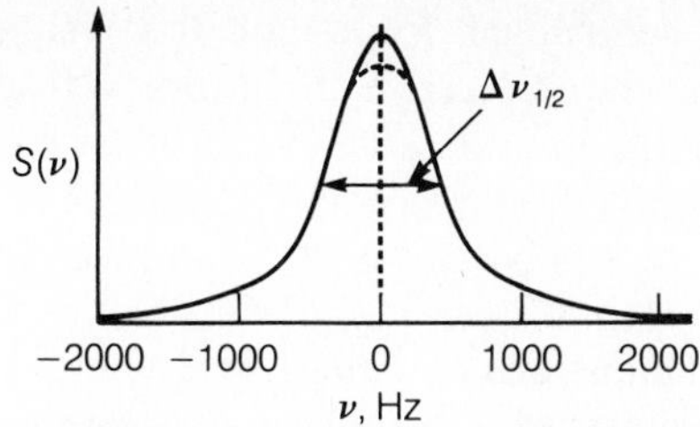

FIGURE 15-10
Power spectral density for the scattering from casein micelles. The solid curve is a Lorentzian function (Eq. 15-21), whose time constant τ_0 is related to the width at half-height ($\Delta\nu_{1/2}$) by Equation 15-22. (From Lin et al., 1971. Reprinted with permission from *Biochemistry 10,* 4788. Copyright by the American Chemical Society.)

zian shape was interpreted by the authors as arising from variation in the size, and hence in the diffusion coefficient, of the particles. The average diffusion coefficient, obtained from the width of the spectral density curve at half-height, is

$$D = \frac{\pi \, \Delta\nu_{1/2}}{2K^2}$$

an equation which is obtained by combining Equation 15-17 with Equation 15-22. Analysis of the curve in Figure 15-10 yields an average diffusion coefficient of 3.23×10^{-8} cm^2 sec^{-1}, corresponding to a sphere of radius 767Å (Lin et al., 1971), using Stokes' law as discussed in Section 15-10.

○ 15–8 Sedimentation Velocity in the Ultracentrifuge

The ultracentrifuge, whose use to determine molecular weights by equilibrium sedimentation was described in Section 7-17, also can be used to measure the velocity of molecular motion under a centrifugal force. As given by Equation 7-79, a particle located a distance r from the axis and moving at angular velocity ω experiences a force F in the rotating frame,

$$F = m\omega^2 r\phi_2$$

in which m is the particle mass and $\phi_2 = (\partial\rho/\partial c_2)_\mu$ is the buoyancy correction, equal to the density increment per gram of solute (called Component 2) at constant chemical potential of the dialyzable (solvent) components (see Sec. 7-18 for a discussion of this term). Using Equation 15-3, the transport velocity is the force divided by the frictional coefficient,

$$v = \frac{M\omega^2 r\phi_2}{N_A f} \tag{15-23}$$

in which m has been replaced by the equivalent M/N_A.

The velocity depends on experimental quantities such as the speed of rotation of the ultracentrifuge, so it is convenient to define a quantity s called the *sedimentation coefficient,*

$$s = \frac{v}{\omega^2 r} \tag{15-24}$$

which is the velocity divided by the centrifugal force per unit mass; s depends only on molecular and solution parameters. Substituting v from Equation 15-23, we see that

$$s = \frac{M\phi_2}{N_A f} \qquad (15\text{-}25)$$

According to Equation 15-24, s has the dimensions of time. Since values in the range of 10^{-13} sec are commonly encountered, s usually is expressed in multiples of 1×10^{-13} sec, a unit called a *Svedberg, S,* after the Swedish inventor of the ultracentrifuge.

Equation 15-25 is analogous to Equation 15-10 in that it relates a macroscopic quantity, the sedimentation coefficient, to the frictional coefficient, f. Measurement of s and the density increment, ϕ_2, allows calculation of the frictional coefficient, assuming that the molecular weight is known from another experiment such as light scattering or sedimentation equilibrium.

EXERCISE 15-9

The molecular weight of lysozyme is 1.47×10^4 g mol^{-1}, the sedimentation coefficient is 1.87 S, and the density increment is 0.312. Calculate the frictional coefficient.

ANSWER

$f = M\phi_2/N_A s = 1.47 \times 10^4 \text{ g mol}^{-1} \times 0.312/(6.022 \times 10^{23} \text{ mol}^{-1} \times 1.87 \times 10^{-13} \text{ sec}) = 4.07 \times 10^{-8} \text{ g sec}^{-1}$.

Sedimentation coefficients can be measured by a *boundary* or a *band* technique. Figure 15-11 shows these two methods in a schematic diagram. In 15-11(*a*) the solution is initially homogeneous, and as the high molecular weight material sediments, a boundary is formed between solute and solution. As time increases, diffusion causes spreading of the boundary. In the band technique [Fig. 15-11(*b*)], the solution is layered on a more dense solvent, which is usually stabilized against convection by a *density gradient,* in which the concentration of a solute such as sucrose or CsCl is higher at the bottom (large r) than at the top (small r) of the cell. (Convection does not occur because heavy material at the bottom of the cell would have to mix with lighter material above it.) Again, diffusion causes spreading of the initially sharp band. The contribution is given quantitatively by Equation 15-15.

The sedimentation coefficient is measured by determining the rate of movement of the boundary or band midpoint, r_b (Fig. 15-11). Substituting $v = dr_b/dt$ in Equation 15-24 yields

$$s = \frac{dr_b/dt}{\omega^2 r_b}$$

Since $dr/r = d \ln r$, we may rearrange this equation to obtain the relationship

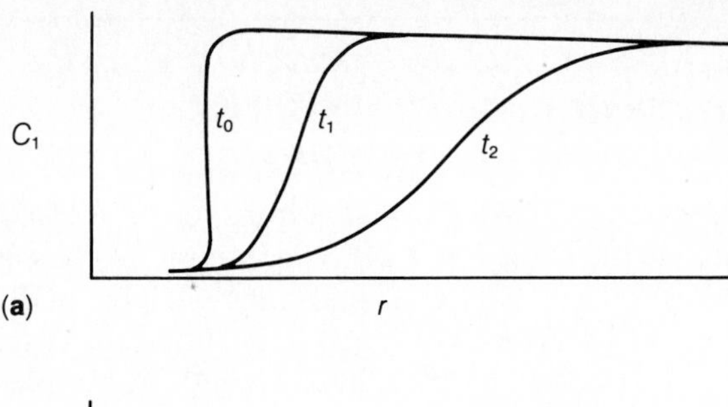

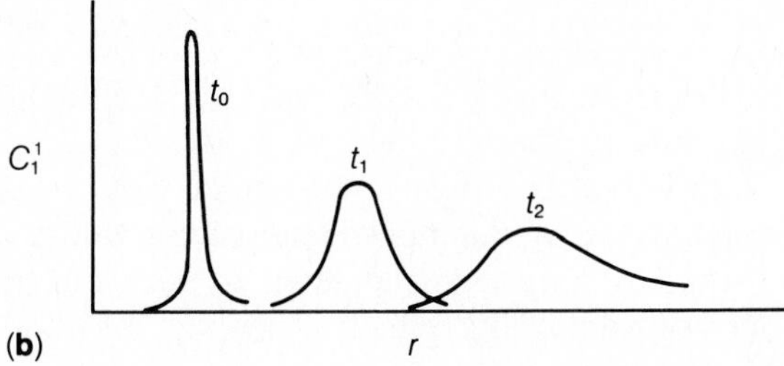

FIGURE 15-11
Schematic diagrams of the boundary **(a)** and band **(b)** sedimentation techniques. In **(a)**, the macromolecules in a homogeneous solution sediment towards the bottom (right) of the cell. Their depletion from the top of the cell causes a boundary to be formed between solution and pure solvent. The spreading of the boundary is caused by diffusion and also possibly by heterogeneity of the macromolecules, which therefore move at different rates. In **(b)**, a layer of solution at the top of the cell moves towards the bottom. The spreading of the band results from diffusion and other factors as in **(a)**.

$$\frac{d \ln r_b}{dt} = \omega^2 s \tag{15-26}$$

Therefore, if $\ln r_b$ is plotted against t, the result should be a straight line with slope $\omega^2 s$, from which s is readily calculated because the angular velocity ω is known ($\omega = 2\pi \times$ revolutions/sec).

EXERCISE 15-10

The boundary position of a solution of bovine serum albumin is found to move from $r_b = 6.15$ cm to $r_b = 6.83$ cm in a time interval 157 min, at a rotor speed of 45,000 rpm. Estimate the sedimentation coefficient (20°C).

ANSWER

By Equation 15-26, $\Delta(\ln r_b)/(\omega^2 \Delta t) = s$, or

$$s = \frac{\ln (6.83/6.15)}{\left(2\pi \times \dfrac{45{,}000 \text{ rpm}}{60 \text{ sec/min}}\right)^2 \left(157 \text{ min} \times 60 \dfrac{\text{sec}}{\text{min}}\right)}$$
$$= 5.01 \times 10^{-13} \text{ sec}$$
$$= 5.01\ S$$

15–9 Diffusion and Sedimentation Velocity Measurements Can Be Combined to Determine Molecular Weights

Both sedimentation and diffusion measure transport of molecules against the frictional resistance developed by their own motion. In the case of sedimentation the driving force depends on molecular weight, and in the case of diffusion it does not. Therefore you might expect that the two measurements can be used to determine the molecular weight. A simple way to show that this is true is to solve Equations 15-10 and 15-25 for the frictional coefficient

$$f = \frac{k_B T}{D} \qquad f = \frac{M\phi_2}{N_A s}$$

If the two experiments are done under identical conditions of temperature and solvent and the results are extrapolated to zero macromolecule concentration to remove nonideality effects, then the two f values must be equal. Hence

$$\frac{k_B T}{D} = \frac{M\phi_2}{N_A s}$$

or

$$M = \frac{RT}{\phi_2}\frac{s}{D} \tag{15-27}$$

This equation provides an important method for determining molecular weights from transport measurements. It requires three experimental techniques: sedimentation, s, diffusion, D, and density measurements to determine the density increment, ϕ_2 (see Eq. 7-80).

EXERCISE 15-11

The diffusion coefficient of bovine serum albumin is 7×10^{-7} cm²/sec at 20°C, and the density increment is 0.266. Use the results of Exercise 15-10 to estimate the molecular weight.

$$M = \frac{8.314 \times 10^7 \text{ erg K}^{-1}\text{ mol}^{-1} \times 293\text{ K}}{(0.266)} \times \frac{5.01 \times 10^{-13}\text{ sec}}{7 \times 10^{-7}\text{ cm}^2\text{ sec}^{-1}}$$
$$= 65{,}500 \text{ g mol}^{-1}$$

(Remember that 1 erg has the dimensions of g cm^2 sec^{-2}.)

15–10 The Frictional Coefficient Depends on Molecular Size and Shape

So far in this chapter we have seen that measurement of either the sedimentation coefficient or the diffusion coefficient (both macroscopic transport quantities) allows us to calculate the molecular frictional coefficient. Such measurements would be more informative if we could use them to draw conclusions about the size and shape of molecules in solution, but this requires that we know how the frictional coefficient f varies with molecular properties.

Theoretical work directed at calculating the frictional properties of macromolecules has relied on idealized models for representing their shape. The principal models include spheres, elongated or flattened spheres called ellipsoids of revolution, rods, and flexible polymer chains.

Stokes' Law Gives the Frictional Coefficient of a Sphere

The value of the frictional coefficient of a sphere (f_0) was calculated theoretically by Stokes about 1850. He found that

$$f_0 = 6\pi\eta R \tag{15-28}$$

in which R is the radius of the sphere and η is a macroscopic quantity called the *solvent viscosity,* which measures the resistance to flow provided by the fluid (see Sec. 15-12). Stokes' law (Eq. 15-28) states that the frictional coefficient of a sphere varies linearly with its radius. This relationship has been tested extensively using macroscopic spheres, whose radii and rates of fall through a fluid can readily be measured.

Ellipsoids of Revolution are Models for Elongated or Flattened Molecules

Few molecules are perfectly spherical, so we make use of idealized models, which allow the molecule to be elongated or flattened from spherical shape. A frequently used shape is the *ellipsoid of revolution,* which is the solid region swept out when an ellipse is rotated about one of its axes. Notice from Figure 15-12 that an ellipse can be rotated about either its short or its long

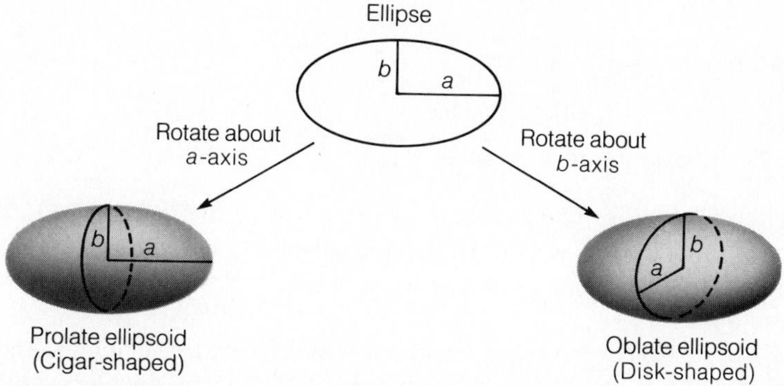

FIGURE 15-12
A *prolate* ellipsoid is swept out when an ellipse is rotated about its long axis and an *oblate* ellipsoid results when the ellipse is rotated about its short axis.

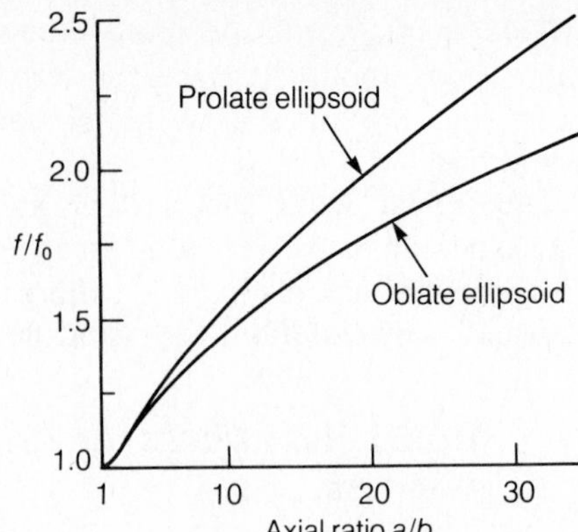

FIGURE 15-13
Ratio of the frictional coefficient of an ellipsoid of revolution (f) to the frictional coefficient (f_0) of a sphere having the same volume as the ellipsoid. (From Tanford, 1961. Reprinted with permission of John Wiley & Sons, Inc.)

axis, producing two classes of ellipsoid. When the ellipse is rotated about its longer axis, a cigar-shaped object results, which is called a *prolate* ellipsoid of revolution. The other rotation, about the shorter axis, creates an *oblate* ellipsoid of revolution, shown in Figure 15-12 as a disk-shaped body.

Ellipsoids of revolution are characterized by their *axial ratio,* the ratio of the lengths of their long and short axes. The longer radius of the ellipse (half the length of the longer axis) is given the symbol a, and the shorter radius is b. The axial ratio is defined as the ratio of a to b:

$$\text{Axial ratio} = a/b \tag{15-29}$$

The frictional coefficient of ellipsoids of revolution was calculated by Perrin in 1936. His results are expressed in terms of the frictional coefficient of the ellipsoid, divided by the frictional coefficient of a sphere of the same volume as the ellipsoid. The rather complicated equations for f/f_0 (see Tanford, 1961) are

$$\frac{f}{f_0} = \frac{(a/b)^{2/3}(1 - b^2/a^2)^{1/2}}{\ln\left\{\dfrac{1 + (1 - b^2/a^2)^{1/2}}{b/a}\right\}} \quad \text{(prolate)} \tag{15-30a}$$

for prolate ellipsoids, and

$$\frac{f}{f_0} = \frac{(a^2/b^2 - 1)^{1/2}}{(a/b)^{2/3}\tan^{-1}(a^2/b^2 - 1)^{1/2}} \quad \text{(oblate)} \tag{15-30b}$$

for oblate ellipsoids. These functions are graphed in Figure 15-13. You can see that f is not very sensitive to shape when the volume is held constant, since a factor change of 20 in the axial ratio a/b results in a factor change of only about 2 in f.

Equation 15-30 can be used to estimate the frictional coefficient, and hence the sedimentation or diffusion coefficient of a macromolecule whose shape can be approximated by an ellipsoid of revolution. The following exercises illustrate the procedure for short, rod-like pieces of DNA.

EXERCISE 15-12

A rod of length L and diameter d can be approximated by a prolate ellipsoid of major radius $L/2$. In order to have the same volume as the rod, it can be shown that the ellipsoid must have an axial ratio equal to $(a/b) = (\frac{2}{3})^{1/2}L/d$. Use Equation 15-30$a$ to calculate the frictional coefficient ratio f/f_0 for a DNA molecule 680Å long (200 base pairs, molecular weight 135,000) and 20Å in diameter.

ANSWER

The axial ratio of the equivalent ellipsoid of revolution is $a/b = (\frac{2}{3})^{1/2}$ $(\frac{680}{20}) = 27.8$. Therefore

$$\frac{f}{f_0} = \frac{(27.8)^{2/3}[1 - (1/27.8)^2]^{1/2}}{\ln(\{1 + [1 - (1/27.8)^2]^{1/2}\}/27.8)} = 2.28$$

EXERCISE 15-13

Use the result of Exercise 15-12 to calculate the frictional coefficient f of the DNA molecule. Use $\eta = 0.010$ g cm^{-1} sec^{-1} for the solvent viscosity.

ANSWER

We need to calculate f_0, the frictional coefficient of a sphere of the same volume as the rod, for which we will use Stokes' law. The volume of the rod is $\pi Ld^2/4 = \pi \times 6.8 \times 10^{-6}$ cm $\times (2 \times 10^{-7}$ cm$)^2/4 = 2.14 \times 10^{-19}$ cm^3. This corresponds to a sphere of volume $(\frac{4}{3})\pi R^3$, or

$$R = \left(\frac{3 \times 2.14 \times 10^{-19}}{4\pi}\right)^{1/3} = 3.71 \times 10^{-7} \text{ cm}$$

Therefore $f_0 = 6\pi\eta R = 6\pi \times 0.01$ g cm^{-1} sec^{-1} $\times 3.71 \times 10^{-7}$ cm $=$ 6.99×10^{-8} g sec^{-1}. Using $f/f_0 = 2.28$, we calculate $f = 2.28 \times 6.99 \times 10^{-8}$ g sec^{-1} $= 1.59 \times 10^{-7}$ g sec^{-1}.

EXERCISE 15-14

Calculate the expected sedimentation coefficient of the DNA molecule in Exercise 15-12, using the results of Exercises 15-12 and 15-13, and the density increment $\phi_2 = 0.49$.

ANSWER

By Equation 15-25, $s = M\phi_2/N_A f = 1.35 \times 10^5$ g $\times 0.49/(6.022 \times 10^{23} \times 1.59 \times 10^{-7}$ g sec$^{-1}) = 6.91 \times 10^{-13}$ sec^{-1} or 6.91 S. The experimental sedimentation coefficient of DNA in this size range is approximately 6 S.

Exercises 15-12 through 15-14 show that the sedimentation coefficient of a rod is predicted with reasonable accuracy by Perrin's equations for the frictional coefficient. Notice that the frictional coefficient predicted for the same mass of DNA packed in a sphere would be 6.91 (f/f_0) or 15.8, which is much larger than the measured value. The small value of the sedimentation coefficient implies that the DNA molecule has a large axial ratio, as we know to be the case for a double helix.

Hydration Affects the Frictional Coefficient

Unfortunately, frictional coefficient measurements cannot give a precise determination of macromolecular axial ratios, because the axial ratio is not the only factor that determines the frictional coefficient. All macromolecules have water molecules bound to them. Some are so tightly bound that they move with the macromolecule and must be considered to be a part of it for the purpose of calculating the frictional coefficient. Therefore, bound water molecules can increase the effective volume of the macromolecule, increasing its frictional coefficient and decreasing its sedimentation coefficient. We conclude that increased hydration has the same influence on S values as does an increase in the axial ratio.

The ambiguity in interpretation of friction coefficients is illustrated by the data for ribonuclease (Tanford, 1961). The frictional coefficient of this enzyme is found to be 1.14 times as large as the calculated value for a sphere of volume equal to that of the macromolecule. This could mean on one hand that the molecule is a sphere with a layer of water (0.35 g/g of protein) around it, increasing its radius by a factor 1.14, or alternatively that the molecule is not hydrated but that its axial ratio is increased to 2.1. Probably some combination of the two influences is responsible for the observations. Without other information it is not possible to decide how much of the frictional coefficient increase is due to hydration, and how much results from a nonspherical shape of the protein. A more modern application of frictional coefficient data is to use the shape of the protein as determined from x-ray diffraction methods in combination with frictional coefficient measurements to calculate the extent of macromolecular hydration.

A Rod also can be Modeled as a Chain of Beads

Macromolecules can be represented by more than one idealized model. Because there are a number of macromolecules such as DNA which are stiff chains, rod-like models are of particular interest. One such model is a string of touching beads, each of diameter δ (see Fig. 15-14). The rod length is $N\delta$, and the axial ratio is N. The frictional coefficient calculated for this model by Kirkwood and Riseman (1956) is

$$f = \frac{3\pi\eta N\delta}{\ln N} \tag{15-31}$$

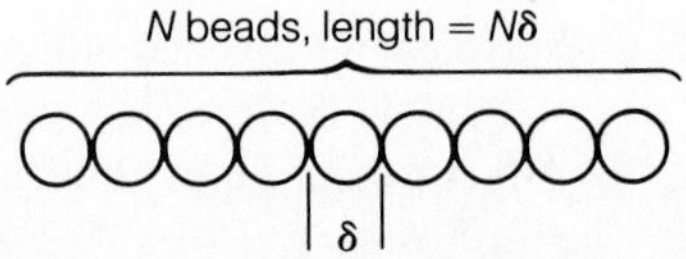

N beads, length = Nδ

δ

FIGURE 15-14
A rod-like molecule can be modeled by a chain of N spherical beads of diameter δ.

The important point to notice about this equation is that when the axial ratio N is large, f varies nearly linearly with the length $N\delta$ of the rod. The mathematical reason is that ln N is a slowly varying function of N when N is large. Thus, to a good approximation the frictional coefficient of a rod depends simply on its length, as is intuitively reasonable if the rod is long enough so that the ends are unimportant. This observation will help us understand the sedimentation behavior of short, rod-like pieces of DNA.

EXERCISE 15-15

Calculate the volume and number of beads required to construct a string-of-beads model with the same volume and length as the DNA molecule in Exercises 15-12 through 15-14.

ANSWER

Length $= N\delta = 6.8 \times 10^{-6}$ cm. Volume $= (\frac{4}{3})\pi(\delta/2)^3 N = 2.14 \times 10^{-19}$ cm^3 (Ex. 15-13). Divide the equation for the volume by the equation for the length to get $(4\pi/24)\delta^2 = (2.14 \times 10^{-19}/6.8 \times 10^{-6})$ cm^2. Solving this equation yields $\delta = 2.45 \times 10^{-7}$ cm (24.5Å). Dividing the length by δ gives $N = 27.7$.

EXERCISE 15-16

Calculate the frictional coefficient and the sedimentation coefficient for the model in Exercise 15-15, using $\eta = 0.01$ g cm^{-1} sec^{-1} and $\phi_2 = 0.49$.

ANSWER

$f = 3\pi\eta N\delta/\ln N = 3\pi \times 0.01$ g cm^{-1} sec^{-1} $\times$ 6.8×10^{-6} cm/ln (27.7) $=$ 1.93×10^{-7} g sec^{-1}. The sedimentation coefficient, calculated as in Exercise 15-14, is $s = 5.69$ S, which is close to the measured value.

Flexible Chains can be Assigned an Effective Hydrodynamic Radius

So far we have considered the frictional coefficient only for solid bodies. A long flexible polymer chain (see Chap. 14) can be thought of as a sphere of effective hydrodynamic radius R_e, with frictional coefficient

$$f = 6\pi\eta R_e \tag{15-32}$$

which follows by direct analogy with Equation 15-28 for a hard sphere. Equation 15-32 serves as a definition of the effective hydrodynamic radius of a flexible chain, but leaves us with the problem of relating R_e to other molecular properties. It turns out, as seems physically reasonable, that R_e is, to a good approximation, proportional to the root mean square end-to-end distance of the polymer chain (see Sec. 14-12),

$$R_e = \langle \gamma L^2 \rangle^{1/2} \tag{15-33}$$

in which γ is a proportionality constant independent of molecular weight (to a good approximation; see Kirkwood and Riseman, 1956).

A relationship between frictional coefficient and polymer chain length can be obtained by combining Equation 14-60, which states that $\langle L^2 \rangle^{1/2} \propto N^{(1+\epsilon)/2}$, with Equations 15-32 and 15-33, yielding

$$f \propto \eta N^{(1+\epsilon)/2} \tag{15-34}$$

in which N is the chain length and $\epsilon \geq 0$ is the chain expansion parameter that takes account of excluded volume effects (Sec. 14-12). Equation 15-34 predicts that the frictional coefficient of a flexible polymer should increase with the square root or higher power of the chain length, depending on the solvent, which determines ϵ.

15–11 The Sedimentation Behavior of DNA

The observed sedimentation properties of DNA samples of varying molecular weight provide an interesting application of the frictional coefficient relations discussed in the previous section. First we note that Equation 15-25 states that the sedimentation coefficient is inversely proportional to the frictional coefficient. Our problem is to determine how the sedimentation coefficient should vary with molecular weight M. To a good approximation the density increment ϕ_2 is independent of M, so the basic proportionality is, from Equation 15-25,

$$s \propto M/f \tag{15-35}$$

There are two simple limiting cases to consider. At low molecular weights ($M \leq 10^5$), when the DNA molecules are shorter than the Kuhn statistical segment length (see Chap. 14), DNA can be considered to be a rigid rod. In this limit, Equation 15-31 states that the frictional coefficient should increase approximately linearly with chain length or M. Hence

$$s \propto M/M = \text{constant} \qquad \text{(rigid rod)} \tag{15-36}$$

and the sedimentation coefficient of DNA should become roughly independent of M when M is below 10^5.

The other limit is reached when the molecule is long enough to contain many statistical segment lengths, so that it may be considered to be a flexible chain. In this case Equation 15-34 applies, and we obtain, by combining it with 15-35 and noting that $M \propto N$,

$$s \propto \frac{M}{M^{(1+\epsilon)/2}} \propto M^{(1-\epsilon)/2} \qquad \text{(flexible chain)} \tag{15-37}$$

This equation reveals that s should increase with the $(\frac{1}{2} - \epsilon/2)$ power of M.

Figure 15-15 shows experimental data for the sedimentation coefficient of DNA samples of varying molecular weight. The plot shows log s versus log M, so that the slope gives the exponent in the variation of s with M. The

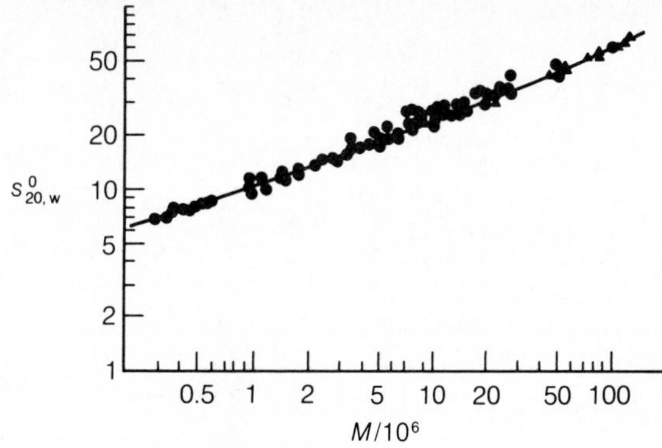

FIGURE 15-15
Variation of the sedimentation coefficient of DNA with molecular weight. The symbol $s^0_{20,w}$ means that the sedimentation coefficient has been extrapolated to zero concentration, and is expressed in terms of the value it would have in a solvent with the viscosity of water at 20°C. Notice the slight curvature toward higher slope at high M on the log-log plot of s against M. (Reproduced with permission from J. Eigner and P. Doty, *J. Mol. Biol.* 12:549 (1965). Copyright by Academic Press Inc.)

limiting slope for large M has been estimated to be 0.45 (see Crothers and Zimm, 1965), so that $(\frac{1}{2} - \epsilon/2) = 0.45$ or $\epsilon = 0.1$. Other estimates of ϵ for DNA range from 0 to 0.13. The sedimentation data do not extend to low enough molecular weight to test rigorously the zero power dependence of s on M predicted in the rigid rod limit (Eq. 15-36). However, the curvature of the log s-log M plot toward a smaller slope at low M is small but unmistakable, so that the prediction of a smaller length dependence for s at low M is verified.

EXERCISE 15-17
Estimate the molecular weight dependence of the diffusion coefficient of DNA at low and high molecular weight.

ANSWER
Since $D = k_BT/f$ (Eq. 15-10), $D \propto M^{-1}$ in the rigid rod limit, and $D \propto M^{-(1+\epsilon)/2}$ at high M. With $\epsilon = 0.1$, this result implies that $D \propto M^{-0.55}$.

○ 15–12 Viscosity is a Measure of the Resistance to Bulk Flow

The qualitative concept of *viscosity* is simple: liquids that flow slowly, such as honey, molasses, or heavy oils, are relatively viscous. In order to make this concept useful, however, we must place it on a firmer quantitative basis. Suppose a liquid is contained between two parallel plates (Fig. 15-16). Let the lower plate be stationary and the upper move with velocity u_{max}. The liquid in contact with the lower plate is stationary, and that in contact with

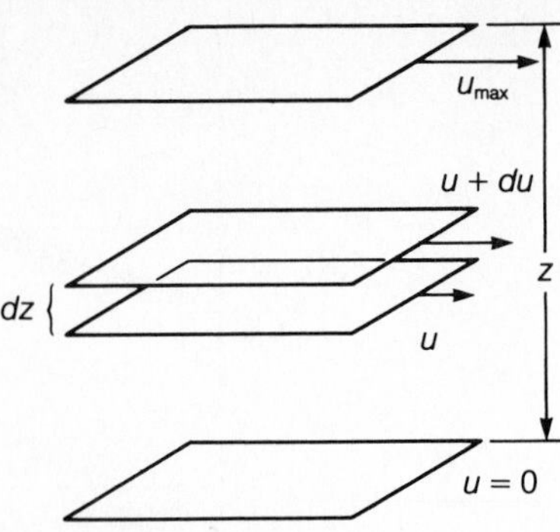

FIGURE 15-16
Two planes moving at velocity u_{max} (upper) and $u = 0$ (lower) are separated by fluid. The two imaginary planes separated by distance dz are moving at velocities u and $u + du$ respectively. The force per unit area (F/A) required to maintain motion against the frictional resistance is $\eta(du/dz)$, which serves to define the coefficient η, called the *fluid viscosity*.

the upper plate moves with it at velocity u_{max}. A force is required to maintain motion because of the frictional resistance of the fluid layers to sliding past each other. The work done by the force heats the solution.

The magnitude of the force required to move the upper plate at fixed velocity depends on the viscosity. Consider the two imaginary planes of unit area drawn in Figure 15-16. The upper fluid plane, a distance dz from the lower, moves relative to it at a velocity du. This motion of one part of the fluid with respect to another is called *shear*. The frictional force, F, per unit area, A, opposing shear is

$$\frac{F}{A} = \eta\left(\frac{du}{dz}\right) \tag{15-38}$$ ◀

(a relationship given by Newton in the seventeenth century). The frictional coefficient η is called the viscosity. The higher the viscosity, the greater the force required to maintain a given *shear rate* (du/dz). Similarly, with a constant force, such as flow under gravity, the greater the viscosity the lower the shear rate and hence the slower the liquid flow. (Liquid flow by definition involves shear.) The force per unit area (F/A) applied to the liquid is called the *shear stress*. Viscosity has, according to Equation 15-38, the cgs dimensions g cm^{-1} sec^{-1}, a unit called a *poise*. The viscosity of water at room temperature is roughly 0.01 poise, or 1 centipoise, which varies strongly with temperature (see Table 15-1). The dimensions of viscosity are such that when η is multiplied by the square of the shear rate, $(du/dz)^2$ in sec^{-2}, the result is the rate of energy dissipation per unit volume of fluid, in ergs cm^{-3} sec^{-1}. This energy, applied as the work required to cause the shear process in flow, appears as heat in the fluid. At constant shear rate, the energy dissipation is proportional to the viscosity.

Viscosity can be conveniently determined by measuring the rate of flow through a cylindrical tube or capillary. The relation between flow rate U (in cm^3 per second) and radius (a), pressure drop across the capillary (P), capillary length (l), and viscosity (η) is called *Poiseuille's law:*

$$U = \frac{\pi P a^4}{8\eta l} \tag{15-39}$$

Note the very strong (fourth power) dependence of volume flow on capillary radius.

TABLE 15-1 *The Viscosity of Water at Various Temperatures*

T/°C	η/CENTIPOISE
0	1.794
5	1.519
10	1.310
15	1.145
20	1.009
25	0.895
30	0.800
35	0.721
40	0.654
45	0.597
50	0.549

EXERCISE 15-18

The flow of blood through capillaries is controlled in part by constriction of the size of the small vessels called *arterioles,* which feed the capillaries. Use Equation 15-39 to calculate the fractional change in flow rate U produced when the arteriole radius is decreased by a factor 0.8.

ANSWER

$U_{\text{final}}/U_{\text{initial}} = (a_{\text{final}}/a_{\text{initial}})^4$, assuming P, η, and l are unchanged. Therefore $U_{\text{final}}/U_{\text{initial}} = (0.8)^4 = 0.41$. Notice that the fourth power dependence of U on a makes variation of the capillary radius an effective mechanism for regulating blood flow.

Flow can also be Turbulent

Equation 15-38 is, like all the equations that describe transport, a linear form that applies rigorously only when the driving force is small. There are two important exceptions to Equation 15-38. First, it describes only what is called *laminar flow,* which means that adjacent planes in the fluid flow parallel to each other, as in Figure 15-16. The contrary is *turbulent* flow, which involves complex and time-dependent motions in the fluid. Liquid which streams rapidly through an orifice is turbulent, as is the flow in a mountain stream. If two streams merge in laminar flow, mixing is very slow since the fluid layers continue on parallel, nonintersecting paths. Rapid mixing requires turbulence.

For flow through cylindrical tubes, there is a dimensionless quantity called the *Reynolds' number,* $\boldsymbol{R_N}$, which can be used to estimate the onset of turbulence. It is given by

$$R_N = 2\rho a u/\eta$$

in which ρ is the fluid density, and u is the mean *linear* fluid velocity in a tube of radius a. As the flow velocity increases, inertial forces due to fluid motion (proportional to u) become increasingly important relative to the damping forces of viscosity (proportional to η). When the Reynolds' number exceeds a critical value of about 2000, flow generally is observed to be turbulent.

EXERCISE 15-19

Show that the Reynolds' number is dimensionless.

EXERCISE 15-20

Calculate the Reynolds' number for blood flowing at 3×10^{-2} cm sec^{-1} through a capillary of radius 10^{-3} cm. Use $\rho = 1\,\text{g cm}^{-3}$ and a blood viscosity of $\eta = 0.04$ poise.

ANSWER

$$R_N = \frac{2 \times 1\text{ g cm}^{-3} \times 10^{-3}\text{ cm} \times 3 \times 10^{-2}\text{ cm sec}^{-1}}{0.04\text{ g cm}^{-1}\text{ sec}^{-1}}$$
$$= 1.5 \times 10^{-3}$$

Consequently, the flow of blood in the capillary (as in all the small blood vessels) is laminar, not turbulent.

EXERCISE 15-21

During vigorous physical activity, the flow of blood from the heart through the aorta is about 25 liters min^{-1}. Using the quantities in Exercise 15-20, and a 1 cm radius for the aorta, calculate the Reynolds' number.

ANSWER

The linear flow rate u is the volume flow rate U divided by the cross-sectional area of the vessel, or

$$u = \frac{25 \times 10^3\text{ cm}^3\text{ min}^{-1}}{60\text{ sec min}^{-1} \times \pi\text{ cm}^2}$$
$$= 132\text{ cm sec}^{-1}$$

Therefore the Reynolds' number is

$$R_N = \frac{2 \times 1\,\text{g cm}^{-3} \times 1\,\text{cm} \times 132\,\text{cm sec}^{-1}}{0.04\text{ g cm}^{-1}\text{ sec}^{-1}} = 6600$$

We conclude that blood flow through the aorta is turbulent during physical stress, since the critical Reynolds' number of 2000 is exceeded. The resting flow rate is about 5 liters min^{-1}, which should produce laminar flow.

The other exception to Equation 15-38 occurs when η depends on the shear rate (du/dz). This often happens for solutions of polymers because rapid flow deforms the flexible polymer and alters (reduces) its frictional resistance. Such flow is said to be *non-Newtonian*. At small enough shear rates, all polymer solutions are Newtonian.

Dissolved Polymers Increase Solution Viscosity

The addition of a polymer to a solution causes the viscosity to rise. The reason is that the polymer extends across the flow planes (see Fig. 15-16) and increases the frictional resistance to shear. The increment in the viscosity is very sensitive to both polymer size and shape. We define the viscosity increment in terms of the *specific viscosity* η_{sp}:

$$\eta_{sp} = \frac{\eta - \eta_0}{\eta_0} \tag{15-40}$$

in which η is the polymer solution viscosity and η_0 is the viscosity of the solvent without polymer. The *relative viscosity* η_{rel} is η/η_0, and from Equation 15-40 you can see that

$$\eta_{sp} = \eta_{rel} - 1 \tag{15-41}$$

so η_{sp} is the increase in the relative viscosity of the solution beyond the value 1 for pure solvent.

For low concentrations of added polymer, the viscosity increment η_{sp} is proportional to the concentration. We define the limiting ratio as $[\eta]$

$$[\eta] = \lim_{c \to 0} \left(\frac{\eta_{sp}}{c} \right) \tag{15-42}$$

in which $[\eta]$ is called the *intrinsic viscosity,* and c is the polymer concentration in wt/unit volume. Intrinsic viscosity has the units concentration^{-1}, or vol g^{-1}.

The Viscosity Increment Depends on Polymer Size and Shape

In 1906 Einstein showed that for spheres the specific viscosity is

$$\eta_{sp} = \tfrac{5}{2}\phi \tag{15-43}$$ ◄

in which ϕ is the fraction of the volume of the solution that is occupied by the spheres. Equation 15-43 is the analogue in viscosity studies of Stokes' law (Eq. 15-28) in sedimentation. It states that the viscosity increment produced by a given volume fraction of spheres does not depend on how large they are. As the spheres are made larger, their individual frictional contribution increases, but there are fewer of them and the two effects exactly cancel.

We can examine the influence of particle shape on the viscosity by employing the more general version of Equation 15-43 introduced by Simha (1940):

$$\eta_{sp} = \nu\phi \tag{15-44}$$

The *asymmetry factor* ν is 2.5 for spheres, but increases for asymmetric bodies. Simha derived the equations (Tanford, 1961)

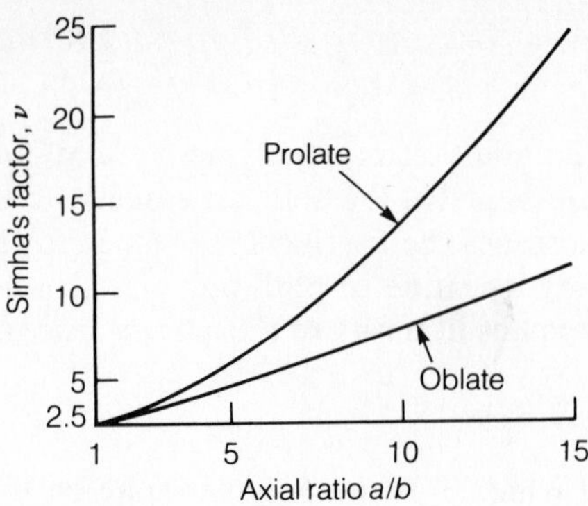

FIGURE 15-17
Variation of the Simha factor ν with axial ratio. Compare this figure with Figure 15-13 to see that viscosity is much more sensitive to macromolecular shape than is the sedimentation coefficient. (From Tanford, 1961. Reprinted with permission of John Wiley & Sons, Inc.)

$$\nu = \frac{(a/b)^2}{15\left[\ln\left(\frac{2a}{b}\right) - \frac{3}{2}\right]} + \frac{(a/b)^2}{5\left[\ln\left(\frac{2a}{b}\right) - \frac{1}{2}\right]} + \frac{14}{15} \qquad \text{(Prolate ellipsoid)} \quad (15\text{-}45)$$

$$\nu = \frac{16}{15}\,\frac{(a/b)}{\tan^{-1}(a/b)} \qquad \text{(Oblate ellipsoid)} \quad (15\text{-}46)$$

for the ellipsoids of revolution shown in Figure 15-12. The relations 15-45 and 15-46 are accurate only when $a/b > 10$. The variation of ν with a/b is shown in Figure 15-17. You can see by a comparison of Figures 15-17 and 15-13 that the viscosity increment η_{sp} (at constant volume fraction ϕ) is much more sensitive to particle shape than is the frictional coefficient (at constant particle volume).

EXERCISE 15-22

Calculate the asymmetry factor ν for the DNA molecule in Exercise 15-12 (200 base pairs long), with $a/b = 27.8$.

ANSWER

$$\nu = \frac{(27.8)^2}{15[\ln(55.5) - 1.5]} + \frac{(27.8)^2}{5[\ln(55.5) - 0.5]} + \frac{14}{15}$$
$$= 65.2$$

EXERCISE 15-23

Use the results of Exercise 15-22 to calculate the intrinsic viscosity of the DNA molecule in the usual units of deciliters g^{-1}. Calculate the volume fraction ϕ using the partial specific volume $\overline{v_2} = 0.51$.

ANSWER

The volume fraction is $\phi = \overline{v_2}$ ml $g^{-1} \times c$ g $dl^{-1}/100$ dl $ml^{-1} = 0.51 \times c/100$ when c is in g/dl and $\overline{v_2}$ is in ml/g. The intrinsic viscosity is $[\eta] = \nu\phi/c = \nu \times 0.51/100 = 0.33$ dl g^{-1}. The experimental intrinsic viscosity of DNA molecules in this size range has not been measured accurately, but can be estimated to lie between 0.3 and 0.4 dl g^{-1} from data for higher molecular weight samples. Notice that $[\eta]$ for DNA of this size would be smaller by a factor $\nu/2.5 = 65.2/2.5 = 26.1$ if the molecule were spherical. As indicated earlier, viscosity is very sensitive to molecular shape.

Flexible Polymers have an Effective Hydrodynamic Radius for Viscosity

The intrinsic viscosity of spheres provides a useful starting point for rationalizing the viscosity of flexible polymer chains, just as we began with Stokes' law and the equivalent hydrodynamic radius of a polymer in our discussion of the sedimentation of flexible polymers (Secs. 15-10 and 15-11). A flexible polymer chain can be thought of as a sphere of effective hydrodynamic radius R_e, whose viscosity is given by Equation 15-43 as $\eta_{sp} = \frac{5}{2}\phi_e$, in which ϕ_e is the effective volume fraction occupied by the spheres. (Much of this effective hydrodynamic sphere is occupied by the solvent, of course.) We can express ϕ_e as the product of an effective hydrodynamic specific volume v_e (v_e is the hydrodynamic volume divided by the polymer mass) times the weight concentration of the polymer, c:

$$\phi_e = cv_e$$

Therefore, Equation 15-43 becomes

$$\eta_{sp} = \tfrac{5}{2}cv_e \tag{15-47}$$

Using Equation 15-42 for $[\eta]$ we obtain

$$[\eta] = \tfrac{5}{2}v_e \tag{15-48}$$

This equation states that the intrinsic viscosity is proportional to the effective hydrodynamic volume of the polymer divided by its mass. Notice that $[\eta]$ has units of volume/mass, as determined by the choice of units for c and v_e, whose product ϕ_e is dimensionless.

The effective hydrodynamic radius (R_e) is related to v_e by the equation v_e = volume/mass, or

$$v_e = \tfrac{4}{3}\pi R_e^3/(M/N_A)$$

Using Equation 15-33, which states that R_e is proportional to the root-mean-square end-to-end distance, $\langle L^2\rangle^{1/2}$, and combining with Equation 15-48, we find that

$$[\eta] \propto v_e \propto R_e^3/M \propto \langle L^2\rangle^{3/2}/M \tag{15-49}$$

Introducing the chain length dependence $\langle L^2\rangle \propto M^{(1+\epsilon)}$ (Equation 14-60, with $M \propto N$), yields

$$[\eta] \propto M^{1/2+3\epsilon/2} \tag{15-50}$$

Notice that excluded volume effects ($\epsilon > 0$) cause $[\eta]$ to increase more rapidly with M than would be the case for a theta solvent ($\epsilon = 0$), in which $[\eta]$

would be proportional to $\sqrt{M}$. This result should be compared with Equation 15-37 for the sedimentation coefficient, in which excluded volume effects cause the exponent α in $s \propto M^{\alpha}$ to decrease by $\epsilon/2$ from the theta solvent value of $\frac{1}{2}$.

The rigid-rod limiting viscosity is also of interest for stiff polymers such as DNA. It can be shown (see Kirkwood and Riseman, 1956), that

$$[\eta] \propto M^{1.8} \tag{15-51}$$

for a rod.

EXERCISE 15-24

Derive an expression in terms of fractional powers of s and $[\eta]$ which is linear in M and in which the chain expansion parameter ϵ is eliminated.

ANSWER

It is apparent from Equations 15-50 and 15-37 that multiplication of s^3 by $[\eta]$ eliminates ϵ:

$$s^3[\eta] \propto M^2$$

or

$$(s[\eta]^{1/3})^{3/2} \propto M$$

The full form of this equation, including the proportionality constants, can be shown to be:

$$M = \left(\frac{s[\eta]^{1/3}\eta_0 N_A}{10^{13}\beta\phi_2} \right)^{3/2} \tag{15-52}$$

in which β is a parameter that depends only slightly on polymer and solvent. All values of β are within the range $(2.4 \pm 0.4) \times 10^6$ when s is in Svedbergs, $[\eta]$ in dl g^{-1}, and η_0 is the solvent viscosity in poise. Equation 15-52, called the *Flory-Mandelkern-Scheraga* equation, is useful for estimating molecular weights from sedimentation-viscosity data.

EXERCISE 15-25

Calculate the parameter β for the DNA molecule of length 200 base pairs, $M = 135{,}000$ (see Ex. 15-14 and 15-23), for which the ellipsoid of revolution model predicted $s = 6.91\ S$ and $[\eta] = 0.33$ dl g^{-1}. Use $\phi_2 = 0.49$, $\eta_0 =$ 0.01 poise.

ANSWER

Rearrange Equation 15-52 to read

$$\beta = \frac{s[\eta]^{1/3}\eta_0 N_A}{10^{13}\phi_2 M^{2/3}} = \frac{6.91 \times (0.33)^{1/3} \times 0.01 \times 6.02 \times 10^{23}}{10^{13} \times 0.49 \times (135{,}000)^{2/3}}$$

$= 2.23 \times 10^6$, which is well within the usual range of values.

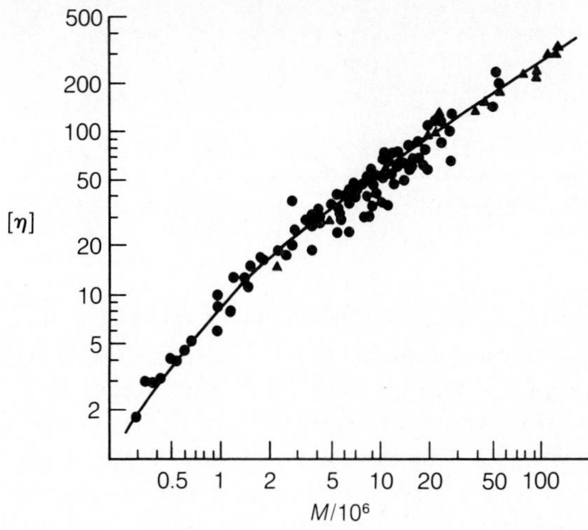

FIGURE 15-18
Dependence of the intrinsic viscosity, [η], of DNA on molecular weight. The slope of the log-log plot becomes smaller at high molecular weight, corresponding to the conversion from rigid rod to flexible coil behavior as molecular weight increases. This figure should be compared to the analogous results for the sedimentation coefficient in Figure 15-15. (Reproduced with permission from J. Eigner and P. Doty, *J. Mol. Biol.* 12:549 (1965). Copyright by Academic Press Inc.)

The Viscosity of DNA Depends Strongly on Molecular Weight

The intrinsic viscosity of double helical DNA changes from that expected for a stiff chain at low molecular weight (Fig. 15-18) to the behavior of a flexible chain with excluded volume effects at high M. The limiting slope of the log-log plot in Figure 15-18 is about 0.665, implying $\epsilon = 0.1$. The same value is consistent with the sedimentation results of Figure 15-15, discussed in Section 15-11. At low molecular weights, the log-log plot curves to a steeper slope, tending toward the expected value of 1.8 for the rigid rod limit. Note the opposite curvatures of the log $[\eta]$-log M and log s-log M plots, resulting from the very different molecular-weight dependence of s and $[\eta]$ in the rigid-rod limit.

15–13 Polymer Normal Modes: The Retardation Time is Sensitive to Molecular Weight

A flexible polymer molecule changes in shape in solution. Zimm (1960) showed that the motion can be described by a set of *normal modes,* which are analogous to the vibrational modes of a set of coupled springs, or of the atoms in a molecule (Chap. 13). Figure 15-19 shows schematically the first

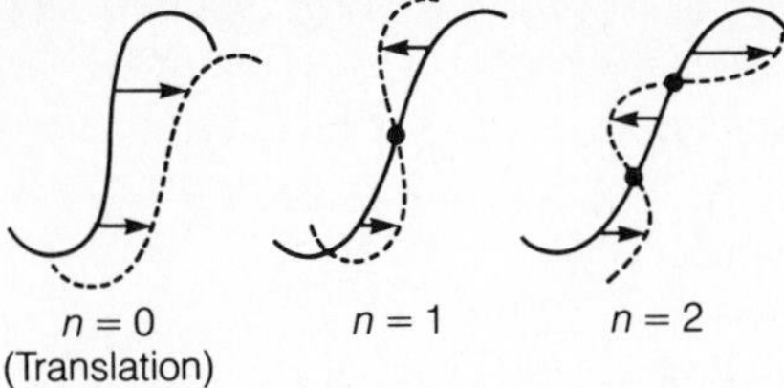

FIGURE 15-19
Normal modes of a polymer molecule. The mode $n = 0$ corresponds to translational motion, and $n = 1$ is the longest internal relaxation mode, which can be described as movement of the ends with a stationary point or node at the center. The mode $n = 2$ has 2 nodes, etc. (From Zimm, 1960.)

three normal modes of a polymer chain. The lowest ($n = 0$) describes translational motion, with no change in shape. The higher modes, $n \geq 1$, describe vibrational motion. For vibrational modes, the integer n is the number of *nodes* in the vibration, which are stationary positions where the direction of vibration changes sign. The lowest frequency mode is $n = 1$, which describes the slowest relaxational motion that would result from stretching the chain by pulling the two ends in opposite directions, and then releasing them.

The total motion of the polymer under any conditions can be obtained from a linear sum of the individual normal modes. If the molecule is held under equilibrium or steady-state conditions, the motion consists of vibrational fluctuation about the equilibrium or steady-state average conformation, coupled with translation and rotation. However, if a perturbation is applied and then removed, the polymer structure will relax with a sum of exponential decays of the normal modes, just as discussed in Section 6-14 for relaxation kinetics.

Measurement of the largest relaxation times ($n = 1$ in Fig. 15-19) is the basis for determination of the *retardation time* for DNA molecules (Klotz and Zimm, 1972). A DNA solution is placed between two concentric cylinders (Fig. 15-20), and a force is applied to rotate the inner cylinder. The shear forces stretch the DNA molecules, accompanied by displacement of the

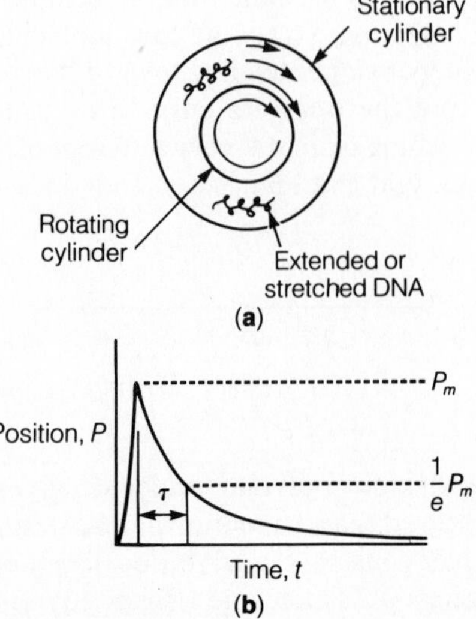

FIGURE 15-20
Measurement of the retardation time of DNA. **(a)** The motion of an inner cylinder separated from a stationary outer cylinder by the DNA solution stretches the DNA molecules. **(b)** When the driving force is removed, the inner cylinder coasts to a stop, then reverses direction and moves toward an equilibrium position. The reversal is caused by the elastic force of the stretched DNA molecules. The time constant is determined by the largest normal mode (n = 1) relaxation time.

$n = 1$ vibrational mode from its equilibrium position. Next, the force is removed from the inner cylinder, which coasts to a halt and then reverses itself and turns in the other direction (Fig. 15-20). The force that causes reversal is the elasticity of the polymer, and the exponential decay time required to reach a final resting position is a measure of the largest normal mode relaxation time.

Klotz and Zimm (1972) showed that the retardation time τ_r depends on molecular weight according to

$$\tau_r \propto M^{3(1+\epsilon)/2} \tag{15-53}$$

For several reasons this is an excellent method for determining the molecular weight of very large DNA molecules in solution. Unlike viscosity, one does not need to know the DNA concentration, and the method does not seem to be subject to the anomalies that limit the use of sedimentation methods for very large molecules (Levin and Hutchinson, 1973). Finally, the high sensitivity to molecular weight (Eq. 15-53) enhances the accuracy of the technique.

15–14 Electrophoresis is Transport Driven by an Electric Field

Electrophoresis is superficially analogous to sedimentation. A charged molecule placed between electrodes of opposite sign is subject to a force, given by

$$F = zeE^* \tag{15-54}$$

where E^* is the electrostatic field (the electrostatic potential difference between the two electrodes divided by the distance between them), z the number of charges on the molecule, and e the protonic charge. As was the case for sedimentation, the driving force should be balanced by the viscous drag, or

$$fv = zeE^*$$

in which f is the frictional coefficient and v the velocity. The analog of the sedimentation coefficient is the *electrophoretic mobility, u,* which is the velocity divided by the field,

$$u = \frac{v}{E^*} = \frac{ze}{f} \tag{15-55}$$

For several reasons, the theoretical analysis of electrophoresis is much more difficult than sedimentation, so that the latter has been the more reliable tool for study of macromolecular structure. One immediate problem in electrophoresis is the binding of counterions to the charged macromolecule, reducing the effective charge below the maximum possible value ze. Also, the ion atmosphere surrounding the polymer (see discussion of Debye-Hückel theory in Sec. 8-10) will screen it from the applied field, reducing the effective field experienced by the particle. In addition, there are complications in the transport itself. Besides simple frictional resistance, described for spheres by Stokes' law, motion of the transported particle is retarded by an *electrophoretic effect* and an *asymmetry effect*. The electrophoretic effect is due to the transport of oppositely charged ions toward the other electrode, producing a current of ions and associated solvent molecules that slows down movement of the macromolecule. The asymmetry effect arises from the inability of the ion atmosphere to adjust instantaneously to the motion of the transported ion. Oppositely charged ions tend to accumulate in an ion's wake (Fig. 15-21), producing an electric field that opposes the applied field. For small electrolytes (such as Na^+Cl^-) this problem was solved by Onsager in the limit of dilute solutions, but there remain serious theoretical obstacles for macroions.

To illustrate the difficulty of making correct predictions about the electrophoretic mobility of macroions, we can, by analogy with our treatment of sedimentation and viscosity, consider the molecular weight dependence of DNA mobility. For a rigid rod, we expect the frictional coefficient to increase linearly with the molecular length (Eq. 15-31), and the charge ze also should increase linearly as molecular weight increases because of the increase in z. Hence, from Equation 15-55, u should be independent of molecular weight. However, as the molecule approaches the flexible chain limit, the frictional coefficient should rise less than linearly with molecular weight (Eq. 15-34). Thus, if ze continues to rise linearly, the electrophoretic mobility should increase as molecular weight increases, just as the sedimentation coefficient does. The experimental result is that the electrophoretic mobility of DNA in solution is independent of molecular weight over the range from 0.26×10^6 to 130×10^6 [in 0.01 M NaCl (B. M. Olivera et al., 1964)]. In effect, DNA seems to remain "free draining" in electrophoresis, which means that each polymer segment retains its rigid-rod frictional resistance as though solvent could flow freely through the domain that the molecule occupies. This means

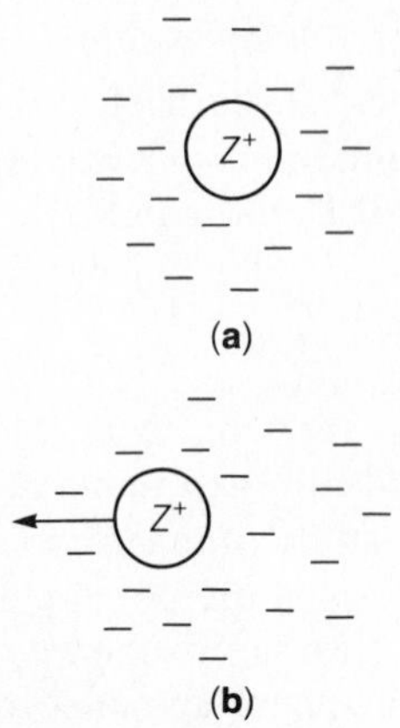

Z^+

(a)

Z^+

(b)

FIGURE 15-21
When an ion Z^+ is static **(a)**, its counterion distribution is spherically symmetric. When it moves under an electrostatic field **(b)**, the ion atmosphere does not adjust rapidly enough to remain spherically symmetric. Instead, the counterions accumulate in the wake of the transported ion, producing an electrostatic force which retards the motion of the ion.

that there is no effective hydrodynamic radius, and Equation 15-32 cannot be used. The book by Bloomfield et al. (1974) contains a discussion of the theoretical basis for the electrophoretic mobility of DNA.

Gel Electrophoresis is Exceptionally Useful for Separation of Polymers

Even though the theoretical basis for electrophoretic mobilities is only imperfectly understood, a variant of electrophoresis is widely used for analytical and preparative separation of both proteins and nucleic acids. This technique, called *gel electrophoresis,* is illustrated in Figure 15-22. The difference between gel and conventional electrophoresis is the presence of a polymeric gel through which electrophoresis occurs. Polyacrylamide gels are frequently used for this purpose: a solution of acrylamide, $CH_2{=}CH{-}CONH_2$, in buffer is mixed with a small amount of bis-acrylamide ($CH_2{=}CH{-}CONH{-}CH_2{-}NH{-}CO{-}CH{=}CH_2$), which serves as a crosslinking agent. Polymerization is initiated by a catalyst which produces free radicals. The product of the polymerization reaction is a crosslinked polymer chain,

```
· · · —CH₂—CH—CH₂—CH— · · ·
           |       |
           CO      CONH₂
           |
           NH
           |
           CH₂
           |
           NH
           |
           CO      CONH₂
           |       |
· · · —CH₂—CH—CH₂—CH— · · ·
```

which forms a gel, reminiscent of a gelatin pudding.

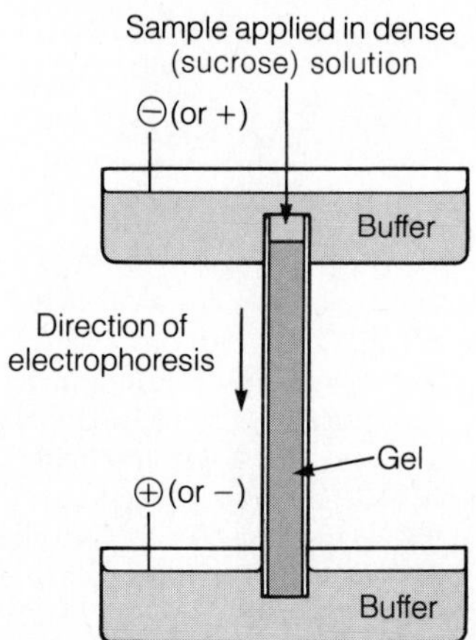

FIGURE 15-22
Schematic diagram of a gel electrophoresis apparatus. The tube or slab containing the gel is connected to buffer reservoirs at different electrical potentials. The sample is applied to the top of the gel in a dense solution to prevent mixing with the buffer. Electrophoresis runs from top to bottom of the gel.

The plates containing the slab gel are placed in contact with a buffer reservoir at each end, and an electric field is applied across the gel (Fig. 15-22). The macromolecules, added to the top of the gel in a dense solution containing sucrose or similar material to prevent mixing with the buffer solution, enter the gel under the force exerted by the field. Each component migrates a distance determined by its mobility in the gel.

It is nearly always found that distance migrated by a macromolecule depends on the logarithm of its molecular weight. Figure 15-23(*a*) shows the gel migration pattern observed for a series of single stranded DNA fragments, and Figure 15-23(*b*) illustrates the linear relationship between log M and distance migrated in the gel. Once a standard curve of log M against distance is established, this technique is a simple and rapid method for measuring DNA molecular weight.

The electrophoretic mobility of nucleic acids in *solution* is independent of molecular weight, but Figure 15-23 shows that the mobility in gel electrophoresis is strongly dependent on molecular size. The qualitative reason for this difference is the difficulty encountered by large molecules in finding their way between the crosslinked polymer strands of the gel. The result is a large dependence of the effective frictional coefficient on molecular weight, an effect sometimes referred to as *molecular sieving*.

The mobility of macromolecules in gel electrophoresis also depends on molecular shape. For example, 5S ribosomal RNA can be separated on gels

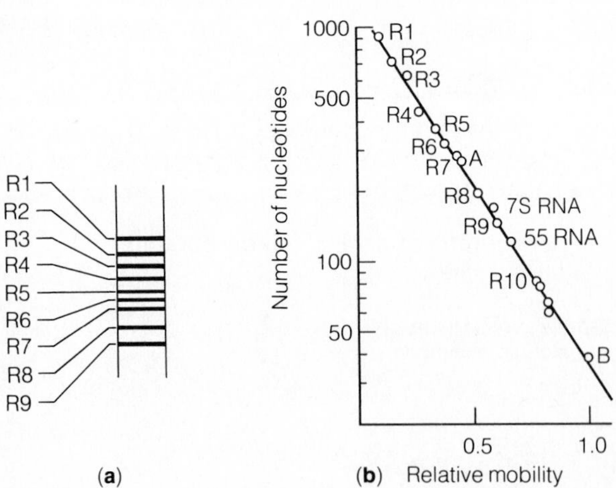

FIGURE 15-23
(a) Migration pattern of a series of DNA fragments on a 5% polyacrylamide gel which contains formamide to separate the strands of the double helix. The DNA molecules contain radioactive ^{32}P which causes exposure of a strip of x-ray film wherever a DNA band occurs. **(b)** Dependence of mobility in the gel of **(a)** on the logarithm of nucleic acid size, expressed in terms of the number of nucleotides in the chain. Samples of both DNA and RNA are included. The migration positions of two dyes (A = xylene cyanol and B = bromophenol blue) are also shown. (From Maniatis et al., 1975, *Biochemistry 14,* 3787. Reprinted with permission from *Biochemistry 14,* 3787. Copyright by the American Chemical Society.)

into two conformational isomers of identical molecular weight and base sequence. More compact, spherically shaped molecules generally move differently through the gel than do elongated forms.

Unlike nucleic acids, proteins do not have a constant ratio of charge to chain length, so the mobility of native proteins in gel electrophoresis is not a simple test for molecular weight. However, a constant ratio of charge to size can be achieved by performing the electrophoresis in the presence of sodium dodecyl sulfate (SDS), $CH_3(CH_2)_{10}CH_2OSO_3^-Na^+$. This detergent unfolds proteins into separated polypeptide chains by surrounding them with a sheath of negatively charged alkyl sulfate ions. The resulting polypeptide chains are negatively charged, to an extent that is independent of amino acid sequence. Therefore, the electrophoretic mobility of proteins on gels containing SDS depends only on the molecular weight of the individual polypeptide chains that make up the protein.

15–15 *Onsager's Generalization of the Phenomenological Transport Equations*

All the equations describing transport velocity or flux are of the general form of Equation 15-4, which says that the flow or flux is the product of a coefficient times an effective force. Examples are 15-13 for diffusion, 15-23 for sedimentation, and 15-38 for viscosity. These equations seem to bear little further relation to each other. However, in 1931 Onsager provided the theoretical basis for writing consistent and general phenomenological equations for transport. Furthermore, his treatment includes the influence of coupling between different transport phenomena. For example, suppose that heat transport and diffusion are occurring simultaneously. Let J_1 be the flux of matter and J_2 the flux of heat. The phenomenological equations take the form

$$\begin{aligned} J_1 &= L_{11}X_1 + L_{12}X_2 \\ J_2 &= L_{21}X_2 + L_{22}X_2 \end{aligned} \tag{15-56}$$

in which X_1 and X_2 are force terms appropriate to diffusion and heat conduction, respectively, and L_{ij} are called the phenomenological coefficients. In Onsager's formulation, J_i is the time derivative of a state function, and furthermore J_iX_i must have the units of entropy production per unit time (energy $\deg^{-1}$ time^{-1}). For example, for diffusion

$$J = \text{flux (matter/unit area/unit time)}$$

which is the same as J in Equation 15-13. However,

$$X = \left(\frac{\partial\mu}{\partial x}\right)$$

where μ is the chemical potential and x is distance. This definition is required in order to give XJ the proper units of entropy production per unit time. Hence Equation 15-13 would read (all other fluxes equal 0)

$$J_1 = L_{11}\left(\frac{\partial \mu}{\partial x}\right)$$
$$= L_{11}\left(\frac{\partial \mu}{\partial c}\right)\left(\frac{\partial c}{\partial x}\right) \tag{15-57}$$

from which we see by comparing Equations 15-13 and 15-57 that

$$L_{11} = D\Big/\left(\frac{\partial \mu}{\partial c}\right) \tag{15-58}$$

The off-diagonal terms (L_{12} and L_{21}) in Equation 15-56 characterize an important physical concept flux of one kind is influenced by other nonconjugate force terms in the system. Onsager was able to show for systems close to equilibrium, using the principle of microscopic reversibility (Chap. 6) that these terms obey the relation

$$L_{ij} = L_{ji} \tag{15-59}$$

implying a mutually symmetric coupling effect between two fluxes and their conjugate forces. Equation 15-59, called the *Onsager reciprocal relation,* further implies (by arguments we will not develop here) that a transport system close to equilibrium will relax to equilibrium with a set of exponential decay times. The reciprocal relation has been verified in several experimental systems, and exponential relaxation, with Lorentzian power spectral density, is always observed for systems close to equilibrium. Hence the Onsager theory provides a secure theoretical basis for the phenomenological equations contained in this chapter.

QUESTIONS FOR REVIEW

1. Define the transport velocity.
2. What is the frictional coefficient?
3. What determines the steady-state velocity?
4. What is the Maxwell distribution?
5. How can an effective force be obtained from an expression for the chemical potential?
6. Write Fick's first and second laws of diffusion.
7. Define the flux in a transport experiment.
8. How is the diffusion coefficient related to the frictional coefficient?
9. What is Einstein's relationship between time and average distance moved by a particle undergoing Brownian motion?
10. Define the sedimentation coefficient. How is it related to the frictional coefficient?
11. How can molecular weight be determined by a combination of sedimentation and diffusion experiments?

12. Describe the influence of axial ratio on the frictional coefficient when molecular volume is held constant.
13. Define viscosity by an equation.
14. Define shear rate and shear stress.
15. What is the Reynold's number?
16. Define specific viscosity and intrinsic viscosity.
17. How does the axial ratio of a particle affect its specific viscosity?
18. Explain why the intrinsic viscosity and sedimentation coefficient of a random-flights polymer chain vary with the square root of chain length.
19. Define the retardation time and explain why it is useful for measuring DNA molecular weight.
20. What relationship is generally found between DNA molecular weight and distance traveled in gel electrophoresis?

PROBLEMS

1. Are the following statements true or false?
   (a) The steady-state velocity is reached very rapidly compared to the time scale of most transport experiments.
   (b) Flux can have the units, moles cm^{-2} sec^{-1}.
   (c) A particle takes four times as long to move by diffusion a distance $2x$ as to move a distance x.
   (d) The sedimentation coefficient of a macromolecule usually varies linearly with its molecular weight.
   (e) The sedimentation coefficient of a macromolecule decreases if it is hydrated.
   (f) The SI unit of viscosity is the poise.
   (g) The viscosity increment produced by a given volume fraction of spheres is independent of their radius.

Answer: (a) T. (b) T. (c) T. (d) F. (e) T. (f) F. (g) T.

2. Add the appropriate word or words.
   (a) The proportionality constant between force and velocity is called the ________ .
   (b) According to Fick's first law, the flux is proportional to the ________ .
   (c) The ratio of sedimentation velocity to the centrifugal force on a unit mass is called the ______ .
   (d) The frictional coefficient of a sphere is given by ________ law.
   (e) A cigar-shaped object is called a ________ ellipsoid of revolution.

(f) The proportionality constant between shear rate and shear stress is called the ______.

(g) When the Reynolds' number is exceeded, flow becomes ______.

Answer: (a) Frictional coefficient. (b) Concentration gradient and diffusion coefficient. (c) Sedimentation coefficient. (d) Stokes'. (e) Prolate. (f) Viscosity. (g) Turbulent.

PROBLEMS RELATED TO EXERCISES

3. (Exercise 15-1) Calculate the diffusion constant of ribonuclease at 1°C, assuming the frictional coefficient is unchanged from the 20°C value (3.6×10^{-8} g sec^{-1}).

4. (Exercise 15-3) Calculate the time required for a protein molecule of diffusion constant 10^{-6} cm^2 sec^{-1} to move by diffusion through a distance of 10 μm.

5. (Exercise 15-9) A protein has a molecular weight of 14,000, a sedimentation coefficient of 1.87S and a diffusion coefficient of 1.33×10^{-6} cm^2 sec^{-1} at 300 K. Calculate the density increment.

6. (Exercise 15-12) Calculate the frictional coefficient ratio (f/f_0) for a DNA molecule 340Å long and 20Å in diameter, using the model of a prolate ellipsoid.

7. (Exercise 15-13) Calculate the frictional coefficient of the molecule in Problem 6, assuming $\eta = 0.01$ poise.

8. (Exercise 15-14) Calculate the sedimentation coefficient of the molecule in Problems 6 and 7, assuming $\phi_2 = 0.49$.

9. (Exercise 15-20) Calculate the Reynolds' number for water with viscosity $\eta = 0.01$ poise flowing at 10 cm sec^{-1} through a capillary of radius 10^{-3} cm.

OTHER PROBLEMS

10. A solution of ribonuclease has a concentration gradient of $10^{-6}M$/cm at a point at which the concentration is $10^{-5}M$. Using $D = 1.1 \times 10^{-6}$ cm^2/sec at 20°C, calculate the transport velocity.

11. The sedimentation coefficient of a virus of molecular weight 2×10^8 is 500S. Calculate the time required for the virus to move the 10^{-4} cm length of a bacterial cell by diffusion at $T = 37$°C. Use $\phi_2 = 0.35$.

12. Begin with Equation 15-48, substitute for v_e the specific volume corresponding to the hydrodynamic radius $R_e^{(\eta)}$, and show that $R_e^{(\eta)} = (30[\eta]M/[\pi N_A])^{1/3}$ in which $[\eta]$ is in dl g^{-1}, and $R_e^{(\eta)}$ is in cm. Use this result to calculate the hydrodynamic radius of a DNA molecule of molecular weight 20×10^6 and intrinsic viscosity 100 dl g^{-1}.

Answer: 3.17×10^{-5} cm, or 3160Å

13. Use the Flory-Mandelkern-Scheraga equation to calculate the sedimentation coefficient expected for the DNA molecule in Problem 12, which has $[\eta] = 100$ dl g^{-1} and $M = 20 \times 10^6$. Use $\phi_2 = 0.49$, $\eta = 0.01$ poise, $\beta = 2.4 \times 10^6$.

14. Combine Equations 15-25 and 15-28 to show that the effective hydrodynamic radius for sedimentation, $R_e^{(S)}$, is

$$R_e^{(S)} = \frac{10^{13} M \phi_2}{6\pi N_A \eta_0 s}$$

in which s is expressed in Svedbergs. Use this result to calculate the effective hydrodynamic radius of the DNA molecule in Problems 12 and 13. (Answer: 2.78×10^{-5} cm). Comment on the comparison with the result of Problem 12.

15. Let α be the ratio of effective hydrodynamic radii for viscosity and sedimentation

$$\alpha = R_e^{(\eta)}/R_e^{(S)}$$

Use the results of Problems 12 and 14 to show that

$$\alpha = \beta/(2.11 \times 10^6)$$

in which β is the parameter that appears in the Flory-Mandelkern-Scheraga equation (Eq. 15-52).

16. Begin with Equation 15-51 and show that the effective hydrodynamic volume of a rod of length L increases with $L^{2.8}$.

17. Material at concentration C_0 is initially contained in a one-dimensional box of length L. The ends of the box are continuously washed so that the concentration C is zero when distance x is 0 and L. Show that the diffusion of material out of the box is described by the equation

$$C = \frac{4C_0}{\pi} \sum_{j=0}^{\infty} \frac{1}{(2j+1)} \sin\left\{\frac{(2j+1)\pi x}{L}\right\} \exp\left\{-\left[\frac{(2j+1)\pi}{L}\right]^2 Dt\right\}$$

by showing that this expression satisfies Fick's second law. Show that the time required for diffusion depends on L^2. Sketch the concentration profile in the box at various times.

● 18. Sketch the autocorrelation function for the following.

(a) The y-component of the magnetization of a nuclear spin precessing at the Larmor frequency ν about the z-axis.

(b) The average daily temperature (recorded for many years) at a weather station in North America.

19. Sedimentation coefficients are usually expressed in terms of the value ($s_{20,w}$) that would be obtained in a solvent of the viscosity of water at 20°C. However, water at 20°C is not often a convenient solvent, because many macromolecules are unstable in the absence of added ions, or may denature at temperatures much above 0°C. Suppose that the sedimentation coefficient of lysozyme is found to be 1.41 S at 10°C in a solvent whose viscosity relative to water at 10°C is 1.02. Use the data in Table 15-1 to show that $s_{20,w} = 1.87\ S$. (*Hint:* Correct first to the viscosity of water at 10°C, then correct for the viscosity change of water between 10°C and 20°C.)

20. The partial specific volume of hemoglobin is $\overline{v}_2 = 0.749$ cm^3 g^{-1} and molecular weight is 6.45×10^4. Calculate:

(a) The radius of a sphere which has the same volume as hemoglobin.

(b) The frictional coefficient expected for such a sphere when $\eta = 1.00$ centipoise.

(c) The diffusion coefficient expected for the sphere at 20°C.

(d) The sedimentation coefficient of the sphere; solvent density equals 1.

21. The measured sedimentation coefficient of hemoglobin is $s_{20,w} = 4.31\ S$. Use also the information in Problem 20 to:

(a) Calculate the amount of water (δ_1) that would have to be hydrated by hemoglobin (in g per g) to produce a sphere of sedimentation coefficient 4.31 S. Assume that the volume of the sphere is

$$V = \frac{M}{N_A}(\overline{v_2} + \delta_1\overline{v_1})$$

in which $\overline{v_1} = 1.00\ \text{cm}^3\ \text{g}^{-1}$ is the partial specific volume of water.

(b) Calculate the axial ratio of hemoglobin that would be required to produce a prolate ellipsoid of the observed sedimentation coefficient if there is no water of hydration. (*Hint:* Use Figure 15-14 to make a preliminary estimate, then insert trial values of a/b into Equation 15-30a until the observed f/f_0 is obtained.)

22. A collagen sample of molecular weight 3.45×10^5 has $D_{20,w} = 0.69 \times 10^{-7}\ \text{cm}^2\ \text{sec}^{-1}$ and partial specific volume $\overline{v_2} = 0.695$. Calculate the axial ratio, assuming a prolate ellipsoid and no water of hydration. $\eta = 0.01$ poise.

23. A T7 DNA sample in 1 M NaCl has $[\eta]$ 112 dl g^{-1} and $s_{20,w} = 31.5\ S$. The same DNA in 1 M NaCl at pH 12 has $[\eta] = 17.3$ dl g^{-1} and $s_{20,w} = 37.0\ S$. Estimate the fractional change in the molecular weight. (*Hint:* Take the *ratio* of the Flory-Mandelkern-Scheraga equation expressions for the two DNAs, assuming ϕ_2, η_0, and β are unchanged.)

FURTHER READING

Berne, B. J., and Pecora, R. 1974. "Laser Light Scattering from Liquids." *Ann. Rev. Phys. Chem. 25,* 233. A review of the field of laser light scattering.

Davidovitz, P. 1975. *Physics in Biology and Medicine.* Englewood Cliffs, N. J.: Prentice-Hall. Contains an excellent, brief discussion of the flow of blood in the circulatory system, along with a number of other interesting topics.

Rainer-Maurer, H., and De Cruyter, W. 1971. *Disc Electrophoresis.* New York: Berlinard. A useful manual on the background and procedures of gel electrophoresis.

Tanford, C. 1961. *Physical Chemistry of Macromolecules.* New York: John Wiley & Sons. A classic work with a thorough discussion of macromolecular transport.

OTHER REFERENCES

Broersma, S. 1959. *J. Chem. Phys. 32,* 1626.

Crothers, D. M., and Zimm, B. H. 1965. *J. Mol. Biol. 12,* 525.

Eigner, J., and Doty, P. 1965. *J. Mol. Biol. 12,* 549.

Kirkwood, J. G., and Riseman, J. 1956. In Eirich, F., ed. *Rheology*, vol. 1. New York: Academic Press.

Klotz, L., and Zimm, B. H. 1972. *J. Mol. Biol. 72,* 779.

Levin, D., and Hutchinson, F. 1973. *J. Mol. Biol. 75,* 455.

Lin, S. H. C., Dewan, R. K., Bloomfield, V. A., and Morr, C. V. 1971. *Biochemistry 10,* 4788.

Maniatis, T., Jeffrey, A., and van de Sande, H. 1975. *Biochemistry 14,* 3787.

Olivera, B. M., Baine, P., and Davidson, N. 1964. *Biopolymers 2,* 245.

Zimm, B. 1960. In Eirich, F., ed. *Rheology,* vol. 3. New York: Academic Press.

PART FIVE

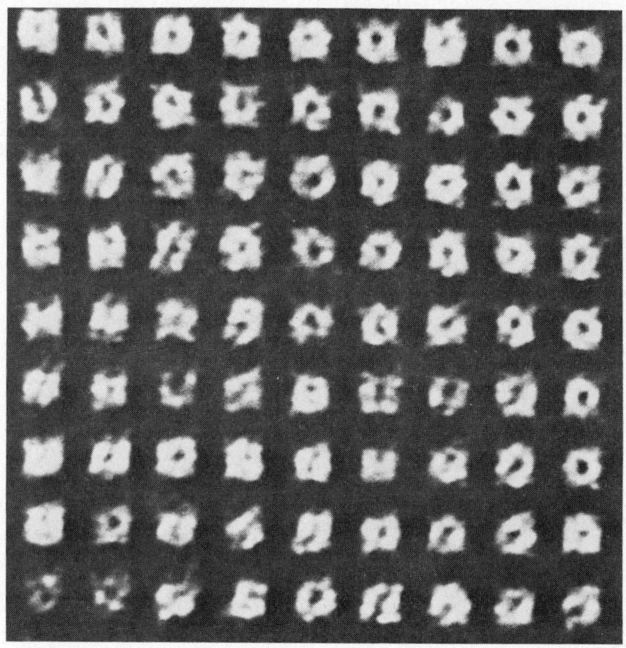

Symmetry and Molecular Structure

CHAPTER 16

Symmetry

If one searches for a theme that runs through the sciences and the arts, it is difficult to find a more pervasive one than symmetry. In biology, one of the first observations must have been that organisms are symmetric. We are most familiar with the apparent mirror (left-right) symmetry of vertebrates, but some lower forms of life exhibit much more elaborate symmetries. Examples of the highly symmetric skeletons of several species of the microscopic animals, *Radiolaria,* are shown in Figure 16-1. In biochemistry, recent work shows that many viruses and macromolecules are symmetric and that their symmetry is a property essential to their formation. We will discuss these symmetries and the related phenomenon of *self-assembly* in Sections 16-2 and 16-4.

Physical scientists also are concerned with symmetry. Crystals, which have been studied avidly by chemists and geologists during the past hundred years, are symmetric not only in their surface faces, but also in their microscopic structure. They are formed by the translational repetition of molecules in three dimensions, as we will discuss in Section 16-6. We will return to crystals in Chapter 17 for what they can reveal about molecular structure.

Even more basic to chemistry are the electronic orbitals that determine the chemical properties of atoms—these are classified by their symmetry, which is their most essential characteristic. Similarly, in physics many of the most basic relationships of quantum mechanics, nuclear structure, and relativity are expressed in terms of symmetry. Mathematicians have developed a branch of their science, *group theory,* for describing symmetry. We will touch on some aspects of group theory in Section 16-3.

Symmetry also is an essential element in the arts. This is most apparent in architecture. Many public buildings, including the Capitol and most cathedrals, have *mirror* symmetry. Some, such as the Pentagon and the Leaning Tower of Pisa, have *cyclic* symmetries of the sort we will discuss in Section 16-2. Still another symmetry commonly found in buildings is the *translational* symmetry of the pillars and arches in a colonnade. Music also is highly symmetric if we think of repetition in time, like that in space, as translation.

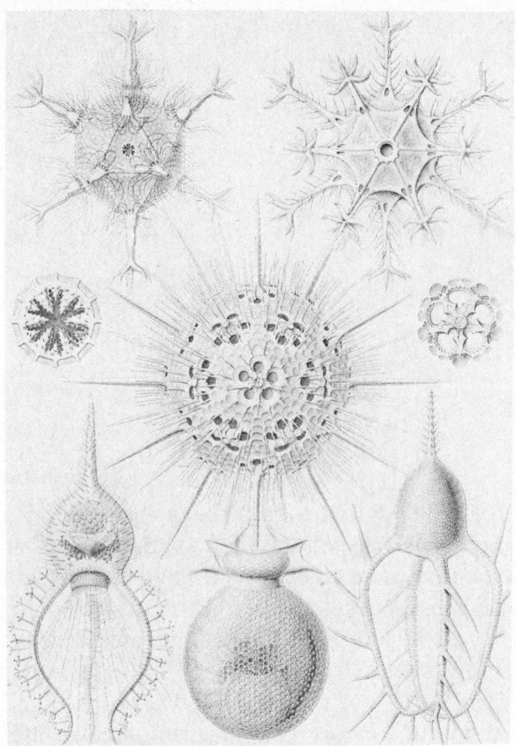

FIGURE 16-1
Drawings of skeletons of microscopic *Radiolaria,* a type of marine protozoa. (Reproduced from Haeckel, 1974.)

The rhythm of music constitutes an underlying symmetry. Some melodies are symmetric, such as the voices in a fugue, or the AABA pattern of popular songs. Symmetric designs have been employed in graphic and fabric art for at least six millennia. Often the symmetry is a simple repetition, such as repeated figures on a Greek vase, but some artists, such as the twentieth century Dutch printmaker M. C. Escher, have employed much more complicated symmetries in their work (see Fig. 16-2).

Symmetry sometimes appears even in literature. For example, a *palindrome* is a word, phrase, or sentence that reads the same forwards as backwards. "Able was I ere I saw Elba" is a palindrome reputed to have been uttered by Napoleon as he set out of Elba to re-establish his empire. (Why Napoleon chose at that moment to speak in English or to concern himself with symmetry is unexplained.) Today, chemists who study DNA are interested in sequences of the nucleic acid which have palindromic symmetry (see Prob. 16-18). On a grander scale, some literary works are designed with symmetric structure. One example is Lawrence Durrell's *Alexandria Quartet,* which describes a course of events from the points of view of four individuals.

We begin this chapter with a consideration of the meaning of the term "symmetry" and with examples of the four basic types of symmetry: rotation,

FIGURE 16-2
A symmetric drawing by M. C. Escher. (The woodcut by M. C. Escher is reproduced by permission of the Escher Foundation, Haags Gemeentemuseum, The Hague. © The Escher Foundation, 1965. Reproduction rights arranged courtesy of the Vorpal Galleries: New York, Chicago, San Francisco, Laguna Beach.)

translation, reflection, and inversion. In Section 16-2 we ignore translations and consider only the possible symmetries about a point—these are the symmetries that a molecule can possess. In Section 16-5 we consider translation along a line, and the types of helices that can be formed. Finally, in Section 16-6, we consider the types of symmetries that arise when translations are permitted in two and three dimensions. These are the symmetries found in graphic designs and crystals.

○ 16–1 Symmetry Operations and Elements

Symmetry is Defined by the Manipulations that Superimpose an Object on Itself

An object is symmetric if some spatial manipulation of it results in an indistinguishable object. (Said a friend of the authors, "If you can't tell when an object has been turned, it has symmetry.") Another way of saying this is that a symmetric object can be superimposed on itself by some operation. The operation that leads to superimposition of an object on itself is a *symmetry operation*. We can see several symmetry operations in Figure 16-2 if we imagine that the figure extends infinitely on all sides by further repetitions. One symmetry operation is that of rotation by π radians (180°). Suppose that there is an axis perpendicular to the drawing at a point where the noses of two white lizards in the Escher drawing meet; then by rotating the drawing π radians about the axis, the pattern can be superimposed on itself. In other words, this operation yields a drawing that is indistinguishable from the drawing in its original orientation. The imaginary axis about which the rotation has been made is one example of a *symmetry element*. In general, a symmetry element is a geometrical entity (such as a line, a point, or a plane) about which a symmetry operation is carried out.

EXERCISE 16-1

Find three other symmetry operations of rotation that characterize Figure 16-2.

ANSWER

Rotations by π radians about three other axes superimpose the drawing on itself: these axes are at the points where the noses of the black lizards meet, where the left hindpaws meet, and where the right forepaws meet.

A different sort of symmetry possessed by Figure 16-2 is *translation*. Suppose again that the figure is infinite in extent. Then if the drawing is translated to the right by placing the junction of the noses of two white lizards on the identical point to the right, the pattern is superimposed on itself. All crystals, and repeating two-dimensional designs like those on wallpaper and tile floors, are characterized by such translational symmetries.

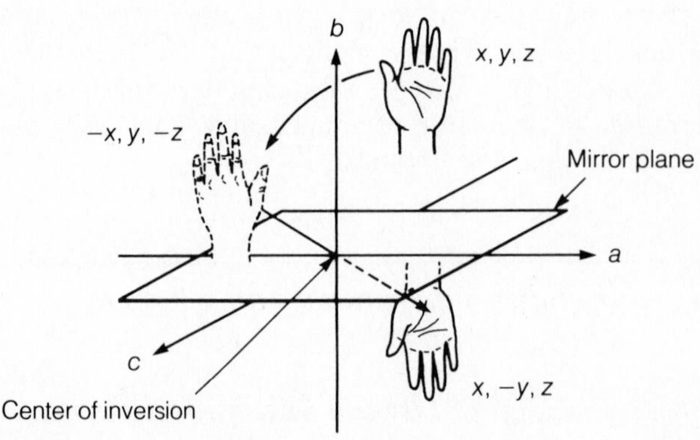

FIGURE 16-3
The symmetry operations of reflection and inversion, as described in the text. Note that the coordinate x measures displacement along the a axis, y along b, and z along c. The curved arrow indicates a 2-fold rotation. (Redrawn from Glusker and Trueblood, 1972.)

EXERCISE 16-2

Find at least two other translational symmetry operations in Figure 16-2.

Rotation and translation are two of the four main types of symmetry operations. The other two are *reflection* and *inversion*. These are illustrated in Figure 16-3. The two hands with solid outlines in this figure are related to each other by the operation of reflection. The symmetry element in this case is a *mirror* plane that lies midway between them. Note that this element acts as though it were a real mirror, changing the left hand above the plane to a right hand below. The operation of inversion may be thought of as a reflection through a point: it transforms each coordinate $(-x, y, -z)$ of the dashed left hand of Figure 16-3 to its negative $(x, -y, z)$ giving the solid right hand. The *center of inversion* lies midway between the hands. Notice that both inversion and reflection change left to right, but rotation and translation do not.

EXERCISE 16-3

What symmetry operation relates the two hands above the mirror?

ANSWER

An axis of rotation of π radians around b.

EXERCISE 16-4

Why is it that biological molecules such as enzymes may contain axes of rotation, but never contain the symmetry elements of reflection or inversion?

ANSWER

The operations of reflection and inversion require for each left-handed object a corresponding right-handed object. Left and right on the molecular level correspond to absolute configurations about optically active carbon atoms. Thus each carbon atom of L-configuration must be paired with a corresponding atom of D-configuration. With few exceptions, such as in bacterial cell walls, organisms contain amino acids with only the L-configuration at the α carbon. Therefore, there are no corresponding D-amino acids, and the operations of reflection and inversion are never found in protein molecules.

16–2 Symmetry about a Point

Point Groups Describe Symmetry about a Point

The symmetry of any object can be described by some combination of rotation, translation, reflection, and inversion. In this section we will restrict our attention to the symmetry of an object around a single point; thus we can

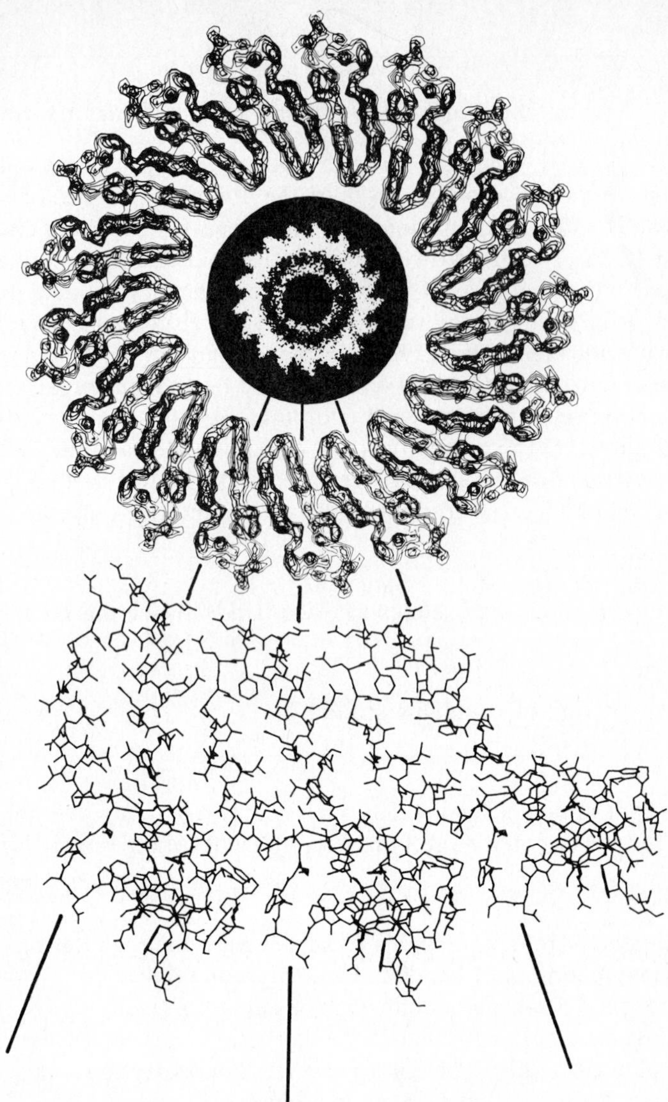

FIGURE 16-4
The 17-subunit disk structure of tobacco mosaic virus, displayed at three resolutions. The structure was determined by the methods of electron microscopy and x-ray diffraction (see Chap. 17). Shown in the center is the disk structure, as determined at low resolution by electron microscopy. Each protein subunit is seen as a white comma. Towards the inside of the protein subunits is a trough, in which RNA is threaded in the virus.

Surrounding the electron micrograph is the structure as determined at low resolution by x-ray diffraction. Each of the 17 subunits is a polypeptide chain (seen here at a resolution too coarse to show amino acids, much less atoms). The polypeptide is represented by contours of electron density (from Champness et al., 1976).

At the bottom, portions of the polypeptide chains of three of the subunits are shown. The conformation of the polypeptide was determined by high-resolution x-ray diffraction (diagram from Dr. A. C. Bloomer). The *resolution* is the smallest separation of two points that can be distinguished. It is about 3Å for the x-ray work that yielded the polypeptide conformation, and about 25Å for the electron micrograph at the center.

neglect translations for the moment. The other three types of symmetry operations are sufficient to describe the symmetry of any molecule, or in fact of any finite object.

The collection of symmetry operations that describe the symmetry of an object about a point is called a *point group*. This term is used because the collection of symmetry operations exhibits all the properties of a mathematical group. Rather than take the abstract approach of group theory to this subject, we will attempt to develop a feeling for symmetry by considering examples of point groups. Two systems of notation are used for point groups: (1) the *Schoenflies symbol,* which is a capital letter and is favored by spectroscopists, and (2) the *Hermann-Mauguin symbol,* which is an explicit list of the symmetry elements and is favored by crystallographers. We will use both in this chapter, citing the Schoenflies symbol first and following it with the Hermann-Mauguin symbol in parentheses.

Cyclic Point Groups Describe Head-To-Tail Arrangements Around a Circle

The simplest point groups are the cyclic groups. They contain a single axis of rotation, and also may contain one or more planes of reflection. The axis is called an n-fold axis of rotation, if its operation is a rotation of $2\pi/n$ radians. Thus a 3-fold axis of rotation is associated with the operation of rotation by $2\pi/3$ radians $= 120°$. Point groups containing only an n-fold axis of rotation are given the symbol n in the Hermann-Mauguin system and C_n in the Schoenflies notation, where C stands for cyclic. We can think of an n-fold axis as relating n equivalent points in the symmetric object (such as the tips of the index fingers in the hands above the mirror in Fig. 16-3).

Examples of point groups C_1 (1), C_2 (2), C_3 (3), and C_{17} (17) are given in Table 16-1. Point group C_1 consists of only a 1-fold axis; thus a rotation of 2π radians (360°) superimposes the object with this symmetry upon itself. Of course, this is true of any object no matter how asymmetric, so that point group C_1 describes the "symmetry" of asymmetric objects.

Point group C_2 (2) consists of a 2-fold axis of rotation (sometimes called a *diad*). An object with this point-group symmetry is superimposed upon itself if rotated by π radians (180°) about the axis. The hemoglobin molecule possesses a single 2-fold axis of rotation, and thus has point group C_2 (see Table 16-1). Molecules that have point-group symmetry of the form C_n in which n is greater than 2 are unusual. One example is the double-disk aggregate of protein from tobacco mosaic virus that is formed at low ionic strength and pH values just below 7. This aggregate is known (Durham et al., 1971) to have C_{17} symmetry (see Fig. 16-4). Thus, the aggregate can be superimposed on itself if it is rotated by $2\pi/17$ radians about its axis.

Some cyclic point groups also contain one or more mirrors. Because proteins and other biological molecules generally cannot contain planes of reflection, we will consider only a few of these. Three examples are given in Table 16-1. The notation for these point groups is straightforward. If the

TABLE 16-1 *Examples of Cyclic Point Groups*

POINT GROUP[a] S	H-M	NUMBER OF ASYMMETRIC UNITS	CONTACT TYPE[b]	SYMMETRY ELEMENTS	EXAMPLES[c]	
C_1	1	1	none	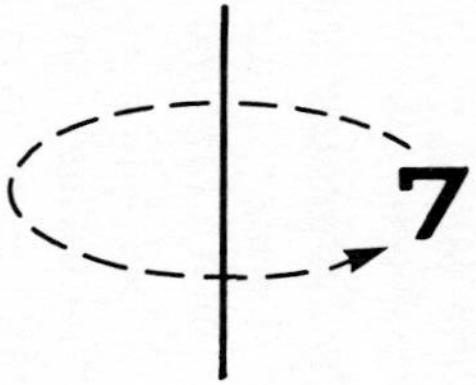	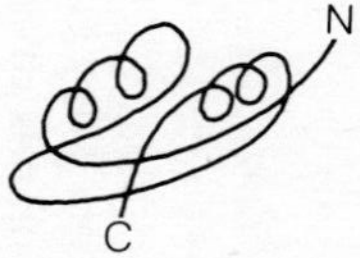 N, C — Any polypeptide chain	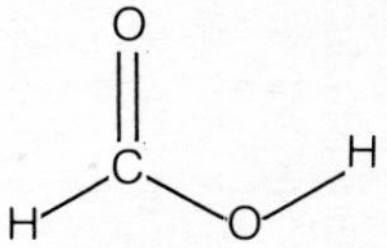 O, C, H, O, H — Formic acid
C_2	2	2	i	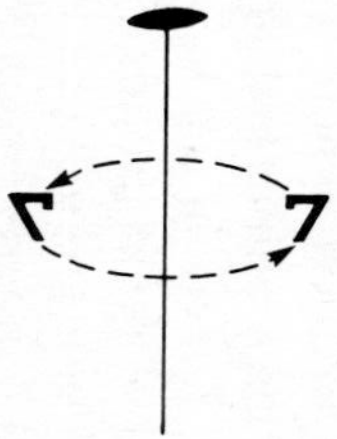	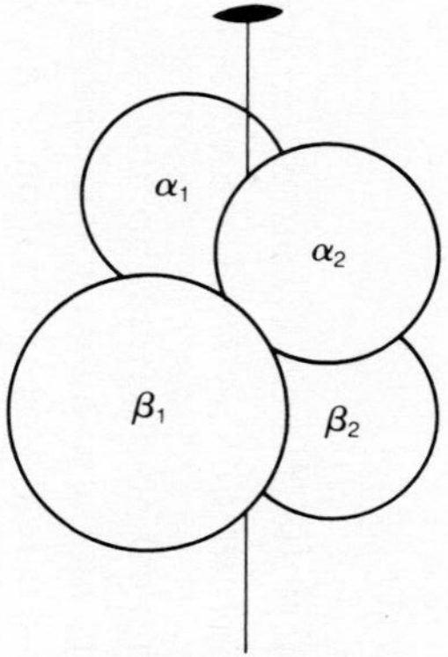 α_1, α_2, β_1, β_2 — Hemoglobin	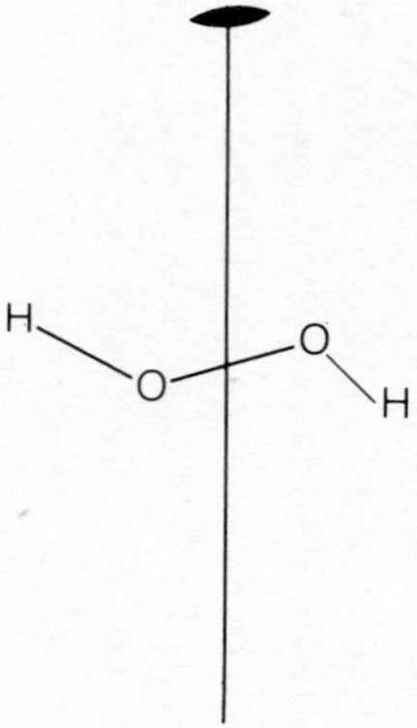 H, O, O, H — Hydrogen peroxide

C_3	3 	3 	*h*	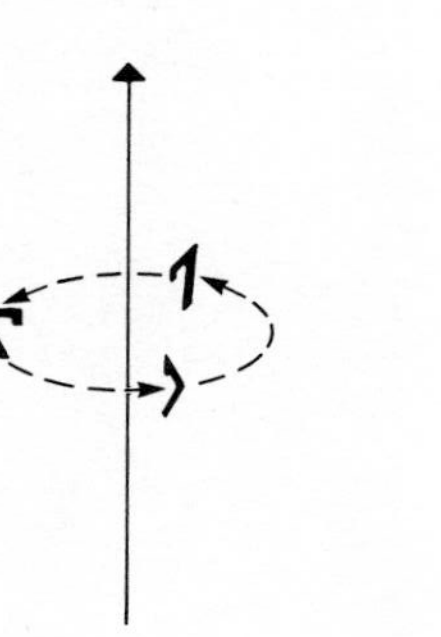 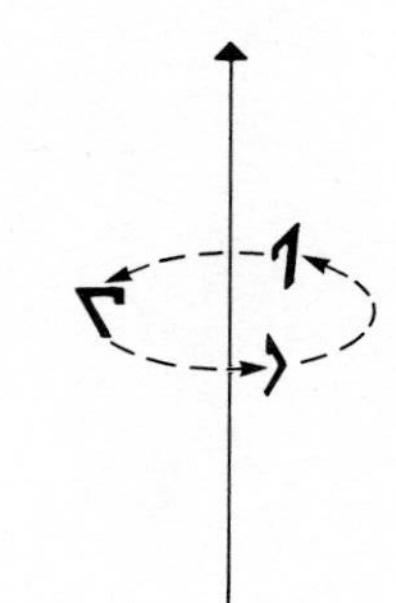	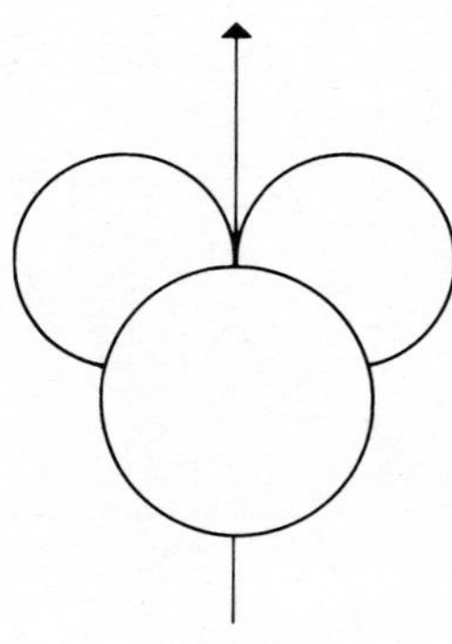 Aldolase from *Pseudomonas putida*	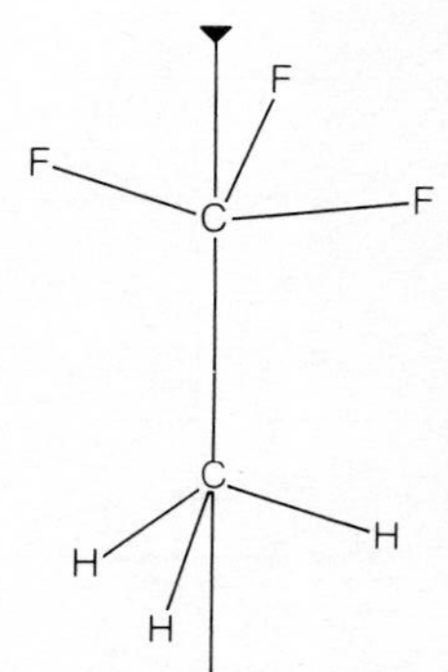 1,1,1-trifluoroethane
C_{17}	17	17	*h*	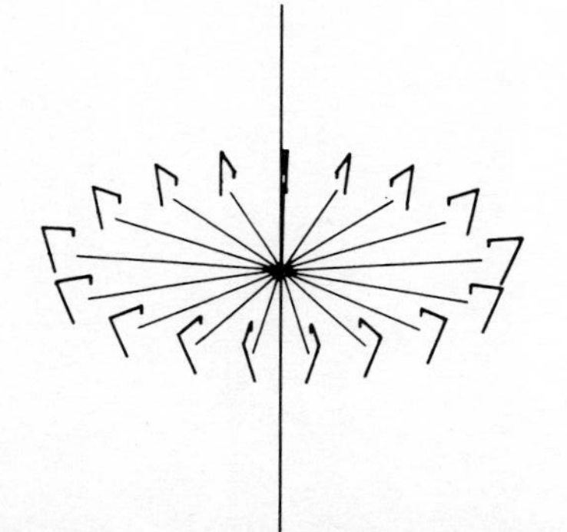		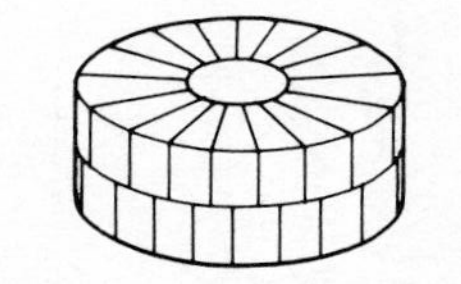 Double disk of protein from tobacco mosaic virus

(Continued)

TABLE 16-1 *Examples of Cyclic Point Groups (Continued)*

POINT GROUP[a] S	POINT GROUP[a] H-M	NUMBER OF ASYMMETRIC UNITS	CONTACT TYPE[b]	SYMMETRY ELEMENTS	EXAMPLES[c]
C_{2v}	$2mm$	4	—		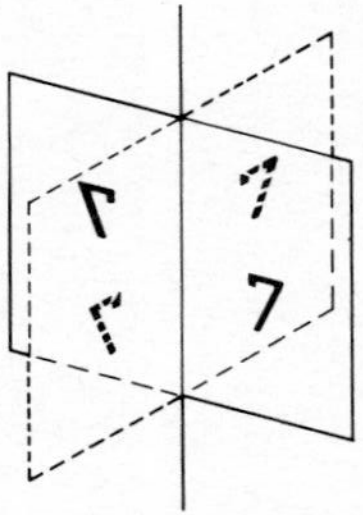 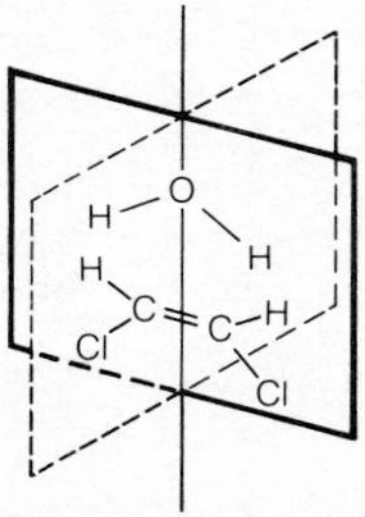
C_{1h} or C_s	$1/m$ or m	2	—		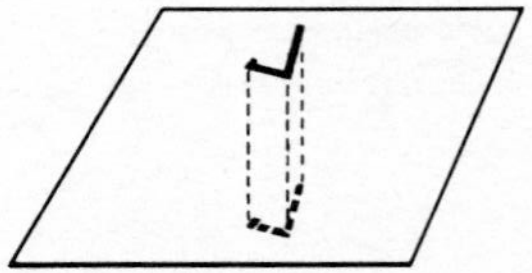 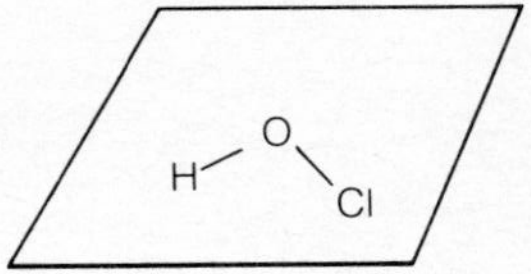
C_{2h}	$2/m$	4	—		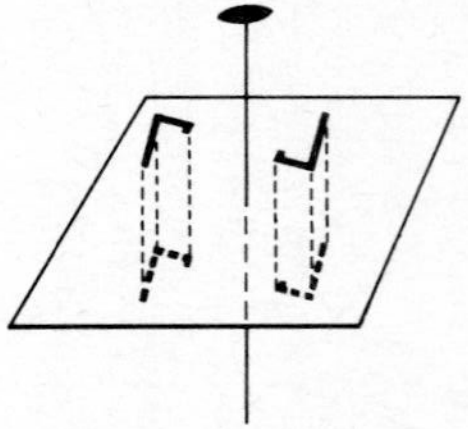 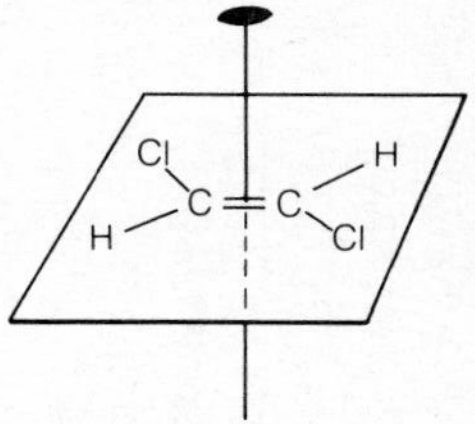

[a] Columns S and H-M give the Schoenflies and Hermann-Mauguin notation, respectively.
[b] *i* indicates isologous and *h* indicates heterologous; see text.
[c] The protein examples are highly schematic.

n-fold axis of rotation is perpendicular to the mirror, the mirror is said to be horizontal, and the Schoenflies notation is C_{nh} (h stands for horizontal). The Hermann-Mauguin notation is n/m, in which the slash indicates that the n-fold axis is perpendicular to the mirror plane of reflection. From Table 16-1 you can see that the molecule *trans*-1,2-dichloroethene belongs to point group C_{2h} ($2/m$) because all of its atoms lie in a plane and a 2-fold axis is perpendicular to the mirror. If, instead, the n-fold axis lies *in* a mirror, the mirror is said to be vertical, and the Schoenflies notation is C_{nv}, in which v stands for vertical. The corresponding Hermann-Mauguin notation is nmm if n is even, and nm if n is odd. The example of point group C_{2v} ($2mm$) is given in Table 16-1. Water and *cis*-1,2-dichloroethane are examples of molecules that belong to this point group.

EXERCISE 16-5

Which point groups best characterize the symmetries of each of the following: (a) the Pentagon building in Arlington, Va., (b) the Golden Gate Bridge, (c) a bridge table, (d) a propeller with three curved blades?

ANSWERS

(a) C_5 (5). (b) C_{2v} ($2mm$). (c) C_{4v} ($4mm$). (d) C_3 (3).

Some Terminology and Notation

A term that we will use often in this chapter and the following one is *asymmetric unit;* it is the part of the symmetric object from which the whole is built up by repeats. In formal terms, it is the smallest unit from which the object can be generated by the symmetry operations of its point group. In every example in the fourth column of Table 16-1, the asymmetric unit is the number 7. Thus in point group C_2 (2), the 2-fold axis of rotation can be thought of as generating a second 7 from the first (which is the asymmetric unit). The asymmetric unit in multisubunit enzymes usually is a single polypeptide chain or *subunit.* This is not always the case; in hemoglobin it is one $\alpha\beta$ pair of polypeptide chains (see Table 16-1). Protein chemists sometimes use the term *protomer* for an asymmetric unit, and *oligomer* for the entire multisubunit enzyme. Thus the protomer of hemoglobin is an $\alpha\beta$ pair.

Terms for the contacts between protomers were introduced by Monod et al. (1965) and are used in the literature of protein chemistry. A *binding set* is the region of a protomer (likely to be several amino acid side chains) that binds to one other protomer. The region that encloses the complementary binding sets that link two protomers is called a *domain of binding.* Monod et al. defined an *isologous association* of two protomers as one in which the domain of binding involves two identical binding sets. An equivalent but simpler definition is that an isologous association of protomers contains a 2-fold axis. Monod et al. defined a *heterologous association* as one in which the domain of binding involves two different binding sets. For example, a protomer related to others by a 4-fold axis of rotation has two distinct binding sets in its domain of binding and thus forms heterologous associations with neighboring protomers.

Conventional symbols are used to denote the most common axes of rotation. If the axis is more or less perpendicular to the plane of the paper the symbols ⬮ ▲ ◆ ⬢ are used to denote 2-fold, 3-fold, 4-fold, and 6-fold axes of rotation, respectively. Some of these symbols are used in Table 16-1. If a 2-fold axis of rotation lies in the plane of the paper, it is represented by an arrow ⟶.

⬡ *Symmetry can be Represented Algebraically*

Up to this point, we have been describing symmetry in geometric terms of rotations around axes and so forth. We also can describe symmetry operations algebraically, which is useful in many types of calculations. We do so by thinking of a symmetry operation as a transformation of the coordinates of a point. Suppose a point, such as the tip of the little finger of the solid hand above the plane in Figure 16-3, has coordinates x, y, z. Then the action of some symmetry operations on the hand moves this point to some *equivalent position* x', y', z'. We may represent this as

$$S{:}\quad x, y, z \rightarrow x', y', z'$$

The actual values of the new coordinates x', y', z' depend on the nature of the symmetry operation S. If S is the 2-fold rotation of the solid hand of Figure 16-3 into the dashed hand, then $y' = y$, because the tip of the little finger is not shifted along b (the y direction) during the rotation. But the rotation around the b-axis moves the little finger as far in the negative x direction as it was initially in the positive x direction; thus $x' = -x$. The same is true for z: $z' = -z$. Therefore, we can summarize the entire 2-fold rotation by

$$2{:}\quad x, y, z \rightarrow -x, y, -z$$

For an n-fold rotation about the b-axis, the algebraic representation of the first application of the symmetry operation is

$$n{:}\quad x, y, z \rightarrow \left(\cos\frac{2\pi}{n}\right)x - \left(\sin\frac{2\pi}{n}\right)z, \quad y, \quad \left(\sin\frac{2\pi}{n}\right)x + \left(\cos\frac{2\pi}{n}\right)z \qquad (16\text{-}1)$$

Repeating this operation will produce each of the equivalent positions in turn. After n repeats it will produce the original coordinate x, y, z.

EXERCISE 16-6

For a 4-fold rotation about the b-axis, determine the three positions equivalent to x, y, z, by repeated applications of Equation 16-1. Plot the four equivalent positions in the x-z plane.

ANSWER

$x, y, z; \; -z, y, x; \; -x, y, -z; \; z, y, -x$

EXERCISE 16-7

Derive the form of Equation 16-1 for a 3-fold rotation about the b-axis.

ANSWER

$$3: \quad x, y, z \rightarrow -\frac{1}{2}x - \frac{\sqrt{3}}{2}z, \qquad y, \qquad \frac{\sqrt{3}}{2}x - \frac{1}{2}z.$$

The symmetry operations of reflections and inversion are easier to represent algebraically. Again referring to Figure 16-3, note that the mirror is perpendicular to the y direction. The operation of reflection across the mirror will transform a point y to $-y$; that is, it is reflected as far in the negative direction as it was initially in the positive direction. However, the coordinates x and z are unchanged during reflection. Thus the operation of reflection across a mirror perpendicular to the y direction can be expressed

$$m_y: \quad x, y, z, \rightarrow x, -y, z \tag{16-2}$$

From inspection of Figure 16-3, you can see that the operation of inversion through a point involves changes in sign of all coordinates:

$$i: \quad x, y, z, \rightarrow -x, -y, -z \tag{16-3}$$

Dihedral Point Groups have Axes of Rotation at Right Angles to Each Other

The dihedral point groups are the ones to which the symmetries of most multisubunit enzymes belong. Structures with dihedral symmetry may be thought of as two-layered structures, each layer having cyclic symmetry, with the two layers joined back-to-back. The definition of the dihedral point group denoted D_n in the Schoenflies notation is an n-fold axis of rotation

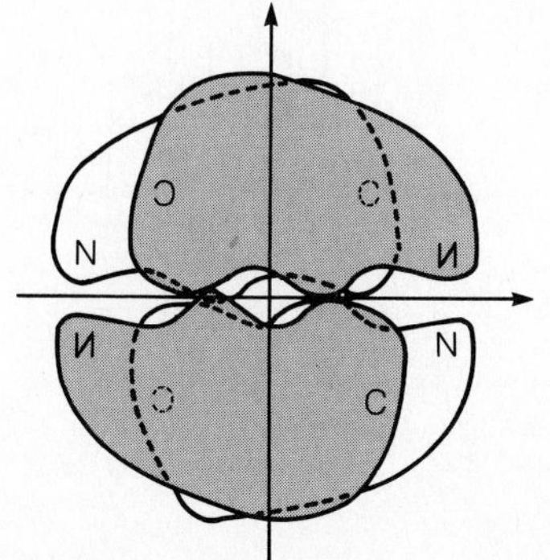

FIGURE 16-5
A hypothetical enzyme with D_2 (222) symmetry. The asymmetric unit represents a polypeptide chain. The symbols show the N and C termini. The shaded polypeptide is nearer. Dashed lines are seen through the unshaded polypeptide.

perpendicular to n 2-fold axes of rotation. For example, point group D_2 (222 in the Hermann-Mauguin notation) has a 2-fold axis perpendicular to two 2-fold axes. That is, it consists of three mutually perpendicular 2-fold axes. As shown in Figure 16-5, this point group contains four asymmetric units. The upper (shaded layer) consists of two asymmetric units related by the 2-fold axis perpendicular to the paper; so does the lower layer. The two 2-fold axes in the plane of the paper transform the upper layer on the lower, and vice versa. Several enzymes have this symmetry, the asymmetric unit being a single polypeptide chain.

EXERCISE 16-8

Draw an arrangement of four figure 7s that has D_2 (222) symmetry.

ANSWER

One possibility:

where the light and dark 7s are seen from the opposite side. Draw another arrangement.

An enzyme of dihedral point-group symmetry D_3 (32) is aspartate transcarbamylase. This enzyme consists of six copies each of a catalytic polypeptide chain, C, and of a regulatory polypeptide chain, R, so that the structure can be represented $(RC)_6$. The asymmetric unit of the point group is a single

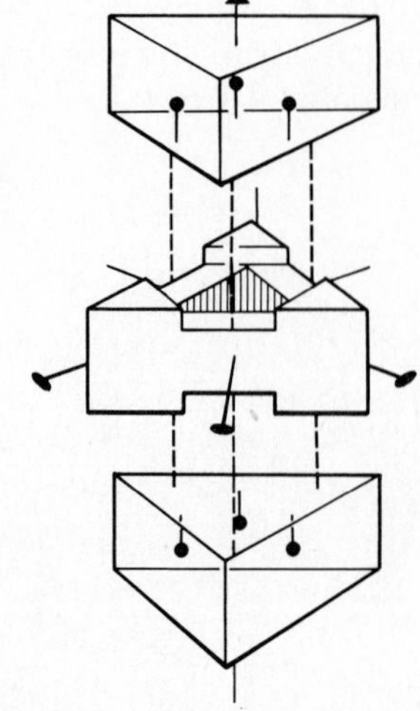

FIGURE 16-6
Exploded view of the aspartate transcarbamylase molecule, giving a schematic representation of three catalytic polypeptide chains above and three below a central region made up mainly of the six regulatory chains. The 3-fold axis is shown perpendicular to the three 2-fold axes. The molecule surrounds a central, solvent-filled cavity 25Å along the 3-fold direction and 50Å along the 2-fold direction. Dots within the catalytic polypeptide chains represent sites where a mercurial compound binds, presumably near the active site of the enzyme. These sites are accessible from the central cavity. The overall molecular dimensions are about 90Å along the 3-fold axis and 110Å along each 2-fold axis. (Adapted from Evans et al., 1973. *Science* 179, 683–685. Copyright 1973 by the American Association for the Advancement of Science.)

RC unit. Figure 16-6 shows schematically the gross structure of this enzyme as determined by an x-ray diffraction study of its atomic structure (see Chap. 17) at a preliminary stage. At this stage the symmetry of the molecule is evident, and the relative positions of the *C* and *R* chains can be seen.

Other dihedral point groups are illustrated in Table 16-2. Notice that the number of asymmetric units in the point group D_n is $2n$. The types of contacts present in an oligomer of each symmetry also are listed in the table.

Dihedral point groups, like cyclic ones, can include mirrors, although these elements are never found in biological structures such as enzymes. The notation D_{nh} signifies that a mirror is perpendicular to the n-fold axis. This mirror, when combined with the rotation axes, generates horizontal mirror planes. For example, the benzene molecule belongs to point group D_{6h} (6/*mmm*); its symmetry elements include a 6-fold axis, six perpendicular 2-fold axes, one horizontal mirror plane perpendicular to the 6-fold axis, and 6 vertical mirror planes.

Cubic Point Groups Describe the Highest Symmetries, Have the Largest Number of Asymmetric Units, and are Best Suited for the Shell-Like Structures of Viruses

The only remaining ways of arranging objects symmetrically about a point are described by the cubic point groups. Their essential characteristic is four 3-fold axes arranged as the four body diagonals (lines connecting opposite corners) of a cube. We will discuss the three cubic point groups that can be occupied by biological molecules; there are several others that also contain mirrors.

The simplest cubic point group is denoted *T* (for tetrahedral) in the Schoenflies system and 23 in the Hermann-Mauguin system. It consists of the four 3-fold axes that run from a vertex to the center of the opposite face of a tetrahedron, plus the three 2-fold axes that connect opposite edges (see Table 16-3); *T* contains 12 asymmetric units. The molecule neopentane belongs to this point group (Table 16-3) and so may the enzyme aspartate-β-decarboxylase.

EXERCISE 16-9

What is the asymmetric unit of the neopentane molecule?

ANSWER

One twelfth of the central C atom, one third of a side chain C, and one H atom.

The point group *O* (for octahedral; 432 in the Hermann-Mauguin notation) contains 24 asymmetric units. Its symmetry elements are all the rotation axes of a cube: three 4-fold axes perpendicular to the cube faces, four 3-fold axes along the body diagonals, and six 2-fold axes through opposite edges. The

TABLE 16-2 *Examples of Dihedral Point Groups*

POINT GROUP[a] S	POINT GROUP[a] H-M	NUMBER OF ASYMMETRIC UNITS	CONTACTS[b] MAXIMUM	CONTACTS[b] MINIMUM	SYMMETRY ELEMENTS[c]	EXAMPLES
D_2	222	4	*iii*	*ii*	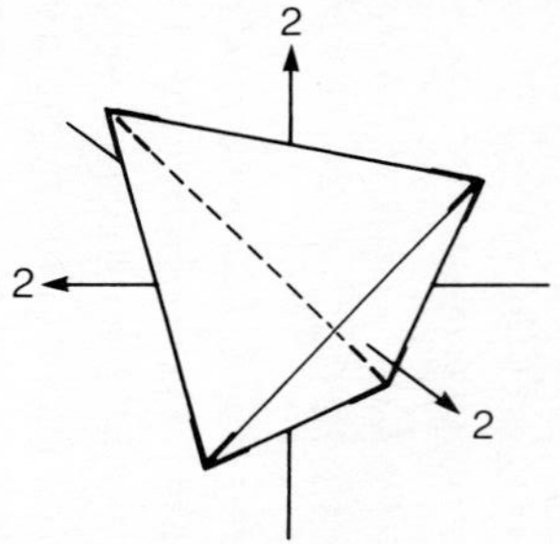 3 2-fold	Aldolase from rabbit muscle Concanavalin A Glyceraldehyde 3-phosphate dehydrogenase Lactate dehydrogenase
D_3	32	6	*hii*	*hi* or *ii*	1 3-fold 3 2-fold	Aspartate transcarbamylase
D_4	422	8	*hii*	*hi* or *ii*	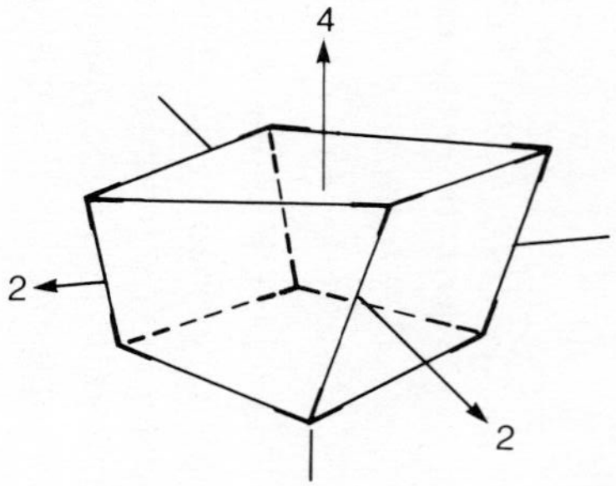 1 4-fold 4 2-fold	Ribulose bisphosphate carboxylase/oxygenase

D_6	622	12	*hii*	*hi* or *ii*		1 6-fold 6 2-fold	Glutamine synthetase Hexamino benzene[d]
D_{6h}	6/*mmm*	12				1 6-fold 6 2-fold 1 *h* mirror 6 *v* mirrors	Benzene

[a] Column S gives Schoenflies notations; column H-M gives Hermann-Mauguin notations.
[b] Contacts between one protomer and its neighbors. The symbol *i* stands for isologous, *h* for heterologous. The minimum contacts are those required to hold the oligomer together. See also Matthews and Bernhard (1973).
[c] Drawings from Klug (1969).
[d] From Bernal et al. (1972). Note that all amino groups have the same tilt relative to the aromatic ring.

TABLE 16-3 *Examples of Cubic Point Groups*

POINT GROUP[a] S	POINT GROUP[a] H-M	NUMBER OF ASYMMETRIC UNITS	CONTACTS[b] MAXIMUM	CONTACTS[b] MINIMUM	SYMMETRY ELEMENTS	EXAMPLES
T	23	12	*hi*	*hi*	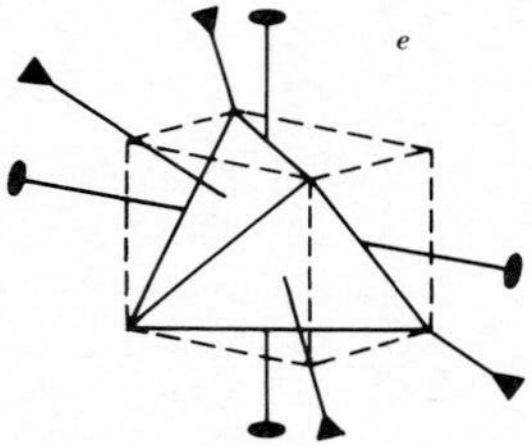 [e] 3 2-fold 4 3-fold	Aspartate-β-decarboxylase[c] Neopentane[d] 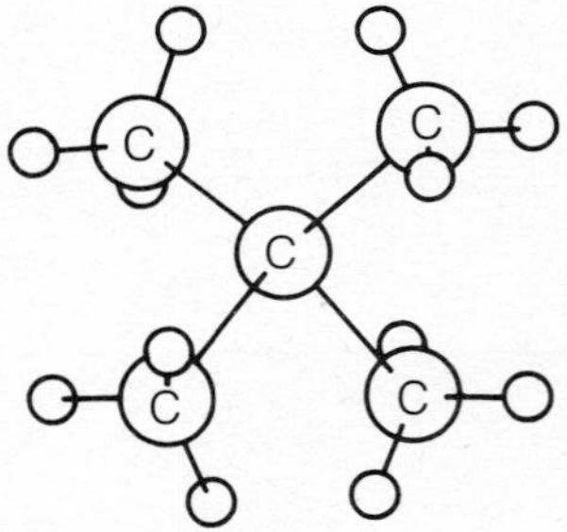
0	432	24	*hii*	*hh* or *hi*	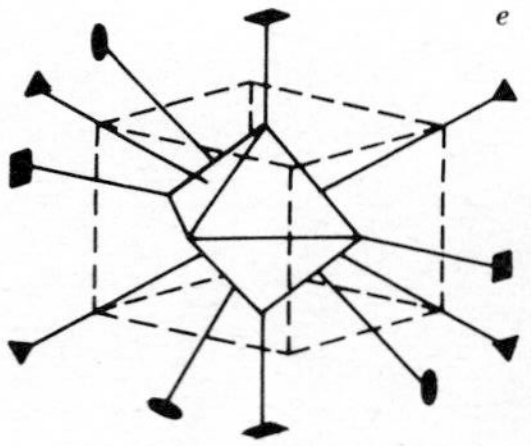 [e] 6 2-fold 4 3-fold 3 4-fold	Dihydrolipoyl transsuccinylase[f] 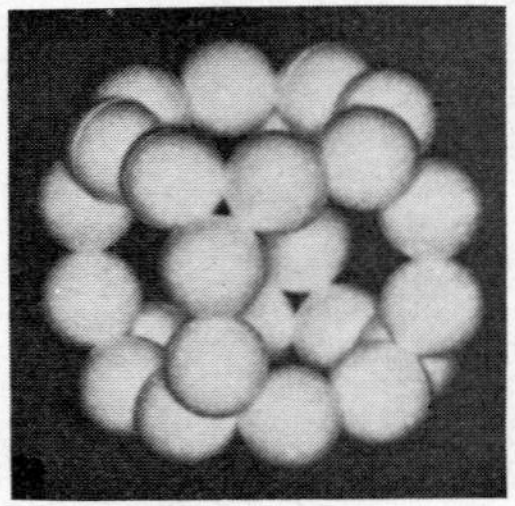 Spherical subunit model 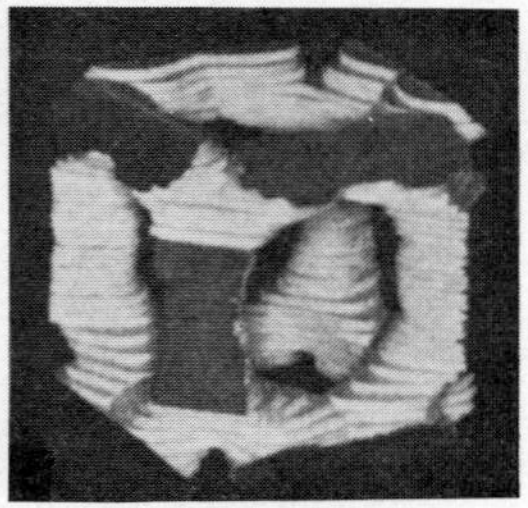 Electron density at low resolution

I	532	60	*hii*	*hh* or *ii*	[e] 15 2-fold 10 3-fold 6 5-fold

Spherical viruses, with asymmetric units related as shown below [g]

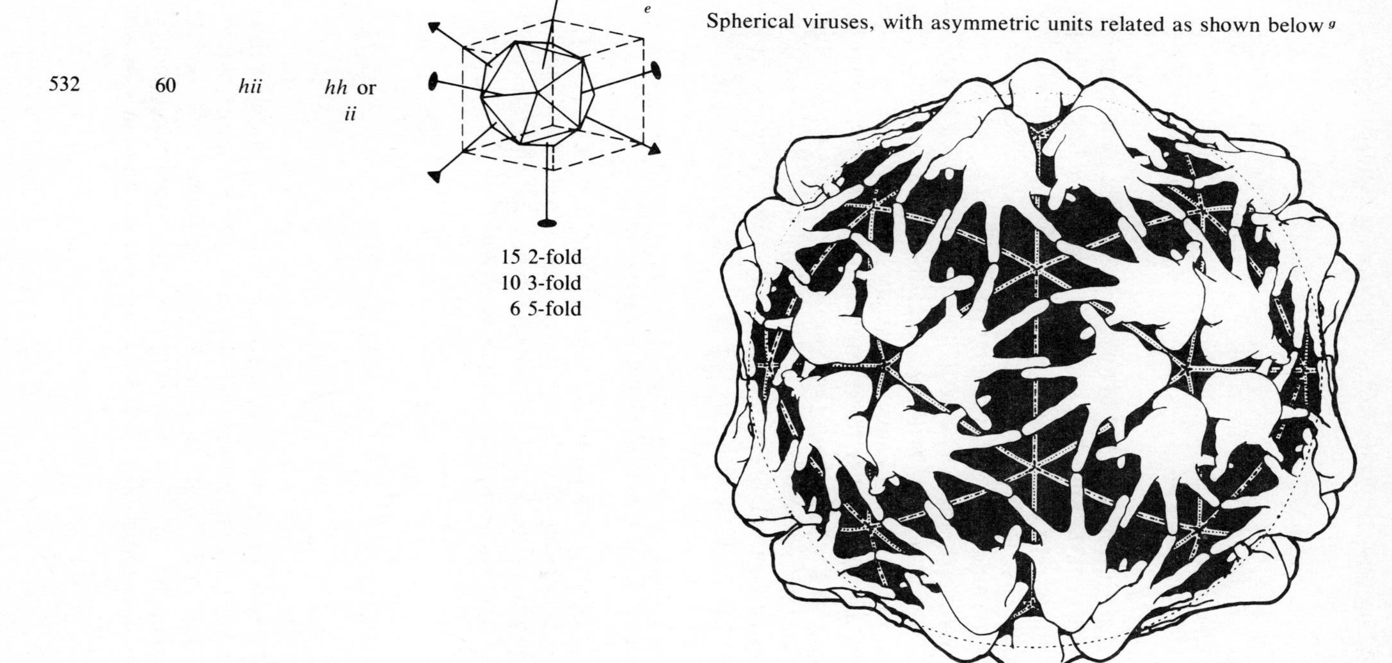

[a] Column S gives Schoenflies notation; column H-M gives Hermann-Mauguin notation.
[b] See corresponding footnote in Table 16-2.
[c] Haschemeyer (1970).
[d] From Bernal et al. (1972).
[e] From Wilson (1966).
[f] From DeRosier (1969).
[g] From Caspar (1964).

enzyme dihydrolipoyl transsuccinylase with 24 identical subunits belongs to this point group.

EXERCISE 16-10

Sketch a cube as seen from one side, from one vertex and from one edge. Draw in axes of symmetry for each sketch.

The point group **I** (for icosohedral; 532 in the Hermann-Mauguin notation) is the symmetry upon which the so-called spherical viruses are built. The 60 asymmetric units of this group are arranged about six 5-fold axes, ten 3-fold axes, and fifteen 2-fold axes. Table 16-3 shows this symmetry, depicting a left hand as the asymmetric unit. Note that the 60 hands form a closed shell, suitable for enclosing other matter. The ϕX174 and polio viruses belong to this point group. It is not certain, however, that their protein shells are comprised of only 60 polypeptide chains. They may be composed of some multiple of 60 chains, and in this case the asymmetric unit is more than one chain. In other words, not every chain necessarily occupies an equivalent position. They may instead occupy *quasi-equivalent* positions, as do protein chains in other spherical viruses. We will return to the topic of virus structure in Section 16-4.

Why There are no Other Point Groups

Students often ask why there are no other point groups. The reason is that there are no other ways to place objects in equivalent positions. Consider for example a 1.9-fold rotation axis, perpendicular to the page. If this axis operates twice on the number 7, it produces first the ′7 and then the ″7 shown here.

It is clear that these three 7s do not have identical environments, and thus together do not constitute an object with point-group symmetry [other than C_1 (1)]. Thus a 1.9-fold axis is not a valid symmetry element. A similar non-equivalence occurs for various combinations of symmetry elements, such as a 5-fold axis perpendicular to a 6-fold axis. Such combinations are therefore not point groups.

JOHANNES KEPLER

1571–1630

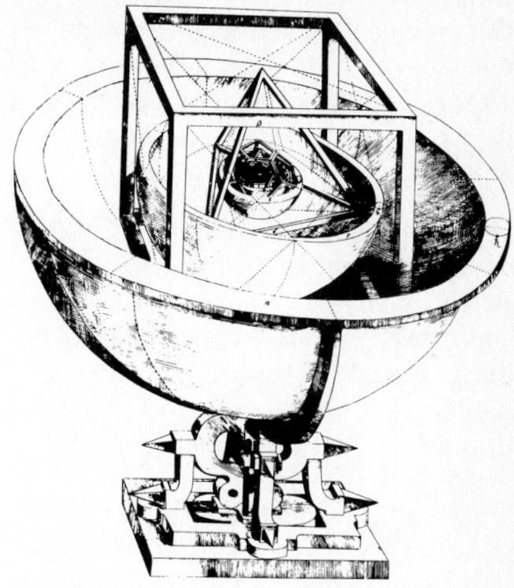

Kepler's model of the solar system involving six known planets and five regular polyhedra.

From Pythagoras to de Broglie, scientists have described the world in terms of symmetry. Doubtless the most bizarre of these descriptions was that of Kepler, who believed that the shapes of geometric solids dictate the structure of our universe. Perhaps even stranger was that, in his determination to prove correct his notions on symmetry, Kepler started the Scientific Revolution.

Kepler's three laws of planetary motion have been called the pillars on which Newton built the modern universe. They were the first natural laws in the modern sense: precise statements, based on observation, about universal relations between physical quantities. Though in discovering his laws, Kepler was one of the first modern scientists, he also retained many viewpoints of medieval science. He was convinced that God created the world in accordance with principles of mathematical harmony, and much of his scientific life was a search for the harmonic relations of the heavenly bodies. However, he did come increasingly to recognize the importance of describing phenomena with mathematics, and of checking and improving old observations with new ones. His analysis of movements of the planets, along with the dynamics developed by his contemporary Galileo, were elements necessary for Newton to formulate his mechanics. "If I have been able to see farther," Newton said, "it was because I stood on the shoulders of the giants."

Kepler was a gloomy, neurotic genius, whose life was filled with all manner of personal woes. He was born in Germany to a middle-class family that had fallen on hard times. Three generations of Keplers lived crowded in a narrow house, the scene of much disease and chaos. Kepler described his father as, "vicious, inflexible, quarrelsome, and doomed to a bad end." He "treated my mother extremely ill, went finally into exile and died." His mother he called, "small, thick, swarthy, gossiping, and quarrelsome, of a bad disposition." Others must have seen her the same way and even worse because as an elderly woman she was charged with witchcraft and brought to the point of torture to confess. Kepler devoted many months to her successful defense. He once wrote that he viewed himself as a wily lap dog who greedily grabs whatever he can, and who continually seeks the goodwill of his masters but barks at and bites others. He reproached himself for his "inconsistency, thoughtlessness, lack of discipline, and rashness."

Despite this poor view of himself, Kepler was an excellent student at school and at the University of Tübingen, where he received both bachelor's and master's degrees. Though he studied for a degree in theology, he was introduced to Copernicus's idea that the planets revolve about the sun rather than the earth. The event that launched him on a career as an astronomer was the death of the teacher of mathematics at the Lutheran school in Graz in southern Austria. Kepler, then 22 years old, was recommended as a replacement. One of his duties in Graz was to issue an annual calendar with astrological forecasts.

While teaching at Graz, Kepler hit on what he considered to be the key to understanding the solar system. What is so fascinating about Kepler's theory is that while it was basically mystical and aimed at answering what we would consider today a question without meaning, it nevertheless brought him to contemplate answerable problems of astronomy. Kepler was interested at first in the question of why there were six, rather than twenty or a hundred, planets and why they have their observed spacings from the sun. (Of course we now know of nine planets, and believe that the questions of their number and spacings are beyond the scope of current scientific theories.) While searching for an answer to his question, Kepler one day drew for his class a circle inscribed within a triangle, and another circle circumscribed about the triangle. It occurred to him that if the circles represented planetary orbits, their relative radii might be determined by the condition that both orbits touch the triangle. But why six planets? Then Kepler recalled that there are exactly five regular three-dimensional polyhedra: the tetrahedron, cube, octrahedron, dodecahedron, and iscosahedron. What if each planet moves on a sphere that is either inscribed or circumscribed about one of the regular solids? This, he believed, could account for there being six planets. By trial and error, Kepler found he could nest the solids in an order that could account remarkably well for the observed planetary radii (see Kepler's figure).

This theory, with its mystical reliance on the concept of symmetry, was the substance of Kepler's first book, *Mystery of the Cosmos,* published in 1596. Though medieval in overall conception, the book contained aspects of modern science. It was the first work to accept openly Copernicus's picture of the solar system. In it Kepler also asked important questions of mechanics, such as why planets nearer to the sun travel more rapidly in their orbits than farther planets. He suggested that there must be some force emanating from the sun that they feel more strongly. This foreshadowed his third law (that the squares of the periods of the planets are proportional to the cubes of their distances to the sun) and Newton's law of gravitation. How often has this happened in the history of science: a person searching for the answer for one question turns up a more important question? Kepler noted this later in life: "The roads by which men arrive at their insights into celestial matters seem to me almost as worthy of wonder as those matters in themselves."

Kepler was soon forced to leave his position in Graz when the Counter Reformation of the Hapsburg monarchs closed Lutheran schools. This caused him many financial and family hardships, but it gave him the opportunity to seek more accurate astronomical observations in order to check and improve his model of the solar system. The man who possessed the most accurate data was Tycho Brahe, a Danish astronomer with a passion for exact observations, who had recently become Imperial Mathematicus in

Prague. Kepler set out to join Tycho's staff and to gain access to his treasure of observations on January 1, 1600.

Tycho was of noble birth and had developed into an eccentric in the grand style. As a student he had become obsessed with astronomy, against the advice of his stepfather, who wanted him to study subjects more fitting for a nobleman. At one point he fought a duel with another noble Danish youth over which of them was the better mathematician. The bridge of Tycho's nose was sliced off and he had a silversmith fashion a replacement from an alloy of gold and silver (or perhaps copper—when Tycho's grave was opened in 1900, his skull contained green stains from copper.) Tycho carried a snuffbox filled with an "ointment of glutinous composition" that he rubbed on his silver nose, especially while trying to disconcert opponents during debate. His contribution to astronomy was his insistence on the need for precise and continuing observational data. He made his reputation at the age of 26 by noting a supernova star. This discovery was of significance in the sixteenth century because it contradicted the accepted doctrine that all change is confined to the immediate vicinity of the earth. After this, the King of Denmark was so anxious to keep Tycho in the country that he granted Tycho a beautiful island, the funds to construct a house and observatory, and an annual income among the largest in Denmark.

After twenty years on his island paradise, Tycho left because his high-handed ways had alienated the King as well as others. He traveled about Europe looking for an equally grand setting for his work. He took with him a party of twenty, including family, assistants, servants, and his jester, a dwarf. They brought along a library, furniture, observational instruments, and even a printing press. At length the Emperor Rudolph II appointed him Imperial Mathematicus, and gave him a princely salary and a castle near Prague, and it was there that Kepler arrived in 1600.

It might be expected that the high-living, self-assured Tycho and the impoverished, sensitive Kepler would not hit it off, and in fact they soon fell to quarreling. Kepler wanted to delve into Tycho's accumulated observations, but Tycho instead assigned Kepler to observe the orbit of Mars and to write disparaging tracts about Brahe's scientific enemies. Tycho would tantalize Kepler by revealing important astronomical facts one at a time during the course of a meal. Nevertheless, they stuck it out with one another because Kepler knew he needed Brahe's data, and Tycho seemed to realize that among his followers only Kepler could make a grand synthesis from his observations. Then some 18 months after Kepler arrived in Prague, Tycho died. He had been a guest at a baronial banquet in Prague, and in Kepler's words, "Imbibing somewhat liberally, he felt as if his bladder were going to burst, but he placed civility ahead of health." Modern authorities believe Tycho had prostate, not bladder, trouble. In any case, after eleven painful days he died. He was buried with great pomp, and Kepler, as his successor as Imperial Mathematicus, came into possession of his records of 38 years of observations.

Much of the remaining 29 years of Kepler's life was consumed in completion, analysis, and publication of Tycho's observations. Out of these studies emerged numerous books on astronomy and optics, and many ideas, including the three laws of planetary motions. One of Kepler's works that should be mentioned in this chapter on symmetry is the pamphlet, *On the Six-Cornered Snowflake*. Kepler was the first person to ponder the reason why all snowflakes are hexagonal, and more importantly, the first person to analyze the periodic arrangements of basic units that can give rise to regular macroscopic shapes.

The Meaning of Point-Group Symmetry: A Summary

We have discussed point-group symmetry in terms of symmetry operations that superimpose a symmetric object on itself. There are equivalent ways to describe symmetry that are helpful to have in mind as we discuss the biological applications of the following sections. One is that every asymmetric unit of the point group has an environment that is identical to the environment of every other asymmetric unit. This is true because each asymmetric unit is situated relative to the symmetry element in the same way as every other. As an example, consider the enzyme aspartate transcarbamylase, depicted in Figure 16-6. If you could shrink yourself to atomic scale and examine the view from one of the six active sites, you would have exactly the same view from each of the other five.

We also can think of point-group symmetry in terms of the chemical forces that cause the asymmetric units to adhere to each other, thereby forming a symmetric aggregate. These "bonds" or "binding sets" must have the symmetry of the point group. In fact, in enzymes, where the asymmetric units are polypeptide chains, it is the bonds that cause the aggregate to form and have the symmetry that it does. Thus the structure of an aggregate can be specified in one asymmetric unit by specifying the relative angle of the binding sets. This is the basis of the biological phenomenon of *self-assembly* that we discuss in the following section.

In using the concepts of symmetry, we must remember that real objects never possess perfect mathematical symmetry. In fact, the closer we look, the finer the resolution of our study, the less likely is an object to retain its apparent symmetry. A man appears to have mirror symmetry until we note he parts his hair on the left, or until we x-ray him and find he has a single heart on the left with no mirror image on the right. Similarly, hemoglobin molecules appear to have C_2 (2) symmetry until we apply some spectroscopic technique that is able to detect different thermal vibrations of the heme groups in the different asymmetric units. In the same vein, we speak loosely when we say that a virus has I (532) symmetry because the strand of nucleic acid enclosed within the protein coat does not share this symmetry.

16–3 Group Theory Describes Symmetry Concisely

The symmetry operations of a molecule when considered together have some interesting properties. It turns out that they constitute a *group* in the sense used by mathematicians. These properties can be illustrated by the example of the NH_3 molecule.

The NH_3 molecule contains a 3-fold axis of rotation, and three vertical mirror planes. These are shown in Figure 16-7. Let us consider these symmetry elements to be fixed in space. Then as symmetry operations act on the

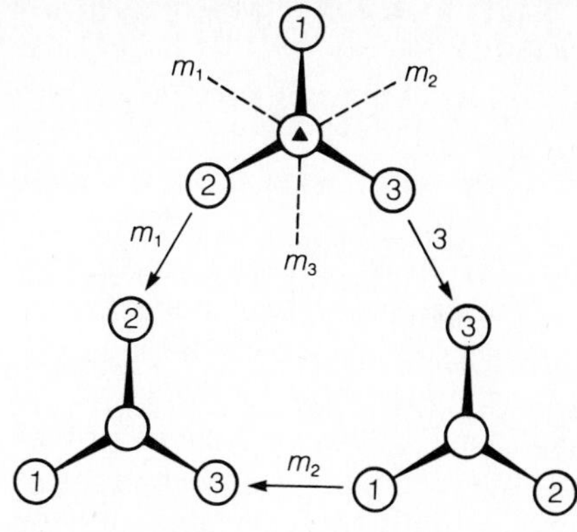

FIGURE 16-7
Symmetry operations of the NH_3 molecule. The N atom is bonded to three H atoms which are labeled for identification. The view of the pyramidal molecule is down the 3-fold axis. The symmetry elements are fixed in place; they are shown with the upper molecule. The lower molecules are produced by symmetry operations on the upper molecule.

molecule, the molecule moves. This is illustrated in the figure for the counterclockwise rotation about the 3-fold axis and for reflection across mirrors m_1 and m_2.

What is the result of carrying out two successive operations on the NH_3 molecule? First we apply the rotation (always counterclockwise) and then we apply the reflection across mirror m_2. By convention we write this process

$$m_2 \times 3$$

with the first operation on the right. You can see from Figure 16-7 that the resulting configuration is the same as would result from the operation m_1 alone. We can express this by the equation

$$m_2 \times 3 = m_1$$

in which the operation m_1 is called the *product* of operations m_2 and 3.

EXERCISE 16-11

Determine the product of operating on NH_3 first with m_2 and then with 3. Write an equation to express the sequence of operations.

ANSWER

$3 \times m_2 = m_3$. Notice that symmetry operations do not necessarily commute. They do, however, obey the associative law of multiplication.

Some sequences of operations take the molecule into a configuration which is identical to its original configuration. The result is called the *identity operation* and is denoted E. An example is two successive operations of any given mirror:

$$m_1 \times m_1 = m_2 \times m_2 = m_3 \times m_3 = E$$

Another example is three successive operations of the 3-fold axis:

$$3 \times 3 \times 3 = 3 \times 3^2 = 3^2 \times 3 = E$$

Here two successive operations of the 3-fold axis are represented by 3^2, which is considered to be an operation itself.

From these examples you can see that for any symmetry operation of the NH_3 molecule, there is another operation, called the *inverse,* that returns the molecule to its original configuration. The inverse of m_1 is m_1, and the inverse of 3 is 3^2.

The symmetry operations of a molecule such as NH_3 can be summarized conveniently in a multiplication table. More generally, the multiplication table summarizes the operations of the point group to which the molecule belongs, in this case C_{3v} ($3m$). The table is arranged in the following way: if an operation given by a symbol in the left-hand column is followed by an operation given by a symbol in the top row, the product of the two operations is given by the symbol at their intersection. The multiplication table for point group C_{3v} ($3m$) is the following:

E	3	3^2	m_1	m_2	m_3
3	3^2	E	m_3	m_1	m_2
3^2	E	3	m_2	m_3	m_1
m_1	m_2	m_3	E	3	3^2
m_2	m_3	m_1	3^2	E	3
m_3	m_1	m_2	3	3^2	E

EXERCISE 16-12

Confirm with the aid of Figure 16-7 that this multiplication table is correct.

We are now ready to see that the symmetry operations of a molecule constitute a mathematical group. A *group* is defined as a set of operations with the following properties:

1. The product of any two operations is equivalent to a single operation which also belongs to the set.
2. The operations obey the associative law of multiplication.
3. The set includes the identity operation, E.
4. The set includes the inverse of every operation.

We have seen that the symmetry operations E, 3, 3^2, m_1, m_2, and m_3 have these properties. They thus comprise a group, point group C_{3v} ($3m$). While this example shows little more than the beautiful symmetry of a point group, it turns out that group theory is a powerful tool in both spectroscopy and x-ray crystallography in describing motions. For further material on applications of group theory in chemistry you can consult Brand and Speakman (1975) or Cotton (1963).

16–4 Self-Assembly and the Structure of Spherical Viruses

Self-Assembling Structures Usually are Symmetric

A fundamental discovery of molecular biology is that many biological structures assemble themselves from simpler precursors. This means that their assembly is achieved without the aid of templates, or assembly enzymes, by the spontaneous decrease in free energy as the units bind to one another. An example is the enzyme aldolase that consists of four polypeptide chains of molecular weight 40,000, arranged in point-group symmetry D_2 (222). When exposed to 4M urea or acid solution of pH 2, the enzyme separates into subunits (chains) and the chains assume random-coil configurations. But when this denatured protein is dialyzed against a renaturing buffer, as much as 70% of its original activity returns. Apparently the amino acid sequence of the polypeptide chain determines the three-dimensional structure of the native subunit, and the subunit determines the structure of the oligomeric enzyme by having binding groups at the proper angles on its surface. Similar *in vitro* demonstrations of self-assembly have been achieved for numerous multisubunit enzymes and viruses.

How can the three-dimensional structure of an oligomer be determined by its constituent subunits? The simplest way is for the binding groups on the surface of the subunit to be oriented to yield a symmetric structure. The argument for this is as follows. The binding groups give a substantial decrease in free energy when in contact with complementary groups on another subunit. The maximum decrease in free energy occurs only when all the binding groups are paired. The complete saturation of binding groups occurs most easily when all bonds are equivalent. Arguments of Section 16-2 show that structures with equivalent bonds are symmetric. The symmetry can be one of the point-groups, or it can be translational symmetry. Linear oligomers with translational symmetry are not suitable for globular enzymes or for the shells of spherical viruses. Therefore, natural selection has chosen point-group symmetry for these biological structures.

The Protein Subunits of Spherical Viruses have Quasi-Equivalent Positions

Let us consider the structure of "spherical" viruses in greater detail. These viruses consist of a core of genetic material (either DNA or RNA) and a protective protein coat composed of subunits. A shell built from many subunits rather than from a single, long polypeptide chain is in some cases a biological necessity, because the genetic material is too limited to code for a very large protein. Crick and Watson (1956) were the first to note that there are very few ways a spherical shell can be built up from subunits; they suggested that the coats most likely belong to the cubic point groups.

However, a problem arises when one considers the very large spherical shells required by most viruses. To cover a shell 280 Å (28 nm) in diameter, about 150 to 250 subunits of molecular weight 20,000 are required. Since many viruses are of this size, it is clear that to enclose the genetic material, often many more than 60 protein subunits are required. However we saw in the preceding section that no point group (other than those with mirrors) has more asymmetric units than the 60 of point group *I* (532). Therefore, subunits in viruses with more than 60 units in the shell cannot be in equivalent positions, and self-assembly by the simple device of saturating identical bonds does not seem possible.

Caspar and Klug found an elegant solution to the problem (which had been adopted by viruses even before the appearance of their paper in 1962). They discovered ways in which a multiple of 60 subunits can be placed in nearly equivalent (they called them *quasi-equivalent*) positions in a closed shell. Because the quasi-equivalent positions are nearly identical, nearly identical bonds form, and self-assembly occurs. The small distortions of bonds from the truly equivalent bonding pattern of point group *I* (532) lower the free energy of interaction. This is known because the shell that forms is the larger one with quasi-equivalent bonding, rather than the smaller equivalent shell with only 60 subunits. If the equivalent bonds were of lower free energy, the smaller shell would form.

What are the quasi-equivalent arrangements of subunits that Caspar and Klug proposed? They all are based on the icosahedron of point group *I* (532), shown in Table 16-3. The icosahedron has 20 triangular faces, and an object belonging to point group *I* (532) can be thought of as having one of its 60 asymmetric units (subunits in this case) in a corner of a triangle. Quasi-equivalent arrangements are based on polyhedra with $20T$ triangular faces. T is the *triangulation number,* which can have only certain values, as explained below. Figure 16-8(*a*) shows polyhedra corresponding to values of T of 1 (the icosahedron), 3, 4, and 7. The number of subunits in a structure characterized by the triangulation number T is $60\ T$: a structure with $T = 1$, has the 60 equivalent positions of point group *I* (532); a structure with $T = 3$ has 180 quasi-equivalent positions. Turnip yellow mosaic virus is a structure of the $T = 3$ type, and is represented in Figure 16-8(*b*).

EXERCISE 16-13

The shell structure of Figure 16-8(*b*) retains point group symmetry *I* (532), although the asymmetric unit is not a single subunit. What is the asymmetric unit?

ANSWER

Point group *I* (532) has 60 asymmetric units. Turnip yellow mosaic virus contains 180 subunits. Therefore there are $180/60 = 3$ subunits in the asymmetric unit. One asymmetric unit is shaded in Figure 16-8(*b*).

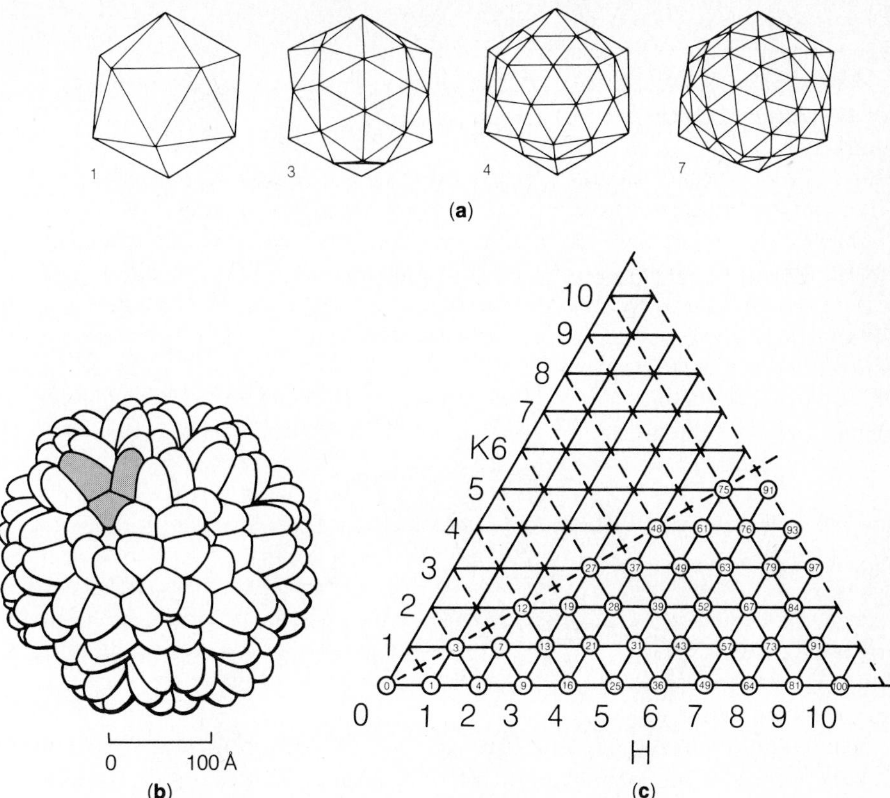

FIGURE 16-8
The theory of quasi-equivalence. (**a**) Polyhedra with 20*T* triangular faces. The *T* number of each is given in the lower left corner. The left panel is the icosahedron ($T = 1$). (From Caspar and Klug, 1962.) (**b**) A drawing of the protein shell of turnip yellow mosaic virus. The shell is formed from 180 identical subunits in quasi-equivalent positions; the structure thus is $T = 3$. (Original drawing by Caspar.) (**c**) The allowed *T* numbers, plotted on the lattice points of a plane triangular net. They are derived by considering the possible ways of folding the triangular net into closed polyhedra, such as those of panel (**a**). This process is described in the text. (From Caspar and Klug, 1963.)

There is a simple way in which you can visualize the arrangements of subunits that are quasi-equivalent. First you note that a plane surface covered by a net of equilateral triangles may be thought of as a set of equivalent 6-fold vertices [Fig. 16-9(*a*)]. Then you imagine that you cut the net along the line $\overline{CA}$, and that you superimpose two adjacent triangles, placing point *C* on point *D*. This produces a 5-fold vertex at *A*, and slightly curves the net. You notice that a subunit placed next to *B* of Figure 16-9(*b*) will not have quite the same environment as a subunit next to *A*; they are quasi-equivalently related. When twelve 5-fold vertices have been created by similar cuts, the net will form a closed surface [Fig. 16-9(*c*)]. The *T* numbers describe the various types of closed surfaces with triangular faces that can be formed in this way.

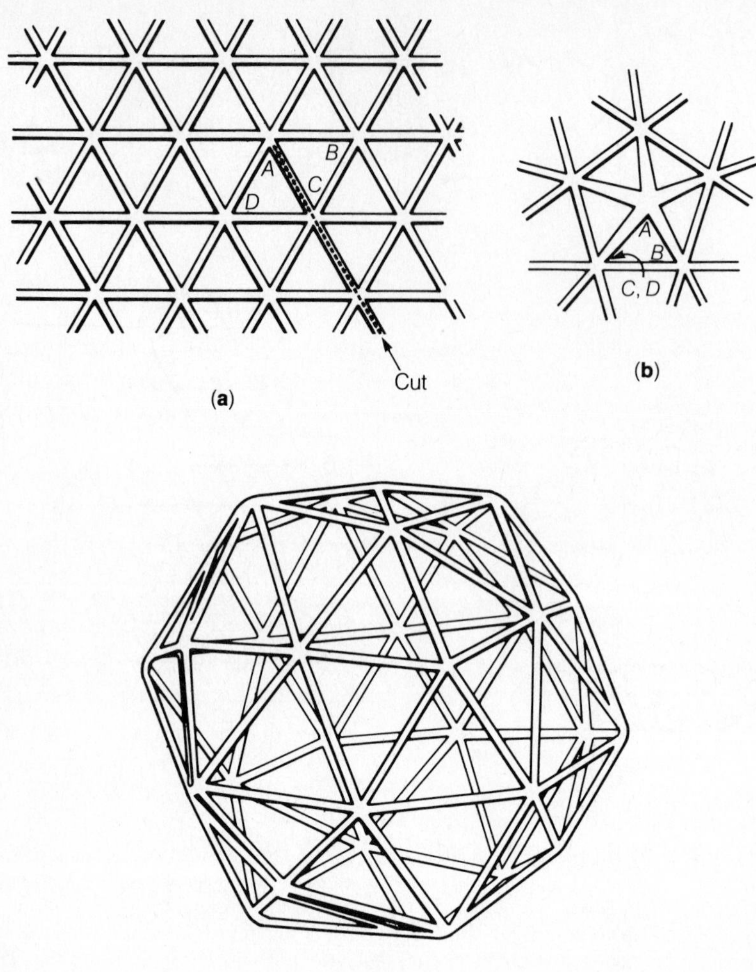

FIGURE 16-9
The folding of a plane hexagonal net (a) into a closed surface with quasi-equivalent positions. To fold the net, a cut is made in (a) along the line that joins A to C. By placing the point C on top of point D, the 6-fold vertex at A is replaced by a 5-fold vertex (b), and some curvature is produced in the net. When twelve similar cuts have been made, the surface shown in (c) results. If one protein subunit is placed at each corner of the 60 equilateral triangles of the surface, a $T = 3$ structure is formed. Note that subunits at a 5-fold vertex will not have exactly the same environment as subunits at a 6-fold vertex. (Redrawn from Wilson, 1966.)

Possible Values for T

The values that T can assume are given by $T = H^2 + HK + K^2$, where H and K are non-negative integers. These values are determined by considering the possible ways to fold a triangular plane net into a closed polyhedron. Suppose at the origin of the net of Figure 16-8(*c*) you form a 5-fold vertex, as described above. If you move to the adjacent lattice point and form a second 5-fold vertex, and then repeat this procedure until you have 12 5-fold vertices, you will produce an icosahedron [Fig. 16-8(*a*)]. This is the $T = 1$ structure, and is represented in Figure 16-8(*c*) by the 1 in the circle on the lattice point adjacent to the origin. Suppose instead that you move two lattice points from the origin before making the second 5-fold vertex, and that you repeat this procedure until you form 12 5-fold vertices. The structure you produce now is the $T = 4$ polyhedron [Fig. 16-8(*a*)]. On the net of Figure 16-8(*c*) this is represented by the 4 at the lattice point two away from the origin. If instead you move two lattice points from the origin and then one to the left (like the Knight's move in chess) you produce the $T = 7$ structure.

In this way you can see that division of the icosahedron into polyhedra containing $60T$ quasi-equivalent positions on their surfaces is described by the T numbers on the lattice points of the net in Figure 16-8(*c*). These are given by the equation $T = H^2 + HK + K^2$, in which H and K are the coordinates on the axes. It is assumed in this figure that there is a 5-fold vertex at the origin, and that the nearest 6-fold lattice point of the net to be converted to another 5-fold vertex is the one with coordinates H, K.

The Caspar-Klug theory also is able to relate electron microscope pictures of viruses to the underlying molecular structure. These pictures, capable of resolving units about 20 Å, show clumps of matter on the virus surface, but cannot show atoms. The clumps are referred to as *morphological units.* Caspar and Klug noted that structures of the same T number could exhibit different morphological units, depending on how the subunits are grouped. Consider the $T = 3$ structure shown in Figure 16-8(*b*). In this structure, the subunits are grouped into 32 morphological units, 20 containing six subunits and 12 containing five subunits. An electron microscope photograph would show only 32 clumps on the surface. The 180 subunits could instead be clustered into 90 dimers (as they are in bushy stunt virus, another $T = 3$ structure), or they could be clustered into 60 trimers. Thus viruses that appear very different in electron micrographs may be built on the same basic symmetry.

16–5 Symmetry along a Line: Helices

Screw Axes are Symmetry Elements that Combine Translation with Rotation

Translation is the symmetry operation of shifting an object a given distance, say d, in a given direction, say the b direction. The group of hands in Figure

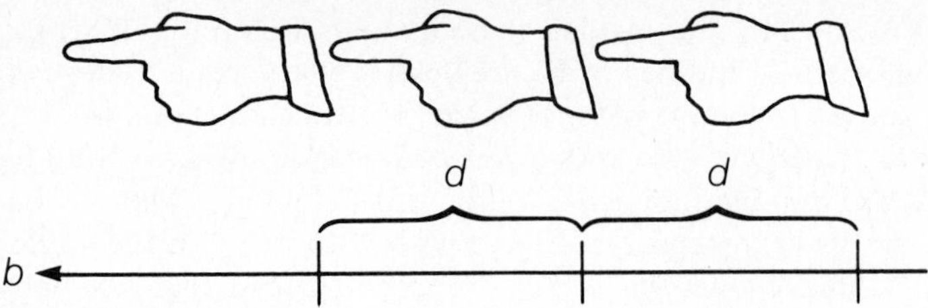

FIGURE 16-10
The symmetry operation of translation by d in the b direction.

16-10 can be superimposed on itself if it is shifted in the b direction by the translation d. We can express this algebraically by

$$T: \quad y' = y + td \tag{16-4}$$

in which y is the coordinate of some point in an asymmetric unit (say the fingertip) and y' is the coordinate of the equivalent point t asymmetric units away. If we consider the hands of Figure 16-10 to constitute a one-dimensional crystal, then each hand occupies one unit cell and the distance d is the *unit-cell edge* of the crystal.

Combining the symmetry operations of translation and rotation produces a *helix*. Let us first consider special helices called *screw axes*. These are symmetry elements of crystals that are helices with an integral number of asymmetric units per turn of helix. Then we will consider more general helices, of the types that describe the structures of fibrous proteins, DNA, and viruses.

A *screw axis*, n_m, is the symmetry element describing rotation by $2\pi/n$ radians about an axis and then translation along the axis of m/n of the unit-cell edge. Figure 16-11 illustrates 2_1, 4_1, and 4_2 screw axes. The action of a 2_1 axis is rotation of the asymmetric unit by π radians about the axis and then translation of it by half the unit-cell edge. A repeat of this action places an asymmetric unit one unit cell above the first. The screw axes found in crystals include 2_1, 3_1, 3_2, 4_1, 4_2, 4_3, 6_1, 6_2, 6_3, 6_4, and 6_5. All these screw axes share the property of an integral number of asymmetric units (n) in one turn of the helix.

EXERCISE 16-14

From the definition of a 4_1 axis, show that it describes the symmetry of the hands in Figure 16-11(b).

EXERCISE 16-15

From the definition of a 4_2 axis, show that it describes the hands in Figure 16-11(c).
Hint: Notice that an asymmetric unit and one related to it by translation are equivalent.

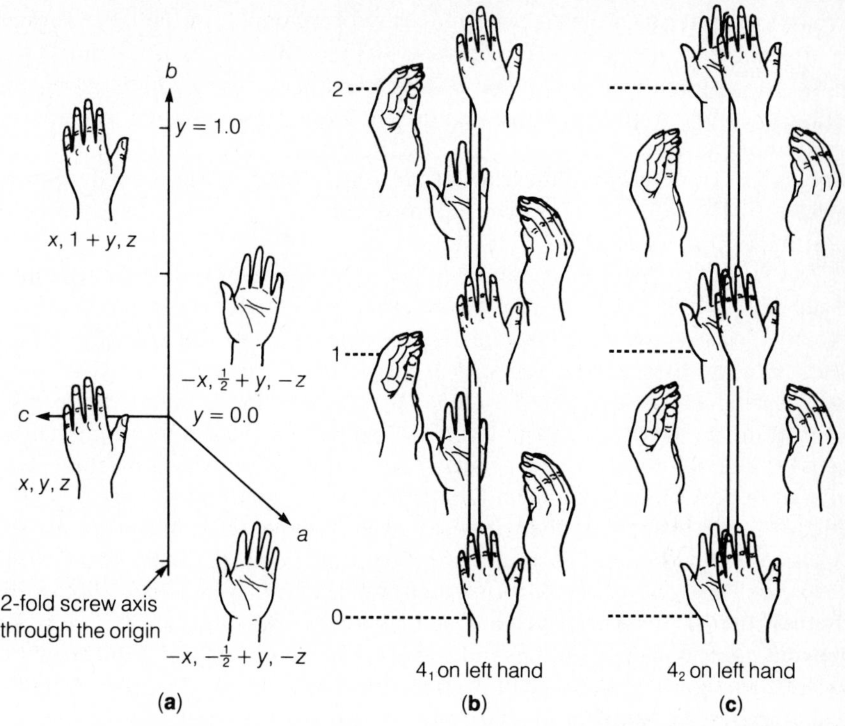

FIGURE 16-11
Some screw axes, with the asymmetric unit being one left hand. **(a)** 2_1; **(b)** 4_1; **(c)** 4_2. (Redrawn from *Crystal Structure Analysis: A Primer* by Jenny Pickworth Glusker and Kenneth N. Trueblood. Copyright © 1972 by Jenny Pickworth Glusker and Kenneth N. Trueblood. Reprinted by permission of Oxford University Press, Inc.)

To get an algebraic expression for an n_m screw axis we combine Equations 16-1 and 16-4. If the screw axis is parallel to b, we obtain

$$n_m\colon\ x, y, z \rightarrow \left(\cos\frac{2\pi}{n}\right)x - \left(\sin\frac{2\pi}{n}\right)z, \quad y + \frac{m}{n}b, \quad \left(\sin\frac{2\pi}{n}\right)x + \left(\cos\frac{2\pi}{n}\right)z \qquad (16\text{-}5)$$

in which b is the unit-cell edge.

Many Biological Structures are Helical

Helices which are symmetry elements of crystals are restricted to having an integral number of asymmetric units per turn. This restriction does not apply to helices in general. Nearly all extended polymers are helical, because a helix is the general form for an extended, repeated structure. Helices can be self-assembling, because they are formed by repetition of one type of bond.

In describing general helices, it is convenient to use the number of units per turn, n^*, which is the ratio of the rise along the helix axis per complete turn, P (the pitch), to the rise per unit, p. For example, in the helix formed by protein subunits of tobacco mosaic virus (TMV) (Fig. 16-12 and Table 16-4), each subunit is a distance $p = 1.4$ Å above its neighbor along the helical axis. The total rise, P, for one turn of helix is 23 Å, so that the number of subunits per turn is $23/1.4 = 16.3$. In other words, after a single turn, there is no subunit related by pure translation to a subunit in the preceding turn. For the TMV helix to have two subunits related by translation, the helix must make three turns (Table 16-4).

TMV was the first virus found to have the property of self-assembly. When its protein subunits and RNA are mixed under proper conditions, infective viruses are formed. This is possible because the subunits of the helix, like the asymmetric units of a point group, occupy equivalent positions. Each has the same environment as all the others, so that if one forms a bond to the adjacent subunit in the helix, so will all the others. Thus helical viruses, like spherical viruses, have their structures encoded by the placement of bond-forming groups on the surface of their individual subunits. The helical viruses differ from the spherical viruses in that their size is not limited by the number of subunits required to form a spherical shell. What is the factor that prevents TMV from forming infinitely long helices? This is one function of the RNA that is enclosed by the protein subunits. The RNA strand is some 6300 nucleotides in length; each protein subunit binds to three nucleotides, limiting the virus to 2100 subunits in length, a total rise along the helical axis of about 3000 Å (300 nm).

The parameters for three other biological helices are given in Table 16-4. DNA is not a true helix if the identity of the four nucleic acid side chains is considered, for then there is no repetition of a single asymmetric unit. If we consider only the sugar-phosphate backbone, however, then each strand of

FIGURE 16-12
A drawing of part of the helical structure of TMV. Each shoe-shaped protein subunit is bound to three RNA nucleotides. Part of the RNA chain is shown stripped of its protein subunits in a configuration it could not maintain without them. Each turn of 16.3 protein subunits is closely related to the disk structure of Figure 16-4. (From Klug and Caspar, 1960.)

TABLE 16-4 *Properties of Some Helices*

HELIX	UNITS/TURN OF HELIX $n^* = u/t$	RISE ALONG HELIX AXIS/UNIT p	PITCH = RISE/TURN $P = pn^*$	NO. UNITS/ REPEAT u	TURNS/ REPEAT t	TRANSLATIONAL REPEAT (RISE/REPEAT) $c = Pt = pu$	FRAC-TIONAL TWIST OF HELIX/ UNIT $\left(= \frac{1}{n^*}\right)$
4_1 axis	4	$c/4$	c	4	1	c	0.25 = 90°
n_m screw axis	n	$\frac{c\dagger}{n}$	c	n	1	c	$\frac{360°}{n}$
α-helix	3.6	1.5 Å	5.4 Å	18	5	27 Å	0.278 = 100°
β-protein	2	3.5 Å	7 Å	2	1	7 Å	0.5 = 180°
DNA (B form)	10	3.4 Å	34 Å	10	1	34 Å	0.1 = 36°
TMV	16.3	1.4 Å	23 Å	49	3	69 Å	0.061 = 22°

† Average rise per unit. Actual rise per unit = mc/n.

DNA is a helix, and the two intertwined strands are related by a 2-fold axis of rotation perpendicular to the helical axes. Structural proteins also are helical. Many of these are formed from two basic helical structures, the α helix, and the β structure that forms pleated sheets (Table 16-4).

16–6 Symmetry in a Plane and in Three Dimensions: Designs and Crystals

Any Periodic Structure may be Viewed as the Combination of a Motif with a Lattice

Designs are periodic structures in two dimensions, and crystals are periodic structures in three dimensions. The symmetries of all structures fall within a limited number of types: these are the 17 two-dimensional plane groups and the 230 three-dimensional space groups. For the biological substances with no reflection or inversion operations, only five plane groups and 65 space groups are possible. The limitation on the number of possible symmetries arises in the same way as the limitation of possible point groups. It is that all other repetition schemes place asymmetric units in non-equivalent positions. Problem 16-15, which demonstrates that crystals cannot contain 5-fold axes of symmetry, gives some insight into these limitations.

Any periodic structure can be generated by placing a *motif* at every point of a *lattice*. A lattice is a rule for translation, and a motif is the object that is translated. In both Figures 16-10 and 16-13, the motif is a right hand. The lattice in Figure 16-10 is one-dimensional and that in Figure 16-13 is two-dimensional. A three-dimensional lattice with the same motif also could be constructed. However, it is not often that the motif is an asymmetric object such as the right hand. More frequently the motif itself has the symmetry of one of the point groups, or one of the screw axes that we discussed in preceding sections. In this case, the periodic structure contains translational symmetry plus rotational (or reflection or inversion) symmetry. Then we can think of the periodic structure as being built up in two steps. First, a motif is generated from the asymmetric unit by the symmetry operations of the point group. Second, the structure is generated from the motif by the translational symmetry operations of the lattice:

$$\text{Asymmetric unit} \xrightarrow[\text{symmetry}]{\text{point-group}} \text{motif} \xrightarrow[\text{symmetry}]{\text{lattice}} \text{structure}$$

The symmetry of the structure is described by a *plane group* if it is two-dimensional or by a *space group* if it is three-dimensional. The plane groups and space groups can be generated by combining all types of lattice with all

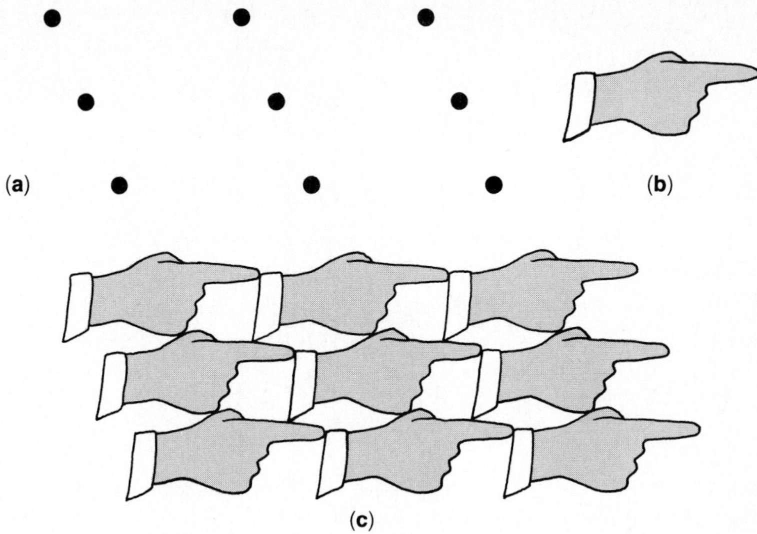

FIGURE 16-13
A lattice (a) specifies the translation of a motif (b) to give a structure (c). Here the lattice is primitive; thus, there is one motif per unit cell. The motif has no internal symmetry in this example. (Redrawn from Eisenberg, 1970.)

types of motif symmetry. We will consider a few examples here; for a more detailed discussion, you should turn to an indispensable guide on symmetry, *The International Tables for Crystallography* (1969).

Plane Groups Describe all Periodic Two-Dimensional Designs

The five two-dimensional lattices, sometimes called *nets,* are shown in Figure 16-14. Each has the characteristic that all its lattice points have the same environment. There are no other arrays of points in two dimensions with this property. Each lattice is characterized by three numbers: the lengths of the two unit-cell edges, a and b; and the angle between them, γ. A *unit cell* is outlined in each of the lattices; it is defined as the smallest unit (having the full available symmetry) from which the lattice can be generated by translations. The phrase in parentheses is necessary to permit the centered lattice, as we shall see.

Except for the centered rectangle lattice, all the lattices are *primitive*—that is, they have lattice points only at their corners. The symbol for the primitive lattice is P; that for the centered lattice is C. It is possible to divide the centered lattice into a primitive lattice with smaller, parallelogram-shaped unit cells, but the smaller cells do not have the full symmetry of the rectangle lattice with its right angle. This point will become clearer after we discuss the combination of motifs with lattices.

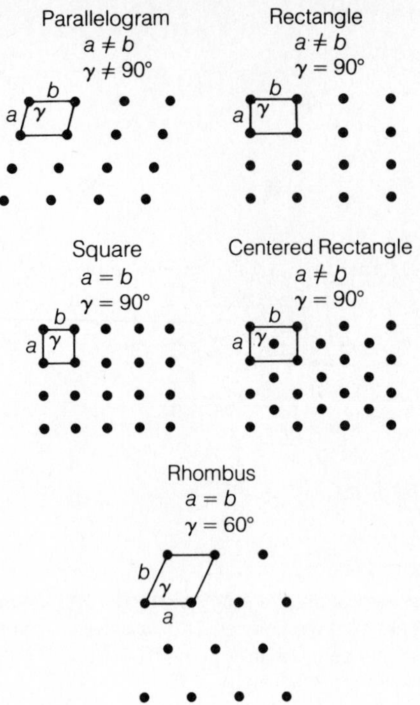

FIGURE 16-14
The five two-dimensional lattices.

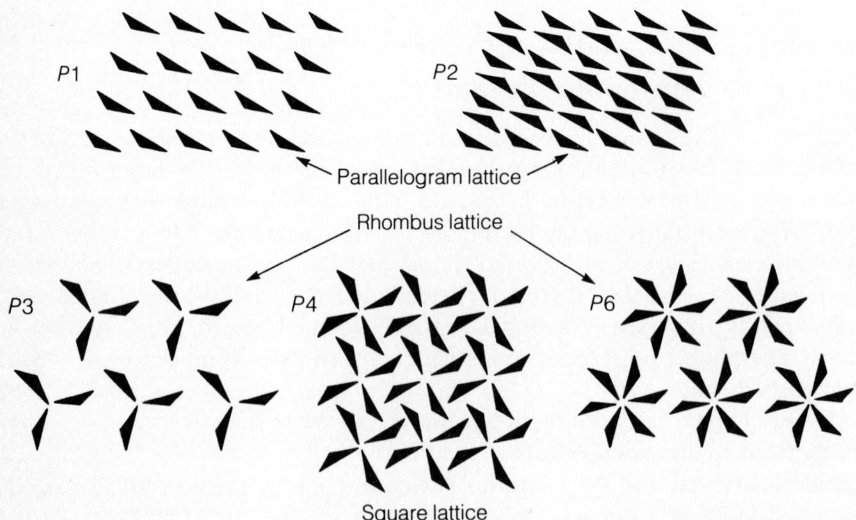

FIGURE 16-15
The five two-dimensional plane groups for biological structures. The asymmetric unit is a triangle in every case; the motif is some group of triangles (1, 2, 3, 4, or 6). (Adapted from Brand and Speakman, 1975.)

EXERCISE 16-16

How many lattice points per unit cell are there in each of the two-dimensional lattices?

ANSWER

In the primitive lattices, each cell is bounded by four lattice points. Each of these points is shared by the four cells around it. Thus there is $4 \times \frac{1}{4} = 1$ lattice point per primitive cell. The centered cell has one lattice point from its corners, plus one from its center, for a total of two.

EXERCISE 16-17

Can a triangular unit cell be selected for the rhombus lattice?

ANSWER

No, because the lattice cannot be generated purely by translations of this unit.

We now can form a two-dimensional periodic design by placing a motif on each lattice point. The *plane group* describes the symmetry of the design and is denoted by the lattice symbol followed by the Hermann-Mauguin symbol for the point group of the motif. The design in Figure 16-13 is built on a primitive (parallelogram) lattice; the motif is a hand and belongs to point group C_1 (1). Therefore the plane group of the entire design is $P1$.

A second example is given by Figure 16-2. If you study this figure you will see that it is built on a primitive lattice. Although you may be tempted to select a lattice with $a = b$ and $\gamma = 90°$, there is no need to do so because no symmetry of the unit cell is lost by selecting a parallelogram lattice. The motif is a group of two white lizards and two black lizards. If you count parts of lizards, you will find that there is one motif per unit cell, as there must be in a primitive lattice. The motif symmetry is C_2 (2), so that the plane group is $P2$. The asymmetric unit of the design is one white lizard plus one black lizard. You can think of a second pair of lizards being generated from the first by the 2-fold axis.

Now let us suppose that all lizards are the same shade. This causes the symmetry of the motif to increase: the points where the right forepaws and the left hindpaws meet are 4-fold axes. Thus the new motif symmetry is C_4 (4), the new asymmetric unit is a single lizard, and the plane group becomes $P4$. Notice that if the lattice points are taken at either of the sets of 4-fold axes, the result is a square lattice. This illustrates an important point. Not all motif symmetries can be associated with all lattices. A motif of point group C_4 (4) demands a square lattice. In fact, the square lattice is the result of the motif having the same length on its two edges, *and it has the same length because of the 4-fold axis*. A parallelogram lattice is no longer correct because its unit cell does not have the full symmetry. Similarly, 3- and 6-fold axes

TABLE 16-5 *The 65 "Biological" Space Groups*

CRYSTAL SYSTEM	LATTICE	MINIMUM SYMMETRY OF UNIT CELL	UNIT CELL EDGES AND ANGLES[a]	DIFFRACTION PATTERN SYMMETRY[b]	SPACE GROUPS[c]
Triclinic	P	None	$a \neq b \neq c$	$\bar{1}$	$P1$
			$\alpha \neq \beta \neq \gamma$		
Monoclinic	P	2-fold axis parallel to **b**	$a \neq b \neq c$	$2/m$	$P2$, $P2_1$
	C		$\alpha = \gamma = 90°$		$C2$
			$\beta \neq 90°$		
Orthorhombic	P	3 mutually perpendicular 2-fold axes	$a \neq b \neq c$	mmm	$P222$, $P2_12_12_1$, $P222_1$, $P2_12_12$
	C		$\alpha = \beta = \gamma = 90°$		$C222$, $C222_1$
	I				[$I222$, $I2_12_12_1$]
	F				$F222$
Tetragonal	P	4-fold axis parallel to **c**	$a = b \neq c$	$4/m$	$P4$, ($P4_1$, $P4_3$), $P4_2$
	I		$\alpha = \beta = \gamma = 90°$		$I4$, $I4_1$
				$4/mmm$	$P422$, ($P4_122$, $P4_322$), $P4_222$
					$P42_12$, ($P4_12_12$, $P4_32_12$), $P4_22_12$
					$I422$, $I4_122$
Trigonal/rhombohedral	R^d	3-fold axis parallel to **c**	$a = b = c$	$\bar{3}$	$R3$
	P^d		$\alpha = \beta = \gamma \neq 90°$		$P3$, ($P3_1$, $P3_2$)
				$\bar{3}m$	$R32$
					[$P321$, $P312$]
					[($P3_121$, $P3_221$), ($P3_112$, $P3_212$)]

Hexagonal	P	6-fold axis parallel to **c**	$a = b \neq c$	$6/m$	$P6$, $(P6_1, P6_5)$
			$\alpha = \beta = 90°$		$P6_3$, $(P6_2, P6_4)$
			$\gamma = 120°$	$6/mmm$	$P622$, $(P6_122, P6_522)$
					$P6_322$, $(P6_222, P6_422)$
Cubic	P	3-fold axes along cube diagonals	$a = b = c$	$m3$	$P23$
			$\alpha = \beta = \gamma = 90°$		$P2_13$
	I				$[I23, I2_13]$
	F				$F23$
				$m3m$	$P432$, $(P4_132, P4_332)$
					$P4_222$
					$I432$, $I4_132$
					$F432$, $F4_132$

[a] a, b, and c are lengths of unit cell edges; α, β, and γ are the angles between b and c, c and a, and a and b, respectively.
[b] Symbols: number with an overbar, a rotary inversion axis; m, a mirror plane; $2/m$, a mirror plane perpendicular to a 2-fold axis; and $6/m$, a mirror plane perpendicular to a 6-fold axis.
[c] Pairs of space groups in parentheses differ from each other only in that they are enantiomorphs. Space groups enclosed in brackets (and also those in parentheses) cannot be distinguished from one another by systematic extinctions of reflections in the diffraction pattern. All other space groups can be assigned on the basis of the diffraction pattern.
[d] The rhombohedral system is often regarded as a subdivision of the hexagonal system, and unit cells in this system may be chosen on either hexagonal or rhombohedral axes.

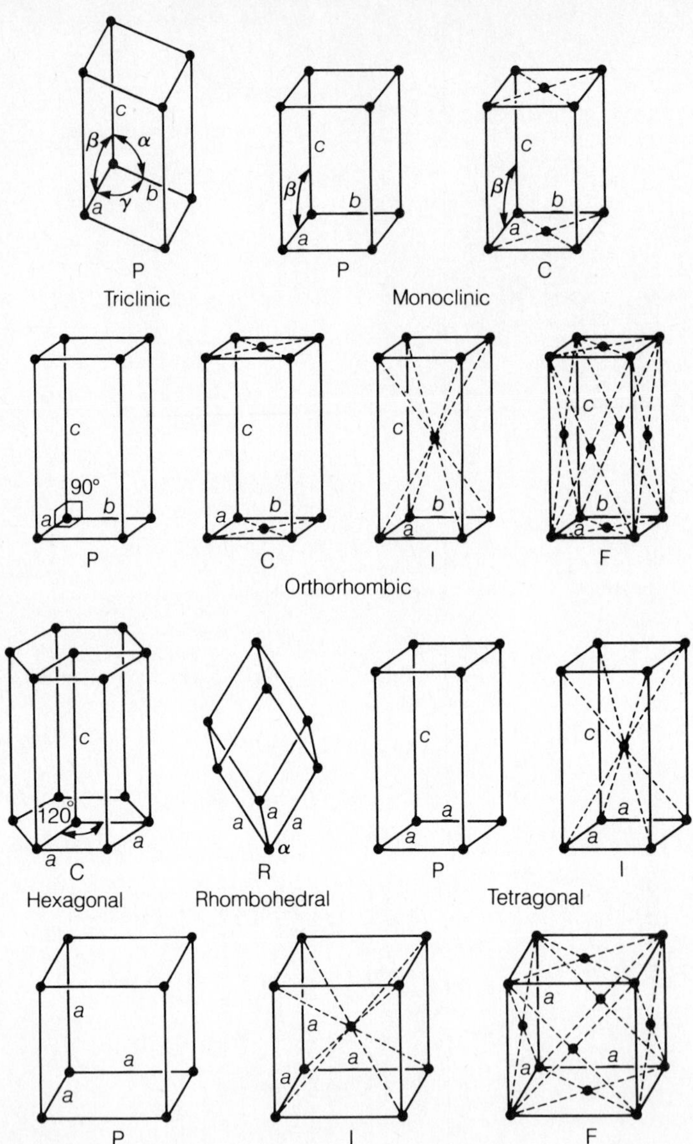

FIGURE 16-16
The 14 three-dimensional Bravais lattices. (Adapted from Wilson, 1966.)

demand a rhombus lattice. It turns out that, because of such restrictions, the only two-dimensional plane groups for "biological" structures (those with no mirrors or centers of inversion) are *P*1, *P*2, *P*3, *P*4, and *P*6. These are illustrated in Figure 16-15. The rectangle lattices are found only with motifs having mirrors as symmetry elements.

Space Groups Describe the Symmetries of all Possible Crystals

The principles of space groups are exactly the same as those of the plane groups. In three dimensions, however, there are 14 possible lattices, and many more motif symmetries, thereby leading to a total of 65 space groups for biological structures. The 14 three-dimensional lattices, called *Bravais lattices* after the scientist who first determined them, are shown in Figure 16-16. As in the two-dimensional lattices, all points in a lattice have the same environment, and there are no other arrays of points satisfying this requirement. Six of the lattices are primitive, having $8 \times \frac{1}{8} = 1$ lattice point, and thus one motif, per unit cell. The others are various centered lattices, having 2, 3, or 4 lattice points per cell. Restrictions on cell edges and angles are given in Table 16-5.

In three dimensions, the motifs may have cyclic, dihedral, or cubic-point symmetry, or may have the symmetry of one of the screw axes, or some combination of screw axes. When all possible motif symmetries are combined with the lattices, subject to the restrictions of the sort we discussed above for the two-dimensional lattices, the 65 biological space groups of Table 16-5 are the result. Any crystal of a biological substance must be characterized by one of these symmetries. A crystal's space group can be determined by the x-ray diffraction methods described in the following chapter.

QUESTIONS FOR REVIEW

1. Define *symmetry operation* and *symmetry element*. Describe inversion, rotation, translation, and reflection.
2. Describe the cyclic point groups. Give the Schoenflies and Hermann-Mauguin notation for these groups.
3. Define *asymmetric unit, protomer* and *oligomer*.
4. Draw the symbols for 2-, 3-, 4-, and 6-fold axes.
5. Write algebraic expressions for reflection, inversion, and 2-fold rotation.
6. Describe the dihedral point groups. Give the Schoenflies and Hermann-Mauguin notations for these groups.
7. What are the three cubic point groups for biological molecules? Why is point group 532 suitable for shell-like structures?
8. Define a group in the mathematical sense. Describe how the symmetry operations of a point group satisfy this definition.
9. Define self-assembly, quasi-equivalence, *T*-number, and morphological unit. Outline the Caspar-Klug theory of virus structure.
10. What is a helix? Why are helices common in biological structures? What parameters are used to describe helical symmetry?

11. Define lattice, motif, unit cell, and primitive. What is the difference between the lattice symmetry and the plane (or space) group?

12. How is a structure built up from the asymmetric unit by symmetry operations?

13. Why is there a limited number of point groups and plane groups?

PROBLEMS

1. Are the following statements true or false?

   (a) Both the operations of inversion and reflection take a molecule into its enantiomorph.

   (b) *Symmetry operation* and *symmetry element* are different terms for exactly the same concept.

   (c) All objects with dihedral point group symmetry contain more than two 2-fold axes.

   (d) All objects with cubic point group symmetry contain at least one 4-fold axis.

   (e) Only cubic point groups include both 2-fold and 4-fold axes of rotation.

   (f) The symmetry operations of a point group necessarily commute.

   (g) All products of symmetry operations of a point group are operations of the group.

   (h) A spherical virus containing more than 60 identical protein subunits does not have point group symmetry.

   (i) A spherical virus containing more than 60 identical protein subunits does not have all subunits in equivalent positions.

   (j) The number of asymmetric units per turn of a helix equals the ratio of its pitch to its rise per asymmetric unit along the helix axis.

   (k) A screw axis of the type found in crystals must have an integral number of asymmetric units per translational repeat.

   (l) Two motifs related by translation do not necessarily have the same orientation.

   (m) There are five two-dimensional plane groups for "biological molecules."

Answers: (a) T. (b) F. (c) T. (d) F, but all contain 3-fold axes. (e) F. (f) F. (g) T. (h) F. (i) T. (j) T. (k) T. (l) F. (m) T.

2. Add the appropriate word.

   (a) A symmetry ________ is the geometric entity about which a symmetry operation takes place.

   (b) A 5-fold rotation is the symmetry operation in which a rotation of ________ radians is made about an axis.

(c) A 2-fold rotation followed by a reflection across a mirror perpendicular to the axis of rotation is equivalent to ______.

(d) ______ point groups contain n 2-fold axes perpendicular to an n-fold axis.

(e) Among the properties of a mathematical group are that both the ______ and ______ operations are members of the group.

(f) Two protomers that have nearly but not exactly the same environment are said to be ______.

(g) An aggregate formed from several protomers is an ______.

(h) The ______ of a helix is its rise along the axis per turn.

(i) In the two-dimensional plane groups, a 4-fold axis of symmetry demands a ______ lattice.

(j) A lattice with more than one lattice point per unit cell is said to be ______.

(k) A unit cell is the smallest unit, having the full available symmetry, from which the structure can be generated by ______ symmetry operations.

(l) Proteins and other biological molecules are not characterized by symmetry operations of ______ and ______.

Answers: (a) Element. (b) $2\pi/5 = 72°$. (c) Inversion. (d) Dihedral. (e) Inverse, identity. (f) Quasi-equivalent. (g) Oligomer. (h) Pitch. (i) Square. (j) Centered. (k) Translational. (l) Reflection and inversion.

PROBLEMS RELATED TO EXERCISES

3. (Exercise 16-3) Show that L-alanine is related to D-alanine by a mirror and by a center of inversion. Alanine is $^+H_3N—CH(CH_3)—COO^-$.

4. (Exercise 16-4) Which point-groups best characterize the following objects?

   (a) A snowflake.

   (b) The DNA double-helical oligomers of Figure 4-8(*c*).

   (c) The glutamine synthetase molecule from *E. coli*, top and side views of which are shown in Figure 5-5.

   (d) The *Radiolaria* in Figure 16-1, assuming they are biological (non-enantiomorphic) structures.

5. (Section 16-2) Figure 16-3 shows that the operation of reflection across a mirror perpendicular to b is the equivalent of a 2-fold rotation about b followed by inversion. Prove this algebraically. (The operation of an n-fold rotation followed by inversion is sometimes called n-fold rotary-inversion, and is denoted $\bar{n}$.)

6. (Section 16-2) The plant protein concanavalin A exhibits symmetry D_2 (222). Each of the four protomers binds one ion of Mn^{2+}. Suppose one of these ions sits at coordinates x, y, z, with respect to an origin at the intersection of the 2-fold axes. What are the coordinates of the three other Mn^{2+} ions?

7. (Exercise 16-6) The symmetry elements of point group C_{3v} (3*m*) are shown in Figure 16-7. Suppose that the *x*-axis is horizontal and that the *y*-axis is vertical. Write algebraic expressions for the following symmetry operations:

   (a) Reflection across m_3.

   (b) 3-fold rotation.

   (c) Reflection across m_2.

8. (Exercise 16-8) Using the number 7 as an asymmetric unit, sketch an object with D_3 (32) symmetry. Indicate the top view of the 7 by a light gray line and the bottom by a heavy black line.

9. (Exercise 16-12) For point group C_4 (4),

   (a) Write algebraic expressions for the symmetry operations E, 4, 4^2, and 4^3.

   (b) Determine the multiplication table.

10. (Exercise 16-15) For 6_1, 6_3, and 6_5 screw axis,

   (a) Sketch each screw axis.

   (b) Explain how the 6_1 and 6_5 screw axes are related.

   (c) Which screw axes contain a 3-fold axis?

11. (Exercise 16-17) Why is there no centered square lattice?

OTHER PROBLEMS

12. Glyceraldehyde-3-phosphate dehydrogenase from muscle normally is a four subunit enzyme. However, at high ionic strength it dissociates into dimers. Does this suggest that the native molecule belongs to point group D_2 (222) or to C_4 (4)? Why?

13. Suppose you study the various subunit dissociation products of aspartate transcarbamylase under different solvent conditions. What different combinations of subunits might you expect to find?

14. Each protein subunit of tobacco mosaic virus binds lead, when the virus is soaked in a solution of lead acetate. Suppose that the perpendicular distance from the Pb atom in one subunit to the helical axis is 25Å. Then, by selecting a coordinate system in which the helical axis is *z*, and the Pb atom lies on the *x*-axis, the coordinates of the Pb atom are 25 Å, 0, 0. Using the helical parameters for TMV given in Table 16-4, determine the coordinates of the Pb atoms in the six adjacent subunits surrounding the one with the Pb atom at 25 Å, 0, 0.

15. Prove that a 5-fold rotation axis cannot be a symmetry element of a plane group (it also is forbidden in space groups). *Hint:* Assume it is a symmetry element and show that equivalent points cannot all lie on a lattice.

16. What is the highest *n*-fold rotation axis that can lie perpendicular to the axis of a helix?

17. Suppose that you have grown a crystal of a globular protein.

(a) Can the asymmetric unit of the crystal contain less than one polypeptide chain? Explain.

(b) Can the asymmetric unit contain less than one molecule? More than one molecule? Explain.

(c) Can the unit cell contain less than one molecule? More than one molecule? Explain.

18. Repressors are proteins that bind to DNA, preventing a gene from expressing its genetic message. Two of the most thoroughly studied repressors, *lac* and λ, each are believed to bind to DNA in tetrameric form. H. M. Sobell [*Proc. Natl. Acad. Sci.*, *69*, 2483 (1972)] noted that if these proteins belong to point group C_4 (4), then the region of DNA to which they bind also will have *4*-fold symmetry. Although it is not immediately obvious how a region of double-stranded DNA might acquire this symmetry, Sobell postulated a way: a region of about 14 complementary base pairs, in which the sequence has two-fold symmetry $\left(\text{say, } \begin{matrix} 3' \ldots \text{TAACGGT ACCGTTA} \ldots 5' \\ 5' \ldots \text{ATTGCCA TGGCAAT} \ldots 3' \end{matrix}\right)$ undergoes a tandem genetic duplication, giving rise to two contiguous runs of the same sequence, containing three *2*-fold axes. Then, rearrangement of hydrogen bonds can give the needed *4*-fold symmetry to the local region of DNA. Draw this region of DNA in its *4*-fold configuration. (Notice that the sequence above is a palindrome.)

QUESTIONS FOR DISCUSSION

19. Suppose that Academic Protection Inc. hires you as an expert in symmetry to design an office lock that will accept its key either right side up or upside down so that professors returning to their offices in darkened corridors can enter them quickly. What point group do you recommend for lock and key?

20. What are the characteristics of a biological assembly process, in terms of energy changes, that dictate whether
    (i) Self-assembly is possible.
    (ii) Enzymatic control is required.
    (iii) Control of a template is required.

21. What is the point group of furniture in the room you are in? What is the plane group of the wallpaper or the tile floor?

22. Notice that the arrangement of leaves on plant stems is often symmetric. Why might this be so?

REFERENCES

Brand, J. C. D. and Speakman, J. C. 1975. *Molecular Structure: The Physical Approach*, 2nd ed., revised by J. K. Tyler and J. C. Speakman. London: Edward Arnold.

Caspar, D. L. D., and Klug, A. 1962. *Cold Spring Harbor Symposium 27*, 1.

Caspar, D. L. D. and Klug, A. 1963. *Virus, Nucleic Acids and Cancer* (17th M. D. Anderson Symposium). Williams & Williams Co., 27.

Champness, J. N., Bloomer, A. C., Bricogne, G., Buttler, P. J. G., and Klug, A. 1976. *Nature 259,* 20.

Cotton, F. A. 1964. *Chemical Applications of Group Theory.* New York: Interscience.

Crick, F. H. C., and Watson, J. D. 1956. *Nature 177,* 780.

Crowther, R. A. 1971. *Endeavour XXX,* 124.

DeRosier, D. J. 1971. *Cold Spring Harbor Symposia on Quantitative Biology XXXVI, 199.*

Durham, A. C. H., Finch, J. T., and Klug, A. 1971. *Nature New Biology 229,* 37.

Eisenberg, D. 1970. *The Enzymes,* ed. P. D. Boyer, *1,* 1. New York: Academic Press.

Evans, D. R., Warren, S. G., Edwards, B. F. P., McMurray, C. H., Bethge, P. H., Wiley, D. C., and Lipscomp, W. N. 1973. *Science, 179,* 683.

Haeckel, E. 1974. *Art Forms in Nature.* New York: Dover Publications, Inc. (Reproductions of plates from Haeckel's book of 1904.)

Haschemeyer, R. H. 1970. *Adv. Enzymol. 33,* 71.

Klug, A. 1969. In *Nobel Symposium, 11,* 425, ed. A. Engström and B. Strandberg. New York: Wiley.

Klug, A., and Caspar, D. L. D. 1960. *Adv. Virus Res. 7,* 225.

Monod, J., Wyman, J., and Changeux, J-P. 1965. *J. Mol. Biol. 12,* 88.

FURTHER READING

Bernal, I., Hamilton, W. C., and Ricci, John S. 1972. *Symmetry.* San Francisco: W. H. Freeman and Co. The symmetries of numerous plane groups and point groups illustrated with stereoscopic drawings.

Caspar, D. L. D. 1965. "Design Principles in Virus Particle Construction," in *Viral and Rickettsial Infections of Man,* 4th ed., eds. F. Horsfall and I. Tamm. New York: Lippincott. A beautifully illustrated, fairly elementary explanation of virus structure and self-assembly.

Dickerson, R. E. 1964. "X-ray Analysis of Protein Structure", in *The Proteins,* ed. H. Neurath. New York: Academic Press. An excellent summary of work on helical structures.

Glusker, J. P., and Trueblood, K. N. 1972. *Crystal Structure Analysis: A Primer.* New York: Oxford University Press. A very clear introduction to symmetry and x-ray diffraction.

MacGillavry, C. H. 1965. *Symmetry Aspects of M. C. Escher's Periodic Drawings.* Utrecht: International Union of Crystallography. Escher's periodic drawings with comments by a crystallographer.

Matthews, B. W. and Bernard, S. A. 1973. *Ann. Rev. Biophys. Bioeng. 2,* 257. A recent summary of work on multisubunit enzymes.

Thompson, D. W. 1961. *On Growth and Form.* Cambridge: Cambridge University Press. A fascinating treatise on symmetry in organismic biology.

Weyl, H. 1952. *Symmetry.* Princeton, N.J.: Princeton University Press. A masterful essay on symmetry in nature and the arts.

Wilson, H. R. 1966. *Diffraction of X-rays by Proteins, Nucleic Acids, and Viruses.* London: Edward Arnold. A very good introduction to symmetry and x-ray diffraction on about the level of this text.

CHAPTER 17

X-ray Diffraction and the Determination of Molecular Structure

X-ray, neutron, and electron diffraction are the most powerful tools for determining the structures of molecules. They can provide information on the arrangements of atoms in gases, liquids, glasses, and fibers, but the information is most complete when crystals of the molecules can be studied. In this case, the method is called *crystallography,* and it is mainly x-ray crystallography that we will consider here. The related techniques are sufficiently similar that once the principles of x-ray crystallography have been learned the others can be understood readily.

X-ray and neutron crystallography are the only techniques that presently reveal the relative positions of all, or nearly all, the atoms in a protein or virus. This is called a *structure determination.* It has been achieved to date for about 100 proteins and for the protein coats of several viruses. In the preliminary stages of structure determination, x-ray crystallography often gives information about the molecular weight and the symmetry of a protein or a virus. Thousands of structure determinations of salts, metals, minerals, and organic compounds also have been made. These form the foundation of modern chemistry and biochemistry.

○ *17–1 Outline of the Determination of Protein Structure by X-ray Crystallography*

Suppose that you wish to determine the structure of a protein. First you must grow a crystal of the material that is at least 0.2 mm on an edge. This often can be done by adding ammonium sulfate to a concentrated solution of the pure protein. Figure 17-1 shows crystals of the enzyme aldolase, which were grown by this procedure over a period of about six weeks.

FIGURE 17-1
Crystals of the enzyme aldolase from rabbit muscle. The crystals are about 1 mm long.

Protein and virus crystals differ from crystals of small molecules in that they contain large amounts of liquid of crystallization. This solvent occupies the large spaces that remain, even when the molecules are packed as tightly as possible next to one another. The fraction of the crystal occupied by solvent usually is between 30% and 60%, but in the extraordinary case of the muscle protein tropomyosin it is 95%. Thus protein crystals are really ordered solutions. If the solvent is allowed to evaporate, crystals lose much of their regular periodicity and consequently their diffraction pattern; thus the crystals become useless for diffraction studies. For this reason, protein crystallographers routinely mount crystals in thin-walled glass capillaries and seal the ends to prevent evaporation (see the biographical sketch on J. D. Bernal in Chap. 5).

To record diffraction data, the crystal is placed in a beam of x-rays (Fig. 17-2). The diffracted x-rays emerge from the crystal at different angles and have different intensities. The angles and intensities can be recorded on a piece of photographic film. Each diffracted x-ray makes a spot or "reflection" where it intersects the x-ray film. The entire pattern of diffraction can be thought of as a three-dimensional lattice of spots. Crystallographers refer to this lattice as the *reciprocal lattice*.

As shown in Figure 17-2(*a*), the reciprocal lattice can be recorded on x-ray film in undistorted form, each film showing one section of the three-dimensional lattice. The actual method of recording the undistorted lattice on film takes some background to understand. The method involves a precession motion of the crystal, and a screen with a transparent circle between the crystal and film which allows only the diffracted rays for one section of the lattice to pass. However, it is easier to understand other methods of recording x-ray data, such as a simple rotation of the crystal in the x-ray beam which gives a distorted picture of the reciprocal lattice, as shown in Figure 17-2(*b*). In Section 17-2 you will learn to interpret the appearance of such x-ray photographs.

The intensities of reflections are the primary data of x-ray crystallography. Each reflection is identified by three indices, h, k, and l, that specify its place in the reciprocal lattice [see Fig. 17-2(*a*)]. The intensity of reflection h, k, l is written as $I(hkl)$. Since it is the square root of the intensity (the amplitude of

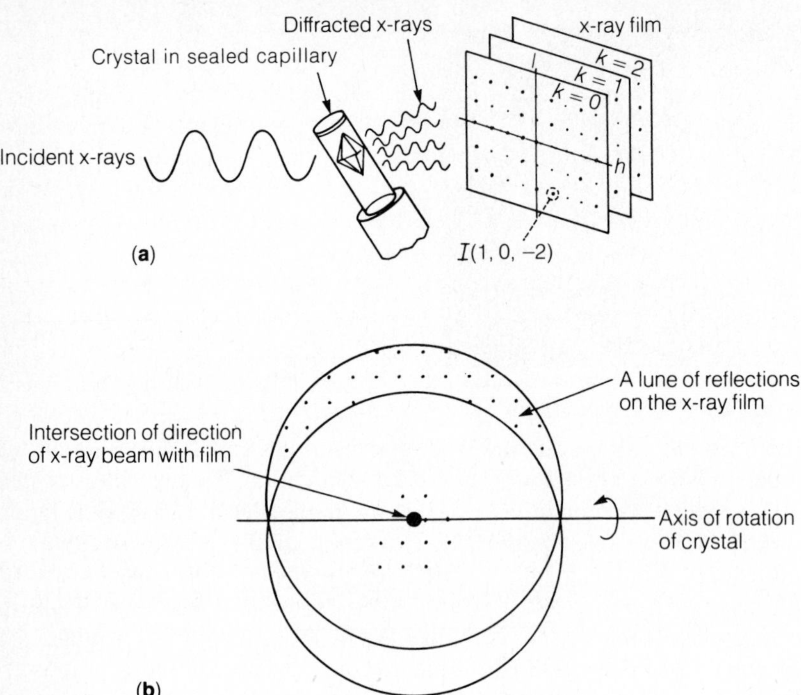

FIGURE 17-2
Recording of x-ray data. A beam of x-rays is incident on a crystal. Diffracted x-rays emerge and can be detected on photographic film. Each spot or "reflection" on the film arises from the intersection of a diffracted ray with the film. The pattern of diffraction spots may be thought of as a three-dimensional lattice, as shown in **(a)**.

If the crystal is rotated in the x-ray beam (with the axis of rotation perpendicular to the x-ray beam) the pattern on the film consists of the more complicated *lunes* shown in **(b)**. To record the diffraction pattern in the simple form of the lattice shown in **(a)**, the crystal must make a precession motion. The reasons for this are described in Glusker and Trueblood (1972) and other books on x-ray crystallography. In **(a)** you can see that each spot is assigned an index *hkl* giving its position on the three-dimensional lattice, and that each spot has a characteristic blackness or intensity, represented by $I(hkl)$. The 1, 0, −2 reflection is circled in **(a)**. With either the precession method of **(a)** or the rotation method of **(b)**, the main x-ray beam must be stopped by a lead beam trap to prevent fogging of the x-ray film.

the diffracted ray) that is characteristic of the molecular structure, the intensity of each reflection is converted to an amplitude $F(hkl)$:

$$F(hkl) = \sqrt{cI(hkl)} \tag{17-1}$$

$F(hkl)$ is called the *structure-factor magnitude* (or amplitude) of reflection *hkl*. The constant in this equation, c, is characteristic of the method of data collection, and corrects for irrelevant experimental factors. The structure of the molecule in the crystal is determined by manipulation of these $F(hkl)$s.

The end product of a crystallographic analysis of molecular structure is a plot of the electron density, $\rho(xyz)$, of one unit cell of the crystal as a function of the three coordinate axes of the unit cell. This plot can be displayed on a stack of plexiglass sheets, each showing contours of electron density in one section through the unit cell. In studies of small molecules this function usually is detailed enough to reveal the arrangement of atoms, even when there is no prior knowledge of the structure. However, in protein crystallography it almost always is necessary to know the amino acid sequence of the protein to identify each feature of the electron-density plot with a chemical grouping. Once this process of identification is complete, the structure has been determined. Figure 17-3 shows the fitting of a model (which represents the amino acid sequence of ribonuclease *S*) to an electron-density map determined by x-ray crystallography.

How are the structure-factor magnitudes, $F(hkl)$, manipulated to give the desired electron density, $\rho(xyz)$? In Section 17-5 we will find that the structure factor magnitudes actually comprise only half the information needed to determine $\rho(xyz)$. The other half consists of a "phase" for each reflection, which is denoted $\alpha(hkl)$. We shall see that $F(hkl)$s and $\alpha(hkl)$s can be thought of respectively as the amplitudes and phases of component waves that are summed to describe the crystal structure. Phases of x-ray reflections cannot be measured by direct experiment.

Once the $F(hkl)$s have been measured for a crystal, determining the molecular structure reduces to figuring out the phase for each reflection. Crystallographers have glorified this difficult step by the term *The Phase Problem.* For small molecules the phases often can be determined from information within the $F(hkl)$, using a device called a *Patterson map*, which we will describe in Section 17-6. For proteins, additional information is required. This information is obtained by binding an atom of high atomic number to each protein molecule within the crystal and by measuring a set of $F(hkl)$s for the modified crystal. By combining the $F(hkl)$s from several modified crystals with the original $F(hkl)$s, the phase of each reflection, hence the electron density, can be determined. This is the method of *isomorphous replacement,* which is described in Section 17-8.

Often the structure of a protein is solved in stages of increasing *resolution.* The resolution of the structure is the minimum separation of two groups in the electron-density plot that can be distinguished from one another. Thus, as the fineness of resolution increases, finer detail can be seen in the electron-density plot. At 3 Å resolution the path of the polypeptide backbone of a protein usually can be traced in an electron-density map. At 2 Å resolution most amino acid side chains can be positioned accurately. And at 1.5 Å resolution (achieved so far for only a few proteins) many resolved atoms can be seen. It is the number of available $F(hkl)$s and $\alpha(hkl)$s that determines the resolution of the electron-density plot: as terms with increasing values of h, k, and l are added, the resolution increases. In the crystal structure analysis of the protein myoglobin, which has a molecular weight of 17,800, about 400

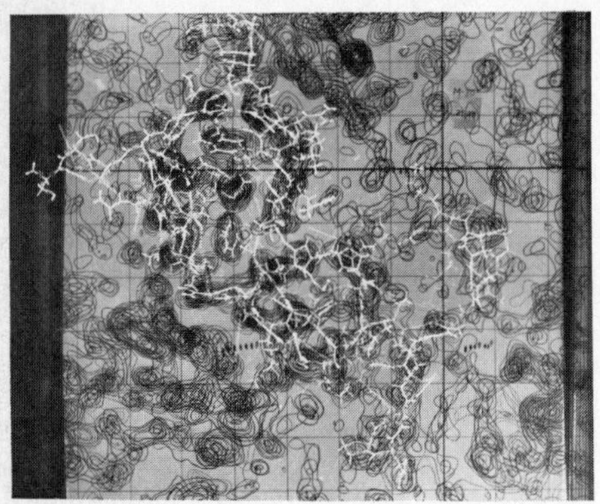
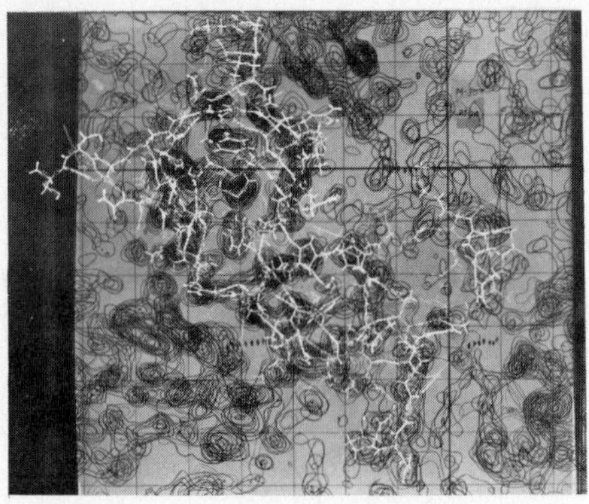

FIGURE 17-3 (a)
A stereopair photograph of a portion of the structure of ribonuclease S, shown by optical superposition of eight sections of the 2.0Å resolution electron density map and the skeletal molecular model derived from the map. The grid squares are 5Å on a side, and the sections are separated by 7.02Å. You can see this figure in stereo if you look through a stereo viewer. Some people can achieve the same effect by allowing their eyes to drift apart until they see the left image with the left eye and the right image with the right eye, and then focusing on the superimposed image in the center.

reflections were required for 6 Å resolution, 9,600 for 2 Å resolution, and more than 25,000 reflections for 1.4 Å resolution.

○ *17–2 Diffraction Occurs by Interference of Waves*

Diffraction is defined as the non-straight-line propagation of light, or of other particle-waves. It occurs when a beam of light is scattered into wavelets and the wavelets then recombine, or *interfere* with each other. Interference of waves was discussed briefly in Section 10-1, and it may be helpful to review Figure 10-2 before going on.

You can observe diffraction by looking at a street light through a fine lace curtain, or at a candle through a nylon stocking or fine handkerchief. You will see a pattern like that of Figure 17-4(*i*), in which the single beam from the light source has been diffracted into many secondary beams. To observe sharp diffraction patterns, such as those of Figure 17-4, it is necessary to replace the candle or street light by a laser beam, and to replace the irregular holes of cloth with small holes in an opaque material (and for safety, to replace your eye with a camera). Each diffraction pattern on the left of Figure 17-4 was formed by light passing through the corresponding set of circular apertures shown on the right. You can see from the first row of the figure that the diffraction pattern of a circular aperture is a set of circular

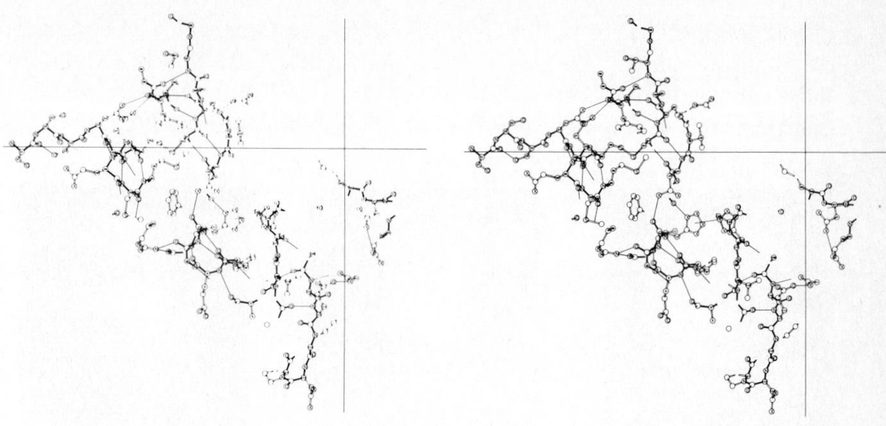

FIGURE 17-3 (b)
A computer-drawn stereopair picture of the same region of the molecule. Bonds connecting main chain atoms are dark and hydrogen bonds are light. The residue number is printed on α- and β-carbon atoms of each residue. Uncertain atoms are labeled *X*. (Reproduced from Wyckoff et al., 1970.)

fringes. Also you can see from panel (*d*) that the diffraction pattern of two circular apertures is the pattern of a single aperture [compare (*d*) to (*a*)] but crossed by a set of vertical fringes. For two vertical apertures [panel (*g*)] the crossing fringes are horizontal.

Beyond these basic facts, Figure 17-4 illustrates four general characteristics of diffraction:

1. *Dimension*. The dimensions of a diffraction pattern are inversely related to the dimensions of the diffracting object. For example, notice that the smallest of the three holes in the top row gives rise to the broadest set of circular fringes. Also notice that the greater the separation of the diffracting object is, the finer is the separation of the fringes of the diffraction pattern. This is illustrated in the bottom row of Figure 17-4.
2. *Sharpness*. The greater the number of diffracting objects, the sharper the diffraction pattern. This is illustrated by the sequence of panels (*d*), (*e*), and (*f*), in which the number of horizontal apertures is increased, and in which the sharpness of the vertical fringes increases. The basic reason for this is that destructive interference is more complete for greater numbers of repeats.
3. *Sampling or convolution property*. If the diffracting object is a repeat of a basic unit, its diffraction pattern is the product of the diffraction pattern of the basic unit with the diffraction pattern of the repeating function. For example, we just noted that the diffraction of two apertures (*d*) is that of a single aperture crossed by vertical fringes. In other words, (*d*) is the product of (*a*) and of vertical fringes. The vertical fringes are just the

diffraction pattern for two points. The diffracting object may be thought of as a single hole repeated on two points. Thus by this principle, its diffraction pattern (*d*) is a product. In (*d*) the diffraction pattern of an aperture (*a*) is said to be *sampled* by the diffraction pattern of two horizon-

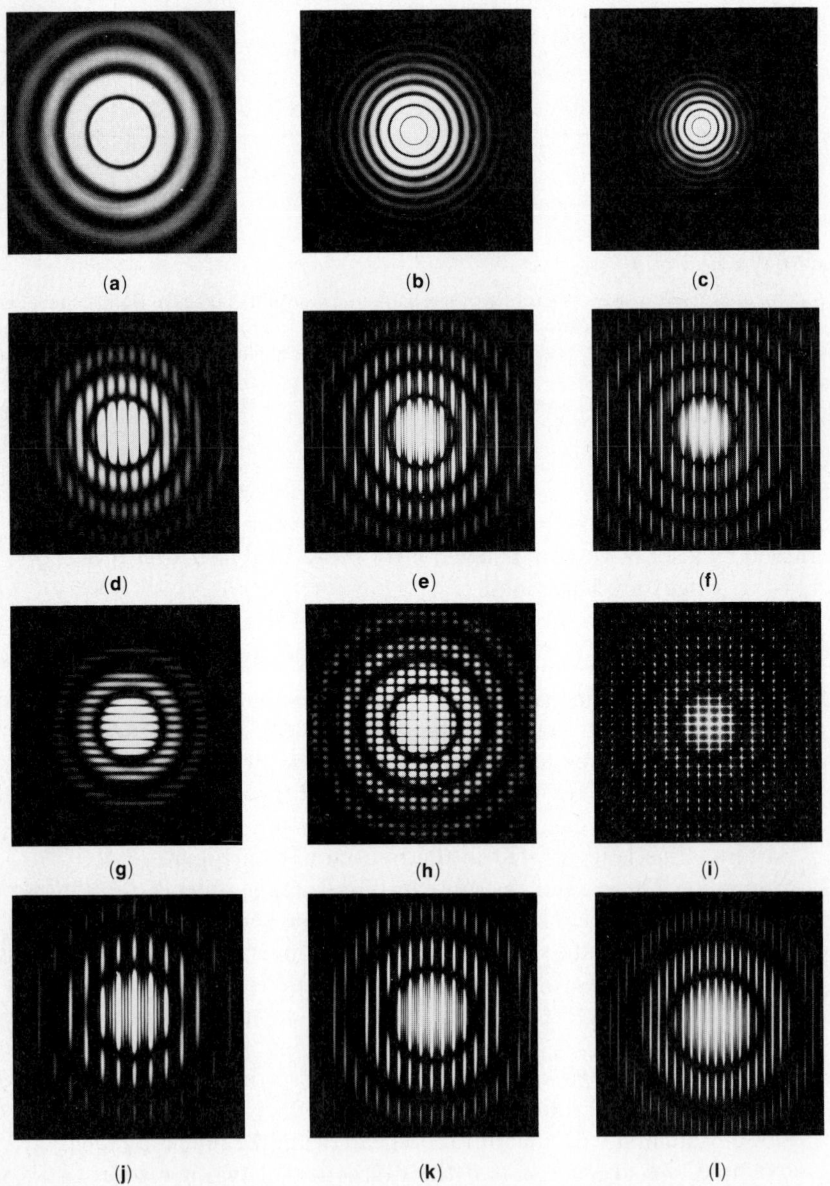

FIGURE 17-4
For legend see opposite page.

tal points (a set of vertical fringes). The technical term for a repeat of one function on another is a *convolution*.

An important point emerges from the sampling property: the distribution of diffracted intensities is determined by the nature of the repeated unit (an aperture in every example of Fig. 17-4), but location of the diffraction spots is determined by how the units are repeated. This is the reason that we can determine the unit cell dimensions of a crystal from the spacing of reflections in its diffraction pattern (Sec. 17-3), but that we must measure the intensities of the reflections to determine the molecular structure. In this section we discuss the factors that determine the spacings or *geometry* of diffraction, and in Section 17-4 we discuss the factors that determine the *intensity* of diffraction.

EXERCISE 17-1

Find another example of the sharpness characteristic in Figure 17-4.

ANSWER

The diffraction pattern of panel (i) is sharper than that of (k) because it is formed from 100 apertures rather than four.

EXERCISE 17-2

Describe the diffraction pattern 17-4(h) in terms of the sampling or convolution principle.

FIGURE 17-4

Optical diffraction patterns formed by a laser beam passing through holes in opaque copper sheets. The diffracted light has been photographed with the camera facing the oncoming beam. The arrangement of holes for each diffraction pattern is shown below. The patterns were photographed by Dr. T. S. Baker.

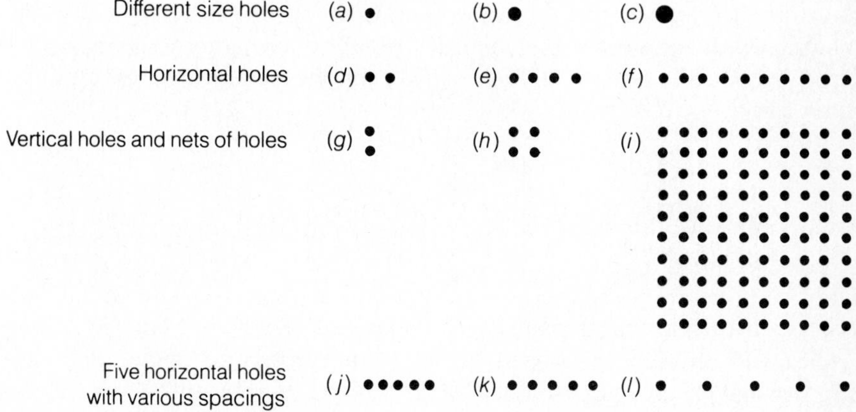

ANSWER

The diffracting object ⁚ ⁚ may be thought of as a combination (convolution) of • • with ⁚ , and thus the diffraction pattern of this object is the product of the diffraction patterns of • • and ⁚ [Fig. 17-4(*d*) and 17-4(*g*)].

---

4. *Spacing*. The greater is the separation of the diffraction objects, the finer is the separation of the fringes of the diffraction pattern. This is illustrated for five apertures in the bottom row of Figure 17-4.

The Origin of Diffraction is in Interference

Why does light passing through several point holes experience diffraction? The answer is that each point acts as a source of a light wavelet, and that the emerging wavelets interfere with each other. In some directions the interference is constructive and light is observed, and in other directions the interference is destructive, and there is darkness. This is illustrated in Figure 17-5(*a*).

Why does light passing through a single hole experience diffraction? This is illustrated in Figure 17-5(*b*). Light wavelets are in phase as they enter the hole, and in the forward direction (A) they remain in phase. Consequently they interfere constructively. But in another direction (say, C) the wavelets from the hole are partly out of phase because of their change in direction. Consequently, they suffer some destructive interference. The variation of interference with angle gives rise to the diffraction pattern of Figure 17-4(*a*).

The origin of x-ray diffraction from a crystal is basically the same, except that instead of small holes the scattered wavelets arise from atoms, and the incident beam of light is a beam of x-rays. When the x-rays fall on the crystal, the atoms scatter secondary x-rays spherically. These secondary x-rays then interfere with each other to produce the diffraction pattern. In short, with both light and x-rays, diffraction is simply scattering, followed by interference.

Laue's Description of X-ray Diffraction

X-ray diffraction was first observed in 1912 at the University of Munich. Until that time it was not certain that x-rays were electromagnetic radiation, nor was it certain that crystals were composed of atoms. Max von Laue, a 33-year-old physicist, reasoned by the same analogy to optical diffraction that we have just discussed that if x-rays are waves and if crystals are periodic arrangements of atoms, then crystals should diffract x-rays. Disregarding the advice of the distinguished theorist Professor Sommerfeld, Laue and two associates looked for diffraction from a crystal of copper sulfate.

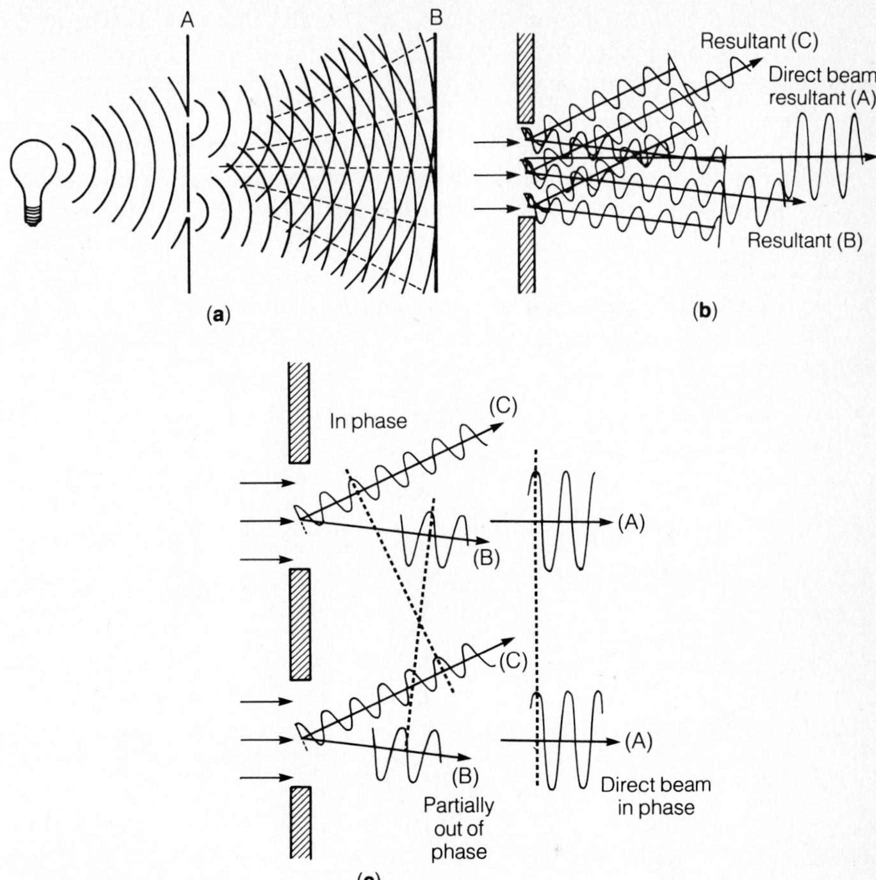

FIGURE 17-5
(a) The origin of diffraction by two holes. The light source produces spherical waves. These pass through the point holes in screen A, and interference takes place. The dashed lines indicate directions of constructive interference, where light spots will be observed on a film or screen at B. **(b)** The origin of diffraction by a single hole. Light passing through the hole behaves as if it is scattered in wavelets which emerge in various directions. These wavelets are in phase in the direction of the direct beam, and form resultant wave A. In another direction they are largely out of phase, and sum to give resultant wave C. **(c)** Light passing through two holes, in greater detail. The resultant wave in a given direction depends on the interference effect shown in **(a)** and the diffraction of a single hole shown in **(b)**. The result is the diffraction pattern from a single hole sampled by the fringes produced by two points. This is illustrated in Figure 17-4**d**. [**(b)** and **(c)** are from *Crystal Structure Analysis: A Primer* by Jenny Pickworth Glusker and Kenneth N. Trueblood. Copyright © 1972 by Jenny Pickworth Glusker and Kenneth N. Trueblood. Reprinted by permission of Oxford University Press, Inc.]

Their positive result must be counted among the most successful experiments of all time. It at once confirmed the wave nature of x-rays and the atomic basis of crystal structure, and it brought Laue the Nobel Prize in 1914.

Laue derived a set of three equations to describe the diffraction of x-rays by a crystal. He started by treating a one-dimensional crystal, a row of point atoms each a distance a from the next, as shown in Figure 17-6. The incoming x-ray beam impinges on the crystal from the direction indicated by $\mathbf{s}_0$, and secondary wavelets are scattered spherically. Laue realized that if x-rays scattered in the direction $\mathbf{s}$ are to interfere constructively, thereby forming a diffracted ray, then the difference in lengths of paths traversed by rays scattered from successive atoms must equal a whole number of wavelengths. You can see from Figure 17-6(a) that this condition can be stated as

$$\Delta(\text{path}) = \overline{A_1C} - \overline{A_2B} = h\lambda \tag{17-2}$$

in which λ is the wavelength of x-rays and h is an integer.

The lengths in Equation 17-2 can be expressed in terms of angles and the interatomic separation a:

$$\overline{A_1C} = a \cos \alpha \qquad \overline{A_2B} = a \cos \alpha_0 \tag{17-3}$$

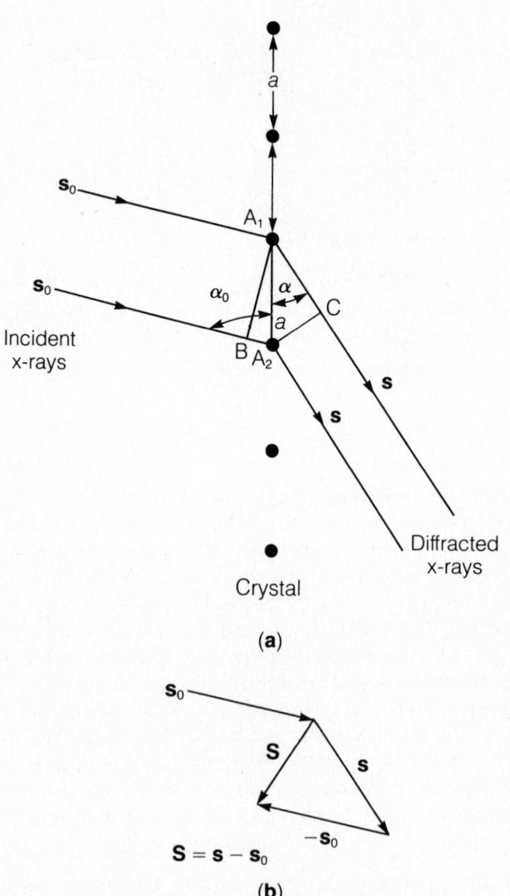

FIGURE 17-6
(a) Derivation of Laue's equation. The dots represent point atoms, equally separated by the distance a. The vectors $\mathbf{s}_0$ and $\mathbf{s}$ are both of length $1/\lambda$, and specify the directions of the incident and diffracted x-rays, respectively. **(b)** The vector $\mathbf{S}$ bisects the angle between $\mathbf{s}_0$ and $\mathbf{s}$.

Substituting Equation 17-3 into Equation 17-2, we get

$$a(\cos\alpha - \cos\alpha_0) = h\lambda \qquad (17\text{-}4a)$$

To describe the x-ray scattering from a three-dimensional crystal with unit cell lengths, a, b, and c, Equation 17-4a must be supplemented by two other equations:

$$b(\cos\beta - \cos\beta_0) = k\lambda \qquad (17\text{-}4b)$$
$$c(\cos\gamma - \cos\gamma_0) = l\lambda \qquad (17\text{-}4c)$$

in which β_0 and γ_0 are the angles formed by the incident beam with the unit cell edges b and c, β and γ are the angles formed by the diffracted ray with the cell edges b and c, and k and l are integers.

Equations 17-4 mean that there are two ways in which x-ray diffraction can be produced from a crystal, depending on whether the x-ray beam contains many values of λ or just one. The first is named for Laue because his historical experiment was performed in this way. An x-ray beam containing many wavelengths impinges on a crystal with unit-cell edges a, b, and c, at the angles α_0, β_0, and γ_0. Diffracted rays emerge from the crystal at the particular angles α, β, and γ that produce simultaneous solutions to Equations 17-4. The second method was used first by W. H. and W. L. Bragg and has become the standard method. In it, a monochromatic beam of x-rays of wavelength λ is used, and Equations 17-4 do not have a solution for every orientation of the crystal. The crystal must be rotated in the x-ray beam, thereby varying α_0, β_0, and γ_0 until Equations 17-4 are satisfied and a diffracted ray flashes out. The physical meaning of the Laue equations is not easy to visualize, so it is fortunate that Bragg developed a much simpler picture of diffraction. You have probably guessed from the notation of Equations 17-4 that the reflection that flashes out is the *hkl* reflection.

Before considering Bragg's idea, let us put Laue's equations into the simpler vector notation that will be useful to us later. The directions of the incoming and diffracted rays in Figure 17-6(b) are defined by the vectors $\mathbf{s}_0$ and $\mathbf{s}$ both of which are defined as having length $1/\lambda$. Thus we can write

$$(\cos\alpha_0)(1/\lambda)a = \mathbf{a}\cdot\mathbf{s}_0 \quad \text{and} \quad (\cos\alpha)(1/\lambda)a = \mathbf{a}\cdot\mathbf{s} \qquad (17\text{-}5)$$

in which $\mathbf{a}$ is a vector defining the length and direction of the a unit-cell edge. Substituting Equation 17-5 into Equation 17-4a, we obtain

$$\mathbf{a}\cdot\mathbf{s} - \mathbf{a}\cdot\mathbf{s}_0 = h \qquad (17\text{-}6)$$

or $$\mathbf{a}\cdot(\mathbf{s} - \mathbf{s}_0) = h \qquad (17\text{-}7)$$

This equation can be simplified further by defining the *diffraction vector* $\mathbf{S} \equiv \mathbf{s} - \mathbf{s}_0$, a quantity that we shall discuss later. Figure 12-6(b) shows that $\mathbf{S}$ bisects the angle formed by the incoming and diffracted rays. Substituting the definition of $\mathbf{S}$ into Equation 17-7, we get

$$\mathbf{a}\cdot\mathbf{S} = h \qquad (17\text{-}8a)$$

Analogous substitutions into Equations 17-5*b* and 17-5*c* yield the other two Laue equations:

$$\mathbf{b} \cdot \mathbf{S} = k \qquad (17\text{-}8b)$$
$$\mathbf{c} \cdot \mathbf{S} = l \qquad (17\text{-}8c)$$

Bragg's Description of Diffraction is of Reflections from Sets of Parallel Lattice Planes

In 1912, a simpler description of diffraction was given by W. L. Bragg, a 22-year-old researcher at Cambridge University. Bragg's interest in x-rays came from his father, who was a professor of physics at Leeds. With the younger Bragg's law of x-ray diffraction, and with monochromatic x-rays and an ionization counter constructed by the elder Bragg, the father-and-son team rapidly determined the atomic structures of zinc blende, fluorspar, sodium nitrate, potassium chloride, sodium chloride, diamond, and other materials. This concentrated period of research, which gave the first glimpse of the atomic arrangement of matter, was brought to a close by the outbreak of World War I. In fact, it was while serving on the front that W. L. Bragg heard that he and his father had been awarded the Nobel Prize. After the war Bragg resumed his research in crystal structure analysis, turning first to simple minerals, then to the phase problem, then to silicates, then to metals, and then to proteins. He played a part in nearly every advance of the science until his death in 1971, and nearly every crystallographer was a student of his, or a student's student, or some further descendant in an academic line. In an obituary by Perutz in 1971, Bragg was compared to Giotto, the great artist of the Italian Renaissance: "It is as though he had himself invented three-dimensional representation, and then lived through all the styles of European painting from the Renaissance to the present day, to be finally confronted by computer art."

Bragg's idea was that diffraction can be considered as *reflection* of x-rays from planes of atoms within a crystal. To see this we must first consider the conventional nomenclature for planes within a crystal, then derive Bragg's law for the geometry of reflection of x-rays. Then we will see how Bragg's law is related to Laue's equations.

Lattice Planes and Their Miller Indices

Lattice planes are imaginary parallel planes within crystals which are described by their *Miller indices*. The three Miller indices of a plane are the reciprocals of the intercepts, in units of cell edge lengths, that the plane makes with the axes of the unit cell. Consider the unit cell shown in Figure 17-7(*a*), and the plane that intersects its a, b, and c axes at a', b', and c'. Then the Miller indices of this plane, conventionally denoted h, k, and l, are

$$h = \frac{a}{a'} \qquad k = \frac{b}{b'} \qquad l = \frac{c}{c'} \qquad (17\text{-}9) \blacktriangleleft$$

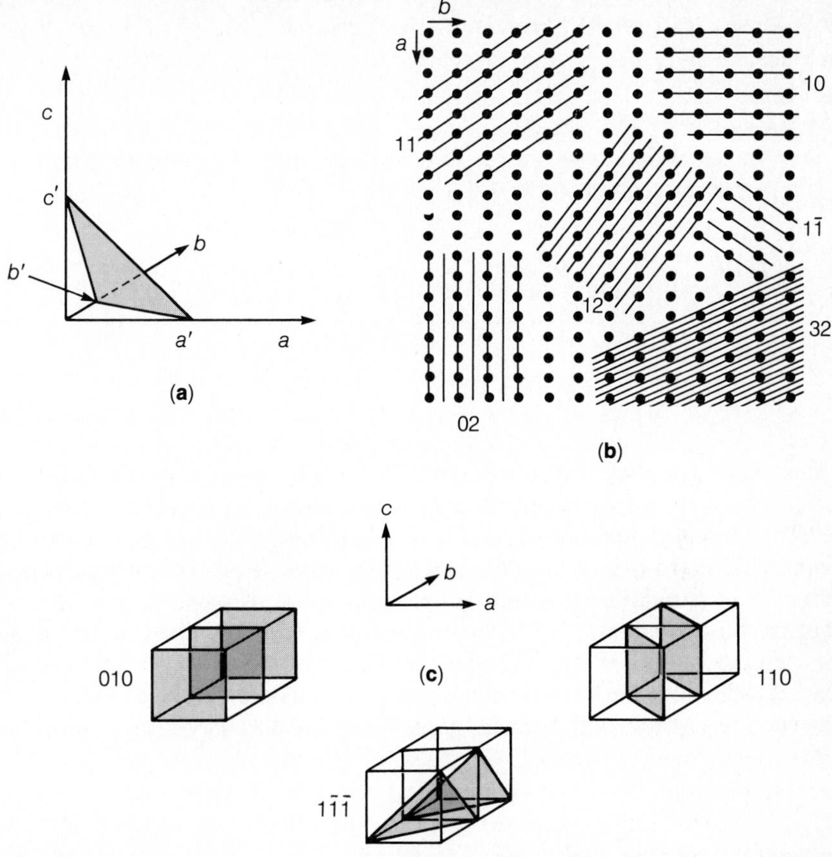

FIGURE 17-7
Miller indices of lattice planes in a crystal. (a) A lattice plane with intercepts a', b', and c' along the a, b, and c axes. (b) Lattice planes in a two-dimensional lattice. (After Bunn, 1946.) (c) Lattice planes in a three-dimensional lattice.

When a plane is parallel to a unit-cell axis, its intercept on that axis is at infinity, hence its Miller index is zero. Planes with negative intercepts are indicated by a bar over the intercept, for example, $\bar{1} = -1$. Some examples of lattice planes and their Miller indices are shown in Figure 17-7(*b*) and (*c*). It is important to note that Miller indices specify a set of parallel planes in the crystal, because a crystal is periodic, extending over many unit cells.

EXERCISE 17-3

Draw the 1,1 planes in a two-dimensional square lattice. How are they related to the $\bar{1}$,1 planes?

ANSWER

The two sets are perpendicular.

EXERCISE 17-4

How are the $\bar{1}$,1 planes related to the 1,$\bar{1}$ planes?

ANSWER

They are the same set. In general the *hkl* and $\overline{hkl}$ planes are identical.

EXERCISE 17-5

Sketch the 2,0,0 planes in a three-dimensional primitive cubic lattice. How are these planes related to the 1,0,0 planes?

ANSWER

They are parallel to them but with half the spacing of the 1,0,0 planes.

Bragg's Law

At this point, you may have surmised that there is some connection between the indices *hkl* of an x-ray reflection on a film and the Miller indices *hkl* of a set of parallel planes in a crystal. Bragg's law makes this connection. Bragg treated diffraction from a crystal as though it arose from reflection of x-rays from a set of parallel planes with Miller indices *hkl*. The construction for his proof is shown in Figure 17-8(*a*). The incident beam is reflected from the planes, as though from a mirror, with the angle of incidence, θ, equal to the angle of reflection. For rays reflected from two successive planes to interfere constructively, their path lengths must differ by one wavelength. In other words,

$$\overline{PQ} + \overline{QR} = \lambda \tag{17-10}$$

Examining the figure, you can see that

$$\overline{PQ} = \overline{QR} = d \sin \theta \tag{17-11}$$

in which d is the spacing between two successive planes of the *hkl* set of planes. Substituting Equation 17-11 into Equation 17-10, we obtain the usual form of Bragg's law:

$$2d \sin \theta = \lambda \tag{17-12}$$ ◀

The physical interpretation of Bragg's law is as follows. A crystal is rotated in a monochromatic beam of x-rays of wavelength λ. When a set of lattice planes, spaced at distance d from one another, forms an angle θ with the incident beam, a diffracted ray flashes out. Suppose that the set of planes has Miller indices *hkl*. Then the spot on the x-ray film caused by the diffracted ray is reflection *hkl* of the reciprocal lattice. Anticipating our discussion of the *intensity* of diffraction in Section 17-4, we may say that if many atoms in the structure happen to lie on lattice planes *hkl*, and if the planes midway between are not heavily populated, then the reflection will be intense. With simple structures such as sodium chloride, all atoms (or ions in this case) lie on *hkl* planes with low indices. Therefore it was a straightforward matter for Bragg to observe the strong reflections in the diffraction pattern of sodium chloride and to deduce the packing of the ions. Such a

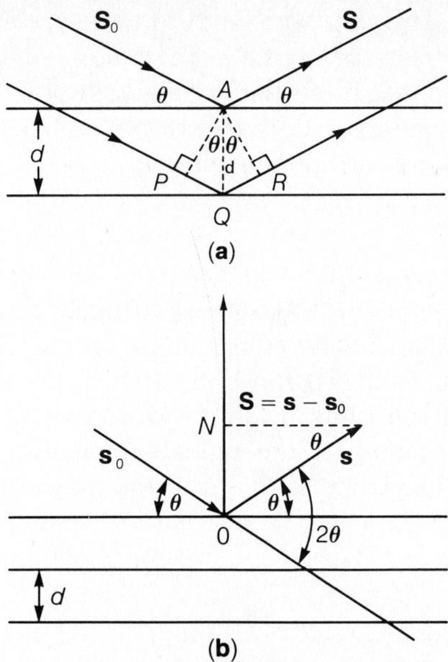

FIGURE 17-8
Bragg's law of x-ray reflection. **(a)** X-rays being "reflected" from the *hkl* set of lattice planes at the angle θ. **(b)** The diffraction vector **S** is perpendicular to the lattice planes that are reflecting x-rays.

procedure is hopeless for much larger molecules, because there are many more atoms and many more reflections in the diffraction pattern; hence we do not consider it here.

Another important point about Bragg's law is that it shows that angle 2θ between incident and diffracted rays depends both on the wavelength of x-rays and on the separation of the *hkl* planes giving rise to the reflection. In protein crystallography, copper-target x-ray tubes ($\lambda = 1.54$ Å) are almost always employed. Small-molecule crystallographers often use molybdenum targets ($\lambda = 0.71$ Å), which for a unit cell of the same size, give a more tightly spaced diffraction pattern. Examination of Bragg's law also shows that for small *d* spacings (large values of *hkl*), 2θ is large. Since it is the planes with small *d* spacings that contain the fine details of molecular structure, higher-angle data reveal this detail. Thus higher-angle data are added to increase the resolution of a crystal structure analysis.

EXERCISE 17-6

What is the theoretical lower limit to the measured spacings of a crystal structure if copper radiation is employed?

ANSWER

From Bragg's law we get

$$d_{\text{min}} = \frac{\lambda}{2(\sin \theta)_{\text{max}}}$$

where $(\sin \theta)_{max} = 1$. This corresponds to the scattering of x-rays back towards the incident beam ($2\theta = 180°$). Then $d_{min} = \lambda/2 = 0.77$ Å for copper radiation. Note that in protein crystallography this theoretical limit is never achieved, because the crystals are not sufficiently periodic to exhibit a diffraction pattern to this resolution.

The diffraction vector **S** that we discussed in connection with Laue's equations has a simple interpretation, according to Bragg's law. Recall from Figure 17-6(*b*) that **S** bisects the angle formed by the incident and diffracted rays. Then, from Figure 17-8(*b*), **S** is perpendicular to the *hkl* planes that are in the reflecting position. Moreover, the length of **S** is the reciprocal of *d*, the spacing of the reflecting planes. This can be shown from Figure 17-8(*b*). The length $\overline{ON}$ is half the magnitude of **S**. Therefore

$$\frac{|\mathbf{S}|/2}{|\mathbf{s}|} = \sin \theta$$

Since the magnitude of **s** is $1/\lambda$, we get

$$|\mathbf{S}| = \frac{2 \sin \theta}{\lambda} \tag{17-13}$$

Comparing Equation 17-13 with Bragg's law (Equation 17-12) we find

$$|\mathbf{S}| = \frac{1}{d} \tag{17-14}$$ ◀

In summary, the diffraction vector **S** is normal to the *hkl* planes and its length is the reciprocal of the spacing of planes which give rise to reflection *hkl*.

⬢ *The Sphere of Reflection*

One additional concept knits together all that we have said about the geometry of diffraction and enables us to analyze any diffraction experiment. This concept is a construction called the *sphere of reflection,* which is a geometric combination of Bragg's law and the reciprocal lattice.

Before making this construction we must say a little more about the reciprocal lattice. In the preceding section we found that when the *hkl* planes are in the reflecting position, the diffraction vector **S** is perpendicular to them, with a magnitude equal to the reciprocal of their separation, *d*. If we think of the vector **S** as being drawn from the origin of the reciprocal lattice to the reciprocal lattice point *hkl,* we have a rule for constructing the reciprocal lattice that corresponds to any given crystal lattice: for any set of crystal lattice planes *hkl,* we draw a vector perpendicular to them with length $1/d$. At the tip of this vector we place the reciprocal lattice point (or reflection) *hkl*. By doing this for all sets of crystal planes, we generate the reciprocal lattice.

This procedure is shown in Figure 17-9 for a two-dimensional lattice. Consider the 1,0 set of crystal planes that are horizontal. The **S** vector for the

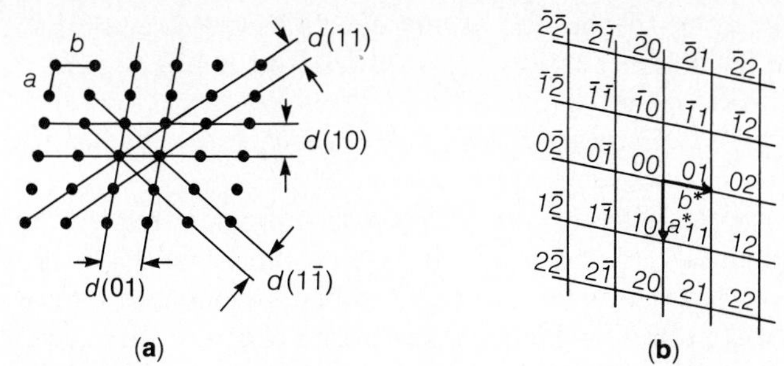

FIGURE 17-9
Relationship between **(a)** the direct (crystal) lattice, and **(b)** the reciprocal lattices. (From Brand and Speakman, 1975.)

planes is perpendicular to them and of length $1/d$; we place its tail at the origin of the reciprocal lattice in Figure 17-9(*b*), and place the reciprocal lattice point 1,0 at its head. We repeat this procedure for all sets of lattice planes in Figure 17-9(*a*), thereby generating the complete reciprocal lattice of Figure 17-9(*b*). Note that the longer crystal lattice dimension *b* corresponds to the shorter reciprocal lattice dimension b^*. Note also that the b^* of the reciprocal axis is perpendicular to the *a*-axis of the direct crystal lattice, and that a^* is perpendicular to *b*.

In constructing the reciprocal lattice corresponding to a three-dimensional lattice, one proceeds in the same way. The $\mathbf{a}^*$-axis of the reciprocal lattice is

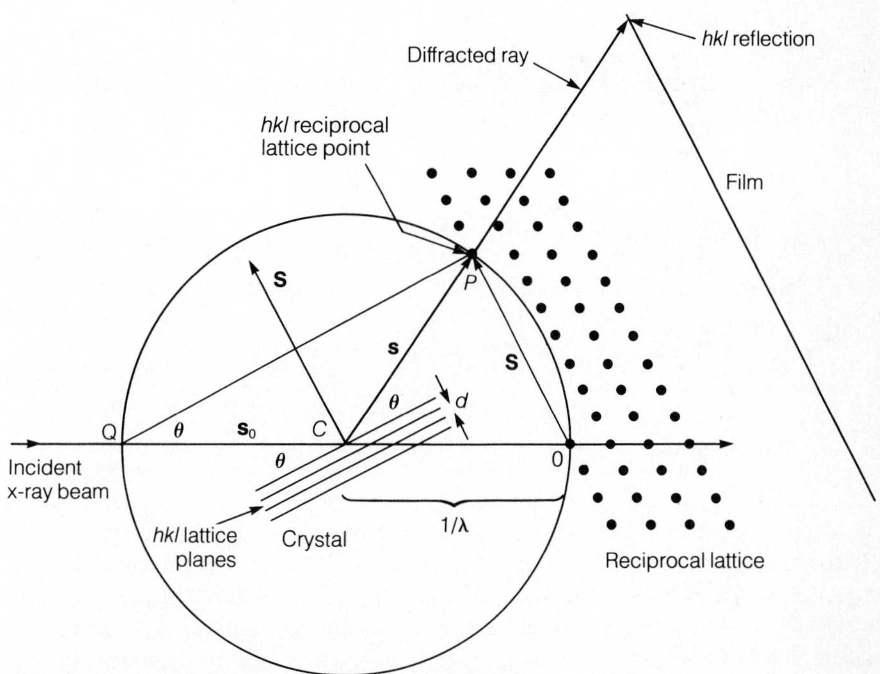

FIGURE 17-10
The sphere of reflection, a geometrical illustration of Bragg's law; see text for discussion.

then perpendicular to both the **b**- and **c**-axes of the direct lattice, and similarly for the **b***- and **c***-axes. We can express this mathematically as follows.

$$\begin{aligned} \mathbf{a}\cdot\mathbf{a}^* &= \mathbf{b}\cdot\mathbf{b}^* = \mathbf{c}\cdot\mathbf{c}^* = 1 \\ \mathbf{a}\cdot\mathbf{b}^* &= \mathbf{a}\cdot\mathbf{c}^* = \mathbf{b}\cdot\mathbf{c}^* = \mathbf{b}\cdot\mathbf{a}^* = \mathbf{c}\cdot\mathbf{a}^* = \mathbf{c}\cdot\mathbf{b}^* = 0 \end{aligned} \qquad (17\text{-}15)$$

Now we are ready to consider the construction of the sphere of reflection, shown in Figure 17-10. An imaginary sphere of radius $1/\lambda$ is drawn around the crystal, which is at point C. The x-ray beam enters the sphere at point Q, and leaves it at point O, which is taken as the origin of the reciprocal lattice. Now the crystal is rotated, and because the orientation of the reciprocal lattice is determined by that of the crystal lattice, the reciprocal lattice rotates in concert with the crystal. The main point of this construction is that when the reciprocal lattice point *hkl* touches the sphere, the *hkl* lattice planes are exactly in the reflecting position, thereby forming an angle θ with both the incoming and reflected beams. The diffracted ray emerges from the crystal at C, passes through the sphere at P (where the reciprocal lattice point is making contact), and continues to the film where it forms the spot *hkl*. Note that the vector **S** is perpendicular to the *hkl* crystal planes. This is so because two vectors with the same length and direction are identical.

EXERCISE 17-7

Prove that Bragg's law is satisfied when the reciprocal lattice point *hkl* is touching the sphere of reflection. (*Hint:* Note that then the triangle *OPQ* is a right triangle with the angle *OQP* equal to θ.)

ANSWER

Focusing on triangle *QPO* and recalling that the diameter of the sphere is $2/\lambda$, we get

$$\sin\theta = \frac{|\mathbf{S}|}{2/\lambda}$$

or, since by Equation 17-14, $|\mathbf{S}| = 1/d$,

$$2d\sin\theta = \lambda$$

which is Bragg's law.

○ *Summary*

Some of the terminology of this chapter is summarized in Table 17-1. Molecules are said to be in crystal space (sometimes called "direct" or "real" space), whereas the diffraction pattern is said to be in reciprocal space. Crystal space is described by the function $\rho(xyz)$, which is periodic and continuous. The positions of atoms are given by coordinates X, Y, and Z, or equivalently by a vector **r**, which extends from the origin to the atom. Reciprocal space is described by the function **F**(*hkl*), which we soon will see is

TABLE 17-1 *Some Terminology of X-ray Diffraction*

SPACE:	CRYSTAL (OR DIRECT OR REAL)	RECIPROCAL
Contents	Molecules	Diffraction pattern
Described by function	$\rho(XYZ)$ electrons Å^{-3}	$F(hkl)$ electrons
Nature of function	Periodic, continuous, real	Discrete, complex
Coordinates	X, Y, Z; fractions	h, k, l; integers
Variable	$\mathbf{r}(XYZ)$	$\mathbf{S}(hkl)$
Unit-cell dimensions	a, b, c	$a^* \propto \frac{1}{a}, b^* \propto \frac{1}{b}, c^* \propto \frac{1}{c}$
Relationship	← Fourier transform →	

complex. The magnitudes of these $\mathbf{F}(hkl)$s are the $F(hkl)$s of Equation 17-1. Points in reciprocal space are described by the coordinates h, k, l, or by the vector $\mathbf{S}$, which extends from the origin to point hkl. The letters hkl also are used to denote sets of parallel planes in the crystal. The reflections of x-rays from the hkl set of crystal planes gives rise to the reflection hkl on the x-ray film. The unit-cell dimensions of crystal space a, b, and c are inversely proportional to the unit-cell dimensions of reciprocal space a^*, b^*, and c^*. As we shall see in Section 17-5, crystal space and reciprocal space are related by a mathematical operation known as a *Fourier transform*.

○ 17–3 Determination of Crystal Parameters, Molecular Weight, and Molecular Symmetry

The first step in any crystal structure analysis is to decide if the unit cell dimensions and diffraction pattern of the crystals warrant a structural determination. This usually can be settled by examining several precession x-ray photographs of sections of the reciprocal lattice. Moreover, from the same photographs it often is possible to extract some information about the molecular weight and molecular symmetry of the substance in the crystals. The steps used to extract this information are as follows.

Determination of Unit-Cell Dimensions and Volume

Each of the reciprocal unit-cell dimensions a^*, b^*, and c^* can be found by measuring the separation between appropriate reflections on x-ray films. This separation must be scaled to take account of the wavelength of x-rays, and of the magnifying effect of the separation of the crystal and film, G. If $\Delta(a^*)$ is the separation on the film of reflections along the a^* axis, then

$$a^* = \frac{\Delta(a^*)}{G\lambda} \tag{17-16}$$

When the unit-cell angles are 90°, the unit-cell dimensions are simply the reciprocals of the reciprocal cell dimensions, and the volume of the unit cell is $V = abc$.

EXERCISE 17-8

Figure 17-11 is an x-ray photograph of a tetragonal crystal of the enzyme lysozyme taken by the precession method. It is the $h0l$ section of the reciprocal lattice. The $\mathbf{a}^*$ axis is in the vertical direction (h increasing), and the $\mathbf{c}^*$ axis is in the horizontal direction (l increasing). The reflection $h = 0$, $k = 0$, $l = 0$ is at the center of the film, covered by the shadow of the beam stop, which prevents the incident x-ray beam from fogging the film. This photograph was made with copper x-radiation ($\lambda = 1.5418$ Å) with a film-to-crystal distance of 60.0 mm. Recalling that in a tetragonal unit cell $a = b$, calculate a, b, c, and V.

ANSWER

For horizontal rows (perpendicular to a^*), $\Delta(\text{row}) = 1.17$ mm; thus $a = (60 \text{ mm})(1.542 \text{ Å})/1.17 \text{ mm} = 79.1 \text{ Å} = b$. For vertical rows, $\Delta(\text{row}) = 2.44$ mm; thus $c = (60 \text{ mm})(1.542 \text{ Å})/2.44 \text{ mm} = 37.9$ Å. Therefore, $V = (79.1 \text{ Å})^2(37.9 \text{ Å}) = 2.37 \times 10^5 \text{ Å}^3$.

Determination of the Space Group

The symmetry elements of the crystal (Sec. 16-6) lead to a symmetry in the intensities of the reciprocal lattice, and often to special groups of reflections with zero intensity. By consulting the *International Tables for X-Ray Crystallography* (Henry and Lonsdale, 1969) it is a simple matter to determine the space group, or at least to narrow the possibilities to a small number. An example of a group of reflections in Figure 17-11 with zero intensities are those on the row $h = 0$, with $l/4$ not an integer. This observation indicates that the crystal contains a 4_1 or 4_3 screw axis along **c**. When coupled with similar observations for the $hk0$ section of the reciprocal lattice, which indicate a 2_1 axis along **a**, this indicates that the space group is $P4_12_12$ (or $P4_32_12$).

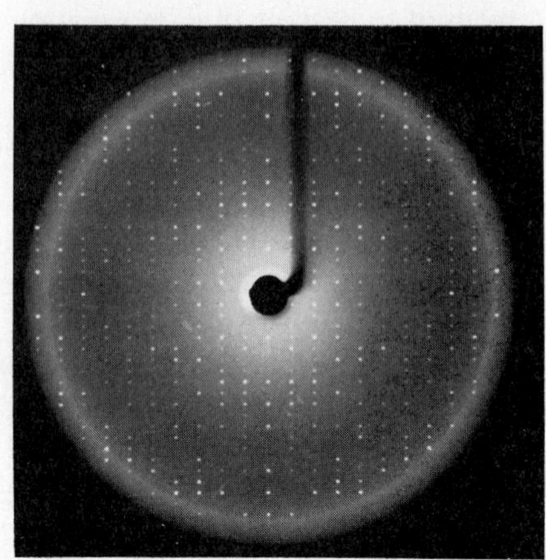

FIGURE 17-11
The $h0l$ section of the reciprocal lattice of hen egg lysozyme. The c^*-axis is horizontal and the a^*-axis vertical. The L-shaped shadow is of the main beam stop. (Photograph taken by Mr. L. Weissman.)

Determination of the Number of Molecules Per Unit Cell and the Molecular Weight

The molecular weight of the substance in a crystal is related to the unit-cell volume and other measurable properties. If the unit cell contained only one protein molecule, the molecular mass would be $V\rho_c$, in which V is the unit-cell volume and ρ_c is the density of the crystal. Similarly, the molecular weight of the protein would be $N_A V\rho_c$. In fact, only the fraction X_P of this weight is protein; the rest is solvent. There may be more than one molecule per unit cell, so that

$$nM = N_A V \rho_c X_P \qquad (17\text{-}17)$$

in which n is the number of molecules per unit cell. The crystal density can be determined by measuring the density of a mixture of nonpolar liquids in which the crystal is exactly buoyant. The weight fraction of protein in the crystal, X_P, can be measured by weighing a group of crystals, evaporating the solvent from them, and reweighing them. If salt is present, a correction for its mass must be applied.

At first glance, Equation 17-17 does not seem useful for determining M because it yields only the product of n and M. But often the space group restricts n so that its value is clear, and M can be calculated. In other cases, an approximate molecular weight is known, so that an approximate n can be calculated from Equation 17-17. Then n can be rounded to the nearest integer and resubstituted into Equation 17-17 to calculate a better value for M. In these cases the accuracy of M is limited by the determination of X_P.

EXERCISE 17-9

The density of the lysozyme crystals of Exercise 17-8 was found to be 1.242 g cm^{-3} and X_P was found to be 0.665. Suppose that it is known from gel electrophoresis that lysozyme consists of a single polypeptide chain of molecular weight between 10,000 and 20,000. Calculate a more accurate value, given that there are 8 asymmetric units per unit cell in space group $P4_12_12$. (The asymmetric unit is the smallest unit from which the crystal can be generated by the symmetry operations of the space group.)

ANSWER

The asymmetric unit would be one polypeptide chain or some integral multiple, m. Thus there are 8 m lysozyme molecules per unit cell. Therefore 8 mM = $(6.022 \times 10^{23}\ mol^{-1})(2.37 \times 10^5\ Å^3)(10^{-24}\ cm^3/Å^3)(1.242\ g\ cm^{-3})(0.665)$. For $m = 1$, $M = 1.47 \times 10^4\ g\ mol^{-1}$. Since only for $m = 1$ does M lie between 10,000 and 20,000, M must be 1.47×10^4.

Determination of Molecular Symmetry

When oligomeric enzymes or viruses crystallize, one or more of the symmetry elements of the molecule often is a symmetry element of the crystal. In such a case, information on the molecular symmetry can be deduced merely

TABLE 17-2 *Molecular Symmetry of Proteins Deduced from Diffraction Patterns*

PROTEIN	SPACE GROUP	n' (A.U./UNIT CELL)	n (MOLECULE/ UNIT CELL)	n/n' (MOLECULE/ A.U.)	MINIMUM SYMMETRY OF MOLECULE
Alcohol dehydrogenase	$C222_1$	8	4	$\frac{1}{2}$	2-fold axis
Aspartate transcarbamylase	$P422$ or $P4_22_12$	8	4	$\frac{1}{2}$	2-fold axis[a]
Aspartate transcarbamylase	$P321$	6	2	$\frac{1}{3}$	3-fold axis[a]
Hemoglobin					
(oxy, horse)	$C2$	4	2	$\frac{1}{2}$	2-fold axis
(deoxy, horse)	$C222_1$	8	4	$\frac{1}{2}$	2-fold axis
(oxy, ox)	$P4_132$	24	12	$\frac{1}{2}$	2-fold axis
Immunoglobin	$C2$	4	2	$\frac{1}{2}$	2-fold axis
L-Lactate dehydrogenase	$I422$	16	4	$\frac{1}{2}$	4-fold axis or three mutually perpendicular 2-fold axes[c]
Ribulose bisphosphate carboxylase-oxygenase	$I422$[b]	16	2	$\frac{1}{8}$	4-fold axis and perpendicular 2-fold axes

[a] Both 2-fold and 3-fold axes are presumed to be present, but only one is a crystallographic axis in each space group.
[b] Low-resolution space group.
[c] The latter symmetry has been confirmed from a Fourier map.

from the space group and the number of molecules per unit cell. A test for whether such information is available is the following. Each space group has a certain number of asymmetric units per unit cell (see Sec. 16-6). This number, n', is given in the *International Tables for X-Ray Crystallography*. The ratio of the number of molecules per cell, n, to n' is the number of molecules per asymmetric unit. When n/n' is less than 1, some symmetry element of the space group is a symmetry element of the molecule, and the information of symmetry can be deduced by referring to the *International Tables*. Several examples of this procedure are given in Table 17-2.

EXERCISE 17-10

Does the diffraction pattern of lysozyme contain information on the symmetry of the molecule?

ANSWER

For space group $P4_12_12$, n' is 8 (see answer to Exercise 17-9). From that exercise we also know that $n/n' = 1$. Since this ratio is not less than 1, there is no symmetry information.

17–4 The Intensity of Diffraction

In this section we will see how the intensities of diffraction can be computed from a knowledge of the atomic positions within the unit cell. In practice, chemists and biochemists normally wish to proceed in the opposite direction, from intensities to atomic positions. But nearly every structural determination involves calculation of intensities at some stage, and the concepts of this section are fundamental to understanding the remaining steps in crystal structure determination.

Review of Complex Numbers

The structure factors, which we shall see shortly describe the diffraction from a crystal, are quantities that contain two numbers, one for an amplitude and one for a phase. Physical quantities that contain two numbers are often described conveniently by complex numbers. Throughout our discussion of complex numbers, it should be remembered that there is nothing inherently "complex" or "imaginary" about them, and that they always can be thought of as a combination of an amplitude and a phase, both of which are real numbers. The quantity $i = \sqrt{-1}$ is simply a bookkeeping device for keeping two components separated.

Any complex number can be written as

$$\mathbf{F} = A + iB \tag{17-18}$$

in which A and B are real numbers and i is the square root of -1. The complex conjugate of $\mathbf{F}$, denoted $\mathbf{F}^*$, is defined as $A - iB$. We use the same symbols for complex numbers as for vectors because a complex number can be represented as a vector in the complex plane [Fig. 17-12(*a*)]. The component along the real axis is A and the component along the imaginary axis is B. Let us call the angle between the positive real axis and the vector the *phase angle* or the *phase* of the complex number, and let us measure it in radians. By examining Figure 17-12(*a*), you see that

$$A = F \cos \alpha \qquad \text{and} \qquad B = F \sin \alpha \tag{17-19}$$

in which F is the length of the vector $\mathbf{F}$ and α is its phase. The length F often is called the magnitude or *modulus* of $\mathbf{F}$. From Figure 17-12(*a*) you also can see that

$$F = |\mathbf{F}| = \sqrt{A^2 + B^2} = \sqrt{\mathbf{FF}^*} \tag{17-20}$$ ◀

and

$$\alpha = \tan^{-1} \frac{B}{A} \tag{17-21}$$ ◀

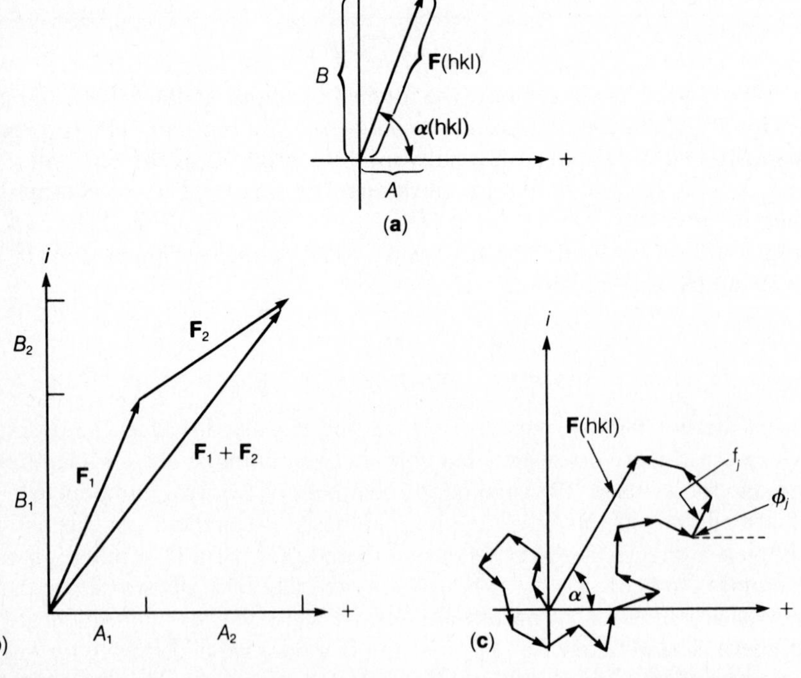

FIGURE 17-12
Representation of structure factors by vectors in the complex plane. **(a)** The structure factor magnitude **F**(*hkl*) is represented by the length of a vector in the complex plane. The phase angle $\alpha(hkl)$ is given by the angle, measured counterclockwise, between the positive real axis and the vector **F**. **(b)** Complex numbers can be added by adding their real and complex components. **(c)** The structure factor for a reflection may be thought of as the vector sum of the x-ray scattering contributions from many atoms. Each of the j contributions may be represented as a vector in the complex plane, with amplitude f_j and phase ϕ_j.

Combining Equations 17-18 and 17-19, we can write

$$\mathbf{F} = F \cos \alpha + iF \sin \alpha \tag{17-22}$$

and substituting Euler's expansion (see inside back cover)

$$e^{ix} = \cos x + i \sin x \tag{17-23}$$

we can convert Equation 17-22 into the simpler exponential form

$$\mathbf{F} = Fe^{i\alpha} \tag{17-22a}$$ ◀

Equations 17-18, 17-22, and 17-22*a* are different ways of expressing the complex number **F**. As shown in Figure 17-12(*b*), we can add complex numbers by adding their real and complex components:

$$\mathbf{F}_1 + \mathbf{F}_2 + \mathbf{F}_3 = (A_1 + A_2 + A_3) + i(B_1 + B_2 + B_3) \tag{17-24}$$

The Structure Factor Describes the Scattering of all Atoms of the Unit Cell for a Given Reflection

Now we are ready to ask why the reflections in an x-ray diffraction pattern differ in intensity. The reason is mainly that for different reflections the contributing atoms scatter with different phases, hence the sum of wavelets is different for each reflection. The situation is exactly analogous to optical diffraction from two holes as described in Figure 17-5.

Each diffracted ray, or reflection, is described by a *structure factor,* $\mathbf{F}(hkl)$. It is a complex number whose magnitude can be determined from the intensity of the *hkl* reflection, according to Equation 17-1. Each structure factor may be regarded as a sum of the contributions of the x-rays scattered from all atoms within the unit cell. Thus we can write

$$\begin{aligned}\mathbf{F}(hkl) = F(hkl)e^{i\alpha(hkl)} &= \sum_{j=1}^{N'} \mathbf{f}_j(hkl) \\ &= \sum_{j=1}^{N'} f_j(hkl)e^{i\phi_j(hkl)}\end{aligned} \tag{17-25}$$ ◀

in which the sum is over all N' atoms within one unit cell. The quantity $\mathbf{f}_j(hkl)$ is called the *atomic scattering factor* of the *j*th atom in the unit cell for the *hkl* reflection. It has an amplitude $f_j(hkl)$ and a phase $\phi_j(hkl)$. The amplitude sometimes is called the *atomic form factor*. This factor is a number related to the number of electrons in the scattering atom, and is tabulated in the *International Tables for X-ray Crystallography*. The phase depends on the position of the atom in the unit cell in a way we now will consider. Once we know how to compute the phase $\phi(hkl)$, we can look up the atomic scattering amplitude $f(hkl)$ in the *International Tables,* and then compute the structure factor for each reflection from Equation 17-25. Note that ϕ_j is the phase for the scattering of the *j*th atom and is distinct from $\alpha(hkl)$, which is the phase for the *hkl* reflection. Their relationship is shown in Figure 17-12(*c*).

The physical interpretation of Equation 17-25 is that the structure factor describes an electromagnetic wave diffracted by the unit cell. h, k, and l describe the direction of the wave, F gives its amplitude, and α gives its phase. This wave is a combination of the wavelets scattered by each atom, and these in turn are described by the atomic scattering factors, $\mathbf{f}_j$. Each atomic scattering factor contains a direction, an amplitude, and a phase. We could describe all these waves by sine and cosine functions, but this is more cumbersome than using complex numbers.

To find how the phase ϕ_j depends on the position of the jth atom, consider Figure 17-13(a), which shows four unit cells of a hypothetical crystal containing a diatomic molecule. We can think of this structure as being composed of two interpenetrating lattices, one for the dark atom and one for the light.

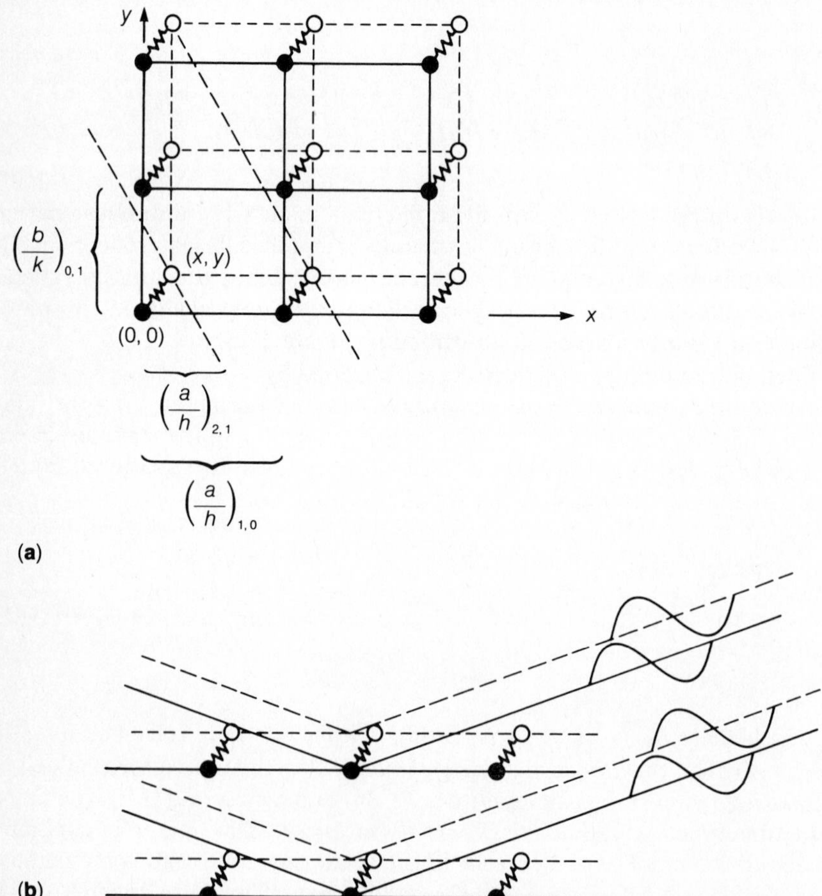

FIGURE 17-13
(a) A two-dimensional crystal containing a "diatomic molecule," represented by ●∿○, showing that different reflections differ in intensity (see text). **(b)** Waves scattered from the lattice of dark atoms are in phase with each other, as are waves scattered from the lattice of light atoms. However, the waves from the two lattices are not in phase with one another.

Though our example contains only two atoms, the discussion is completely general, because we can consider a structure with N' atoms in the unit cell as N' interpenetrating lattices. When we consider the *geometry* of diffraction it is clear that the reflections for the lattice of the dark atoms will be superimposed exactly on the reflections for the lattice of the light atoms. This is so because, by Bragg's law, the angles of reflections depend only on the spacings between crystal planes, and not on their absolute positions, and the spacings for the dark and light atoms are identical.

However, when we consider the intensities of reflections we find that relative positions of the atoms affect the intensities. Consider first the 0,1 reflection, which is determined by the 0,1 set of crystal planes [the horizontal lines in Figure 17-13(*a*)]. A glance at Figure 17-13(*b*) reveals that when Bragg's law is satisfied for the 0,1 planes, wavelets of all the dark atoms are in phase with one another, and that all the light atoms are in phase with one another; but the wavelets scattered by dark atoms are not in phase with wavelets scattered from the light atoms. Thus the intensity for the 0,1 reflection will not be simply the algebraic sum of the contributions from dark and light atoms; a phase difference must be taken into account. If the 2,1 planes now are considered [diagonal lines in Figure 17-13(*a*)], it can be seen that again there is a phase difference between dark and light atoms, but the phase difference is different than for the 0,1 planes. Thus the 2,1 reflection will have a different intensity than will the 0,1 reflection, and in general the intensities of the various reflections depend on the distribution of atoms within the unit cell. When many atoms lie on the *hkl* set of planes, the *hkl* reflection is intense.

Expression for the Structure Factor in Terms of Atomic Coordinates

Now we can derive an expression for the phase difference that arises by displacing the light atoms from the dark atoms. Consider the 0,1 planes shown in Figure 17-13(*a*). If the light atom were moved b/k along b from the dark atom, the phase of its scattering for the *hkl* reflection would shift by 2π radians relative to the dark atom, and hence would be exactly in phase. But it is shifted only a distance y. Therefore, the phase difference for the shift in the b direction is $\phi_y/2\pi = y/(b/k)$, or $\phi_y = 2\pi ky/b$. Similarly, the phase difference for a shift of a light atom by a distance x in the direction of a is given by $\phi_x = 2\pi hx/a$. For a three-dimensional unit cell, the total phase difference contributed to the reflection *hkl* by moving an atom distance x, y, and z from the unit-cell origin is

$$\phi = 2\pi\left(\frac{hx}{a} + \frac{ky}{b} + \frac{lz}{c}\right) = 2\pi(hX + kY + lZ) \tag{17-26}$$

Here X, Y, and Z are coordinates in fractions of unit cell edges. We now can combine Equations 17-26 and 17-25 to get an expression for the structure factor of reflection *hkl* in terms of atomic positions within the unit cell and the atomic scattering factors:

$$\mathbf{F}(hkl) = \sum_{j=1}^{N'} f_j \exp\left[2\pi i \left(\frac{hx_j}{a} + \frac{ky_j}{b} + \frac{lz_j}{c}\right)\right] \tag{17-27}$$ ◀

in which x_j, y_j, and z_j are the coordinates of the *j*th atom in the unit cell. This expression, along with Equation 17-1, allows us to compute the diffraction pattern from atomic positions.

The physical significance of the structure factor equation is that every atom in the unit cell contributes to each *hkl* reflection. The atoms that contribute most strongly are those with the same separation as that of the *hkl* set of crystal planes.

EXERCISE 17-11

A cubic unit cell contains two atoms, each with an atomic scattering factor of 10. One atom is at the origin and the other at $X = Y = Z = \frac{1}{2}$; the capital letters indicate coordinates in fraction of a unit cell edge.

(a) Draw four pictures of the unit cell showing its two atoms and their positions relative to the sets of lattice planes: 1,0,0; 2,0,0; 1,1,0; 1,1,1.
(b) Calculate the intensities of diffraction for the reflections: 1,0,0; 2,0,0; 1,1,0; 1,1,1. (Assume $c = 1$ in Eq. 17-1).
(c) Explain the calculated intensities in terms of the positions of the atoms on the Bragg planes of (a).

ANSWER

(b) $F(hkl) = 10[\exp 2\pi i(0) + \exp 2\pi i(h/2 + k/2 + l/2)]$
$= 10[1 + \exp 2\pi i(h/2 + k/2 + l/2)]$

For 1,0,0: $F(1,0,0) = 10[1 + \exp(2\pi i/2)] = 10[1 - 1] = 0$;
$I(1,0,0) = 0$.

For 2,0,0: $F(2,0,0) = 10[1 + \exp 2\pi i(1)] = 10[1 + 1] = 20$.
$I(2,0,0) = 20^2 = 400$.

For 1,1,0: $F(1,1,0) = 10[1 + \exp 2\pi i(1)] = 20$
$I(1,1,0) = 20^2 = 400$.

For 1,1,1: $F(1,1,1) = 10[1 + \exp 2\pi i(\frac{3}{2})] = 10[1 - 1] = 0$
$I(1,1,1) = 0$.

(c) For 2,0,0: the 2,0,0 lattice planes all contain atoms and hence $I(2,0,0)$ is strong. The same is true for 1,1,0.

For 1,0,0: the 1,0,0 lattice planes contain atoms, but there are an equal number of identical atoms midway between, and these scatter exactly out of phase to interfere destructively. The same is true for 1,1,1.

Other Ways of Writing the Structure-Factor Equation

The structure-factor equation (17-27) can be simplified by putting it into vector form. Instead of specifying a reflection by h, k, and l, let us use the vector **S**, which runs from the origin of the reciprocal lattice to the reciprocal lattice point *hkl*. We can write this vector as

$$\mathbf{S} = h\mathbf{a}^* + k\mathbf{b}^* + l\mathbf{c}^* \tag{17-28}$$

in which h, k, and l are the components of **S** along the reciprocal axes $\mathbf{a}^*$, $\mathbf{b}^*$, and $\mathbf{c}^*$. We also can express the atomic positions in terms of the vector $\mathbf{r}_j$, which extends from the origin of the unit cell to the jth atom. We can write **r** as

$$\mathbf{r} = X\mathbf{a} + Y\mathbf{b} + Z\mathbf{c} \tag{17-29}$$

in which $X_j = x_j/a$ is the x coordinate in units of the unit cell length, and similarly for Y_j and Z_j. Now the scalar product of **S** and **r** has a particularly simple form:

$$\mathbf{r} \cdot \mathbf{S} = hX + kY + lZ \tag{17-30}$$

because cross terms of the form $\mathbf{a} \cdot \mathbf{b}^*$ are zero and self terms of the form $\mathbf{a} \cdot \mathbf{a}^*$ are unity, as shown in Equation 17-15. Thus the vector form of the structure-factor equation is

$$\mathbf{F}(\mathbf{S}) = \sum_{j=1}^{N'} f_j\,[\exp 2\pi i(\mathbf{r}_j \cdot \mathbf{S})] \tag{17-31}$$

A more general form of the structure-factor equation, which is closely analogous to Equation 17-31, treats the electron density of the diffracting object as continuous, rather than as a sum of discrete atoms:

$$\mathbf{F}(\mathbf{S}) = \int \rho(r)\,[\exp 2\pi i(\mathbf{r} \cdot \mathbf{S})]\,dV \tag{17-32}$$ ◀

in which the integral is over the volume of the object.

17–5 *Fourier Showed That Periodic Objects can be Represented as a Sum of Waves*

In the preceding section we saw that the diffraction pattern of a crystal may be regarded as a sum of waves. In this section we shall find that the crystal also can be considered to be a sum of waves, and that the amplitudes of these waves are the structure-factor magnitudes obtained from the diffraction pattern.

In the early nineteenth century the French mathematician Fourier showed that any well-behaved periodic function can be represented as a superposition of sinusoidal waves. For example, the periodic function at the bottom of Figure 17-14 can be represented as a sum of the five sinusoidal waves above it, plus a constant. Each of these waves is characterized by a wave number n (the number of oscillations made by the wave within one period of the function), a phase $\alpha(n)$ (the horizontal displacement of the wave maximum from a common origin), and an amplitude $A(n)$ (one-half the vertical distance from crest to trough).

Since the electron density of a crystal is a periodic function it also can be represented as a sum of component waves, each of which has the proper amplitude and phase. Two one-dimensional "crystals" are shown in Figure 17-15. One contains a "diatomic molecule" and the other a "triatomic

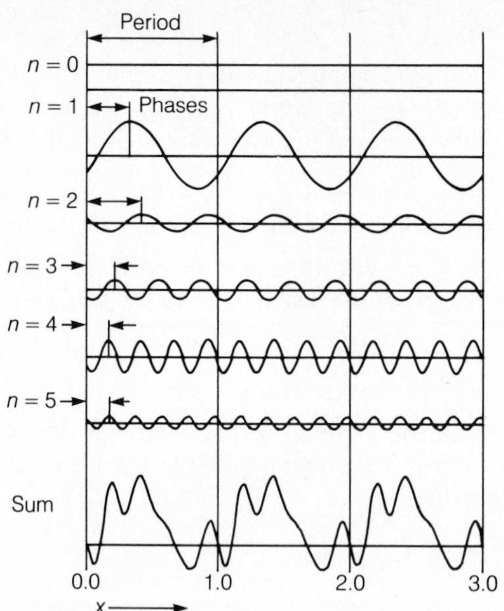

FIGURE 17-14
Superposition of sinusoidal waves to yield a periodic function. The wave numbers and phases of the component waves are shown. (From Waser, 1968.)

molecule," yet they both are constructed from the same waves. It is the differences in the amplitudes and phases of the waves in the two crystals that result in the different distributions of "electron density." Note that the waves of small wave number define the locations of the molecules within the unit cell, whereas the waves of higher wave number define the molecular details. In other words, adding higher terms increases the resolution of the structure.

We can express the sum of waves in mathematical terms as follows.

$$\rho(X) = \sum_n A(n)\{\cos[2\pi nX + \alpha(n)] + i \sin[2\pi nX + \alpha(n)]\} \quad (17\text{-}33)^1$$

Using Euler's expansion (Eq. 17-23), we find

$$\rho(X) = \sum_n A(n) \exp[i\alpha(n)] \exp(2\pi inX) = \sum_n \mathbf{A}(n) \exp(2\pi inX) \quad (17\text{-}34)$$

The relationship of the amplitude $A(n)$ to the structure factor can be determined by inserting this expression for $\rho(X)$ given in Equation 17-34 into Equation 17-32. If this is done, it is found that $\mathbf{A}(n) = \mathbf{F}(-h)/L$, in which L is the length of the period, and F is a structure factor for the "one-dimensional

[1] For this equation to hold, n must run from $-\infty$ to ∞, rather than from 0 to ∞ as suggested in Figure 17-14. An alternative expression for this sum with terms only from 0 to ∞ is given in Problem 17-27, but is not commonly used in crystallography.

crystal'' of Figure 17-14. In other words, Equation 17-33 for the ''electron density'' of the one-dimensional crystal can be written

$$\rho(X) = \frac{1}{L} \sum_h F(h) \exp i\alpha(h) \exp(-2\pi i h X) \tag{17-35}$$ ◀

In x-ray diffraction experiments, it is observed that $F(h) = F(-h)$. This is called *Friedel's law.* One would expect this equality since $F(h)$ and $F(-h)$ describe reflections from opposite sides of the same set of lattice planes. The summation of exponentials in Equation 17-35 is called a *Fourier synthesis.*

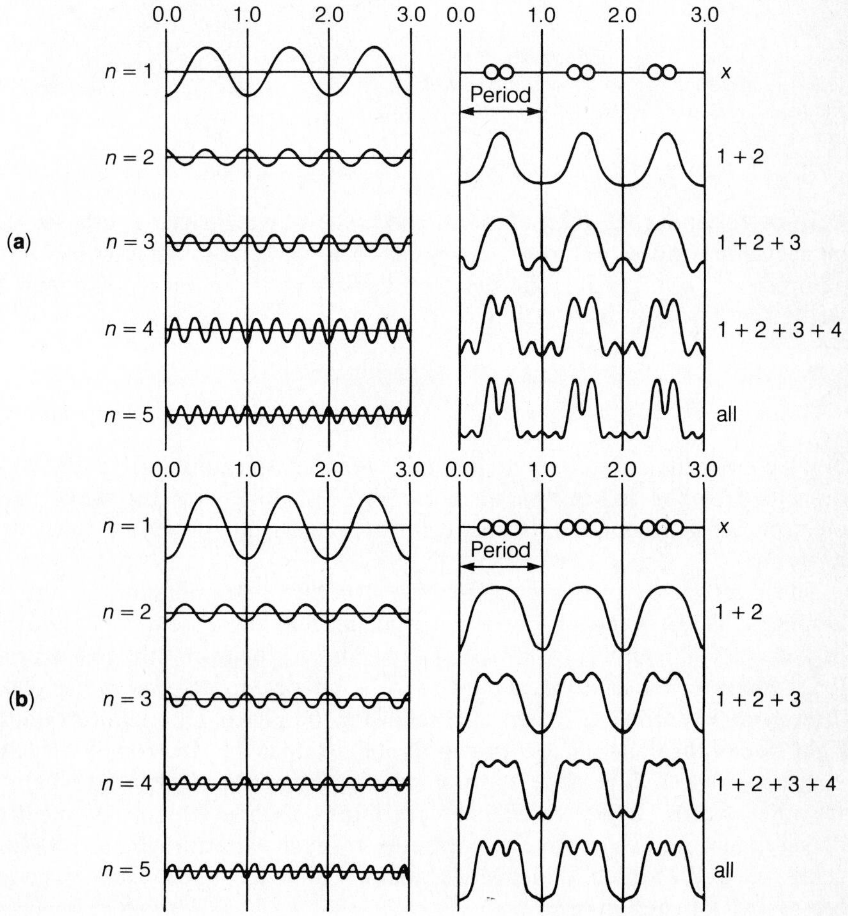

FIGURE 17-15
(a) Representation of the electron density of a one-dimensional "crystal" by a superposition of waves. The crystal is formed by a periodic repetition of a diatomic molecule, as shown at the top of the right-hand column. The component waves, each with proper phase and amplitude, are on the left. The curves on the right show the successive superposition of the five waves on the left. (From Waser, 1968.) **(b)** Representation of another one-dimensional crystal, this one containing a triatomic molecule. Note that this crystal is built up from the same waves as the crystal of **(a)**; only the amplitudes and phases have been changed. (From Waser, 1968.)

(Continued)

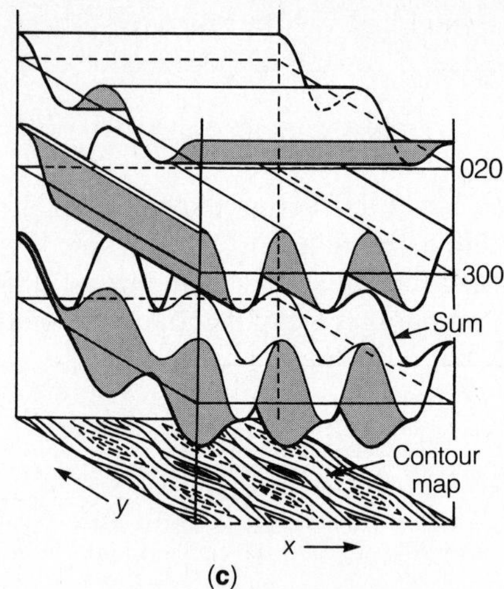

FIGURE 17-15 *(Continued)*
(c) The summation of two-dimensional waves to produce a two-dimensional "electron density." (From Jeffery, 1972.)

To represent a real, three-dimensional crystal by a Fourier synthesis, one must include waves in three dimensions. The wave number now has three components: h, k, and l. The electron density at the point X, Y, Z can be written by analogy to Equation 17-35 as

$$\rho(X,Y,Z) = \frac{1}{V}\sum_h\sum_k\sum_l F(hkl)\exp[i\alpha(hkl)] \exp[-2\pi i(hX + kY + lZ)] \quad (17\text{-}36)$$

in which V is the volume of the unit cell, $F(hkl)$ is the amplitude of the wave described by the indices hkl, and $\alpha(hkl)$ is the phase of the wave. This electron-density equation is the central relationship in crystal structure analysis.

The significance of this equation for structure determination is that the amplitude $F(hkl)$ is the structure-factor magnitude for reflection hkl, and can be determined in an x-ray diffraction experiment. The term "the phase problem" refers to the fact that as yet there is no way to determine the phase directly by experiment. Note that except for the phase, all quantities on the right side of the equation are known. Note also that the electron density at a single point (X,Y,Z) is made up from contributions from all structure factors. In other words, every reflection contributes to every point of electron density, just as every atom contributes to each structure factor. In this sense, as well as in cell dimensions, reciprocal space and crystal space are reciprocals of each other.

Figure 17-15(c) is an aid in visualizing the meaning of Equation 17-36. It shows a two-dimensional Fourier synthesis of two waves. The resulting surface (marked "sum") shows $\rho(X,Y)$. It is also represented in the lowest level of the figure as a contour map. A three-dimensional Fourier synthesis shows $\rho(X,Y,Z)$. It can be represented [as in Fig. 17-3(a)] by a stack of sections, each a slice through the unit cell at some constant value of Z, showing contours of $\rho(X,Y,Z)$.

EXERCISE 17-12

Show that Equation 17-35 can be simplified to the form

$$\rho(X) = \frac{2}{L}\sum_{h=1}^{\infty} F(h)\cos\left[\alpha(h) - 2\pi ihX\right] + \frac{F(0)}{L}$$

where the term for $h = 0$ is not included in the summation where it would be multiplied by 2. (*Hint:* Use Equation 17-23 to expand the exponentials, then pair terms for h and $-h$.)
Note that $F(h) = F(-h)$ and that $\alpha(h) = -\alpha(-h)$.

ANSWER

$$\rho(X) = \frac{1}{L}\sum_{h=-\infty}^{\infty} F(h)\exp i\alpha(h)\exp(-2\pi ihX)$$

$$= \frac{1}{L}\sum_{h=-\infty}^{\infty} F(h)\exp\left[i\alpha(h) - 2\pi ihX\right]$$

$$= \frac{1}{L}\sum_{h=-\infty}^{\infty} F(h)\{\cos\left[\alpha(h) - 2\pi hX\right] + i\sin\left[\alpha(h) - 2\pi hX\right]\}$$

Including explicitly terms of negative h within the sum, we get

$$\rho(X) = \frac{1}{L}\sum_{h=1}^{\infty} F(h)\{\cos\left[\alpha(h) - 2\pi hX\right] + \cos\left[\alpha(-h) + 2\pi hX\right]$$
$$+ i\sin\left[\alpha(h) - 2\pi hX\right] + i\sin\left[\alpha(-h) + 2\pi hX\right]\} + \frac{F(0)}{L}$$

Recalling that $\cos(-x) = \cos x$, and that $\sin(-x) = -\sin x$, and noting that $F(h) = F(-h)$ and that $\alpha(h) = -\alpha(-h)$, we obtain

$$\rho(X) = \frac{2}{L}\sum_{h=1}^{\infty} F(h)\cos\left[\alpha(h) - 2\pi hX\right] + F(0)/L \qquad (17\text{-}37)$$ ◀

JEAN BAPTISTE JOSEPH FOURIER

1768–1830

By virtue of intelligence and administrative skill, Fourier rose from humble beginnings in the Burgundy region of France to a position of power and influence during the time of Napoleon. He also achieved lasting fame for the analytical methods that bear his name. For modern scientists, the wonder of his career is that he was able to make profound discoveries in between revolutions, insurrections, expeditions in Egypt, stints in jails, and service in high political offices.

Fourier was born the son of a poor tailor, and was orphaned by the time he was eight. He was placed by the town's archbishop in the local military school where he displayed both literary and mathematical talent. At age 12 he was writing sermons that were used in Paris (reputedly with great effect), and by the time he was 21 he delivered his first mathematical paper before the Academy of Sciences.

Fourier selected the military for a career and applied for admission to the artillery, with a strong recommendation from Legendre. He was rejected, however, with the statement, "Fourier, not being of noble birth, cannot enter the artillery, not even if he is a second Newton." Instead he took a place teaching mathematics at his old school.

Having been barred from his intended career by the archaic social order, it is not surprising that Fourier embraced the cause of the French Revolution. He participated, among other ways, as a member of the Citizens' Committee on Surveillance. In this position he defended victims of the Terror, so that soon he himself was arrested and jailed. An appeal to Robespierre, head of the Terror, was unsuccessful, but Fourier was released after the turn of events that led to Robespierre's own arrest and execution.

In 1795, when the École Polytechnique was organized as an elite institute of higher education, Fourier was selected as a teaching assistant to the great mathematicians Lagrange and Monge. Once again he was arrested (this time as a supposed supporter of Robespierre's), and again released. Then in 1798 he was selected to join Napoleon's Egyptian campaign. During three years in Egypt Fourier organized factories for the army, led scientific expeditions, was in charge of receiving complaints from the Egyptian population, collected material for the famous *Description of Egypt,* and in collaboration with Napoleon wrote the introduction to this work. This writing so enhanced Fourier's literary reputation that eventually he won membership in the French Academy.

Returning to France in 1801, Fourier wished to resume mathematical work at the École Polytechnique, but Napoleon had become so impressed with Fourier's administrative skills that he appointed him prefect (governor) of the department of Isere, centered at Grenoble. During the next 15 years, his achievements in

Isere included extensive road building and drainage of marshlands. For these Napoleon named him Baron of the Empire in 1808.

During this period Fourier developed the infinite series and the integral transforms that are named for him. These mathematical techniques, now used in the most diverse applications, were stimulated by Fourier's study of the conduction of heat. His paper before the Academy of Sciences on this subject in 1807 was accepted by Laplace and others, but was opposed by Lagrange, who did not accept the idea of trigonometric series. To induce Fourier to extend and improve his ideas, the Academy offered a prize in 1812 for a mathematical theory of the propagation of heat. Fourier rewrote his paper, introducing his integral transform. The judges—Laplace, Lagrange, and Legendre—awarded the prize to Fourier, but the paper was still criticized on grounds of lack of "rigor and generality."

Fourier was thrown from political office during the events that followed Napoleon's return from Elba. Despite political realities, Fourier remained loyal to Napoleon, and with the Emperor's final defeat, Fourier was out of office and nearly penniless. This is where his story takes a heartwarming turn—at least for teachers: a former student of Fourier's from the École Polytechnique, by this time prefect of the department of the Seine, appointed Fourier director of its Bureau of Statistics. This position paid an adequate salary, but was not demanding in time, so that he could continue his mathematical research. Then as the years passed, Fourier's intellectual achievements were rewarded by memberships in both the French Academy of Sciences and the French Academy (of letters). Several times, however, these rewards were delayed because of his past political associations. He died at the age of 62 from complications of an illness that dated from his service in Egypt.

⬡ *Which are More Important, Phases or Intensities? The Example of DNA*

In the following three sections we will discuss methods for determining phases from intensity data. This might lead one to suppose that the amplitudes are much more important in determining the electron density than are phases. In fact, the phases may be more important, as the following example indicates.

Our example is taken from a discussion (Donohue, 1969) of methods used to determine the structure of DNA. Figure 17-16(*a*) shows the electron density at 2 Å resolution of an adenine-thymine pair. The electron density was calculated by Equation 17-36, using amplitudes and phases from a model, via Equation 17-27. As one would expect at this resolution, the contours of electron density show a somewhat blurred representation of the model (which is indicated by the ball-and-stick drawing embedded in the electron density). The model is, of course, reflected in the electron density, since the density is computed from phases and amplitudes that both come from the model. Figure 17-16(*b*) shows a conceivable alternative mode of hydrogen bonding of the A-T pair. The electron density of the alternative was calculated from the alternative model in the same way. The question arises, "Can we determine which pairing is the correct one if we have only the measured

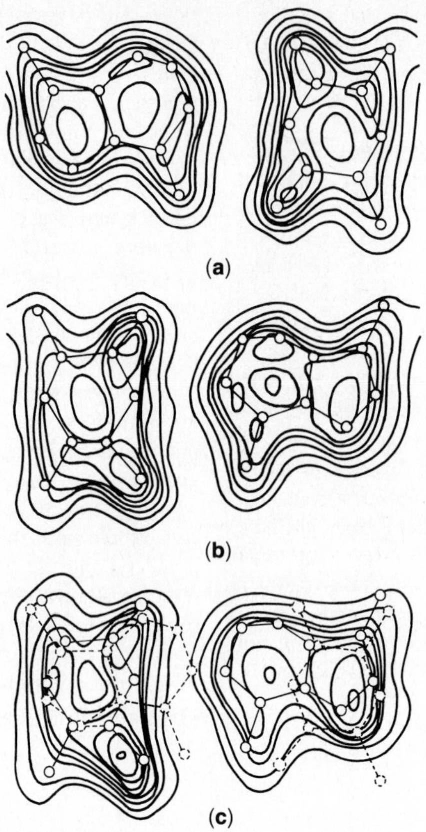

FIGURE 17-16
Calculations of the electron density of adenine-thymine pairs, illustrating the relative importance of phases and amplitudes. **(a)** Electron density of an adenine-thymine pair in one possible configuration, calculated with the amplitudes and phases of this pair. **(b)** Electron density of an alternative configuration, calculated with the amplitudes and phases of this pair. **(c)** Electron density calculated with the amplitudes of the pairing scheme in **(a)** (dotted circles) and the phases of the pairing scheme in **(b)** (full circles). Notice that the phases determine the electron density to a far greater extent than do the amplitudes. (From Donohue, J., 1969, *Science 165,* 1091. Copyright 1969 by the American Association for the Advancement of Science.)

amplitudes?'' Figure 17-16(*c*) suggests that it may be difficult to do so. It shows the electron density calculated by combining the amplitudes from Figures 17-16(*a*) with the phases from Figure 17-16(*b*), and vice versa. Apparently the electron density resembles the model from which the phases were taken much more closely than the model from which the amplitudes were taken.

Since it is the amplitudes that are available from experiment and the phases that must be discovered, this result might be taken as an indication that determination of the phases from the amplitudes is impossible. In fact, the ingenuity of crystallographers during the past fifty years has produced a variety of methods that work. We shall consider three of the most useful in the following three sections.

17–6 The Patterson Function

One of the most potent weapons in the crystallographer's arsenal for assault on the phase problem is the *Patterson function*. This function is a map, similar to the electron-density map, that shows the vectors between all atoms in a unit cell, rather than the atoms themselves. It has the distinct advantage over the electron-density map in that it can be computed from observed amplitudes alone. For simple structures, the Patterson map often can be unraveled to yield atomic positions. Even for larger structures this function is often helpful. For protein studies it yields the locations of bound heavy atoms, which in turn yield the phases.

The Patterson function is a Fourier synthesis, similar to Equation 17-36, in which the coefficients $F(hkl)\exp[i\alpha(hkl)]$ are replaced by $[F(hkl)]^2$:

$$P(U,V,W) = \frac{1}{V}\sum_h\sum_k\sum_l F^2(hkl)\exp\left[-2\pi i(hU + kV + lW)\right] \quad (17\text{-}38)$$

The coordinates U, V, and W are used in place of X, Y, and Z to emphasize that the Patterson vector map is distinct from the Fourier electron-density map. Notice that once the amplitudes $F(hkl)$ have been measured, the function $P(U,V,W)$ can be completed.

EXERCISE 17-13

Express the Patterson function for a one-dimensional crystal in the simplified cosine form of Equation 17-37.

ANSWER

$$P(U) = \frac{2}{L}\sum_{h=1}^{\infty} F^2(h)\left[\cos 2\pi hU\right] + F^2(0)/L \quad (17\text{-}39)$$

You can understand the meaning of the Patterson function by considering a one-dimensional crystal which has only two atoms per cell at fractional coordinates X_1 and X_2. The structure factor for reflection h from this crystal is

$$\mathbf{F}(h) = f_1 e^{2\pi i(hX_1)} + f_2 e^{2\pi i(hX_2)} \quad (17\text{-}40)$$

By Equation 17-20 and the definition of the complex conjugate, we can express $F^2(h)$ as follows:

$$F^2(h) = \mathbf{F}(h)\,\mathbf{F}^*(h) = \mathbf{F}(h)\,\mathbf{F}(\bar{h}) \quad (17\text{-}41)$$

in which $\bar{h}$ is crystallographic notation for $-h$. Substituting Equation 17-40 into Equation 17-41, we obtain

$$\begin{aligned} F^2(h) &= [f_1 e^{2\pi i(hX_1)} + f_2 e^{2\pi i(hX_2)}]\cdot[f_1 e^{-2\pi i(hX_1)} + f_2 e^{-2\pi i(hX_2)}] \\ &= f_1^2 + f_2^2 + f_1 f_2 e^{2\pi i[h(X_2-X_1)]} + f_1 f_2 e^{2\pi i[h(X_1-X_2)]} \end{aligned} \quad (17\text{-}42)$$

Examination of this equation shows that it is the structure-factor expression for a hypothetical "crystal" containing four "atoms" in the unit cell. Of

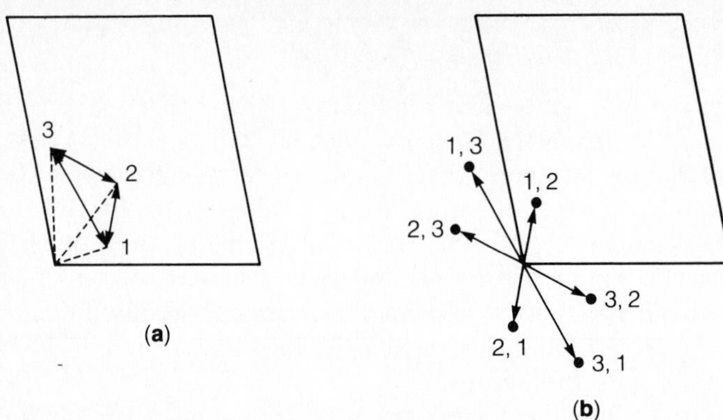

FIGURE 17-17
(a) A structure consisting of three-point atoms. **(b)** The Patterson function corresponding to the structure. (From Buerger, 1970.)

these "atoms," two with atomic form factors f_1^2 and f_2^2 are located at the origin, and two with form factors f_1f_2 are located at $(X_2 - X_1)$ and $(X_1 - X_2)$. The positions of these "atoms" are in fact the vectors between the two original atoms (including the self-vectors between each atom and itself). The self-vectors have zero length, and accordingly are at the origin. Thus the Patterson function of Equation 17-39 may be thought of as a map that shows not atoms at X_1 and X_2, but the vectors between all pairs of atoms. Buerger (1970) shows that this result may be extended from two atoms to any number of atoms in the unit cell.

An example of a Patterson map corresponding to a structure of three points is shown in Figure 17-17. Notice that if there are N atoms in a structure, there are N^2 peaks in the corresponding Patterson map, N of which represent the "self-vectors" at the origin. Thus, whereas it is trivial to derive the locations of the three atoms of Figure 17-17(*a*) from the Patterson map of Figure 17-17(*b*), it is hopeless to derive the locations of the, say, 10^4 atoms in the unit cell of a protein from the 10^8 peaks of its corresponding Patterson map.

Nevertheless, the Patterson function is an indispensable tool in protein crystallography. We shall see in Section 17-8 that the phases of reflections from a protein crystal can be determined by studying at least two *isomorphous heavy-atom derivatives.* Each of these is a crystal identical to a crystal of the native protein, except that it contains a heavy atom bound at the same point in every unit cell. A preliminary step in determining phases is to find the coordinates of the heavy atom within the unit cell. This can be done once the structure-factor amplitudes have been measured for both the native crystal and for a crystal containing a heavy atom bound to the protein. Let us denote these amplitudes $F(hkl)$ and $F_H(hkl)$, respectively. Then the *difference Patterson function,*

$$P(UVW) = \frac{1}{V}\sum_h\sum_k\sum_l [F_H(hkl) - F(hkl)]^2 \times \exp[-2\pi i(hU + kV + lW)] \quad (17\text{-}43)$$

will show the vectors between heavy atoms in the unit cell. Since there usually are relatively few heavy atoms, their positions can often be determined, just as the atomic positions of a simple crystal structure can be determined from its Patterson function. In Section 17-8 we will see how the heavy-atom positions can be used to determine the phases of reflections.

17–7 The Heavy-Atom and Fourier Methods

The *heavy-atom method* can solve the phase problem for molecules of molecular weight as great as about 1,000. The method has been used to determine structures for many molecules, both with and without biological significance.

In the heavy-atom method, attention is focused initially on an atom of high atomic number in the structure to be solved. The atom may be a naturally occurring ligand, or it may have been inserted chemically by a crystallographer. The first step is to determine the coordinates of the heavy atom within the unit cell. This often can be achieved with a Patterson map, since the weight with which terms contribute to the Patterson function (as shown by Equation 17-42) is proportional to the product of the atomic scattering factors of the two atoms separated by a given vector. If you recall that the atomic scattering factor is related closely to the number of electrons of an atom, then it will be clear that the ratio of the contribution of the vectors between heavy atoms to one between light atoms is proportional to the ratio of the *squares* of their atomic numbers. In short, if an atom is heavy enough, it dominates the Patterson map, and its position in the unit cell can be determined from the map.

Once the heavy atom's position is known, the positions of other atoms can be determined by a procedure of "bootstrapping." The basis for this bootstrapping is that the diffraction pattern is dominated by the heavy atom. This situation is illustrated in Figure 17-18, in which the structure factor of a crystal with a heavy atom is represented as a sum of atomic scattering factors. When the heavy atom has an atomic scattering factor far larger than

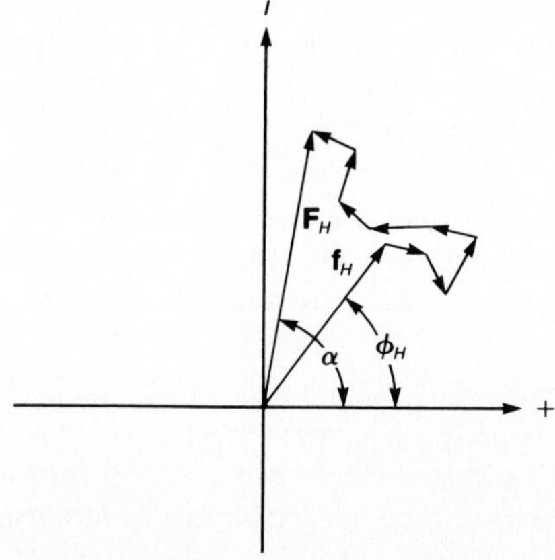

FIGURE 17-18
The phase α for a reflection of a compound containing a heavy atom is likely to be near the phase ϕ_H for the heavy atom contribution.

that of any light atom, the phase of the heavy atom $\phi_H(hkl)$ is likely not to be very different from the phase of the reflection, $\alpha(hkl)$. Thus, as a first approximation it is reasonable to take the ϕ_H for the unknown $\alpha(hkl)$. The heavy-atom phase can be computed from Equation 17-27, with the aid of Equation 17-21 (once the coordinates of the heavy atom have been determined from the Patterson map).

Assuming that $\alpha(hkl) = \phi_H(hkl)$, one can compute the electron density by Equation 17-36, using the observed structure-factor amplitudes, $F(hkl)$. This trial electron-density map certainly will show the heavy atom, since its phases are used in computing the map. The map is likely to show some other atoms too, because experimental amplitudes are used in the map, and these contain information on the positions of the other atoms. The positions of these atoms can be used, along with the position of the heavy atom, to determine a new set of trial phases. Then these phases can be combined with the experimental amplitudes to yield an improved electron-density map with still more atoms. This procedure of iterative bootstrapping (by successive phase and electron-density calculations) from a knowledge of one atom to a knowledge of the entire structure is called the *Fourier method.*

EXERCISE 17-14

The combined heavy-atom and Fourier method works best when the square of the atomic number of the heavy atom approximately equals the sum of the squares of the light atoms in the structure—that is, when

$$(Z_H)^2 \simeq \sum_{j=1}^{N'} Z_i^2 \qquad \text{(17-44)}$$

Using this equation, make an order-of-amplitude calculation to show that the heavy-atom method is hopeless for a protein of molecular weight 28,000.

ANSWER

The mass of the protein arises mainly from oxygen with 8 electrons, nitrogen with 7 electrons, and carbon with 6 electrons. Let us then represent the x-ray scattering as that from N' nitrogen atoms, where N' is roughly $2.8 \times 10^3/14 = 2 \times 10^3$. Thus

$$(Z_H)^2 \simeq \sum_{j=1}^{2000} 7^2 = 2 \times 10^3 \times 49 = 9.8 \times 10^4$$

and $Z_H \simeq 3.1 \times 10^2$. Unfortunately, the periodic table ends just above $Z = 1 \times 10^2$.

Recalling from the illustration of DNA that the phases rather than the amplitudes dominate an electron-density map, you may wonder if there is danger in this bootstrapping method. Could you bootstrap yourself into an imaginary structure, with the phases of the imagined structure dominating the experimental intensities? The answer is that such a disaster is always possible, and that the crystallographer must appraise his maps with an objective eye. One check on reliability is that the final structure is chemically reasonable. A more quantitative measure of correctness is the *discrepancy index, R,* defined by

$$R = \frac{\sum_{hkl} \left| |F_{\text{obs}}| - |F_{\text{calc}}| \right|}{\sum_{hkl} |F_{\text{obs}}|} \tag{17-45}$$

where F_{obs} is the observed amplitude (from Eq. 17-1) and F_{calc} is the amplitude calculated from the structure by Equation 17-27. Clearly R is a measure of the deviation of the model structure from the true structure. For a careful determination of the structure of a small molecule, R is often between 0.02 and 0.08. For a random arrangement of atoms within the unit cell, R is 0.59. For molecules as large as proteins, R is seldom smaller than about 0.2.

17–8 Isomorphous Replacement

The principal method of phase determination used in solving protein structures is *isomorphous replacement*. In this method the phase of each reflection is found by comparing its structure-factor amplitude to that of corresponding reflection of modified crystals, called *heavy-atom derivatives,* to which heavy atoms have been added. The heavy atom may be covalently bonded to the protein (such as an organomercurial bound to a protein sulfhydryl) or it may be held in place by noncovalent forces (such as a $PtCl_4^{2-}$ group coordinated to a sulfur atom of a methionine residue). All that matters is that the modified crystal is the same as the parent crystal, except that the heavy-atom group replaces some solvent.

The presence of the heavy atom in a derivative crystal alters the intensities of diffraction slightly. If we represent the structure factor of the derivative crystal for reflection *hkl* by $F_H(hkl)$, then we may write

$$\mathbf{F}_H(hkl) = \mathbf{F}(hkl) + \mathbf{f}_H(\text{hkl}) \tag{17-46}$$

where $\mathbf{f}_H(hkl)$ is the atomic scattering factor for the heavy atom. This equation is a special case of Equation 17-27, in which the contribution of the heavy atom to the scattering is shown separately from the contributions of all other atoms of the unit cell. A geometric representation of Equation 17-46 is shown in Figure 17-19(a), where $\mathbf{F}_H$ is seen to be the vector sum of $\mathbf{F}$ and $\mathbf{f}_H$.

When we have recorded the diffraction patterns of the parent and derivative crystals, we initially know only the amplitudes of $\mathbf{F}$ and $\mathbf{F}_H$. Referring to Figure 17-19(b), we see that knowing the amplitude of a reflection is equivalent to knowing the radius of a circle in the complex plane. However, it is possible to find both the amplitude and phase of the heavy-atom contribution, $\mathbf{f}_H$. This is done by first determining the coordinates of the heavy atom in the unit cell, $X_H Y_H Z_H$, from the difference Patterson function, Equation

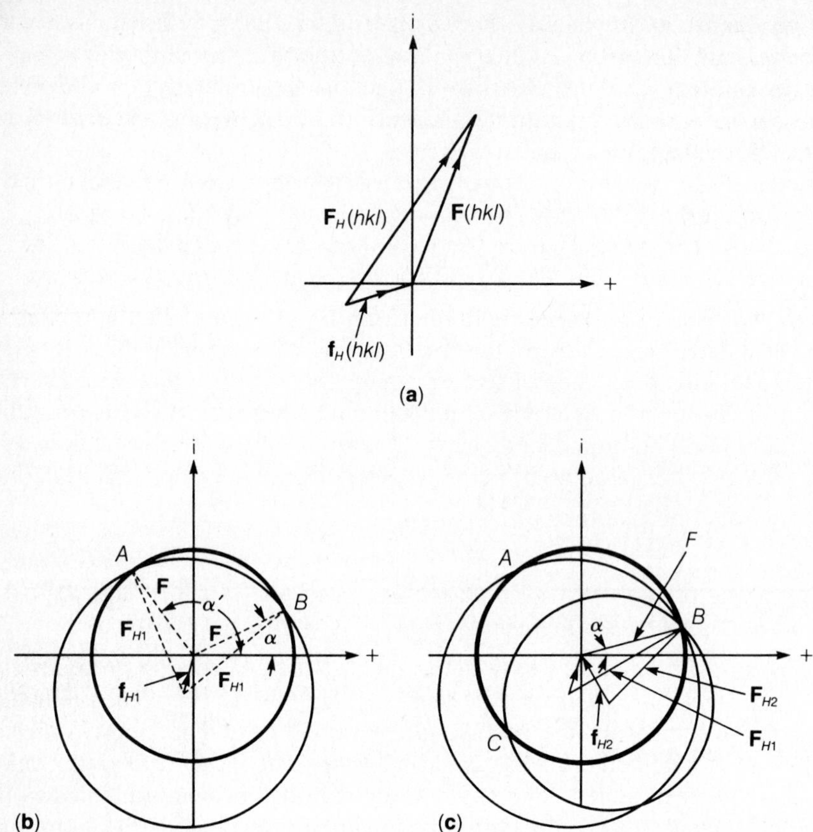

FIGURE 17-19
Phase determination by the method of isomorphous replacement. **(a)** The structure factor for any reflection of a heavy atom derivative can be represented as a vector sum of the structure factors of the native protein and the heavy atom. **(b)** A single heavy atom derivative gives two possible values (α and α') for the phase of each reflection, as explained in the text. The phase circle for the native protein crystal is indicated by a heavy line. **(c)** Data from a second heavy atom derivative indicate that α is the phase angle for this reflection.

17-43. Then inserting these coordinates into the structure-factor equation, we can write $\mathbf{f}_H$ as

$$\mathbf{f}_H = f_H \exp[2\pi i(hX_H + kY_H + lZ_H)] \tag{17-47}$$

We are now in a position to determine the phase of reflection *hkl*. First we represent the heavy-atom structure factor $\mathbf{f}_H$ as a vector in the complex plane, with its head at the origin as shown in Figure 17-19(*b*). Because we know the amplitude of **F**, we can represent by a circle of radius *F* the locus of points where the head of **F** may lie. Similarly, we can represent by a circle of radius F_H, the locus of points where the head of $\mathbf{F}_H$ may lie. Let us place the center of the second circle at the tail of the vector $\mathbf{f}_H$. Then the two intersections of the circles [points *A* and *B* in Figure 17-19(*b*)] represent the only possible points where the heads of vectors **F** and $\mathbf{F}_H$ may lie and simultaneously satisfy Equation 17-46, which states that $\mathbf{F}_H$, **F**, and f_H must form a triangle. In other words, the phase of reflection *hkl* is limited to two possible values, α and α' in Figure 17-19(*b*). These values correspond respectively to intersections *B* and *A* of the two circles.

Which of the two intersections is the true phase angle for reflection *hkl*? We can select the correct phase by adding a second heavy-atom derivative to the analysis. The coordinates of the second heavy atom can be found from a difference Patterson map. These lead via Equation 17-47 to $\mathbf{f}_{H_2}$, which can be plotted in the complex plane as shown in Figure 17-19(*c*). Then by centering a circle of radius F_{H_2} at the tail of the vector $\mathbf{f}_{H_2}$, we will find that the circle intersects the native protein circle at two points [*B* and *C* in Fig. 17-19(*c*)]. One of these points (*B*) lies near to one of the two intersections of the circle for the first derivative. This common intersection is the phase angle $\alpha(hkl)$. Thus in the method of isomorphous replacement, the phase angle is determined directly from experiment.

The successes of protein crystallography stem directly from the demonstration by Max Perutz in 1954 that isomorphous replacement could be used to determine phases of protein diffraction patterns. Owing to difficulties in preparing the heavy-atom derivatives and in carrying out the necessary calculations before our present era of powerful computers, the first protein structure was not determined until 1960, when Kendrew and his co-workers solved the structure of myoglobin. Perutz and Kendrew shared the Nobel Prize for chemistry in 1962 for this monumental achievement.

QUESTIONS FOR REVIEW

1. What is a *reflection?* What are its *indices?* How is the intensity of a reflection related to the magnitude of the corresponding structure factor?
2. Describe diffraction from a hole and from a pair of holes. Why does diffraction occur?
3. Describe each of the following properties of diffraction: dimension, sharpness, and sampling. Give an illustration of each.
4. In what ways is x-ray diffraction from a crystal like optical diffraction from an array of holes?
5. What is Laue's equation for diffraction? What is the meaning of **S**, the diffraction vector?
6. What are lattice planes and their Miller indices? Give some examples.
7. What is Bragg's law of diffraction? How is the diffraction vector **S** related to the lattice planes?
8. How is information on molecular weight and molecular symmetry derived from diffraction patterns?
9. Define the structure factor in terms of atomic properties. Why do different reflections have different intensities?

10. Write the structure-factor equation in three ways.

11. Write the expression for a periodic structure as a sum of waves and define all symbols. How is the structure factor related to the waves?

12. What is the Patterson function? What is its interpretation in terms of atomic positions? What is its usefulness?

13. Describe two ways in which a heavy atom can be used to determine the phases of x-ray reflections.

PROBLEMS

1. Are the following statements true or false?

   (a) Each x-ray reflection corresponds to a diffracted ray in a given direction.

   (b) Each x-ray reflection originates from a single atom.

   (c) The spacing of the diffraction pattern is inversely related to the spacing of the diffracting object.

   (d) The reciprocal lattice vector a^* must be perpendicular to the direct lattice vectors b and c.

   (e) The higher are the indices hkl of an x-ray reflection, the more narrowly spaced are the lattice planes from which it is reflected.

   (f) The sharpness of a diffraction pattern decreases as the number of diffracting objects increases.

   (g) Every atom in the unit cell contributes to the intensity of a given reflection.

   (h) Phases are more important than amplitudes for defining electron density.

   (i) The Patterson function is a Fourier synthesis in which the coefficients are the structure factor amplitudes, $F(hkl)$.

   (j) Once the position of a heavy atom is known, its contribution to the structure factor can be computed.

   Answers: (a) T. (b) F. (c) T. (d) T. (e) T. (f) F. (g) T. (h) T. (i) F, they are $F^2(hkl)$. (j) T.

2. Add the appropriate words.

   (a) Diffraction is the ________ propagation of light.

   (b) The diffraction pattern arising from a row of horizontal holes is a set of ________ fringes.

   (c) The diffraction vector **S** is ________ to the set of Bragg planes of spacing $d = 1/|\mathbf{S}|$ when the set is in the reflecting position.

   (d) A complex number differs from its complex conjugate by the ________ of its imaginary component.

(e) When many atoms lie on the *hkl* set of lattice planes and there are an equal number of atoms on planes midway in between, the reflection from the *hkl* set is ________.

(f) ________ law states that $F(h) = F(-h)$.

(g) A Patterson map shows all ________ between pairs of atoms in a structure.

(h) In the method of isomorphous replacement, a single isomorphous heavy atom derivative limits the phase of a reflection to ________ possible values.

Answers: (a) Non-straight line. (b) Vertical. (c) Perpendicular. (d) Sign. (e) Absent. (f) Friedel's. (g) Vectors. (h) Two.

PROBLEMS RELATED TO EXERCISES

3. (Exercise 17-1) An application of optical diffraction to modern biology is the study of periodic structures seen in electron microscope pictures (micrographs). The micrograph is used as a diffracting object for visible (laser) light. Often the diffraction spots have large areas. Why is this so?

4. (Exercise 17-2) How is the diffraction pattern of light passing through two small holes related to the diffraction pattern from two slightly larger holes?

5. (Exercise 17-3) For a two-dimensional square lattice, sketch the $2,\bar{1}$ planes and the $\bar{2},1$ planes.

6. (Exercise 17-4) Sketch the $\bar{1},\bar{1},\bar{1}$ planes in a three-dimensional primitive cubic lattice.

7. (Exercise 17-6) Suppose you wish to record 2 Å resolution x-ray data ($d_{\text{min}} = 2$ Å) on photographic film with copper K_α radiation ($\lambda = 1.542$ Å). With square sheets of film 12.5 cm on an edge, what is the maximum distance the film can be placed from the crystal to receive all desired data? Assume the film is perpendicular to the beam of x-rays.

8. (Exercise 17-7) What changes would you have to make in Figure 17-10 for a different set of *hkl* lattice planes with greater *d* spacing than the set shown?

9. (Exercise 17-8) Suppose you have recorded an x-ray precession $h0l$ photograph of a tetragonal crystal of the enzyme glutamine synthetase. You have used copper x-radiation with film-to-crystal distance of 60.0 mm. In measuring the photograph you find the spacing of reflections perpendicular to a^*, $\Delta(a^*) = 0.717$ mm and for reflections perpendicular to c^* that $\Delta(c^*) = 0.532$ mm. What is the volume of the unit cell?

10. (Exercise 17-9) Suppose the density of the crystals of Problem 9 is 1.242 g cm^{-3} and that their weight fraction of protein is 0.55. How many molecules of molecular weight 600,000 are contained in the unit cell?

11. (Exercise 17-10) The space group of the crystal described in Problems 9 and 10 was determined to be $P4_2$, which contains four asymmetric units per unit-cell. What information about the symmetry of the glutamine synthetase molecule can you deduce?

12. (Exercise 17-11) A heavy atom of atomic scattering factor $f = 80$ is located in a unit cell at coordinates (in fractions of a unit cell edge) of 0, $\frac{1}{2}$, $\frac{1}{4}$. Calculate its contribution in phase and amplitude to the following reflections: 1,0,0; 0,1,0; 0,2,0; 0,1,2; 1,1,1. Plot each of these contributions in the complex plane.

● 13. (Exercise 17-12) In Exercise 17-12 it is stated that $\alpha(-h) = -\alpha(h)$. Prove that this is so (*Hint:* Consider Equations 17-25 and 17-26).

● 14. (Exercise 17-13) Express Equation 17-38 for the three-dimensional Patterson function in the cosine form of Equation 17-37.

● 15. (Exercise 17-14) What is about the largest molecular weight of a compound whose structure can be determined by the heavy atom–Fourier method, using an attached mercury atom?

OTHER PROBLEMS

16. For lattice planes spaced by 2.5 Å, what is the length of the corresponding scattering vector **S**, and what is the angle 2θ between the incident and scattered x-rays with Cu K_α radiation ($\lambda = 1.542$ Å)?

17. Consider a protein P which can be crystallized. P has an estimated molecular weight (determined, say, by sedimentation equilibrium) of 1.9×10^5. The weight fraction of P in its crystals is 0.375. The unit cell volume of the crystals is 2.40×10^6 Å^3. The density of the crystals is 1.111 g cm^{-3}.

    (a) How many molecules of P are there in one unit cell of the crystal?

    (b) Estimate the molecular weight of P from these data.

18. What are the maximum and minimum values of the real component of $e^{2\pi i x}$? What is the physical significance in x-ray diffraction of these limits?

19. W. H. and W. L. Bragg solved the structures of simple crystals essentially by guessing the structure and then by showing that calculated structure-factor amplitudes for reflections agreed with observed structure-factor amplitudes. By this method, they found that NaCl has a face-centered cubic structure, with Cl^- ions at the following coordinates (in fractions of a unit cell edge):

    $$0,0,0;\ \tfrac{1}{2},\tfrac{1}{2},0;\ \tfrac{1}{2},0,\tfrac{1}{2};\ 0,\tfrac{1}{2},\tfrac{1}{2}$$

    and with Na^+ ions at the following coordinates:

    $$0,0,\tfrac{1}{2};\ 0,\tfrac{1}{2},0;\ \tfrac{1}{2},0,0;\ \tfrac{1}{2},\tfrac{1}{2},\tfrac{1}{2}$$

    Calculate $|\mathbf{F}|$, α, and I for the 1,1,1, 2,0,0, and 1,0,0 reflections assuming that

    (a) The magnitude of the atomic scattering factor f for an ion is equal to its number of electrons.

    (b) $I = \mathbf{F} \cdot \mathbf{F}^*$, where $\mathbf{F}^*$ is the complex conjugate of $\mathbf{F}$.

20. Why is the intensity of the 1,1,1 reflection of the KCl crystal very weak compared to that of NaCl, even though these salts have the same face-centered cubic structure?

21. Suppose that the motif in a one-dimensional crystal possesses a center of inversion (that is, for every atom at x, there is another atom of the same type at $-x$). Show that all structure factors for the crystal are real numbers. [*Hint:* expand the exponential in the structure-factor equation by Euler's relationship ($\exp ix = \cos x + i \sin x$). Then, pair terms for atoms at x and $-x$.]

22. For structures that contain a center of inversion, the method of isomorphous replacement requires only a single heavy-atom derivative. Draw figures analogous to Figure 17-19(*b*) to show how the method works in this case.

23. In space group $P2_1$, the motif symmetry is a 2-fold screw axis parallel to the b crystal axis. Show that all reflections of the form $0k0$ are absent (zero intensity) where k is odd. (Hint: start by deriving the equivalent positions for a 2_1 axis parallel to b. Then insert these equivalent positions in the structure-factor equation.)

24. Explain why an x-ray rotation photograph [Fig. 17-2(*b*)] displays the moon-shaped region of reflections.

25. Deduce the form of the diffraction pattern of a helix, given the following two hints: (1) Figure 17-4(*f*) shows that the diffraction pattern of a linear array of points is a set of parallel lines, each perpendicular to the array. The converse is also true: the diffraction pattern of a set of parallel lines is a linear array of points perpendicular to the lines. (2) The structure of a helix can be represented crudely by two sets of parallel line segments:

26. Using the definition of the *Fourier transform*, **F(S)**, given by Equation 17-32, compute the Fourier transform of the "slit function," $\rho(x) = A$, $-a/2 \leq x \leq a/2$; $\rho(x) = 0$, $x > a/2$, and $x < -a/2$.

27. An alternative to Equation 17-33 for expressing a Fourier series is

$$\rho(x) = a(0) + \sum_{n=1}^{\infty} a(n) \cos 2\pi nX + b(n) \sin 2\pi nX$$

Show that $A(n)$ of Equation 17-33 is given by

$$A(n) = \tfrac{1}{2}[a(n) - ib(n)] \text{ and } A(-n) = \tfrac{1}{2}[a(n) + ib(n)].$$

QUESTIONS FOR DISCUSSION

28. (a) Musical tones are functions periodic in time, rather than space. How can it be that two tones of the same pitch and loudness have different qualities? For example, why is it that a violin and a clarinet both playing middle C sound quite different?

(b) Devise a theory of how your brain allows you to recognize the voice of a friend.

REFERENCES

Brand, J. C. D., and Speakman, J. C. 1960. *Molecular Structure*. London: Edward Arnold.

Buerger, M. J. 1970. *Contemporary Crystallography*. New York: John Wiley & Sons. The mathematical development of Section 17-6 closely follows this reference.

Bunn, C. W. 1946. *Chemical Crystallography*. Oxford: The Clarendon Press.

Donohue, J. 1969. *Science 165*, 1092.

Henry, N. F. M., and Lonsdale, K., eds. 1969. *International Tables for X-ray Crystallography*. Birmingham: The Kynoch Press.

Jeffrey, J. W. 1971. *Methods in X-ray Crystallography*. New York: Academic Press.

Lipson, H., and Taylor, C. A. 1958. *Fourier Transforms and X-ray Diffraction*. London: Bell.

Moore, W. J. 1972. *Physical Chemistry*, 4th ed. Englewood Cliffs, N. J.: Prentice Hall.

Perutz, M. F. 1971. *Nature 233*, 74.

Waser, J. 1968. *J. Chem. Ed. 45*, 446.

Wykoff, H. W., et al. 1970. *J. Biol. Chem. 245*, 305.

FURTHER READINGS

Blundell, T. L., and Johnson, L. N. 1976. *Protein Crystallography*. New York: Academic Press. An indispensable guide to the practice of this art.

Cold Spring Harbor Symposia on Quantitative Biology XXXVI. 1971. A collection of papers on current research in enzyme and virus crystallography.

Dickerson, R. E. 1964. In *The Proteins*, ed. H. Neurath, Vol. 2, 603. New York: Academic Press. A superb, elementary introduction to diffraction of crystals and fibers.

Dickerson, R. E., and Geis, J. 1969. *The Structure and Action of Proteins*. New York: Harper & Row. A beautifully illustrated summary of structural studies of proteins.

Glusker, J. P., and Trueblood, K. N. 1972. *Crystal Structure Analysis: A Primer*. New York: Oxford University Press. Highly recommended to those who want to read further about x-ray crystallography.

Sherwood, D. 1976. *Crystals, X-rays and Proteins*. New York: John Wiley & Sons. A very complete, understandable treatment of the subject.

SELECTED REFERENCES FOR BIOGRAPHICAL SKETCHES

J. Willard Gibbs

Wheeler, L. P. 1951. *Josiah Willard Gibbs*. New Haven, Conn.: Yale University Press.

Benjamin Thompson, Count Rumford

Sanborn, C. Brown. 1962. *Count Rumford, Physicist Extraordinaire*. Garden City, N.Y.: Doubleday Anchor.

William Thomson, Baron Kelvin

Thompson, S. P. 1910. *The Life of William Thomson, Baron Kelvin of Largs*. London: MacMillan & Co., Ltd.

Buchwald, J. Z. 1976. *Dictionary of Scientific Biography.* Vol. XIII, p. 374.

Percy W. Bridgman

Newitt, D. M. 1962. *Biographical Memoirs of Fellows of the Royal Society.* Vol. VIII, p. 27.

Kemble, E. C., F. Birch, and G. Holton. 1970. *Dictionary of Scientific Biography.* Vol. II, p. 457.

Trueblood, K. N. Personal communication.

J. D. Bernal and Max Delbrück

Ewald, P. P., ed. 1962. *Fifty Years of X-ray Diffraction*. Utrecht: N. V. A. Oosthoek's Uitgeversmaatschappij.

Stent, G. 1963. *Molecular Biology of Bacterial Viruses*. San Francisco: W. H. Freeman and Co.

Cairns, J., G. Stent, and J. D. Watson. 1966. *Phage and the Origins of Molecular Biology.* Long Island, N.Y.: Cold Springs Harbor Laboratory.

Hodgkin, D. C., and D. P. Riley. 1968. *Structural Chemistry and Molecular Biology,* ed. by N. Davidson and A. Rich. San Francisco: W. H. Freeman and Co.

Watson, J. D. 1968. *The Double Helix*. New York: Atheneum Publishing Co.

Delbrück, M. 1970. *Science 168,* 1312. (Delbrück's Nobel Prize acceptance speech.)

Crowther, J. G. 1971. "John Desmond Bernal—an appreciation," *New Scientist and Science Journal*. September 23, p. 666.

Mackay, A. 1971. *Physics Today.* December, p. 65.

Delbrück, M. personal communication.

Marya Sklodovska—Marie Curie

Curie, E. 1937. *Madame Curie*. Garden City, N.Y.: Doubleday, Doran & Co., Inc.

Weill, A. R. 1971. *Dictionary of Scientific Biography.* Vol. III, p. 497.

Reid, R. 1974. *Marie Curie*. New York: Saturday Review Press/E. P. Dutton & Co.

Hermann von Helmholtz

Margenau, H. 1954. Introduction to the Dover Publications, New York, edition of Helmholtz' book *Sensations of Tone*.

Turner, R. Steven. 1972. *Dictionary of Scientific Biography.* Vol. VI, p. 241.

Michael Faraday

Williams, L. Pearce. 1971. *Dictionary of Scientific Biography.* Vol. IV, p. 527.

Max Planck and Albert Einstein

Born, M. 1948. *Obituary Notices of Fellows*. Vol. 6, p. 141. London: Royal Society. (Obituary of Max Planck.)

Clark, R. W. 1971. *Einstein: The Life and Times*. New York and Cleveland: The World Publishing Co.

J. Robert Oppenheimer

Bethe, H. 1968. *Biographical Memoirs*. London: Royal Society. Vol. 14, p. 391.

Peierls, R. 1974. *Dictionary of Scientific Biography.* Vol. X, p. 213.

John Strutt, Lord Rayleigh

Strutt, R. J. 1968. *Life of John William Strutt, Third Baron Rayleigh*. Madison, Wis.: University of Wisconsin Press.

Linsay, R. B. 1979. *Dictionary of Scientific Biography.* Vol. XIII, p. 100.

Louis Pasteur

Dubos, René. 1950. *Louis Pasteur: Free Lance of Science*. Boston: Little Brown and Co.

Dubos, René. 1961. *Pasteur and Modern Science*. London: Heinemann.

Geison, Gerald L. 1974. *Dictionary of Scientific Biography.* Vol. X, pp. 350–416.

Dubos, René. 1978. Personal communication.

Leo Szilard

Szilard, L. 1968. "Reminiscences" (ed. by G. W. Szilard and K. R. Winsor) in *Perspectives in American History,* ed. by Donald Fleming and Bernard Bailyn. Vol. II, p. 94. Cambridge: Harvard University Press.

Feld, B. T., and G. W. Szilard, eds. 1972. *The Collected Works of Leo Szilard*. Cambridge: MIT Press.

Feld, B. T. 1976. *Dictionary of Scientific Biography.* Vol. XIII, p. 226.

Johannes Kepler

Koestler, A. 1960. *The Watershed: A Biography of Johannes Kepler.* Garden City, N.Y.: Anchor Books.

Kepler, J. 1966. *The Six-Cornered Snowflake*. Oxford: The Clarendon Press.

Gingerich, Owen. 1973. *Dictionary of Scientific Biography.* Vol. VII, p. 289.

Gingerich, Owen. 1977. *Harvard Magazine*. March–April, p. 53.

Jean-Baptiste Joseph Fourier

Hutchins, M., ed. 1952. *Great Books of the Western World*. Vol. 15, p. 193.

Ravetz, J. R., and I. Grattan-Guiness. 1972. *Dictionary of Scientific Biography.* Vol. V, p. 93.

INDEX